ELECTROMAGNETIC FIELDS AND WAVES

Second Edition

ELECTROMAGNETIC FIELDS AND WAVES

Paul Lorrain
University of Montreal
Dale R. Corson
Cornell University

W. H. FREEMAN AND COMPANY
New York

Printed in the United States of America

International Standard Book Number: 0-7167-0331-9

Library of Congress Catalog Card Number: 72-94872

12 13 14 15 16 17 19 20 MP 0 8 9 8 7 6 5 4

Contents

3

ELECTROSTATIC FIELDS II 91
Dielectric Materials

4

ELECTROSTATIC FIELDS III 138

General Methods for Solving Laplace's and Poisson's Equations

5

RELATIVITY I 193

The Basic Concepts

6

RELATIVITY II 247

The Electric and Magnetic Fields of Moving Electric Charges

7

MAGNETIC FIELDS I 292

Steady Currents and Nonmagnetic Materials

8

MAGNETIC FIELDS II 332

Induced Electromotance and Magnetic Energy

10

MAXWELL'S EQUATIONS 422

11

PROPAGATION OF ELECTROMAGNETIC WAVES I 459

Plane Waves in Infinite Media

12

PROPAGATION OF ELECTROMAGNETIC WAVES II 504

Reflection and Refraction

13

PROPAGATION OF ELECTROMAGNETIC WAVES III 557

Guided Waves

14

RADIATION OF ELECTROMAGNETIC WAVES 595

Preface

This is a revised version of *Introduction to Electromagnetic Fields and Waves* by the same authors. The general level is unchanged, despite the fact that the phrase "Introduction to" has been omitted in the title.

Like the first edition, this book is intended primarily for students of Physics or Electrical Engineering at the junior or senior levels, although some schools will prefer to use it with first-year graduate students. The book should also be useful for scientists and engineers who wish to review the subject.

The major change has been the addition of two chapters on relativity. *Those who are pressed for time may omit these without losing continuity.* The other chapters have been largely rewritten, with many additions and deletions. There are about 100 new figures; as in the first edition, three-dimensional objects and phenomena are represented as such, and the field maps (such as those in Chapter 4) have been plotted on a computer. Most of the 140 examples and most of the 413 problems are new.

The aim of this book is to give the reader a working knowledge of the basic concepts of electromagnetism. Indeed, as Alfred North Whitehead stated, half a century ago, "Education is the acquisition of the art of the *utilization* of knowledge." This explains the relatively large number of examples and problems. It also explains why we have covered fewer subjects more thoroughly. For instance, Laplace's equation is solved in rectangular and in spherical coordinates, but not in cylindrical coordinates.

CONTENTS

So as to reduce the mathematical requirements, we have included a chapter on vectors (Chapter 1), a discussion of Legendre's differential equation

(Section 4.5), an appendix on the technique that involves replacing $\cos \omega t$ by $\exp j\omega t$, and an appendix on wave propagation.

After the introductory chapter on vectors, Chapters 2, 3, and 4 describe electrostatic fields, both in a vacuum and in dielectrics. All of Chapter 4 is devoted to the solution of Laplace's and of Poisson's equations.

Chapters 5 and 6 discuss relativity. They may be omitted without losing continuity. Indeed, this book presents both the conventional and the relativistic approaches to electromagnetism, which are, in a sense, complementary. Chapter 5 is a short exposition of the basic concepts of special relativity, with little reference to electric charges. It requires nothing more, in the way of mathematics, than elementary differential calculus and the vector analysis of Chapter 1. Chapter 6 contains a demonstration of Maxwell's equations that is based on Coulomb's law and on the Lorentz transformation and which is valid only for the case where the charges move at constant velocities.

Chapters 7 and 8 deal with the conventional approach to the magnetic fields associated with constant and with variable currents. Here, as elsewhere, references to Chapter 6 may be disregarded.

Chapter 9 contains a discussion of magnetic materials that parallels, to a certain extent, that of Chapter 3 on dielectrics.

In Chapter 10, the Maxwell equation for the curl of $\boldsymbol{B}$ is rediscovered, without using relativity. This is followed by a discussion of the four Maxwell equations, as well as of some of their more general implications. The point of view is different from that of Chapter 6, and there is essentially no repetition.

The last four chapters, 11 to 14, concern various applications of Maxwell's equations: plane waves in infinite media in Chapter 11, reflection and refraction in Chapter 12, guided waves in Chapter 13, and radiation in Chapter 14. The only three media considered in Chapters 11 and 12 are perfect dielectrics, good conductors, and low-pressure ionized gases. Similarly, Chapter 13 is limited to the two simplest types of guided wave, namely the TEM mode in coaxial lines and the $TE_{1,0}$ mode in rectangular guides. Chapter 14 discusses electric and magnetic dipoles and quadrupoles, as well as the essential ideas concerning the half-wave antenna, antenna arrays, and the reciprocity theorem.

For a basic and relatively simple course on electromagnetism, one could study only Chapters 2, 3 (less Sections 3.3, 3.4, 3.8, 3.9, and 3.10), 4 (less Sections 4.4 and 4.5), 7, 8, 9 (less Section 9.3 but conserving the equation $\nabla \cdot \boldsymbol{B} = 0$), and 10. For a rather advanced course, on the other hand, Chapters 2, 3, 4, 5, 7, 8, and 9 could be reviewed briefly using the summaries at the end of each chapter. One would then start with Chapter 6, and then go on to Chapter 10 and the following chapters. There are, of course, many other possibilities.

In Chapter 12, Sections 12.3 and 12.7 could be dispensed with. They involve the application of Fresnel's equations to particular cases and are not essential for the remaining chapters. Chapter 13 is instructive, both because of the insight it provides into the propagation of electromagnetic waves and because of its engineering applications, but it is not required for understanding Chapter 14. Finally, Chapter 14 is based on Chapter 10 and on the first two sections of Chapter 11.

PROBLEMS

This second edition contains an extensive collection of problems, many of which are drawn from the current literature of Physics and Engineering. Such problems should serve as a clear indication that electromagnetism has a wealth of applications in a wide variety of fields. They should favor the transferability of what has been learned, and they should develop creativity. Few persons will spend the time and energy required for solving all of the longer problems. Those left unsolved should provide some profitable reading.

Double daggers (‡) indicate those problems for which hints are given.

It is suggested that the problems be solved on a computer when this is feasible.

UNITS AND NOTATION

The units used are those of the *Système International d'Unités* (designated SI in all languages), adopted in Paris at the 1960 Eleventh General Conference on Weights and Measures. This system, which is based on the meter, kilogram, second, ampere, kelvin, and candela, is essentially the same as the Giorgi, or MKSA, system in the field of electromagnetism.

The exponential function for periodic phenomena can be either $\exp(j\omega t)$ or $\exp(-j\omega t)$, since the real part of both of these functions is $\cos \omega t$. We have used the positive exponent, which is preferable at this level; otherwise, an impedance becomes $R - jX$ instead of the conventional $R + jX$.

ACKNOWLEDGMENTS

The authors owe much to Gilles Cliche, to Gaétan Marchand, and to Paul Carrière who performed the numerical calculations and assisted in the design of the figures for the first edition. Several other persons assisted in the preparation of this second edition. François Lorrain, Luce Gauthier-Labonté, and Gilles Labonté all checked the complete manuscript and proposed innumerable improvements. François Lorrain and Guy Basque performed the numerical calculations required for the new figures. Jean and Michel Barrette checked the answers of many of the problems, Roland Savage drew the one-hundred-odd new figures, and Angèle Elias typed over a thousand pages of text with admirable care. Despite the invaluable assistance of all these persons, the undersigned is clearly responsible for any errors or inaccuracies that may remain.

The authors are grateful to the many students who proposed corrections and modifications to the first edition. They are especially grateful to the many persons who took the trouble to send their comments and lists of errata. Among these it is a pleasure to mention Professors C. Rutherford Fischer of Adelphi College; Gaston Fischer of the University of Montreal; R. H. Good of California State College, Hayward; W. M. Nunn, Jr., of the Florida Institute of Technology; Frank E. Rose of Flint College; Paul A. Smith of Coe College; Herschel Weil of the University of Michigan and R. E. Worley of the University of Nevada.

The number of bits of information in a book such as this is truly enormous, and errors are bound to occur. Further feedback from readers will therefore be appreciated.

Finally, I wish to thank the Universidad de Madrid, where I was on sabbatical leave during the final stages of the preparation of the manuscript, and to thank Maximino Rodríguez Vidal and his group for many fruitful discussions.

I note with regret that Dale Corson's duties as Provost, and later as President of Cornell University prevented him from participating in the preparation of this second edition.

May 1969 PAUL LORRAIN

List of Symbols

SPACE, TIME, MECHANICS

Element of length or distance	dl, ds, dr
Total length or distance	l, L, s, r
Element of area	da
Total area	S
Element of volume	$d\tau$
Total volume	τ
Solid angle	Ω
Normal to a surface	$\boldsymbol{n}$
Wave length	λ
Wave length in free space	λ_0
Wave length of a guided wave	λ_g
Radian length	$\lambda\!\!\!^{-} = \lambda/2\pi = 1/k_r$
Wave number	$k = k_r - jk_i$
Attenuation constant	k_i
Attenuation distance	$\delta = 1/k_i$
Time	t
Period	$T = 1/f$
Frequency	$f = 1/T$
Angular frequency	$\omega = 2\pi f$
Velocity	$\boldsymbol{u}$, $\boldsymbol{v}$, $\mathcal{V}$, $\gamma = \dfrac{1}{\{1 - (v/c)^2\}^{1/2}}$
Acceleration	$\boldsymbol{a} = \partial \boldsymbol{u}/\partial t$
Angular velocity	$\boldsymbol{\omega}$
Angular acceleration	$\boldsymbol{\alpha} = \partial \boldsymbol{\omega}/\partial t$
Mass	m
Density	ρ
Curvilinear coordinate	q
Four-vector	$\mathbf{r}$
Momentum	$\boldsymbol{p}$
Four-momentum	$\mathbf{p}$
Moment of inertia	$\boldsymbol{I}$
Force	$\boldsymbol{F}$
Torque	$\boldsymbol{\Gamma}$
Pressure	p
Energy	W

ELECTRICITY AND MAGNETISM

Quantity of electricity	Q
Velocity of light	$c = 2.998 \times 10^8$ meters/second
Volume charge density	ρ
Surface charge density	σ

Linear charge density	ν, λ
Electric potential, scalar potential	V
Induced Electromotance, Voltage	$\mathcal{V}$
Electric field intensity	$\boldsymbol{E}$
Electric displacement	$\boldsymbol{D}$
Permittivity of vacuum	$\epsilon_0 = 8.854 \times 10^{-12}$ farad/meter
Relative permittivity	ϵ_r
Permittivity	$\epsilon = D/E$
Q of a medium	$\mathcal{Q}$
Electric dipole moment	$\boldsymbol{p}$
Electric polarization, electric dipole moment per unit volume	$\boldsymbol{P}$
Molecular polarizability	α
Electric susceptibility	χ_e
Electric quadrupole moment	q
Electric current	I
Volume current density	$\boldsymbol{J}$
Four-current density	$\boldsymbol{J}$
Surface current density	$\boldsymbol{\lambda}$
Avogadro's number	$N_A = 6.023 \times 10^{23}$/mole
Molecular weight	M
Boltzmann constant	$k = 1.381 \times 10^{-23}$ joule/kelvin
Electronic charge	$e = 1.602 \times 10^{-19}$ coulomb
Planck's constant divided by 2π	$\hbar = 1.055 \times 10^{-34}$
Vector potential	$\boldsymbol{A}$
Four-potential	$\boldsymbol{A}$
Magnetic induction	$\boldsymbol{B}$
Magnetic field intensity	$\boldsymbol{H}$
Magnetic flux	Φ
Permeability of vacuum	$\mu_0 \equiv 4\pi \times 10^{-7}$ henry/meter
Relative permeability	μ_r
Permeability	$\mu = B/H$
Magnetic dipole moment per unit volume	$\boldsymbol{M}$
Magnetic susceptibility	χ_m
Magnetic dipole moment	$\boldsymbol{m}$
Resistance	R
Capacitance	C
Self-inductance	L
Mutual inductance	M
Resistivity	ρ
Conductivity	σ
Poynting vector	$\mathbf{S}$

MATHEMATICAL SYMBOLS

Approximately equal to	$\approx$
Proportional to	$\propto$
Factorial n	$n!$
Exponential of x	e^x, $\exp(x)$
Real part of z	Re z
Imaginary part of z	Im z
Modulus of z	$\lvert z \rvert$
Decadic log of x	$\log x$
Natural log of x	$\ln x$

Arc tangent x	arc tan x
Complex conjugate of z	z^*
Vector	$\boldsymbol{E}$
Gradient	$\boldsymbol{\nabla}$
Divergence	$\boldsymbol{\nabla}\cdot$
Curl	$\boldsymbol{\nabla}\times$
Quad	$\square$
Laplacian	$\boldsymbol{\nabla}^2$
Unit vectors in Cartesian coordinates	$\boldsymbol{i}, \boldsymbol{j}, \boldsymbol{k}$
Unit vectors in cylindrical coordinates	$\boldsymbol{\rho}_1, \boldsymbol{\varphi}_1, \boldsymbol{z}_1$
Unit vectors in spherical coordinates	$\boldsymbol{r}_1, \boldsymbol{\theta}_1, \boldsymbol{\varphi}_1$
Unit vector along $\boldsymbol{r}$	$\boldsymbol{r}_1$
Field point	x, y, z
Source point	x', y', z'
Average	$\overline{\mathrm{E}}$

ELECTROMAGNETIC FIELDS AND WAVES

CHAPTER 1

VECTORS

We shall discuss electric and magnetic phenomena in terms of the *fields* of electric charges and currents. For example, we shall consider the force between two electric charges to be due to an interaction between either one of the charges and the field of the other.

It is therefore essential that you acquire at the very outset a thorough understanding of the mathematical methods required to deal with fields. This is the purpose of the present chapter on vectors. Note that the concept of field and the mathematics of vectors are essential not only to electromagnetic theory but also to most of present-day physics.

We shall assume that you are not familiar with vectors and that a thorough discussion is required.

If, on the contrary, you are already quite familiar with vector analysis, you may wish to omit Sections 1.1 to 1.8 and concentrate your attention on Section 1.9, which deals with curvilinear coordinates.

Mathematically, a field is a function that describes a physical quantity at all points in space. In *scalar fields* this physical quantity is completely specified by a single number for each point. Temperature, density, and electric potential are examples of scalar quantities that can vary from one point to another in space. For *vector fields* both a number and a direction are required. Wind velocity, gravitational force, and electric field intensity are examples of such vector quantities.

Vector quantities will be indicated by ***boldface*** type; *lightface* type will indicate either a scalar quantity or the magnitude of a vector quantity.

We shall follow the usual custom of using *right-hand Cartesian coordinate*

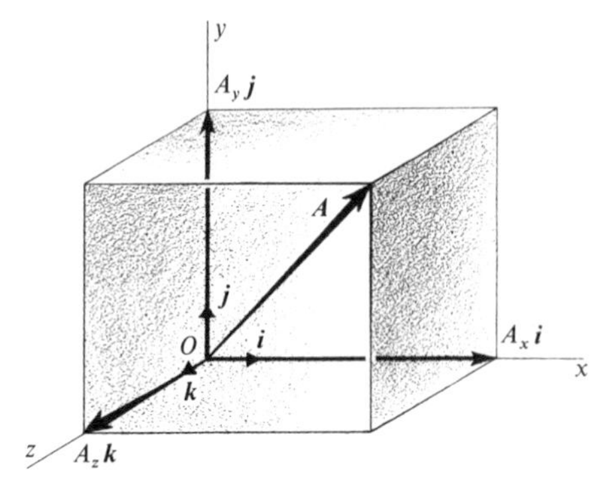

Figure 1-1. A vector $\boldsymbol{A}$ and the three vectors $A_x\boldsymbol{i}$, $A_y\boldsymbol{j}$, $A_z\boldsymbol{k}$, which, when placed end-to-end, are equivalent to $\boldsymbol{A}$.

systems as in Figure 1-1: the positive z direction is the direction of advance of a right-hand screw rotated in the sense that turns the positive x-axis into the positive y-axis through the 90° angle.

1.1 VECTOR ALGEBRA

A vector can be specified by its *components* along any three mutually perpendicular axes. In the Cartesian coordinate system of Figure 1-1, for example, the vector $\boldsymbol{A}$ has components A_x, A_y, A_z.

The vector $\boldsymbol{A}$ can be uniquely expressed in terms of its components through the use of unit vectors $\boldsymbol{i}$, $\boldsymbol{j}$, $\boldsymbol{k}$, which are defined as vectors of unit magnitude in the positive x, y, z directions, respectively:

$$\boldsymbol{A} = A_x\boldsymbol{i} + A_y\boldsymbol{j} + A_z\boldsymbol{k}. \tag{1-1}$$

The vector $\boldsymbol{A}$ is the sum of three vectors of magnitude A_x, A_y, A_z, parallel to the x-, y-, z-axes, respectively.

The magnitude of $\boldsymbol{A}$ is

$$A = (A_x^2 + A_y^2 + A_z^2)^{1/2}. \tag{1-2}$$

The sum of two vectors is obtained by adding their components:

$$\boldsymbol{A} + \boldsymbol{B} = (A_x + B_x)\boldsymbol{i} + (A_y + B_y)\boldsymbol{j} + (A_z + B_z)\boldsymbol{k}. \tag{1-3}$$

Subtraction is simply addition with one of the vectors changed in sign:

$$\boldsymbol{A} - \boldsymbol{B} = \boldsymbol{A} + (-\boldsymbol{B}) = (A_x - B_x)\boldsymbol{i} + (A_y - B_y)\boldsymbol{j} + (A_z - B_z)\boldsymbol{k}. \tag{1-4}$$

We shall use two types of multiplication: the scalar, or dot product; and the vector, or cross product. The *scalar*, or *dot product*, is the scalar quantity obtained on multiplying the magnitude of the first vector by the magnitude

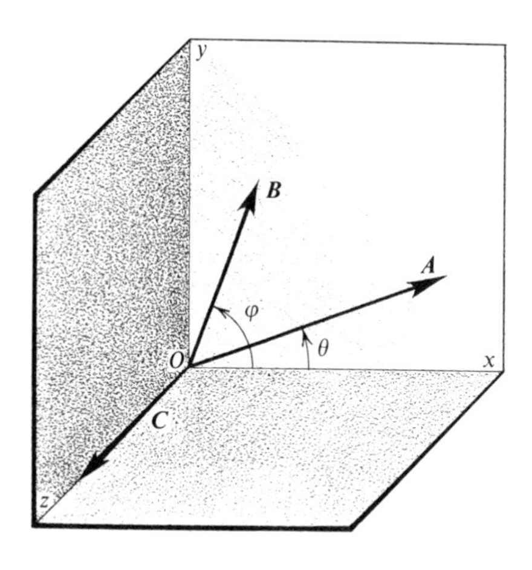

Figure 1-2. Two vectors $\boldsymbol{A}$ and $\boldsymbol{B}$ in the xy-plane. The vector $\boldsymbol{C}$ is their vector product $\boldsymbol{A} \times \boldsymbol{B}$.

of the second and by the cosine of the angle between the two vectors. In Figure 1-2, for example,

$$\boldsymbol{A}\cdot\boldsymbol{B} = AB\cos(\varphi - \theta) \tag{1-5}$$

It follows from this definition that the usual commutative and distributive rules of ordinary arithmetic multiplication apply to the scalar product:

$$\boldsymbol{A}\cdot\boldsymbol{B} = \boldsymbol{B}\cdot\boldsymbol{A}, \tag{1-6}$$

and

$$\boldsymbol{A}\cdot(\boldsymbol{B} + \boldsymbol{C}) = \boldsymbol{A}\cdot\boldsymbol{B} + \boldsymbol{A}\cdot\boldsymbol{C}. \tag{1-7}$$

From the definition of the scalar product it follows that

$$\boldsymbol{i}\cdot\boldsymbol{i} = 1, \quad \boldsymbol{j}\cdot\boldsymbol{j} = 1, \quad \boldsymbol{k}\cdot\boldsymbol{k} = 1, \tag{1-8}$$

$$\boldsymbol{j}\cdot\boldsymbol{k} = 0, \quad \boldsymbol{k}\cdot\boldsymbol{i} = 0, \quad \boldsymbol{i}\cdot\boldsymbol{j} = 0. \tag{1-9}$$

Then

$$\boldsymbol{A}\cdot\boldsymbol{B} = (A_x\boldsymbol{i} + A_y\boldsymbol{j} + A_z\boldsymbol{k})\cdot(B_x\boldsymbol{i} + B_y\boldsymbol{j} + B_z\boldsymbol{k}), \tag{1-10}$$

$$= A_xB_x + A_yB_y + A_zB_z. \tag{1-11}$$

It is easy to check that this result is correct for two vectors in a plane, as in Figure 1-2:

$$\boldsymbol{A}\cdot\boldsymbol{B} = AB\cos(\varphi - \theta) = AB\cos\varphi\cos\theta + AB\sin\varphi\sin\theta, \tag{1-12}$$

$$= A_xB_x + A_yB_y. \tag{1-13}$$

Example | A simple physical example of the scalar product is the work W done by a force $\boldsymbol{F}$ acting through a displacement $\boldsymbol{s}$: $W = \boldsymbol{F}\cdot\boldsymbol{s}$.

The *vector product*, or *cross product*, of two vectors is a vector whose direction is perpendicular to the plane containing the two initial vectors and

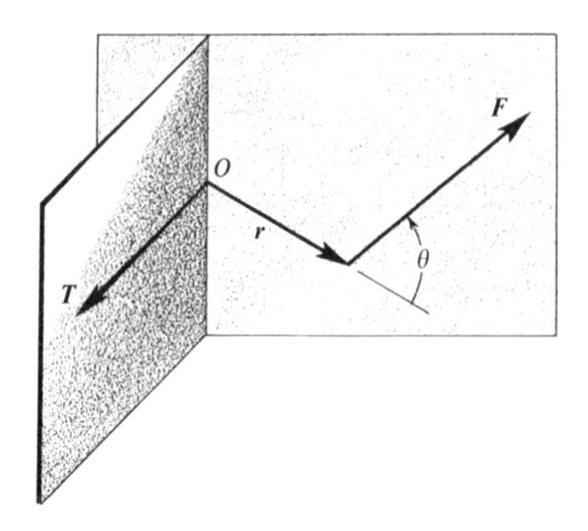

Figure 1-3. An example of vector multiplication. The torque $\boldsymbol{T}$ of the force $\boldsymbol{F}$ about the point O is $\boldsymbol{r} \times \boldsymbol{F}$. This vector has a magnitude of $rF \sin \theta$ and is oriented as shown.

whose magnitude is the product of the magnitudes of those vectors and the sine of the angle between them. We indicate the vector product thus:

$$\boldsymbol{A} \times \boldsymbol{B} = \boldsymbol{C}, \tag{1-14}$$

where the magnitude of $\boldsymbol{C}$ is

$$C = |AB \sin (\varphi - \theta)|, \tag{1-15}$$

with φ and θ defined as in Figure 1-2. The direction of $\boldsymbol{C}$ is given by the right-hand screw rule: it is the direction of advance of a right-hand screw whose axis, held perpendicular to the plane of $\boldsymbol{A}$ and $\boldsymbol{B}$, is rotated in the sense that rotates the first-named vector ($\boldsymbol{A}$) into the second-named ($\boldsymbol{B}$) through the smaller angle.

The commutative rule is *not* followed for the vector product, since inverting the order of $\boldsymbol{A}$ and $\boldsymbol{B}$ inverts the direction of $\boldsymbol{C}$:

$$\boldsymbol{A} \times \boldsymbol{B} = -(\boldsymbol{B} \times \boldsymbol{A}). \tag{1-16}$$

The distributive rule, however, is followed:

$$\boldsymbol{A} \times (\boldsymbol{B} + \boldsymbol{C}) = (\boldsymbol{A} \times \boldsymbol{B}) + (\boldsymbol{A} \times \boldsymbol{C}). \tag{1-17}$$

This will be shown in Problem 1-7.

From the definition of the vector product it follows that

$$\boldsymbol{i} \times \boldsymbol{i} = 0, \quad \boldsymbol{j} \times \boldsymbol{j} = 0, \quad \boldsymbol{k} \times \boldsymbol{k} = 0, \tag{1-18}$$

and, for the usual right-handed coordinate systems, such as that of Figure 1-1,

$$\boldsymbol{i} \times \boldsymbol{j} = \boldsymbol{k}, \quad \boldsymbol{j} \times \boldsymbol{k} = \boldsymbol{i}, \quad \boldsymbol{k} \times \boldsymbol{i} = \boldsymbol{j}, \quad \boldsymbol{j} \times \boldsymbol{i} = -\boldsymbol{k}, \text{ and so on.} \tag{1-19}$$

Writing out the vector product of $\boldsymbol{A}$ and $\boldsymbol{B}$ in terms of the components, we have

$$\boldsymbol{A} \times \boldsymbol{B} = (A_x\boldsymbol{i} + A_y\boldsymbol{j} + A_z\boldsymbol{k}) \times (B_x\boldsymbol{i} + B_y\boldsymbol{j} + B_z\boldsymbol{k}), \tag{1-20}$$

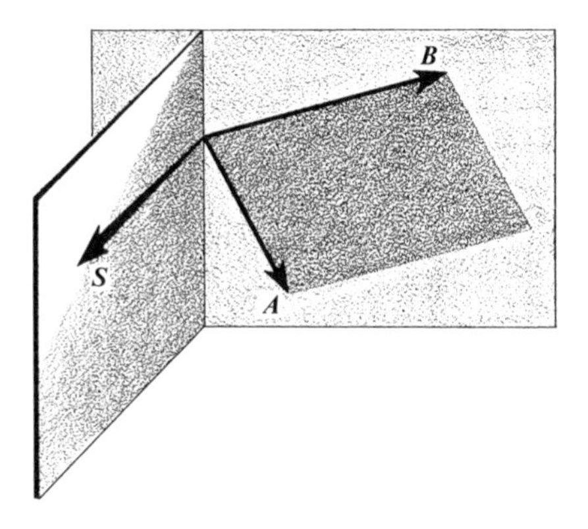

Figure 1-4. Another example of vector multiplication. The area of the parallelogram is $\boldsymbol{A} \times \boldsymbol{B} = \mathbf{S}$. The vector $\mathbf{S}$ is normal to the parallelogram.

$$\boldsymbol{A} \times \boldsymbol{B} = (A_y B_z - A_z B_y)\boldsymbol{i} + (A_z B_x - A_x B_z)\boldsymbol{j} + (A_x B_y - A_y B_x)\boldsymbol{k}, \quad (1\text{-}21)$$

$$= \begin{vmatrix} \boldsymbol{i} & \boldsymbol{j} & \boldsymbol{k} \\ A_x & A_y & A_z \\ B_x & B_y & B_z \end{vmatrix}. \quad (1\text{-}22)$$

We can check this result for the two vectors of Figure 1-2 by expanding $\sin(\varphi - \theta)$ and noting that the vector product is in the positive z direction.

Examples | A good physical example of the vector product is the torque $\boldsymbol{T}$ produced by a force $\boldsymbol{F}$ acting with a moment arm $\boldsymbol{r}$ about a point O, as in Figure 1-3, where $\boldsymbol{T} = \boldsymbol{r} \times \boldsymbol{F}$.

A second example is the area of a parallelogram, as in Figure 1-4, where the area $\mathbf{S} = \boldsymbol{A} \times \boldsymbol{B}$. The area is thus represented by a vector perpendicular to the surface.

Many other multiplication operations could be defined. The scalar and the vector product operations as defined here are unique, however, in that they are useful in describing real physical quantities.

1.2 INVARIANCE

You have probably noticed that we have *two* definitions for the scalar product: Equations 1-5 and 1-11.

The first definition clearly does not refer to a particular coordinate system and depends solely on the magnitudes of the two vectors and on the angle between them. A quantity that is independent of the coordinate system is said to be *invariant*.

The second definition, Eq. 1-11, involves the components of $\boldsymbol{A}$ and of $\boldsymbol{B}$

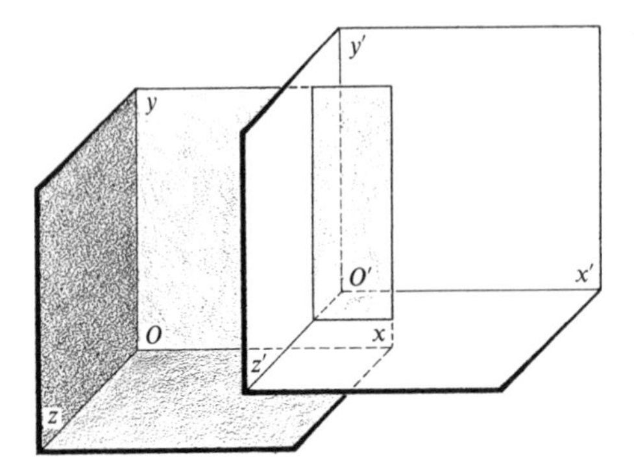

Figure 1-5. Two coordinate systems. The x', y', z' system is obtained from x, y, z by a translation. Corresponding coordinate axes, such as Ox and $O'x'$ are parallel, but the origins O and O' do not coincide.

in some particular coordinate system. Since this second definition is deduced from the first one, it should also be invariant. For example, given a force $\boldsymbol{F}$ and a displacement $\boldsymbol{s}$, this second definition should always give the correct numerical result for any right-handed Cartesian coordinate system. If it does not, then it is of no interest to us.

We shall show that the second definition, Eq. 1-11, really is invariant. First, let us consider two right-handed coordinate systems x, y, z and x', y', z', as in Figure 1.5, one system being displaced, but not rotated, with respect to the other. It is quite obvious that the components of $\boldsymbol{F}$ and of $\boldsymbol{s}$ are the same in the two systems and that Eq. 1-11 therefore leads to identical results in the two systems.

Figure 1-6 shows a more general type of relation between the x, y, z and x', y', z' systems, involving both a translation and a rotation.

Since we have already disposed of translation, we need only consider rotation. Now any rotation of a coordinate system can be decomposed into

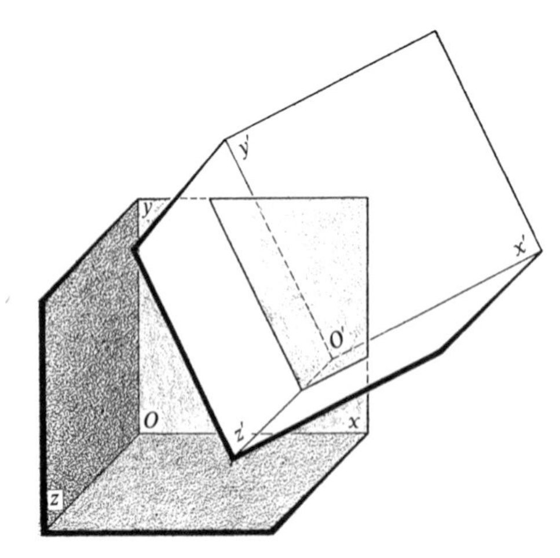

Figure 1-6. The x', y', z' system is obtained from x, y, z by a translation *and* a rotation. The two origins O and O' do not coincide and corresponding axes are not parallel.

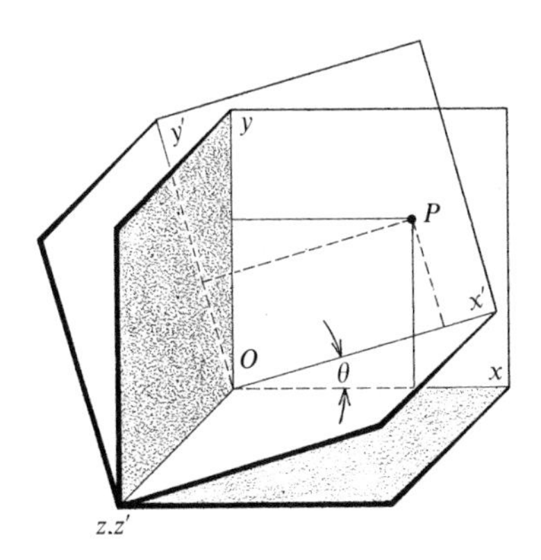

Figure 1-7. The x', y', z' system is obtained from x, y, z by a rotation about the z-axis.

three rotations performed successively about three different axes in space. It will therefore suffice to investigate whether Eq. 1-11 is invariant under only one such rotation, namely the rotation illustrated in Figure 1-7, where the x'- and y'-axes are tilted at an angle θ with respect to the x- and y-axes. For this condition the coordinates x, y, z and x', y', z' of a point P are related as follows:

$$x' = x\cos\theta + y\sin\theta, \tag{1-23}$$

$$y' = -x\sin\theta + y\cos\theta, \tag{1-24}$$

$$z' = z, \tag{1-25}$$

$$x = x'\cos\theta - y'\sin\theta, \tag{1-26}$$

$$y = x'\sin\theta + y'\cos\theta, \tag{1-27}$$

$$z = z', \tag{1-28}$$

and the components of the vector

$$\boldsymbol{A} = A_x\boldsymbol{i} + A_y\boldsymbol{j} + A_z\boldsymbol{k} = A_{x'}\boldsymbol{i}' + A_{y'}\boldsymbol{j}' + A_{z'}\boldsymbol{k}' \tag{1-29}$$

obey similar equations of transformation:

$$A_{x'} = A_x\cos\theta + A_y\sin\theta, \tag{1-30}$$

$$A_{y'} = -A_x\sin\theta + A_y\cos\theta, \tag{1-31}$$

$$A_{z'} = A_z, \tag{1-32}$$

$$A_x = A_{x'}\cos\theta - A_{y'}\sin\theta, \tag{1-33}$$

$$A_y = A_{x'}\sin\theta + A_{y'}\cos\theta, \tag{1-34}$$

$$A_z = A_{z'}. \tag{1-35}$$

Similar equations apply to the vector $\boldsymbol{B}$. If we use the equations of transformation in

$$\boldsymbol{A}\cdot\boldsymbol{B} = A_xB_x + A_yB_y + A_zB_z, \tag{1-36}$$

we should obtain a similar equation in terms of the primed components, if the expression on the right is invariant. We find indeed that

$$\boldsymbol{A}\cdot\boldsymbol{B} = A_xB_x + A_yB_y + A_zB_z = A_{x'}B_{x'} + A_{y'}B_{y'} + A_{z'}B_{z'}. \tag{1-37}$$

We have therefore shown that the scalar product, as defined in Eq. 1-11, is an invariant under both a translation and a rotation of the coordinate axes.

Now what about the vector product? It has also been defined in two different ways, first in Eq. 1-14 and then in Eq. 1-22.

From the first definition, the vector product of two vectors is an invariant since it depends only on the magnitude of the two vectors and on the angle between them. Is the vector product as defined in Eq. 1-22 also an invariant? It should be, since Eq. 1-22 follows directly from Eq. 1-14. We have again the same equations of transformation 1-30 to 1-35, and the unit vectors $\boldsymbol{i}$, $\boldsymbol{j}$, $\boldsymbol{k}$ must also be transformed to $\boldsymbol{i}'$, $\boldsymbol{j}'$, $\boldsymbol{k}'$ according to the following rule:

$$\boldsymbol{i} = \cos\theta\boldsymbol{i}' - \sin\theta\boldsymbol{j}', \tag{1-38}$$

$$\boldsymbol{j} = \sin\theta\boldsymbol{i}' + \cos\theta\boldsymbol{j}', \tag{1-39}$$

$$\boldsymbol{k} = \boldsymbol{k}'. \tag{1-40}$$

We find that

$$\boldsymbol{A}\times\boldsymbol{B} = \begin{vmatrix} \boldsymbol{i} & \boldsymbol{j} & \boldsymbol{k} \\ A_x & A_y & A_z \\ B_x & B_y & B_z \end{vmatrix} = \begin{vmatrix} \boldsymbol{i}' & \boldsymbol{j}' & \boldsymbol{k}' \\ A_{x'} & A_{y'} & A_{z'} \\ B_{x'} & B_{y'} & B_{z'} \end{vmatrix}. \tag{1-41}$$

The cross product as defined in Eq. 1-22 is therefore also invariant.

1.3 THE TIME DERIVATIVE

We shall often be concerned with the rates of change of scalar and vector quantities with both time and space coordinates, and thus with the time and space derivatives.

The time derivative of a vector quantity is straightforward. In a time Δt, a vector $\boldsymbol{A}$, as in Figure 1-8, may change by $\Delta\boldsymbol{A}$, which in general represents a change both in magnitude and in direction. Since $\Delta\boldsymbol{A}$ has components ΔA_x, ΔA_y, and ΔA_z,

$$\Delta\boldsymbol{A} = \Delta A_x\boldsymbol{i} + \Delta A_y\boldsymbol{j} + \Delta A_z\boldsymbol{k}. \tag{1-42}$$

On dividing $\Delta\boldsymbol{A}$ by Δt and taking the limit in the usual way, we arrive at the definition of $d\boldsymbol{A}/dt$:

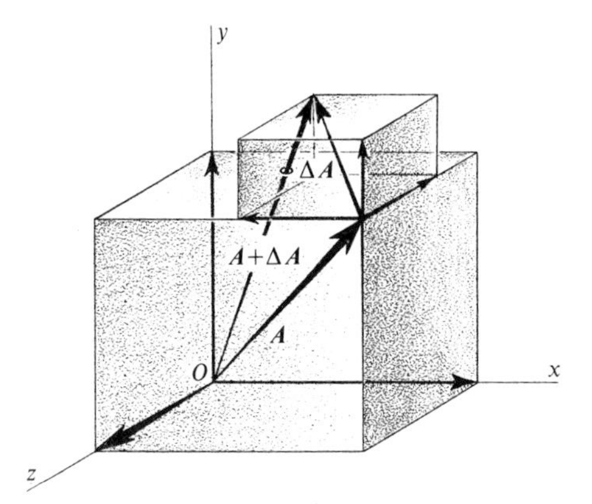

Figure 1-8. A vector A, its increment ΔA, and their components.

$$\frac{dA}{dt} = \lim_{\Delta t \to 0} \frac{A(t + \Delta t) - A(t)}{\Delta t}, \tag{1-43}$$

$$= \lim_{\Delta t \to 0} \frac{\Delta A_x \boldsymbol{i} + \Delta A_y \boldsymbol{j} + \Delta A_z \boldsymbol{k}}{\Delta t}, \tag{1-44}$$

$$= \frac{dA_x}{dt}\boldsymbol{i} + \frac{dA_y}{dt}\boldsymbol{j} + \frac{dA_z}{dt}\boldsymbol{k}. \tag{1-45}$$

The time derivative of a vector is thus equal to the vector sum of the time derivatives of its components.

Example | The time derivative of the position $\boldsymbol{r}$ of a point is its velocity $\boldsymbol{v}$, and the time derivative of $\boldsymbol{v}$ is the acceleration $\boldsymbol{a}$.

We again have two definitions, Eq. 1-43, which is necessarily invariant, and Eq. 1-45, which refers to a given coordinate system. Is the right-hand side of Eq. 1-45 invariant? It is not affected by a translation of the coordinate system, since this changes neither the unit vectors $\boldsymbol{i}, \boldsymbol{j}, \boldsymbol{k}$ nor the components A_x, A_y, A_z. We need only check whether or not Eq. 1-45 is affected by a rotation of the coordinate axes. If we express Eq. 1-45 in terms of the primed coordinates, using the equations of transformation 1-33 to 1-35 for A_x, A_y, A_z, and 1-38 to 1-40 for $\boldsymbol{i}, \boldsymbol{j}, \boldsymbol{k}$, we find that

$$\frac{dA}{dt} = \frac{dA_x}{dt}\boldsymbol{i} + \frac{dA_y}{dt}\boldsymbol{j} + \frac{dA_z}{dt}\boldsymbol{k} = \frac{dA_{x'}}{dt}\boldsymbol{i}' + \frac{dA_{y'}}{dt}\boldsymbol{j}' + \frac{dA_{z'}}{dt}\boldsymbol{k}'. \tag{1-46}$$

The derivative of a vector with respect to time, as defined in Eq. 1-45, is therefore an invariant.

1.4 THE GRADIENT

We shall be interested in one particular function of the space derivatives of a scalar quantity (the gradient) and in two particular functions of the space derivatives of a vector quantity (the divergence and the curl). Again, many other such functions could be defined, but these particular ones are unique in that they are useful to describe certain physical quantities.

Let us consider a scalar quantity that is a continuous and differentiable function of the coordinates and has the value f at a certain point, as in Figure 1-9. We wish to know how f changes over the distance $\boldsymbol{dl}$ measured from that point. We know that

$$df = \frac{\partial f}{\partial x}\,dx + \frac{\partial f}{\partial y}\,dy. \tag{1-47}$$

Now df is the scalar product of two vectors $\boldsymbol{A}$ and $\boldsymbol{dl}$ with

$$\boldsymbol{A} = \frac{\partial f}{\partial x}\,\boldsymbol{i} + \frac{\partial f}{\partial y}\,\boldsymbol{j}, \tag{1-48}$$

$$\boldsymbol{dl} = dx\,\boldsymbol{i} + dy\,\boldsymbol{j}. \tag{1-49}$$

The vector $\boldsymbol{A}$, whose components are the rates of change of f with distance along the coordinate axes, is called the *gradient* of the scalar quantity f. Gradient is commonly abbreviated as "grad," and the operation on the scalar f defined by the term gradient is indicated by the symbol $\boldsymbol{\nabla}$, called "del." Thus

$$\boldsymbol{A} = \text{grad} f \equiv \boldsymbol{\nabla} f. \tag{1-50}$$

In three dimensions the operator $\boldsymbol{\nabla}$ is defined as

$$\boldsymbol{\nabla} = \boldsymbol{i}\frac{\partial}{\partial x} + \boldsymbol{j}\frac{\partial}{\partial y} + \boldsymbol{k}\frac{\partial}{\partial z}. \tag{1-51}$$

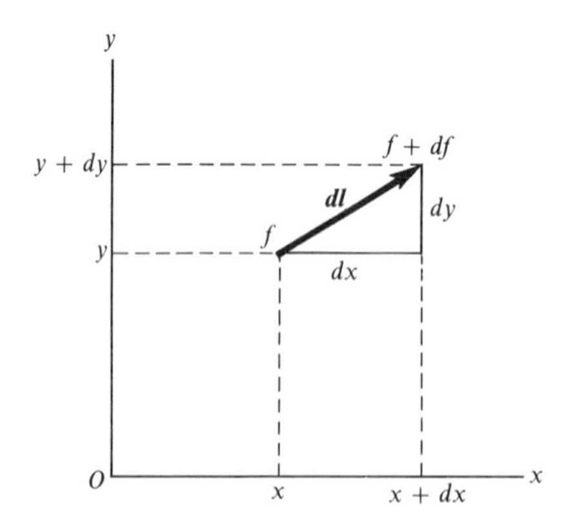

Figure 1-9. The quantity f, a function of position, changes from f to $f + df$ over the distance $\boldsymbol{dl}$.

The partial differentiations indicated are to be carried out on whatever scalar quantity stands to the right of the $\boldsymbol{\nabla}$ symbol. Thus

$$\boldsymbol{\nabla} f = \frac{\partial f}{\partial x}\boldsymbol{i} + \frac{\partial f}{\partial y}\boldsymbol{j} + \frac{\partial f}{\partial z}\boldsymbol{k}, \tag{1-52}$$

and

$$|\boldsymbol{\nabla} f| = \left[\left(\frac{\partial f}{\partial x}\right)^2 + \left(\frac{\partial f}{\partial y}\right)^2 + \left(\frac{\partial f}{\partial z}\right)^2\right]^{1/2}. \tag{1-53}$$

The magnitude of df can also be written as

$$df = \boldsymbol{\nabla} f \cdot \boldsymbol{dl} = |\boldsymbol{\nabla} f||\boldsymbol{dl}| \cos\theta, \tag{1-54}$$

where θ is the angle between the vectors $\boldsymbol{\nabla} f$ and $\boldsymbol{dl}$.

We can now ask what direction we should choose for $\boldsymbol{dl}$ in order that df shall be maximum. The answer is: the direction in which $\cos\theta = 1$ or $\theta = 0$, that is, the direction of $\boldsymbol{\nabla} f$. The gradient of f is thus a vector whose magnitude and direction are those of the maximum space rate of change of f.

To summarize, the gradient of a scalar function is a vector having the following properties.

(1) Its components at any point are the rates of change of the function along the directions of the coordinate axes at that point.

(2) Its magnitude at the point is the maximum rate of change of the function with distance.

(3) Its direction is that of the maximum rate of change of the function.

(4) It points toward *larger* values of the function.

The gradient is thus a vector point-function derived from a scalar point-function.

Example As an example of the gradient, consider Figure 1-10, in which E, the elevation above sea level, is a function of the x- and y- coordinates measured on a horizontal plane. Points of constant elevation may be joined together by contour lines. The gradient of the elevation E at a given point then has the following properties: (a) it is perpendicular to the contour line at that point, (b) its magnitude is equal to the maximum rate of change of elevation with displacement measured in a horizontal plane at that point, and (c) it points toward an increase in elevation.

Again we have two definitions: the gradient is a vector whose magnitude and direction are those of the maximum space rate of change of f, and it is also the vector of Eq. 1-52.

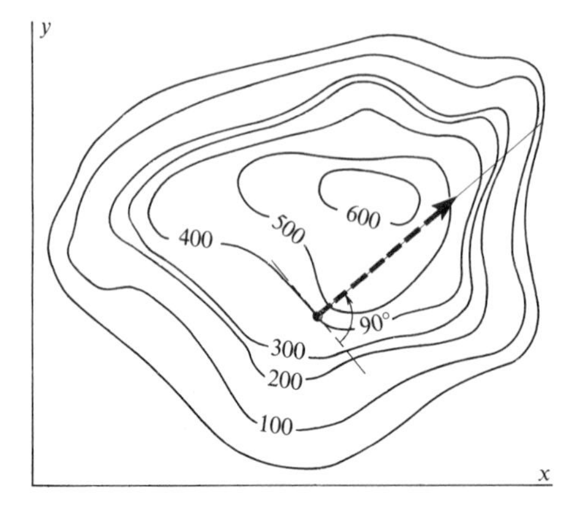

Figure 1-10. Topographic map of a hill. The numbers shown give the elevation E in meters. The gradient of E is the slope of the hill at the point considered, and it points toward an *in*crease in elevation: $\boldsymbol{\nabla} E = (\partial E/\partial x)\boldsymbol{i} + (\partial E/\partial y)\boldsymbol{j}$. The arrow shows $\boldsymbol{\nabla} E$ at one point where the elevation is 400 meters.

The latter definition will be invariant if the operator $\boldsymbol{\nabla}$ is invariant, that is, if

$$\boldsymbol{\nabla} = \frac{\partial}{\partial x}\boldsymbol{i} + \frac{\partial}{\partial y}\boldsymbol{j} + \frac{\partial}{\partial z}\boldsymbol{k} = \frac{\partial}{\partial x'}\boldsymbol{i}' + \frac{\partial}{\partial y'}\boldsymbol{j}' + \frac{\partial}{\partial z'}\boldsymbol{k}'. \tag{1-55}$$

The scalar quantity f is of course invariant.

Let us check whether $\boldsymbol{\nabla}$ is invariant under a rotation of axes. We have

$$\frac{\partial f}{\partial x} = \frac{\partial f}{\partial x'}\frac{\partial x'}{\partial x} + \frac{\partial f}{\partial y'}\frac{\partial y'}{\partial x} + \frac{\partial f}{\partial z'}\frac{\partial z'}{\partial x}, \tag{1-56}$$

or, from Eqs. 1-23 to 1-25,

$$\frac{\partial f}{\partial x} = \cos\theta\frac{\partial f}{\partial x'} - \sin\theta\frac{\partial f}{\partial y'}. \tag{1-57}$$

Thus

$$\frac{\partial}{\partial x} = \cos\theta\frac{\partial}{\partial x'} - \sin\theta\frac{\partial}{\partial y'}. \tag{1-58}$$

Similarly,

$$\frac{\partial}{\partial y} = \sin\theta\frac{\partial}{\partial x'} + \cos\theta\frac{\partial}{\partial y'}, \tag{1-59}$$

$$\frac{\partial}{\partial z} = \frac{\partial}{\partial z'}. \tag{1-60}$$

Using these last three equations, together with the equations of transformation for the unit vectors $\boldsymbol{i}, \boldsymbol{j}, \boldsymbol{k}$, we do find Eq. 1-55.

We have shown that the $\boldsymbol{\nabla}$ operator is invariant and therefore that the gradient $\boldsymbol{\nabla} f$ is also invariant.

1.5 FLUX AND DIVERGENCE. THE DIVERGENCE THEOREM

It is often necessary to calculate the flux of a vector through a surface. The flux $d\Phi$ of a vector $\boldsymbol{A}$ through an infinitesimal surface $\boldsymbol{da}$ is defined as

$$d\Phi = \boldsymbol{A}\cdot\boldsymbol{da}, \tag{1-61}$$

where the vector $\boldsymbol{da}$, which represents the element of area, is normal to its surface. The flux $d\Phi$ is the component of the vector normal to the surface, multiplied by da. For a finite surface we find the total flux by integrating $\boldsymbol{A}\cdot\boldsymbol{da}$ over the entire surface:

$$\Phi = \int_S \boldsymbol{A}\cdot\boldsymbol{da}. \tag{1-62}$$

It should be obvious that this integral is invariant since the scalar product $\boldsymbol{A}\cdot\boldsymbol{da}$ is invariant.

For a closed surface bounding a finite volume, the vector $\boldsymbol{da}$ is taken to point outward.

Example Let us consider fluid flow. We define a vector $\rho\boldsymbol{v}$, ρ being the fluid density and $\boldsymbol{v}$ the fluid velocity at a point. The flux of $\rho\boldsymbol{v}$ through any closed surface is the net rate at which mass leaves the volume bounded by the surface. In an incompressible fluid this flux is always equal to zero.

The outward flux of a vector through a closed surface can be calculated either from the above equation or as follows. Let us consider an infinitesimal volume dx, dy, dz and a vector $\boldsymbol{A}$, as in Figure 1-11, whose components A_x, A_y, A_z are functions of the coordinates x, y, z. We consider an infinitesimal volume and first-order variations of the vector $\boldsymbol{A}$.

The value of A_x at the center of the right-hand face can be taken to be the average value over the entire face. Through the right-hand face of the volume element, the outgoing flux is

$$d\Phi_R = \left(A_x + \frac{\partial A_x}{\partial x}\frac{dx}{2}\right) dy\, dz, \tag{1-63}$$

since the normal component of $\boldsymbol{A}$ at the right-hand face is the x-component of $\boldsymbol{A}$ at that face.

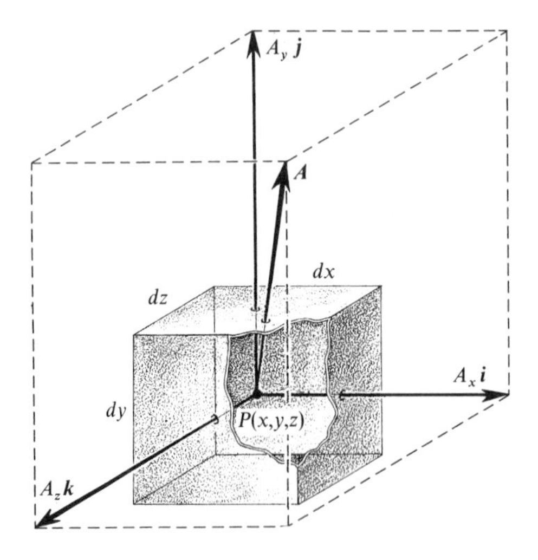

Figure 1-11. Element of volume dx, dy, dz around a point P, where the vector $\boldsymbol{A}$ has the value illustrated by the arrow.

At the left-hand face,

$$d\Phi_L = -\left(A_x - \frac{\partial A_x}{\partial x}\frac{dx}{2}\right) dy\, dz. \tag{1-64}$$

The minus sign before the parenthesis is necessary here because, $A_x\boldsymbol{i}$ being inward at this face and $\boldsymbol{da}$ being outward, the cosine of the angle between the two vectors is -1. The net outward flux through the two faces is then

$$d\Phi_R + d\Phi_L = \frac{\partial A_x}{\partial x} dx\, dy\, dz = \frac{\partial A_x}{\partial x} d\tau, \tag{1-65}$$

where $d\tau$ is the volume of the infinitesimal element.

If we calculate the net flux through the other pairs of faces in the same manner, we find the total outward flux for the element of volume $d\tau$ to be

$$d\Phi_{\text{tot}} = \left(\frac{\partial A_x}{\partial x} + \frac{\partial A_y}{\partial y} + \frac{\partial A_z}{\partial z}\right) d\tau. \tag{1-66}$$

Suppose now that we have two adjoining infinitesimal volume elements and that we add the flux through the bounding surface of the first volume to the flux through the bounding surface of the second. At the common face the fluxes are equal in magnitude but opposite in sign, and they cancel. The sum then, of the flux from the first volume plus that from the second is the flux through the bounding surface of the combined volumes. The total outward flux is

$$d\Phi_{1+2} = \left(\frac{\partial A_x}{\partial x} + \frac{\partial A_y}{\partial y} + \frac{\partial A_z}{\partial z}\right)_1 d\tau_1 + \left(\frac{\partial A_x}{\partial x} + \frac{\partial A_y}{\partial y} + \frac{\partial A_z}{\partial z}\right)_2 d\tau_2. \tag{1-67}$$

To extend this calculation to a finite volume, we sum the individual fluxes for each of the infinitesimal volume elements in the finite volume and the total outward flux is

$$\Phi_{\text{tot}} = \int_\tau \left(\frac{\partial A_x}{\partial x} + \frac{\partial A_y}{\partial y} + \frac{\partial A_z}{\partial z}\right) d\tau. \tag{1-68}$$

At any given point in the volume, the quantity

$$\left(\frac{\partial A_x}{\partial x} + \frac{\partial A_y}{\partial y} + \frac{\partial A_z}{\partial z}\right)$$

is thus the *outgoing* flux per unit volume. We call this the *divergence* of the vector $\boldsymbol{A}$ at the point.

The divergence of a vector point-function is a scalar point-function.

According to the rules for the scalar product, we write the divergence of $\boldsymbol{A}$ as

$$\boldsymbol{\nabla}\cdot\boldsymbol{A} = \frac{\partial A_x}{\partial x} + \frac{\partial A_y}{\partial y} + \frac{\partial A_z}{\partial z} \tag{1-69}$$

where the operator $\boldsymbol{\nabla}$ is as defined in Eq. 1-51.

The divergence of a vector as defined above must be invariant since, as we have shown, both the operator $\boldsymbol{\nabla}$ and the scalar product are invariant. We can also check the invariance of $\boldsymbol{\nabla}\cdot\boldsymbol{A}$ by substituting primed quantities for A_x, A_y, A_z, $\partial/\partial x$, $\partial/\partial y$, $\partial/\partial z$, using the transformation equations 1-33 to 1-35 and 1-58 to 1-60. This demonstration is formally identical to that of the invariance of the scalar product.

The operator $\boldsymbol{\nabla}\cdot$ has physical meaning, not by itself, but only as it operates on a function standing to the right of it. The symbol $\boldsymbol{\nabla}$ is not a vector symbol, of course, but it is convenient to use the notation of the scalar product to indicate the operation that is to be carried out.

In Eq. 1-68 the total outward flux is also equal to the surface integral of the normal outward component of $\boldsymbol{A}$, thus

$$\int_S \boldsymbol{A}\cdot d\boldsymbol{a} = \int_\tau \left(\frac{\partial A_x}{\partial x} + \frac{\partial A_y}{\partial y} + \frac{\partial A_z}{\partial z}\right) d\tau = \int_\tau \boldsymbol{\nabla}\cdot\boldsymbol{A}\, d\tau. \tag{1-70}$$

This is the *divergence theorem*. Note that the left-hand side involves only the values of $\boldsymbol{A}$ on the surface S, whereas the right-hand side involves the values of $\boldsymbol{A}$ throughout the volume τ enclosed by S. This is also called *Green's theorem*.

If the volume τ is allowed to shrink sufficiently, so that $\boldsymbol{\nabla}\cdot\boldsymbol{A}$ does not vary appreciably over it, then

$$\int_S \boldsymbol{A}\cdot d\boldsymbol{a} = (\boldsymbol{\nabla}\cdot\boldsymbol{A})\tau, \tag{1-71}$$

and the divergence can therefore be defined as

$$\boldsymbol{\nabla}\cdot\boldsymbol{A} = \lim_{\tau\to 0}\frac{1}{\tau}\int_S \boldsymbol{A}\cdot d\boldsymbol{a}. \tag{1-72}$$

As we have seen, the divergence is the *outward* flux per unit volume as the volume τ approaches zero.

Examples In an incompressible fluid, $\boldsymbol{\nabla}\cdot(\rho\boldsymbol{v})$ is everywhere equal to zero since the outward mass flux per unit volume is zero.

Within an explosion, $\boldsymbol{\nabla}\cdot(\rho\boldsymbol{v})$ is positive.

1.6 LINE INTEGRAL AND CURL

The integrals

$$\int_a^b \boldsymbol{A}\cdot d\boldsymbol{l}, \qquad \int_a^b \boldsymbol{A}\times d\boldsymbol{l}, \qquad \text{and} \qquad \int_a^b f\, d\boldsymbol{l},$$

evaluated from the point a to the point b on some specified curve, are examples of line integrals.

In the first one, which is especially important, each element of length $d\boldsymbol{l}$ on the curve is multiplied by the local value of $\boldsymbol{A}$ according to the rule for the scalar product. These products are then summed to obtain the value of the integral.

Example The work W done by a force $\boldsymbol{F}$ acting from a to b along some specified path is

$$W = \int_a^b \boldsymbol{F}\cdot d\boldsymbol{l}, \tag{1-73}$$

where both $\boldsymbol{F}$ and $d\boldsymbol{l}$ must of course be known functions of the coordinates if the integral is to be evaluated analytically. Let us calculate the work done by a force $\boldsymbol{F}$, which is in the y direction and has a magnitude proportional to y, as it moves around the circular path from a to b in Figure 1-12. Since

$$\boldsymbol{F} = ky\,\boldsymbol{j} \qquad \text{and} \qquad d\boldsymbol{l} = dx\,\boldsymbol{i} + dy\,\boldsymbol{j},$$

$$W = \int_a^b \boldsymbol{F}\cdot d\boldsymbol{l} = \int_0^r ky\,dy = \frac{kr^2}{2}. \tag{1-74}$$

A vector field $\boldsymbol{A}$ is said to be *conservative* if the line integral of $\boldsymbol{A}\cdot d\boldsymbol{l}$ around any closed curve is zero:

$$\oint \boldsymbol{A}\cdot d\boldsymbol{l} = 0 \tag{1-75}$$

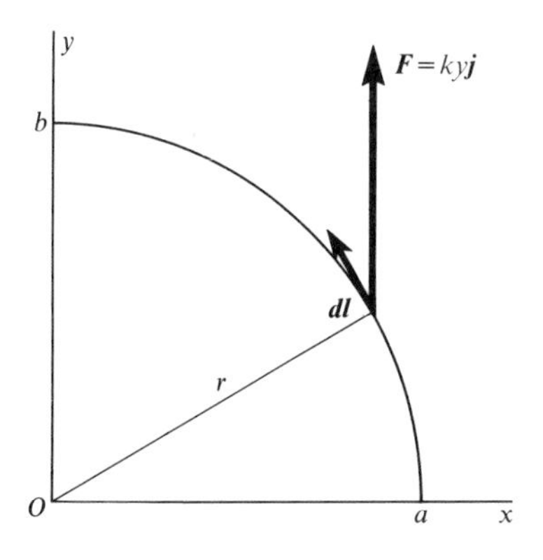

Figure 1-12. The force $\boldsymbol{F}$ is proportional to y, and its point of application moves from a to b. The work done is given by the line integral of $\boldsymbol{F}\cdot\boldsymbol{dl}$ over the circular path.

The circle on the integral sign indicates that the path of integration is closed.

Let us calculate the value of the above integral in the more general case where it is not equal to zero.

For an infinitesimal element of path $\boldsymbol{dl}$ in the xy-plane, and from the definition of the scalar product,

$$\boldsymbol{A}\cdot\boldsymbol{dl} = A_x\,dx + A_y\,dy. \tag{1-76}$$

Thus, for any closed path in the xy-plane and for any $\boldsymbol{A}$,

$$\oint \boldsymbol{A}\cdot\boldsymbol{dl} = \oint A_x\,dx + \oint A_y\,dy. \tag{1-77}$$

Now consider the infinitesimal path in Figure 1-13. There are two contributions to the first integral on the right-hand side of the above equation, one at $y - (dy/2)$ and one at $y + (dy/2)$:

$$\oint A_x\,dx = \left(A_x - \frac{\partial A_x}{\partial y}\frac{dy}{2}\right)dx - \left(A_x + \frac{\partial A_x}{\partial y}\frac{dy}{2}\right)dx. \tag{1-78}$$

There is a minus sign before the second term because the path element at $y + (dy/2)$ is in the negative x direction. Therefore

$$\oint A_x\,dx = -\frac{\partial A_x}{\partial y}\,dy\,dx. \tag{1-79}$$

Similarly,

$$\oint A_y\,dy = \frac{\partial A_y}{\partial x}\,dx\,dy, \tag{1-80}$$

and

$$\oint \boldsymbol{A}\cdot\boldsymbol{dl} = \left(\frac{\partial A_y}{\partial x} - \frac{\partial A_x}{\partial y}\right)dx\,dy \tag{1-81}$$

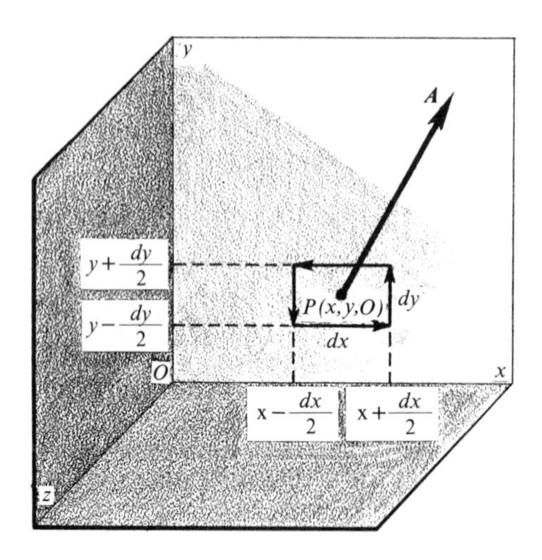

Figure 1-13. Closed, rectangular path in the xy-plane, centered on the point $P(x, y, 0)$ where the vector $\boldsymbol{A}$ has the value illustrated by the heavy arrow. The integration around the path is performed in the direction of the light arrows.

for the infinitesimal path of Figure 1-13.

If we set

$$g_3 = \frac{\partial A_y}{\partial x} - \frac{\partial A_x}{\partial y}, \tag{1-82}$$

then

$$\oint \boldsymbol{A} \cdot d\boldsymbol{l} = g_3\, da, \tag{1-83}$$

where $da = dx\, dy$ is the area enclosed on the xy-plane by the infinitesimal path. Note here that this is correct only if the line integral is evaluated in the positive direction in the xy-plane, that is, in the direction in which one would have to turn a right-hand screw to make it advance in the positive direction along the z-axis.

Let us now consider g_3 and the other two symmetrical quantities as the components of a vector

$$\left(\frac{\partial A_z}{\partial y} - \frac{\partial A_y}{\partial z}\right) \boldsymbol{i} + \left(\frac{\partial A_x}{\partial z} - \frac{\partial A_z}{\partial x}\right) \boldsymbol{j} + \left(\frac{\partial A_y}{\partial x} - \frac{\partial A_x}{\partial y}\right) \boldsymbol{k}, \tag{1-84}$$

which may be written as

$$\boldsymbol{\nabla} \times \boldsymbol{A} = \begin{vmatrix} \boldsymbol{i} & \boldsymbol{j} & \boldsymbol{k} \\ \dfrac{\partial}{\partial x} & \dfrac{\partial}{\partial y} & \dfrac{\partial}{\partial z} \\ A_x & A_y & A_z \end{vmatrix}. \tag{1-85}$$

We shall call this vector the *curl* of $\boldsymbol{A}$. The quantity g_3 is then its z component.

If we now consider the element of area as a vector $\boldsymbol{da}$ pointing in the direction of advance of a right-hand screw turned in the direction chosen for the line integral, then

$$\boldsymbol{da} = da\,\boldsymbol{k} \tag{1-86}$$

and

$$\oint \boldsymbol{A}\cdot\boldsymbol{dl} = (\boldsymbol{\nabla}\times\boldsymbol{A})\cdot\boldsymbol{da}. \tag{1-87}$$

This means that the line integral of $\boldsymbol{A}\cdot\boldsymbol{dl}$ around the edge of an element of area $\boldsymbol{da}$ is equal to the scalar product of the curl of $\boldsymbol{A}$ by this element of area, as long as we observe the above sign convention.

We have arrived at this result for an element of area $dx\,dy$ in the xy plane. Is this result general? Does it apply to any small area, whatever the orientation with respect to the coordinate axes? It does *if* it is invariant. We have already seen that the scalar product is invariant. Thus the above line integral is also invariant. We have also seen that the $\boldsymbol{\nabla}$ operator and the vector product are invariant. If you wish, you can check to see whether $\boldsymbol{\nabla}\times\boldsymbol{A}$ really is invariant. Equation 1-87 does apply to any element of area $\boldsymbol{da}$ and

$$(\boldsymbol{\nabla}\times\boldsymbol{A})_n = \lim_{S\to 0}\frac{1}{S}\oint \boldsymbol{A}\cdot\boldsymbol{dl}: \tag{1-88}$$

the component of the curl of a vector normal to a surface S is equal to the line integral of the vector around the boundary of the surface divided by the area of the surface when this area approaches zero.

Example Let us consider a fluid stream in which the velocity $\boldsymbol{v}$ is proportional to the distance from the bottom. We set the z-axis parallel to the direction of flow, and the x-axis perpendicular to the stream bottom, as in Figure 1-14. Then

$$v_x = 0, \qquad v_y = 0, \qquad v_z = cx. \tag{1-89}$$

We shall calculate the curl from Eq. 1-88. For $(\boldsymbol{\nabla}\times\boldsymbol{v})_x$ we choose a path parallel to the yz-plane. In evaluating

$$\oint \boldsymbol{v}\cdot\boldsymbol{dl}$$

around such a path we note that the contributions are equal and opposite on the parts parallel to the z-axis, hence $(\boldsymbol{\nabla}\times\boldsymbol{v})_x = 0$. Likewise, $(\boldsymbol{\nabla}\times\boldsymbol{v})_z = 0$.

For the y-component we choose a path parallel to the xz-plane and evaluate the integral around it in the sense that would advance a right-hand screw in the positive y direction. On the parts of the path parallel to the x-axis, $\boldsymbol{v}\cdot\boldsymbol{dl} = 0$ since $\boldsymbol{v}$ and $\boldsymbol{dl}$ are perpendicular.

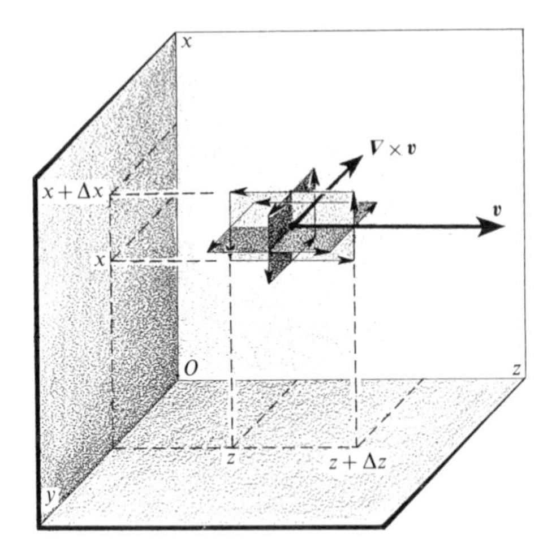

Figure 1-14. The velocity $\boldsymbol{v}$ in a viscous fluid is assumed to be in the direction of the z-axis and proportional to the distance x from the bottom. Then $\boldsymbol{\nabla} \times \boldsymbol{v} = -c\boldsymbol{j}$.

On the bottom part of the path, at a distance x from the yz-plane,

$$\int_z^{z+\Delta z} \boldsymbol{v} \cdot \boldsymbol{dl} = cx\,\Delta z, \tag{1-90}$$

whereas at $(x + \Delta x)$

$$\int_z^{z+\Delta z} \boldsymbol{v} \cdot \boldsymbol{dl} = -c(x + \Delta x)\,\Delta z. \tag{1-91}$$

For the whole path,

$$\oint \boldsymbol{v} \cdot \boldsymbol{dl} = -c\,\Delta x\,\Delta z, \tag{1-92}$$

and the y-component of the curl is

$$(\boldsymbol{\nabla} \times \boldsymbol{v})_y = \lim_{S \to 0} \frac{\oint \boldsymbol{v} \cdot \boldsymbol{dl}}{S} = \frac{-c\,\Delta x\,\Delta z}{\Delta x\,\Delta z} = -c. \tag{1-93}$$

Calculating $\boldsymbol{\nabla} \times \boldsymbol{v}$ directly from Eq. 1-85,

$$\boldsymbol{\nabla} \times \boldsymbol{v} = \begin{vmatrix} \boldsymbol{i} & \boldsymbol{j} & \boldsymbol{k} \\ \dfrac{\partial}{\partial x} & \dfrac{\partial}{\partial y} & \dfrac{\partial}{\partial z} \\ 0 & 0 & cx \end{vmatrix} = -c\boldsymbol{j}, \tag{1-94}$$

which is the same result as above.

We have used the operator $\boldsymbol{\nabla}$ for the gradient of a scalar point-function and for the divergence and curl of a vector point-function. In all three cases, $\boldsymbol{\nabla}$ is defined by a single expression, and we obtain the gradient, the divergence, or the curl by performing the appropriate multiplication. This relatively simple situation is peculiar to the Cartesian coordinate system. As we shall see later on, other coordinate systems do not permit a single definition for the operator

∇ but lead to more complicated expressions for the gradient, the divergence, and the curl.

1.7 STOKES'S THEOREM

Equation 1-87 is true only for a path so small that $\nabla \times \boldsymbol{A}$ can be considered constant over the surface $\boldsymbol{da}$ bounded by the path. What if the path is so large that this condition is not met? The equation can be extended readily to arbitrary paths. We divide the surface—any surface bounded by the path of integration in question—into elements of area $\boldsymbol{da}_1$, $\boldsymbol{da}_2$, and so forth, as in Figure 1-15. For any one of these small areas,

$$\oint \boldsymbol{A} \cdot \boldsymbol{dl}_i = (\nabla \times \boldsymbol{A}) \cdot \boldsymbol{da}_i. \tag{1-95}$$

We add the left-hand sides of these equations for all the $\boldsymbol{da}$'s, and then we add all the right-hand sides. The sum of the left-hand sides is the line integral around the external boundary, since there are always two equal and opposite contributions to the sum along every common side between adjacent $\boldsymbol{da}$'s. The sum of the right-hand sides is merely the integral of $(\nabla \times \boldsymbol{A}) \cdot \boldsymbol{da}$ over the finite surface. Thus, for an *arbitrary* path,

$$\oint \boldsymbol{A} \cdot \boldsymbol{dl} = \int_S (\nabla \times \boldsymbol{A}) \cdot \boldsymbol{da}. \tag{1-96}$$

This is *Stokes's theorem.* It relates a line integral to a surface integral over any surface bounded by the line integral path.

Figure 1-15 illustrates the sign convention.

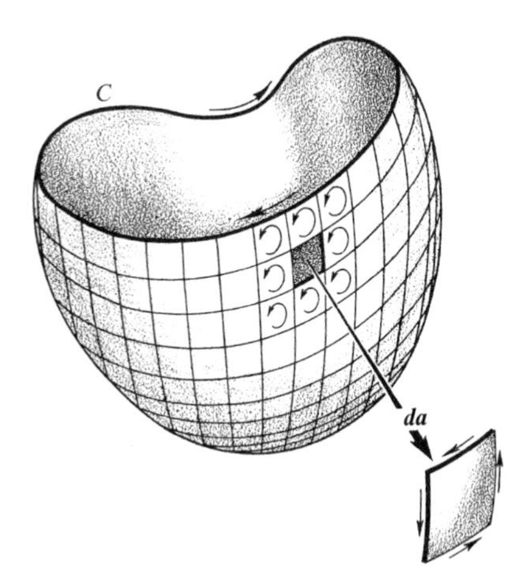

Figure 1-15. An arbitrary surface bounded by the curve C. The sum of the line integrals around the curvilinear squares shown is equal to the line integral around C.

Example Under what condition is a vector field conservative? In other words, under what condition is the line integral of $\boldsymbol{A}\cdot d\boldsymbol{l}$ around an arbitrary closed path zero? From Stokes's theorem, the line integral of $\boldsymbol{A}\cdot d\boldsymbol{l}$ around an arbitrary closed path is zero if $\boldsymbol{\nabla}\times\boldsymbol{A} = 0$ everywhere. This condition can be met if

$$\boldsymbol{A} = \boldsymbol{\nabla} f. \tag{1-97}$$

For then

$$A_x = \frac{\partial f}{\partial x}, \qquad A_y = \frac{\partial f}{\partial y}, \qquad A_z = \frac{\partial f}{\partial z}, \tag{1-98}$$

and

$$(\boldsymbol{\nabla}\times\boldsymbol{A})_x = \frac{\partial A_z}{\partial y} - \frac{\partial A_y}{\partial z} = \frac{\partial^2 f}{\partial y\,\partial z} - \frac{\partial^2 f}{\partial z\,\partial y} \equiv 0, \tag{1-99}$$

and so on for the other components of the curl.

The field of $\boldsymbol{A}$ is therefore conservative if $\boldsymbol{A}$ can be expressed as the gradient of some scalar function f.

1.8 THE LAPLACIAN

The divergence of the gradient is of great importance in electromagnetic theory, as well as in many other parts of physics. Since

$$\boldsymbol{\nabla} f = \frac{\partial f}{\partial x}\boldsymbol{i} + \frac{\partial f}{\partial y}\boldsymbol{j} + \frac{\partial f}{\partial z}\boldsymbol{k}, \tag{1-100}$$

then

$$\boldsymbol{\nabla}\cdot\boldsymbol{\nabla} f = \boldsymbol{\nabla}^2 f = \frac{\partial^2 f}{\partial x^2} + \frac{\partial^2 f}{\partial y^2} + \frac{\partial^2 f}{\partial z^2}. \tag{1-101}$$

The divergence of the gradient is the sum of the second derivatives with respect to the rectangular coordinates. The quantity $\boldsymbol{\nabla}\cdot\boldsymbol{\nabla} f$ is abbreviated to $\boldsymbol{\nabla}^2 f$, and is called the *Laplacian* of f. The operator $\boldsymbol{\nabla}^2$ is called the *Laplace operator*.

The Laplacian is invariant because it is the result of two successive invariant operations.

Example We shall show in the next chapter that the Laplacian of the electrostatic potential is equal to zero in regions where there is zero space charge.

We have defined the Laplacian of a scalar point-function f. It is also useful to define the Laplacian of a vector point-function $\boldsymbol{A}$:

$$\boldsymbol{\nabla}^2\boldsymbol{A} = \boldsymbol{\nabla}^2 A_x\,\boldsymbol{i} + \boldsymbol{\nabla}^2 A_y\,\boldsymbol{j} + \boldsymbol{\nabla}^2 A_z\,\boldsymbol{k}, \tag{1-102}$$

The Laplacian of a vector is also invariant.

1.9 CURVILINEAR COORDINATES

It is frequently convenient, because of the symmetries that exist in certain fields, to use coordinate systems other than the rectangular one. Of all the other possible coordinate systems, we shall restrict our discussion to cylindrical and spherical polar coordinates, the two most commonly used.

We could calculate the gradient, the divergence, and so on, directly in both cylindrical and spherical coordinates, but it is less tedious and more useful to introduce first the idea of orthogonal curvilinear coordinates.

Consider the equation

$$f(x, y, z) = q, \tag{1-103}$$

in which q is a constant. This equation determines a family of surfaces in space, each member of the family being characterized by a particular value of the parameter q. An obvious example is $x = q$, which determines the surfaces parallel to the yz-plane in Cartesian coordinates.

Consider now the three equations

$$\begin{aligned} f_1(x, y, z) &= q_1, \\ f_2(x, y, z) &= q_2, \\ f_3(x, y, z) &= q_3, \end{aligned} \tag{1-104}$$

which are chosen so that the three families of surfaces are mutually perpendicular, or orthogonal. A point in space can then be defined as the intersection of three of these surfaces, one of each family; the point is completely defined if we state the values of q_1, q_2, q_3 corresponding to these three surfaces. The variables q_1, q_2, q_3 are called the *curvilinear coordinates* of the point, as in Figure 1-16.

Let us call dl_1 an element of length *perpendicular to the surface* q_1. This

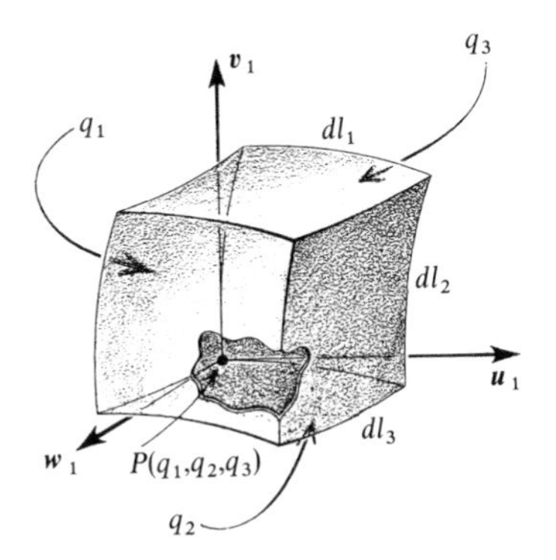

Figure 1-16. Element of volume in curvilinear coordinates. The unit vectors $\boldsymbol{u}_1$, $\boldsymbol{v}_1$, $\boldsymbol{w}_1$ are respectively normal to the q_1, q_2, q_3 surfaces at the point $P(q_1, q_2, q_3)$. They are mutually perpendicular and are oriented in such a way that $\boldsymbol{u}_1 \times \boldsymbol{v}_1 = \boldsymbol{w}_1$.

element of length is the distance between the surfaces q_1 and $q_1 + dq_1$ in the infinitesimal region considered. The element of length dl_1 is related to dq_1 by the equation

$$dl_1 = h_1\, dq_1, \tag{1-105}$$

in which h_1 is, in general, a function of the coordinates q_1, q_2, q_3. Similarly,

$$dl_2 = h_2\, dq_2, \tag{1-106}$$

$$dl_3 = h_3\, dq_3. \tag{1-107}$$

In the Cartesian system of coordinates, h_1, h_2, h_3 are all unity.

The *unit vectors* $\boldsymbol{u}_1, \boldsymbol{v}_1, \boldsymbol{w}_1$, are of unit length, normal respectively to the q_1, q_2, q_3 surfaces, and oriented toward increasing values of these coordinates. They are chosen so that $\boldsymbol{u}_1 \times \boldsymbol{v}_1 = \boldsymbol{w}_1$.

The orientation of the three unit vectors depends, in general, on the point of space considered. Only in rectangular coordinates do they all have fixed directions.

The volume element is

$$d\tau = dl_1\, dl_2\, dl_3 = h_1\, h_2\, h_3 (dq_1\, dq_2\, dq_3). \tag{1-108}$$

We can now find the q's, the h's, the elements of length, and the elements of volume for cylindrical and spherical coordinates.

1.9.1 Cylindrical Coordinates

In cylindrical coordinates, as in Figure 1-17, the position of any point P in space is specified by ρ, φ, z. The coordinate ρ is the perpendicular distance from the z-axis; φ is the azimuth angle of the plane containing P and the

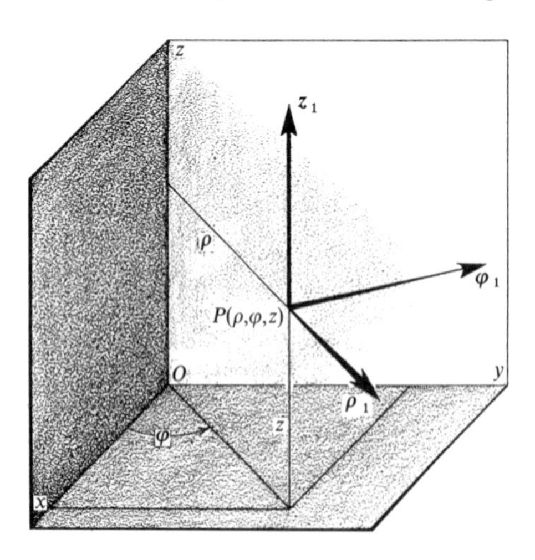

Figure 1-17. Cylindrical coordinate system.

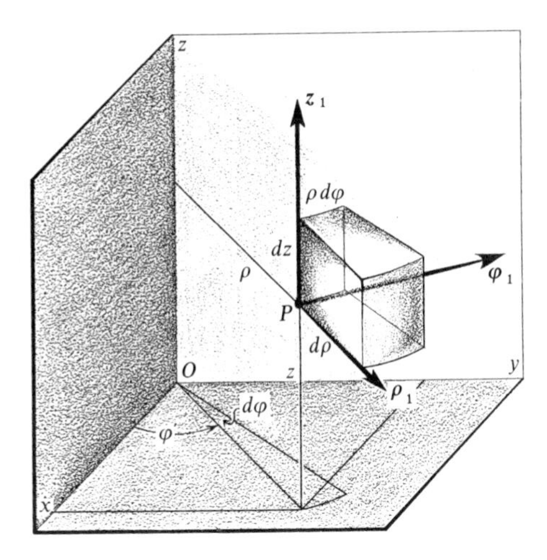

Figure 1-18. Element of volume in cylindrical coordinates.

z-axis, measured from the xz-plane in the right-hand screw sense; and z is the distance from the xy-plane. The q's are thus ρ, φ, z in this case.

At the point P there are three mutually orthogonal directions specified by three unit vectors: $\boldsymbol{\rho}_1$ in the direction of the perpendicular from the z-axis extended through P; $\boldsymbol{\varphi}_1$ perpendicular to the plane containing the z-axis and P in the direction corresponding to increasing φ; and $\boldsymbol{z}_1$ in the positive z direction. These unit vectors do *not* maintain the same directions in space as the point P moves about, but they always remain mutually orthogonal.

The vector describing the position of P is

$$\boldsymbol{r} = \rho\,\boldsymbol{\rho}_1 + z\,\boldsymbol{z}_1 \tag{1-109}$$

Note that the angle φ does not appear explicitly on the right-hand side; it is given by the orientation of the unit vector $\boldsymbol{\rho}_1$.

Elements of length corresponding to infinitesimal changes in the coordinates of a point are important. If the coordinates φ and z of the point P are kept constant while ρ is allowed to increase by $d\rho$, P is displaced by $d\boldsymbol{r} = d\rho\,\boldsymbol{\rho}_1$. On the other hand, if ρ and z are held constant while φ is allowed to increase by $d\varphi$, then P is displaced by $d\boldsymbol{r} = \rho d\varphi\,\boldsymbol{\varphi}_1$. Finally, if ρ and φ are held constant while z is allowed to increase by dz, then $d\boldsymbol{r} = dz\,\boldsymbol{z}_1$. For arbitrary increments $d\rho$, $d\varphi$, dz,

$$d\boldsymbol{r} = d\rho\,\boldsymbol{\rho}_1 + \rho d\varphi\,\boldsymbol{\varphi}_1 + dz\,\boldsymbol{z}_1 \tag{1-110}$$

and

$$dr = [(d\rho)^2 + (\rho\,d\varphi)^2 + (dz)^2]^{1/2}. \tag{1-111}$$

Figure 1-18 shows the volume element whose edges are the elements of length corresponding to infinitesimal increments in the coordinates at the point P of Figure 1-17. The infinitesimal volume is

$$d\tau = \rho\, d\rho\, d\varphi\, dz. \tag{1-112}$$

1.9.2 Spherical Coordinates

In spherical coordinates the position of a point P is specified by r, θ, φ, where r is the distance from the origin, θ the angle between the z-axis and the radius vector, and φ the azimuthal angle. At the point P the unit vectors are shown in Figure 1-19: $\boldsymbol{r}_1$ is in the direction of the radius vector extended through P, $\boldsymbol{\theta}_1$ is perpendicular to the radius vector in the plane containing the z-axis and the radius vector, and $\boldsymbol{\varphi}_1$ is perpendicular to that plane. Again, these unit vectors do not maintain the same directions in space as the point P moves about.

The vector $\boldsymbol{r}$ describing the position of P is now simply $\boldsymbol{r} = r\boldsymbol{r}_1$, the coordinates θ and φ being determined by the orientation of the unit vector $\boldsymbol{r}_1$.

The distance element $\boldsymbol{dl}$ corresponding to arbitrary increments of the coordinates is

$$\boldsymbol{dl} = dr\, \boldsymbol{r}_1 + rd\theta\, \boldsymbol{\theta}_1 + r \sin\theta d\varphi\, \boldsymbol{\varphi}_1, \tag{1-113}$$

$$dl = [(dr)^2 + r^2(d\theta)^2 + r^2 \sin^2\theta(d\varphi)^2]^{1/2}. \tag{1-114}$$

The volume element at point P is shown in Figure 1-20, and

$$d\tau = r^2 \sin\theta\, dr\, d\theta\, d\varphi. \tag{1-115}$$

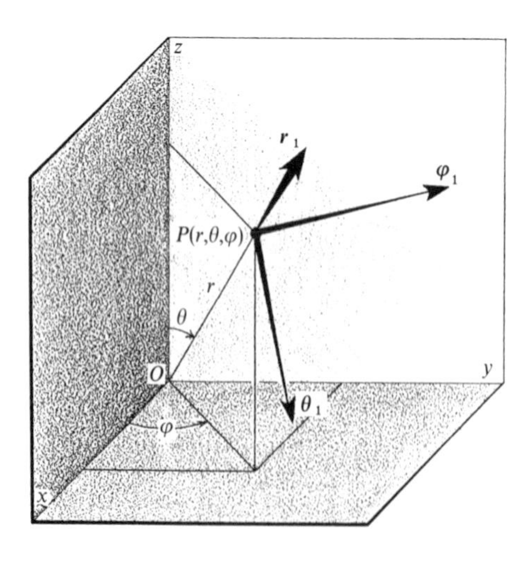

Figure 1-19. Spherical coordinate system.

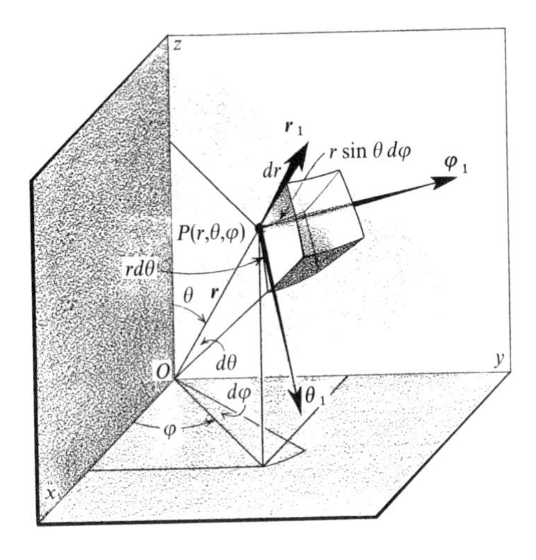

Figure 1-20. Element of volume in spherical coordinates.

The table below shows the correspondence of curvilinear coordinates to Cartesian, cylindrical, and spherical coordinates.

Curvilinear	*Cartesian*	*Cylindrical*	*Spherical*
q_1	x	ρ	r
q_2	y	φ	θ
q_3	z	z	φ
h_1	1	1	1
h_2	1	ρ	r
h_3	1	1	$r \sin\theta$
$\boldsymbol{u}_1$	$\boldsymbol{i}$	$\boldsymbol{\rho}_1$	$\boldsymbol{r}_1$
$\boldsymbol{v}_1$	$\boldsymbol{j}$	$\boldsymbol{\varphi}_1$	$\boldsymbol{\theta}_1$
$\boldsymbol{w}_1$	$\boldsymbol{k}$	$\boldsymbol{z}_1$	$\boldsymbol{\varphi}_1$

It should be noted that the angle φ in both cylindrical and spherical coordinates is *un*defined for points on the z-axis.

We are now in a position to find the gradient, divergence, and curl operators in curvilinear coordinates. Once these are found, it will be a simple matter to find the operators in cylindrical and spherical coordinates, for then we need only substitute the appropriate values of q_1, q_2, q_3, h_1, h_2, h_3, $\boldsymbol{u}_1$, $\boldsymbol{v}_1$, $\boldsymbol{w}_1$.

1.9.3 The Gradient

To find the form of the gradient, we require the rate of change of a scalar function f in each of the three coordinate directions:

$$\nabla f = \frac{\partial f}{\partial l_1} \boldsymbol{u}_1 + \frac{\partial f}{\partial l_2} \boldsymbol{v}_1 + \frac{\partial f}{\partial l_3} \boldsymbol{w}_1, \tag{1-116}$$

$$= \frac{1}{h_1} \frac{\partial f}{\partial q_1} \boldsymbol{u}_1 + \frac{1}{h_2} \frac{\partial f}{\partial q_2} \boldsymbol{v}_1 + \frac{1}{h_3} \frac{\partial f}{\partial q_3} \boldsymbol{w}_1, \tag{1-117}$$

where $\boldsymbol{u}_1$, $\boldsymbol{v}_1$, $\boldsymbol{w}_1$ are the unit vectors as in Figure 1-16.

In cylindrical coordinates,

$$\nabla f = \frac{\partial f}{\partial \rho} \boldsymbol{\rho}_1 + \frac{1}{\rho} \frac{\partial f}{\partial \varphi} \boldsymbol{\varphi}_1 + \frac{\partial f}{\partial z} \boldsymbol{z}_1, \tag{1-118}$$

while, in spherical coordinates,

$$\nabla f = \frac{\partial f}{\partial r} \boldsymbol{r}_1 + \frac{1}{r} \frac{\partial f}{\partial \theta} \boldsymbol{\theta}_1 + \frac{1}{r \sin \theta} \frac{\partial f}{\partial \varphi} \boldsymbol{\varphi}_1. \tag{1-119}$$

On the z-axis, ρ and $\sin \theta$ are both zero and neither of these expressions is valid.

We must ask ourselves again whether or not these expressions are invariant. Let us just check the one for cylindrical coordinates.

Let us first check whether it is invariant to a change from cylindrical coordinates to Cartesian coordinates. You can do this yourself by using the transformation equations

$$x = \rho \cos \varphi, \qquad y = \rho \sin \varphi, \qquad z = z, \tag{1-120}$$

$$\boldsymbol{\rho}_1 = \cos \varphi\, \boldsymbol{i} + \sin \varphi\, \boldsymbol{j}, \qquad \boldsymbol{\varphi}_1 = -\sin \varphi\, \boldsymbol{i} + \cos \varphi\, \boldsymbol{j}, \qquad \boldsymbol{z}_1 = \boldsymbol{k}, \tag{1-121}$$

You will find that ∇f in cylindrical coordinates does transform into the proper expression for ∇f in Cartesian coordinates, so that ∇f is invariant to a change from cylindrical to Cartesian coordinates. The inverse is equally true.

Now what about invariance to a change from one system of cylindrical coordinates S_1 to another system of cylindrical coordinates S_2? The gradient is also invariant under this transformation for the following reason. Call C_1 and C_2 the corresponding Cartesian systems. We have just shown that we have invariance in going from S_1 to C_1. Now we also have invariance in going from C_1 to C_2, as we showed earlier in Section 1.4, and invariance in going from C_2 to S_2. It follows that the gradient is invariant if one changes from S_1 to S_2.

The gradient expressed in spherical coordinates is similarly invariant to a change from a spherical system of reference to either a Cartesian system, or to a cylindrical system, or to a different spherical system.

1.9.4 The Divergence

To find the divergence, we consider the volume element of Figure 1-21. The quantity A_1 is the q_1 component of the vector $\boldsymbol{A}$ at the center of the volume

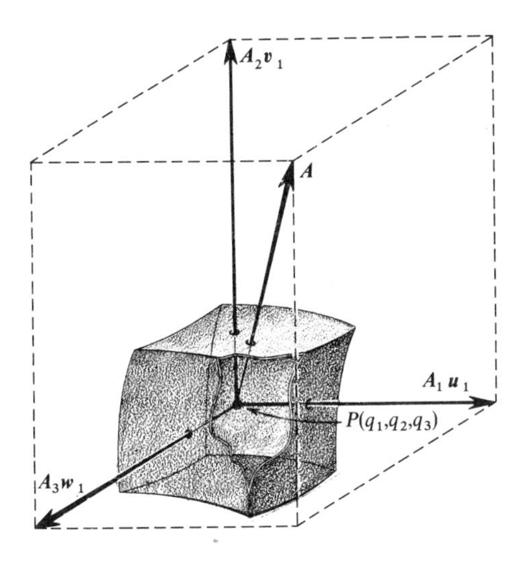

Figure 1-21. Element of volume in curvilinear coordinates centered on the point $P(q_1, q_2, q_3)$, where the vector A has the value shown by the arrow.

element; h_1, h_2, h_3 are the h values at the center. Then the outward flux through the left-hand face is

$$d\Phi_L = -A_{1L}\, h_{2L}\, h_{3L}\, dq_2\, dq_3, \tag{1-122}$$

$$= -\left(A_1 - \frac{\partial A_1}{\partial q_1}\frac{dq_1}{2}\right)\left(h_2 - \frac{\partial h_2}{\partial q_1}\frac{dq_1}{2}\right)\left(h_3 - \frac{\partial h_3}{\partial q_1}\frac{dq_1}{2}\right) dq_2\, dq_3. \tag{1-123}$$

It must be remembered that h_2 and h_3 may be functions of q_1, as well as A_1. If we neglect differentials of order higher than the third, then

$$d\Phi_L = -A_1 h_2 h_3\, dq_2\, dq_3 + \frac{\partial}{\partial q_1}(A_1 h_2 h_3)\frac{dq_1}{2}\, dq_2\, dq_3. \tag{1-124}$$

By a similar argument,

$$d\Phi_R = A_1 h_2 h_3\, dq_2\, dq_3 + \frac{\partial}{\partial q_1}(A_1 h_2 h_3)\frac{dq_1}{2}\, dq_2\, dq_3 \tag{1-125}$$

for the right-hand face. The net flux through this pair of faces is then

$$d\Phi_{LR} = \frac{\partial}{\partial q_1}(A_1 h_2 h_3)\, dq_1\, dq_2\, dq_3. \tag{1-126}$$

The same calculation can be repeated for the other pairs of faces to find the net outward flux through the bounding surface. Dividing by the volume of the element then gives the divergence. Since we have considered only differentials up to the third order, we have already passed to the limit $\tau \to 0$, and

$$\boldsymbol{\nabla}\cdot\boldsymbol{A} = \frac{1}{h_1 h_2 h_3}\left[\frac{\partial}{\partial q_1}(A_1 h_2 h_3) + \frac{\partial}{\partial q_2}(A_2 h_3 h_1) + \frac{\partial}{\partial q_3}(A_3 h_1 h_2)\right]. \tag{1-127}$$

Substituing the h's and q's gives the divergence in cylindrical coordinates:

$$\nabla \cdot A = \frac{1}{\rho}\frac{\partial}{\partial \rho}(\rho A_\rho) + \frac{1}{\rho}\frac{\partial A_\varphi}{\partial \varphi} + \frac{\partial A_z}{\partial z}, \tag{1-128}$$

$$= \frac{A_\rho}{\rho} + \frac{\partial A_\rho}{\partial \rho} + \frac{1}{\rho}\frac{\partial A_\varphi}{\partial \varphi} + \frac{\partial A_z}{\partial z}, \tag{1-129}$$

and, in spherical coordinates,

$$\nabla \cdot A = \frac{1}{r^2 \sin\theta}\left[\frac{\partial}{\partial r}(r^2 \sin\theta A_r) + \frac{\partial}{\partial \theta}(r \sin\theta A_\theta) + \frac{\partial}{\partial \varphi}(r A_\varphi)\right], \tag{1-130}$$

$$= \frac{2}{r}A_r + \frac{\partial A_r}{\partial r} + \frac{A_\theta}{r}\cot\theta + \frac{1}{r}\frac{\partial A_\theta}{\partial \theta} + \frac{1}{r\sin\theta}\frac{\partial A_\varphi}{\partial \varphi}. \tag{1-131}$$

These expressions are again not valid on the z-axis where ρ and $\sin\theta$ are zero.

1.9.5 The Curl

From the fundamental definition given in Eq. 1-88,

$$(\nabla \times A)_1 = \lim_{S \to 0} \frac{1}{S} \oint A \cdot dl, \tag{1-132}$$

where the path of integration must lie in the surface defined by q_1 = constant, and where the direction of integration must be related to the direction of the unit vector u_1 by the right-hand screw rule. For the paths labeled a, b, c, d in Figure 1-22, we have the following contributions to the line integral:

(a) $$-A_3 h_3\, dq_3,$$

(b) $$\left(A_3 + \frac{\partial A_3}{\partial q_2} dq_2\right)\left(h_3 + \frac{\partial h_3}{\partial q_2} dq_2\right) dq_3,$$

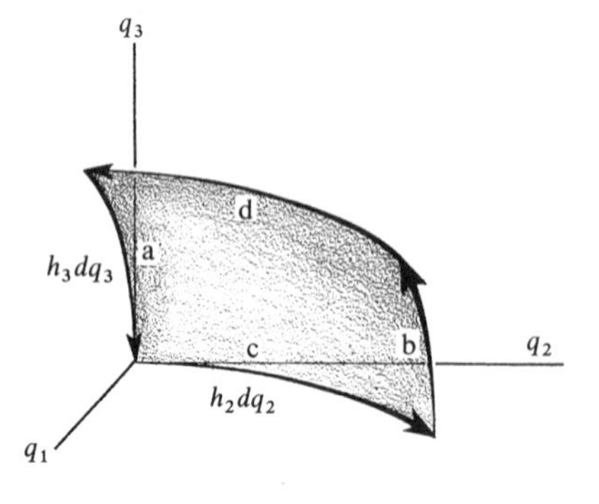

Figure 1-22. Path of integration for component 1 of the curl in curvilinear coordinates.

(c) $$+A_2 h_2\, dq_2,$$

(d) $$-\left(A_2 + \frac{\partial A_2}{\partial q_3} dq_3\right)\left(h_2 + \frac{\partial h_2}{\partial q_3} dq_3\right) dq_2.$$

Then, adding up these four terms and neglecting the higher order differentials because we are interested in the limit $S \to 0$,

$$(\nabla \times A)_1 = \frac{1}{h_2 h_3\, dq_2\, dq_3}\left[\frac{\partial}{\partial q_2}(A_3 h_3)\, dq_2\, dq_3 - \frac{\partial}{\partial q_3}(A_2 h_2)\, dq_2\, dq_3\right], \quad (1\text{-}133)$$

$$= \frac{1}{h_2 h_3}\left[\frac{\partial}{\partial q_2}(A_3 h_3) - \frac{\partial}{\partial q_3}(A_2 h_2)\right]. \quad (1\text{-}134)$$

Corresponding expressions for the other two components of the curl can be found either by proceeding again as above or by proper rotation of the indices. Finally,

$$\nabla \times A = \frac{1}{h_1 h_2 h_3}\begin{vmatrix} h_1 \boldsymbol{u}_1 & h_2 \boldsymbol{v}_1 & h_3 \boldsymbol{w}_1 \\ \dfrac{\partial}{\partial q_1} & \dfrac{\partial}{\partial q_2} & \dfrac{\partial}{\partial q_3} \\ h_1 A_1 & h_2 A_2 & h_3 A_3 \end{vmatrix}. \quad (1\text{-}135)$$

For cylindrical coordinates,

$$\nabla \times A = \frac{1}{\rho}\begin{vmatrix} \boldsymbol{\rho}_1 & \rho \boldsymbol{\varphi}_1 & \boldsymbol{z}_1 \\ \dfrac{\partial}{\partial \rho} & \dfrac{\partial}{\partial \varphi} & \dfrac{\partial}{\partial z} \\ A_\rho & \rho A_\varphi & A_z \end{vmatrix}. \quad (1\text{-}136)$$

whereas, for spherical coordinates,

$$\nabla \times A = \frac{1}{r^2 \sin\theta}\begin{vmatrix} \boldsymbol{r}_1 & r\boldsymbol{\theta}_1 & r\sin\theta \boldsymbol{\varphi}_1 \\ \dfrac{\partial}{\partial r} & \dfrac{\partial}{\partial \theta} & \dfrac{\partial}{\partial \varphi} \\ A_r & rA_\theta & r\sin\theta A_\varphi \end{vmatrix}. \quad (1\text{-}137)$$

These definitions are valid everywhere except on the z-axis.

1.9.6 The Laplacian

We calculate the Laplacian of a scalar function f in curvilinear coordinates by combining the expressions for the divergence and for the gradient:

$$\nabla^2 f = \nabla \cdot \nabla f. \quad (1\text{-}138)$$

$$= \frac{1}{h_1 h_2 h_3}\left[\frac{\partial}{\partial q_1}\left(\frac{h_2 h_3}{h_1}\frac{\partial f}{\partial q_1}\right) + \frac{\partial}{\partial q_2}\left(\frac{h_3 h_1}{h_2}\frac{\partial f}{\partial q_2}\right) + \frac{\partial}{\partial q_3}\left(\frac{h_1 h_2}{h_3}\frac{\partial f}{\partial q_3}\right)\right]. \quad (1\text{-}139)$$

For cylindrical coordinates,

$$\nabla^2 f = \frac{1}{\rho}\frac{\partial f}{\partial \rho} + \frac{\partial^2 f}{\partial \rho^2} + \frac{1}{\rho^2}\frac{\partial^2 f}{\partial \varphi^2} + \frac{\partial^2 f}{\partial z^2}, \tag{1-140}$$

except on the z-axis and, for spherical coordinates,

$$\nabla^2 f = \frac{1}{r^2}\frac{\partial}{\partial r}\left(r^2 \frac{\partial f}{\partial r}\right) + \frac{1}{r^2 \sin\theta}\frac{\partial}{\partial \theta}\left(\sin\theta \frac{\partial f}{\partial \theta}\right) + \frac{1}{r^2 \sin^2\theta}\frac{\partial^2 f}{\partial \varphi^2}, \tag{1-141}$$

$$= \frac{2}{r}\frac{\partial f}{\partial r} + \frac{\partial^2 f}{\partial r^2} + \frac{\cot\theta}{r^2}\frac{\partial f}{\partial \theta} + \frac{1}{r^2}\frac{\partial^2 f}{\partial \theta^2} + \frac{1}{r^2 \sin^2\theta}\frac{\partial^2 f}{\partial \varphi^2}, \tag{1-142}$$

again except on the z-axis.

The divergence, the curl, and the Laplacian are all invariants.

We have already seen in Section 1.8 that the Laplacian of a vector $\boldsymbol{A}$ in Cartesian coordinates is itself a vector whose components are the Laplacians of A_x, A_y, A_z. It will be shown in Problem 1-28 that

$$\boldsymbol{\nabla} \times \boldsymbol{\nabla} \times \boldsymbol{A} = \boldsymbol{\nabla}(\boldsymbol{\nabla} \cdot \boldsymbol{A}) - \nabla^2 \boldsymbol{A} \tag{1-143}$$

is then an identity in Cartesian coordinates. In the more general case of curvilinear coordinates, $\nabla^2 \boldsymbol{A}$ is defined as the vector whose components are those of $\boldsymbol{\nabla}(\boldsymbol{\nabla} \cdot \boldsymbol{A}) - \boldsymbol{\nabla} \times \boldsymbol{\nabla} \times \boldsymbol{A}$, and *not* as a vector whose components are the Laplacians of A_1, A_2, A_3.

1.10 SUMMARY

In Cartesian coordinates a *vector* quantity is written in the form

$$\boldsymbol{A} = A_x \boldsymbol{i} + A_y \boldsymbol{j} + A_z \boldsymbol{k}, \tag{1-1}$$

where $\boldsymbol{i}$, $\boldsymbol{j}$, $\boldsymbol{k}$ are *unit vectors* directed along the x, y, z axes respectively.

The *magnitude* of the vector $\boldsymbol{A}$ is the scalar

$$A = (A_x^2 + A_y^2 + A_z^2)^{1/2}. \tag{1-2}$$

Vectors can be added and subtracted:

$$\boldsymbol{A} + \boldsymbol{B} = (A_x + B_x)\boldsymbol{i} + (A_y + B_y)\boldsymbol{j} + (A_z + B_z)\boldsymbol{k}, \tag{1-3}$$

$$\boldsymbol{A} - \boldsymbol{B} = (A_x - B_x)\boldsymbol{i} + (A_y - B_y)\boldsymbol{j} + (A_z - B_z)\boldsymbol{k}. \tag{1-4}$$

The *scalar product* (Figure 1-2)

$$\boldsymbol{A} \cdot \boldsymbol{B} = AB\cos(\varphi - \theta), \tag{1-5}$$

$$= A_x B_x + A_y B_y + A_z B_z, \tag{1-11}$$

obeys the commutative and distributive rules:

$$\boldsymbol{A} \cdot \boldsymbol{B} = \boldsymbol{B} \cdot \boldsymbol{A}, \tag{1-6}$$

$$\boldsymbol{A} \cdot (\boldsymbol{B} + \boldsymbol{C}) = \boldsymbol{A} \cdot \boldsymbol{B} + \boldsymbol{A} \cdot \boldsymbol{C}. \tag{1-7}$$

The *vector product* (Figure 1-2)

$$\boldsymbol{A} \times \boldsymbol{B} = \boldsymbol{C}, \tag{1-14}$$

$$C = AB \sin (\varphi - \theta), \tag{1-15}$$

follows the distributive rule

$$\boldsymbol{A} \times (\boldsymbol{B} + \boldsymbol{C}) = \boldsymbol{A} \times \boldsymbol{B} + \boldsymbol{A} \times \boldsymbol{C}, \tag{1-17}$$

but *not* the commutative rule:

$$\boldsymbol{A} \times \boldsymbol{B} = -\boldsymbol{B} \times \boldsymbol{A}. \tag{1-16}$$

The *time derivative* of $\boldsymbol{A}$ is

$$\frac{d\boldsymbol{A}}{dt} = \frac{dA_x}{dt}\boldsymbol{i} + \frac{dA_y}{dt}\boldsymbol{j} + \frac{dA_z}{dt}\boldsymbol{k}. \tag{1-45}$$

The *del operator* is defined as follows:

$$\boldsymbol{\nabla} = \frac{\partial}{\partial x}\boldsymbol{i} + \frac{\partial}{\partial y}\boldsymbol{j} + \frac{\partial}{\partial z}\boldsymbol{j}, \tag{1-51}$$

and the *gradient* of a scalar function f is written

$$\boldsymbol{\nabla} f = \frac{\partial f}{\partial x}\boldsymbol{i} + \frac{\partial f}{\partial y}\boldsymbol{j} + \frac{\partial f}{\partial z}\boldsymbol{k}. \tag{1-52}$$

The gradient gives the maximum rate of change of f with distance at the point considered, and it points toward larger values of f.

The *flux* Φ of a vector $\boldsymbol{A}$ through a surface S is

$$\Phi = \int_S \boldsymbol{A} \cdot \boldsymbol{da}. \tag{1-62}$$

For a closed surface the vector $\boldsymbol{da}$ is chosen to point outward.

The *divergence* of $\boldsymbol{A}$,

$$\boldsymbol{\nabla} \cdot \boldsymbol{A} = \frac{\partial A_x}{\partial x} + \frac{\partial A_y}{\partial y} + \frac{\partial A_z}{\partial z}, \tag{1-69}$$

is the outward flux of $\boldsymbol{A}$ per unit volume at the point considered.

The *divergence theorem* states that

$$\int_\tau \boldsymbol{\nabla} \cdot \boldsymbol{A}\, d\tau = \int_S \boldsymbol{A} \cdot \boldsymbol{da}, \tag{1-70}$$

where S is the surface bounding the volume τ.

The *line integral*

$$\int_a^b \boldsymbol{A}\cdot d\boldsymbol{l}$$

over a specified curve is the sum of the terms $\boldsymbol{A}\cdot d\boldsymbol{l}$ for each element $d\boldsymbol{l}$ of the curve between the points a and b and, for a closed curve C which bounds a surface S, we have *Stokes's theorem:*

$$\oint_C \boldsymbol{A}\cdot d\boldsymbol{l} = \int_S (\boldsymbol{\nabla}\times\boldsymbol{A})\cdot d\boldsymbol{a}, \qquad (1\text{-}96)$$

where

$$\boldsymbol{\nabla}\times\boldsymbol{A} = \begin{vmatrix} \boldsymbol{i} & \boldsymbol{j} & \boldsymbol{k} \\ \frac{\partial}{\partial x} & \frac{\partial}{\partial y} & \frac{\partial}{\partial z} \\ A_x & A_y & A_z \end{vmatrix} \qquad (1\text{-}85)$$

is the *curl* of the vector function $\boldsymbol{A}$. The above surface integral is evaluated over *any* surface bounded by the curve C.

The *Laplacian* is the divergence of the gradient:

$$\boldsymbol{\nabla}\cdot\boldsymbol{\nabla}f = \boldsymbol{\nabla}^2 f = \frac{\partial^2 f}{\partial x^2} + \frac{\partial^2 f}{\partial y^2} + \frac{\partial^2 f}{\partial z^2}. \qquad (1\text{-}101)$$

The Laplacian of a vector in Cartesian coordinates is defined as

$$\boldsymbol{\nabla}^2\boldsymbol{A} = \boldsymbol{\nabla}^2 A_x\,\boldsymbol{i} + \boldsymbol{\nabla}^2 A_y\,\boldsymbol{j} + \boldsymbol{\nabla}^2 A_z\,\boldsymbol{k}. \qquad (1\text{-}102)$$

In *cylindrical coordinates*, Figures 1-17 and 1-18,

$$\boldsymbol{r} = \rho\boldsymbol{\rho}_1 + z\boldsymbol{z}_1, \qquad (1\text{-}109)$$

$$d\boldsymbol{r} = d\rho\,\boldsymbol{\rho}_1 + \rho\,d\varphi\,\boldsymbol{\varphi}_1 + dz\,\boldsymbol{z}_1, \qquad (1\text{-}110)$$

$$d\tau = \rho\,d\rho\,d\varphi\,dz, \qquad (1\text{-}112)$$

$$\boldsymbol{\nabla}f = \frac{\partial f}{\partial\rho}\boldsymbol{\rho}_1 + \frac{1}{\rho}\frac{\partial f}{\partial\varphi}\boldsymbol{\varphi}_1 + \frac{\partial f}{\partial z}\boldsymbol{z}_1, \qquad (1\text{-}118)$$

$$\boldsymbol{\nabla}\cdot\boldsymbol{A} = \frac{A_\rho}{\rho} + \frac{\partial A_\rho}{\partial\rho} + \frac{1}{\rho}\frac{\partial A_\varphi}{\partial\varphi} + \frac{\partial A_z}{\partial z}, \qquad (1\text{-}129)$$

$$\boldsymbol{\nabla}\times\boldsymbol{A} = \frac{1}{\rho}\begin{vmatrix} \boldsymbol{\rho}_1 & \rho\boldsymbol{\varphi}_1 & \boldsymbol{z}_1 \\ \frac{\partial}{\partial\rho} & \frac{\partial}{\partial\varphi} & \frac{\partial}{\partial z} \\ A_\rho & \rho A_\varphi & A_z \end{vmatrix}, \qquad (1\text{-}136)$$

$$\boldsymbol{\nabla}^2 f = \frac{1}{\rho}\frac{\partial f}{\partial\rho} + \frac{\partial^2 f}{\partial\rho^2} + \frac{1}{\rho^2}\frac{\partial^2 f}{\partial\varphi^2} + \frac{\partial^2 f}{\partial z^2}. \qquad (1\text{-}140)$$

These last four equations are meaningless on the z-axis, where $\rho = 0$.

In *spherical coordinates*, Figures 1-19 and 1-20,

$$\boldsymbol{r} = r\boldsymbol{r}_1,$$

$$\boldsymbol{dr} = dr\,\boldsymbol{r}_1 + rd\theta\,\boldsymbol{\theta}_1 + r\sin\theta\,d\varphi\,\boldsymbol{\varphi}_1, \tag{1-113}$$

$$d\tau = r^2 \sin\theta\,dr\,d\theta\,d\varphi, \tag{1-115}$$

$$\boldsymbol{\nabla} f = \frac{\partial f}{\partial r}\boldsymbol{r}_1 + \frac{1}{r}\frac{\partial f}{\partial \theta}\boldsymbol{\theta}_1 + \frac{1}{r\sin\theta}\frac{\partial f}{\partial \varphi}\boldsymbol{\varphi}_1, \tag{1-119}$$

$$\boldsymbol{\nabla}\cdot\boldsymbol{A} = \frac{2}{r}A_r + \frac{\partial A_r}{\partial r} + \frac{A_\theta}{r}\cot\theta + \frac{1}{r}\frac{\partial A_\theta}{\partial \theta} + \frac{1}{r\sin\theta}\frac{\partial A_\varphi}{\partial \varphi}, \tag{1-131}$$

$$\boldsymbol{\nabla}\times\boldsymbol{A} = \frac{1}{r^2\sin\theta}\begin{vmatrix} \boldsymbol{r}_1 & r\boldsymbol{\theta}_1 & r\sin\theta\boldsymbol{\varphi}_1 \\ \frac{\partial}{\partial r} & \frac{\partial}{\partial \theta} & \frac{\partial}{\partial \varphi} \\ A_r & rA_\theta & r\sin\theta A_\varphi \end{vmatrix}, \tag{1-137}$$

$$\boldsymbol{\nabla}^2 f = \frac{2}{r}\frac{\partial f}{\partial r} + \frac{\partial^2 f}{\partial r^2} + \frac{\cot\theta}{r^2}\frac{\partial f}{\partial \theta} + \frac{1}{r^2}\frac{\partial^2 f}{\partial \theta^2} + \frac{1}{r^2\sin^2\theta}\frac{\partial^2 f}{\partial \varphi^2}. \tag{1-142}$$

These last four equations are also meaningless on the z-axis, where $\sin\theta = 0$.

In other than Cartesian coordinates, $\boldsymbol{\nabla}^2\boldsymbol{A}$ is *defined* as follows:

$$\boldsymbol{\nabla}^2\boldsymbol{A} = \boldsymbol{\nabla}(\boldsymbol{\nabla}\cdot\boldsymbol{A}) - \boldsymbol{\nabla}\times\boldsymbol{\nabla}\times\boldsymbol{A}.$$

We shall have occasion to use the following operations:*

$$\boldsymbol{\nabla}(fg) = f\boldsymbol{\nabla} g + g\boldsymbol{\nabla} f;$$

$$\boldsymbol{\nabla}(\boldsymbol{A}\cdot\boldsymbol{B}) = (\boldsymbol{B}\cdot\boldsymbol{\nabla})\boldsymbol{A} + (\boldsymbol{A}\cdot\boldsymbol{\nabla})\boldsymbol{B} + \boldsymbol{B}\times(\boldsymbol{\nabla}\times\boldsymbol{A}) + \boldsymbol{A}\times(\boldsymbol{\nabla}\times\boldsymbol{B});$$

$$\boldsymbol{\nabla}\cdot(f\boldsymbol{A}) = (\boldsymbol{\nabla} f)\cdot\boldsymbol{A} + f(\boldsymbol{\nabla}\cdot\boldsymbol{A});$$

$$\boldsymbol{\nabla}\cdot(\boldsymbol{A}\times\boldsymbol{B}) = \boldsymbol{B}\cdot(\boldsymbol{\nabla}\times\boldsymbol{A}) - \boldsymbol{A}\cdot(\boldsymbol{\nabla}\times\boldsymbol{B});$$

$$\boldsymbol{\nabla}\times(f\boldsymbol{A}) = (\boldsymbol{\nabla} f)\times\boldsymbol{A} + f(\boldsymbol{\nabla}\times\boldsymbol{A});$$

$$\boldsymbol{\nabla}\times(\boldsymbol{A}\times\boldsymbol{B}) = (\boldsymbol{B}\cdot\boldsymbol{\nabla})\boldsymbol{A} - (\boldsymbol{A}\cdot\boldsymbol{\nabla})\boldsymbol{B} + (\boldsymbol{\nabla}\cdot\boldsymbol{B})\boldsymbol{A} - (\boldsymbol{\nabla}\cdot\boldsymbol{A})\boldsymbol{B};$$

$$\boldsymbol{\nabla}\times\boldsymbol{\nabla}\times\boldsymbol{A} = \boldsymbol{\nabla}(\boldsymbol{\nabla}\cdot\boldsymbol{A}) - \boldsymbol{\nabla}^2\boldsymbol{A} \text{ (see Section 1.9.6)};$$

$$(\boldsymbol{A}\cdot\boldsymbol{\nabla})\boldsymbol{B} = \left[A_x\frac{\partial B_x}{\partial x} + A_y\frac{\partial B_x}{\partial y} + A_z\frac{\partial B_x}{\partial z}\right]\boldsymbol{i} + \left[A_x\frac{\partial B_y}{\partial x} + A_y\frac{\partial B_y}{\partial y} + A_z\frac{\partial B_y}{\partial z}\right]\boldsymbol{j} + \left[A_x\frac{\partial B_z}{\partial x} + A_y\frac{\partial B_z}{\partial y} + A_z\frac{\partial B_z}{\partial z}\right]\boldsymbol{k};$$

$$\boldsymbol{\nabla}'\left(\frac{1}{r}\right) = \frac{\boldsymbol{r}_1}{r^2},$$

* See J. Van Bladel, *Electromagnetic Fields* (McGraw-Hill, New York, 1964), Appendixes 1 and 2, for an extensive collection of vector identities and theorems.

where $\boldsymbol{\nabla}'\left(\frac{1}{r}\right)$ is calculated at the source point (x', y', z') and $\boldsymbol{r}_1$ is the unit vector from the source point (x', y', z') to the field point (x, y, z);

$$\boldsymbol{\nabla}\left(\frac{1}{r}\right) = -\frac{\boldsymbol{r}_1}{r^2},$$

when the gradient is calculated at the field point, with the same unit vector;

$$\int_\tau \boldsymbol{\nabla} f \, d\tau = \int_S f \, \boldsymbol{da};$$

$$\int_\tau (\boldsymbol{\nabla} \times \boldsymbol{A}) \, d\tau = -\int_S \boldsymbol{A} \times \boldsymbol{da},$$

where S is the surface that bounds the volume τ.

PROBLEMS

NOTE: Double daggers (‡) indicate problems for which hints are given at the end of the Problems section.

1-1. Show that the two vectors $\boldsymbol{A} = 9\boldsymbol{i} + \boldsymbol{j} - 6\boldsymbol{k}$ and $\boldsymbol{B} = 4\boldsymbol{i} - 6\boldsymbol{j} + 5\boldsymbol{k}$ are perpendicular to each other.

1-2. Show that the angle between the two vectors $\boldsymbol{A} = 2\boldsymbol{i} + 3\boldsymbol{j} + \boldsymbol{k}$ and $\boldsymbol{B} = \boldsymbol{i} - 6\boldsymbol{j} + \boldsymbol{k}$ is 130.5°.

1-3. The vectors $\boldsymbol{A}, \boldsymbol{B}, \boldsymbol{C}$ are coplanar. Show graphically that

$$\boldsymbol{A}\cdot(\boldsymbol{B} + \boldsymbol{C}) = \boldsymbol{A}\cdot\boldsymbol{B} + \boldsymbol{A}\cdot\boldsymbol{C}$$

1-4. If $\boldsymbol{A}$ and $\boldsymbol{B}$ are adjacent sides of a parallelogram, $\boldsymbol{C} = \boldsymbol{A} + \boldsymbol{B}$ and $\boldsymbol{D} = \boldsymbol{A} - \boldsymbol{B}$ are the diagonals, and θ is the angle between $\boldsymbol{A}$ and $\boldsymbol{B}$, show that $(C^2 + D^2) = 2(A^2 + B^2)$ and that $(C^2 - D^2) = 4AB\cos\theta$.

1-5. Let $\boldsymbol{a}$ and $\boldsymbol{b}$ be two unit vectors lying in the xy-plane. Let α be the angle $\boldsymbol{a}$ makes with the x-axis, and let β be the angle $\boldsymbol{b}$ makes with the x-axis, so that $\boldsymbol{a} = \cos\alpha\,\boldsymbol{i} + \sin\alpha\,\boldsymbol{j}$ and $\boldsymbol{b} = \cos\beta\,\boldsymbol{i} + \sin\beta\,\boldsymbol{j}$.

Show that the trigonometric relations for the sine and cosine of the sum and difference of two angles follow from the interpretation of $\boldsymbol{a}\cdot\boldsymbol{b}$ and $\boldsymbol{a} \times \boldsymbol{b}$.

1-6. Show that the magnitude of $(\boldsymbol{A} \times \boldsymbol{B})\cdot\boldsymbol{C}$ is the volume of a parallelepiped whose edges are $\boldsymbol{A}, \boldsymbol{B}, \boldsymbol{C}$, and show that $(\boldsymbol{A} \times \boldsymbol{B})\cdot\boldsymbol{C} = \boldsymbol{A}\cdot(\boldsymbol{B} \times \boldsymbol{C})$.

1-7. Show that the distributive rule applies to the vector product (Eq. 1-17).‡

1-8. Show that $\boldsymbol{a} \times (\boldsymbol{b} \times \boldsymbol{c}) = \boldsymbol{b}(\boldsymbol{a}\cdot\boldsymbol{c}) - \boldsymbol{c}(\boldsymbol{a}\cdot\boldsymbol{b})$.

1-9. If $\boldsymbol{r}\cdot(d\boldsymbol{r}/dt) = 0$, show that $r = \text{constant}$.

1-10. A gun fires a bullet at a velocity of 500 meters/second and at an angle of 30° with the horizontal. Find the position vector $\mathbf{r}$, the velocity vector $\mathbf{v}$, and the acceleration vector $\mathbf{a}$ of the bullet, t seconds after the gun is fired. Draw a sketch of the trajectory and show the three vectors for some time t.

1-11. Let $\mathbf{r}$ be the radius vector from the origin of coordinates to any point, and let $\mathbf{A}$ be a constant vector. Show that $\nabla(\mathbf{A}\cdot\mathbf{r}) = \mathbf{A}$.

1-12. The vector $\mathbf{r}$ is directed from $P'(x', y', z')$ to $P(x, y, z)$.

(a) If the point P is fixed and the point P' is allowed to move, show that the gradient of $(1/r)$ under these conditions is

$$\nabla'\left(\frac{1}{r}\right) = \frac{\mathbf{r}_1}{r^2},$$

where $\mathbf{r}_1$ is the unit vector along $\mathbf{r}$. Show that this is the maximum rate of change of $1/r$.

(b) Show similarly that, if P' is fixed and P is allowed to move,

$$\nabla\left(\frac{1}{r}\right) = -\frac{\mathbf{r}_1}{r^2}.$$

1-13. (a) Show that $\nabla\cdot\mathbf{r} = 3$.

(b) What is the flux of $\mathbf{r}$ through a spherical surface of radius r?

1-14. Show that

$$\nabla\cdot(f\mathbf{A}) = f\nabla\cdot\mathbf{A} + \mathbf{A}\cdot\nabla f,$$

where f is a scalar function and $\mathbf{A}$ is a vector function.

1-15. The vector $\mathbf{A} = 3x\mathbf{i} + y\mathbf{j} + 2z\mathbf{k}$, and $f = x^2 + y^2 + z^2$.

(a) Show that $\nabla\cdot(f\mathbf{A})$ at the point (2, 2, 2) is 120, by first calculating $f\mathbf{A}$ and then calculating its divergence.

(b) Make the same calculation by first finding ∇f and $\nabla\cdot\mathbf{A}$, and then using the identity of Problem 1-14 above.

(c) If x, y, z are measured in meters, what are the units of $\nabla\cdot(f\mathbf{A})$?

1-16. It is suggested that $\nabla\cdot(\mathbf{A}\times\mathbf{r}/r^3)$ is identically equal to zero as long as the vector $\mathbf{A}$ is a constant.

Is this correct?

1-17. The components of a vector $\mathbf{A}$ are

$$A_x = y\frac{\partial f}{\partial z} - z\frac{\partial f}{\partial y}, \qquad A_y = z\frac{\partial f}{\partial x} - x\frac{\partial f}{\partial z}, \qquad A_z = x\frac{\partial f}{\partial y} - y\frac{\partial f}{\partial x},$$

where f is a function of x, y, z. Show that

$$\mathbf{A} = \mathbf{r}\times\nabla f, \qquad \mathbf{A}\cdot\mathbf{r} = 0, \qquad \text{and} \qquad \mathbf{A}\cdot\nabla f = 0.$$

1-18. Show that $\nabla\times(f\mathbf{A}) = (\nabla f)\times\mathbf{A} + f(\nabla\times\mathbf{A})$, where f is a scalar function and $\mathbf{A}$ is a vector function.

1-19. Show that $\nabla\cdot(\mathbf{A}\times\mathbf{D}) = \mathbf{D}\cdot(\nabla\times\mathbf{A}) - \mathbf{A}\cdot(\nabla\times\mathbf{D})$, where $\mathbf{A}$ and $\mathbf{D}$ are any two vectors.

1-20. One of the four Maxwell equations states that $\nabla \times \boldsymbol{E} = -\partial \boldsymbol{B}/\partial t$, where $\boldsymbol{E}$ and $\boldsymbol{B}$ are respectively the electric field intensity in volts/meter and the magnetic induction in teslas at a point.

Show that

$$\left| \oint \boldsymbol{E} \cdot d\boldsymbol{l} \right|$$

is 20.0 microvolts over a square 10.0 centimeters on the side when $\boldsymbol{B}$ is $2.00 \times 10^{-3} t$ tesla and normal to the square.

1-21. Show that

$$\nabla \cdot (\nabla^2 \boldsymbol{A}) = \nabla^2 (\nabla \cdot \boldsymbol{A})$$

in Cartesian coordinates.

1-22. Show, by differentiating the appropriate expressions for $\boldsymbol{r}$, that the velocity $\dot{\boldsymbol{r}}$ in cylindrical coordinates is

$$\dot{\rho} \boldsymbol{\rho}_1 + \rho \dot{\varphi} \boldsymbol{\varphi}_1 + \dot{z} \boldsymbol{z}_1,$$

while in spherical coordinates it is

$$\dot{r} \boldsymbol{r}_1 + r \dot{\theta} \boldsymbol{\theta}_1 + r \sin \theta \dot{\varphi} \boldsymbol{\varphi}_1.$$

1-23. The vector $\boldsymbol{A}$ is everywhere perpendicular to, and directed away from, a given straight line; that is, in cylindrical coordinates, $A_z = A_\varphi = 0$. Calculate the net outgoing flux for a volume element, and show that

$$\nabla \cdot \boldsymbol{A} = \frac{A_\rho}{\rho} + \frac{\partial A_\rho}{\partial \rho}.$$

1-24. A fluid rotates with an angular velocity ω about the z-axis. The direction of rotation is related to that of the z-axis by the right-hand screw rule.

(a) Find the velocity $\boldsymbol{v}$ of a point in the fluid, and show that

$$\nabla \times \boldsymbol{v} = 2\omega \boldsymbol{k}.$$

(b) If now ω is a function of the radius r, show that $\nabla \times \boldsymbol{v} = 0$ if $\omega = \text{Constant}/r^2$.

1-25. The gravitational forces exerted by the Sun on the planets are always directed toward the Sun and depend only on the distance r. This type of field is called a *central force field.*

Find the potential energy at a distance r from a center of attraction when the force varies as $1/r^2$. Set the potential energy equal to zero at infinity.

1-26. The azimuthal force exerted on an electron in a certain betatron is proportional to $r^{0.4}$.

Find the curl of this force in cylindrical coordinates, and show that the force is nonconservative.

1-27. A vector field is defined by $\boldsymbol{A} = f(r)\boldsymbol{r}$.

(a) Show that $f(r) = \text{Constant}/r^3$ if $\nabla \cdot \boldsymbol{A} = 0$.

(b) Show that $\nabla \times \boldsymbol{A}$ is always equal to zero.

1-28. Show that

$$\boldsymbol{\nabla} \times (\boldsymbol{\nabla} \times \boldsymbol{A}) = \boldsymbol{\nabla}(\boldsymbol{\nabla} \cdot \boldsymbol{A}) - \boldsymbol{\nabla}^2 \boldsymbol{A}$$

in Cartesian coordinates, but not in the general case of curvilinear coordinates, *if* $\boldsymbol{\nabla}^2 \boldsymbol{A}$ is taken to be the vector whose components are the Laplacians of A_1, A_2, A_3.‡

1-29. Show that

$$\int_\tau \boldsymbol{\nabla} f \, d\tau = \int_S f \, \boldsymbol{da},$$

where f is a function of (x, y, z) and S is the surface bounding the volume τ.‡

1-30. Show that

$$\int_\tau (\boldsymbol{\nabla} \times \boldsymbol{A}) \, d\tau = -\int_S \boldsymbol{A} \times \boldsymbol{da},$$

where $\boldsymbol{A}$ is an arbitrary vector and S is the surface bounding the volume τ.‡

Hints

1-7. Write out the components in Cartesian coordinates.

1-28. You can solve this problem without having to write out the equations in full.

1-29. Multiply both sides by a constant vector (scalar product) and use the identity of Problem 1-14.

1-30. Multiply both sides by a constant vector (scalar product) and use the identities of Problems 1-6 and 1-19.

CHAPTER 2

ELECTROSTATIC FIELDS I

Electrostatic Fields in a Vacuum

We begin our study of electromagnetic fields by investigating those fields that originate from stationary electric charges. This will occupy us during the next three chapters.

We shall start with Coulomb's law for electrostatic forces because it is fundamental. We shall deduce from it the basic concepts and laws of electrostatics: the electric field intensity and the electric potential, Gauss's law and Poisson's equation, and, finally, the electrostatic energy density, which is related to the electrostatic forces.

2.1 COULOMB'S LAW

It has been found experimentally that the force between two stationary electric *point* charges Q_a and Q_b (a) acts along the line joining the two charges, (b) is proportional to the product Q_aQ_b, and (c) is inversely proportional to the square of the distance r separating the charges.

If the charges are extended, the situation is more complicated in that the "distance between the charges" has no definite meaning. Moreover, the presence of Q_b can modify the charge distribution within Q_a, and vice versa, leading to a complicated variation of force with distance.

We thus have *Coulomb's law* for stationary point charges:

$$\boldsymbol{F}_{ab} = K\frac{Q_aQ_b}{r^2}\,\boldsymbol{r}_1, \tag{2-1}$$

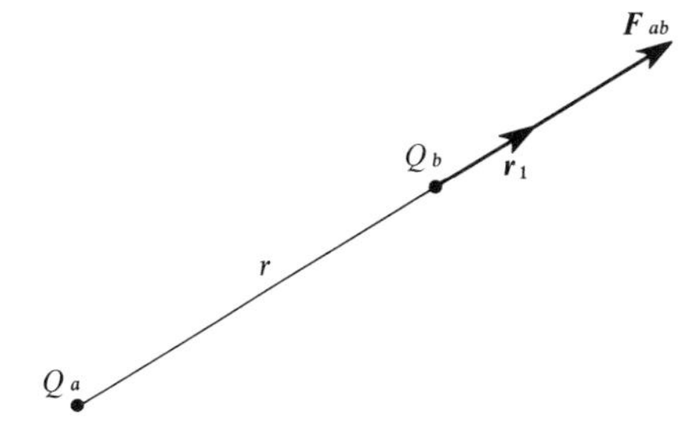

Figure 2-1. Charges Q_a and Q_b separated by a distance r. The force exerted on Q_b by Q_a is $\mathbf{F}_{ab}$ and is in the direction $\mathbf{r}_1$ along the line joining the two charges.

where $\mathbf{F}_{ab}$ is the force exerted *by* Q_a *on* Q_b, K is a constant of proportionality, and $\mathbf{r}_1$ is a unit vector pointing in the direction *from* Q_a *to* Q_b, as in Figure 2-1. The force is repulsive if Q_a and Q_b are of the same sign; it is attractive if they are of different signs.

The magnitude of the constant of proportionality K depends on the units that are used for the measurement of force, charge, and distance. In the rationalized *MKSA* (Meter, Kilogram, Second, Ampere) system, which we shall use throughout this book,

$$\mathbf{F}_{ab} = \frac{1}{4\pi\epsilon_0}\frac{Q_a Q_b}{r^2}\mathbf{r}_1, \tag{2-2}$$

where the force $\mathbf{F}_{ab}$ is measured in newtons; the charges Q_a and Q_b, in coulombs (the magnitude of which will be defined in terms of magnetic interactions in Chapters 6 and 7); and the distance r, in meters. The quantity 4π appears in explicit form so as to simplify other equations that are used much more extensively than Coulomb's law. The constant ϵ_0 is called the *permittivity of free space:*

$$\epsilon_0 = 8.8542 \times 10^{-12} \text{ farad/meter.}$$

Coulomb's law applies to a pair of point charges situated in a vacuum. It also applies in dielectrics and conductors if $\mathbf{F}_{ab}$ is taken to be the direct force between Q_a and Q_b, irrespective of the forces arising from other charges within the medium.

Substituting the value of ϵ_0 into Coulomb's law,

$$\mathbf{F}_{ab} = 9 \times 10^9 \frac{Q_a Q_b}{r^2}\mathbf{r}_1, \tag{2-3}$$

where the factor of 9 is accurate to about one part in 1000; it should really be 8.988.

Example | The Coulomb forces in nature are enormous when compared with the gravitational forces, which are given by

$$F = 6.67 \times 10^{-11} \frac{m_a m_b}{r^2} \mathbf{r}_1. \tag{2-4}$$

One good example is the following. The gravitational force on a proton at the surface of the Sun (Mass = 2.0×10^{30} kilograms, Radius = 7.0×10^8 meters) is equal to the electric force between a proton and one microgram of electrons, separated by a distance equal to the Sun's radius.

It is remarkable indeed that we should not be conscious of these enormous forces in everyday life. The positive and negative charges carried respectively by the proton and the electron are very nearly, if not exactly, the same. Recent experiments have shown them to be equal within about one part in 10^{22}.* Ordinary matter is thus neutral, and the enormous Coulomb forces prevent the accumulation of any appreciable quantity of charge of either sign. You will be able to show in Problem 2-35 that a sphere of protons about one foot in diameter and weighing about 20 kilograms would have enough energy to pulverize the Moon!

2.2 THE ELECTRIC FIELD INTENSITY

We think of the interaction between the point charges Q_a and Q_b in Coulomb's law as being an interaction between Q_a and the *field* of Q_b, or vice versa. We define the *electric field intensity* $\mathbf{E}$ to be the force per unit charge exerted on a test charge in the field. Thus the electric field intensity due to the point charge Q_a is

$$\mathbf{E}_a = \frac{\mathbf{F}_{ab}}{Q_b} = \frac{Q_a}{4\pi\epsilon_0 r^2} \mathbf{r}_1. \tag{2-5}$$

The electric field intensity is measured in volts/meter. The electric field intensity due to the point charge Q_a is the same whether the test charge Q_b is in the field or not, even if Q_b is large compared to Q_a.

When the electric field is produced by a charge distribution that is disturbed by the introduction of a finite test charge Q', we can define $\mathbf{E}$ to be the limiting force per unit charge as the magnitude of the test charge Q' tends to zero:

$$\mathbf{E} = \lim_{Q' \to 0} \frac{\mathbf{F}}{Q'}. \tag{2-6}$$

* John G. King: private communication.

If the electric field is produced by more than one charge distribution, each one produces its own field, and the resultant $\boldsymbol{E}$ is simply the vector sum of all the individual $\boldsymbol{E}$'s:

$$\boldsymbol{E} = \frac{1}{4\pi\epsilon_0}\int_{\tau'} \frac{\rho \boldsymbol{r}_1}{r^2}\, d\tau'. \tag{2-7}$$

This is the *principle of superposition*. In this integral, ρ is the electric charge density at the source point (x', y', z'), $\boldsymbol{r}_1$ is a unit vector pointing from the source point to the field point (x, y, z) where $\boldsymbol{E}$ is calculated, r is the distance between these two points, and $d\tau'$ is the element of volume $dx'\,dy'\,dz'$. If there exist surface distributions of charge, then we must add a similar integral with ρ replaced by σ and τ' by A'.

2.3 THE ELECTRIC POTENTIAL

Consider a test point charge Q' that can be moved about in an electric field. The work W required to move it at a constant speed from a point P_1 to a point P_2 along a given path is

$$W = -\int_{P_1}^{P_2} \boldsymbol{E}Q' \cdot \boldsymbol{dl}. \tag{2-8}$$

The negative sign is required to obtain the work done against the field. Here again, we assume that Q' is so small that the charge distributions are not appreciably disturbed by its presence.

If the path is closed, the total work done is

$$W = -\oint \boldsymbol{E}Q' \cdot \boldsymbol{dl}. \tag{2-9}$$

Let us evaluate this integral. To simplify matters, we first consider the electric field produced by a single point charge Q. Then

$$\oint \boldsymbol{E}Q' \cdot \boldsymbol{dl} = \frac{QQ'}{4\pi\epsilon_0}\oint \frac{(\boldsymbol{r}_1 \cdot \boldsymbol{dl})}{r^2}. \tag{2-10}$$

Now the term under the integral on the right is simply dr/r^2 or $-d(1/r)$. The sum of the increments of $(1/r)$ over a closed path is zero, since r has the same value at the beginning and at the end of the path. Then the line integral is zero, and the net work done in moving a point charge Q' around any closed path in the field of a point charge Q, which is fixed, is zero.

If the electric field is produced not by a single point charge Q but by some fixed charge distribution, the line integrals corresponding to each individual charge of the distribution are all zero. Thus, for any distribution of fixed charges,

$$\oint \mathbf{E} \cdot d\mathbf{l} = 0. \tag{2-11}$$

An electrostatic field is therefore conservative (Section 1.6). This important property follows simply from the fact that the Coulomb force is a central force.

Then, from Stokes's theorem (Section 1.6), at all points in space,

$$\boldsymbol{\nabla} \times \mathbf{E} = 0, \tag{2-12}$$

and we can write that

$$\mathbf{E} = -\boldsymbol{\nabla} V, \tag{2-13}$$

where V is a scalar point function, since $\boldsymbol{\nabla} \times \boldsymbol{\nabla} V \equiv 0$. We can thus describe the field completely by means of the function $V(x, y, z)$, which is called the *electric potential*. The negative sign is required in order that the electric field intensity $\mathbf{E}$ can point toward a *decrease* in potential, according to the usual convention. It is important to note that V is not uniquely defined; we can add to it any quantity that is independent of the coordinates without affecting $\mathbf{E}$ in any way.

We can now show that the work done in moving a test charge at a constant speed from a point P_1 to a point P_2 is independent of the path. Let a and b be any two different paths leading from P_1 to P_2. Then these two paths together form a closed curve and the work done in going from P_1 to P_2 along a, and then from P_2 back to P_1 along b is zero. Then the work done in going from P_1 to P_2 is the same along a as it is along b.

We must remember that we are dealing here with electro*statics*. If there were moving charges present, $\boldsymbol{\nabla} \times \mathbf{E}$ would not necessarily be zero, and $\boldsymbol{\nabla} V$ would then describe only part of the electric field intensity $\mathbf{E}$. We shall investigate these more complicated phenomena later on.

According to Eq. 2-13,

$$\mathbf{E} \cdot d\mathbf{l} = -\boldsymbol{\nabla} V \cdot d\mathbf{l} = -dV. \tag{2-14}$$

Then

$$V_2 - V_1 = -\int_1^2 \mathbf{E} \cdot d\mathbf{l} = \int_2^1 \mathbf{E} \cdot d\mathbf{l}, \tag{2-15}$$

as in Figure 2-2.

Note that the electric field intensity $\mathbf{E}$ determines only *differences* between the potentials at two different points. When we wish to speak of the electric

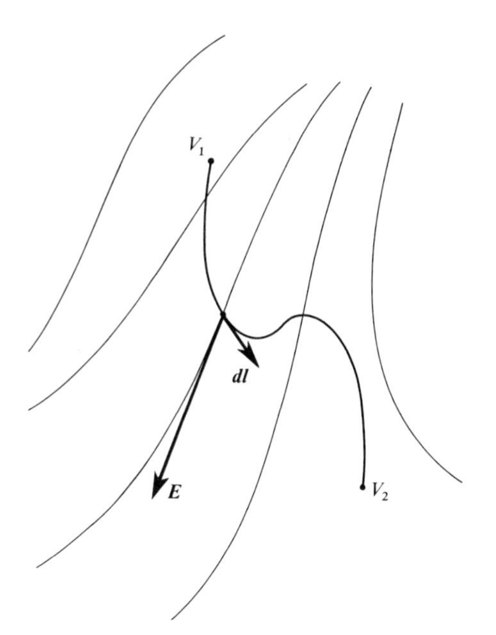

Figure 2-2. The potential difference $V_1 - V_2$ between two points is given by the line integral of $\boldsymbol{E} \cdot \boldsymbol{dl}$ from 1 to 2, where $\boldsymbol{E}$ is the electric field intensity and $\boldsymbol{dl}$ is an element of the path along which the integral is to be calculated. The light lines represent lines of force.

potential at a given point, we must therefore arbitrarily define the potential in a given region of space to be zero. It is usually convenient to choose the potential at infinity to be zero. Then the potential V at the point 2 is

$$V = \int_2^\infty \boldsymbol{E} \cdot \boldsymbol{dl}. \tag{2-16}$$

The work W required to bring a charge Q' from a point at which the potential is defined to be zero to the point considered is VQ'. Thus V is W/Q' and can be defined to be the work per unit charge. The potential V is expressed in joules/coulomb, or in volts.

When the field is produced by a single point charge Q,

$$V = \int_r^\infty \frac{Q}{4\pi\epsilon_0} \frac{dr}{r^2} = \frac{Q}{4\pi\epsilon_0 r} = 9 \times 10^9 \frac{Q}{r}. \tag{2-17}$$

It will be observed that the sign of the potential V is the same as that of Q.

The principle of superposition applies to the electric potential V as well as to the electric field intensity $\boldsymbol{E}$.

The set of points in space that are at a given potential defines an *equipotential surface*. For example, the equipotential surfaces about a point charge are concentric spheres. We can see from Eq. 2-13 that the electric field intensity $\boldsymbol{E}$ is everywhere normal to the equipotential surfaces (Section 1.4).

If we join end-to-end infinitesimal vectors representing $\boldsymbol{E}$, we get a curve in space—called a *line of force*—that is everywhere normal to the equipotential surfaces. The vector $\boldsymbol{E}$ is everywhere tangent to a line of force.

2.4 THE ELECTRIC FIELD OUTSIDE AND INSIDE MACROSCOPIC BODIES

We have so far been thinking in terms of point charges. Now electric charges are usually distributed over macroscopic bodies, which are composed of positively charged nuclei and negative electrons. This brings up two questions.

(a) First of all, can we calculate the field *outside* an electrically charged body by assuming that the charge distribution inside the body is continuous? If our assumption is valid, we can calculate the field by integrating over the charge distribution. If our assumption is not valid, then we must find some other form of calculation.

It is, in fact, usually appropriate to treat the discrete charges carried by nuclei and electrons within macroscopic bodies as though they were continuously distributed within these bodies. Even the largest nuclei have diameters that are only of the order of 10^{-14} meter. Nuclei and electrons are therefore so small and so closely packed compared to the dimensions of ordinary macroscopic objects that we may define an average electric charge density ρ, measured in coulombs/meter^3, as $\Delta Q/\Delta\tau'$, where ΔQ is the net charge within $\Delta\tau'$.

The volume $\Delta\tau'$ may be assumed large enough to render the fluctuations in ΔQ with time, or from one $\Delta\tau'$ to a neighboring one, negligible. At the same time the volume $\Delta\tau'$ may be assumed small enough to permit the use of integral calculus.

Thus, instead of summing the potentials of a large number of individual charges, we may integrate over a continuous distribution $\rho(x', y', z')$. An element of charge $\rho\, d\tau'$ contributes at a point $P(x, y, z)$ outside $d\tau'$ a potential

$$dV = \frac{1}{4\pi\epsilon_0}\frac{\rho\, d\tau'}{r}, \tag{2-18}$$

where r is the distance from $d\tau'$ to P, and

$$V = \frac{1}{4\pi\epsilon_0}\int_\infty \frac{\rho\, d\tau'}{r}. \tag{2-19}$$

The integral is evaluated over all regions where the net charge density ρ is not zero.

The above integral is not always applicable, however, because in some cases it diverges. In such cases you can calculate $\boldsymbol{E}$ and then deduce a potential V by integration.

(b) Now what about the electric field *inside* a charged body? It is obvious that the electric field intensity in the immediate neighborhood of a nucleus or of an electron must be enormous. It is also obvious that this electric field must change rapidly with time, since the charges are never perfectly stationary. It is not useful for our purposes to look at the electric field as closely as that. We shall be satisfied if we can calculate average values of $\boldsymbol{E}$ and V inside a charged body by assuming a continuous distribution of charge. This should be meaningful, since nuclei and electrons are so exceedingly small.

But is it then really possible to define the electric field at a point P inside a continuously distributed charge? It appears at first sight that the dV contributed by the charge element $\rho\, d\tau'$ at P is infinite, since r is zero. In fact, it is not infinite.

Consider a volume element $d\tau'$ that is a spherical shell of thickness dr and radius r centered on P. The charge in this shell contributes at P a dV of $(\rho/4\pi\epsilon_0)(4\pi r^2\, dr/r)$. Another shell of smaller radius contributes a smaller dV because, as r decreases, the $4\pi r^2\, dr$ in the numerator decreases more rapidly than does the r in the denominator. The electric potential V therefore converges, and the integral is finite. A similar argument shows that the electric field intensity $\boldsymbol{E}$ also converges.

The electric fields of real charge distributions can therefore be calculated with the usual techniques of the integral calculus, both inside and outside the distributions.

2.5 GAUSS'S LAW

Gauss's law relates the flux of $\boldsymbol{E}$ through a closed surface to the total charge enclosed within the surface. Consider Figure 2-3, in which a point charge Q is located inside a closed surface S at a point P'. We can calculate the flux of the electric field intensity $\boldsymbol{E}$ through the closed surface as follows. The flux of $\boldsymbol{E}$ through the element of area $\boldsymbol{da}$ is

$$\boldsymbol{E}\cdot\boldsymbol{da} = \frac{Q}{4\pi\epsilon_0}\frac{\boldsymbol{r}_1\cdot\boldsymbol{da}}{r^2} \tag{2-20}$$

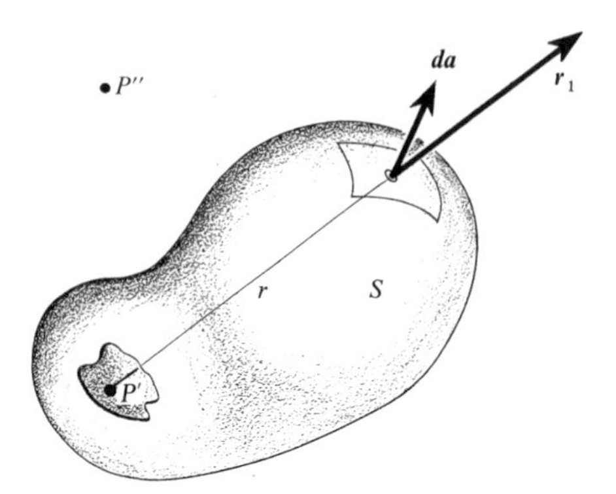

Figure 2-3. A point charge Q located at P' inside a closed surface S. The total flux of the electric field intensity $\boldsymbol{E}$ through the surface S is Q/ϵ_0.

where $\boldsymbol{r}_1 \cdot \boldsymbol{da}$ is the projection of $\boldsymbol{da}$ on a plane normal to $\boldsymbol{r}_1$. Then

$$\boldsymbol{E} \cdot \boldsymbol{da} = \frac{Q}{4\pi\epsilon_0} \, d\Omega, \tag{2-21}$$

where $d\Omega$ is the element of solid angle subtended by $\boldsymbol{da}$ at the point P'.

To find the total flux of $\boldsymbol{E}$, we integrate over the whole surface S:

$$\int_S \boldsymbol{E} \cdot \boldsymbol{da} = \frac{Q}{\epsilon_0}. \tag{2-22}$$

If the point charge Q were outside the surface at some point P'', the solid angle subtended by the surface S at P'' would be zero.

The situation remains unchanged if the surface is convoluted in such a way that a line drawn outward from P' cuts the surface at more than one point. The total solid angle subtended by the closed surface is still 4π at an inside point P', and is still zero at an outside point P''.

If more than one point charge resides within S, the fluxes add algebraically, and the total flux of $\boldsymbol{E}$ leaving the volume is equal to the total enclosed charge divided by ϵ_0. This is *Gauss's law*.

Gauss's law provides us with a powerful method for calculating the electric field intensity $\boldsymbol{E}$ of simple charge distributions.

If the charge enclosed by the surface S' is distributed over a finite volume, the total enclosed charge is

$$Q = \int_{\tau'} \rho \, d\tau', \tag{2-23}$$

where ρ is the charge density and τ' is the volume enclosed by the surface S'. Then

$$\int_{S'} \boldsymbol{E} \cdot \boldsymbol{da}' = \frac{1}{\epsilon_0} \int_{\tau'} \rho \, d\tau'. \tag{2-24}$$

This is *Gauss's law stated in integral form.*

Applying the divergence theorem to the left-hand side,

$$\int_{\tau'} \nabla \cdot \boldsymbol{E}\, d\tau' = \frac{1}{\epsilon_0} \int_{\tau'} \rho\, d\tau'. \tag{2-25}$$

Since this equation is valid for any closed surface in the field, the integrands must be equal; at every point in the field,

$$\nabla \cdot \boldsymbol{E} = \frac{\rho}{\epsilon_0}. \tag{2-26}$$

This is *Gauss's law stated in differential form*. It concerns the *derivatives* of $\boldsymbol{E}$ with respect to the coordinates, and not $\boldsymbol{E}$ itself.

Example

THE AVERAGE POTENTIAL OVER A SPHERICAL SURFACE

As an illustration of Gauss's law, we shall now show that the average potential over any spherical surface (a) is equal to the potential at the center of the sphere if there are no charges *inside* and (b) is equal to $Q/4\pi\epsilon_0 a$, where a is the radius of the sphere and Q is the enclosed charge, if there are no charges *outside*.

Let us first think of a spherical shell of radius a carrying a uniform surface charge density σ and a total charge Q, as in Figure 2-4a. Then, from Gauss's law, the electric field intensity at some point P situated outside at a distance R from the center of the shell is $Q/4\pi\epsilon_0 R^2$ and the potential is

$$V = \frac{Q}{4\pi\epsilon_0 R}. \tag{2-27}$$

Also, according to Coulomb's law,

$$V = \int \frac{\sigma\, da}{4\pi\epsilon_0 r}, \tag{2-28}$$

where the integral is evaluated over the surface of the shell. Equating the two values of V and remembering that σ is $Q/4\pi a^2$, we obtain a purely geometrical relation concerning a sphere of radius a and a point P at a distance R from its center, as can be seen by canceling the Q's:

$$\frac{Q}{4\pi\epsilon_0 R} = \frac{1}{4\pi a^2} \int \frac{Q\, da}{4\pi\epsilon_0 r}. \tag{2-29}$$

We now shift our attention to Figure 2-4b. The left-hand side of this equation is the potential at the center of the sphere due to the charge Q, and the right-hand side is the average potential on the spherical surface.

We have therefore demonstrated that the average potential over a spherical surface due to a point charge situated *outside* is equal

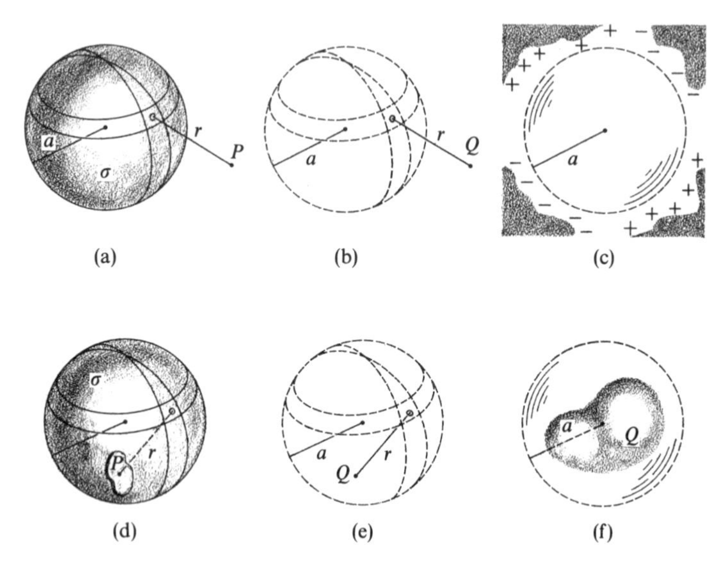

Figure 2-4. (a) Spherical surface carrying a uniform charge distribution σ. (b) Point charge Q outside an imaginary spherical surface of radius a. (c) Imaginary spherical surface in a charge-free region. It is shown that the average potential over the surface is equal to the potential at the center. (d) Spherical surface carrying a uniform charge distribution σ. The point P is situated inside. (e) Point charge Q inside an imaginary spherical surface of radius a. (f) Imaginary spherical surface enclosing a charge Q. It is shown that the average potential over the surface is equal to $Q/4\pi\epsilon_0 a$.

to the potential at the center. This result also applies to any charge distribution situated outside the sphere, as in Figure 2-4c, because of the principle of superposition.

Problem 2-16 illustrates one application of this property of the electric potential to the calculation of electric fields.

Now imagine for a moment that there is a potential maximum at some point O in a region where $\rho = 0$. Then the average potential over some sphere centered on O must be lower than the potential at O, which is contrary to the above result. *Thus we can never have a potential maximum in a charge-free region. For the same reason we can never have a potential minimum either.*

We can proceed in a similar fashion to find the average potential over a spherical surface when the charges are situated *inside*. We start again with a charge Q distributed uniformly over the surface as in Figure 2-4d. At any point P *inside*, the electric field intensity must be zero for the following reason. The symmetry of the field requires that E_θ and E_φ be zero. To find E_r we apply Gauss's law to a concentric spherical surface having some radius smaller than a and we find that E_r is also zero. Then $\boldsymbol{E}$ is zero inside, and the potential V at the point P inside must be equal to that at the surface, namely $Q/4\pi\epsilon_0 a$. Using now Coulomb's law as previously,

$$V = \frac{Q}{4\pi\epsilon_0 a} = \int \frac{\sigma\, da}{4\pi\epsilon_0 r} = \frac{1}{4\pi a^2} \int \frac{Q\, da}{4\pi\epsilon_0 r}. \tag{2-30}$$

But the last term is just the average potential over the spherical surface of Figure 2-4e.

Therefore the average potential over a spherical surface of radius a containing a point charge Q is equal to $Q/4\pi\epsilon_0 a$, whatever the position of Q inside the surface. The same applies if the charge Q is distributed over a finite volume as in Figure 2-4f.

2.6 THE EQUATIONS OF POISSON AND OF LAPLACE

If we replace $\boldsymbol{E}$ by $-\boldsymbol{\nabla} V$ in Eq. 2-26,

$$\nabla^2 V = -\frac{\rho}{\epsilon_0}. \tag{2-31}$$

This is *Poisson's equation.* In a region of the field where the charge density ρ is zero,

$$\nabla^2 V = 0, \tag{2-32}$$

which is *Laplace's equation.*

The general problem of finding the electric potential V corresponding to a given charge distribution amounts to finding a solution of either Laplace's or Poisson's equation that will satisfy the given boundary conditions.

REVIEW. At this stage it is worthwhile reviewing briefly the material that we have studied since the beginning of the chapter.

Coulomb's law states that the force between two stationary electric charges (a) is proportional to the magnitudes of each one of the two charges, (b) is directed along the line joining the two charges, and (c) varies inversely as the square of the distance between them. We interpreted this force as the product of the magnitude of either one of the charges by the electric field intensity of the other at the point occupied by the first one.

The principle of superposition follows from the observation that each charge produces its own field independently of all the other charges and that the resultant electric field intensity is therefore the vector sum of all the individual fields.

It follows from (b) and from the principle of superposition that the electrostatic field is conservative, and that we can describe the electric field intensity in terms of a potential V: $\boldsymbol{E} = -\boldsymbol{\nabla} V$.

Finally, (a), (b), (c), and the principle of superposition led us to Gauss's law $\boldsymbol{\nabla} \cdot \boldsymbol{E} = \rho/\epsilon_0$. When we expressed this in terms of the electric potential V, we found Poisson's equation $\boldsymbol{\nabla}^2 V = -\rho/\epsilon_0$, and Laplace's equation $\boldsymbol{\nabla}^2 V = 0$ for the special case where $\rho = 0$.

2.7 CONDUCTORS

A conductor can be defined as a material inside which electric charges can flow freely. Since we are dealing with electrostatics, we assume *a priori* that the charges have reached their equilibrium positions and are fixed in space. Then, inside a conductor, there is zero electric field and all points are at the same potential.

Thus, when a conductor is charged, the charges arrange themselves so that the net electric field due to *all* the charges can be zero inside the conductor. If a conductor is placed in an electric field, charges flow temporarily within it so as to produce a second field that, added to the first, will give a net field equal to zero.

It is therefore impossible to have net charge in one region of a conductor without having another set of charges, somewhere, to make the net electric field intensity within the conductor equal to zero. For example, it is impossible to have a charge density on one surface of an isolated plane conducting plate without having an identical charge density on the other side so as to produce within the plate two fields of equal intensity but of opposite direction. Or, if one plate of a parallel-plate capacitor is charged on the inside face, an equal and opposite charge must exist on the inside face of the opposite plate in order that $\boldsymbol{E}$ inside the plates can be zero.

Coulomb's law applies within conductors, even though the *net* field is zero. Then Gauss's law $\boldsymbol{\nabla} \cdot \boldsymbol{E} = \rho/\epsilon_0$, which is a consequence of Coulomb's law, must also be valid within conductors. Now, under static conditions, $\boldsymbol{E} = 0$ inside a conductor, and hence the charge density ρ must be zero.

As a corollary, any net static charge on a conductor must reside on its surface.

At the surface of a conductor, the electric field intensity $\boldsymbol{E}$ must be normal, for if there were a tangential component of $\boldsymbol{E}$, charges would flow along the surface, which would be contrary to our hypothesis. Then, according to Gauss's law, just outside the surface, $E = \sigma/\epsilon_0$, where σ is the surface charge density.

It is paradoxical that we should be able to express the electric field intensity at the surface of a conductor in terms of the local surface charge density σ alone, despite the fact that the field is of course due to *all* the charges, whether they are on the conductor or elsewhere.

Example A hollow conductor has a charge on its inner surface that is equal in magnitude and opposite in sign to any charge that may be enclosed within the hollow. This is readily demonstrated by considering a Gaussian surface that lies within the conductor and that encloses the hollow. Since $\boldsymbol{E}$ is everywhere zero on this surface, the total enclosed charge must be zero.

2.8 CALCULATION OF THE ELECTRIC FIELD PRODUCED BY A SIMPLE CHARGE DISTRIBUTION

A problem often encountered in electrostatics is that of finding the electric field intensity $\boldsymbol{E}$ produced by a given charge distribution. The vector $\boldsymbol{E}$ can be found in several different ways.

(a) We can use Eq. 2-7 and integrate over the whole charge distribution. We must of course keep in mind that the electric field intensity is a vector quantity.

(b) Another way consists in writing down the potential produced by an element of charge, integrating over the whole charge to obtain V, and then calculating the electric field intensity from $\boldsymbol{E} = -\boldsymbol{\nabla} V$. The potential calculation is generally simpler because the potential is a scalar quantity that can be integrated more easily than the vector $\boldsymbol{E}$.

(c) In some cases we can integrate either Laplace's or Poisson's equation, depending on whether the charge density ρ is zero or not, if the geometry is simple enough, and if we have sufficient information about the field to determine the constants of integration.

(d) We can use Gauss's law if the charge distribution possesses a symmetry that ensures constancy of the electric field intensity over certain imaginary surfaces in the field. When this method is applicable, it is usually the easiest one.

(e) We shall find still another method in Section 4.7.

These five types of calculation are often inadequate, and we must then use other methods, described in Chapter 4 and in Appendix B. Even then, only quite simple charge distributions can be treated by analytical means. The fields of more complex charge distributions are calculated on electronic computers.

Example

FIELD OF A UNIFORM SPHERICAL CHARGE DISTRIBUTION

As an illustration, we shall calculate the electric field intensity $\boldsymbol{E}$, both inside and outside a uniform spherical charge distribution, using the methods (a), (b), (c), and (d). Our calculation will be quite detailed and will serve as a review of all the material we have discussed so far.

The distribution has a radius R and a uniform electric charge density ρ, as in Figure 2-5. Our problem is to find the electric field intensity $\boldsymbol{E}$ as a function of the distance r from the center O of the sphere to the point P. It should be obvious, by symmetry, that $\boldsymbol{E}$ is independent of the two other spherical coordinates θ and φ.

We shall use the index o to indicate that we are dealing with the field outside the charge distribution, and the index i inside.

Field $\boldsymbol{E}_o$ at an External Point

Calculation of $\boldsymbol{E}_o$ Using Coulomb's Law. Let us first calculate $\boldsymbol{E}_o$ directly from Eq. 2-7. We can find the contribution to $\boldsymbol{E}_o$ due to the charge $\rho\, d\tau'$ in the element of volume $d\tau'$ and then integrate the resultant expression over the whole sphere. It is convenient to use spherical coordinates, since the charge has spherical symmetry. Then the volume element is $dr'\, r'\, d\theta'\, r' \sin\theta'\, d\varphi'$. The charge in this volume produces a field at the point P which is directed away from the volume element if ρ is positive, and toward the volume element if ρ is negative. Its magnitude is

$$d^3e_o = \frac{1}{4\pi\epsilon_0}\frac{\rho r'^2 \sin\theta'\, dr'\, d\theta'\, d\varphi'}{s^2}, \tag{2-33}$$

where s is the distance from the volume element to the point P. The axis along which $\theta = 0$ can be taken to be the line OP. The element of electric field intensity is written d^3e_o, since it is a third-order differential.

It should be obvious, from the symmetry of the charge distribution, that $\boldsymbol{E}_o$ must be radial. For example, while the charge element shown in Figure 2-5 produces an electric field intensity $\boldsymbol{d^3e_o}$ that is not along the radius OP, there is another symmetrically placed element that produces a symmetrically oriented field of the same magnitude, and the net result is a field along OP.

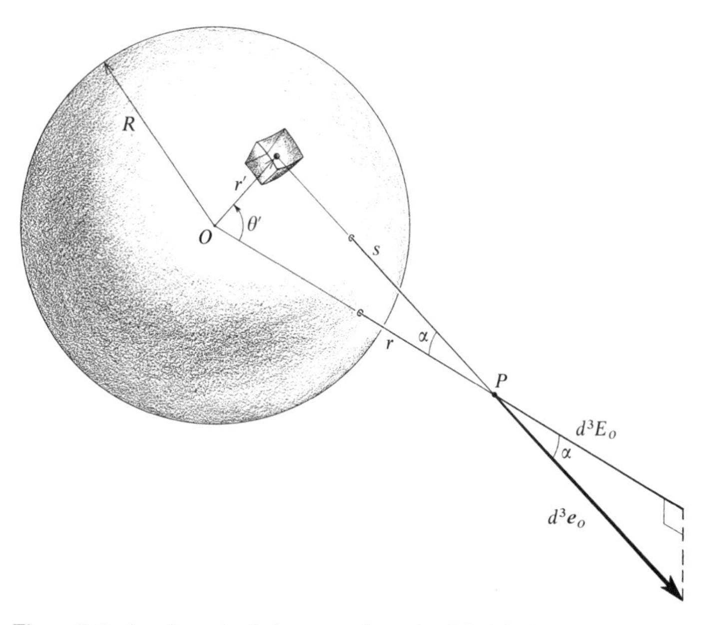

Figure 2-5. An element of charge at the point (r', θ) inside a uniform, spherical charge distribution produces an element of electrostatic field intensity $\boldsymbol{d^3E_0'}$ at a point P outside the sphere. The projection of $\boldsymbol{d^3E_o'}$ on the axis joining P and the center of the sphere is d^3E_o.

We therefore consider only the radial component of $\boldsymbol{e}_o$ and

$$d^3E_o = \frac{1}{4\pi\epsilon_0}\frac{\rho r'^2 \sin\theta' \, dr' \, d\theta' \, d\varphi'}{s^2} \cos\alpha. \qquad (2\text{-}34)$$

The integration over the azimuth angle φ' around OP is straightforward, and the angle φ' varies from 0 to 2π. We can carry out the other two integrations by using r' and s as independent variables. To do this, we eliminate α with the aid of the cosine law:

$$\cos\alpha = \frac{s^2 + r^2 - r'^2}{2sr}. \qquad (2\text{-}35)$$

Similarly,

$$\cos\theta' = \frac{r'^2 + r^2 - s^2}{2r'r}. \qquad (2\text{-}36)$$

Now we wish to eliminate $\sin\theta' \, d\theta'$ from the expression for d^3E_o. We can find $\sin\theta' \, d\theta'$ in terms of r', r, s, by differentiating the above equation. Here we must remember that r is a constant and that r' is taken as a constant when we integrate Eq. 2-34 with respect to θ'. Thus we must differentiate Eq. 2-36, taking both r and r' as constants, and thus

$$\sin\theta' \, d\theta' = \frac{s \, ds}{r'r}. \qquad (2\text{-}37)$$

If we substitute Eqs. 2-35 and 2-37 into Eq. 2-34, and integrate,

$$E_o = \frac{\rho\pi}{4\pi\epsilon_0 r^2}\int_{r'=0}^{r'=R}\int_{s=r-r'}^{s=r+r'} r'\left(1+\frac{r^2-r'^2}{s^2}\right) ds\, dr', \quad (2\text{-}38)$$

$$\mathbf{E}_o = \frac{1}{4\pi\epsilon_0}\frac{(4/3)\pi R^3\rho}{r^2}\mathbf{r}_1 = \frac{1}{4\pi\epsilon_0}\frac{Q}{r^2}\mathbf{r}_1, \quad (2\text{-}39)$$

where Q is the total charge $(4/3)\pi R^3\rho$, and where $\mathbf{r}_1$ is the radial unit vector directed outward.

The vector $\mathbf{E}_o$ is directed outward along OP if Q is positive, and inward along OP if Q is negative.

This result is the same as if the total charge Q were concentrated at the center of the sphere.

Calculation of $\mathbf{E}_o$ *Using the Potential.* To compute the electric field intensity $\mathbf{E}_o$ from the potential V_o we use the same element of charge as above. Then, from the definition of V_o,

$$d^3V_o = \frac{1}{4\pi\epsilon_0}\frac{\rho r'^2 \sin\theta'\, dr'\, d\theta'\, d\varphi'}{s}. \quad (2\text{-}40)$$

There is no $\cos\alpha$ term now, since V_o is a scalar. To carry out the integration, we eliminate θ' and integrate over φ', s, and r' as we did before. The result is

$$V_o = \frac{1}{4\pi\epsilon_0}\frac{Q}{r}. \quad (2\text{-}41)$$

The electric potential V_o, like E_o, is the same as if the total charge Q were concentrated at the center of the sphere.

To find $\mathbf{E}_o$, we now calculate $\boldsymbol{\nabla} V_o$. By symmetry, $\mathbf{E}_o$ must be radial, hence

$$\mathbf{E}_o = -\boldsymbol{\nabla} V_o = -\frac{\partial V_o}{\partial r}\mathbf{r}_1 = \frac{1}{4\pi\epsilon_0}\frac{Q}{r^2}\mathbf{r}_1, \quad (2\text{-}42)$$

as previously.

Calculation of $\mathbf{E}_o$ *Using Laplace's Equation.* By hypothesis, $\rho = 0$ outside the sphere, and

$$\nabla^2 V_o = 0. \quad (2\text{-}43)$$

Now, by symmetry, V_o is independent of both θ and φ. Therefore

$$\frac{1}{r^2}\frac{\partial}{\partial r}\left(r^2\frac{\partial V_o}{\partial r}\right) = 0, \quad (2\text{-}44)$$

$$\frac{\partial V_o}{\partial r} = \frac{A}{r^2}, \quad (2\text{-}45)$$

$$E_o = -\frac{A}{r^2}, \quad (2\text{-}46)$$

where A is a constant of integration. We shall be able to determine its value later on, after we have found E_i.

Calculation of $\mathbf{E}_o$ *Using Gauss's Law.* The simplest way to compute the electric field intensity in this case is to use Gauss's law. Consider an imaginary sphere of radius $r > R$ concentric with the charged sphere. We know that $\mathbf{E}_o$ must be radial. Then, according to Gauss's law,

$$4\pi r^2 E_o = \frac{Q}{\epsilon_0}, \tag{2-47}$$

and again

$$\mathbf{E}_o = \frac{1}{4\pi\epsilon_0}\frac{Q}{r^2}\mathbf{r}_1. \tag{2-48}$$

If the charge were not distributed with spherical symmetry, $\mathbf{E}_o$ would be a function of θ and of φ, and it would not be constant over the imaginary sphere. Gauss's law would then only give the average value of the normal component of $\mathbf{E}_o$ over the imaginary sphere.

Field $\mathbf{E}_i$ at an Internal Point

Let us now calculate the electric field intensity at a point P within the charge distribution, as shown in Figure 2-6. We can proceed as in the case of the external point, first writing down the contribution from an element of charge either to $\mathbf{E}_i$ or to V_i, and then integrating over the whole charge distribution. However, since the integration is difficult to carry out, we shall simplify the problem by dividing it into two distinct parts.

We draw an imaginary sphere of radius r through the point P, as in Figure 2-6, to divide the charge distribution into two parts. We then calculate the electric field intensity due to the charge contained within the sphere of radius r and that due to the charge contained within the hollow outer sphere of inner and outer radii r and R, respectively. Then, by the principle of superposition, the resultant field intensity for the two charge systems must be the vector sum of the two component field intensities.

The separation of the charge into two systems is especially advantageous in this case because, as we shall see, the field produced by the hollow outer sphere at a point on its inner surface, or at any point within the hollow, is zero. This can be demonstrated as follows, without integrating.

We draw a small cone of solid angle $d\Omega$, having its vertex at the point P and extending in both directions as in Figure 2-6, and consider the volumes that these small cones intercept within a spherical shell of radius r' and thickness dr' concentric with the sphere. The distances between these volumes and P are s_1 and s_2.

On the left, the volume element is

$$d^2\tau_L = \frac{s_1^2\, d\Omega}{\cos\alpha}\, dr', \tag{2-49}$$

whereas on the right it is

$$d^2\tau_R = \frac{s_2^2\, d\Omega}{\cos\alpha}\, dr'. \tag{2-50}$$

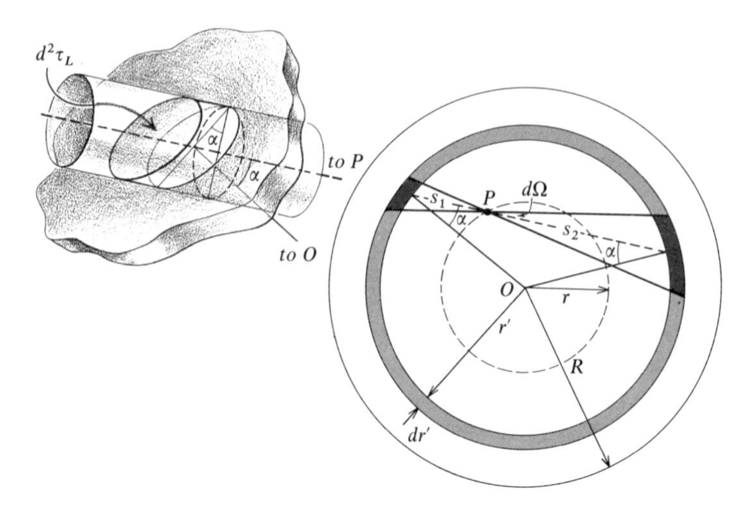

Figure 2-6. To find the electrostatic field intensity $\boldsymbol{E}_i$ at a point P inside a uniform spherical charge distribution, we divide the sphere into a shell and a core by means of an imaginary sphere of radius r. Then any pair of volume elements such as those shown in the shell produces equal and opposite fields at P. The field at P is thus due solely to the charges in the core. The exploded view shows one of the volume elements in detail.

The charge in the left-hand volume element contributes at P a field of magnitude

$$d^2E_i = \frac{\rho}{4\pi\epsilon_0 s_1^2}\frac{s_1^2\,d\Omega}{\cos\alpha}\,dr', \tag{2-51}$$

which is directed toward the right if ρ is positive. Similarly, the charge on the right contributes an identical field, opposite in direction, with the result that the two fields cancel. Since this result is valid for any $d\Omega$ and any dr', the field due to the hollow sphere at a point on its inner surface, or at any point within the hollow, is zero.

A much simpler way of showing that the field due to the hollow sphere is zero anywhere inside the hollow is to use Gauss's law. Imagine a concentric spherical surface within the hollow. According to Gauss's law, the average radial electric field intensity over this surface must be zero, since there is no enclosed charge. Now the symmetry of the problem requires that the electric field, if it exists, be radial and be the same over all the surface of the sphere. Hence the electric field intensity must be zero at every point on any spherical surface within the hollow.

Calculation of $\boldsymbol{E}_i$ Using Coulomb's Law. With the electric field of the hollow sphere thus disposed of, we can calculate the contribution to $\boldsymbol{E}_i$ that is due to the inner sphere of radius r, just as we did in the case of the external point:

$$\boldsymbol{E}_i = \frac{1}{4\pi\epsilon_0}\frac{(4/3)\pi r^3\rho}{r^2}\,\boldsymbol{r}_1 = \frac{\rho r}{3\epsilon_0}\,\boldsymbol{r}_1. \tag{2-52}$$

The electric field intensity thus increases linearly with r inside the spherical charge distribution.

Calculation of $\mathbf{E}_i$ *Using the Potential.* We can arrive at the same result by first calculating the potential V_i as a function of r within the charge distribution. To do this, we could proceed by direct integration. However, it will again be easier and more instructive to divide the charge distribution into two parts as above.

Let us first consider the hollow shell. We have seen that there is no electric field inside a hollow sphere of charge. Then all points within the hollow must be at the same potential and, instead of calculating the potential at a point on the interior surface of the shell, we can calculate the potential at the center of the shell, where the integration is more easily performed. We choose our elementary volume to be a thin shell of radius r' and of thickness dr'. Thus the part of V_i that is due to the hollow sphere is

$$\frac{1}{4\pi\epsilon_0}\int_r^R \frac{\rho 4\pi r'^2\, dr'}{r'} = \frac{\rho}{2\epsilon_0}(R^2 - r^2). \tag{2-53}$$

Next we compute the potential due to the interior sphere of radius r. The calculation is the same as for an external point, and we can use Eq. 2-41. This term is

$$\frac{1}{4\pi\epsilon_0}\frac{\rho(4/3)\pi r^3}{r} = \frac{\rho}{3\epsilon_0} r^2. \tag{2-54}$$

Adding these two contributions, we obtain the potential V_i at a radius r inside the spherical charge distribution:

$$V_i = \frac{\rho}{\epsilon_0}\left(\frac{R^2}{2} - \frac{r^2}{6}\right). \tag{2-55}$$

The potential V_i can also be written as

$$V_i = \frac{Q}{4\pi\epsilon_0 R^3}\frac{(R^2 - r^2)}{2} + \frac{Q}{4\pi\epsilon_0 R}, \tag{2-56}$$

where the second term is the potential at the surface of the sphere, and where the first term is the increase above the surface value for the interior points.

Then

$$\mathbf{E}_i = -\boldsymbol{\nabla} V_i = -\frac{\partial V_i}{\partial r}\mathbf{r}_1 = \frac{\rho r}{3\epsilon_0}\mathbf{r}_1. \tag{2-57}$$

Calculation of $\mathbf{E}_i$ *Using Poisson's Equation.* We now have, inside the charge distribution,

$$\nabla^2 V_i = -\frac{\rho}{\epsilon_0}, \tag{2-58}$$

$$\frac{1}{r^2}\frac{\partial}{\partial r}\left(r^2\frac{\partial V_i}{\partial r}\right) = -\frac{\rho}{\epsilon_0}, \tag{2-59}$$

$$\frac{\partial}{\partial r}\left(r^2 \frac{\partial V_i}{\partial r}\right) = -\frac{\rho}{\epsilon_0} r^2, \tag{2-60}$$

$$r^2 \frac{\partial V_i}{\partial r} = -\frac{\rho r^3}{3\epsilon_0} + B, \tag{2-61}$$

$$E_i = \frac{\rho r}{3\epsilon_0} - \frac{B}{r^2}, \tag{2-62}$$

where B is a constant of integration.

It is intuitively obvious that E_i cannot become infinite at the center of a uniform spherical charge distribution; B must therefore be zero, and

$$\mathbf{E}_i = \frac{\rho r}{3\epsilon_0} \mathbf{r}_1. \tag{2-63}$$

We are now in a position to find the value of the constant of integration A, which we found when calculating $\mathbf{E}_o$ with the equation of Laplace. Should not the two values we have found for the electric field intensity $\mathbf{E}$, one valid inside and one valid outside (Eqs. 2-39 and 2-52), be equal at the surface of the charge distribution at $r = R$? How could they be different? According to Gauss's law, they could be different if we had a surface charge density as on the surface of a charged conductor. But, by assumption, the charged sphere has a uniform volume density ρ out to the radius R and there can be no

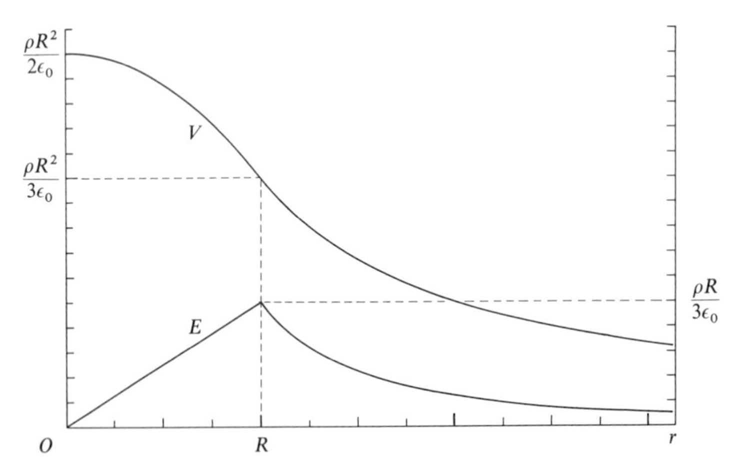

Figure 2-7. In a uniform spherical charge distribution, the electric field intensity E rises linearly from the center to the surface of the sphere and falls off as the inverse square of the distance outside the sphere. On the other hand, the electric potential falls from a maximum at the center in parabolic fashion inside the sphere, and as the inverse first power of the distance outside.

discontinuity in $E = -\partial V/\partial r$ at the surface. Thus our two values of E must be equal at the surface:

$$E_{r=R} = -\frac{A}{R^2} = \frac{\rho R}{3\epsilon_0}, \tag{2-64}$$

so that

$$A = -\frac{\rho R^3}{3\epsilon_0} = -\frac{Q}{4\pi\epsilon_0}, \tag{2-65}$$

and Eq. 2-46 does give the correct value for E_o.

Calculation of $\mathbf{E}_i$ *Using Gauss's Law.* To calculate the electric field intensity at the internal point from Gauss's law, we draw an imaginary sphere of radius r through the point P. Symmetry requires that the electric field intensity be radial; thus

$$4\pi r^2 E_i = \frac{\rho(4/3)\pi r^3}{\epsilon_0}, \tag{2-66}$$

$$\mathbf{E}_i = \frac{\rho r}{3\epsilon_0}\mathbf{r}_1, \tag{2-67}$$

as previously.

Figure 2-7 shows E and V for our spherical charge distribution of radius R as a function of the radial distance r.

2.9 THE ELECTRIC DIPOLE

The electric dipole shown in Figure 2-8 is one type of charge distribution that is encountered frequently. We shall return to it later in this chapter and also in Chapters 3 and 14.

The electric dipole consists of two charges, a positive and a negative charge of the same magnitude, separated by a distance s, which is small compared to the distance r to the point P at which we require the electric potential V and the electric field intensity $\mathbf{E}$.

At P,

$$V = \frac{Q}{4\pi\epsilon_0}\left(\frac{1}{r_b} - \frac{1}{r_a}\right), \tag{2-68}$$

where

$$r_a^2 = r^2 + \left(\frac{s}{2}\right)^2 + rs\cos\theta, \tag{2-69}$$

$$\frac{r}{r_a} = \left[1 + \left(\frac{s}{2r}\right)^2 + \frac{s}{r}\cos\theta\right]^{-1/2}, \tag{2-70}$$

$$= 1 - \frac{1}{2}\left(\frac{s^2}{4r^2} + \frac{s}{r}\cos\theta\right) + \frac{3}{8}\left(\frac{s^2}{4r^2} + \frac{s}{r}\cos\theta\right)^2 - \cdots, \tag{2-71}$$

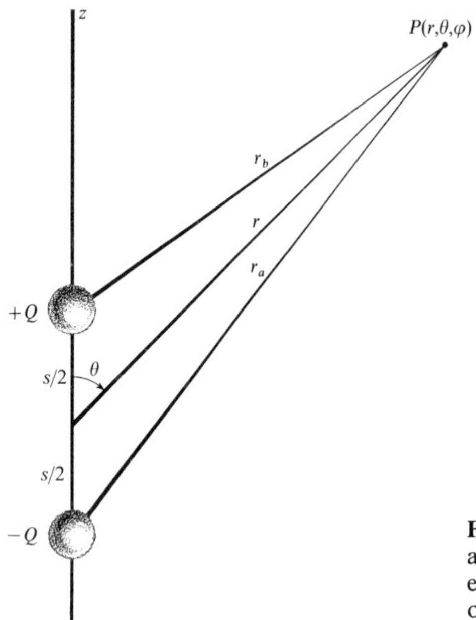

Figure 2-8. The two charges $+Q$ and $-Q$ form a dipole. The electric potential at P is the sum of the potentials due to the individual charges.

or, if we neglect terms of order higher than $\frac{s^2}{r^2}$,

$$\frac{r}{r_a} = 1 - \frac{s}{2r}\cos\theta + \frac{s^2}{4r^2}\,\frac{3\cos^2\theta - 1}{2}. \tag{2-72}$$

Similarly,

$$\frac{r}{r_b} = 1 + \frac{s}{2r}\cos\theta + \frac{s^2}{4r^2}\,\frac{3\cos^2\theta - 1}{2}, \tag{2-73}$$

$$V = \frac{Qs}{4\pi\epsilon_0 r^2}\cos\theta \qquad (r^3 \gg s^3). \tag{2-74}$$

It is interesting to note that the potential due to a dipole falls off as $1/r^2$, whereas the potential from a single point charge varies only as $1/r$. This comes from the fact that the charges of a dipole appear close together for an observer some distance away, and that their fields cancel more and more as the distance r increases.

We define the *dipole moment* $\mathbf{p} = Q\mathbf{s}$ as a vector whose magnitude is Qs and which is directed from the negative to the positive charge. Then

$$V = \frac{\mathbf{p}\cdot\mathbf{r}_1}{4\pi\epsilon_0 r^2}. \tag{2-75}$$

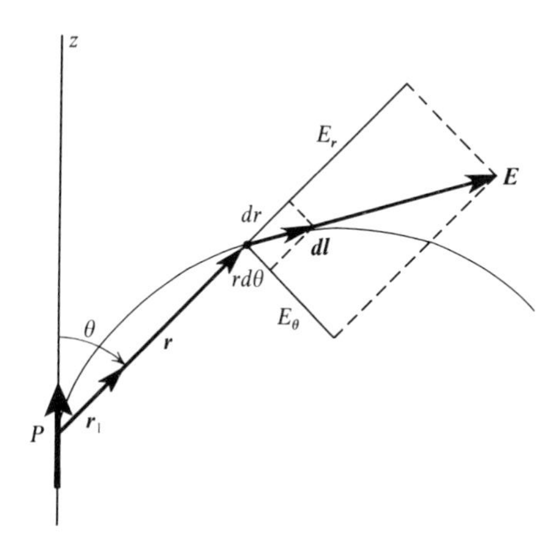

Figure 2-9. A dipole $\boldsymbol{p}$ produces an electric field intensity $\boldsymbol{E}$ with components E_r and E_θ at the point (r, θ). The element $\boldsymbol{dl}$ of the line of force is parallel to $\boldsymbol{E}$ at the point.

The components of $\boldsymbol{E}$ in spherical coordinates can be computed from the gradient of V:

$$E_r = -\frac{\partial V}{\partial r} = \frac{1}{4\pi\epsilon_0}\frac{2p}{r^3}\cos\theta, \tag{2-76}$$

$$E_\theta = -\frac{1}{r}\frac{\partial V}{\partial \theta} = \frac{1}{4\pi\epsilon_0}\frac{p}{r^3}\sin\theta, \tag{2-77}$$

$$E_\varphi = -\frac{1}{r\sin\theta}\frac{\partial V}{\partial \varphi} = 0. \tag{2-78}$$

The electric field intensity of a dipole thus falls off as the *cube* of the distance.

The equation for the lines of force of a dipole can be found by considering Figure 2-9, which shows an element $\boldsymbol{dl}$ of a line of force. Since the two vectors are parallel, the components of $\boldsymbol{E}$ and of $\boldsymbol{dl}$ are proportional and

$$\frac{r\,d\theta}{dr} = \frac{E_\theta}{E_r} = \frac{\sin\theta}{2\cos\theta}, \tag{2-79}$$

$$\frac{dr}{r} = \frac{2d(\sin\theta)}{\sin\theta}, \tag{2-80}$$

$$r = A\sin^2\theta. \tag{2-81}$$

This equation determines the family of lines of force for an electric dipole. The constant A is a parameter that varies from one line of force to another.

Figure 2-10 shows lines of force and equipotential lines for an electric dipole. Equipotential surfaces are generated by rotating equipotential lines around the vertical axis.

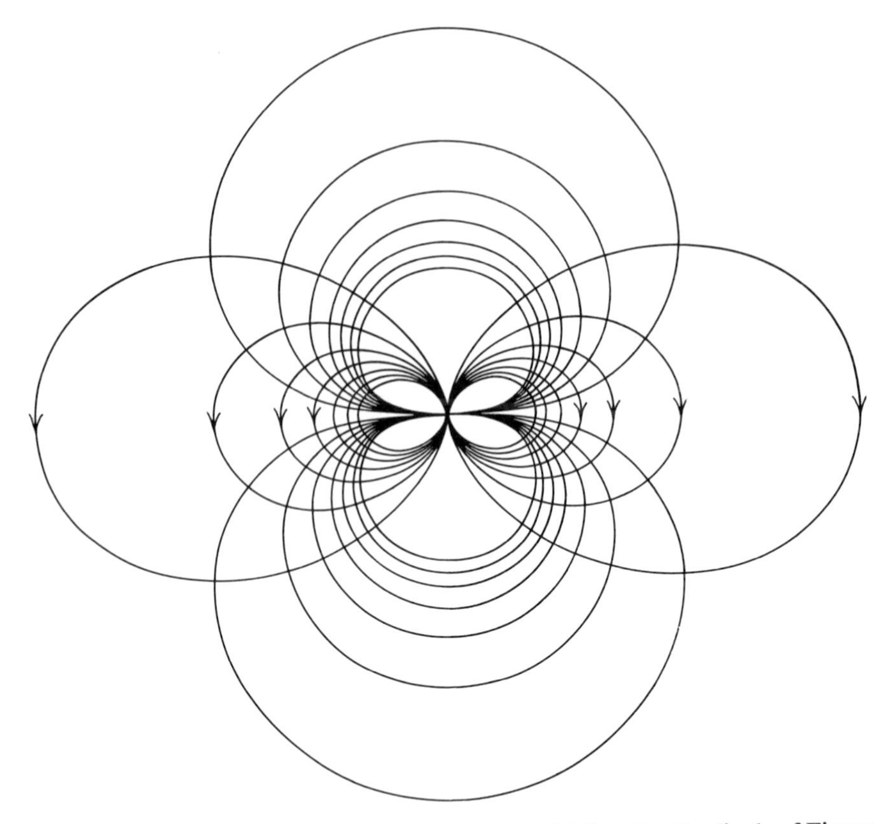

Figure 2-10. Lines of force (arrows) and equipotential lines for the dipole of Figure 2-8. The dipole is in the vertical position at the center of the figure with the positive charge very close to and above the negative charge. In the central region the lines come too close together to be shown.

2.10 THE LINEAR ELECTRIC QUADRUPOLE

The linear electric quadrupole is an arrangement of three charges as in Figure 2-11. The separation s of the charges is again assumed to be small compared to the distance r to the point P. At P,

$$V = \frac{1}{4\pi\epsilon_0}\left(\frac{Q}{r_a} - \frac{2Q}{r} + \frac{Q}{r_b}\right), \tag{2-82}$$

$$= \frac{Q}{4\pi\epsilon_0 r}\left(\frac{r}{r_a} + \frac{r}{r_b} - 2\right). \tag{2-83}$$

The ratios r/r_a and r/r_b can be expanded as previously, except that s re-

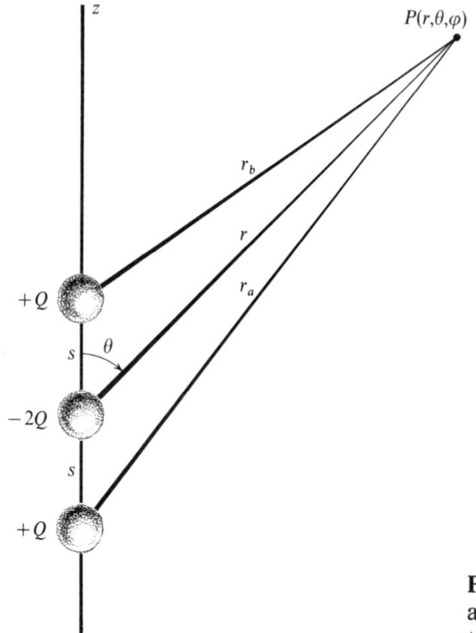

Figure 2-11. Charges $+Q$, $-2Q$, and $+Q$ arranged along a line to form an axial quadrupole.

places $s/2$. Thus, neglecting terms of order higher than s^2/r^2,

$$\frac{r}{r_a} = 1 - \frac{s}{r}\cos\theta + \frac{s^2}{r^2}\frac{(3\cos^2\theta - 1)}{2}, \tag{2-84}$$

$$\frac{r}{r_b} = 1 + \frac{s}{r}\cos\theta + \frac{s^2}{r^2}\frac{(3\cos^2\theta - 1)}{2}, \tag{2-85}$$

and

$$V = \frac{2Qs^2}{4\pi\epsilon_0 r^3}\frac{(3\cos^2\theta - 1)}{2} \qquad (r^3 \gg s^3). \tag{2-86}$$

The electric potential V due to a linear electric quadrupole thus varies as $1/r^3$, whereas the electric field intensity $\boldsymbol{E}$, calculated as for the dipole, varies as $1/r^4$. The fields of the three charges $+Q$, $-2Q$, $+Q$ cancel almost completely for $r \gg s$.

2.11 ELECTRIC MULTIPOLES

It is possible to extend the dipole and quadrupole concepts to larger numbers of positive and negative charges located at small distances from each other. Such charge arrangements are known as *multipoles* and are defined as follows. A single point charge is called a *monopole*. A *dipole* is obtained by displacing a monopole through a small distance $\mathbf{s}_1$ and replacing the original monopole by another of the same magnitude but of opposite sign. Likewise, a *quadrupole* is obtained by displacing a dipole by a small distance $\mathbf{s}_2$ and then replacing the original dipole by one of equal magnitude but of opposite sign. For the linear quadrupole, $\mathbf{s}_2 = \mathbf{s}_1$.

The multipole concept can be continued indefinitely. For example, the quadrupole may be displaced by a small distance $\mathbf{s}_3$, and the original quadrupole replaced by one in which the signs of all the charges have been changed. This produces an *octupole*. In general, we can produce *2^l-poles*, where l is the number of independent displacements $\mathbf{s}_1$, $\mathbf{s}_2$, $\cdots$ required to specify the arrangement.

We have seen that the dipole potential varies as $1/r^2$ and that the quadrupole potential varies as $1/r^3$. For the general multipole, characterized by the letter l, the potential varies as $1/r^{l+1}$ and the field intensity as $1/r^{l+2}$.

2.12 THE ELECTRIC FIELD OUTSIDE AN ARBITRARY CHARGE DISTRIBUTION

Let us consider an arbitrary charge distribution of density $\rho(x', y', z')$ occupying a volume τ' and extending to a maximum distance $r'_{\max}$ from the origin of coordinates O. We select O either within the volume or close to it. Such a distribution is illustrated in Figure 2-12.

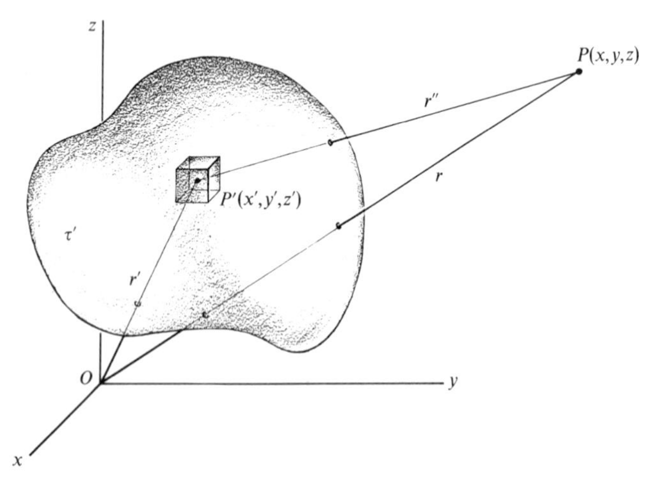

Figure 2-12. An arbitrary charge distribution of density ρ within a volume τ' produces an electric potential V at a point P (x, y, z) outside τ'. It is assumed here for the calculation of V that $r > r'_{\max}$.

We wish to find the electric potential V at some point $P(x, y, z)$ such that $r > r'_{\max}$. This is

$$V = \int_{\tau'} \frac{\rho \, d\tau'}{4\pi\epsilon_0 r''}, \tag{2-87}$$

where r'' is the distance between the point of observation P and the position $P'(x', y', z')$ of the element of charge $\rho \, d\tau'$:

$$r'' = [(x - x')^2 + (y - y')^2 + (z - z')^2]^{1/2}. \tag{2-88}$$

The point of observation $P(x, y, z)$ is taken to be fixed; thus r'' is a function of x', y', z'.

Since r'' is a function of x', y', z', let us expand $1/r''$ as a Taylor series near the origin:

$$\frac{1}{r''} = \frac{1}{r} + \left[x' \frac{\partial}{\partial x'} + y' \frac{\partial}{\partial y'} + z' \frac{\partial}{\partial z'}\right]_0 \left(\frac{1}{r''}\right)$$
$$+ \frac{1}{2!} \left[x' \frac{\partial}{\partial x'} + y' \frac{\partial}{\partial y'} + z' \frac{\partial}{\partial z'}\right]_0^2 \left(\frac{1}{r''}\right) + \cdots, \tag{2-89}$$

where the subscripts 0 indicate that the derivatives are evaluated at the origin. In squaring the second bracket, we consider the *factors* x', y', z' to be constants. The same applies to all the terms that follow.

Now

$$\frac{\partial}{\partial x'}\left(\frac{1}{r''}\right) = -\frac{1}{r''^2}\frac{\partial r''}{\partial x'} = \frac{x - x'}{r''^3} \tag{2-90}$$

and

$$\left[\frac{\partial}{\partial x'}\left(\frac{1}{r''}\right)\right]_0 = \frac{x}{r^3} = \frac{l}{r^2}, \tag{2-91}$$

where $l = x/r$ is the cosine of the angle between the vector $\boldsymbol{r}$ and the x-axis. The other first derivatives with respect to y' and z' are given by similar expressions.

The second derivatives of the third term on the right-hand side of Eq. 2-89 are calculated similarly, the substitution of $x' = 0$, $y' = 0$, $z' = 0$, $r'' = r$ being made only at the end.

The result of the calculation is that at the point P

$$V = \int_{\tau'} \frac{1}{r} \frac{\rho \, d\tau'}{4\pi\epsilon_0} + \int_{\tau'} \frac{1}{r^2} (lx' + my' + nz') \frac{\rho \, d\tau'}{4\pi\epsilon_0}$$
$$+ \int_{\tau'} \frac{1}{r^3} [3mny'z' + 3nlz'x' + 3lmx'y'$$
$$+ \tfrac{1}{2}(3l^2 - 1)x'^2 + \tfrac{1}{2}(3m^2 - 1)y'^2 + \tfrac{1}{2}(3n^2 - 1)z'^2] \frac{\rho \, d\tau'}{4\pi\epsilon_0} + \cdots, \tag{2-92}$$

where l, m, n are the direction cosines of the line joining the origin to the point of observation P.

Let us now examine the various terms in succession. The first term is merely the electric potential that we would have at P if the whole charge were concentrated at the origin. It is called the *monopole term* and is zero only if the total net charge is zero. If the charges are all of the same sign, then it is the most important term of the series, since it decreases only as $1/r$.

The second term varies as $1/r^2$, as does the electric potential of a dipole. Let us find the value of this term when the charge distribution is simply a dipole, as in Figure 2-8. The integral must then be evaluated over the two charges Q and $-Q$ situated at $z' = s/2$ and $z' = -s/2$ respectively. The result is the value of V for the dipole, as in Eq. 2-74. The second term V_2 of Eq. 2-92 can thus be taken to be a *dipole term*. In fact,

$$V_2 = \frac{\boldsymbol{p}\cdot\boldsymbol{r}_1}{4\pi\epsilon_0 r^2} \tag{2-93}$$

if we set

$$\boldsymbol{p} = \int_{\tau'} (x'\boldsymbol{i} + y'\boldsymbol{j} + z'\boldsymbol{k})\,\rho\,d\tau' = \int_{\tau'} \boldsymbol{r}'\rho\,d\tau', \tag{2-94}$$

$\boldsymbol{r}_1$ being the unit vector along $\boldsymbol{r}$ in the direction of P:

$$\boldsymbol{r}_1 = l\boldsymbol{i} + m\boldsymbol{j} + n\boldsymbol{k}. \tag{2-95}$$

The quantity $\boldsymbol{p}$ is the dipole moment of the charge distribution. It is a vector whose components are

$$p_x = \int_{\tau'} x'\rho\,d\tau', \qquad p_y = \int_{\tau'} y'\rho\,d\tau', \qquad p_z = \int_{\tau'} z'\rho\,d\tau'. \tag{2-96}$$

Therefore V_2, the second term of Eq. 2-92, is the same as if we had at the origin a small dipole with dipole moment $\boldsymbol{p}$, as in Eq. 2-94.

The dipole moment of an extended charge distribution can also be defined as

$$\boldsymbol{p} = Q\bar{\boldsymbol{r}}', \tag{2-97}$$

$$p_x = Q\bar{x}', \qquad p_y = Q\bar{y}', \qquad p_z = Q\bar{z}', \tag{2-98}$$

where Q is the total net charge in the distribution, and where $\bar{\boldsymbol{r}}'$ (the vector distance from the origin to the center of charge) is defined in a manner analogous to the center of mass in mechanics:

$$\bar{\boldsymbol{r}}' = \frac{\int_{\tau'} \boldsymbol{r}'\rho\,d\tau'}{\int_{\tau'} \rho\,d\tau'} = \frac{\int_{\tau'} \boldsymbol{r}'\rho\,d\tau'}{Q}. \tag{2-99}$$

If Q is zero, then $\bar{\boldsymbol{r}}' \to \infty$ from the above equation, and $\boldsymbol{p}$ is indeterminate in Eq. 2-97. Equation 2-94, however, always determines $\boldsymbol{p}$ unambiguously. When $Q = 0$, the dipole moment is independent of the choice of origin (see Problem 2-23).

If the net charge is *not* zero, the dipole moment of the distribution can always be made zero by choosing the origin at the center of charge, for then $\bar{\boldsymbol{r}}' = 0$.

Let us now consider the third term V_3 of Eq. 2-92. It involves a $1/r^3$ factor as the V of the linear quadrupole of Section 2-10. If we calculate V_3 for the linear quadrupole with charges $Q, -2Q, Q$ at $z = s$, $z = 0$, $z = -s$ respectively, we find that it gives the electric potential V of Eq. 2-86.

Then the third term V_3 of Eq. 2-92 is the same as if we had a small quadrupole at the origin. Let us write out its quadrupole moment. This third term can be rewritten as follows:

$$V_3 = \frac{1}{4\pi\epsilon_0 r^3}\left[3mn\int_{\tau'} y'z'\rho\,d\tau' + 3nl\int_{\tau'} z'x'\rho\,d\tau' + 3lm\int_{\tau'} x'y'\rho\,d\tau' + \frac{1}{2}(3l^2-1)\int_{\tau'} x'^2\rho\,d\tau' + \frac{1}{2}(3m^2-1)\int_{\tau'} y'^2\rho\,d\tau' + \frac{1}{2}(3n^2-1)\int_{\tau'} z'^2\rho\,d\tau'\right]. \tag{2-100}$$

Note that these integrals, just like the integrals of Eq. 2-94, depend solely on the distribution of charge within the volume τ', and not on the coordinates (x, y, z) of the point of observation P.

These six integrals specify a quantity called the quadrupole moment of the charge distribution:

$$p_{xx} = \int_{\tau'} x'^2\rho\,d\tau' = Q\overline{x'^2}, \tag{2-101}$$

$$p_{yy} = \int_{\tau'} y'^2\rho\,d\tau' = Q\overline{y'^2}, \tag{2-102}$$

$$p_{zz} = \int_{\tau'} z'^2\rho\,d\tau' = Q\overline{z'^2}, \tag{2-103}$$

$$p_{yz} = p_{zy} = \int_{\tau'} y'z'\rho\,d\tau' = Q\overline{y'z'}, \tag{2-104}$$

$$p_{zx} = p_{xz} = \int_{\tau'} z'x'\rho\,d\tau' = Q\overline{z'x'}, \tag{2-105}$$

$$p_{xy} = p_{yx} = \int_{\tau'} x'y'\rho\,d\tau' = Q\overline{x'y'}, \tag{2-106}$$

where the bars indicate average values. Note that the quadrupole moment has six components. Thus

$$V_3 = \frac{1}{4\pi\epsilon_0 r^3}\left[3mn\, p_{yz} + 3nl\, p_{zx} + 3lm\, p_{xy} + \frac{1}{2}(3l^2 - 1)\, p_{xx} + \frac{1}{2}(3m^2 - 1)\, p_{yy} + \frac{1}{2}(3n^2 - 1)\, p_{zz}\right]. \quad (2\text{-}107)$$

If the charge distribution displays cylindrical symmetry about the z-axis, the elements of charge at the points (x', y', z'), $(-x', y', z')$, $(x', -y', z')$, $(-x', -y', z')$ are all equal; their contributions to the integrals for p_{xy}, p_{yz}, p_{zx} cancel; and $p_{xy} = p_{yz} = p_{zx} = 0$. Also, by symmetry, $p_{xx} = p_{yy}$. In such cases it is convenient to define a single quantity q, often called the quadrupole moment of the charge distribution:

$$q = 2\,(p_{zz} - p_{xx}), \quad (2\text{-}108)$$

$$= \int_{\tau'} (3\, z'^2 - r'^2)\rho\, d\tau', \quad (2\text{-}109)$$

$$= Q\,\overline{(3z'^2 - r'^2)}. \quad (2\text{-}110)$$

For a charge distribution that has cylindrical symmetry about the z-axis, the quadrupole potential at the point (r, θ) is

$$V_3 = \frac{q}{4\pi\epsilon_0}\frac{(3\cos^2\theta - 1)}{4r^3}. \quad (2\text{-}111)$$

Note that, in general, the various multipole terms depend on the position of the origin, except for the monopole term that is proportional to the net charge Q of the system. The dipole term is independent of the position of the origin in the special case where $Q = 0$.

In summary, the potential due to an arbitrary charge distribution at a point outside the distribution such that $r > r'_{\max}$ is the same as: (a) the potential of a point charge equal to the net charge of the distribution, plus (b) that of a point dipole with a dipole moment equal to the dipole moment of the distribution, plus (c) that of a point quadrupole with a quadrupole moment equal to that of the distribution, and so on, all located at the origin.

Similarly, the electric field intensity $\boldsymbol{E} = -\boldsymbol{\nabla} V$ outside an arbitrary charge distribution is the sum of the electric field intensities of the above point charge, dipole, quadrupole, etc.

2.13 THE AVERAGE ELECTRIC FIELD INTENSITY INSIDE A SPHERE CONTAINING AN ARBITRARY CHARGE DISTRIBUTION

In discussing dielectrics in the next chapter, we shall require the average electric field intensity within a spherical surface enclosing an arbitrary charge

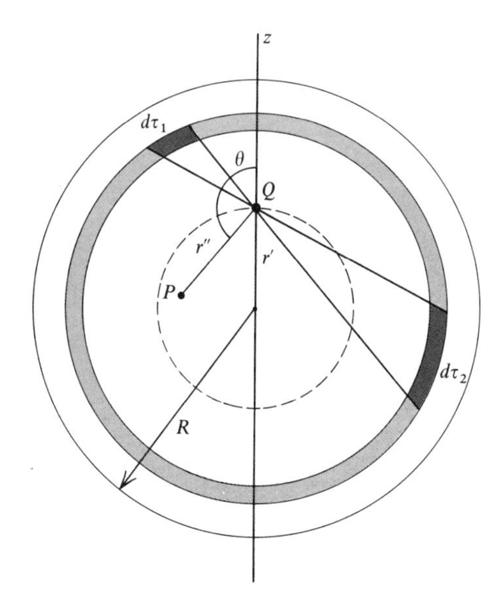

Figure 2-13. A point charge Q located at a distance r' from the center of a sphere of radius R. The average field intensity which Q produces within the sphere is proportional to the dipole moment of Q, is inversely proportional to the volume of the sphere, and is in the direction opposite to the dipole moment of Q. Figure 2-6 shows the elements of volume in detail.

distribution. Let us calculate the average electric field intensity at the interior point directly. To do this we shall first calculate the average $\boldsymbol{E}$ for a single point charge.

Figure 2-13 shows a point charge Q located at a distance r' from the center of a sphere of radius R. The z-axis is taken to be along the line joining the center of the sphere and the charge Q. By symmetry, the average field over the volume of the sphere must be along the z-axis.

We wish then to calculate

$$\overline{E}_z = \frac{\int_\tau E_z\, d\tau}{\int_\tau d\tau} = \frac{1}{\tau}\int_\tau E_z\, d\tau, \tag{2-112}$$

where τ is the volume of the sphere. This integral can be separated into two parts, one over the spherical shell between the radii r' and R, and one over the sphere of radius r'.

We can show that the integral over the outer volume is zero by using an argument somewhat analogous to that used on pages 57 to 61 to find the electric field inside a uniform spherical charge distribution. In Figure 2-13 the solid-angle element $d\Omega$ intercepts the volume elements $d\tau_1$ and $d\tau_2$ in the shell. Since the value of E_z *decreases* as the square of the distance from Q, whereas $d\tau$ *increases* as the square of the distance (see Eq. 2-49), their product remains constant. However, at $d\tau_1$, E_z is positive, whereas at $d\tau_2$ E_z is negative; thus the two contributions to the integral cancel. The same is true for the whole outer shell, the contributions to the integral canceling in pairs.

To calculate the integral of $E_z\,d\tau$ over the inner volume, which is equal to the same integral over the complete volume τ, we shall use spherical polar coordinates and set the origin at the position of the charge Q. Then, at the point P,

$$E_z = \frac{Q}{4\pi\epsilon_0 r''^2}\cos\theta, \tag{2-113}$$

and

$$\int_\tau E_z\,d\tau = \int_{\varphi=0}^{\varphi=2\pi}\int_{\theta=\pi/2}^{\theta=\pi}\int_{r''=0}^{r''=-2r'\cos\theta}\frac{Q}{4\pi\epsilon_0 r''^2}\cos\theta\,(r''^2\sin\theta\,dr''\,d\theta\,d\varphi). \tag{2-114}$$

In this integral, r'' varies from zero to $-2r'\cos\theta$, which is the distance from Q to the surface of the inner sphere in the direction θ. The negative sign comes from the fact that $\cos\theta$ is negative. Then θ sweeps from $\pi/2$ to π and φ from 0 to 2π. Thus

$$\int_\tau E_z\,d\tau = -\frac{Qr'}{\epsilon_0}\int_{\pi/2}^{\pi}\cos^2\theta\sin\theta\,d\theta, \tag{2-115}$$

$$= \frac{Qr'}{\epsilon_0}\left[\frac{\cos^3\theta}{3}\right]_{\pi/2}^{\pi} = -\frac{Qr'}{3\epsilon_0}, \tag{2-116}$$

and

$$\bar{E}_z = -\frac{Qr'}{3\epsilon_0}\frac{3}{4\pi R^3} = -\frac{Qr'}{4\pi\epsilon_0 R^3} \tag{2-117}$$

or, since Qr' is the dipole moment of the charge Q (Equation 2-94),

$$\bar{\boldsymbol{E}} = -\frac{\boldsymbol{p}}{4\pi\epsilon_0 R^3}. \tag{2-118}$$

This average electric field intensity has been calculated for a single charge Q on the z-axis. For an arbitrary charge distribution the field is the superposition of the fields due to the different charges, according to the theorem of superposition. Then $\bar{\boldsymbol{E}}$ is again as above, with $\boldsymbol{p}$ equal to the dipole moment of the arbitrary charge distribution within the sphere of radius R.

2.14 POTENTIAL ENERGY OF A CHARGE DISTRIBUTION

Let us first consider a set of small charges distributed in space as in Figure 2-14. We shall assume that the charge configuration is of finite extent and that no other charges are present. Each individual charge is situated in the electric field produced by the other charges, at a point where the potential V due to

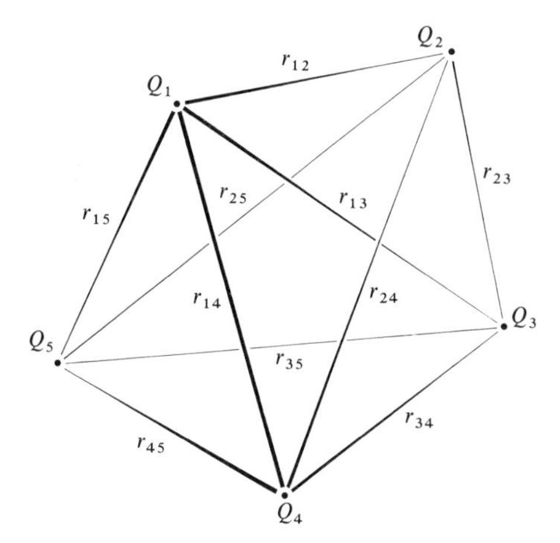

Figure 2-14. Set of point charges $Q_1, Q_2, \ldots, Q_5$ separated by distances $r_{12}, \ldots, r_{45}$.

the other charges has some definite value. Each charge thus has associated with it a definite potential energy, either positive or negative, and the system as a whole has a potential energy that we shall calculate.

We assume that the charges remain in equilibrium under the action of the electric forces and of restraining mechanic forces.

Let $Q_1, Q_2, Q_3, \cdots, Q_N$ be the N charges; let r_{12} be the distance between Q_1 and Q_2, r_{13} the distance between Q_1 and Q_3, and so on, as in Figure 2-14. Now let Q_1 recede to infinity slowly, so that the electric and the mechanical forces are always in equilibrium. In this way there is no acceleration and, therefore, no kinetic energy involved. The other charges remain fixed.

The decrease W_1 in the electric potential energy of the charge is equal to the work done by the electric forces. This is the product of the charge Q_1 and the potential V_1 produced by the other charges at the original position of Q_1. Then

$$W_1 = \frac{Q_1}{4\pi\epsilon_0}\left(\frac{Q_2}{r_{12}} + \frac{Q_3}{r_{13}} + \cdots + \frac{Q_N}{r_{1N}}\right). \tag{2-119}$$

Note that all the charges except Q_1 are represented in the series of terms within the parentheses.

Now that Q_1 has been removed, let Q_2 recede to some point infinitely distant from Q_1. The decrease W_2 in the potential energy of Q_2 is

$$W_2 = \frac{Q_2}{4\pi\epsilon_0}\left(\frac{Q_3}{r_{23}} + \frac{Q_4}{r_{24}} + \cdots + \frac{Q_N}{r_{2N}}\right). \tag{2-120}$$

This series has only $N - 2$ terms within the parentheses, the Q_1 and Q_2 terms being absent. We continue the process for all the remaining charges, there being progressively fewer terms in the series, until finally the Nth charge can be removed without any change in energy, since it is left in a zero field once all the other charges have been removed.

The total potential energy of the original charge distribution is then

$$W = W_1 + W_2 + W_3 + \cdots + W_N, \tag{2-121}$$

$$\begin{aligned} &= \frac{Q_1}{4\pi\epsilon_0}\left(0 + \frac{Q_2}{r_{12}} + \frac{Q_3}{r_{13}} + \frac{Q_4}{r_{14}} + \cdots + \frac{Q_N}{r_{1N}}\right) \\ &+ \frac{Q_2}{4\pi\epsilon_0}\left(\qquad 0 + \frac{Q_3}{r_{23}} + \frac{Q_4}{r_{24}} + \cdots + \frac{Q_N}{r_{2N}}\right) \\ &+ \frac{Q_3}{4\pi\epsilon_0}\left(\qquad\qquad 0 + \frac{Q_4}{r_{34}} + \cdots + \frac{Q_N}{r_{3N}}\right) \\ &\cdots\cdots\cdots\cdots\cdots\cdots\cdots\cdots\cdots\cdots \\ &+ \frac{Q_N}{4\pi\epsilon_0}\left(\qquad\qquad\qquad\qquad 0 \right). \end{aligned} \tag{2-122}$$

Let us now rewrite the array of terms within the parentheses, adding in, to the left of and below the diagonal line of zeros, terms that are equal to their counterparts on the other side of the diagonal. Then every term of the series appears twice, and

$$\begin{aligned} 2W &= \frac{Q_1}{4\pi\epsilon_0}\left(0 + \frac{Q_2}{r_{12}} + \frac{Q_3}{r_{13}} + \frac{Q_4}{r_{14}} + \cdots + \frac{Q_N}{r_{1N}}\right) \\ &+ \frac{Q_2}{4\pi\epsilon_0}\left(\frac{Q_1}{r_{21}} + 0 + \frac{Q_3}{r_{23}} + \frac{Q_4}{r_{24}} + \cdots + \frac{Q_N}{r_{2N}}\right) \\ &+ \frac{Q_3}{4\pi\epsilon_0}\left(\frac{Q_1}{r_{31}} + \frac{Q_2}{r_{32}} + 0 + \frac{Q_4}{r_{34}} + \cdots + \frac{Q_N}{r_{3N}}\right) \\ &\cdots\cdots\cdots\cdots\cdots\cdots\cdots\cdots\cdots\cdots \\ &+ \frac{Q_N}{4\pi\epsilon_0}\left(\frac{Q_1}{r_{N1}} + \frac{Q_2}{r_{N2}} + \frac{Q_3}{r_{N3}} + \frac{Q_4}{r_{N4}} + \cdots + 0\right). \end{aligned} \tag{2-123}$$

Thus the first line is Q_1V_1, the second line is Q_2V_2, and so forth, so that

$$2W = Q_1V_1 + Q_2V_2 + Q_3V_3 + \cdots + Q_NV_N, \tag{2-124}$$

and the potential energy of the original charge distribution is

$$W = \frac{1}{2}\sum_{i=1}^{N} Q_iV_i, \tag{2-125}$$

where V_i is the potential *in the undisturbed system* due to all the charges *except* Q_i at the point occupied by Q_i.

We have *not* taken into account the self energy of the individual charges, that is, the energy that would be liberated if each one were allowed to expand to an infinite volume.

The reason the factor $\frac{1}{2}$ appears in the above equation should be clear from the reasoning we have used to arrive at it. It is that the potential at the position of a given charge at the time it is removed to infinity is less, in general, than the potential at the same point in the original charge distribution. On the average, the potential at the time of removal is just one-half the potential in the original charge distribution.

You will note that this energy W, which does *not* include the energy required to assemble the individual charges themselves, can be either positive or negative. It is the energy required to assemble already existing point charges. For example, if we have two positive point charges Q_1 and Q_2, then the positive charge Q_1 is situated at a position where the potential V_1, due to the other charge Q_2, is positive, so that Q_1V_1 is positive. The product Q_2V_2 is also positive and the potential energy of the charge distribution $Q_1Q_2/4\pi\epsilon_0 r^2$ is positive. However, if we have a positive charge Q_1 and a negative charge Q_2, then V_1 is negative and Q_1V_1 is negative. The other product Q_2V_2 is similarly negative, and W is negative. Note also that, for a single point charge, the above W is zero.

For a continuous electric charge distribution of density $\rho(x', y', z')$, we simply replace Q_i by $\rho\, d\tau'$ and the summation by an integration:

$$W = \frac{1}{2}\int_{\tau'} V\rho\, d\tau'. \tag{2-126}$$

This operation is misleading, however, because we have now included the energies required to assemble the individual macroscopic charges. In fact, as we shall see in the next section, this integral is always positive. It is evaluated over any arbitrary volume τ' that contains all the charges in the system.

If we have conducting bodies carrying surface charge densities σ, then the stored energy is

$$W = \frac{1}{2}\int_{S'} \sigma V\, da', \tag{2-127}$$

where S' is the sum of the areas of all the conductors.

Imagine now a conductor that is situated a long distance away from any other body. Because of the repulsion between similar charges, energy must be expended to charge the conductor. Accordingly, its potential rises as charge is added, the magnitude of the charge in potential being proportional to the amount of charge added and depending on the geometrical configuration of the conductor. According to the above equation, the stored energy is

$$W = \tfrac{1}{2} QV, \tag{2-128}$$

where Q is the net charge carried by the conductor and V is its potential.

The ratio Q/V is called the *capacitance* of the isolated conductor:

$$C = \frac{Q}{V}, \tag{2-129}$$

and thus its stored energy can also be written as

$$W = \frac{1}{2} CV^2 = \frac{1}{2}\frac{Q^2}{C}. \tag{2-130}$$

The capacitance is the electric charge that must be added per unit increase in potential.

One might well ask why the capacitance C should be a constant for a given conductor, as we have implied above. This can be shown as follows. The isolated conductor carries a charge Q, and the electric potential in the region surrounding it is $V(x, y, z)$. At the surface of the conductor the charge density is $\epsilon_0 E$ (Section 2-7), or $-\epsilon_0$ times the rate of change of V in the direction normal to the surface. Since the potential V satisfies Laplace's equation $\nabla^2 V = 0$ (Eq. 2-32), and since this equation is linear, any multiple of V is also a solution. Let us therefore increase V everywhere by some factor a. Then, since $\mathbf{E} = -\nabla V$, the surface charge density will also increase by a, and the total charge Q on the conductor will increase likewise. The charge on an isolated conductor is therefore proportional to the potential, and C is a constant.

Imagine now that we have two uncharged isolated conductors. If a charge Q is transferred from one to the other, a potential difference ΔV is established between them and, by definition, the *capacitance* C between the conductors is $Q/\Delta V$.

The unit of capacitance is the coulomb/volt, or *farad*.

2.15 ENERGY DENSITY IN AN ELECTRIC FIELD

The potential energy W can be related to the electric field of the charge distribution as follows.

According to Poisson's equation,

$$\rho = -\epsilon_0 \nabla^2 V \tag{2-131}$$

at every point in the field. If we substitute this into Eq. 2-126, then

$$W = -\frac{\epsilon_0}{2}\int_{\tau'} V\nabla^2 V\, d\tau', \tag{2-132}$$

where τ' is again any volume containing all the charges in the system.

We may simplify this equation through the vector identity

$$\nabla \cdot f\boldsymbol{A} = f(\nabla \cdot \boldsymbol{A}) + \boldsymbol{A} \cdot \nabla f, \tag{2-133}$$

where f and $\boldsymbol{A}$ are scalar and vector functions respectively. Let $f = V$, and $\boldsymbol{A} = \nabla V$. Then, rearranging,

$$V \nabla^2 V = \nabla \cdot (V \nabla V) - (\nabla V)^2, \tag{2-134}$$

and

$$W = -\frac{\epsilon_0}{2}\left[\int_{\tau'} \nabla \cdot (V \nabla V)\, d\tau' - \int_{\tau'} (\nabla V)^2\, d\tau'\right]. \tag{2-135}$$

We can transform the first term on the right into an integral over the surface S bounding the volume τ' by using the divergence theorem. Thus

$$W = -\frac{\epsilon_0}{2}\left[\int_{S'} (V \nabla V) \cdot d\boldsymbol{a}' - \int_{\tau'} (\nabla V)^2\, d\tau'\right]. \tag{2-136}$$

Now τ' can be any volume that includes all the charges in the system. We are thus free to choose the bounding surface S' at a large distance from the charge distribution. Then, in the first integral, V falls off at least as fast as $1/r$, since the monopole term in the potential of a charge distribution falls off as $1/r$, whereas the dipole, quadrupole, and higher terms fall off as $1/r^2$, $1/r^3$ and so forth, as we saw in Section 2.12. Then ∇V falls off at least as fast as $1/r^2$. Since the surface area increases as r^2, the whole integral decreases at least as fast as $1/r$ and can be made arbitrarily small by choosing the surface S' sufficiently far off.

We are then left with

$$W = \frac{\epsilon_0}{2}\int_{\tau} E^2\, d\tau', \tag{2-137}$$

since $E = |\nabla V|$. The volume τ' need not be infinite now, but only large enough to include all regions where E differs from zero. This integral takes into account the self-energies of the charges; it is always positive, and it is obviously not zero for a single charge.

The potential energy W of Eq. 2-126 is equal to the W we have deduced from it above and is therefore also positive for all charge configurations.

Equation 2-137 shows that the energy associated with a charge distribution, that is, the energy required to assemble it, starting with a configuration in which the charge is spread over an infinite volume, may be calculated by associating with each point of the field an energy density

$$\frac{dW}{d\tau} = \frac{\epsilon_0}{2} E^2 \text{ joules/meter}^3. \tag{2-138}$$

In the next section we shall arrive at this same result by another method.

We can therefore calculate the energy stored in an electric charge distribution, either in terms of the charge density ρ and the potential V, or in terms of the electric field intensity E.

Example An isolated spherical conductor of radius R carries a surface charge density σ. Then the potential energy is

$$W = \frac{1}{2}\int_S \sigma V\, da = \frac{1}{2} QV, \tag{2-139}$$

where

$$V = \frac{Q}{4\pi\epsilon_0 R}. \tag{2-140}$$

Thus

$$W = \frac{Q^2}{8\pi\epsilon_0 R}. \tag{2-141}$$

We can also write that

$$W = \frac{\epsilon_0}{2}\int_\tau E^2\, d\tau, \tag{2-142}$$

$$= \frac{\epsilon_0}{2}\int_R^\infty \left(\frac{Q}{4\pi\epsilon_0 r^2}\right)^2 4\pi r^2\, dr = \frac{Q^2}{8\pi\epsilon_0 R}, \tag{2-143}$$

which is the same result as above. We have integrated from R to infinity, since there is zero E inside the sphere.

2.16 FORCES ON CONDUCTORS

An element of charge $\sigma\, da$ on the surface of a conductor experiences the electric field of all the other charges in the system and is therefore subject to an electric force. In a static field, this force must be perpendicular to the surface of the conductor, for otherwise the charges would move along the surface. Since the charge $\sigma\, da$ is bound to the conductor by internal forces, the force acting on $\sigma\, da$ is transmitted to the conductor itself.

To calculate the magnitude of this force, let us consider a conductor with a surface charge density σ and a field $\boldsymbol{E}$ at the surface. From Gauss's law, the electric field intensity just outside the conductor is σ/ϵ_0 and is perpendicular to the conducting surface. Now the force on the element of charge $\sigma\, da$ is *not* $E\sigma\, da$, since the field that acts on $\sigma\, da$ is the field due only to the *other* charges in the system.

Let us first calculate the electric intensity produced by $\sigma\, da$ itself. We can do this by using Gauss's law. The total flux emerging from $\sigma\, da$ must be $\sigma\, da/\epsilon_0$, half of it inwards and half outwards, as in Figure 2-15. Then the

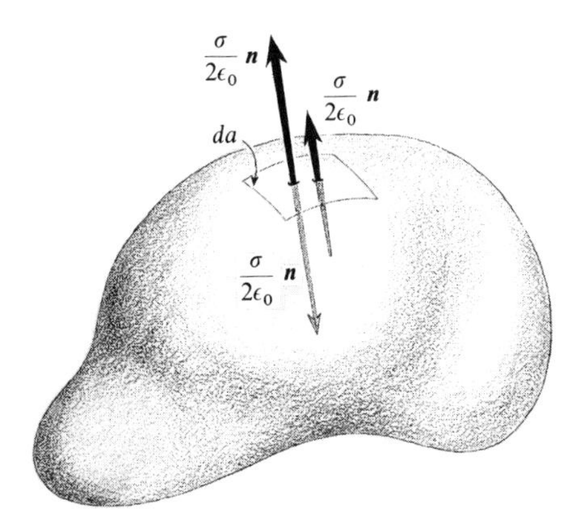

Figure 2-15. The local electric charge density at the surface of a conductor gives rise to oppositely directed electric field intensities $\sigma/2\epsilon_0$ as shown by the two arrows on the left; the other charges on the conductor give rise to the field $\sigma/2\epsilon_0$ shown by the arrow on the right. The net result is σ/ϵ_0 outside, and zero inside. The vector $\boldsymbol{n}$ is a unit vector normal to the conductor surface, and it points outward.

electric field intensity due to the local surface chage density must be $\sigma/2\epsilon_0$ inwards and $\sigma/2\epsilon_0$ outward.

The element of charge $\sigma\, da$ itself therefore produces exactly half the total field at a point outside, arbitrarily close to the surface. This is reasonable, for the nearby charge is more effective than the rest.

Now if $\sigma\, da$ produces half the field, then all the other charges must produce the other half, and the electric field intensity acting on $\sigma\, da$ must be $\sigma/2\epsilon_0$.

The force dF on the element of area da of the conductor is therefore given by the product of its charge $\sigma\, da$ multiplied by the field of all the other charges:

$$dF = \frac{\sigma}{2\epsilon_0}\,\sigma\, da, \tag{2-144}$$

and the force per unit area is

$$\frac{dF}{da} = \frac{\sigma^2}{2\epsilon_0} = \epsilon_0\,\frac{E^2}{2}. \tag{2-145}$$

Now this force is just equal to the energy density that we found in the preceding section. We should really have expected this for the following reason. Let us imagine that the conducting bodies in the field are disconnected from their power supplies. Then, if the electric forces are allowed to perform mechanical work, they must do so at the expense of the electric energy stored in the field. Now imagine that a small area a of a conductor is allowed to be pulled into the field by a small distance x. The mechanical work performed is ax times the force per unit area. It is also equal to the energy lost by the field, which is ax times the energy density. Then the force per unit area must be equal to the energy density.

If we use the same type of argument for a compressed gas, we find that its energy density is equal to its pressure.

We have therefore found another and a more simple demonstration of the fact that we can calculate the energy stored in an electric field by integrating $\epsilon_0 E^2/2$.

This second demonstration is in some ways preferable to that of Section 2.15. First, it is purely local; it does not require that the field be integrated over an infinite volume. As you will remember, we had to perform such an operation to eliminate the first integral in Eq. 2-136. Second, it relies solely on Gauss's law and on the fact that the force on a charge Q situated in an electric field $\boldsymbol{E}$ is $Q\boldsymbol{E}$. Both of these postulates are valid even if the field is not constant. The proof of Section 2.15, on the other hand, is valid only for electrostatic fields.

It is important to note that the electric force on a conductor always tends to pull the conductor into the field. In other words, an electric field exerts a negative pressure on a conductor.

Example

Electric forces are never very large. For example, even for $E = 10^6$ volts/meter, the electric force is only 4.4 newtons/meter^2, or 0.45 kilogram/meter^2, or about 2 ounces/foot^2. See Problem 3-23.

Example

FORCES ON A PARALLEL-PLATE CAPACITOR

As a simple example, let us calculate the forces on the plates of a parallel-plate capacitor of area S and plate separation s carrying a charge density $+\sigma$ on one plate and $-\sigma$ on the other, as in Figure 2-16. We shall assume that the separation of the plates is small compared to their linear extent, in order that the parallel-plate capacitor approximation be valid.

The force per unit area on the plates is equal to the energy density in the field $\epsilon_0 E^2/2$, and the force of attraction between the plates is therefore $\epsilon_0 E^2 S/2$.

Let us now calculate this force from the change in energy corresponding to a slight *de*crease in the spacing s. The energy associated with the electric field between the plates is

$$W = \frac{\epsilon_0}{2} E^2 S s. \tag{2-146}$$

If the plates are insulated, so that the charges remain constant, and

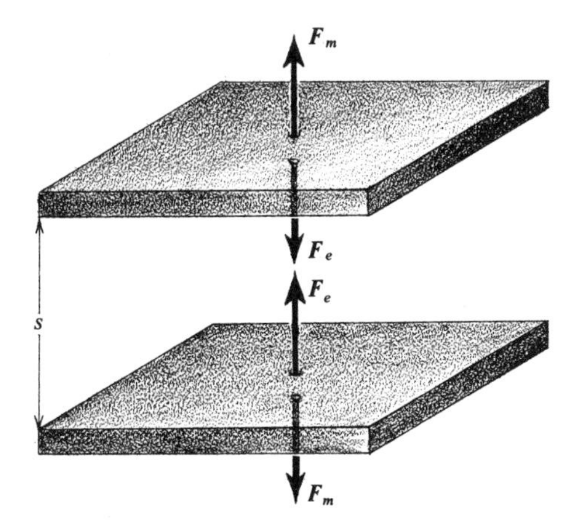

Figure 2-16. A charged parallel-plate capacitor with its plates insulated. The plates are held in equilibrium by mechanical forces $\boldsymbol{F}_m$ which act on each plate in a direction tending to increase the separation s, and by electric forces $\boldsymbol{F}_e$ tending to decrease the separation.

if the plates are moved closer together by a small distance ds, the electric field intensity E remains unchanged, according to Gauss's law. The volume between the plates decreases, however, and hence the energy in the field also decreases. Since we must have conservation of energy, the energy lost by the field has gone into the work $F_m\, ds$ against the mechanical forces F_m holding the plates apart. Then

$$\frac{\epsilon_0}{2} E^2 S\, ds = F_m\, ds, \tag{2-147}$$

and

$$F_m = \frac{\epsilon_0}{2} E^2 S = \frac{\sigma^2}{2\epsilon_0} S = \frac{Q^2}{2\epsilon_0 S}. \tag{2-148}$$

Since the electric force $\boldsymbol{F}_e$ is in equilibrium with the mechanical force, it must be of equal magnitude and oriented so as to pull the plates together.

2.17 SUMMARY

It is found empirically that the force exerted *by* a point charge Q_a *on* a point Q_b is

$$\boldsymbol{F}_{ab} = \frac{1}{4\pi\epsilon_0} \frac{Q_a Q_b}{r^2} \boldsymbol{r}_1, \tag{2-2}$$

where r is the distance between the charges and $\boldsymbol{r}_1$ is a unit vector pointing from Q_a to Q_b. This is *Coulomb's law*. We consider the force $\boldsymbol{F}_{ab}$ as being the product of Q_b by the *electric field intensity* due to Q_a,

$$\boldsymbol{E}_a = \frac{Q_a}{4\pi\epsilon_0 r^2} \boldsymbol{r}_1, \tag{2-5}$$

or vice versa.

According to the *principle of superposition*, two or more electric fields acting at a given point add vectorially.

The electrostatic field is conservative,

$$\nabla \times \boldsymbol{E} = 0, \tag{2-12}$$

hence

$$\boldsymbol{E} = -\nabla V, \tag{2-13}$$

where

$$V = \frac{1}{4\pi\epsilon_0}\int_\infty \frac{\rho\, d\tau'}{r}, \tag{2-19}$$

$\rho\, d\tau'$ is the element of charge contained within the element of volume $d\tau'$, and r is the distance between this element and the point where V is calculated.

Gauss's law follows from Coulomb's law. It can be stated either in integral or in differential form:

$$\int_{S'} \boldsymbol{E}\cdot \boldsymbol{da}' = \frac{1}{\epsilon_0}\int_{\tau'} \rho\, d\tau', \tag{2-24}$$

or

$$\nabla\cdot \boldsymbol{E} = \frac{\rho}{\epsilon_0}. \tag{2-26}$$

Poisson's equation

$$\nabla^2 V = -\frac{\rho}{\epsilon_0} \tag{2-31}$$

then follows from the differential statement of Gauss's law and from the definition of the potential.

When the charge density is zero we have *Laplace's equation*

$$\nabla^2 V = 0. \tag{2-32}$$

The above is summarized schematically in Figure 2-17.

The *electric field produced by simple charge distributions* can be calculated in four different ways: (a) by evaluating $\boldsymbol{dE}$ for an element of charge and integrating over the complete charge distribution, (b) by calculating V in the same manner and then calculating its gradient, (c) by integrating Poisson's or Laplace's equation, or (d) by using Gauss's law. The fields due to more complex charge distributions can be calculated by other methods described in Chapter 4 and in Appendix B.

The *potential produced by an arbitrary charge distribution* at a distance $r > r'_{\max}$ is the same as that produced by a point charge, a point dipole, a point quadrupole, and so forth, all situated at the origin, where the monopole

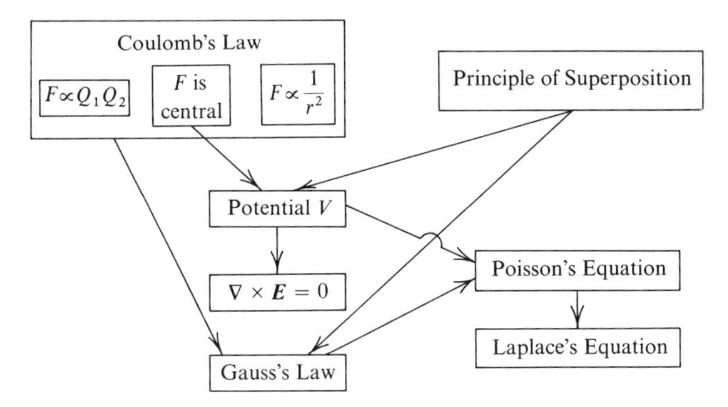

Figure 2-17.

carries the total charge of the distribution, the dipole has the dipole moment of the charge distribution,

$$\boldsymbol{p} = \int_{\tau'} \boldsymbol{r}' \rho \, d\tau', \tag{2-94}$$

the quadrupole has the quadrupole moment of the charge distribution, and so forth.

If Q is the net charge carried by an isolated conductor, and if V is its potential, the ratio Q/V is called its *capacitance.*

The *capacitance between two isolated conductors* is the ratio $Q/\Delta V$, where Q is the charge transferred from one to the other, and ΔV is the resulting potential difference.

The *potential energy* associated with a charge distribution can be written either as

$$W = \frac{1}{2} \int_{\tau'} V \rho \, d\tau' \tag{2-126}$$

or as

$$W = \frac{\epsilon_0}{2} \int_{\tau'} E^2 \, d\tau'. \tag{2-137}$$

In the first integral, τ' must be chosen to include all the charge distribution, and in the second it must include all regions of space where E is nonvanishing. The assignment of an *energy density* $\epsilon_0 E^2/2$ to every point in space therefore leads to the correct potential energy for the whole charge distribution.

Finally, *electric forces* on conductors can be calculated by either of two methods. We can either (a) utilize the fact that the force per unit area is $\sigma^2/2\epsilon_0 = \epsilon_0 E^2/2$, where E is the electric field intensity near the surface of the conductor, or (b) equate the work performed by the electric forces to the work performed against the restraining mechanical forces during an infinitesimal virtual displacement.

PROBLEMS

2-1. Show that ϵ_o is expressed in farads/meter, that E is expressed in volts/meter, and that the volt is a joule/coulomb.

2-2. (a) Calculate the electric field intensity E that would be just sufficient to balance the gravitational force on an electron.
(b) If this electric field were produced by a second electron located below the first one, what would be the distance between the two electrons? (The charge on an electron is -1.6×10^{-19} coulomb and its mass is 9.1×10^{-31} kilogram.)

2-3. It is known that the maximum electric field intensity which can be maintained in air at NTP is about 3×10^6 volts/meter. This determines the maximum surface charge density.
If a spherical body is charged to half this maximum surface charge density and placed in a vertical electric field of 1.5×10^6 volts/meter, what is the largest solid sphere that can be supported by the electric field? Choose any material you wish for the sphere.
Assume that it is nonconducting and that its charge is uniformly distributed over its surface.

2-4. A charge $+Q$ is situated at $(-a, 0, 0)$ and a charge $-2Q$ is situated at $(a, 0, 0)$.
Is there a point in space where $E = 0$?

2-5. A thin infinite conducting plate carries a surface charge density σ. Show that one-half of the electric field intensity E at a point situated z meters from the surface of the plate is due to the charge located on the plate within a circle of radius $\sqrt{3}\, z$.

2-6. Show that the potential due to two parallel line charges of opposite polarity is $(\lambda/2\pi\epsilon_0) \ln (r_1/r_2)$, where r_1 is the distance from the point considered to the negative line, and r_2 is the corresponding distance to the positive line.
If your answer turns out to be 0/0, try another method.

2-7. The electron beam in an oscilloscope is deflected vertically and horizontally by two pairs of parallel deflecting plates that are maintained at appropriate voltages. As the electrons pass between one pair they are accelerated by the electric field, and their kinetic energy increases. Thus, under steady-state conditions, we achieve an increase in the kinetic energy of the electrons without any expenditure of power in the deflecting plates, as long as the beam does not touch the plates.

Could this phenomenon not be used as the basis of a perpetual motion machine?

2-8. In 1906, in the course of a historic experiment which demonstrated the small size of the atomic nucleus, Rutherford observed that an alpha particle ($Q = 2 \times 1.6 \times 10^{-19}$ coulomb) with a kinetic energy of 7.68×10^6 electron-volts ($7.68 \times 10^6 \times 1.6 \times 10^{-19}$ joule) making a head-on collision with a gold nucleus ($Q = 79 \times 1.6 \times 10^{-19}$ coulomb) is repelled.

(a) What is the distance of closest approach at which the electrostatic potential energy is equal to the initial kinetic energy? Express your result in femtometers (10^{-15} meter).
(b) What is the maximum force of repulsion?
(c) What is the maximum acceleration in g's? The mass of the alpha particle is about 4 times that of a proton, or $4 \times 1.7 \times 10^{-27}$ kilogram.

2-9. Electric charge is distributed along an infinite straight line with a density $\lambda = \lambda_0/(|x| + k)$, where k is a positive constant.

(a) Is the total charge infinite?
(b) Can you calculate V at a distance ρ from the wire at $x = 0$? If not, can you show that it is not infinite?

2-10. A flat sheet of copper carries a surface charge density of σ coulombs/meter^2 on each face.

Use Gauss's law to show that the electric field intensity just outside the sheet is σ/ϵ_0 and that it is zero inside.

2-11. (a) Show that the electric field intensity near a flat sheet of charge having a uniform surface density σ is $\sigma/2\epsilon_0$.
(b) Deduce the value of the electric potential.

2-12. An electric field points everywhere in the z direction.

(a) What can you conclude about the values of the partial derivatives of E_z with respect to x, to y, and to z, (*i*) if the charge density ρ is zero, (*ii*) if the charge density ρ is not zero?
(b) Sketch lines of force for some fields that are possible and for some that are not.

2-13. A 1.00-microampere beam of protons is accelerated through a difference of potential of 10,000 volts.

(a) Calculate the charge density once the protons have been accelerated, assuming that the current density is uniform within a diameter of 2.00 millimeters and is zero outside this diameter.
(b) Calculate the radial electric field intensity both inside and outside the beam.
(c) Draw a graph of the radial electric field intensity for values of r ranging from zero to 1.00 centimeter.
(d) Now let the beam be situated along the axis of a grounded cylindrical conducting tube with an inside radius of 1.00 centimeter. Draw a graph of V inside the tube.

2-14. Two infinite parallel plates separated by a distance s are at potentials 0 and V_0.

(a) Use Poisson's equation to find the potential V in the region between the plates where the space charge density is $\rho = \rho_0(x/s)$. The distance x is measured from the plate at zero potential.

(b) What are the charge densities on the plates?

2-15. A two-dimensional electrostatic field varies with the coordinates x and y but is independent of z.

Show that the average value of the potential V on any circle parallel to the xy-plane equals the potential at the center of the circle, provided the charge density in the region is zero.

2-16. You can use the property demonstrated in Problem 2-15 to calculate the potential distribution within a square, two of whose adjacent edges are maintained at 100 volts while the other two are maintained at 0 and at 50 volts.

(a) First draw a square grid of 36 points. Of these, 20 will be on the edges of the square, and 16 will be in the interior.

(b) Guess carefully the potentials at the interior points and write them on your grid. This is your first approximation.

(c) Now you can correct your first approximation as follows. Start at an interior point, near one corner of the square. Let us call this point P. Calculate the sum of the potentials at the four nearest points and subtract four times the potential at P. This number is called the *residual* at P. Show this number on your grid, preferably in a different color, and repeat the calculation for the other interior points.

To obtain a second approximation, select the points where the residuals are the largest and add to your original estimates one quarter of the residuals. Round off the potentials to the nearest volt. This makes the potential distribution satisfy more closely the condition found in Problem 2-15.

(d) Now calculate the new residuals and repeat the operation until the residuals are everywhere less than 2, corresponding to corrections of less than one-half volt.

You might wish to solve this problem on a computer.

(e) Sketch in some equipotentials.

This is a good illustration of the *relaxation method* of calculation, which we largely owe to R. V. Southwell, and which is described in many books on applied mathematics. The method is applicable to a wide variety of problems, and not only to the solution of Laplace's equation.

2-17. A charged conductor has a small-diameter hole in a region where the surface charge density is σ.

Show that the electric field intensity in the hole close to the surface of the conductor is $\sigma/2\epsilon_0$.

2-18. The electric field intensity at the surface of a real conductor falls from its external value $\mathbf{E}_o$ to zero within the conductor in a finite distance δ.

Show that the surface charge density $\sigma = \int_0^\delta \rho(x)\,dx = \epsilon_0 E_o$.

2-19. At an early stage in the development of the atomic theory, J. J. Thomson had proposed an atom consisting of a positive charge Ze, where Z is an integer and e is the fundamental unit of charge (1.6×10^{-19} coulomb), uniformly distributed throughout a sphere of radius a. The electrons, of charge $-e$, were considered to be point charges embedded in the positive charge.

(a) Find the force acting on one electron as a function of its distance r from the center of the sphere. Assume that the other charges are uniformly distributed throughout the sphere.
(b) What type of motion does the electron execute?
(c) What is the frequency for a typical atomic radius of 1 angstrom?
(d) How does this frequency compare with that of the visible light radiated by atoms?

2-20. (a) Calculate the electric field intensity in volts/meter at the surface of an iodine nucleus (53 protons and 74 neutrons).
(b) Find the electric potential at the center of the nucleus.

Assume that the charge density is uniform and that the radius of the nucleus is $1.25 \times 10^{-15} A^{1/3}$ meter, where A is the total number of particles in the nucleus.

2-21. A spherical charge distribution is given by

$$\rho = \rho_0 \left(1 - \frac{r^2}{a^2}\right), \qquad (r \leq a)$$

$$\rho = 0 \qquad (r > a).$$

(a) Calculate the total charge Q.
(b) Find the electric field intensity E and the potential V outside the charge distribution.
(c) Find E and V inside.
(d) Show that the maximum value of E is at $(r/a) = 0.745$.
(e) The above charge distribution applies roughly to light nuclei. Draw graphs showing ρ, E, and V as functions of r/a for calcium (atomic number 20), assuming that $\rho_0 = 5.0 \times 10^{25}$ coulombs/meter^3 and $a = 4.5$ femtometers (1 femtometer $= 10^{-15}$ meter).

2-22. We shall see in Chapter 12 that an electromagnetic wave incident on a conductor exerts a pressure on the conduction electrons. Imagine that you have an insulated copper sphere one micron in diameter and that you suddenly illuminate it with a powerful laser pulse that blasts essentially all the conduction electrons out of the sphere. There is one conduction electron per atom in copper.

(a) What would be the voltage at the surface of the sphere immediately after the irradiation?
(b) What energy would be required in the laser pulse?
(c) Do you think the experiment is feasible?

2-23. Show that the dipole moment of an arbitrary charge distribution is independent of the choice of origin, provided the net charge in the distribution is zero.

2-24. Compute the dipole moment of a spherical shell with a surface charge density $\sigma = \sigma_o \cos\theta$, where θ is the polar angle.

2-25. Calculate the potential for a dipole exactly and indentify the quadrupole and octupole terms.

2-26. A charge Q is uniformly distributed inside a cube of side a. Show that the dipole and quadrupole moments of the charge distribution are zero.

2-27. A charge Q is uniformly distributed along a straight line of length a.
(a) Calculate V by integration, choosing the origin of coordinates at the center of the charge.
(b) Calculate the monopole, dipole, and quadrupole terms in the multipole expansion for V.
(c) For what values of the distance r to the center of the charge is the quadrupole term less than one per cent of the monopole term?
(d) Expand the value of V found in (a) up to terms in $1/r^2$ and compare with (b).

2-28. A parallel-plate capacitor is formed of two parallel conducting plates of area S separated by a distance s in air. Use Gauss's law to show that its capacitance is $\epsilon_0 S/s$ farads. Neglect edge effects.

Can you suggest reasonable dimensions for a 1 picofarad (10^{-12} farad) air-insulated parallel-plate capacitor which is to operate at a potential difference of 500 volts?

2-29. Find the capacitance per unit length of a capacitor consisting of a pair of infinite coaxial cylinders having inner and outer radii a and b, respectively.

2-30. The capacitance of a parallel-plate capacitor is $\epsilon_0 S/s$, where S is the area of the plates and s is their separation, as long as edge effects are negligible.
(a) How is the capacitance affected if an insulated sheet of metal of thickness s' is introduced between the plates, but without touching either plate? (The sheet need not be parallel to the plates.)
(b) What do you think of the suggestion of moving this sheet in order to vary the capacitance without moving the capacitor plates?

2-31. Two capacitors of capacitance C_1 and C_2 have charges Q_1 and Q_2, respectively.
(a) Calculate the amount of energy dissipated when they are connected in parallel.
(b) How is this energy dissipated?

2-32. A capacitor consisting of two concentric spheres is arranged so that the outer sphere can be separated and removed without disturbing the charges on either. The radius of the inner sphere is a, that of the outer sphere is b, and the charges are Q and $-Q$ respectively.
(a) If the outer sphere is removed and restored to its original form, find the increase in energy when the two spheres are separated by a large distance.
(b) Where does this extra energy come from?

2-33. Consider an imaginary sphere of radius a centered on a dipole of moment p. Show that the electric energy associated with the region of space *outside* the sphere is $W = (p^2/12\pi\epsilon_0 a^3)$ by integrating the energy density from $r = a$ to $r = \infty$.

2-34. According to the theory of relativity, a particle at rest has an energy m_0c^2, where m_0 is the particle's mass and c is the velocity of light (Section 5.15). Imagine that this is the electrostatic energy of the electron.

Find the radius of the electron (a) if the charge e is distributed uniformly throughout its spherical volume, and (b) if the charge is distributed uniformly over its spherical surface.

2-35. (a) Calculate the electric potential energy of a sphere of radius R carrying a total charge Q uniformly distributed throughout its volume.
(b) Then calculate the gravitational potential energy of a sphere of radius R' of total mass M.
(c) The Moon has a mass of 7.33×10^{22} kilograms and a radius of 1.74×10^6 meters. Calculate its gravitational potential energy.
(d) Imagine that you can assemble a sphere of protons with a density equal to that of water. What would be the radius of this sphere if its electric potential energy were sufficient to blow up the Moon?

2-36. A light spherical balloon is made of conducting material. It is suggested that it could be kept spherical simply by connecting it to a high-voltage source. The balloon has a diameter of 10.0 centimeters, and the maximum breakdown voltage in air is 3×10^6 volts/meter.
(a) What must be the voltage of the source if the electric force is to be as large as possible?
(b) What gas pressure inside the balloon would produce the same effect?

2-37. We have shown in Section 2.16 that the force of attraction between the plates of a parallel-plate capacitor is $(1/2)\epsilon_0 E^2 S$ when the plates are disconnected from the voltage source. Check this result by repeating the calculation for the case where the capacitor is connected to the source. The force should be the same.

You will find that one half of the energy supplied by the source performs mechanical work and that the other half increases the energy in the field. *This is a general result.*

2-38. Find the time required for the plates (mass m_0 per unit area) of a parallel-plate capacitor to come together when released from a separation x_0, (a) when the plates are charged with a charge density σ and then insulated, and (b) when the plates are maintained at a constant potential difference V.

2-39. A variable capacitor consists of two thin coaxial cylinders of radii a and b, with $(b - a) \ll a$, free to move with respect to each other in the axial direction.

(a) Using energy methods, compute the magnitude and direction of the force on the inner cylinder when it is displaced with respect to the outer one.
(b) Explain how this force arises.

2-40. Imagine the following simple-minded high-voltage generator. A parallel-plate capacitor has one fixed plate that is permanently connected to ground, and one plate that is movable. When the plates are close together at the distance s the capacitor is charged by a battery to a voltage V. Then the movable plate is disconnected from the battery and moved out to a distance ns. The voltage on this plate then increases to nV, if we neglect edge effects. Once the voltage has been raised to nV, the plate is discharged through a load resistance.

(a) Verify that there is conservation of energy.
(b) Can you suggest a rough design for such a high-voltage generator with a more convenient geometry?

2-41. Can you suggest a rough design for an electrostatic motor?

Draw a sketch, explain its operation, specify voltages and currents, and make a rough estimate of what its power would be.

Why is it that electric forces should be so much weaker than magnetic forces in such cases?

2-42. Normal seeds can be separated from discolored ones and from foreign objects by means of an electrostatic seed-sorting apparatus that operates as follows. The seeds are observed by a pair of photocells as they fall one by one inside a tube. If the color is not right, voltage is applied to a needle that deposits a charge on the seed. The seeds then fall between a pair of electrically charged plates that deflect the undesired ones into a separate bin. One such machine can sort peas at the rate of 100/second, or about 2 tons per 24-hour day.

(a) If the seeds are dropped at the rate of 100/second, over what distance must they fall if they must be separated by 2 centimeters when they pass between the photocells? Neglect air resistance.
(b) Assuming that the seeds acquire a charge of 1.5×10^{-9} coulomb, that the deflecting plates are parallel and 5 centimeters apart, and that the potential difference between them is 25,000 volts, how long should the plates extend below the charging needle if the charged seeds must be deflected by 4 centimeters on leaving the plates? Assume that the charging needle and the top of the deflecting plates are close to the photocell.

Hint

2-37. Do not forget to take into account the energy supplied either by or to the battery.

CHAPTER 3

ELECTROSTATIC FIELDS II

Dielectric Materials

Dielectrics differ from conductors in that they have no free charges that can move through the material under the influence of an electric field. In dielectrics, all the electrons are bound; the only motion possible in the presence of an electric field is a minute displacement of positive and negative charges in opposite directions. The displacement is usually small compared to atomic dimensions.

A dielectric in which this charge displacement has taken place is said to be *polarized*, and its molecules are said to possess *induced dipole moments*. These dipoles produce their own field, which adds to that of the external charges. The dipole field and the externally applied electric field can be comparable in magnitude.

In addition to displacing the positive and negative charges, an applied electric field can also polarize a dielectric by orienting molecules that possess a permanent dipole moment. Such molecules experience a torque which tends to align them with the field, but collisions arising from the thermal agitation of the molecules tend to destroy the alignment. An equilibrium polarization is thus established in which there is, on the average, a net alignment.

This is, in fact, a simplified view of dielectric behavior because many solids are not made up of molecules but of individual ions which interact with a large number of other ions. Sodium chloride is an example of such a solid. We shall limit most of our discussion to dielectrics that are made up of molecules that are either distorted or oriented, or both distorted and oriented in an electric field.

In a polarized dielectric each molecule acts as an electric dipole of moment

$\boldsymbol{p}$. We shall treat these dipoles both from a molecular and from a macroscopic point of view: we shall consider the fields of individual dipoles, as well as those of large numbers of dipoles.

In the end we shall deal with dielectrics entirely from the macroscopic point of view, but we must justify this procedure by first examining carefully the molecular aspect.

3.1 THE ELECTRIC POLARIZATION $\boldsymbol{P}$

Let us first consider nonpolar dielectrics, that is, dielectrics whose molecules have zero permanent electric dipole moment. We shall not have to consider polar dielectrics until Section 3.9.

The *electric polarization* $\boldsymbol{P}$ is the dipole moment per unit volume at a given point and is a macroscopic quantity. If $\boldsymbol{p}$ is the average electric dipole moment per molecule in a small volume τ, and if N is the number of molecules per unit volume, then

$$\boldsymbol{P} = N\boldsymbol{p}. \tag{3-1}$$

It will be recalled from Problem 2-23 that the dipole moment of a charge distribution is independent of the origin of coordinates, provided the net charge in the distribution is zero, as it is in molecules.

The volume τ over which we define $\boldsymbol{P}$ must be chosen large enough to render the fluctuations in $\boldsymbol{P}$, from one instant to the next or from one τ to a neighboring one, negligible. Although on the average the volume τ contains $N\tau$ molecules, we may expect, from statistical considerations, that there can be fluctuations of $\pm(N\tau)^{1/2}$ in the number $N\tau$. The actual number of molecules per unit volume thus fluctuates between $N - (N/\tau)^{1/2}$ and $N + (N/\tau)^{1/2}$, or by $\pm 100\,(1/N\tau)^{1/2}$ %. This restriction on the size of τ is ordinarily unimportant. Take the case of water, for example, where $N \approx 3 \times 10^{28}$ molecules/meter^3. Even in a submicroscopic cube measuring 100 angstroms or 10^{-8} meter on the side, the statistical fluctuation in $\boldsymbol{P}$ is only about one half of one percent.

3.2 ELECTRIC FIELD AT AN EXTERIOR POINT

Figure 3-1 shows a block of dielectric material with a dipole moment $\boldsymbol{P}$ per unit volume, $\boldsymbol{P}$ being a function of position within the dielectric. Let us calculate the electric potential V which the dipoles in the dielectric produce at a point P outside. This potential can then be added to that produced by all

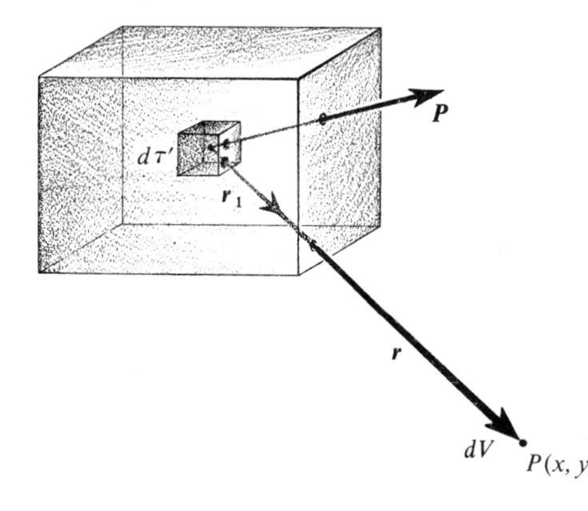

Figure 3-1. Block of dielectric with a dipole moment $\boldsymbol{P}$ per unit volume. The dipoles within the element of volume shown inside the block give rise to an electric potential dV at the point P.

the other charges in the system to give the total potential. The negative gradient of this total potential at the point P is the electric field intensity at that point.

At the point $P(x, y, z)$ the potential dV due to the dipole $\boldsymbol{P}\, d\tau'$ situated at (x', y', z') is

$$dV = \frac{1}{4\pi\epsilon_0} \frac{\boldsymbol{P}\cdot\boldsymbol{r}_1}{r^2}\, d\tau', \tag{3-2}$$

$$= \frac{1}{4\pi\epsilon_0} \left[\boldsymbol{P}\cdot\boldsymbol{\nabla}'\left(\frac{1}{r}\right)\right] d\tau', \tag{3-3}$$

where $\boldsymbol{\nabla}'$ is evaluated at (x', y', z') (Problem 1-12). We have assumed that the distance r is large compared to the dimensions of a molecule so that the dipole approximation can be valid. Integrating over the block of dielectric gives

$$V = \frac{1}{4\pi\epsilon_0} \int_{\tau'} \left[\boldsymbol{P}\cdot\boldsymbol{\nabla}'\left(\frac{1}{r}\right)\right] d\tau'. \tag{3-4}$$

This integration can be carried out analytically only for simple geometries and only if $\boldsymbol{P}$ is a simple function of position within the dielectric. Furthermore, $\boldsymbol{P}$ is never known *a priori*, although we can often deduce it from measurable quantities. We shall nevertheless assume, for the time being, that $\boldsymbol{P}$ is known. This integral can be put into a more interesting form by transforming it through the vector identity

$$\boldsymbol{\nabla}'\cdot f\boldsymbol{A} = f\boldsymbol{\nabla}'\cdot\boldsymbol{A} + \boldsymbol{A}\cdot\boldsymbol{\nabla}'f, \tag{3-5}$$

where f and $\boldsymbol{A}$ are scalar and vector functions respectively. Let $1/r$ be the scalar and $\boldsymbol{P}$ the vector function. Then

$$V = \frac{1}{4\pi\epsilon_0} \int_{\tau'} \left(\boldsymbol{\nabla}'\cdot\frac{\boldsymbol{P}}{r}\right) d\tau' - \frac{1}{4\pi\epsilon_0} \int_{\tau'} \frac{\boldsymbol{\nabla}'\cdot\boldsymbol{P}}{r}\, d\tau'. \tag{3-6}$$

We can use the divergence theorem to express the first term as a surface integral; then

$$V = \frac{1}{4\pi\epsilon_0}\int_{S'} \frac{\boldsymbol{P}\cdot d\boldsymbol{a}'}{r} - \frac{1}{4\pi\epsilon_0}\int_{\tau'} \frac{\boldsymbol{\nabla}'\cdot\boldsymbol{P}}{r}\,d\tau', \tag{3-7}$$

where S' is the surface that bounds the volume τ' of the dielectric and the vector $d\boldsymbol{a}'$ points outward.

Remember that this is the potential due to the dipoles in the dielectric. To obtain the total potential we have to add to this V the V due to the charges that produce the applied electric field.

3.2.1 The Bound Charge Densities ρ_b and σ_b

Now both integrals in Eq. 3-7 involve a $1/r$ dependence and are multiplied by $1/4\pi\epsilon_0$. We thus have the following remarkable result: these two terms are exactly the potentials that would result from surface and volume charge distributions having densitites

$$\sigma_b = \boldsymbol{P}\cdot\boldsymbol{n} \tag{3-8}$$

and

$$\rho_b = -\boldsymbol{\nabla}\cdot\boldsymbol{P} \tag{3-9}$$

respectively, where $\boldsymbol{n}$ is the unit *outward* normal vector at the surface of the dielectric. Thus

$$V = \frac{1}{4\pi\epsilon_0}\int_{S'} \frac{\sigma_b da'}{r} + \frac{1}{4\pi\epsilon_0}\int_{\tau'} \frac{\rho_b d\tau'}{r}. \tag{3-10}$$

We have omitted the prime on $\boldsymbol{\nabla}$ since it is obvious that the divergence must be evaluated at the point where the volume charge density is ρ_b.

The dielectric may therefore be replaced by the bound charge distributions σ_b and ρ_b without affecting the electric field outside the dielectric. We shall see later that we can use the same procedure to find a macroscopic potential at points inside the dielectric.

We can demonstrate with the aid of Figure 3-2 that the bound electric charge densities σ_b and ρ_b represent actual accumulations of charge. Let us first consider the surface density σ_b. We imagine a small element of surface $d\boldsymbol{a}'$ inside the dielectric. Under the action of the field, an average charge separation $\boldsymbol{s}$ is produced in the molecules. Positive charge crosses the surface by moving in the direction of the field; negative charge crosses it by moving in the opposite direction. For the purpose of our calculation, we may consider the positive charge to be in the form of point charges Q and the negative charge to be in the form of point charges $-Q$. Furthermore, we may consider

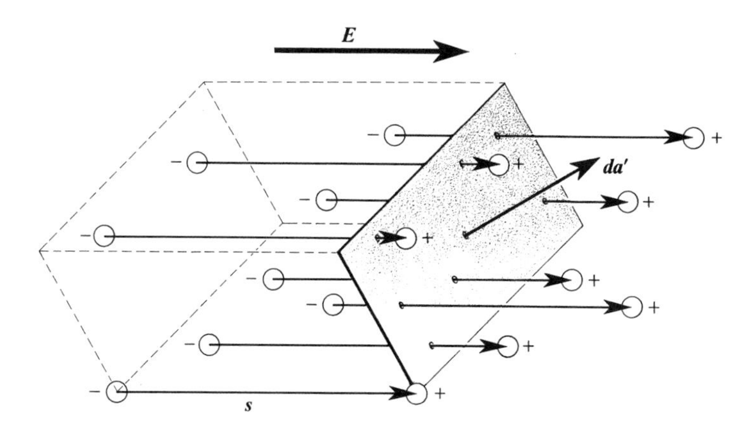

Figure 3-2. Under the action of an electric field $\boldsymbol{E}$, which is the resultant of an external field and of the field of the dipoles within the dielectric, positive and negative charges in the molecules are separated by an average distance $\boldsymbol{s}$. In the process a net charge $dQ = NQ\boldsymbol{s}\cdot\boldsymbol{da}'$ crosses the surface $\boldsymbol{da}'$, N being the number of molecules per unit volume and Q the positive charge in a molecule. The vector $\boldsymbol{da}$ is perpendicular to the shaded surface. The circles indicate the centers of charge for the positive and for the negative charges in one molecule.

the negative charges to be fixed and the positive charges to move a distance $\boldsymbol{s}$. The amount of charge dQ that crosses $\boldsymbol{da}'$ is then just the total amount of *positive* charge within the imaginary parallelepiped shown in Figure 3-2. The volume of this parallelepiped is

$$d\tau' = \boldsymbol{s}\cdot\boldsymbol{da}', \tag{3-11}$$

and

$$dQ = NQ\boldsymbol{s}\cdot\boldsymbol{da}', \tag{3-12}$$

where N is the number of molecules per unit volume and $Q\boldsymbol{s}$ is the dipole moment $\boldsymbol{p}$ of a molecule. Then

$$dQ = \boldsymbol{P}\cdot\boldsymbol{da}'. \tag{3-13}$$

If $\boldsymbol{da}'$ is on the surface of the dielectric material, dQ accumulates there in a layer of thickness $\boldsymbol{s}\cdot\boldsymbol{n}$. Since the thickness of the layer is of the order of the dimensions of a molecule, we may treat the charge as a surface distribution with a density

$$\sigma_b = \frac{dQ}{da} = \boldsymbol{P}\cdot\boldsymbol{n}. \tag{3-14}$$

The bound surface charge density σ_b is thus equal to the normal component of the polarization vector at the surface.

We can show similarly that $-\boldsymbol{\nabla}\cdot\boldsymbol{P}$ represents a volume density of charge. The net charge that flows out of a volume τ' across an element $\boldsymbol{da}'$ of its surface is $\boldsymbol{P}\cdot d\boldsymbol{a}'$, as we found above. The net charge that flows out of the surface S' bounding τ' is thus

$$Q = \int_{S'} \boldsymbol{P}\cdot d\boldsymbol{a}', \tag{3-15}$$

and the net charge that remains within the volume τ' must be $-Q$. If ρ_b is the volume density of the charge remaining within this volume, then

$$\int_{\tau'} \rho_b \, d\tau' = -Q = -\int_{S'} \boldsymbol{P}\cdot d\boldsymbol{a}', \tag{3-16}$$

$$= -\int_{\tau'} (\boldsymbol{\nabla}\cdot\boldsymbol{P}) \, d\tau'. \tag{3-17}$$

Since this equation must be true for all τ', the integrands must be equal at every point, and the bound electric charge density is

$$\rho_b = -\boldsymbol{\nabla}\cdot\boldsymbol{P}. \tag{3-18}$$

We refer to σ_b and ρ_b as either *bound, polarization*, or *induced* charge densities, as distinguished from the *free* or *conductible* charge densities σ_f and ρ_f. Bound charges are those that accumulate through the displacements that occur on a molecular scale in the polarization process. The other charges are called free charges. The conduction electrons in a conductor and the electrons injected into a dielectric with a high-energy electron beam are examples of free charges.*

Coulomb's law applies to any net accumulation of charge, regardless of other matter that may be present.

3.2.2 The Polarization Current Density

When a dielectric is placed in an electric field that is a function of the time, the motion of the bound charges gives rise to a *polarization current*. We can find the polarization current density $\boldsymbol{J}_b$ by using again the fact that charge is conserved.

For any volume τ bounded by a surface S, the rate at which bound charge flows out through S must be equal to the rate of decrease of the bound charge within S:

---

* The concepts of free and bound charges, and even the concept of polarization, are based on the assumption that the dielectric is composed of molecules. If there are no well-defined molecules, the values of σ_b, σ_f, and $\boldsymbol{P}$ become rather arbitrary. See, for example, Edward Purcell, *Electricity and Magnetism*, Berkeley Physics Course Volume 2 (McGraw-Hill, New York, 1965) p. 344.

$$\int_S \boldsymbol{J}_b \cdot d\boldsymbol{a} = -\frac{\partial}{\partial t} \int \rho_b \, d\tau. \tag{3-19}$$

Using the divergence theorem on the left and substituting the value of ρ_b on the right from Eq. 3-18,

$$\int_\tau \boldsymbol{\nabla} \cdot \boldsymbol{J}_b \, d\tau = \frac{\partial}{\partial t} \int_\tau \boldsymbol{\nabla} \cdot \boldsymbol{P} \, d\tau = \int_\tau \boldsymbol{\nabla} \cdot \frac{\partial \boldsymbol{P}}{\partial t} \, d\tau. \tag{3-20}$$

We have put the $\partial/\partial t$ under the integral sign because it is immaterial whether the derivative with respect to the time is calculated first or last. Since τ is any volume, the integrands must be equal and the polarization current density is

$$\boldsymbol{J}_b = \frac{\partial \boldsymbol{P}}{\partial t} \text{ amperes/meter}^2. \tag{3-21}$$

3.3 ELECTRIC FIELD AT AN INTERIOR POINT

It is often necessary to know the electric potential V and the electric field intensity $\boldsymbol{E}$ at points within a dielectric. We may wish to know, for example, the potential difference between two charged conductors separated by a dielectric.

The electric field intensity at a particular time and at a particular point within the dielectric is a rapidly fluctuating quantity, both with regard to position and to time. For example, it reaches enormous magnitudes at the surfaces of nuclei. There are also large variations in its direction: on one side of a nucleus it is in one direction, and on the other side it is in the opposite direction. Thermal agitation of the molecules also produces large time fluctuations at any given point.

The macroscopic electric field intensity $\boldsymbol{E}$, which is what we really wish to know, is the *space-and-time* average of this electric field intensity.

The macroscopic electric field intensity $\boldsymbol{E}$ is a slowly varying function of the coordinates and is independent of the time in the static case. It is this electric field intensity that we shall integrate in order to calculate potential differences.

We shall show that the part of this macroscopic field that originates from the dipoles within the dielectric can be calculated from the bound charge distributions σ_b and ρ_b discussed above, exactly as in Eq. 3-10. This is not intuitively obvious.

The reason the type of calculation used for the exterior point might not be valid at an interior point is that, in the immediate neighborhood of the point considered, there may be large numbers of molecules for which our

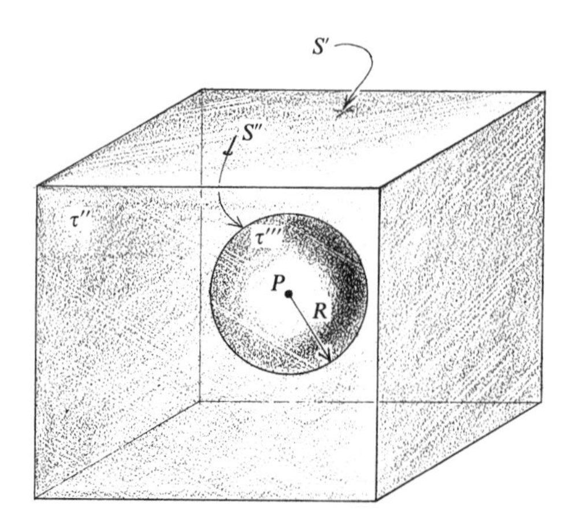

Figure 3-3. Imaginary sphere of radius R, volume τ''' and surface S'' inside a block of dielectric. The remaining volume is τ'', and its outside surface is S'.

expressions for the dipole field are not valid, since Eq. 3-2 requires that the distance between the dipole and the point considered be much larger than the dimensions of the dipole.

To calculate the macroscopic electric field intensity $\boldsymbol{E}$ inside a dielectric, let us consider Figure 3-3, which shows a small imaginary sphere of radius R centered on P. The surface S'' of this sphere divides the dielectric into two volumes, τ'' outside and τ''' inside. The potential at P may be calculated for all the polarized material outside S'' by treating P as an exterior point, as in Section 3.2; however, the molecules within S'' are too close to P to be treated in this fashion. The volume τ''' is macroscopically small, and the macroscopic quantities $\boldsymbol{E}$, $\boldsymbol{P}$, and $\rho_b = -\boldsymbol{\nabla}\cdot\boldsymbol{P}$ do not vary significantly from one side to the other.

Thus the macroscopic electric field intensity that the dipoles of the dielectric produce at the point P is the sum of two terms, which we shall calculate separately:

$$\boldsymbol{E} = \boldsymbol{E}'' + \boldsymbol{E}'''. \tag{3-22}$$

3.3.1 Electric Field Intensity E'' Due to the Distant Dipoles

Since the point P is external to the volume τ'', we can calculate $\boldsymbol{E}''$ from the bound charge (a) on the outer surface S', (b) in the volume τ'', and (c) on the inner surface S''. This latter surface is not real, but the discussion in Section 3.2 applies even when only a portion of the dielectric is under consideration, as it is here.

Each element of polarization charge $\rho_b\, d\tau''$ in τ'' contributes at P an element of field

$$\frac{1}{4\pi\epsilon_0}\frac{\rho_b\, d\tau''}{r^2}\,\boldsymbol{r}_1,$$

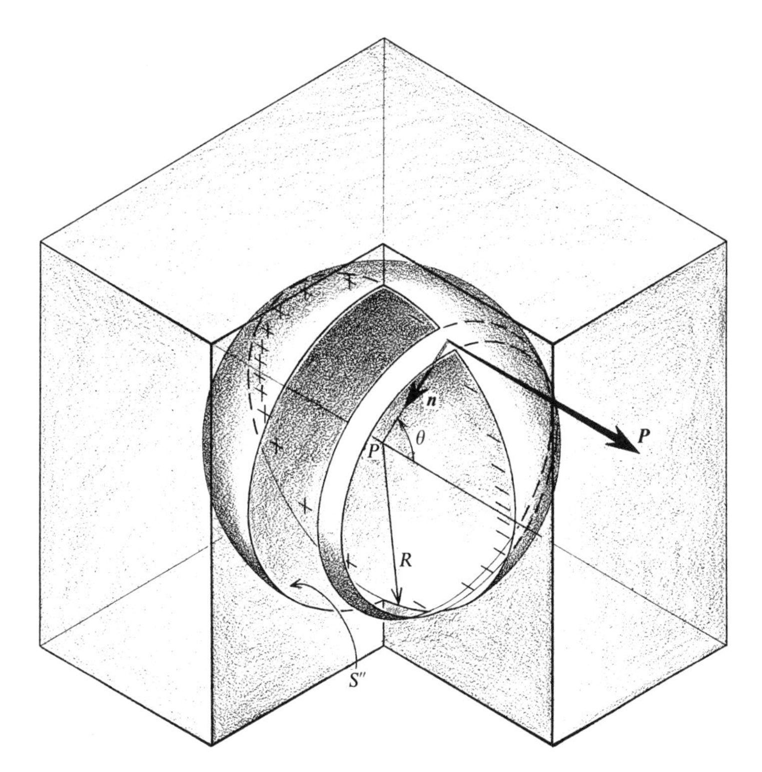

Figure 3-4. Spherical cavity of radius R within a polarized dielectric. The charges shown are due to the polarization $\boldsymbol{P}$.

where $\boldsymbol{r}_1$ is the unit vector in the direction from the element of volume $d\tau''$ to the point P, and r is the distance from $d\tau''$ to P. Proceeding similarly for the surface charges,

$$\boldsymbol{E}'' = \frac{1}{4\pi\epsilon_0}\int_{S'} \frac{\sigma_b\, da'}{r^2}\,\boldsymbol{r}_1 + \frac{1}{4\pi\epsilon_0}\int_{\tau''} \frac{\rho_b\, d\tau''}{r^2}\,\boldsymbol{r}_1 + \frac{1}{4\pi\epsilon_0}\int_{S''} \frac{\sigma_b\, da''}{r^2}\,\boldsymbol{r}_1. \quad (3\text{-}23)$$

Let us just calculate the third term for the moment. Figure 3-4 shows the spherical surface S'' with an axis drawn parallel to the polarization vector $\boldsymbol{P}$ The polarization charge density on the surface S'' is

$$\sigma_b = \boldsymbol{P}\cdot\boldsymbol{n} = -P\cos\theta, \quad (3\text{-}24)$$

and the charge on the annulus of width $R\,d\theta$ contributes an element of field intensity at P of

$$\frac{1}{4\pi\epsilon_0}\frac{P\cos\theta}{R^2}\,2\pi R^2 \sin\theta\, d\theta \cos\theta.$$

The first $\cos\theta$ factor comes from σ_b, and the second comes from taking the component of the electric field intensity in the axial direction, since symmetry requires the resultant field to be in that direction. On performing the integration over the whole surface S'', from $\theta = 0$ to $\theta = \pi$, we find that the third integral of Eq. 3-23 is equal to $P/3\epsilon_0$.

3.3.2 Electric Field Intensity E''' Due to the Near Dipoles

The electric polarization $\boldsymbol{P}$ may be taken to be uniform inside the small sphere S''. Then we may assume that the electric field intensity $\boldsymbol{E}'''$ at the center is equal to the average electric field intensity inside S'' due to this polarization; this is the quantity that we calculated in Section 2-13, and

$$\boldsymbol{E}''' = -\frac{N(4/3)\pi R^3 \boldsymbol{p}}{4\pi\epsilon_0 R^3} = -\frac{\boldsymbol{P}}{3\epsilon_0}, \tag{3-25}$$

where N is the number of molecules per unit volume and $\boldsymbol{p}$ is the average dipole moment per molecule.

Thus the contribution of the molecules within S'' to the macroscopic electric field intensity $\boldsymbol{E}$ at P is equal in magnitude and opposite in direction to the field of the bound charges on the surface S'' (the third integral of Eq. 3-23), and

$$\boldsymbol{E} = \boldsymbol{E}'' + \boldsymbol{E}''' = \frac{1}{4\pi\epsilon_0}\int_{S'} \frac{\sigma_b\, da'}{r^2}\,\boldsymbol{r}_1 + \frac{1}{4\pi\epsilon_0}\int_{\tau''} \frac{\rho_b\, d\tau''}{r^2}\,\boldsymbol{r}_1. \tag{3-26}$$

The second integral is calculated only over that portion of the dielectric that is outside the spherical surface S''. Of course, it would be simpler to calculate the second integral if it extended over the complete volume τ' of the dielectric, but this would add the term

$$\frac{1}{4\pi\epsilon_0}\int_{\tau'''} \frac{\rho_b\, d\tau'''}{r^2}\,\boldsymbol{r}_1.$$

Since the bound charge density does not vary significantly over the small sphere τ''', this integral is merely the electric field intensity at the center of a uniform spherical charge distribution. For every element of charge giving an element of electric field intensity in one direction, there is a symmetrical element of charge giving a field in the opposite direction, and this integral is zero. The second integral of Eq. 3-26 can therefore be extended to the complete volume τ' of the dielectric, and

$$\boldsymbol{E} = \frac{1}{4\pi\epsilon_0}\int_{S'} \frac{\sigma_b\, da'}{r^2}\,\boldsymbol{r}_1 + \frac{1}{4\pi\epsilon_0}\int_{\tau'} \frac{\rho_b\, d\tau'}{r_2}\,\boldsymbol{r}_1, \tag{3-27}$$

$$= \frac{1}{4\pi\epsilon_0}\int_{S'} \frac{\boldsymbol{P}\cdot d\boldsymbol{a}'}{r^2}\,\boldsymbol{r}_1 - \frac{1}{4\pi\epsilon_0}\int_{\tau'} \frac{\boldsymbol{\nabla}\cdot\boldsymbol{P}\, d\tau'}{r^2}\,\boldsymbol{r}_1. \tag{3-28}$$

That part of the electric field intensity that is due to the dielectric itself can therefore be calculated at points inside the dielectric just as it is calculated outside, namely by replacing the dielectric by a space charge distribution $-\nabla \cdot \boldsymbol{P}$ and a surface charge distribution $\boldsymbol{P} \cdot \boldsymbol{n}$.

REVIEW. In the presence of an electric field, a dielectric becomes polarized and acquires a dipole moment $\boldsymbol{P}$ per unit volume. We wish to calculate the electric field produced by the polarized dielectric.

In Section 3.2 we calculated the field at a point situated *outside* the dielectric. We started with the value of the electric potential V for a small dipole $\boldsymbol{P}\, d\tau$ and integrated over the volume of the dielectric. The result (Eq. 3-28) showed that the field is just what one would expect from a volume charge density $\rho_b = -\nabla \cdot \boldsymbol{P}$ and a surface charge density $\sigma_b = \boldsymbol{P} \cdot \boldsymbol{n}$. We then showed that these two quantities do indeed represent real accumulations of charge.

Next we calculated the macroscopic electric field intensity $\boldsymbol{E}$ at a point *inside* the dielectric. This is the space-and-time average of the electric field intensity over a small region of the dielectric.

We divided the dielectric into two regions, a small sphere of volume τ''' centered on the point P, and the remaining volume τ''. The electric field intensity at P due to the dipoles in τ'' comprises three terms arising from (a) the surface charge density on the outside surface S' of the dielectric, (b) the volume charge density in τ'', and (c) the surface charge density on the surface S''. It turns out that this last term just cancels the field at P due to the dipoles within the sphere τ'''. We also found that the second term can be extended without error to the complete volume of the dielectric.

The macroscopic electric field intensity $\boldsymbol{E}$ at a point inside a dielectric can therefore be found by replacing the dielectric with the bound charge densities ρ_b and σ_b in exactly the same manner as for an external point. In other words, we have found that the charge densities ρ_b and σ_b represent actual accumulations of charge, even though bound, and that they produce electric field intensities according to Coulomb's law. This is true inside as well as outside the dielectric.

In calculating the macroscopic field $\boldsymbol{E}$, we have said nothing about the field of the external charge distribution that is responsible for the polarization in the first place. It simply adds at every point to the field of the polarized dielectric, to give the total macroscopic field.

The electric potential V and the electric field intensity $\boldsymbol{E}$ are therefore given correctly by the following integrals, whether or not there are dielectrics, or even conductors, in the field:

$$V = \frac{1}{4\pi\epsilon_0}\int_{\tau'} \frac{\rho_f + \rho_b}{r}\, d\tau' + \frac{1}{4\pi\epsilon_0}\int_{S'} \frac{\sigma_f + \sigma_b}{r}\, da', \tag{3-29}$$

$$\boldsymbol{E} = \frac{1}{4\pi\epsilon_0}\int_{\tau'} \frac{(\rho_f + \rho_b)\boldsymbol{r}_1}{r^2}\, d\tau' + \frac{1}{4\pi\epsilon_0}\int_{S'} \frac{(\sigma_f + \sigma_b)\boldsymbol{r}_1}{r^2}\, da', \tag{3-30}$$

where ρ_f is the free charge density, ρ_b is the bound charge density, r is the distance between the source point where the total charge density is $\rho_f + \rho_b$ and the field point where V and $\boldsymbol{E}$ are calculated, $\boldsymbol{r}_1$ is a unit vector pointing *from* the source point *to* the field point, and τ' is any volume enclosing all the charges.

3.4 THE LOCAL FIELD

Although the above discussion clarifies our understanding of polarization effects in dielectrics, it cannot be used to calculate the macroscopic $\boldsymbol{E}$, since we do not know as yet how to relate the electric polarization $\boldsymbol{P}$ to $\boldsymbol{E}$. We must eventually be able to calculate the electric potential and field intensity at any point, either inside or outside a dielectric, without knowing $\boldsymbol{P}$. We must be able to calculate V and $\boldsymbol{E}$ knowing only the potentials of, or the charges on, the conductors, plus the geometry and characteristics of the conductors and dielectrics in the system.

In order to proceed we must know the average electric field intensity acting on a molecule in the dielectric. We shall call this the *local field* and shall designate it $\boldsymbol{E}_{\text{loc}}$.

In the next section we shall see that the dipole moment $\boldsymbol{p}$ induced in a single molecule is usually proportional to the local field.

We define $\boldsymbol{E}_{\text{loc}}$ in the following way. Imagine that we stop the thermal motion of all the molecules in the dielectric at some particular instant. We remove the molecule in question, keeping all the other molecules frozen in position, and calculate the space-averaged electric field intensity in the cavity previously occupied by the molecule. We replace the molecule, return the system to normal temperature, and repeat the process many times. The resultant space-and-time average of the electric field intensity in the cavity is $\boldsymbol{E}_{\text{loc}}$.

We expect the local field to be larger than the macroscopic $\boldsymbol{E}$, since $\boldsymbol{E}_{\text{loc}}$ does not include the field of the molecule under consideration, whereas the macroscopic field does. In Section 2.13 we found that the average electric field intensity within a sphere containing an arbitrary charge distribution is in the direction opposite to the electric dipole moment of the distribution. Thus

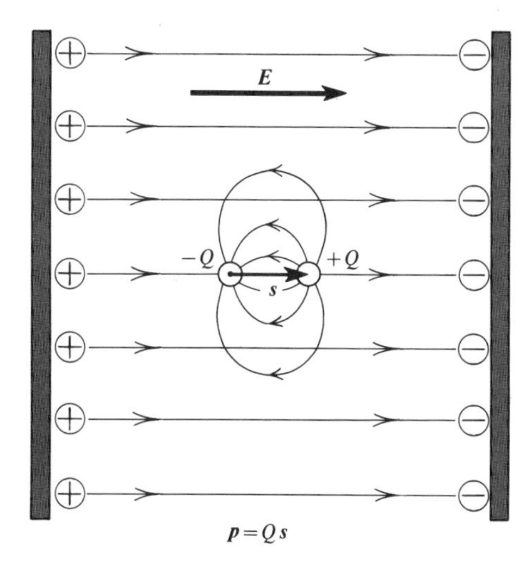

Figure 3-5. The field of an individual dipole in an electric field.

the molecule in question contributes an average field in the direction opposite that of the macroscopic electric field intensity $\boldsymbol{E}$, as indicated in Figure 3-5, at least if the material is isotropic so that the molecules are free to orient their dipole moments according to the direction of $\boldsymbol{E}$.

A numerical calculation of the average electric field intensity due to the molecule would require detailed information about its shape and charge distribution, but we can perform the calculation in general terms, assuming only the shape of the molecule. If we select the simplest shape, namely the sphere, then this average field is (again from Section 2-13)

$$-\frac{\boldsymbol{p}}{4\pi\epsilon_0 R^3} = -\frac{\boldsymbol{P}}{4\pi\epsilon_0 N R^3} = -\frac{\boldsymbol{P}}{3\epsilon_0 N \tau_m}, \tag{3-31}$$

where $\boldsymbol{p}$ is the dipole moment of the molecule and τ_m is its volume.

If we further set $\tau_m \approx 1/N$, then

$$\boldsymbol{E}_{\text{loc}} \approx \boldsymbol{E} + \frac{\boldsymbol{P}}{3\epsilon_0}. \tag{3-32}$$

This expression for the local field is at best approximate, owing to the approximations that have gone into it. Although we assumed a spherical shape for the molecule, we set $\tau_m \approx 1/N$. This means that all of space is filled with molecules, which contradicts the assumption of spherical shape. Nonetheless, dielectric behavior predicted from the above expression for the local field agrees surprisingly well with the experimental data for many substances.

Fortunately, our dielectric theory does not require that the expression for the local field be quantitatively accurate; our conclusions will all be valid if only

$$\boldsymbol{E}_{\text{loc}} = \boldsymbol{E} + b\frac{\boldsymbol{P}}{\epsilon_0}, \tag{3-33}$$

where b is a constant that depends on the nature of the dielectric.

So far we have discussed dielectrics consisting of only one type of molecule. More generally,

$$\boldsymbol{P} = N_1\boldsymbol{p}_1 + N_2\boldsymbol{p}_2 + \cdots, \tag{3-34}$$

where N_1 is the number of molecules of the first type per unit volume, with the average dipole moment $\boldsymbol{p}_1$, N_2 is the number of the second type per unit volume, and so on. The theory is not otherwise modified.

3.5 THE ELECTRIC SUSCEPTIBILITY χ_e

We can now find a relationship between the electric polarization $\boldsymbol{P}$ and the macroscopic electric field intensity $\boldsymbol{E}$. To do this, we consider the electric dipole moment $\boldsymbol{p}$ induced in a single molecule. The magnitude of $\boldsymbol{p}$ depends on the local field $\boldsymbol{E}_{\text{loc}}$: the positive and negative charges are separated under the action of the local field until the restoring force, which is an internal electrostatic force arising from their separation, just balances the local field force.

In most dielectrics the molecular charge separation is directly proportional to, and in the same direction as, the local field. Thus

$$\boldsymbol{p} = \alpha\boldsymbol{E}_{\text{loc}} = \alpha\left(\boldsymbol{E} + b\frac{\boldsymbol{P}}{\epsilon_0}\right), \tag{3-35}$$

where the constant α is known as the *molecular polarizability*.

Dielectrics which show this simple dependence of polarization on local field are said to be *linear* and *isotropic*. Many commercially important dielectrics are homogeneous, as well as linear and isotropic. We shall designate such materials as *Class A dielectrics* and shall confine our attention to them for the time being.

The dipole moment per unit volume in Class A dielectries is

$$\boldsymbol{P} = N\boldsymbol{p} = N\alpha\,\boldsymbol{E}_{\text{loc}} = N\alpha\left(\boldsymbol{E} + b\frac{\boldsymbol{P}}{\epsilon_0}\right), \tag{3-36}$$

where N is again the number of molecules per unit volume, and

$$\boldsymbol{P} = \frac{N\alpha}{1 - (N\alpha b/\epsilon_0)}\boldsymbol{E}, \tag{3-37}$$

$$= \epsilon_0\chi_e\boldsymbol{E}, \tag{3-38}$$

where χ_e is a dimensionless constant known as the *electric susceptibility* of the dielectric. The constant ϵ_0 is included explicitly to make χ_e dimensionless.

For a Class A dielectric, the electric polarization $\boldsymbol{P}$ is thus proportional to the macroscopic electric field intensity $\boldsymbol{E}$ as well as to the local field $\boldsymbol{E}_{\text{loc}}$.

3.6 THE DIVERGENCE OF $\boldsymbol{E}$. THE ELECTRIC DISPLACEMENT $\boldsymbol{D}$

We have found that the electric field due to the atomic and molecular dipoles of a polarized dielectric can be calculated from the bound charge densities and from Coulomb's law, just as for any other charge distribution. Let us investigate the implications of this fact on Gauss's law, which is a direct consequence of Coulomb's law.

Gauss's law relates the flux of the electric field intensity $\boldsymbol{E}$ through a closed surface to the total net charge Q_t enclosed within that surface:

$$\int_S \boldsymbol{E}\cdot d\boldsymbol{a} = \int_\tau \boldsymbol{\nabla}\cdot\boldsymbol{E}\, d\tau = \frac{Q_t}{\epsilon_0}. \tag{3-39}$$

For dielectrics, Q_t must include bound as well as free charges:

$$Q_t = \int_\tau (\rho_f + \rho_b)\, d\tau, \tag{3-40}$$

where the integration is intended to cover both the surface and volume distributions of free and bound charges.

If we substitute this value of Q_t into Gauss's law and equate the integrands of the volume integrals, then

$$\boxed{\boldsymbol{\nabla}\cdot\boldsymbol{E} = \frac{\rho_t}{\epsilon_0}} \tag{3-41}$$

at every point, where $\rho_t = \rho_f + \rho_b$ is the total net charge density. This is *Gauss's law in its more general form. It is one of Maxwell's four fundamental equations of electromagnetism.* It follows from (a) Coulomb's law, (b) the concept of electric field intensity, (c) the principle of superposition, and (d) the fact that the field of the dipoles in a polarized medium can be calculated everywhere from the bound charge densities ρ_b and σ_b.

Since $\boldsymbol{E} = -\boldsymbol{\nabla} V$, it follows that

$$\nabla^2 V = -\frac{\rho_t}{\epsilon_0}. \tag{3-42}$$

This is *Poisson's equation for V*. This equation is valid in dielectrics and is more general than Eq. 2-31. (In Chapter 2 there was no need to distinguish between free and bound charges because all the charges considered were of the former type; the ρ's of Chapter 2 are therefore ρ_f's.)

Now $\rho_b = -\boldsymbol{\nabla}\cdot\boldsymbol{P}$, and, for any dielectric,

$$\boldsymbol{\nabla}\cdot\boldsymbol{E} = \frac{1}{\epsilon_0}(\rho_f - \boldsymbol{\nabla}\cdot\boldsymbol{P}), \tag{3-43}$$

or

$$\boldsymbol{\nabla}\cdot(\epsilon_0\boldsymbol{E} + \boldsymbol{P}) = \rho_f. \tag{3-44}$$

The vector $\epsilon_0\boldsymbol{E} + \boldsymbol{P}$ is therefore such that its divergence depends only on the free charge density ρ_f. This vector is called the *electric displacement* and is designated by $\boldsymbol{D}$:

$$\boldsymbol{D} = \epsilon_0\boldsymbol{E} + \boldsymbol{P}. \tag{3-45}$$

Thus

$$\boldsymbol{\nabla}\cdot\boldsymbol{D} = \rho_f. \tag{3-46}$$

In integral form Gauss's law for $\boldsymbol{D}$ becomes

$$\int_S \boldsymbol{D}\cdot d\boldsymbol{a} = \int_\tau \rho_f \, d\tau, \tag{3-47}$$

and the flux of the electric displacement $\boldsymbol{D}$ through a closed surface is equal to the free charge enclosed by the surface.

From the definition of the electric displacement,

$$\boldsymbol{E} = \frac{\boldsymbol{D}}{\epsilon_0} - \frac{\boldsymbol{P}}{\epsilon_0} \tag{3-48}$$

and the electric field intensity $\boldsymbol{E}$ inside a dielectric is the sum of two fields: $\boldsymbol{D}/\epsilon_0$ associated with the free charges, since

$$\boldsymbol{\nabla}\cdot\left(\frac{\boldsymbol{D}}{\epsilon_0}\right) = \frac{\rho_f}{\epsilon_0}, \tag{3-49}$$

and $-\boldsymbol{P}/\epsilon_0$ associated with the bound charges, since

$$\boldsymbol{\nabla}\cdot\left(-\frac{\boldsymbol{P}}{\epsilon_0}\right) = \frac{\rho_b}{\epsilon_0}. \tag{3-50}$$

Lines of $\boldsymbol{D}$ begin or end only on free charges, whereas lines of $\boldsymbol{E}$ begin or end on either free or bound charges. This does not mean, however, that the magnitude and direction of $\boldsymbol{D}$ at a particular point depend only on the free charges. The only quantity that depends on the free charges alone is $\boldsymbol{\nabla}\cdot\boldsymbol{D}$. To find $\boldsymbol{D}$, we must integrate Eq. 3-46, subject to whatever boundary conditions

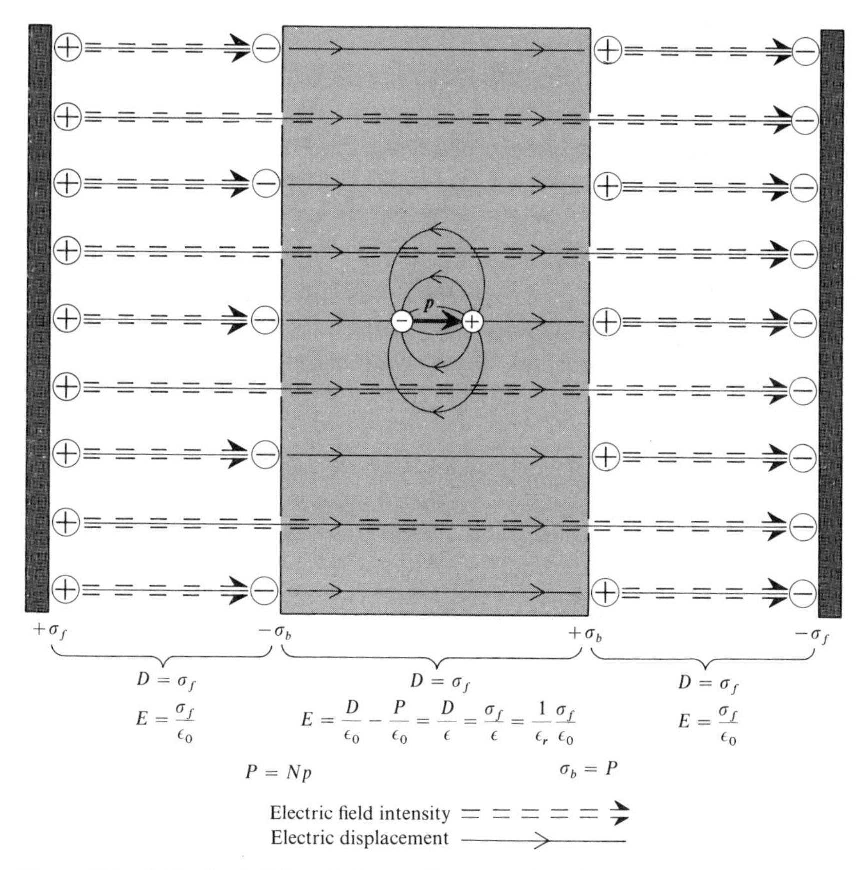

Figure 3-6. A block of dielectric located between the plates of a parallel-plate capacitor. Each molecule of the dielectric has a dipole moment $\boldsymbol{p}$, leading to a polarization $\boldsymbol{P}$. The electric displacement $\boldsymbol{D}$ depends only on the free charges $\pm\sigma_f$ and is the same inside and outside the dielectric. The electric field intensity $\boldsymbol{E}$, on the other hand, is reduced inside the dielectric because the polarization charges $\pm\sigma_b$ produce a field in the opposite direction.

obtain in the particular field. Figure 3-6 shows the vectors $\boldsymbol{p}$, $\boldsymbol{E}$, $\boldsymbol{D}$ and the electric charge densities σ_f and σ_b for a parallel-plate capacitor.

In writing down the expressions for $\boldsymbol{\nabla}\cdot\boldsymbol{E}$ and $\boldsymbol{\nabla}\cdot\boldsymbol{P}$, we have implicity assumed the existence of the space derivatives of $\boldsymbol{E}$ and $\boldsymbol{P}$. These derivatives do not exist, of course, either at a point charge or at the interface between two media. In such cases, we must revert to the integral form of Gauss's law:

$$\epsilon_0 \int_S \boldsymbol{E}\cdot d\boldsymbol{a} = \int_\tau (\rho_f + \rho_b)\, d\tau. \tag{3-51}$$

3.6.1 The Relative Permittivity ϵ_r. Poisson's Equation for Dielectrics

For a Class A dielectric we have, from Eqs. 3-45 and 3-38,

$$\boldsymbol{D} = \epsilon_0(1 + \chi_e)\,\boldsymbol{E} = \epsilon_0\epsilon_r\,\boldsymbol{E} = \epsilon\boldsymbol{E}, \tag{3-52}$$

where

$$\epsilon_r = 1 + \chi_e = \frac{\epsilon}{\epsilon_0} \tag{3-53}$$

is a dimensionless constant known as the *relative permittivity*, or the *dielectric constant* of the material. In a vacuum, $\chi_e = 0$, $\epsilon_r = 1$, and in all Class A dielectrics $\epsilon_r > 1$.

The relative permittivity of dielectrics lies typically between 2 and 5. However, some substances which are not Class A dielectrics can have relative permittivities as high as 10^5 or even higher.

In Class A dielectrics $\boldsymbol{D}$ is proportional to $\boldsymbol{E}$, and, from Eqs. 3-46 and 3-52, if ϵ_r is *not* a function of the coordinates,

$$\boldsymbol{\nabla}\cdot\boldsymbol{E} = \frac{\rho_f}{\epsilon_r\epsilon_0} = \frac{\rho_f}{\epsilon}. \tag{3-54}$$

Then

$$\nabla^2 V = -\frac{\rho_f}{\epsilon_r\epsilon_0} = -\frac{\rho_f}{\epsilon}. \tag{3-55}$$

This is *Poisson's equation for homogeneous Class A dielectrics*. If the dielectric is not Class A, we must write

$$\nabla^2 V = -\frac{(\rho_f + \rho_b)}{\epsilon_0}. \tag{3-56}$$

3.6.2 The Free Charge Density ρ_f and the Bound Charge Density ρ_b

The bound charge density ρ_b is related to the free charge density ρ_f in a Class A dielectric as follows. Since

$$\boldsymbol{P} = \boldsymbol{D} - \epsilon_0\boldsymbol{E} = \frac{\epsilon_r - 1}{\epsilon_r}\boldsymbol{D}, \tag{3-57}$$

and since the dielectric is homogeneous (ϵ_r is independent of the coordinates) by hypothesis,

$$\boldsymbol{\nabla}\cdot\boldsymbol{P} = \left(\frac{\epsilon_r - 1}{\epsilon_r}\right)\boldsymbol{\nabla}\cdot\boldsymbol{D} = \left(1 - \frac{1}{\epsilon_r}\right)\rho_f, \tag{3-58}$$

or

$$-\rho_b = \left(\frac{\epsilon_r - 1}{\epsilon_r}\right)\rho_f. \tag{3-59}$$

The total charge density

$$\rho_t = \rho_f + \rho_b = \rho_f - \boldsymbol{\nabla}\cdot\boldsymbol{P} = \frac{\rho_f}{\epsilon_r} \tag{3-60}$$

is *smaller* than the free charge density ρ_f, since ρ_b cancels part of ρ_f.

At any point in a homogeneous, isotropic, linear dielectric, ρ_b is zero if ρ_f is zero, from Eq. 3-60. Usually, $\rho_f = 0$, $\boldsymbol{\nabla}\cdot\boldsymbol{P} = 0$, and the polarization charges are located only on the surfaces of the dielectric.

3.7 CALCULATION OF ELECTRIC FIELDS INVOLVING DIELECTRICS

We are now in a position to calculate electric fields involving Class A dielectrics.

If the geometry of the field is simple and if the free electric charge density ρ_f is zero, as it often is, we can usually integrate $\boldsymbol{\nabla}\cdot\boldsymbol{D} = 0$ to find the electric displacement $\boldsymbol{D}$. The constants of integration can be determined from the boundary conditions, which can be, for example, the charge densities on the conductors. The integral equation

$$\int_S \boldsymbol{D}\cdot\boldsymbol{da} = \int_\tau \rho_f \, d\tau, \tag{3-61}$$

is also often useful for determining $\boldsymbol{D}$.

Once $\boldsymbol{D}$ is determined, $\boldsymbol{E}$ can be found at once from $\boldsymbol{E} = \boldsymbol{D}/\epsilon_r\epsilon_0$, provided that the relative permittivity ϵ_r is known. It is then a simple step from $\boldsymbol{E}$ to potential difference or to capacitance, or to other physical quantities depending on $\boldsymbol{E}$.

Combining Eqs. 3-29 and 3-60, we also have

$$V = \frac{1}{4\pi}\int_{\tau'} \frac{\rho_f}{\epsilon r}\, d\tau' + \frac{1}{4\pi\epsilon_0}\int_{S'} \frac{\sigma_f + \sigma_b}{r}\, da' \tag{3-62}$$

for linear and isotropic dielectrics. Note that ρ_f/ϵ is equal to ρ_t/ϵ_0, but that no such relation exists for the surface charge densities σ_f and $\sigma_f + \sigma_b$. Indeed, σ_f and σ_b are unrelated. For example, a dielectric carrying zero σ_f acquires a σ_b when it is placed in an electric field. (However, see the second example that follows.)

Example

THE DIELECTRIC-INSULATED PARALLEL-PLATE CAPACITOR

Figure 3-7 shows such a capacitor with the spacing s exaggerated for clarity. We assume that s is small compared to the linear extent of the plates in order that we may neglect fringing effects at the edges. The surface charge densities $+\sigma_f$ on the lower plate and $-\sigma_f$ on the upper plate produce a uniform electric field directed upward. The polarization in the dielectric produces a bound surface charge density $-\sigma_b$ on the lower surface of the dielectric and $+\sigma_b$ on the upper surface as in Figure 3-6. These bound charges produce a uniform electric field directed downward which cancels part of the field of the charges situated on the plates. Since the net field between the plates must remain equal to V/s, where V is the battery voltage, the charge densities $+\sigma_f$ and $-\sigma_f$ on the plates must be larger than when the dielectric is absent. The presence of the dielectric thus has the effect of increasing the charges on the plates at a given value of V, and hence of increasing the capacitance.

To calculate the capacitance, we apply Gauss's law for the displacement $\boldsymbol{D}$ to a cylinder as in Figure 3-7. Then the only flux of $\boldsymbol{D}$ through the Gaussian surface is through the top, and D is numerically equal to σ_f. Also $E = \sigma_f/\epsilon_r\epsilon_0$, the potential difference between the plates is $\sigma_f s/\epsilon_r\epsilon_0$, and the capacitance

$$C = \frac{\sigma_f S}{\sigma_f s/\epsilon_r\epsilon_0} = \epsilon_r\epsilon_0 \frac{S}{s}, \tag{3-63}$$

where S is the area of one plate. The capacitance is therefore increased by a factor ϵ_r through the presence of the dielectric.

The measurement of the capacitance of a suitable capacitor with and without a dielectric provides a convenient method for measuring a relative permittivity ϵ_r.

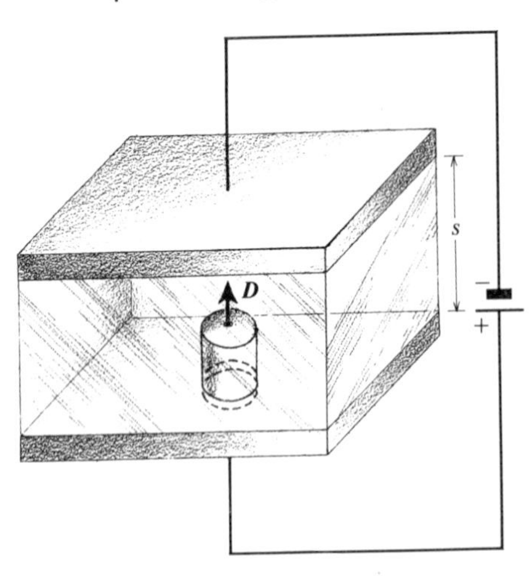

Figure 3-7. Dielectric insulated plane-parallel capacitor.

Example

THE FREE CHARGE DENSITY σ_f, THE BOUND CHARGE DENSITY σ_b, AND THE ELECTRIC DISPLACEMENT $\boldsymbol{D}$ AT A DIELECTRIC-CONDUCTOR BOUNDARY

At the interface between a dielectric and a conductor there is a bound surface charge density on the dielectric and a free surface charge density on the conductor. For example, at the lower plate of the capacitor considered above,

$$\sigma_b = \boldsymbol{P} \cdot \boldsymbol{n} = -P, \tag{3-64}$$

since $\boldsymbol{n}$ is the outward normal to the dielectric surface and points downward, while the electric polarization $\boldsymbol{P}$ points upward like $\boldsymbol{E}$. Thus

$$-\sigma_b = P = D - \epsilon_0 E, \tag{3-65}$$

$$= \frac{\epsilon_r - 1}{\epsilon_r} D = \frac{\epsilon_r - 1}{\epsilon_r} \sigma_f, \tag{3-66}$$

and

$$\sigma_t = \frac{\sigma_f}{\epsilon_r}, \tag{3-67}$$

as in Eq. 3-60.

Since $\epsilon_r > 1$, the bound surface charge density σ_b and the free surface charge density σ_f have opposite signs. The bound charge density is also smaller in magnitude than the free charge density by the factor $(\epsilon_r - 1)/\epsilon_r$. This relationship is always true whenever a Class A dielectric is in contact with a conductor, or carries a free surface charge density.

The free charge density σ_f on a conductor in contact with *any* dielectric is always equal to the electric displacement D just inside the dielectric. This follows from Gauss's law, as in the case of the parallel-plate capacitor. Furthermore, if a conducting surface is introduced into a dielectric so as to coincide with an equipotential surface, then the free surface charge density σ_f on the conductor at a particular point is equal in magnitude to the electric displacement D in the dielectric at that point.

Example

DIELECTRIC SPHERE WITH A POINT CHARGE AT ITS CENTER

Let us consider a Class A dielectric sphere of radius R with a point charge Q embedded at the center, as in Figure 3-8. (It is possible to trap negative charges within dielectrics by bombarding them with high-energy electrons. This example is of little practical interest, but it proves to be an excellent illustration of the behavior of dielectrics.)

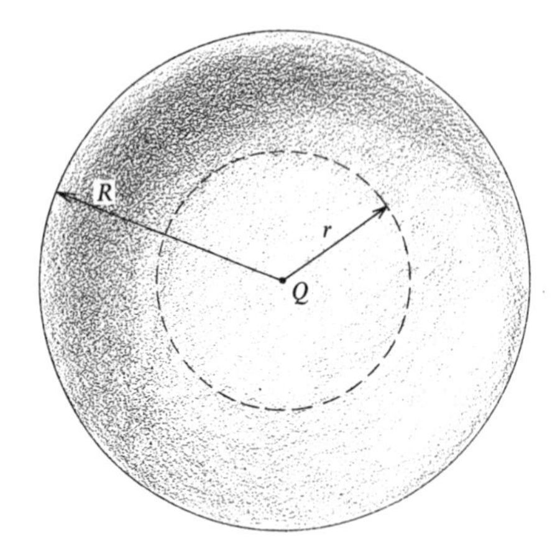

Figure 3-8. Dielectric sphere with a point charge Q at the center.

Again, we can find $\boldsymbol{D}$ from Gauss's law. We draw an imaginary sphere of radius $r < R$ and use it as a Gaussian surface. Then

$$\boldsymbol{D} = \frac{Q}{4\pi r^2}\,\boldsymbol{r}_1, \tag{3-68}$$

$$\boldsymbol{E} = \frac{Q}{4\pi\epsilon_0\epsilon_r r^2}\,\boldsymbol{r}_1. \tag{3-69}$$

$$\boldsymbol{P} = \frac{\epsilon_r - 1}{\epsilon_r}\,\boldsymbol{D} = \frac{\epsilon_r - 1}{\epsilon_r}\,\frac{Q}{4\pi r^2}\,\boldsymbol{r}_1. \tag{3-70}$$

At the outer surface of the sphere,

$$\sigma_b = \boldsymbol{P}\cdot\boldsymbol{n} = P = \frac{\epsilon_r - 1}{\epsilon_r}\,\frac{Q}{4\pi R^2}. \tag{3-71}$$

The total amount of bound charge on the outer surface is thus

$$Q_{bo} = \frac{\epsilon_r - 1}{\epsilon_r}\,Q. \tag{3-72}$$

Since there is no volume density of free charge, there should be no volume density of bound charge, as we saw in the last section. This is correct:

$$\rho_b = -\boldsymbol{\nabla}\cdot\boldsymbol{P} = -\frac{1}{r^2}\frac{\partial}{\partial r}(r^2 P) = 0. \tag{3-73}$$

There is also a bound charge on the surface of the cavity containing the free charge Q at the center. This can be calculated as follows. Let the radius of the cavity be δ. Then, if $-\sigma_{bc}$ is the density of bound charge on the cavity surface, the total bound charge on the cavity is

$$Q_{bc} = -\sigma_{bc}4\pi\delta^2 = -P_\delta 4\pi\delta^2 = -\left(\frac{\epsilon_r - 1}{\epsilon_r}\right) D_\delta 4\pi\delta^2, \tag{3-74}$$

$$= -\left(\frac{\epsilon_r - 1}{\epsilon_r}\right) Q. \tag{3-75}$$

The dielectric as a whole therefore remains neutral, as must be ex-expected. The net charge at the center is then

$$Q_{\text{net}} = Q - \left(\frac{\epsilon_r - 1}{\epsilon_r}\right) Q = \frac{Q}{\epsilon_r} \tag{3-76}$$

and is smaller than the free charge by the factor ϵ_r. This accounts for the reduction of the electric field intensity within the dielectric by the factor ϵ_r, as in Eq. 3-69.

It is instructive to compare the electric field intensity outside the sphere and that inside. Outside, from Gauss's law,

$$\boldsymbol{D} = \frac{Q}{4\pi r^2}\,\boldsymbol{r}_1, \tag{3-77}$$

$$\boldsymbol{E} = \frac{Q}{4\pi\epsilon_0 r^2}\,\boldsymbol{r}_1, \tag{3-78}$$

which is to be compared to Eqs. 3-68 and 3-69 for the field inside the sphere. At the surface of the sphere the electric field intensity is discontinuous, the magnitude just outside the surface being ϵ_r times as large as the magnitude just inside the surface. The difference is due to the bound charges, which produce an opposing field within the dielectric.

Example

THE BAR ELECTRET

An electret is the electric equivalent of a bar magnet. In most dielectrics the polarization disappears immediately when the electric field is removed, but some dielectrics retain their polarization for a very long time. Indeed, several polymers have extrapolated lifetimes of several thousand years at room temperature.

One way of charging a dielectric is to place it in a strong electric field at some high temperature. The bound charge density on the surfaces then builds up slowly as the molecules orient themselves. Free charges are also deposited on the surfaces when sparking occurs

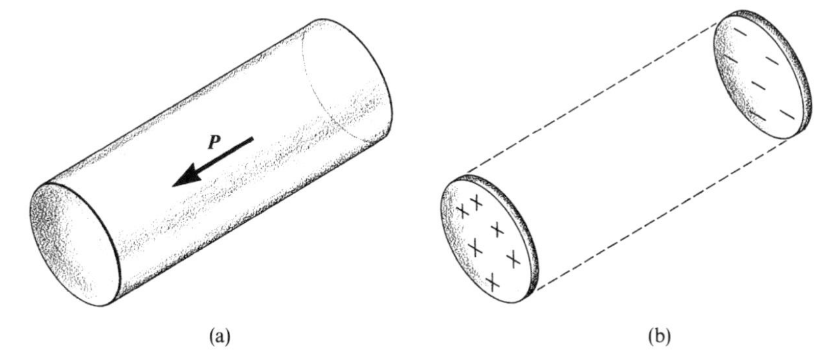

Figure 3-9. (a) Bar electret polarized uniformly parallel to its axis. (b) The $\boldsymbol{E}$ field of the bar electret is the same as that of a pair of circular plates carrying uniform surface charge densities of opposite polarities.

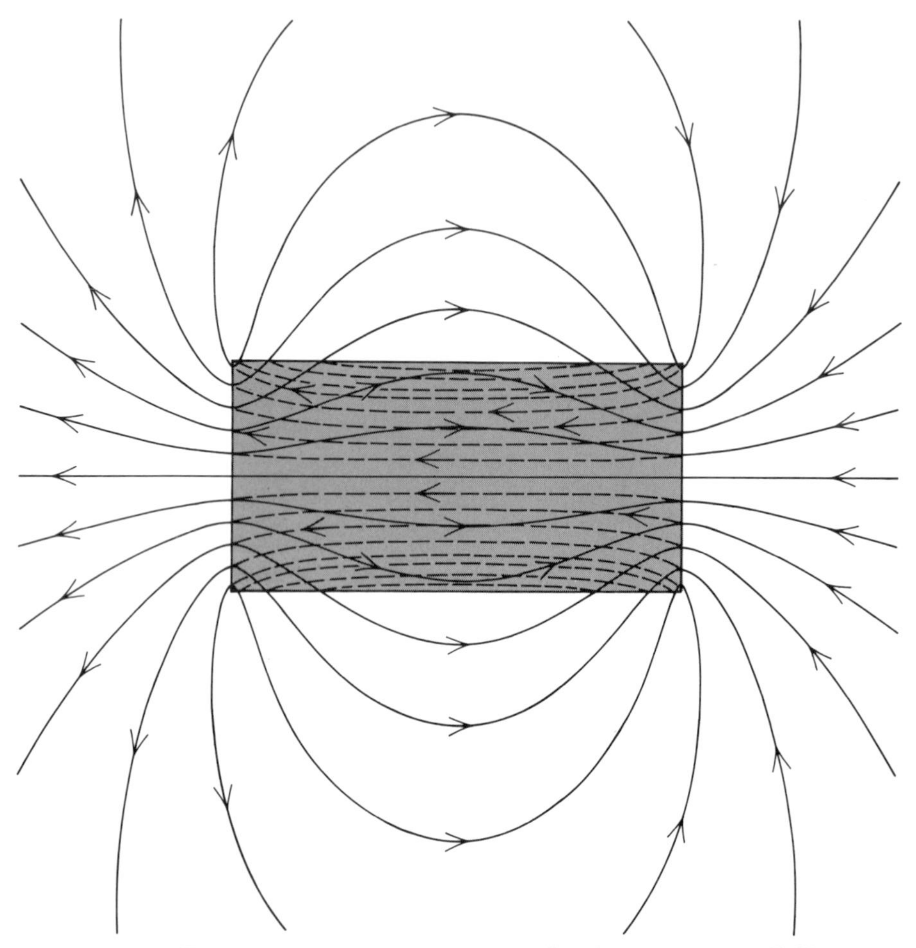

Figure 3-10. Lines of $\boldsymbol{E}$ (solid) for the bar electret of Figure 3-9. Lines of $\boldsymbol{D}$ are shown broken inside the electret; outside, they follow the lines of $\boldsymbol{E}$.

between the electrodes and the dielectic. The free and the bound charges have opposite signs. The sample is then cooled to room temperature without removing the electric field.

In the presence of an external electric field the polarization $\boldsymbol{P}$ inside an electret is related to the total $\boldsymbol{E}$ as follows:

$$\boldsymbol{P} = \chi_e \epsilon_0 \boldsymbol{E} + \boldsymbol{P}_0, \tag{3-79}$$

where $\boldsymbol{P}_0$ is the permanent polarization. Then

$$\boldsymbol{D} = \epsilon_0 \boldsymbol{E} + \boldsymbol{P} = \epsilon_0(1 + \chi_e)\boldsymbol{E} + \boldsymbol{P}_0, \tag{3-80}$$

$$= \epsilon \boldsymbol{E} + \boldsymbol{P}_0. \tag{3-81}$$

Since there are no free charges inside the material, $\rho_b = 0$ (from

Eq. 3-60) and the electric field of an electret can be calculated solely from the surface charges.

As an exercise, let us see how one can calculate the field of the bar electret illustrated in Figure 3-9a. We assume that the polarization $\boldsymbol{P}$ is uniform. We also assume that the surface density of free charges σ_f is zero.

The electric field intensity $\boldsymbol{E}$ *both inside and outside* is the same as that of a pair of parallel circular plates carrying charge densities $\pm P$ as in Figure 3-9b. This field is shown as solid lines in Figure 3-10.

The lines of $\boldsymbol{D}$ are identical to the lines of $\boldsymbol{E}$ outside the electret since $\boldsymbol{D} = \epsilon_0 \boldsymbol{E}$ there. Inside the electret, $\boldsymbol{D}$ is $\epsilon_0 \boldsymbol{E} + \boldsymbol{P}$ with $\boldsymbol{E}$ and $\boldsymbol{P}$ pointing in approximately opposite directions. If the electret were a thin sheet polarized in the direction normal to its surface, the $\boldsymbol{E}$ inside would be $\boldsymbol{P}/\epsilon_0$ and $\boldsymbol{D}$ would be zero. However, inside a bar electret, $\epsilon_0 \boldsymbol{E}$ is smaller than $\boldsymbol{P}$, the result that $\boldsymbol{D}$ points in the direction of $\boldsymbol{P}$, and $\boldsymbol{E}$ and $\boldsymbol{D}$ point in approximately opposite directions, as in Figure 3-10.

3.8 THE CLAUSIUS-MOSSOTTI EQUATION *

Let us return now to the basic polarization mechanism and compute the relative permittivity ϵ_r in terms of the molecular properties. We shall assume the approximate expression of Eq. 3-32 for the local field:

$$\boldsymbol{E}_{\text{loc}} = \boldsymbol{E} + \frac{\boldsymbol{P}}{3\epsilon_0}. \tag{3-82}$$

As we saw, this equation is not necessarily correct, in that the factor multiplying $\boldsymbol{P}/\epsilon_0$ may differ from $\frac{1}{3}$, but $\frac{1}{3}$ is of the right order of magnitude.

If we set $b = \frac{1}{3}$, then from Eq. 3-36,

$$\boldsymbol{P} = N\alpha\left(\boldsymbol{E} + \frac{\boldsymbol{P}}{3\epsilon_0}\right), \tag{3-83}$$

or

$$\epsilon_0(\epsilon_r - 1)\boldsymbol{E} = N\alpha\left(\boldsymbol{E} + \frac{\epsilon_r - 1}{3}\boldsymbol{E}\right), \tag{3-84}$$

and

$$\frac{\epsilon_r - 1}{\epsilon_r + 2} = \frac{N\alpha}{3\epsilon_0}, \tag{3-85}$$

where N is the number of molecules per unit volume and α is the molecular polarizability.

---

* Sections 3-8 to 3-10 can be omitted without losing continuity.

If the dielectric is a compound consisting of a number of different types of molecules or atoms,

$$\frac{\epsilon_r - 1}{\epsilon_r + 2} = \frac{1}{3\epsilon_0} \sum_i N_i \alpha_i, \tag{3-86}$$

where N_i and α_i are the appropriate quantities for the ith type of molecule or atom.

Equation 3-86 is known as the *Clausius-Mossotti* equation. It relates the relative permittivity ϵ_r to the mass density of a material, since

$$N = \frac{\rho}{M} N_A, \tag{3-87}$$

where ρ is the mass density, N_A is Avogadro's number, and M is the molecular weight. In *MKSA* units, ρ is expressed in kilograms/meter^3, N_A is the number of molecules in one kilogram molecular weight, or 6.02×10^{26}, and M is the kilogram molecular weight (for example, the kilogram molecular weight of oxygen is 32 kilograms). Then

$$\frac{M}{\rho} \frac{\epsilon_r - 1}{\epsilon_r + 2} = \frac{N_A}{3\epsilon_0} \alpha = \alpha_M. \tag{3-88}$$

The quantity α_M is called the *molar polarization.*

Now in this equation Avogadro's number N_A and the permittivity of free space ϵ_0 are of course the same for all substances, while both the molecular weight M and the molecular polarizability α depend only on the type of molecule of which the dielectric is made. Then Eq. 3-88 shows that the quantity $(\epsilon_r - 1)/\rho(\epsilon_r + 2)$ is independent of density if our model for the polarization is valid. This prediction of the Clausius-Mossotti equation is borne out in a wide variety of gases and is approximated in nonpolar liquids.

Examination of Eq. 3-88 shows that ϵ_r must increase rapidly and tend to infinity as the number of molecules per unit volume approaches the critical value $3\epsilon_0/\alpha$. Physically, the expression $\epsilon_r \to \infty$ corresponds to infinitely polarizable molecules. In gases and in liquids the polarizabilities are relatively small, with the result that this condition is never approached.

In crystalline solids the interactions between the dipoles produce polarizations that are more complicated than those of our simple model, and the Clausius-Mossotti equation is not valid.

3.9 POLAR DIELECTRICS

Let us now investigate the behavior of dielectrics in which the molecules possess permanent dipole moments. Molecules consisting of two or more

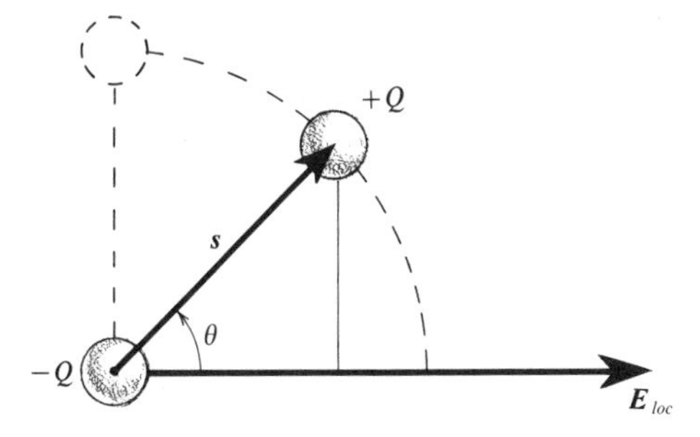

Figure 3-11. Dipole with a permanent dipole moment $\boldsymbol{p} = Q\boldsymbol{s}$ oriented at an angle θ with respect to the local field.

dissimilar atoms exhibit such permanent moments. For example, a molecule composed of two atoms, one of which carries a positive charge and the other a negative charge, is polar. Other mechanisms exist by which other molecular types can also have permanent moments. Diatomic molecules consisting of two similar atoms never have permanent moments because there is no asymmetrical way for them to arrange themselves.

In addition to their permanent dipole moments, polar molecules also exhibit induced moments of the type we have been discussing, but for the time being we shall neglect these induced moments.

Permanent dipole moments are usually of the order of 10^{-30} to 10^{-29} coulomb-meter.

Equation 3-1 obviously does not apply to polar dielectrics, since the individual $\boldsymbol{p}$'s are not necessarily aligned with the field. Our problem is to calculate their net orientation.

A permanent dipole oriented so that its dipole moment $\boldsymbol{p}$ makes an angle θ with the local field, as in Figure 3-11, is subjected to a torque that tends to align it with the field. An unaligned dipole therefore has a greater potential energy than an aligned one.

It is a simple matter to calculate the potential energy of such a dipole. We select arbitrarily the position $\theta = 90°$ to correspond to zero potential energy. The axis of the dipole is then perpendicular to $\boldsymbol{E}_{\text{loc}}$. This is permissible, since only variations in potential energy have physical significance. Then the energy lost by the dipole when its axis rotates from 90° to θ is $E_{\text{loc}}\, Qs \cos\theta$ and the potential energy

$$W = -\boldsymbol{p}\cdot\boldsymbol{E}_{\text{loc}}. \tag{3-89}$$

The potential energy is minimum when $\boldsymbol{p}$ is parallel to $\boldsymbol{E}_{\text{loc}}$.

3.9.1 The Langevin Equation

In gaseous and liquid polar dielectrics, thermal agitation brings about collisions between molecules that tend to destroy any alignment with the local field. However, the local field exerts a restoring force between collisions and, on the average, effects a net alignment, and consequently a net dipole moment $\boldsymbol{P}$ per unit volume.

To compute this net polarization, we consider a unit volume containing N dipoles. In the absence of an external electric field, the dipoles are oriented at random, and, at any instant, dN are oriented at angles lying between θ and $\theta + d\theta$ with respect to a given direction. The fraction dN/N is merely the ratio of the solid angle corresponding to the angular interval $d\theta$ to the total solid angle 4π:

$$\frac{dN}{N} = \frac{2\pi \sin\theta \, d\theta}{4\pi} = \frac{\sin\theta \, d\theta}{2}. \tag{3-90}$$

If the dipoles are subjected to a local field, the solid-angle elements are no longer equally probable, since a dipole has a potential energy W, which depends on its orientation relative to the field. It is shown in statistical mechanics that if a large number of molecules are in statistical equilibrium, the number possessing a particular energy W is proportional to $\exp(-W/kT)$, where k is the Boltzmann constant 1.381×10^{-23} joule/kelvin and T is the absolute temperature in kelvins.

In the present case the dipoles whose axes lie in the range between θ and $\theta + d\theta$, measured from the direction of the local field, all possess an energy $W = -\boldsymbol{p}\cdot\boldsymbol{E}_{\text{loc}} = -pE_{\text{loc}} \cos\theta$. Then the number of dipoles per unit volume lying in the interval θ to $\theta + d\theta$ is

$$dN = C \exp(\boldsymbol{p}\cdot\boldsymbol{E}_{\text{loc}}/kT) \sin\theta \, d\theta = C \exp(u \cos\theta) \sin\theta \, d\theta, \tag{3-91}$$

where we have set

$$u = \frac{pE_{\text{loc}}}{kT}, \tag{3-92}$$

and where the constant C must be chosen so that the total number of molecules in a unit volume is N:

$$N = C \int_0^{\pi} \exp(u \cos\theta) \sin\theta \, d\theta. \tag{3-93}$$

Now the molecules whose moments lie in the angular interval θ to $\theta + d\theta$ possess a total dipole moment in the direction of the local field of

$$dP = p \, dN \cos\theta = pN \frac{\exp(u \cos\theta) \sin\theta \cos\theta \, d\theta}{\int_0^{\pi} \exp(u \cos\theta) \sin\theta \, d\theta}. \tag{3-94}$$

To obtain the dipole moment per unit volume P we now integrate the numerator from $\theta = 0$ to $\theta = \pi$:

$$P = Np \frac{\int_0^\pi \exp(u\cos\theta)\sin\theta\cos\theta\, d\theta}{\int_0^\pi \exp(u\cos\theta)\sin\theta\, d\theta}. \tag{3-95}$$

So as to simplify the integration we set

$$t = \frac{pE_{\text{loc}}}{kT}\cos\theta = u\cos\theta. \tag{3-96}$$

Then

$$P = \frac{\frac{Np}{u}\int_{-u}^{+u}(\exp t)\, t\, dt}{\int_{-u}^{+u}(\exp t)\, dt} = \frac{Np}{u}\frac{[t(\exp t) - (\exp t)]_{-u}^{+u}}{[\exp t]_{-u}^{+u}}, \tag{3-97}$$

$$= Np\left(\coth u - \frac{1}{u}\right) = Np\left(\coth\frac{pE_{\text{loc}}}{kT} - \frac{kT}{pE_{\text{loc}}}\right). \tag{3-98}$$

This is known as the *Langevin equation.*

Figure 3-12 shows P/Np as a function of u. For large values of $u = pE_{\text{loc}}/kT$, that is, for large fields and low temperatures, P approaches Np. The dipoles are then all aligned with the field, and the polarization is maximum.

The region of practical interest is where pE_{loc}/kT is small compared to unity. At room temperature $kT \sim 4 \times 10^{-21}$ joule, whereas a typical dipole moment is of the order of 10^{-30} coulomb-meter. Thus, even with a local field of 10^7 volts/meter, pE_{loc}/kT is only of the order of 2×10^{-3}. We can therefore expand the exponentials in Eq. 3-98 and retain only the terms up to u^3. Then

$$P \approx Np\left[\frac{2 + u^2}{2u[1 + (u^2/6)]} - \frac{1}{u}\right], \tag{3-99}$$

$$\approx Np\left[\frac{1}{2u}(2 + u^2)\left(1 - \frac{u^2}{6}\right) - \frac{1}{u}\right], \tag{3-100}$$

$$\approx \frac{Npu}{3} = \frac{Np^2}{3kT}E_{\text{loc}}. \tag{3-101}$$

Thus, when $pE_{\text{loc}} \ll kT$, the polarization P in a polar dielectric is proportional to the local field, and the dielectric is linear. We found earlier in Eq. 3-36 that the same applies to nonpolar dielectrics. When $pE_{\text{loc}} \ll kT$ the susceptibility of polar dielectric is inversely proportional to the temperature.

From a practical point of view, the only feature that distinguishes a polar from a nonpolar dielectric is the temperature dependence of the susceptibility and, therefore, of the relative permittivity.

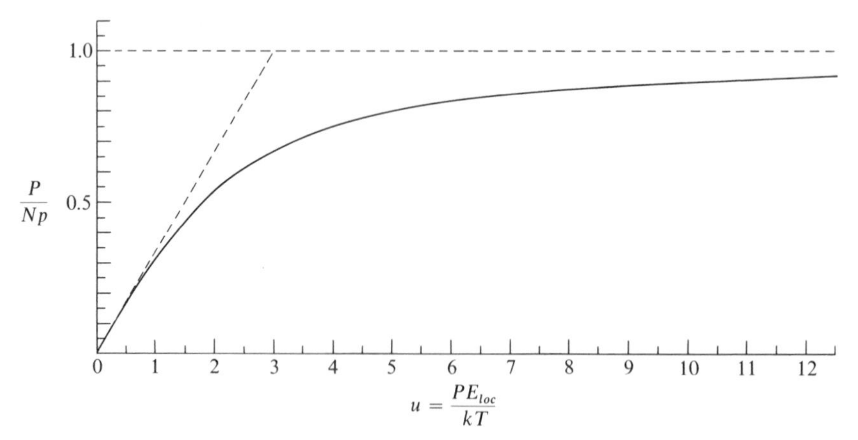

Figure 3-12. The Langevin function.

3.9.2 The Debye Equation

Let us now consider real polar dielectrics in which there are both induced and permanent dipoles. Combining Eqs. 3-36 and 3-101,

$$\boldsymbol{P} = N\left(\alpha + \frac{p^2}{3kT}\right)\boldsymbol{E}_{\text{loc}}, \tag{3-102}$$

where

$$\boldsymbol{E}_{\text{loc}} = \boldsymbol{E} + \frac{\boldsymbol{P}}{3\epsilon_0}, \tag{3-103}$$

as in Eq. 3-32 or, for a linear dielectric (Eqs. 3-38 and 3-53),

$$\boldsymbol{E}_{\text{loc}} = \frac{\epsilon_r + 2}{3}\boldsymbol{E}. \tag{3-104}$$

Then

$$\boldsymbol{P} = \epsilon_0(\epsilon_r - 1)\boldsymbol{E} = N\left(\alpha + \frac{p^2}{3kT}\right)\left(\frac{\epsilon_r + 2}{3}\right)\boldsymbol{E}, \tag{3-105}$$

$$\frac{\epsilon_r - 1}{\epsilon_r + 2} = \frac{N}{3\epsilon_0}\left(\alpha + \frac{p^2}{3kT}\right). \tag{3-106}$$

Multiplying by the molecular weight M and dividing by the mass density ρ, as we did in discussing the Clausius-Mossotti equation, we find a new expression for the molar polarization that is valid for polar dielectrics:

$$\alpha_M = \frac{M}{\rho}\frac{\epsilon_r - 1}{\epsilon_r + 2} = \frac{N_A}{3\epsilon_0}\left(\alpha + \frac{p^2}{3kT}\right) \tag{3-107}$$

This is known as the *Debye equation*. It is similar to Eq. 3-88, except for the term $p^2/3kT$, which comes from the alignment of the polar molecules.

In principle, the Debye equation can be used to determine both the

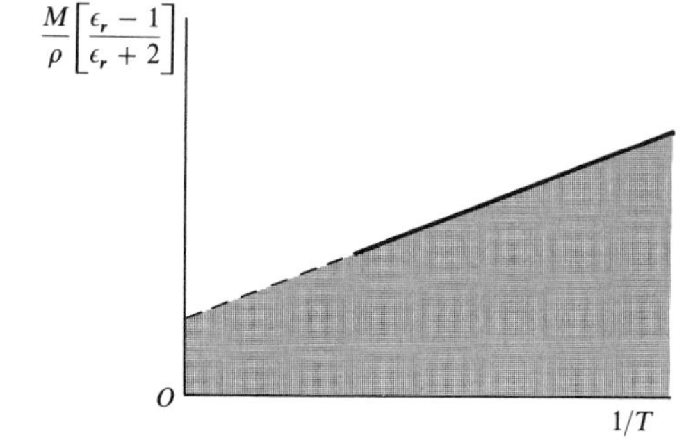

Figure 3-13. The molecular polarizability α and the dipole moment per molecule p can be determined from the intercept on the vertical axis and from the slope of this curve.

molecular polarizability α and the permanent dipole moment p of a molecule. Plotting $(M/\rho)(\epsilon_r - 1)/(\epsilon_r + 2)$ versus $1/T$ gives a straight line, as in Figure 3-13, whose intercept on the vertical axis is $N_A\alpha/3\epsilon_0$ and whose slope is $N_A p^2/9\epsilon_0 k$, where N_A is Avogadro's number and k is the Boltzmann constant. In practice, this equation provides a reliable way of measuring α and p only for gases, where the dielectric coefficient differs only slightly from unity, and for dilute solutions of polar molecules in a nonpolar solvent.

We have made one important assumption in deriving the Debye equation. We have assumed that the field $\boldsymbol{E}_{\text{loc}} = \boldsymbol{E} + (\boldsymbol{P}/3\epsilon_0)$ is responsible both for the induced and the oriented polarization. We saw that the factor $\frac{1}{3}$ is only approximate. Furthermore, there is another more subtle effect that further invalidates the Debye equation. As a permanent dipole turns, part of the local field turns with it and plays no part in orienting the dipole. This comes about because the field of the dipole polarizes the surrounding molecules, and these in turn produce a field at the dipole that is in the same direction as its moment. Furthermore, molecular associations in liquids and in solids complicate the problem. The result is that dielectric behavior in liquids, and especially in crystalline solids, is more complicated than the simple picture we have developed here. The molar polarization of water, for example, shows little temperature dependence, even though water has a large permanent dipole moment of 6.2×10^{-30} coulomb-meter. The basic features of the polarization process are as we have discussed them, but a precise determination of molecular polarizability and of permanent dipole moment must rest on a more sophisticated treatment.

3.10 FREQUENCY DEPENDENCE, ANISOTROPY, AND NONHOMOGENEITY

We have discussed so far two basic polarization processes. We first discussed induced polarization, in which the center of negative charge in a molecule is displaced relative to the center of positive charge when an external field is applied. This type of polarization is also called *electronic*. We then considered *orientational* polarization, in which molecules with a permanent dipole moment tend to be aligned by an external field, the magnitude of the susceptibility being inversely proportional to the temperature. There is also a third basic polarization process, which we may call *ionic*. This process occurs in ionic crystals, in which ions of one sign may move with respect to ions of the other sign when an external field is applied.

Examples Water has a relative permittivity of 81 in an electrostatic field, and of about 1.8 at optical frequencies. The large static value is attributable to the orientation of the permanent dipole moments, but the rotational inertia of the molecules is much too large for any significant response at optical frequencies.

Similarly, the relative permittivity of sodium chloride is 5.6 in an electrostatic field and 2.3 at optical frequencies. The larger static value is attributed to ionic motion, which again is impossible at high frequencies.

For a given magnitude of $\boldsymbol{E}$, both the magnitude and the phase of $\boldsymbol{p}$ are functions of the frequency, first because of the various polarization processes that come into play as the frequency changes, and also because of the existence of resonances. Thus, since ϵ_r is a function of the frequency, ϵ_r is strictly definable only for a pure sine wave.

In many substances the relative permittivity decreases by a large factor as the temperature is lowered through the freezing point.

Example In nitrobenzene, ϵ_r falls from about 35 to about 3 in passing through the freezing point at 279 kelvins. In the solid state, the permanent dipoles of the nitrobenzene molecules are fixed rigidly in the crystal lattice and cannot rotate under the influence of an external field.

Anisotropy is a common departure from the ideal Class A dielectric behavior. Crystalline solids commonly have different dielectric properties in different crystal directions because the charges which constitute the atoms of the crystal are able to move more easily in some directions than in others. The result is that the susceptibility depends on direction. Thus, in general, $\boldsymbol{P}$ is not in the same direction as $\boldsymbol{E}$. For example,

$$P_x = \epsilon_0(\chi_{exx}E_x + \chi_{exy}E_y + \chi_{exz}E_z) \tag{3-108}$$

or, more generally,

$$P_i = \epsilon_0 \sum_j \chi_{eij} E_j, \tag{3-109}$$

where the subscripts i and j represent the three coordinate directions x, y, z. All three components of $\boldsymbol{P}$ depend on all three components of $\boldsymbol{E}$, with different susceptibilities for each. The susceptibility χ_e thus has nine components and is a tensor. Actually, there are only six independent components, and, if the coordinate axes are properly chosen, these six components reduce to three. The relationship between $\boldsymbol{P}$ and $\boldsymbol{E}$ is still linear but is more complicated than for isotropic dielectrics.

For anisotropic dielectrics, Eq. 3-44 still applies:

$$\boldsymbol{\nabla} \cdot (\epsilon_0 \boldsymbol{E} + \boldsymbol{P}) = \rho_f, \tag{3-110}$$

where ρ_f is the free charge density. It is again useful to define the displacement vector as in Eq. 3-45:

$$\boldsymbol{D} = \epsilon_0 \boldsymbol{E} + \boldsymbol{P}, \tag{3-111}$$

except that now $\boldsymbol{D}$, $\boldsymbol{E}$, and $\boldsymbol{P}$ are generally not all in the same direction.

The relation between the displacement $\boldsymbol{D}$ and the electric field intensity $\boldsymbol{E}$ is then

$$D_i = \epsilon_0 \sum_j \epsilon_{rij} E_j, \tag{3-112}$$

where ϵ_{rij} is a tensor which, again, has six independent components. The relationship between $\boldsymbol{D}$ and $\boldsymbol{E}$ is still linear, and the general features of the discussion of Class A dielectrics are valid.

If the relative permittivity is a function of position in the dielectric, then a volume density of bound charge may exist when there is no corresponding volume density of free charge (see Eq. 3-60). Demonstration of this fact will be left as an exercise.

3.11 POTENTIAL ENERGY OF A CHARGE DISTRIBUTION IN THE PRESENCE OF DIELECTRICS

If dielectrics are present in the vicinity of a charge distribution, we may still calculate the electric potential energy of the system in the manner of Section 2.14. We allow each free charge to recede to infinity as before, except that now the potential at a given point in the field depends not only on the charges and on the geometry, but also on the characteristics of the dielectric. Then

$$W = \frac{1}{2} \int_\tau \rho_f V \, d\tau, \tag{3-113}$$

where the volume τ is any volume which includes all the free charges in the system. Now $\rho_f = \boldsymbol{\nabla}\cdot\boldsymbol{D}$, and, if we use the vector identity of Problem 1-14,

$$W = \frac{1}{2}\int_\tau V(\boldsymbol{\nabla}\cdot\boldsymbol{D})\,d\tau, \tag{3-114}$$

$$= \frac{1}{2}\int_\tau (\boldsymbol{\nabla}\cdot V\boldsymbol{D})\,d\tau - \frac{1}{2}\int_\tau (\boldsymbol{D}\cdot\boldsymbol{\nabla}V)\,d\tau. \tag{3-115}$$

Transforming the first term to a surface integral,

$$W = \frac{1}{2}\int_S V\boldsymbol{D}\cdot\boldsymbol{da} + \frac{1}{2}\int_\tau (\boldsymbol{D}\cdot\boldsymbol{E})\,d\tau. \tag{3-116}$$

We now proceed as in Section 2.14 and let τ and S tend to infinity. In the first integral, V falls off at least as fast as $1/r$, D falls off at least as fast as $1/r^2$, and da increases as r^2. The first integral falls off at least as fast as $1/r$ and goes to zero as the surface S recedes to infinity. Thus

$$W = \frac{1}{2}\int_\tau (\boldsymbol{D}\cdot\boldsymbol{E})\,d\tau, \tag{3-117}$$

where τ is any volume that includes all the points where $\boldsymbol{D}$ and $\boldsymbol{E}$ differ from zero. This expression is independent of the type of dielectric present in the system and is completely general.

As for free space, we may define an energy density

$$\frac{dW}{d\tau} = \frac{1}{2}\boldsymbol{D}\cdot\boldsymbol{E}, \tag{3-118}$$

which is useful in calculating the work required to assemble a charge distribution.

For a Class A dielectric, $\boldsymbol{D} = \epsilon_r\epsilon_0\boldsymbol{E}$,

$$W = \frac{\epsilon_0}{2}\int_\tau \epsilon_r E^2\,d\tau, \tag{3-119}$$

and the energy density is $\epsilon E^2/2$.

Example

ENERGY STORED IN A PARALLEL-PLATE CAPACITOR.

Imagine a dielectric-insulated plane parallel capacitor with one plate connected to ground and the other plate at a potential V. According to Eq. 3-113, the stored energy is $QV/2$, or $CV^2/2$. Now the energy density is $\epsilon E^2/2$, and, if each plate has an area S and if the dielectric has a thickness s, the stored energy is

$$\frac{1}{2}\epsilon E^2 Ss = \frac{1}{2}\epsilon\frac{S}{s}(Es)^2 = \frac{1}{2}CV^2 \tag{3-120}$$

as before.

3.12 FORCES ON DIELECTRICS

A dielectric material placed in an electric field is subjected to forces and torques that arise from the interaction of the electric field with the dipoles in the dielectric. Although these forces and torques are ordinarily small, they can also be quite large. See for example Problem 3-19.

A dipole in a *uniform* electric field experiences a torque

$$\boldsymbol{T} = \boldsymbol{p} \times \boldsymbol{E}, \tag{3-121}$$

which tends to align it with the field, but the net force is zero. There can be a net force only if the field is nonuniform, so that one end of the dipole is subjected to a greater force than the other.

Let us calculate the force that a dipole experiences in a *nonuniform* electric field. We select axes as in Figure 3-14 and set the field intensity at the origin to be $\boldsymbol{E}$. Then the x-component of the force on the dipole is

$$F_x = -QE_x + Q\left(E_x + \frac{\partial E_x}{\partial x}\,\boldsymbol{s}\cdot\boldsymbol{i} + \frac{\partial E_x}{\partial y}\,\boldsymbol{s}\cdot\boldsymbol{j} + \frac{\partial E_x}{\partial z}\,\boldsymbol{s}\cdot\boldsymbol{k}\right), \tag{3-122}$$

$$= \frac{\partial E_x}{\partial x}\,p_x + \frac{\partial E_x}{\partial y}\,p_y + \frac{\partial E_x}{\partial z}\,p_z. \tag{3-123}$$

Similarly,

$$F_y = \frac{\partial E_y}{\partial x}\,p_x + \frac{\partial E_y}{\partial y}\,p_y + \frac{\partial E_y}{\partial z}\,p_z, \tag{3-124}$$

$$F_z = \frac{\partial E_z}{\partial x}\,p_x + \frac{\partial E_z}{\partial y}\,p_y + \frac{\partial E_z}{\partial z}\,p_z. \tag{3-125}$$

Then

$$\boldsymbol{F} = \left(p_x\frac{\partial}{\partial x} + p_y\frac{\partial}{\partial y} + p_z\frac{\partial}{\partial z}\right)(E_x\boldsymbol{i} + E_y\boldsymbol{j} + E_z\boldsymbol{k}), \tag{3-126}$$

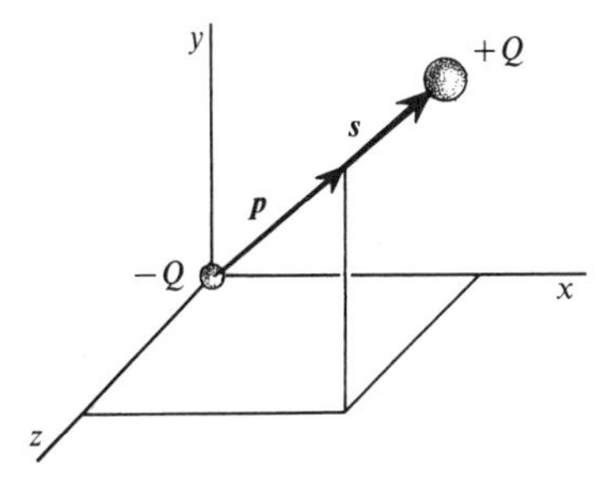

Figure 3-14. Dipole with negative charge at the origin.

$$\mathbf{F} = (\mathbf{p}\cdot\nabla)\mathbf{E}. \tag{3-127}$$

This is the force exerted on a single dipole.

The force per unit volume of the dielectric is N times larger, or $(\mathbf{P}\cdot\nabla)\mathbf{E}$. Since

$$\mathbf{P} = (\epsilon - \epsilon_0)\mathbf{E}, \tag{3-128}$$

the force per unit volume is

$$(\epsilon - \epsilon_0)(\mathbf{E}\cdot\nabla)\mathbf{E} = (\epsilon - \epsilon_0)\left(E_x\frac{\partial}{\partial x} + E_y\frac{\partial}{\partial y} + E_z\frac{\partial}{\partial z}\right)(E_x\mathbf{i} + E_y\mathbf{i} + E_z\mathbf{k}), \tag{3-129}$$

and its x-component is

$$(\epsilon - \epsilon_0)\left[E_x\frac{\partial E_x}{\partial x} + E_y\frac{\partial E_x}{\partial y} + E_z\frac{\partial E_x}{\partial z}\right].$$

Now, in an electrostatic field, $\nabla \times \mathbf{E} = 0$ and

$$\frac{\partial E_x}{\partial y} = \frac{\partial E_y}{\partial x}, \tag{3-130}$$

$$\frac{\partial E_x}{\partial z} = \frac{\partial E_z}{\partial x}, \tag{3-131}$$

so that the x-component is

$$(\epsilon - \epsilon_0)\left[E_x\frac{\partial E_x}{\partial x} + E_y\frac{\partial E_y}{\partial x} + E_z\frac{\partial E_z}{\partial x}\right] = \frac{1}{2}(\epsilon - \epsilon_0)\frac{\partial}{\partial x}E^2 \tag{3-132}$$

and, finally, the force per unit volume is

$$\frac{1}{2}(\epsilon - \epsilon_0)\nabla E^2 = \frac{\epsilon_r - 1}{\epsilon_r}\nabla\left(\frac{1}{2}\epsilon E^2\right) \tag{3-133}$$

inside a homogeneous dielectric. Note that the quantity between parentheses on the right is the electric energy density.

This is the force per unit volume exerted on a dielectric inside which the electric field intensity is E. It is clear that the force is in the direction of an increase in the magnitude of $\mathbf{E}$ and that it is unaffected by a change in the polarity of the field.

This expression is usually valid not only in electrostatic fields but also in alternating electric fields for the following reason. We shall see in Section 8.1.1 that, in the general case, the curl of $\mathbf{E}$ is equal to $\partial\mathbf{B}/\partial t$ where $\mathbf{B}$ is the magnetic induction. For example, the x-component of the curl gives

$$\frac{\partial E_y}{\partial z} - \frac{\partial E_z}{\partial y} = -\frac{\partial B_x}{\partial t}. \tag{3-134}$$

We may therefore set $\nabla \times E = 0$ whenever

$$\left|\frac{\partial E_y}{\partial z}\right| \approx \left|\frac{\partial E_z}{\partial y}\right| \gg \left|\frac{\partial B_x}{\partial t}\right|, \tag{3-135}$$

which is usually the case when we have to deal with dielectrics.

The x-component of the force on the dipole is zero if all three terms on the right of Eq. 3-123 are zero, for example, if $\boldsymbol{p}$ is parallel to the x-axis ($p_y = p_z = 0$) and if E_x does not vary with x ($\partial E_x/\partial x = 0$). This is intuitively quite obvious, and, of course, the same applies to the two other components F_y and F_z. It is a simple matter to discover other conditions where F_x is zero.

Forces on dielectrics can also be calculated using the method of virtual work (Section 2.16).

Example

FORCE PER UNIT VOLUME ON THE INSULATING MATERIAL IN A COAXIAL CABLE

A wire of radius R_1 is insulated with a dielectric of outer radius R_2 that is itself enclosed in a grounded conducting sheath. We wish to calculate the force per unit volume exerted on the dielectric.

We require the electric field intensity E inside the dielectric in terms of the applied voltage V and of the radius ρ. Let the charge per unit length on the wire be λ. From Gauss's law, E is $\lambda/2\pi\epsilon\rho$, the voltage between the inner and outer conductors is

$$V = \int_{R_1}^{R_2} \frac{\lambda}{2\pi\epsilon\rho}\,d\rho = \frac{\lambda}{2\pi\epsilon}\ln\frac{R_2}{R_1}, \tag{3-136}$$

and

$$E = \frac{\lambda}{2\pi\epsilon\rho} = \frac{V}{\rho\ln(R_2/R_1)}. \tag{3-137}$$

Thus, according to Eq. 3-133, the force per unit volume is

$$\frac{1}{2}\epsilon_0(\epsilon_r - 1)\frac{V^2}{\ln^2(R_2/R_1)}\left|\nabla\frac{1}{\rho^2}\right| = -\frac{\epsilon_0(\epsilon_r - 1)V^2}{\ln^2(R_2/R_1)\,\rho^3}. \tag{3-138}$$

The force is radial, and it is directed inward. The reason for this is as follows. Let us assume that the voltage V on the wire is positive. The dipoles then align themselves radically with their negative ends facing the wire, and the inward force on the negative charges is slightly larger than the outward force on the positive charges.

You will be able to show in Problem 3-19 that this force can be larger than the gravitational force by a few orders of magnitude.

3.13 FORCES ON CONDUCTORS IN THE PRESENCE OF DIELECTRICS.

The calculation of the forces between conductors in the presence of dielectrics is best done by the method of virtual work illustrated by the example in Section 2.16.

When the conductors are immersed in a *liquid* dielectric, the forces are always found to be smaller than those in air by the factor ϵ_r if the *charges* are the same in both cases. They are larger than in air by the factor ϵ_r if the electric fields (and hence the voltages) are the same.

It follows (see Section 2.16) that the energy density in a dielectric subjected to an electric field E is $\epsilon E^2/2$, as in Section 3.11.

The case of solid dielectrics will be illustrated in Problem 3-22.

Example FORCES ON A PARALLEL-PLATE CAPACITOR IMMERSED IN A LIQUID DIELECTRIC

We found in the example in Section 2.16 that the force between the plates of an air-insulated parallel-plate capacitor is $(\sigma^2/2\epsilon_0)S$, where σ is the surface charge density and S is the area of one plate. We can perform a similar calculation for a pair of plates immersed in a liquid dielectric. In this case, if the spacing s between the plates decreases by ds, the work done against the mechanical forces F_m holding the plates apart is again $F_m\, ds$ and

$$F_m\, ds = \frac{\epsilon}{2} E^2 S\, ds. \tag{3-139}$$

Then

$$F_m = \frac{\epsilon}{2} E^2 S = \epsilon_r \frac{\epsilon_0}{2} E^2 S, \tag{3-140}$$

and, *for a given electric field strength E*, or for a given voltage difference between the plates, the electric force is ϵ_r times *larger* than in air.

We also have that

$$F_m = \frac{D^2}{2\epsilon} S = \frac{\sigma^2}{2\epsilon} S = \frac{Q^2}{2\epsilon S} = \frac{1}{\epsilon_r} \frac{Q^2}{2\epsilon_0 S} \tag{3-141}$$

and, *for given charges $+Q$ and $-Q$ on the plates*, the electric force is ϵ_r times smaller than in air.

3.14 SUMMARY

When a nonpolar dielectric material is placed in an electric field, the positive and negative charges in the molecules are displaced, one with respect to the other, and the molecules become *polarized.* The induced dipole moment per unit volume $\boldsymbol{P}$ is called the *electric polarization.*

This produces real accumulations of charge that we can use to calculate V and $\boldsymbol{E}$ both inside and outside the dielectric:

$$\sigma_b = \boldsymbol{P}\cdot\boldsymbol{n}, \tag{3-14}$$

$$\rho_b = -\boldsymbol{\nabla}\cdot\boldsymbol{P}, \tag{3-18}$$

where σ_b is the surface density and ρ_b is the volume density of bound charge. The unit vector $\boldsymbol{n}$ is normal to the surface of the dielectric and points outward.

The *polarization current density* is

$$\boldsymbol{J}_b = \frac{\partial \boldsymbol{P}}{\partial t} \text{ amperes/meter}^2. \tag{3-21}$$

The *local electric field intensity* $\boldsymbol{E}_{\text{loc}}$ is the space-and-time average of the electric field intensity acting on a particular molecule:

$$\boldsymbol{E}_{\text{loc}} \approx \boldsymbol{E} + \frac{\boldsymbol{P}}{3\epsilon_0}. \tag{3-32}$$

This result is approximate and, in general,

$$\boldsymbol{E}_{\text{loc}} = \boldsymbol{E} + b\frac{\boldsymbol{P}}{\epsilon_0}, \tag{3-33}$$

where b is a constant that depends on the nature of the dielectric.

In linear and isotropic dielectrics the dipole moment per molecule $\boldsymbol{p}$ is proportional to the local field, the factor of proportionality α being the *molecular polarizability,*

$$\boldsymbol{p} = \alpha \boldsymbol{E}_{\text{loc}}, \tag{3-35}$$

and the electric polarization $\boldsymbol{P}$ is then proportional to $\boldsymbol{E}$:

$$\boldsymbol{P} = \epsilon_0 \chi_e \boldsymbol{E}, \tag{3-38}$$

where χ_e is a dimensionless constant called the *electric susceptibility.*

Gauss's law can be expressed in a form which is valid for dielectrics:

$$\boxed{\boldsymbol{\nabla}\cdot\boldsymbol{E} = \frac{\rho_t}{\epsilon_0},} \tag{3-41}$$

where the total charge density ρ_t is the free charge density ρ_f plus the bound charge density ρ_b. This is one of Maxwell's four fundamental equations of electromagnetism.

The *electric displacement*

$$\boldsymbol{D} = \epsilon_0 \boldsymbol{E} + \boldsymbol{P}, \tag{3-45}$$

$$= \epsilon_0(1 + \chi_e)\boldsymbol{E} = \epsilon_0\epsilon_r\boldsymbol{E} = \epsilon\boldsymbol{E}, \tag{3-52}$$

where ϵ_r is the *relative permittivity*, or the *dielectric constant*, and ϵ is the *permittivity of the medium*. The vector $\boldsymbol{D}$ is related to the free charge density ρ_f:

$$\boldsymbol{\nabla}\cdot\boldsymbol{D} = \rho_f, \tag{3-46}$$

or

$$\int \boldsymbol{D}\cdot d\boldsymbol{a} = \int_\tau \rho_f \, d\tau. \tag{3-47}$$

Poisson's equation for Class A (linear, isotropic, homogeneous) dielectrics is

$$\nabla^2 V = -\frac{\rho_f}{\epsilon}. \tag{3-55}$$

In a Class A dielectric, the bound charge density is related to the free charge density as follows:

$$-\rho_b = \frac{\epsilon_r - 1}{\epsilon_r}\rho_f. \tag{3-59}$$

At a dielectric–conductor boundary the free surface charge density σ_f on the conductor is equal to the electric displacement just inside the dielectric, and the bound surface charge density σ_b is smaller in magnitude by the factor $(\epsilon_r - 1)/\epsilon_r$:

$$-\sigma_b = \frac{\epsilon_r - 1}{\epsilon_r} D = \frac{\epsilon_r - 1}{\epsilon_r}\sigma_f. \tag{3-66}$$

The relative permittivity ϵ_r is related to the mass density ρ and to the molecular polarizability α through the *Clausius-Mossotti equation:*

$$\frac{\epsilon_r - 1}{\epsilon_r + 2} = \frac{N\alpha}{3\epsilon_0} \tag{3-85}$$

and

$$\alpha_M = \frac{M}{\rho}\frac{\epsilon_r - 1}{\epsilon_r + 2} = \frac{N_A}{3\epsilon_0}\alpha, \tag{3-88}$$

where α_M is the *molar polarization.*

For polar molecules we have the *Langevin equation:*

$$P = Np\left(\coth\frac{pE_{\mathrm{loc}}}{kT} - \frac{kT}{pE_{\mathrm{loc}}}\right), \tag{3-98}$$

where p is the permanent dipole moment of the molecule; and for real polar dielectrics, in which there are both induced and permanent dipoles, we have the *Debye equation:*

$$\alpha_M = \frac{M}{\rho}\left(\frac{\epsilon_r - 1}{\epsilon_r + 2}\right) = \frac{N_A}{3\epsilon_0}\left(\alpha + \frac{p^2}{3kT}\right). \tag{3-107}$$

In the presence of dielectrics the potential energy of a charge distribution is given by an integral similar to that for free space:

$$W = \frac{1}{2}\int_\tau \rho_f V\, d\tau, \tag{3-113}$$

$$= \frac{1}{2}\int_\tau (\mathbf{D}\cdot\mathbf{E})\, d\tau. \tag{3-117}$$

And we can define an *energy density*

$$\frac{dW}{d\tau} = \frac{1}{2}(\mathbf{D}\cdot\mathbf{E}), \tag{3-118}$$

which is equal to $(1/2)\epsilon E^2$ for Class A dielectrics.

The *torque* on a dipole is

$$\mathbf{T} = \mathbf{p} \times \mathbf{E}, \tag{3-121}$$

while the *force* is

$$\mathbf{F} = (\mathbf{p}\cdot\nabla)\mathbf{E}. \tag{3-127}$$

The *force per unit volume* is $(\mathbf{P}\cdot\nabla)\mathbf{E}$, or $(\frac{1}{2})(\epsilon - \epsilon_0)\nabla E^2$.

Charged conductors are subjected to electric forces that depend on the nature of the dielectric in which they are immersed. In a liquid dielectric the dielectric decreases the forces by a factor of ϵ_r if the charges are kept constant, and it increases the forces by a factor of ϵ_r if the voltages are kept constant.

PROBLEMS

3-1. A sample of diamond has a density of 3.5 grams/centimeter^3 and a polarization of 10^{-7} coulomb/meter^2.

(a) Compute the average dipole moment per atom.

(b) Find the average separation between centers of positive and negative charge. Carbon has a nucleus with a charge $+6e$, surrounded by 6 electrons.

3-2. Show that the bound charge density at the interface between two dielectrics 1 and 2 that is crossed by an electric field is $(\mathbf{P}_1 - \mathbf{P}_2)\cdot\mathbf{n}$. The polarization in 1 is $\mathbf{P}_1$ and is directed into the interface, while the polarization in 2 is $\mathbf{P}_2$ and points away from the interface. The unit vector $\boldsymbol{n}$ is normal to the interface and points in the direction from 1 to 2.

3-3. A large block of dielectric contains small cavities of various shapes that may be assumed not to disturb appreciably the polarization.

(a) Show that, inside a needle-like cavity parallel to $\boldsymbol{P}$, $\boldsymbol{E}$ is the same as in the dielectric.
(b) Show that, inside a thin crack perpendicular to $\boldsymbol{P}$, $\boldsymbol{E}$ is ϵ_r times larger than in the dielectric.
(c) Show that, at the center of a small spherical cavity, $\boldsymbol{E}$ is $\boldsymbol{P}/3\epsilon_0$.

3-4. The space between the plates of a parallel-plate capacitor with plate separation s and surface area S is partially filled with a dielectric plate of area S and of thickness $t < s$.

Show that the capacitance is

$$\frac{\epsilon_0 S}{s - [(\epsilon_r - 1)t/\epsilon_r]}.$$

3-5. A dielectric sphere of radius R contains a uniform density of free charge ρ_f. Show that the potential at the center is

$$\frac{2\epsilon_r + 1}{2\epsilon_r}\,\frac{\rho_f R^2}{3\epsilon_0}.$$

3-6. Draw graphs of D, E, V as functions of r for a point charge at the center of a dielectric sphere as in the second example on page 111. Set $Q = 1.00 \times 10^{-9}$ coulomb, $R = 2.00$ centimeters, $\epsilon_r = 3.00$.

Show also the curves of D, E, V in the absence of the dielectric sphere.

3-7. A dielectric sphere of radius R is polarized so that $\boldsymbol{P} = (K/r)\boldsymbol{r}_1$, $\boldsymbol{r}_1$ being the unit radial vector.

(a) Calculate the volume and the surface density of bound charge.
(b) Calculate the volume density of free charge.
(c) Calculate the potential inside and outside the sphere.
(d) Sketch a curve of potential versus distance from $r = 0$ to $r = \infty$.

3-8. A conducting wire carrying a charge λ per unit length is embedded along the axis of a circular cylinder of dielectric. The radius of the wire is a; the radius of the cylinder is b.

(a) Show that the bound charge on the outer surface of the dielectric is equal to the bound charge on the inner surface, except for sign.
(b) Show that the net charge along the axis is λ/ϵ_r per unit length.
(c) Show that the volume density of bound charge is zero in the dielectric.

3-9. The relative permittivity of the dielectric between the plates of a parallel-plate capacitor varies linearly from one plate to the other. If ϵ_{r1} and ϵ_{r2} are the values at the two plates, where $\epsilon_{r2} > \epsilon_{r1}$, and if the plate separation is s, show that the capacitance per unit area is

$$\frac{\epsilon_0\,(\epsilon_{r2} - \epsilon_{r1})}{s\,\ln\,(\epsilon_{r2}/\epsilon_{r1})}.$$

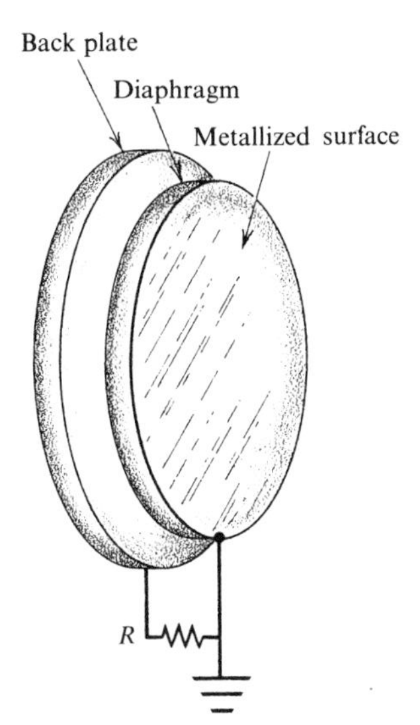

Figure 3-15.

3-10. (a) If the space between two long, coaxial cylindrical conductors were filled with a dielectric, how would the relative permittivity have to depend on the distance ρ from the axis in order that the electric field intensity be independent of ρ?
(b) What would be the volume density of bound charge?

3-11. A capacitor is formed of two concentric spherical conducting shells of radii r_1 and r_3. The space between the shells is filled from r_1 to r_2 with a dielectric of relative permittivity ϵ_{r1}, and from r_2 to r_3 with a dielectric of relative permittivity ϵ_{r2}.

Show that the capacitance is

$$\frac{4\pi\epsilon_0}{\dfrac{1}{\epsilon_{r1}r_1} - \dfrac{1}{\epsilon_{r2}r_3} + \dfrac{1}{r_2}\left(\dfrac{1}{\epsilon_{r2}} - \dfrac{1}{\epsilon_{r1}}\right)}.$$

3-12. An electret has the form of a thin circular sheet of radius R and thickness t, permanently polarized in the direction parallel to its axis. The polarization P is uniform throughout the volume of the disk.

Calculate E and D on the axis, both inside and outside the disk.

3-13. Figure 3-15 shows a microphone in which an electret film is used as diaphragm. The right-hand side of the electret is metallized and serves as one electrode, the other electrode being the back plate.

In actual practice the surface of the back plate is slightly roughened and the diaphragm is placed in physical contact with it. The diaphragm then touches the back plate at a large number of small isolated spots. The back plate is perforated in order to reduce the stiffness of the air cushion and thus increase the sensitivity.

The surfaces of the film carry both free and bound charges, the *free* charges being predominant. The net charge density has a time constant that can be as large as a few thousand years at room temperature.

(a) Calculate the induced charge densities on the electrodes, under steady-state conditions, for the case of a 0.006 millimeter thick Mylar ($\epsilon_r = 3.0$) film and a 0.013 millimeter air gap. Assume constant free charge densities of 10^{-4} coulomb/meter^2 on the film surfaces and neglect permanent bound charges. Neglect edge effects.
(b) Find a differential equation for σ_f when the diaphragm oscillates at an angular frequency ω.

3-14. When a block of insulating material such as Lucite is bombarded with high-energy electrons, the electrons penetrate into the material and remain trapped inside. In one particular instance a 0.1 microampere beam bombarded an area of 25 centimeters^2 of Lucite ($\epsilon_r = 3.2$) for 1 second, and essentially all the electrons were trapped about 6 millimeters below the surface in a region about 2 millimeters thick. The block was 12 millimeters thick.

In the following calculations neglect edge effects and assume a uniform density for the trapped electrons. Assume also that both faces of the Lucite are in contact with grounded conducting plates.

(a) What is the bound charge density in the charged region?
(b) What is the bound charge density at the surface of the Lucite?
(c) Sketch graphs of D, E, V as functions of position inside the dielectric.
(d) Show that the potential at the center of the sheet of charge is about 4 kilovolts.
(e) What is the electric field intensity in the charge-free region?
(f) What is the energy stored in the block? What is the danger that the block will explode?
(g) How would the curve of V be affected if the sheet of electrons were closer to one face of the block than to the other?

3-15. The dipole moment of the H_2O molecule is 6.2×10^{-30} coulomb-meter. Find the maximum electric polarization of water vapor at a temperature of 100°C and at a pressure of 760 millimeters of mercury.

3-16. (a) Show that a nonhomogeneous dielectric can have a volume density of bound charge in the absence of a free charge density.
(b) Calculate ρ_b in this case.

3-17. (a) Show that the maximum energy stored per cubic meter in a parallel-plate capacitor is $\epsilon a^2/2$, where a is the dielectric strength of the insulator (maximum electric field intensity before breakdown).
(b) It is suggested that a small vehicle could be propelled by an electric motor fed by charged capacitors. Comment on this suggestion, assuming that the only problem is one of energy storage. A good dielectric to use would be Mylar which has a dielectric strength of 4000 volts per mil (0.001 inch) and a relative permittivity of 3.2.

3-18. A dipole of moment $\boldsymbol{p}$ is lined up with the z-axis at the origin of coordinates. A second dipole of moment $\boldsymbol{p}$ is centered at the point $(a, 0, a)$ and is pointed toward the origin. Calculate the force on the second dipole.

3-19. (a) Calculate the force per cubic meter on the dielectric of a coaxial cable whose inner conductor has a radius of 1 millimeter and whose outer conductor has an inner radius of 5 millimeters. The dielectric has a relative permittivity of 2.5. The outer conductor is gounded, and the inner conductor is maintained at 25 kilovolts.

(b) Show that the electric force near the inner conductor is about 300 times larger than the gravitational force if the dielectric has the density of water, namely 10^3 kilograms/meter^3.

3-20. In the presence of a nonuniform electric field, a particle of dielectric suspended in a fluid having a lower relative permittivity is submitted to a force that pulls it toward the region where the field is highest. This phenomenon is known as *dielectrophoresis.*

Unlike *electrophoresis*, dielectrophoresis does not require charged particles, but requires a nonuniform field that can be either direct or alternating.

Dielectrophoresis is in general a weak effect that can be used only with strong fields. It has been used to separate living cells from dead ones, carbon particles from a polymer, various minerals and chemicals, etc.

Let us calculate the drift velocity of spherical particles of radius a and relative permittivity p in a fluid of relative permittivity $f < p$, when the field is cylindrical with inner and outer radii ρ_1 and ρ_2 respectively.

First, the polarizing field inside the particles is smaller than the field $\boldsymbol{E}$ in the fluid by the factor

$$\frac{3f}{2f+p},$$

as we shall see in the example on page 173. This factor is strictly valid only for a uniform field, but it is applicable here as long as the particles subtend a small angle at the axis of symmetry. Then the *net* force on the particle is the force on the particle minus the force on the droplet of fluid that it replaces.

(a) You should be able to show that

$$\boldsymbol{F}_{\text{net}} = \frac{2}{3}\pi a^3 \epsilon_0 A \nabla E^2 = -\frac{4\pi}{3}\epsilon_0 a^3 A \frac{V_0^2 \boldsymbol{\rho}_1}{\rho^3 \ln^2(\rho_2/\rho_1)},$$

where V_0 is the voltage difference between the inner and outer electrodes, and

$$A = \frac{f(4f+5) - p(f-1)}{(2f+p)^2}(p-f).$$

(b) Under what conditions is A positive? Sketch a curve of A as a function of p, for a given value of f. Can you explain qualitatively the shape of this curve?

(c) The drift velocity v is given by Stokes's equation:

$$F_{\text{net}} = 6\pi\eta a v,$$

where η is the coefficient of viscosity of the fluid.

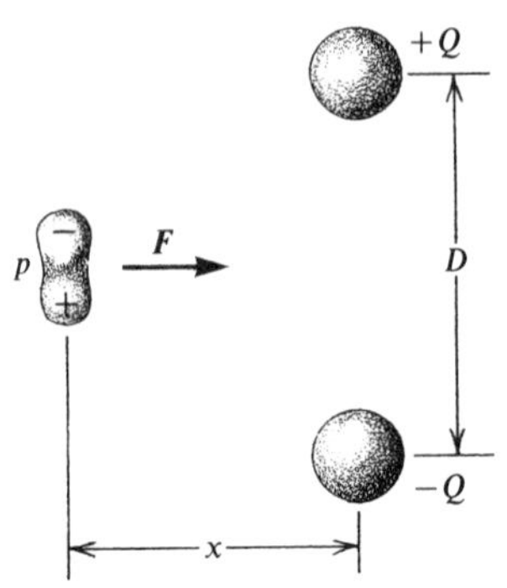

Figure 3-16.

The drift velocity v is therefore proportional to the square of the radius a of the particles and inversely proportional to the cube of the distance ρ from the axis.

Show that $v \approx 3 \times 10^{-4}$ meter/second for particles of a polymer with $p = 4.6$ in benzene ($\mathscr{f} = 2.3$, $\eta = 6.5 \times 10^{-4}$ newton second/meter^2), $V_0 = 4 \times 10^3$ volts, $\rho_2 = 1$ centimeter, $\rho_1 = 0.5$ millimeter, $a = 0.1$ millimeter, and $\rho = 0.5$ centimeter.

3-21. It is possible to accelerate a *neutral* molecule by means of an electric field in the following way. Figure 3-16 shows a pair of spheres carrying charges $+Q$ and $-Q$, respectively, and a molecule whose dipole moment is $\boldsymbol{p}$. It is clear that the molecule is subject to a net force $\boldsymbol{F}$. At the moment the molecule reaches the midpoint between the spheres, the spheres are both grounded, and the molecule continues at a constant velocity.

The purpose of such an accelerator is to study in detail the inelastic processes which occur in molecular collisions.

(a) Show that the kinetic energy of a molecule accelerated in this way is

$$\frac{2pQ}{\pi\epsilon_0 D^2},$$

if $x \gg D$ initially.

Of course the dipole moment p increases as the molecule moves into the electric field between the two spheres, and the kinetic energy is in fact larger.‡

(b) In one particular accelerator the electrodes have a radius of 0.25 millimeter, $D = 1.00$ millimeter, and they are maintained at ± 40 kilovolts. Show that the kinetic energy acquired by a molecule with $p = 2 \times 10^{-29}$ coulomb-meter is approximately 0.01 electron-volt.‡

(c) Since this energy is too low, the accelerator has 700 stages with an average distance of 1.4 centimeters/stage, for a total length of 10 meters. The electrodes are fed at a frequency of 500 kilohertz. Assuming that the total energy is then 700 times larger, show that the equivalent temperature of the accelerated molecules is about 6×10^4 kelvins.

3-22. A capacitor composed of two electrodes of area A separated by a sheet of dielectric of thickness t is connected to a source of voltage V.‡

(a) Calculate the force of attraction between the plates.
(b) In one particular case $S = 68$ square inches, $t = 0.030$ inch, $\epsilon_r = 3.0$, $V = 60$ kilovolts. Calculate the force of attraction in metric tons (1000 kilograms).
(c) The solid dielectric is removed and the plates are submerged in a liquid of the same relative permittivity, so that the space between the plates is always filled with dielectric. Calculate the force of attraction again in metric tons.

3-23. One author states that he can attain fields of 4×10^5 volts/centimeter over a 0.1 inch gap in purified nitrobenzene ($\epsilon_r = 3.5$) and that the resulting electric force on the electrodes is then 36 pounds per square inch, or more than two atmospheres. Is the force really that large?

3-24. Electrostatic clamps are used for holding work pieces while they are being machined. They utilize an insulated conducting plate charged to several thousand volts and covered with a thin insulating sheet. The work piece is placed on the sheet and grounded.

One particular type operates at 3000 volts and is advertised as having a holding power of 30 pounds per square inch. If the insulator is Mylar ($\epsilon_r = 3.2$), what is its thickness?

3-25. It is stated in Section 3.13 that the electric forces on conductors immersed in liquid dielectrics are larger than in air by the factor ϵ_r if the *voltages* are the same, with and without the dielectric, and that they are smaller than in air by the same factor ϵ_r if the *charges* are the same. Can you justify this general statement?

Hints

3-21. (a) Assume that $+Q$ and $-Q$ are point charges, and calculate the force exerted *on them* by the molecule.
(b) You will have to calculate an approximate value for Q.

3-22. (a) Remember that, if the plate separation is increased by ds, the extra field is in *air*.

3-24. See hint for 3-22(a).

CHAPTER 4

ELECTROSTATIC FIELDS III

General Methods for Solving Laplace's and Poisson's Equations

Up to this point our discussion of electrostatic fields has been limited to rather simple charge distributions; we shall now develop methods for calculating more complex fields.

Except for Section 4.1, this chapter deals with the solution of the differential equations $\nabla^2 V = 0$, $\nabla^2 V = -\rho_t/\epsilon_0$, $\nabla^2 \boldsymbol{E} = \nabla \rho_t/\epsilon_0$. The first is of course Laplace's equation, while the other two are Poisson's equations for V and $\boldsymbol{E}$. Now these equations are valid not only for electrostatic fields but also for several other classes of phenomena. For example, the flow of heat in a medium of thermal conductivity K obeys Poisson's equation $\nabla^2 T = -q/K$, where T is the temperature and q is the thermal energy generated per unit volume and per unit time. This chapter therefore has broad applications.*

We shall first discuss the continuity of various quantities at the interface between two different media, and then we shall prove the uniqueness theorem, according to which there is only one physically possible electric field that can satisfy both Poisson's equation and a given set of boundary conditions. Then we shall illustrate the method of images, and finally we shall discuss at considerable length the solution of Laplace's and Poisson's equations in rectangular and in spherical coordinates.

If you are interested in this type of calculation, you can work through Appendix B, which describes a method that is widely used for calculating two-dimensional electrostatic fields when the volume charge density ρ is zero.

---

* Sections 4.1 and 4.2 are fundamental, but the rest of this chapter can be omitted if time is lacking; it is not required for what follows.

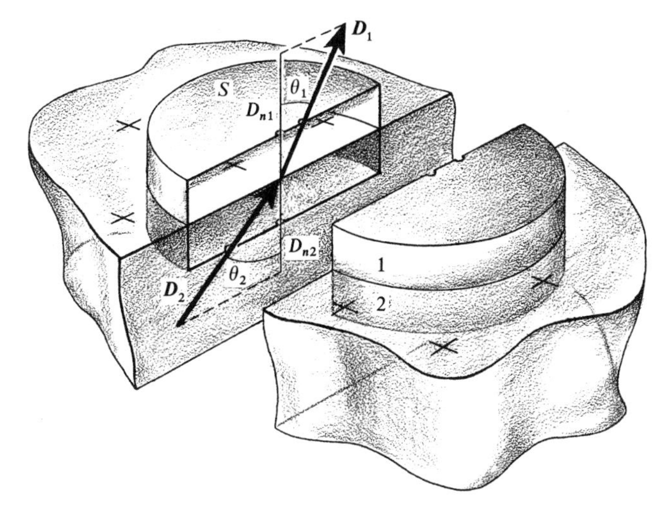

Figure 4-1. Gaussian cylinder on the interface between two different media 1 and 2. The difference $D_{n1} - D_{n2}$ between the *normal* components of $\boldsymbol{D}$ is equal to the surface charge density σ_f.

It should be realized that, once the electrostatic field is known, it becomes possible to deduce many other quantities—such as charge densities and total charges on conductors, potential differences between conductors, capacitances, focusing properties of electrostatic lenses, and so forth.

4.1 CONTINUITY OF V, D_n, E_t AT THE INTERFACE BETWEEN TWO DIFFERENT MEDIA

The solution of Poisson's or Laplace's equation must be consistent with certain boundary conditions.

4.1.1. Potential

At the boundary between two different media the potential V must be continuous, for a discontinuity would imply an infinitely large electric field intensity, which is physically impossible.

The potential must also be zero at infinity if the charge distributions are of finite extent, and it must be constant throughout any conductor as long as the electric charges are at rest.

4.1.2 Normal Component of the Electric Displacement

Consider a short Gaussian cylinder drawn about a boundary surface, as in Figure 4-1. The end faces of the cylinder are parallel to the boundary and

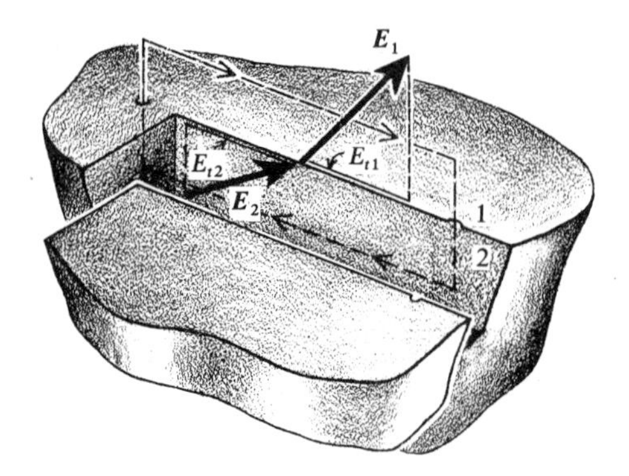

Figure 4-2. Closed path of integration crossing the interface between two different media 1 and 2. Whatever be the surface charge density σ_f the *tangential* components of $\boldsymbol{E}$ on either side of the interface are equal: $E_{t1} = E_{t2}$.

arbitrarily close to it. The boundary carries a free surface charge density σ_f. If the area S is small, D and σ_f do not vary significantly over it, and according to Gauss's law the flux of $\boldsymbol{D}$ emerging from the flat cylinder is equal to the charge enclosed:

$$(D_{n1} - D_{n2})S = \sigma_f S. \tag{4-1}$$

The only flux of $\boldsymbol{D}$ is through the end faces, since the area of the other surface is arbitrarily small. Thus

$$D_{n1} - D_{n2} = \sigma_f. \tag{4-2}$$

At the boundary between two dielectric media the free surface charge density σ_f is generally zero and then D_n is continuous across the boundary. On the other hand, if the boundary is between a conductor and a dielectric, and if the electric field is constant, $\boldsymbol{D} = 0$ in the conductor and $D_n = \sigma_f$ in the dielectric, σ_f being the free charge density on the surface of the conductor.

4.1.3 Tangential Component of the Electric Field Intensity

Consider the path shown in Figure 4-2, with two sides parallel to the boundary and arbitrarily close to it. The other two sides are infinitesmal. If the path is short enough, E_t does not vary significantly over it and the line integral of $\boldsymbol{E}\cdot d\boldsymbol{l}$ is $E_{t_1}L - E_{t_2}L$. Now, according to Stokes's Theorem, this line integral is equal to the integral of $\nabla \times \boldsymbol{E}$ over the surface enclosed by the path. By definition, the enclosed area is zero. So, even when $\nabla \times \boldsymbol{E}$ is not zero, its surface integral is zero and

$$E_{t1}L - E_{t2}L = 0, \tag{4-3}$$

or

$$E_{t1} = E_{t2}. \tag{4-4}$$

The tangential component of $\boldsymbol{E}$ is therefore continuous across the boundary.

If the boundary lies between a dielectric and a conductor, then $\boldsymbol{E} = 0$ in the conductor and $E_t = 0$ in both media. For static fields, $\boldsymbol{E}$ is therefore normal to the surface of a conductor.

4.1.4 Bending of Lines of Force

It follows from the boundary conditions that the $\boldsymbol{D}$ and $\boldsymbol{E}$ vectors change direction at the boundary between two dielectrics. In Figure 4-3,

$$D_1 \cos\theta_1 = D_2 \cos\theta_2, \tag{4-5}$$

or

$$\epsilon_{r1}\epsilon_0 E_1 \cos\theta_1 = \epsilon_{r2}\epsilon_0 E_2 \cos\theta_2, \tag{4-6}$$

and

$$E_1 \sin\theta_1 = E_2 \sin\theta_2. \tag{4-7}$$

Then, dividing the second equation by the third,

$$\epsilon_{r1} \cot\theta_1 = \epsilon_{r2} \cot\theta_2, \tag{4-8}$$

or

$$\frac{\tan\theta_1}{\tan\theta_2} = \frac{\epsilon_{r1}}{\epsilon_{r2}}. \tag{4-9}$$

The larger angle from the normal is in the medium with the larger relative permittivity.

For the case of an electret, see Problem 4-2.

The electric field is therefore subjected to several boundary conditions at an interface, but these conditions do not all have the same degree of generality. The condition that V be continuous is perfectly general. Equation 4-2

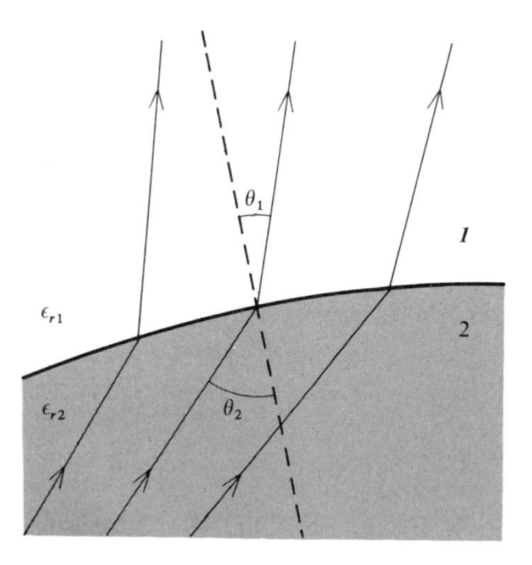

Figure 4-3. Lines of $\boldsymbol{D}$ or of $\boldsymbol{E}$ crossing the interface between two different media 1 and 2. The lines change direction in such a way that $\epsilon_{r1} \tan\theta_2 = \epsilon_{r2} \tan\theta_1$.

is also general. However, the condition $D_n = \sigma_f$ at the interface between a dielectric and a conductor is valid only for static fields, because only then is $\boldsymbol{D} = 0$ in conductors. Similarly, the condition $E_{t1} = E_{t2}$ at any interface, and the bending of lines of force, as above, are also valid only for static fields because they are based on the hypothesis that $\boldsymbol{\nabla} \times \boldsymbol{E} = 0$, which is not general, as we shall see in Chapter 8.

4.2 THE UNIQUENESS THEOREM

We shall demonstrate that a potential V that satisfies both Poisson's equation and the boundary conditions pertinent to a particular field is the only possible potential.

This is an important theorem because it leaves us free to use *any* method, even intuition, to determine an electric field; if we can somehow find a field that satisfies both of the above conditions, then it is the only possible one.

Let us consider a finite region of space that may contain charged conductors at specified potentials, Class A dielectric materials of specified properties, and volume distributions of free charges with specified densities. To demonstrate the uniqueness theorem we shall assume that at each point there are two possible solutions, V_1 and V_2, both of which satisfy Poisson's equation and both of which reduce to the specified potentials on the surfaces of the conductors. This does *not* imply that a given point can be at two different potentials at the same time. Our assumption is that either one or the other of two different fields can exist in the region for which boundary conditions are specified. We shall find that $V_1 \equiv V_2$. This is the uniqueness theorem.

Corresponding to V_1 and V_2, there are two possible electric field intensities

$$\boldsymbol{E}_1 = -\boldsymbol{\nabla} V_1, \qquad \boldsymbol{E}_2 = -\boldsymbol{\nabla} V_2 \tag{4-10}$$

at every point in the field.

We have assumed that Poisson's equation is satisfied by both V_1 and V_2 everywhere. Then

$$\boldsymbol{\nabla} \cdot \boldsymbol{D}_1 = \rho_f, \qquad \boldsymbol{\nabla} \cdot \boldsymbol{D}_2 = \rho_f \tag{4-11}$$

where ρ_f is the free charge density.

We now focus our attention on the difference between the two solutions and call it V_3:

$$V_3 = V_2 - V_1. \tag{4-12}$$

Then the corresponding $\boldsymbol{D}_1$, $\boldsymbol{D}_2$, $\boldsymbol{D}_3$ are such that

$$\mathbf{D}_3 = \mathbf{D}_2 - \mathbf{D}_1, \tag{4-13}$$

and

$$\nabla \cdot \mathbf{D}_3 = \nabla \cdot \mathbf{D}_2 - \nabla \cdot \mathbf{D}_1 = 0 \tag{4-14}$$

at every point. On the surfaces of the conductors $V_3 = 0$ since both V_1 and V_2 reduce to the specified boundary values.

We now use the vector identity

$$\nabla \cdot V_3\mathbf{D}_3 = V_3(\nabla \cdot \mathbf{D}_3) + \mathbf{D}_3 \cdot \nabla V_3. \tag{4-15}$$

Integrating over a volume τ and using the divergence theorem,

$$\int_S V_3\mathbf{D}_3 \cdot d\mathbf{a} = \int_\tau V_3(\nabla \cdot \mathbf{D}_3)\, d\tau + \int_\tau (\mathbf{D}_3 \cdot \nabla V_3)\, d\tau, \tag{4-16}$$

where the surface integral on the left is evaluated over all the surfaces that bound the volume τ. Let us take this volume to be the volume external to the conductors, extending to infinity in all directions.

The surface integral is then to be evaluated over the surfaces of the conductors and over an imaginary sphere of infinite radius. Since the quantity V_3 is zero on all these surfaces, this portion of the integral is zero. To evaluate the integral over the sphere of infinite radius, we consider the integral over a finite sphere and let its surface recede to infinity. Both V_2 and V_1 must fall off as $1/r$ at sufficiently large distances, since all the charge in the system will appear as a point charge from a distance large in comparison to the dimensions of the charge system. Then V_3, the difference between V_2 and V_1, must also fall as $1/r$. Now $\mathbf{D}_3$ must fall off as ∇V_3, or as $1/r^2$. Since the area S over which the integration is performed increases as r^2, the whole integral falls off as $1/r$ and approaches zero at infinity. The left side of the equation is thus zero.

The first term on the right is also zero, since $\nabla \cdot \mathbf{D}_3 = 0$ at every point. We are thus left with the second term on the right, which must be identically equal to zero. Thus

$$\int_\tau (\mathbf{D}_3 \cdot \mathbf{E}_3)\, d\tau = 0. \tag{4-17}$$

In homogeneous, isotropic, linear dielectrics the quantity $\mathbf{D} \cdot \mathbf{E} = \epsilon E^2$ is positive, and the only way in which the integral can be zero is to have $\mathbf{D}_3$ and $\mathbf{E}_3$ equal to zero at every point.

It therefore follows that

$$\nabla V_2 = \nabla V_1, \tag{4-18}$$

or that V_2 can differ from V_1 at most by a constant. Since V_1 and V_2 must be the same on the surfaces of the conductors, they must be the same everywhere. Therefore $V_2 = V_1$, and there is only one possible potential V.

We have therefore shown that the solution of the Poisson equation for given boundary conditions is unique, as long as $\boldsymbol{D}\cdot\boldsymbol{E}$ is positive throughout the dielectric material in the system.

4.3 IMAGES

The method of images involves the conversion of an electric field into another equivalent field that is simpler to calculate. It is particularly useful for point charges near conductors: it is possible in certain cases to replace the conductors by one or more point charges in such a way that the conductor surfaces are replaced by equipotential surfaces at the same potentials. Since the boundary conditions are then conserved, the electric field thus found is the correct one for the region outside the conductors.

Example

POINT CHARGE NEAR AN INFINITE GROUNDED CONDUCTING PLANE

As a first example, consider a point charge Q at a distance D from an infinite conducting plane connected to ground, as in Figure 4-4a. This plane may be taken to be at zero potential. It is clear that if we remove the grounded conductor and replace it by a charge $-Q$ at a distance D behind the plane, then every point of the plane will be equidistant from Q and from $-Q$ and will thus be at zero potential.

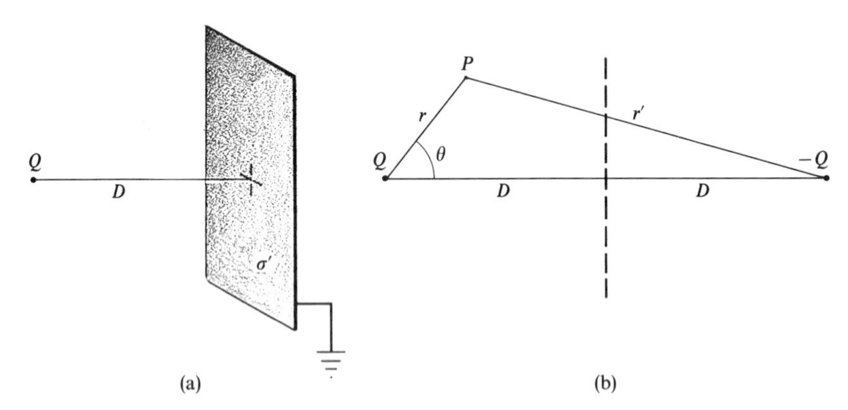

Figure 4-4. (a) Point charge Q near a grounded conducting plane. (b) The conducting plane has been replaced by the image charge $-Q$ to calculate the field at P.

In the region to the left of the plane the two point charges must therefore give the proper solution for the point and the plane. The charge $-Q$ is said to be the *image* of the charge Q in the plane.

The potential V at a point P, whose coordinates are r and θ, as in Figure 4-4b, is given by

$$4\pi\epsilon_0 V = \frac{Q}{r} - \frac{Q}{r'}, \tag{4-19}$$

where

$$r' = \sqrt{r^2 + 4D^2 - 4rD\cos\theta}. \tag{4-20}$$

The components of the electric field intensity at P are given by the components of $\boldsymbol{\nabla} V$:

$$4\pi\epsilon_0 E_r = -4\pi\epsilon_0 \frac{\partial V}{\partial r} = \frac{Q}{r^2} - \frac{Q(r - 2D\cos\theta)}{r'^3}, \tag{4-21}$$

$$4\pi\epsilon_0 E_\theta = -4\pi\epsilon_0 \frac{1}{r}\frac{\partial V}{\partial \theta} = -\frac{2QD\sin\theta}{r'^3}. \tag{4-22}$$

The lines of force and the equipotentials are shown in Figure 4-5.

The induced charge density σ' on the surface of the conducting plane is readily found from the normal component of the electric field intensity at the conductor, since

$$E_n = \sigma'/\epsilon_0. \tag{4-23}$$

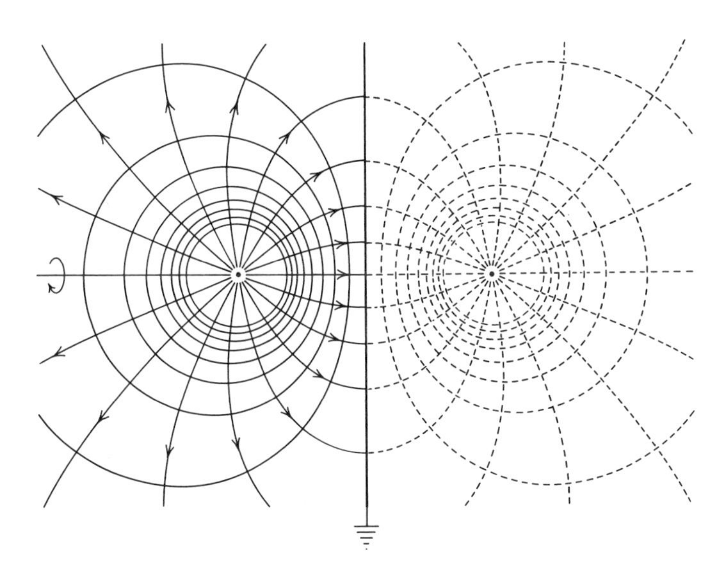

Figure 4-5. Lines of force (identified by arrows) and equipotentials for a point charge near a grounded conducting plane. Equipotentials and lines of force near the charge cannot be shown because they get too close together. Equipotential surfaces are generated by rotating the figure about the axis designated by the curved arrow. The image field to the right of the conducting plane is indicated by broken lines.

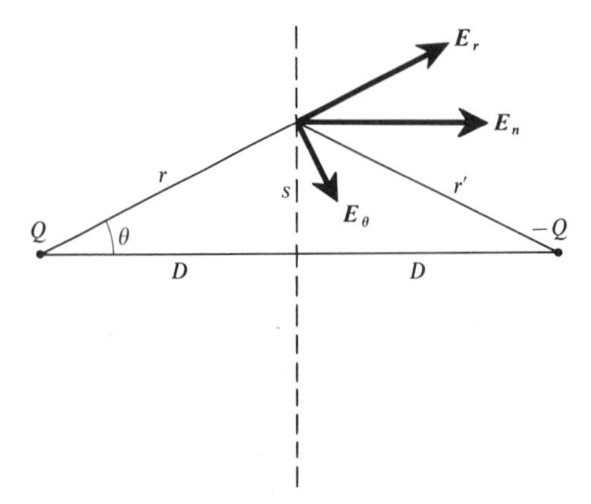

Figure 4-6. The electric field intensity E_n at the surface of the grounded conducting plane is calculated from the fields of Q and of its image $-Q$. It is the vector sum of E_r and E_θ, and is normal to the surface.

In this particular case the surface charge density is negative, and the electric field intensity points to the right at the surface of the conducting plate.

From Figure 4-6, $r = r'$ at all points on the plane and

$$E_n = 2E_r \cos\theta = \frac{2QD}{4\pi\epsilon_0 r^3} = -\frac{\sigma'}{\epsilon_0}, \tag{4-24}$$

$$\sigma' = -\frac{QD}{2\pi r^3}. \tag{4-25}$$

A charge Q induces on the conductor a charge

$$Q' = \int_0^\infty \sigma' 2\pi s\, ds = -\frac{QD}{2}\int_0^\infty \frac{2s\, ds}{(s^2 + D^2)^{3/2}} = -Q, \tag{4-26}$$

as expected.

It will be observed that the image charges are always located at points *outside* the region where we wish to calculate the field. In this case we require the field in the region to the left of the plane, whereas the image is to the right.

Now what is the electrostatic force of attraction between the charge Q and the grounded plate? It is obviously the same as between two charges Q and $-Q$ situated at a distance $2D$ one from the other, since Q cannot tell whether it is in the presence of a point charge $-Q$ or of a grounded plate. The force on Q is thus $Q^2/4\pi\epsilon_0(2D)^2$. This is always true. The force between a charge and a conductor is always given correctly by the Coulomb force between the charge and its image charges.

Example

POINT CHARGE NEAR A GROUNDED CONDUCTING SPHERE

Another case where the image method is applicable is that of a point charge near a grounded conducting sphere, as in Figure 4-7a. We

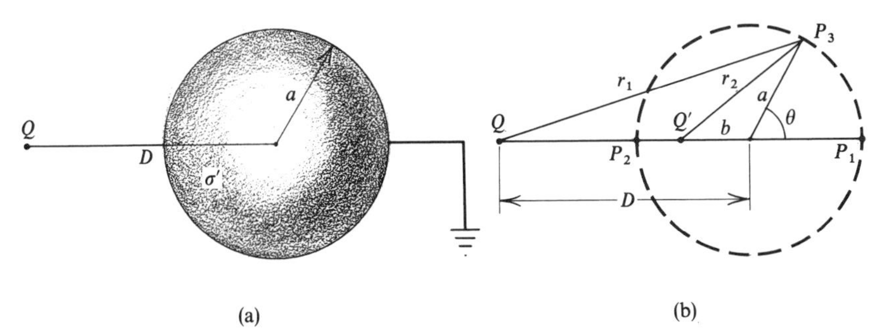

Figure 4-7. (a) Point charge Q at a distance D from the center of a grounded conducting sphere of radius a. When Q is positive, the induced surface charge density σ' is negative. (b) The boundary condition $V = 0$ on the spherical surface is satisfied by the original charge Q and its image $-Q'$.

remove the conductor and try to find the position and magnitude of an "image" charge Q', as in Figure 4-7b, that will make the potential zero on the spherical surface.

It is clear from the symmetry of the problem that if such a charge exists it must lie on the line connecting Q and the center of the sphere. We begin by making the potential zero at the points P_1 and P_2. Then

$$\frac{Q}{(D+a)} + \frac{Q'}{(a+b)} = 0, \tag{4-27}$$

$$\frac{Q}{(D-a)} + \frac{Q'}{(a-b)} = 0. \tag{4-28}$$

Solving these two equations gives

$$Q' = -\frac{a}{D}Q, \qquad b = \frac{a^2}{D}. \tag{4-29}$$

We still have to find whether this charge arrangement will make the potential zero at a general point P_3 on the surface of the sphere. At P_3

$$4\pi\epsilon_0 V = \frac{Q}{r_1} + \frac{Q'}{r_2}, \tag{4-30}$$

where

$$r_1 = \sqrt{(D^2 + a^2 + 2Da\cos\theta)}, \tag{4-31}$$

$$r_2 = \sqrt{(b^2 + a^2 + 2ba\cos\theta)}. \tag{4-32}$$

Taking Q' and b as in Eq. 4-29 does in fact make $V = 0$ at P_3, as required.

Since the original point charge and the image charge Q' satisfy the boundary condition, they must give the correct field at every point in the space outside the conducting sphere.

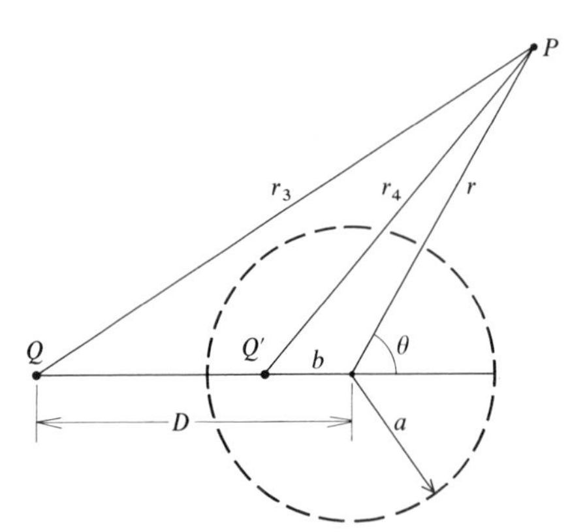

Figure 4-8. The electric potential V at the point $P(r, \theta)$ is calculated from Q and its image $-Q'$.

We can now write down the potential at an arbitrary point $P(r, \theta)$ as in Figure 4-8:

$$4\pi\epsilon_0 V = \frac{Q}{r_3} - \frac{(a/D)Q}{r_4}, \tag{4-33}$$

where

$$r_3 = \sqrt{D^2 + r^2 + 2Dr\cos\theta}, \tag{4-34}$$

$$r_4 = \sqrt{\left(\frac{a^2}{D}\right)^2 + r^2 + 2\frac{a^2}{D}r\cos\theta}. \tag{4-35}$$

Then, to find $\mathbf{E}$, we merely have to calculate $-\nabla V$.

As in the case of the point charge and the plane, we calculate the induced charge density σ' from the value of the electric field intensity on the surface of the sphere. We calculate E_r and evaluate it at $r = a$:

$$4\pi\epsilon_0 E_r = \frac{Q(r + D\cos\theta)}{r_3^3} - \frac{aQ\left(r + \frac{a^2}{D}\cos\theta\right)}{Dr_4^3}, \tag{4-36}$$

and, at $r = a$,

$$\sigma' = \epsilon_0 E_r = -\frac{Q(D^2 - a^2)}{4\pi a(D^2 + a^2 + 2Da\cos\theta)^{3/2}}. \tag{4-37}$$

If we integrate this density over the surface of the sphere, we get the total induced charge:

$$Q' = \int_0^\pi \sigma' 2\pi a^2 \sin\theta\, d\theta = -\frac{a}{D}Q. \tag{4-38}$$

The total induced charge on the real conducting sphere is thus the same as the image charge that replaced the sphere. This must be true because of Gauss's law: if we draw a Gaussian surface just outside the sphere, then the flux of $\mathbf{E}$ through this surface, and hence the enclosed charge, must be the same, no matter whether the conducting sphere is present or whether it is replaced by the image

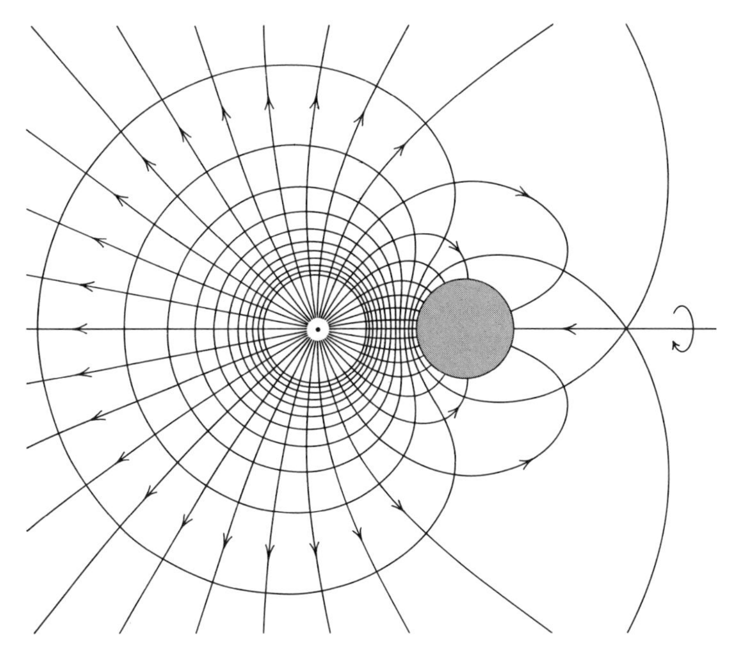

Figure 4-9. Lines of force (identified by arrows) and equipotentials for a point charge near a conducting sphere. Equipotential surfaces are generated by rotating the figure about the axis identified by the curved arrow. Equipotentials and lines of force in the vicinity of the point charge are again not shown because they get too close together.

charge, since the fields outside the sphere are identical in either case.

Figure 4-9 shows lines of force and equipotentials for a point charge in the vicinity of a grounded conducting sphere.

Example

POINT CHARGE NEAR A CHARGED CONDUCTING SPHERE

If the sphere is at a potential other than zero, we may still determine the field by the method of images. We first replace the conducting sphere with an image charge Q', as we did for a grounded sphere. This makes the surface occupied by the sphere an equipotential. We next add a second image charge at the center to raise the spherical surface to the required potential.

If we are given a sphere of radius a with a charge Q_s on it, and if its center is at a distance $D > a$ from a point charge Q, we can replace the sphere by an image charge $Q' = -(a/D)Q$ at a distance $b = a^2/D$ from the center, plus a charge $(Q_s - Q')$ at the center. The surface density of charge is then $\sigma' + \sigma''$, where σ' is the nonuniform distribution calculated from Q and Q', and where σ'' is the uniform distribution calculated from $(Q_s - Q')$.

Example

CHARGED SPHERE NEAR A GROUNDED CONDUCTING PLANE

Some fields may be calculated by the method of images through successive approximations. As an illustration we shall calculate the capacitance of a charged sphere near a grounded conducting plane. We shall replace both the sphere and the plane by a set of point charges that will maintain these surfaces as equipotentials.

First, we put a charge Q_1 at the center of the sphere, as in Figure 4-10b. This makes the sphere, but not the plane, an equipotential. Next we put the image $-Q_1$ of Q_1 to the right of the plane. This makes the plane an equipotential but destroys the spherical equipotential, so we put the image Q_2 of $-Q_1$ inside the sphere. This makes the sphere again an equipotential but upsets the plane. We continue the process, which converges rapidly, until we have the

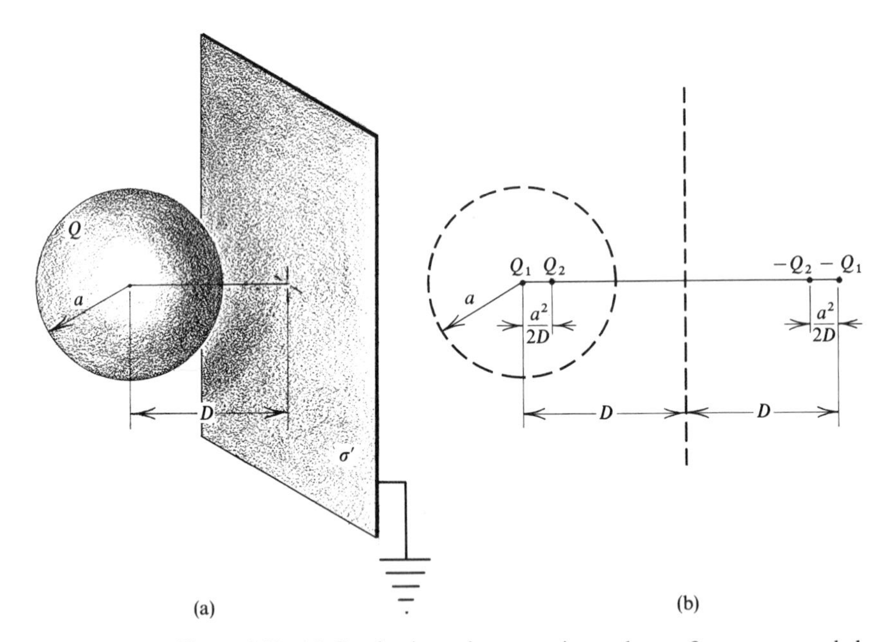

Figure 4-10. (a) Conducting sphere carrying a charge Q near a grounded conducting plane. When Q is positive, the induced surface charge density σ' is negative. The plane is assumed to be infinite. (b) The field outside the sphere and to the left of the conducting plane is calculated by successive approximations by using the image charges Q_1, Q_2, . . . , etc., and $-Q_1$, $-Q_2$, . . . , etc.

required precision. The charges and their locations are shown in Table 4-1.

If we set $(a/2D) = r$, then

$$
\begin{aligned}
Q_1 &= Q_1,\\
Q_2 &= rQ_1,\\
Q_3 &= \frac{r^2}{1 - r^2} Q_1,\\
Q_4 &= \frac{r^3}{(1 - r^2)\left(1 - \dfrac{r^2}{1 - r^2}\right)} Q_1, \qquad (4\text{-}39)\\
Q_5 &= \frac{r^4}{(1 - r^2)\left(1 - \dfrac{r^2}{1 - r^2}\right)\left[1 - \dfrac{r^2}{1 - [r^2/(1 - r^2)]}\right]} Q_1,
\end{aligned}
$$

and so on. The equipotentials for $D = 3a$ are shown in Figure 4-11.

The total charge on the sphere is

$$
Q = Q_1\left(1 + r + \frac{r^2}{1 - r^2} + \cdots\right), \qquad (4\text{-}40)
$$

but only Q_1 contributes to its potential: the charges $-Q_1$ and Q_2 make the potential of the sphere zero, and the same is true of all the following pairs of charges. The potential of the sphere is therefore

$$
V = \frac{Q_1}{4\pi\epsilon_0 a}, \qquad (4\text{-}41)
$$

and the capacitance between the sphere and the plane is

$$
C = \frac{Q}{V} = \frac{Q_1(1 + r + \cdots)}{Q_1/4\pi\epsilon_0 a} = 4\pi\epsilon_0 a(1 + r + \cdots). \qquad (4\text{-}42)
$$

Table 4-1. *Images for the Case of a Charged Sphere Near a Grounded Conducting Plane*

Left of Plane		Right of Plane	
Charge	Distance from Center of Sphere	Charge	Distance from Center of Sphere
Q_1	0		
$Q_2 = (a/2D)Q_1$	$a^2/2D$	$-Q_1$	$2D$
$Q_3 = \dfrac{a}{\left(2D - \dfrac{a^2}{2D}\right)} \dfrac{a}{2D} Q_1$	$\dfrac{a^2}{2D - (a^2/2D)}$	$-Q_2$	$2D - (a^2/2D)$
$= \dfrac{(a/2D)^2}{1 - (a/2D)^2} Q_1$	$= \dfrac{a^2/2D}{1 - (a/2D)^2}$		

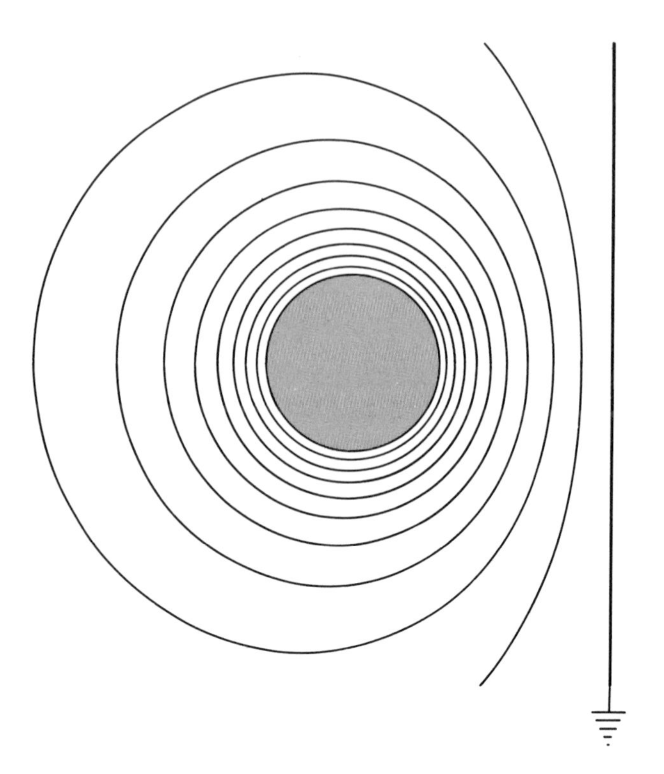

Figure 4-11. Equipotentials for a charged sphere near a grounded conducting plane.

The presence of the plane increases the capacitance of the sphere, as must be expected.

Figure 4-12 shows the ratio C/a as a function of the ratio r. For $r = 0$ the sphere is infinitely distant from the grounded plate, and C is simply the capacitance of an isolated sphere, $4\pi\epsilon_0 a$ farad. For $r \to 1/2$, the sphere comes infinitely close to the conducting plane, and the capacitance tends to infinity.

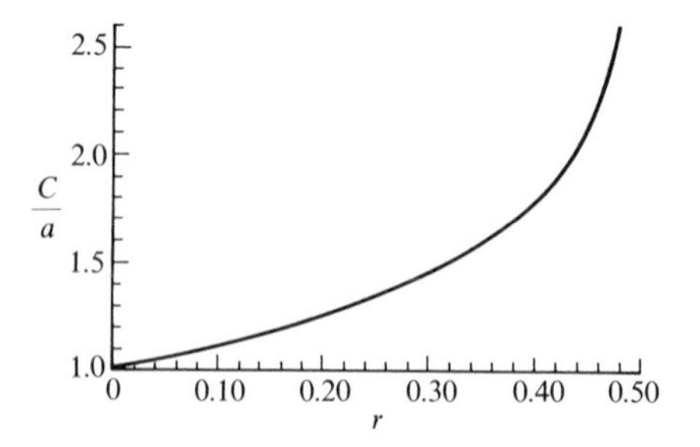

Figure 4-12. The ratio C/a as a function of r.

Example

CHARGE NEAR A SEMI-INFINITE DIELECTRIC

The method of images can also be used to calculate fields that involve dielectrics. As an example, consider a point charge Q at a distance D from a semi-infinite block of Class A dielectric, as in Figure 4-13. The electric field of the charge Q polarizes the dielectric, and a surface charge density σ_b appears on the surface. The resultant electric field at any point can then be computed by replacing the dielectric with this surface charge and treating the problem as if these charges, together with Q, were in free space. The value of σ_b is, of course, a function of the distance s.

We have two different fields, that due to Q and that due to σ_b. Let us consider only the normal components. That part of the normal component of the electric field intensity at the dielectric surface arising from Q is

$$\frac{Q}{4\pi\epsilon_0}\frac{D}{(s^2 + D^2)^{3/2}}.$$

This field is the same, both in magnitude and in direction, just inside and just outside the dielectric surface, as in Figure 4-13.

From Gauss's law, the normal component of the field of σ_b is $\sigma_b/2\epsilon_0$. This field is directed away from the boundary if σ_b is positive (Q negative), and into the boundary if σ_b is negative (Q positive).

Therefore, just inside the dielectric surface, the normal component of the electric field intensity is

$$E_{ni} = -\frac{Q}{4\pi\epsilon_0}\frac{D}{(s^2 + D^2)^{3/2}} - \frac{\sigma_b}{2\epsilon_0}. \tag{4-43}$$

This quantity is positive if $\boldsymbol{E}_{ni}$ points outward.

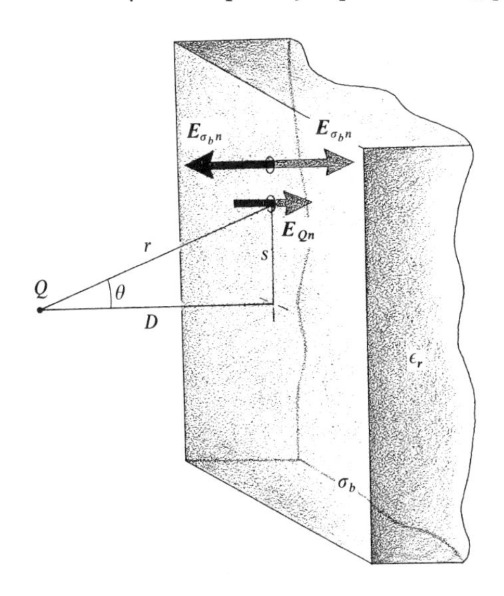

Figure 4-13. Point charge Q near the plane surface of a large block of dielectric. The quantity $\boldsymbol{E}_{\sigma bn}$ is the normal component of $\boldsymbol{E}$ due to the induced surface charge density σ_b; E_{Qn} is the normal component of $\boldsymbol{E}$ due to the charge Q. So as to simplify our calculation, the arrows for $\boldsymbol{E}_{Qn}$ and for $\boldsymbol{E}_{\sigma bn}$ are oriented on the assumption that Q and σ_b are positive. We find, as expected, that if Q is positive σ_b is negative and that the fields $\boldsymbol{E}_{\sigma bn}$ point into the boundary.

Now, from Eqs. 3-8 and 3-51,

$$\sigma_b = \boldsymbol{P} \cdot \boldsymbol{n} = P_n = \epsilon_0(\epsilon_r - 1)E_{ni}, \tag{4-44}$$

where the unit vector $\boldsymbol{n}$ is normal to the surface and points outward, and

$$\sigma_b = -\epsilon_0(\epsilon_r - 1)\left[\frac{Q}{4\pi\epsilon_0}\frac{D}{(s^2 + D^2)^{3/2}} + \frac{\sigma_b}{2\epsilon_0}\right] \tag{4-45}$$

$$= -\frac{(\epsilon_r - 1)QD}{2\pi(\epsilon_r + 1)(s^2 + D^2)^{3/2}}. \tag{4-46}$$

The induced charge density σ_b and Q have opposite signs, as expected.

At this stage we could calculate the electric potential and field intensity at any point, either within the dielectric or in free space, by using Coulomb's law, integrating over the σ_b distribution, and adding the contribution from the point charge Q. But this is not the simplest way to deal with this field. We shall find instead a set of image charges that will satisfy the boundary conditions.

To find these charges, we confine our attention to the boundary and write down the normal components of the resultant electric field intensity for a point just inside the dielectric, E_{ni}, and just outside, E_{no}. From Eqs. 4-43 and 4-46,

$$E_{ni} = -\left[1 - \left(\frac{\epsilon_r - 1}{\epsilon_r + 1}\right)\right]\frac{QD}{4\pi\epsilon_0(s^2 + D^2)^{3/2}} \tag{4-47}$$

$$= -\left(\frac{2}{\epsilon_r + 1}\right)\frac{QD}{4\pi\epsilon_0(s^2 + D^2)^{3/2}}. \tag{4-48}$$

For the outside point,

$$E_{no} = -\left[1 + \left(\frac{\epsilon_r - 1}{\epsilon_r + 1}\right)\right]\frac{QD}{4\pi\epsilon_0(s^2 + D^2)^{3/2}}, \tag{4-49}$$

$$= -\left(\frac{2\epsilon_r}{\epsilon_r + 1}\right)\frac{QD}{4\pi\epsilon_0(s^2 + D^2)^{3/2}}. \tag{4-50}$$

It will be observed that the normal component of $\boldsymbol{D}$ is continuous across the boundary:

$$\epsilon_r E_{ni} = E_{no}. \tag{4-51}$$

This is because there is zero *free* charge density at the surface of the dielectric.

Can we now find a set of image charges that will give these normal field components? We recall that an image charge is always *outside* the region in which the field is to be determined, and that both $\boldsymbol{E}_{ni}$ and $\boldsymbol{E}_{no}$ point outward from the dielectric. We consider first a point just outside the dielectric. We can see from Eq. 4-49 that E_{no} is the same as if we replaced the dielectric by the image charge

$$Q' = -\frac{(\epsilon_r - 1)}{(\epsilon_r + 1)}Q, \tag{4-52}$$

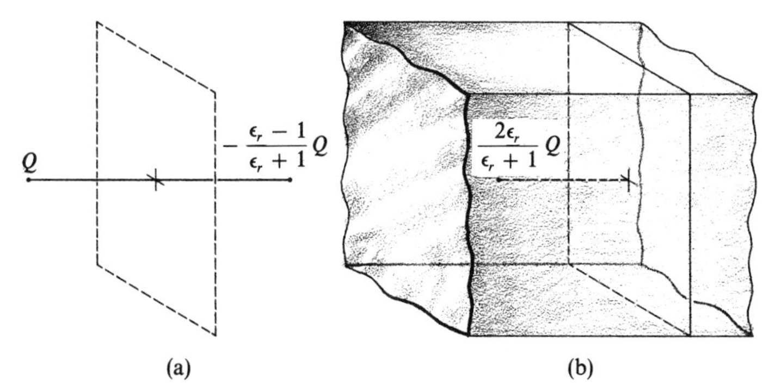

Figure 4-14. (a) If the dielectric is replaced by the image charge $-(\epsilon_r - 1)Q/(\epsilon_r + 1)$, the field is unaffected *outside* the dielectric. (b) If the dielectric is extended to both sides and $2\epsilon_r Q/(\epsilon_r + 1)$ is substituted for Q, the field is unaffected *inside* the dielectric.

located at a distance D behind the boundary, as in Figure 4-14a.

For a point inside the dielectric, there are two sets of point charges that will give the proper normal component of E: (a) the point charge Q together with an image charge

$$Q' = +\frac{(\epsilon_r - 1)}{(\epsilon_r + 1)} Q \tag{4-53}$$

at the image position inside the dielectric and (b) a single charge

$$Q'' = +\frac{2\epsilon_r}{\epsilon_r + 1} Q, \tag{4-54}$$

that replaces Q, the dielectric extending in this case on both sides of the boundary. The first of these must be ruled out because we always want the image charge to be outside the region in which the field is required. The image charge must therefore be located as in Figure 4-14b.

What we have done, then, is the following. In order to find the field outside the dielectric we replaced the dielectric by an image charge Q' located at a distance D behind the boundary, as in Figure 4-14a. In order to find the field inside the dielectric, we replaced Q by a single charge Q'' at the position of Q, and we extended the dielectric to both sides of the boundary, as in Figure 4-14b. Since these combinations of charges satisfy the boundary conditions, we know from the uniqueness theorem that they provide the correct field. The shape of the field is shown in Figure 4-15.

In general, for two media having relative permittivities ϵ_{r1} and ϵ_{r2}, with the point charge Q in medium 1, the point charges Q' and Q'' are the following. The field in medium 1 is the same as if we had

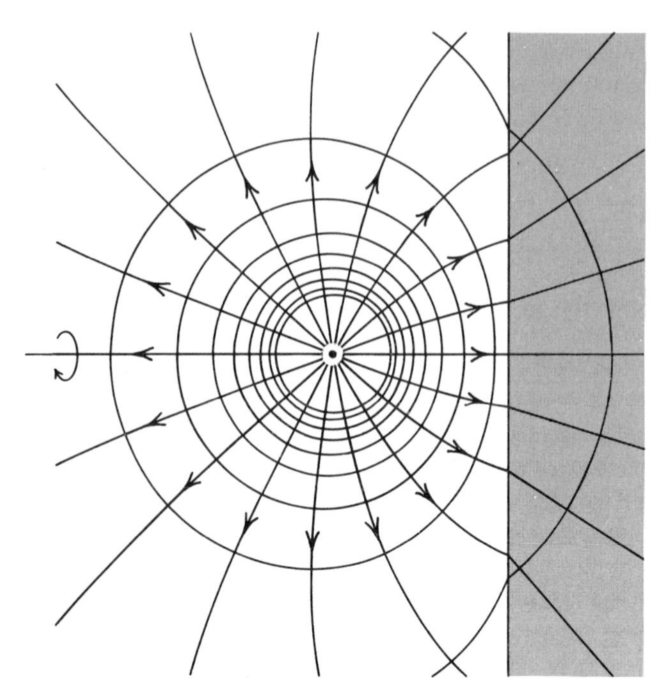

Figure 4-15. Lines of $\boldsymbol{D}$ (identified by arrows) and equipotentials for a point charge near a dielectric. As previously, equipotential surfaces are generated by rotating the figure about the axis indicated by the curved arrow. Equipotentials and lines of $\boldsymbol{D}$ near the point charge are not shown.

the same medium 1 on both sides of the interface, the original charge Q, and a charge

$$Q' = \frac{\epsilon_{r1} - \epsilon_{r2}}{\epsilon_{r1} + \epsilon_{r2}} Q \tag{4-55}$$

at the image position. The field in medium 2 is the same as if medium 2 extended on both sides of the interface and the original charge Q were replaced by

$$Q'' = \frac{2\epsilon_{r2}}{\epsilon_{r1} + \epsilon_{r2}} Q \tag{4-56}$$

4.4 SOLUTION OF LAPLACE'S EQUATION IN RECTANGULAR COORDINATES

The methods that we have considered until now for the calculation of electric fields are useful only in special cases. We must find more general methods for solving Poisson's equation,

$$\nabla^2 V = -\frac{\rho}{\epsilon}. \tag{4-57}$$

To begin with, we shall confine our attention to electric fields with zero space charge density, and we shall deal with Laplace's equation,

$$\nabla^2 V = 0. \tag{4-58}$$

Solutions of Laplace's equation are known as *harmonic functions*. These functions have a number of general properties, of which we shall use the following one. If the functions $V_1, V_2, V_3, \cdots$ are solutions, then any linear combination $A_1V_1 + A_2V_2 + A_3V_3 + \cdots$ of these functions, where the A's are arbitrary constants, is also a solution. This can be readily demonstrated by substitution into the original equation.

It is usually possible to find solutions of Laplace's equation that will satisfy required boundary conditions by separating the variables. For example, in Cartesian coordinates, we can usually find a solution of the form

$$V = X(x)Y(y)Z(z), \tag{4-59}$$

where $X(x)$, $Y(y)$, $Z(z)$ are respectively functions only of x, y, z. We can then fit boundary conditions by adding a series of such solutions multiplied by suitable coefficients. The uniqueness theorem assures us that the solution found in this way is correct.

We can find the form of the functions $X(x)$, $Y(y)$, $Z(z)$ by substituting $V = X(x)Y(y)Z(z)$ into Laplace's equation. Then

$$YZ\frac{d^2X}{dx^2} + ZX\frac{d^2Y}{dy^2} + XY\frac{d^2Z}{dz^2} = 0, \tag{4-60}$$

where we have written total instead of partial derivatives, since each one of the X, Y, Z functions is a function of a single variable. Dividing through by XYZ,

$$\frac{1}{X}\frac{d^2X}{dx^2} + \frac{1}{Y}\frac{d^2Y}{dy^2} + \frac{1}{Z}\frac{d^2Z}{dz^2} = 0. \tag{4-61}$$

Now since the second and third terms are independent of x, and since the three terms must add to zero at all points, the first term must also be independent of x. It is therefore constant in value and

$$\frac{1}{X}\frac{d^2X}{dx^2} = C_1. \tag{4-62}$$

Similarly,

$$\frac{1}{Y}\frac{d^2Y}{dy^2} = C_2, \tag{4-63}$$

$$\frac{1}{Z}\frac{d^2Z}{dz^2} = C_3, \tag{4-64}$$

and

$$C_1 + C_2 + C_3 = 0.$$

The problem then is to solve three *ordinary* differential equations, subject to this condition and to the boundary conditions.

Example

FIELD BETWEEN TWO GROUNDED SEMI-INFINITE PARALLEL ELECTRODES TERMINATED BY A PLANE ELECTRODE AT A POTENTIAL V_0

Figure 4-16 shows two grounded, semi-infinite, parallel electrodes separated by a distance b. At $x = 0$ an electrode is maintained at a potential V_0. The problem is to find the potential V at any point between the plates.

By hypothesis, the potential is independent of z, and the constant C_3 is zero. We must therefore solve two ordinary differential equations:

$$\frac{d^2X}{dx^2} - k^2X = 0, \tag{4-66}$$

$$\frac{d^2Y}{dy^2} + k^2Y = 0, \tag{4-67}$$

We have substituted k^2 for C_1 and $-k^2$ for C_2 so as to eliminate square roots in the solution. The choice between C_1 and C_2 as the negative constant is immaterial; the boundary conditions will force us to the same final solution in either case.

We solve Eq. 4-67 by setting

$$Y = A \sin ky + B \cos ky, \tag{4-68}$$

where A and B are arbitrary constants. This can be easily verified by substitution. Our value of V must satisfy the boundary conditions

$$V = 0 \qquad (y = 0, y = b), \tag{4-69}$$

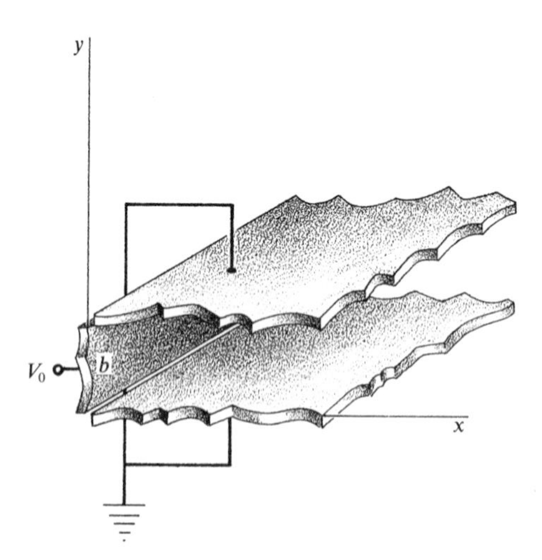

Figure 4-16. Grounded, plane-parallel electrodes terminated by a plane electrode at potential V_0. The electrodes are assumed to be infinite in the direction perpendicular to the paper and are assumed to extend infinitely on the right.

$$V = V_0 \qquad (x = 0), \tag{4-70}$$

$$V \to 0 \qquad (x \to \infty). \tag{4-71}$$

In order to have $V = 0$ at $y = 0$ we must have $B = 0$; and in order to have $V = 0$ at $y = b$ we must have

$$kb = n\pi \qquad (n = 1, 2, \cdots). \tag{4-72}$$

Thus

$$Y = A \sin \frac{n\pi y}{b} \qquad (n = 1, 2, \cdots). \tag{4-73}$$

The value $n = 0$ must be omitted, for it corresponds to zero field.

The X equation is now

$$\frac{d^2X}{dx^2} - \left(\frac{n\pi}{b}\right)^2 X = 0, \tag{4-74}$$

and

$$X = Ge^{n\pi x/b} + He^{-n\pi x/b}, \tag{4-75}$$

where G and H are arbitrary constants. We can again verify this solution by substitution. The condition that $V \to 0$ as $x \to \infty$ requires that $G = 0$.

Altogether then,

$$V'(x, y) = C \sin \frac{n\pi y}{b} e^{-n\pi x/b}, \tag{4-76}$$

where C is another arbitrary constant.

The solution as it is will obviously satisfy the boundary conditions stated in Eqs. 4-69 and 4-71. It will not, however, satisfy Eq. 4-70. We therefore use an infinite sum of such solutions and set

$$V(x, y) = \sum_{n=1}^{\infty} C_n \sin \frac{n\pi y}{b} e^{-n\pi x/b}. \tag{4-77}$$

To evaluate the coefficients C_n, we use the boundary condition at $x = 0$, namely

$$V(0, y) = V_0 = \sum_{n=1}^{\infty} C_n \sin \frac{n\pi y}{b}. \tag{4-78}$$

The expression on the right is called a *Fourier series*. It can be shown that, provided an infinite series of cosine terms is also included, it constitutes a *complete* set of functions. This means that *an arbitrary boundary condition can be satisfied with such an infinite series.*

Using a technique devised by Fourier, we multiply both sides by $\sin [(p\pi y)/b]$, where p is an integer, and integrate from $y = 0$ to $y = b$:

$$\int_0^b V_0 \sin \frac{p\pi y}{b}\, dy = \int_0^b \sum_{n=1}^{\infty} C_n \sin \frac{n\pi y}{b} \sin \frac{p\pi y}{b}\, dy. \tag{4-79}$$

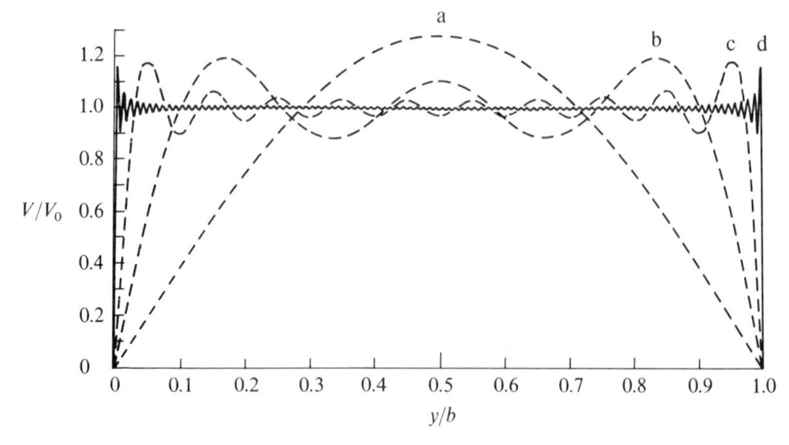

Figure 4-17. The condition $V = V_o$ as satisfied by a Fourier series taking (a) only the first term, (b) the first 3 terms, (c) the first 10 terms, and (d) the first 100 terms. The Fourier series provides an increasingly better approximation as the number of terms is increased.

On the left-hand side,

$$\int_0^b V_0 \sin \frac{p\pi y}{b}\, dy = \begin{cases} \dfrac{2bV_0}{p\pi} & \text{if } p \text{ is odd,} \\ 0 & \text{if } p \text{ is even,} \end{cases} \tag{4-80}$$

whereas, on the right-hand side,

$$\int_0^b C_n \sin \frac{n\pi y}{b} \sin \frac{p\pi y}{b}\, dy = \begin{cases} 0 & \text{if } p \neq n, \\ C_n \dfrac{b}{2} & \text{if } p = n. \end{cases} \tag{4-81}$$

Thus the only term of the infinite series on the right-hand side of Eq. 4-79 that differs from zero is the one for which $n = p$. A sequence of functions possessing this property is said to be *orthogonal.*

Combining Eqs. 4-80 and 4-81,

$$C_n = \begin{cases} \dfrac{4V_0}{n\pi} & \text{if } n \text{ is odd,} \\ 0 & \text{if } n \text{ is even.} \end{cases} \tag{4-82}$$

We can now write down the potential V at any point (x, y, z):

$$V(x, y, z) = \frac{4V_0}{\pi} \sum_{n=1,3,5,\cdots}^{\infty} \frac{1}{n} \sin \frac{n\pi y}{b}\, e^{-n\pi x/b}. \tag{4-83}$$

The successive terms in the series become progressively less important, both because of the factor $(1/n)$ and because of the exponential function. The degree of approximation that is achieved at $x = 0$ with one, three, ten, and one hundred terms of the series is indicated in Figure 4-17. At $y = b$ the first term alone gives a good approximation. The equipotentials are shown in Figure 4-18.

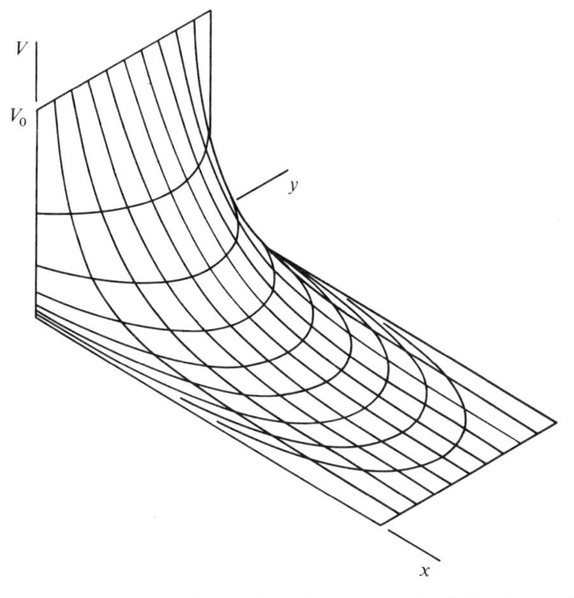

Figure 4-18. Three-dimensional plot of the potential V for the configuration of Figure 4-17. The U-shaped curves are equipotentials; the others show the intersections of the potential surface with planes parallel to the xV-plane.

Example

FIELD BETWEEN TWO GROUNDED PARALLEL ELECTRODES TERMINATED ON TWO OPPOSITE SIDES BY PLATES AT POTENTIALS V_1 AND V_2

In Figure 4-19 two grounded plane parallel electrodes of width a are separated by a distance b and extend to infinity in the other direction. At $x = 0$ a conducting surface is maintained at a potential V_1, and the plane at $x = a$ is occupied by a conductor maintained at a potential V_2. The problem is again to find the electric potential V at any point between the plates.

The field is again independent of z, and the constant C_3 is zero. Since the Y part of the solution is identical with that of the previous example, Eq. 4-73 is valid, as are Eqs. 4-74 and 4-75.

From this point on, the solution differs from that of the previous example, since the boundary conditions are different. Here we have

$$V = V_1 \quad \text{at} \quad x = 0, \tag{4-84}$$

$$V = V_2 \quad \text{at} \quad x = a. \tag{4-85}$$

The most general solution, and the one required to satisfy the boundary conditions, is

$$V(x, y, z) = \sum_{n=1}^{\infty} (A_n e^{-n\pi x/b} + B_n e^{n\pi x/b}) \sin \frac{n\pi y}{b}, \tag{4-86}$$

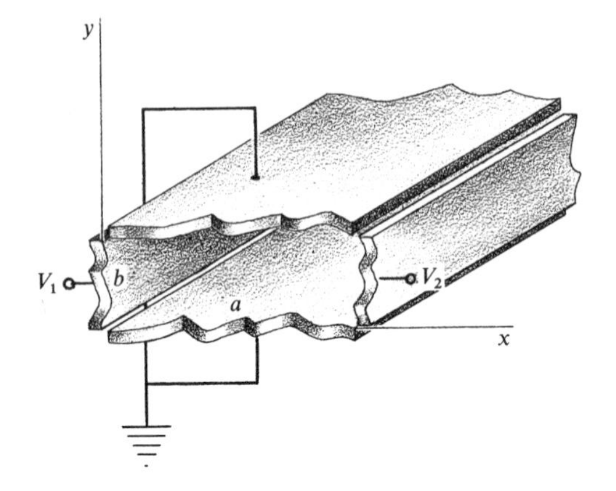

Figure 4-19. Grounded plane-parallel electrodes terminated on two sides with plane electrodes at potentials V_1 and V_2. The electrodes are assumed to be infinite in the direction perpendicular to the paper.

where A_n and B_n are again constants that must be determined from the boundary conditions.

At $x = 0$,

$$V_1 = \sum_{n=1}^{\infty} (A_n + B_n) \sin \frac{n\pi y}{b}. \tag{4-87}$$

The coefficients are evaluated by the same Fourier method used in the previous example. On multiplying by $\sin \frac{p\pi y}{b}$ and integrating from $y = 0$ to $y = b$, we have again, out of the whole infinite series, only one term corresponding to $p = n$:

$$V_1 \int_0^b \sin \frac{n\pi y}{b}\, dy = (A_n + B_n) \frac{b}{2}, \tag{4-88}$$

and

$$A_n + B_n = \begin{cases} \dfrac{4V_1}{n\pi} & \text{if } n \text{ is odd,} \\ 0 & \text{if } n \text{ is even.} \end{cases} \tag{4-89}$$

We can find another relationship between A_n and B_n from the boundary condition at $x = a$: from Eq. 4-86,

$$V_2 = \sum_{n=1}^{\infty} (A_n e^{-n\pi a/b} + B_n e^{n\pi a/b}) \sin \frac{n\pi y}{b}. \tag{4-90}$$

Multiplying by $\sin \frac{p\pi y}{b}$ and integrating from $y = 0$ to $y = b$, as before,

$$A_n e^{-n\pi a/b} + B_n e^{n\pi a/b} = \begin{cases} \dfrac{4V_2}{n\pi} & \text{if } n \text{ is odd,} \\ 0 & \text{if } n \text{ is even,} \end{cases} \tag{4-91}$$

and, from Eqs. 4-89 and 4-91,

$$A_n = \frac{4}{n\pi}\left(\frac{V_1 - V_2 e^{-n\pi a/b}}{1 - e^{-2n\pi a/b}}\right), \tag{4-92}$$

$$B_n = \frac{4e^{-n\pi a/b}}{n\pi}\left(\frac{V_2 - V_1 e^{-n\pi a/b}}{1 - e^{-2n\pi a/b}}\right), \tag{4-93}$$

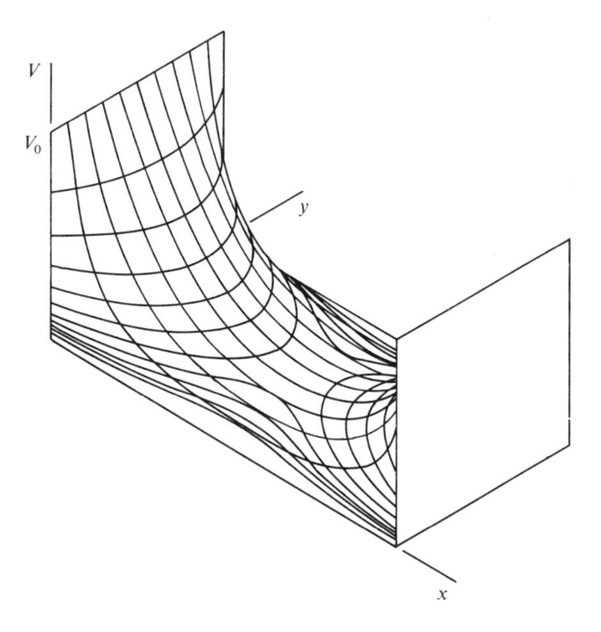

Figure 4-20. A three-dimensional plot of the potential V for the configuration of Figure 4-18 with $V_1 = V_2 = V_o$.

where $n = 1, 3, 5, \cdots$. The potential V at any point (x, y, z) is given by Eq. 4-86 with A_n and B_n as above.

Figure 4-20 shows the equipotentials for $V_1 = V_2 = V_0$.

4.5 SOLUTION OF LAPLACE'S EQUATION IN SPHERICAL COORDINATES. LEGENDRE'S EQUATION. LEGENDRE POLYNOMIALS

Although electrostatic fields can usually be calculated in Cartesian coordinates, certain cases are best treated in spherical polar coordinates. Laplace's equation then takes the form

$$\nabla^2 V = \frac{1}{r^2}\frac{\partial}{\partial r}\left(r^2 \frac{\partial V}{\partial r}\right) + \frac{1}{r^2 \sin\theta}\frac{\partial}{\partial \theta}\left(\sin\theta \frac{\partial V}{\partial \theta}\right) + \frac{1}{r^2 \sin^2\theta}\frac{\partial^2 V}{\partial \varphi^2} = 0. \quad (4\text{-}94)$$

The solutions of this equation are known as *spherical harmonic functions.*

We shall restrict ourselves here to fields with axial symmetry, that is, to fields where V is independent of the angle φ. Then

$$\frac{\partial}{\partial r}\left(r^2 \frac{\partial V}{\partial r}\right) + \frac{1}{\sin\theta}\frac{\partial}{\partial\theta}\left(\sin\theta \frac{\partial V}{\partial\theta}\right) = 0. \tag{4-95}$$

As in Cartesian coordinates, we seek solutions in which the variables are separated and set

$$V(r, \theta) = R(r)\,\Theta(\theta), \tag{4-96}$$

where R is a function of r only and Θ is a function of θ only. Substituting $V = R\Theta$ into Eq. 4-95,

$$\Theta \frac{\partial}{\partial r}\left(r^2 \frac{\partial R}{\partial r}\right) + \frac{R}{\sin\theta}\frac{\partial}{\partial\theta}\left(\sin\theta \frac{\partial\Theta}{\partial\theta}\right) = 0, \tag{4-97}$$

and dividing through by $R\Theta$,

$$\frac{1}{R}\frac{d}{dr}\left(r^2 \frac{dR}{dr}\right) + \frac{1}{\Theta\sin\theta}\frac{d}{d\theta}\left(\sin\theta \frac{d\Theta}{d\theta}\right) = 0. \tag{4-98}$$

We have written total instead of partial derivatives, since R and Θ are each functions of a single variable.

Since the second term is independent of r, the first term must also be independent of r. The first term must therefore be constant:

$$\frac{1}{R}\frac{d}{dr}\left(r^2 \frac{dR}{dr}\right) = k, \tag{4-99}$$

and then

$$\frac{1}{\Theta\sin\theta}\frac{d}{d\theta}\left(\sin\theta \frac{d\Theta}{d\theta}\right) = -k, \tag{4-100}$$

since the sum of the two constants must equal zero.

Let us examine the R equation first. Multiplying both sides by R and differentiating the term between parentheses, we obtain

$$r^2 \frac{d^2R}{dr^2} + 2r\frac{dR}{dr} - kR = 0. \tag{4-101}$$

The solution of this equation is of the form

$$R = Ar^n + \frac{B}{r^{n+1}}. \tag{4-102}$$

On substituting we find that

$$n(n+1) = k. \tag{4-103}$$

Let us now examine Eq. 4-100 for Θ:

$$\frac{d}{d\theta}\left(\sin\theta \frac{d\Theta}{d\theta}\right) + n(n+1)\sin\theta\,\Theta = 0 \tag{4-104}$$

It is convenient to change variables in this equation and to let

$$\mu = \cos\theta. \tag{4-105}$$

Now, for any function $f(\mu)$,

$$\frac{df}{d\theta} = \frac{df}{d\mu}\frac{d\mu}{d\theta} = -\sin\theta\,\frac{df}{d\mu} = -(1-\mu^2)^{1/2}\frac{df}{d\mu}, \tag{4-106}$$

and our equation for Θ becomes

$$\frac{d}{d\mu}\left[(1-\mu^2)\frac{d\Theta}{d\mu}\right] + n(n+1)\Theta = 0. \tag{4-107}$$

This is *Legendre's equation*. When n is an integer, its solutions are polynomials in $\cos\theta$ and are known as *Legendre polynomials*. They are designated by $P_n(\mu)$ or $P_n(\cos\theta)$:

$$\Theta = P_n(\mu) = P_n(\cos\theta), \tag{4-108}$$

where n is called the *degree* of the polynomial. There is a different polynomial for each value of n.

Before proceeding to find solutions of Eq. 4-107, we must point out an important property of Legendre's equation. The number n must satisfy Eq. 4-103, but

$$n' = -(n+1) \tag{4-109}$$

will equally satisfy this equation because

$$n'(n'+1) = n(n+1) = k. \tag{4-110}$$

That is, Eq. 4-107 remains unchanged when the number n' is substituted for n. Hence

$$P_{-(n+1)}(\cos\theta) = P_n(\cos\theta), \tag{4-111}$$

and, for every solution of Laplace's equation of the form

$$V = Ar^nP_n(\cos\theta), \tag{4-112}$$

there is another solution of the form

$$V = \frac{B}{r^{n+1}}P_{-(n+1)}(\cos\theta) = \frac{B}{r^{n+1}}P_n(\cos\theta). \tag{4-113}$$

This also follows from the R function of Eq. 4-102.

Let us now proceed to find the Legendre polynomials $P_n(\cos\theta)$ that are solutions of Legendre's equation. We know from our experience with point charges that

$$V_1 = \frac{C}{r} \tag{4-114}$$

is a solution of Laplace's equation, C being a constant. This is readily verified by substitution in Eq. 4-94. Since we are looking for solutions of the form indicated in Eqs. 4-112 or 4-113, it follows from the latter equation that

$$P_0'(\cos\theta) = 1. \tag{4-115}$$

We use a prime on the symbol P because the polynomials that we derive here differ from the Legendre polynomials by constant factors, as we shall see later. They are nevertheless solutions of Eq. 4-107. Substituting $\Theta = P_0'(\cos\theta) = 1$ and $n = 0$ into Eq. 4-107 does in fact solve it.

Having found $P_0'(\cos\theta)$, how can we find $P_1'(\cos\theta)$ and all the other polynomials corresponding to integral values of the index n in Eq. 4-107? We shall do this starting with V_1, but first we must know that any partial derivative of a solution of Laplace's equation with respect to any of the Cartesian coordinate variables is also a solution. This is easily demonstrated by substituting $\partial V/\partial x$ in Laplace's equation and remembering that the order of differentiation in partial derivatives is immaterial.

Let us therefore find the negative partial derivative of V_1 with respect to z:

$$-\frac{\partial}{\partial z}\frac{C}{r} = \frac{C}{r^2}\frac{\partial r}{\partial z} \tag{4-116}$$

where

$$\frac{\partial r}{\partial z} = \frac{\partial}{\partial z}(x^2 + y^2 + z^2)^{1/2} = \frac{z}{r} = \cos\theta. \tag{4-117}$$

We therefore have a new solution of Laplace's equation:

$$V_2 = C\frac{\cos\theta}{r^2}. \tag{4-118}$$

Comparing once again with Eq. 4-113, we see that

$$P_1'(\cos\theta) = \cos\theta. \tag{4-119}$$

Substitution of $P_1'(\cos\theta) = \cos\theta$ for Θ into Eq. 4-107 shows that it really is a solution when $n = 1$.

Equation 4-112 provides another possible solution for a given $P_n(\cos\theta)$. In addition to V_2, we have another solution:

$$V_2' = DrP_1'(\cos\theta) = Dr\cos\theta. \tag{4-120}$$

To find $P_2'(\cos\theta)$, we differentiate V_2 with respect to z:

$$V_3 = -\frac{\partial}{\partial z}\left(C\frac{\cos\theta}{r^2}\right) = -\frac{\partial}{\partial z}\left(C\frac{z}{r^3}\right) = C\frac{(3\cos^2\theta - 1)}{r^3}. \tag{4-121}$$

Comparing this with Eq. 4-113,

$$P_2'(\cos\theta) = (3\cos^2\theta - 1). \tag{4-122}$$

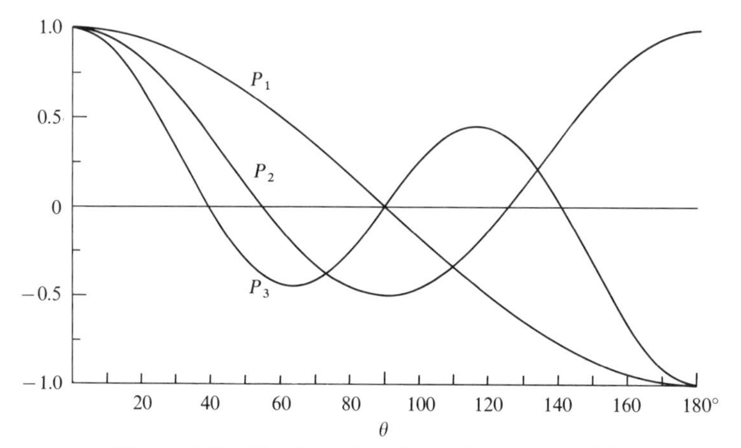

Figure 4-21. The first three Legendre polynomials.

Again there is another solution:

$$V_3' = Fr^2(3\cos^2\theta - 1), \tag{4-123}$$

which corresponds to Eq. 4-112. We shall stop here, but we could continue to find further polynomials in this way by repeated partial differentiations with respect to z.

It is convenient to multiply the above polynomials by normalizing factors to make them all equal to unity at $\cos\theta = 1$. For example, $P_2'(\cos\theta)$ must be multiplied by $\frac{1}{2}$ to make $P_2(\cos\theta) = 1$ when $\cos\theta = 1$. The *general form of the normalized Legendre polynomial* is

$$P_n(\cos\theta) = \frac{1}{2^n n!}\frac{\partial^n}{\partial(\cos\theta)^n}(\cos^2\theta - 1)^n. \tag{4-124}$$

The first six are shown in Table 4-2; those for $n = 1, 2, 3$ are plotted as functions of θ in Figure 4-21.

Table 4-2. *Legendre Polynomials*

n	$P_n(\cos\theta)$
0	1
1	$\cos\theta$
2	$\frac{3}{2}\cos^2\theta - \frac{1}{2}$
3	$\frac{5}{2}\cos^3\theta - \frac{3}{2}\cos\theta$
4	$\frac{35}{8}\cos^4\theta - \frac{15}{4}\cos^2\theta + \frac{3}{8}$
5	$\frac{63}{8}\cos^5\theta - \frac{35}{4}\cos^3\theta + \frac{15}{8}\cos\theta$

Table 4-3. *Solutions of Laplace's Equation in Spherical Polar Coordinates in the Case of Axial Symmetry*

n	$r^n P_n(\cos\theta)$	$r^{-(n+1)} P_n \cos\theta$
0	1	r^{-1}
1	$r\cos\theta$	$r^{-2}\cos\theta$
2	$\frac{1}{2}r^2(3\cos^2\theta - 1)$	$\frac{1}{2}r^{-3}(3\cos^2\theta - 1)$
3	$\frac{1}{2}r^3(5\cos^3\theta - 3\cos\theta)$	$\frac{1}{2}r^{-4}(5\cos^3\theta - 3\cos\theta)$
4	$\frac{1}{8}r^4(35\cos^4\theta - 30\cos^2\theta + 3)$	$\frac{1}{8}r^{-5}(35\cos^4\theta - 30\cos^2\theta + 3)$
5	$\frac{1}{8}r^5(63\cos^5\theta - 70\cos^3\theta + 15\cos\theta)$	$\frac{1}{8}r^{-6}(63\cos^5\theta - 70\cos^3\theta + 15\cos\theta)$

A general solution of Laplace's equation in spherical polar coordinates, assuming axial symmetry, is therefore

$$V = \sum_{n=0}^{\infty} A_n r^n P_n(\cos\theta) + \sum_{n=0}^{\infty} B_n r^{-(n+1)} P_n(\cos\theta). \qquad (4\text{-}125)$$

The lower order terms are given in Table 4-3.

It can be shown that the above functions form a *complete set of functions:* an arbitrary boundary condition with axial symmetry can be satisfied with such an infinite series. Moreover, any function of the polar angle θ can be represented as a series of Legendre polynomials, provided the function is continuous within the range of θ considered. Finally,

$$\int_{-1}^{+1} P_m(\cos\theta)\, P_n(\cos\theta)\, d(\cos\theta) = \begin{cases} 0 & \text{if } m \neq n, \\ \dfrac{2}{2n+1} & \text{if } m = n. \end{cases} \qquad (4\text{-}126)$$

This property of *orthogonality* of the Legendre polynomials is important in evaluating the coefficients A_n and B_n of Eq. 4-125.

Example

CONDUCTING SPHERE IN A UNIFORM ELECTRIC FIELD

Figure 4-22 shows an insulated conducting sphere situated in a uniform electric field $\mathbf{E}_o$. At any point, either inside or outside the sphere, the electric field intensity is $\mathbf{E}_o$ plus that due to the induced charges. We assume that the charges that produce $\mathbf{E}_o$ are so far away that they are unaffected by the presence of the sphere. This is reasonable because, as we shall see, the field is distorted only in the immediate neighborhood of the sphere. The induced charges arrange themselves on the conducting sphere so that the total field is zero *inside*. We shall calculate the field outside the sphere by solving Laplace's equation in three different ways.

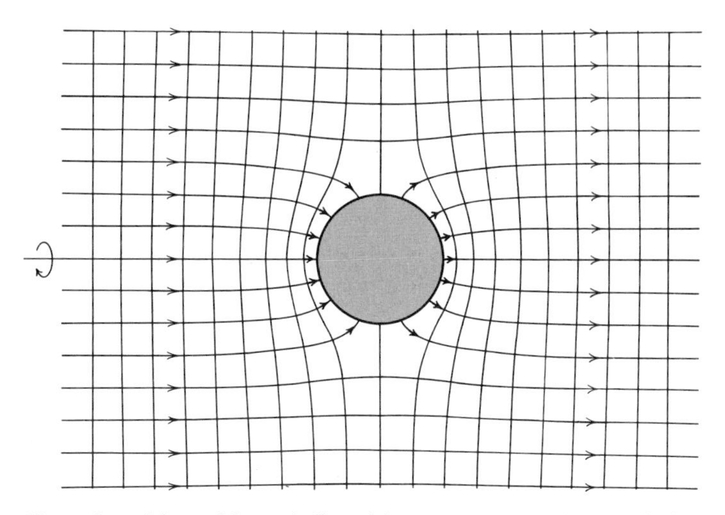

Figure 4-22. Lines of force (indicated by arrows) and equipotentials for a *conducting sphere* in a uniform electric field. The lines of force are normal at the surface of the sphere, and there is zero electric field intensity inside. Observe that the field is hardly disturbed at distances larger than one *radius* from the surface of the sphere. The origin is at the center of the sphere and the polar axis used in the calculation points to the right.

(a) The field is best described in terms of spherical polar coordinates, with the origin at the center of the sphere and the polar axis along $\mathbf{E}_o$. Our boundary conditions are then

$$V = 0 \qquad (r = a), \tag{4-127}$$

$$V = -E_o z = -E_o r \cos\theta \qquad (r = \infty). \tag{4-128}$$

At $r = a$, from Eqs. 4-125 and 4-127,

$$0 = \sum_{n=0}^{\infty} A_n a^n P_n(\cos\theta) + \sum_{n=0}^{\infty} B_n a^{-(n+1)} P_n(\cos\theta). \tag{4-129}$$

The method of evaluating the coefficients A_n and B_n is similar to that which we used for evaluating the C_n's of Eq. 4-78. We multiply both sides of the equation by $P_m(\cos\theta)$ and integrate from $\cos\theta = -1$ to $\cos\theta = +1$:

$$0 = \sum_{n=0}^{\infty} \int_{-1}^{+1} A_n a^n P_n(\cos\theta) P_m(\cos\theta)\, d(\cos\theta)$$

$$+ \sum_{n=0}^{\infty} \int_{-1}^{+1} B_n a^{-(n+1)} P_n(\cos\theta) P_m(\cos\theta)\, d(\cos\theta). \tag{4-130}$$

According to Eq. 4-126, the only nonvanishing terms are those for for which $n = m$ and

$$0 = A_n a^n \int_{-1}^{+1} P_n^2(\cos\theta)\, d(\cos\theta) + B_n a^{-(n+1)} \int_{-1}^{+1} P_n^2(\cos\theta)\, d(\cos\theta), \tag{4-131}$$

$$= A_n a^n \left(\frac{2}{2n+1}\right) + B_n a^{-(n+1)} \left(\frac{2}{2n+1}\right). \tag{4-132}$$

Thus

$$B_n = -A_n a^{2n+1}. \tag{4-133}$$

For $r \to \infty$ the potential V is $-E_o r \cos\theta$. All the terms involving inverse powers of r go to zero, and

$$-E_o r P_1(\cos\theta) = \sum_{n=0}^{\infty} A_n r^n P_n(\cos\theta). \tag{4-134}$$

Inspection of this equation shows that the only term that is not zero on the right-hand side is that for which $n = 1$. We can show this in a formal manner by multiplying both sides by $P_m(\cos\theta)$ and integrating from $\cos\theta = -1$ to $\cos\theta = +1$. By either method

$$A_1 = -E_o, \tag{4-135}$$

and all the other A_n's are zero. Then all the B's are also zero except B_1:

$$B_1 = -A_1 a^3 = E_o a^3. \tag{4-136}$$

Finally, at any point (r, θ),

$$V(r, \theta) = -E_o r \cos\theta + E_o \frac{a^3 \cos\theta}{r^2} = -E_o \left(1 - \frac{a^3}{r^3}\right) r \cos\theta \tag{4-137}$$

and

$$E_r = -\frac{\partial V}{\partial r} = E_o \left(1 + \frac{2a^3}{r^3}\right) \cos\theta, \tag{4-138}$$

$$E_\theta = -\frac{1}{r}\frac{\partial V}{\partial \theta} = -E_o \left(1 - \frac{a^3}{r^3}\right) \sin\theta. \tag{4-139}$$

The surface density of induced charge on the sphere is simply equal to ϵ_0 times E_r at $r = a$:

$$\sigma = 3\epsilon_0 E_o \cos\theta. \tag{4-140}$$

Returning now to Eq. 4-137 we observe that the first term is the potential corresponding to the uniform field $\boldsymbol{E}_o$. The second term has the form of the potential due to a dipole (Section 2.9). In fact, if we replace the sphere by a dipole of moment

$$p = 4\pi\epsilon_0 E_o a^3 \tag{4-141}$$

located at the center, the field outside the surface previously occupied by the sphere remains unchanged. We shall examine the image aspect of this field in a problem at the end of this chapter.

(b) We could also have determined the field quickly from Eq. 4-125 by a less formal method. We must have the term $-E_o r \cos\theta$ to fit the condition at infinity. No other function with positive powers

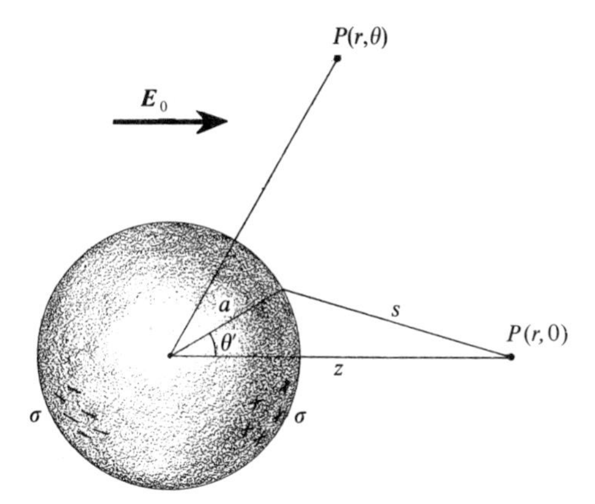

Figure 4-23. The electric potential V at $P(r, \theta)$ can be calculated from the applied electric field intensity $\boldsymbol{E}_o$ and the induced surface charge density σ. We first calculate V on the axis of symmetry at $P(r, 0)$ by a simple integration. This value is then used as a boundary condition to determine the coefficients of the Legendre polynomials in the series for V at any point $P(r, \theta)$.

of r can be included. This one term, however, is inadequate to fit the condition at $r = a$, where V must be independent of θ. We must therefore add another function which also includes the $\cos\theta$ factor in order that the coefficient of $\cos\theta$ can be zero at $r = a$. Then

$$V = -E_o r \cos\theta + \frac{B\cos\theta}{r^2}. \tag{4-142}$$

We finally set $B = E_o a^3$ to make $V = 0$ at $r = a$. Our solution satisfies both Laplace's equation and the boundary conditions; thus, it is the correct solution, according to the uniqueness theorem.

(c) There is still another method of calculating this same field that will add to our understanding and that will further illustrate the use of Legendre polynomials. Consider Figure 4-23. As indicated previously, the potential at any point (r, θ) arises from two charge distributions: (1) that which produces the electric field intensity $\boldsymbol{E}_o$ and which resides on electrodes situated far away, and (2) that which is induced on the surface of the sphere. This latter distribution is unknown, and we denote it by $\sigma(\theta')$. We use a prime on θ to distinguish it from the polar angle for a point (r, θ) outside the sphere. At the general point (r, θ) the total potential from these two sources must be of the form shown in Eq. 4-125.

Now it is possible to compute from Coulomb's law the potential at a point P on the axis $\theta = 0$ at a distance $r = z$ from the center of the sphere:

$$V = -E_o z + \frac{1}{4\pi\epsilon_0}\int_0^{\pi} \frac{\sigma(\theta')2\pi a^2 \sin\theta'\, d\theta'}{s}, \tag{4-143}$$

where a is the radius of the sphere and s is the distance from a point on the sphere to P as in Figure 4-23:

$$s = \sqrt{z^2 + a^2 - 2az\cos\theta'} = z\sqrt{1 + \frac{a^2}{z^2} - \frac{2a}{z}\cos\theta'}. \quad (4\text{-}144)$$

Expanding $1/s$ and grouping terms involving the same power of (a/z),

$$\frac{1}{s} = \frac{1}{z}\left[1 + \frac{a}{z}\cos\theta' + \left(\frac{3}{2}\cos^2\theta' - \frac{1}{2}\right)\frac{a^2}{z^2} + \left(\frac{5}{2}\cos^3\theta' - \frac{3}{2}\cos\theta'\right)\frac{a^3}{z^3} + \cdots\right], \quad (4\text{-}145)$$

$$= \frac{1}{z} + \frac{a}{z^2}P_1(\cos\theta') + \frac{a^2}{z^3}P_2(\cos\theta') + \cdots, \quad (4\text{-}146)$$

as long as $z > a$.

We have already seen that any continuous function of the polar angle θ can be expanded as a series of Legendre polynomials. Thus

$$\sigma(\theta') = b_0 + b_1P_1(\cos\theta') + b_2P_2(\cos\theta') + \cdots, \quad (4\text{-}147)$$

where b_0, b_1, $\cdots$ are constants. Equation 4-143 then becomes

$$V = -E_oz + \frac{a^2}{2\epsilon_0}\int_{-1}^{+1}[b_0 + b_1P_1(\cos\theta') + b_2P_2(\cos\theta') + \cdots] \times\left[\frac{1}{z} + \frac{a}{z^2}P_1(\cos\theta') + \frac{a^2}{z^3}P_2(\cos\theta') + \cdots\right]d(\cos\theta'). \quad (4\text{-}148)$$

The orthogonality property of the Legendre polynomials makes this an easy integral to evaluate, and

$$V = -E_oz + \frac{a^2}{2\epsilon_0}\left(\frac{2b_0}{z} + \frac{2b_1}{3}\frac{a}{z^2} + \frac{2b_2}{5}\frac{a^2}{z^3} + \cdots\right). \quad (4\text{-}149)$$

Thus, when $\theta = 0$ and $r = z$, the general solution for V shown in Eq. 4-125 reduces to the above form, and we can match coefficients term by term to find V at *any* point (r, θ) outside the sphere. On doing this we find that all the A_n's are zero, except for

$$A_1 = -E_o, \quad (4\text{-}150)$$

and that

$$B_n = \frac{b_n}{2n+1}\frac{a^{n+2}}{\epsilon_0}. \quad (4\text{-}151)$$

To evaluate the b_n's, we use the fact that the potential is zero at $r = a$. Substituting the above coefficients into Eq. 4-125 and setting $r = a$,

$$0 = -E_oaP_1(\cos\theta) + \frac{b_0a}{\epsilon_0} + \frac{b_1a}{3\epsilon_0}P_1(\cos\theta) + \frac{b_2a}{5\epsilon_0}P_2(\cos\theta) + \cdots, \quad (4\text{-}152)$$

which must be true for all θ. Thus both the term b_0a/ϵ_0, which is independent of θ, and the coefficients of all the P_n's must be equal to zero:

$$b_0 = 0, \tag{4-153}$$

and

$$-E_o a + \frac{b_1 a}{3\epsilon_0} = 0, \tag{4-154}$$

or

$$b_1 = 3\epsilon_0 E_o. \tag{4-155}$$

All other b_n's are zero.

The potential V at any point (r, θ) is thus given by substituting into Eq. 4-125 $A_1 = -E_o$, as in Eq. 4-150, and

$$B_1 = E_o a^3, \tag{4-156}$$

as in Eqs. 4-151 and 4-155. The field is the same as that found previously.

The surface charge density $\sigma(\theta')$ on the conducting sphere can be obtained from Eq. 4-147 now that the b_n's are known: we find the value previously found in Eq. 4-140.

Example

DIELECTRIC SPHERE IN A UNIFORM ELECTRIC FIELD

We can calculate this field by either of the formal methods discussed above if we write a general solution like Eq. 4-125 for points outside the sphere, and write another solution with different coefficients for points inside the sphere. The coefficients must be chosen in such a way that the boundary conditions are satisfied:

$V \to -E_o r \cos\theta$ $\quad (r \to \infty)$;

V is continuous across the boundary $\quad (r = a)$;

the normal component of $\boldsymbol{D}$ is continuous $\quad (r = a)$.

Instead of following such a formal procedure, however, we shall write down a combination of spherical harmonics that will satisfy all the boundary conditions.

Outside the sphere, we must have $-E_o r \cos\theta$ as one of the terms in the solution to satisfy the condition at $r \to \infty$. Furthermore, this is the only harmonic with a positive power of r that we can permit, for otherwise the condition at $r \to \infty$ would be violated. As regards this condition, all the terms with inverse powers of r are acceptable.

Consider now the solution for points inside the dielectric sphere. No inverse powers at all are permissible here, since such terms would make the potential infinite at the center. This is clearly impossible, since the only charges in the system are those which produce the field E_o and those induced on the surface of the sphere, if we assume a Class A dielectric, with the result that no volume distribution of induced charge exists.

Writing V_o for the potential outside the sphere and V_i for that inside,

$$V_o = -E_o r \cos\theta + \sum_{n=0}^{\infty} B_n r^{-(n+1)} P_n(\cos\theta), \tag{4-157}$$

$$V_i = \sum_{n=0}^{\infty} C_n r^n P_n(\cos\theta). \tag{4-158}$$

We require that V_o be equal to V_i at $r = a$ and that

$$-\left(\frac{\partial V_o(r,\theta)}{\partial r}\right)_{r=a} = -\left(\epsilon_r \frac{\partial V_i(r,\theta)}{\partial r}\right)_{r=a}, \tag{4-159}$$

where ϵ_r is the relative permittivity of the sphere. These are the second and third boundary conditions discussed above. Therefore

$$-E_o a P_1(\cos\theta) + \frac{B_0}{a} + \frac{B_1 P_1(\cos\theta)}{a^2} + \frac{B_2 P_2(\cos\theta)}{a^3} + \cdots$$

$$= C_0 + C_1 a P_1(\cos\theta) + C_2 a^2 P_2(\cos\theta) + \cdots, \tag{4-160}$$

and

$$E_o P_1(\cos\theta) + \frac{B_0}{a^2} + \frac{2B_1 P_1(\cos\theta)}{a^3} + \frac{3B_2 P_2(\cos\theta)}{a^4} + \cdots$$

$$= -\epsilon_r C_1 P_1(\cos\theta) - 2\epsilon_r C_2 a P_2(\cos\theta) + \cdots. \tag{4-161}$$

In order that these two equations be true for all values of θ, the coefficient of each Legendre polynomial on the left must be equal to the coefficient of the same Legendre polynomial on the right. Thus, from Eq. 4-160,

$$\frac{B_0}{a} = C_0, \tag{4-162}$$

$$-E_o a + \frac{B_1}{a^2} = C_1 a, \tag{4-163}$$

$$\frac{B_2}{a^3} = C_2 a^2, \cdots, \tag{4-164}$$

and, from Eq. 4-161,

$$\frac{B_0}{a^2} = 0, \tag{4-165}$$

$$E_o + \frac{2B_1}{a^3} = -\epsilon_r C_1, \tag{4-166}$$

$$\frac{3B_2}{a^4} = -2\epsilon_r C_2 a. \tag{4-167}$$

These sets of equations lead to the following values for the coefficients:

$$B_0 = C_0 = 0, \tag{4-168}$$

$$B_1 = \left(\frac{\epsilon_r - 1}{\epsilon_r + 2}\right) E_o a^3, \tag{4-169}$$

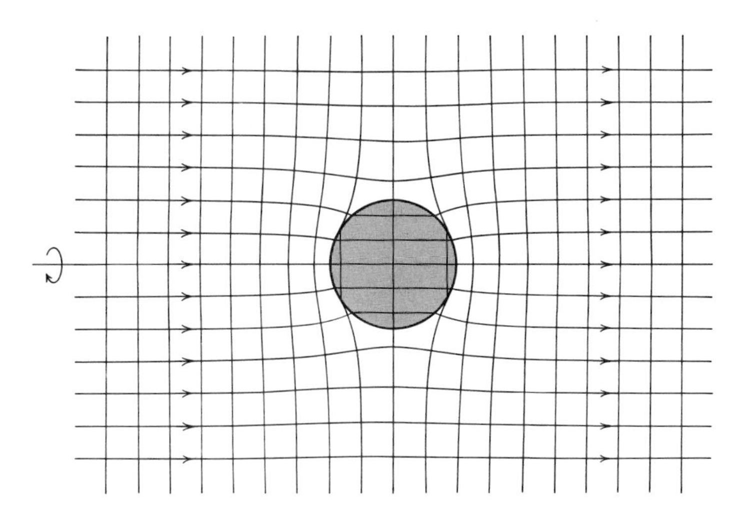

Figure 4-24. The field near a *dielectric sphere* ($\epsilon_r = 3$) in a uniform electric field. The lines of $\boldsymbol{D}$ (indicated by arrows) crowd into the sphere as shown, with the result that D is larger inside than outside. Since there is no free charge at the surface of the sphere, the lines of $\boldsymbol{D}$ neither originate nor terminate there, and they are continuous across the boundary. The equipotentials spread out inside, corresponding to a *lower* electric field intensity E. The electric field intensity E is discontinuous at the surface, and the density of lines of force is lower inside than outside. As in the conducting sphere, the field is hardly disturbed at distances larger than one radius from the surface. The field inside is uniform. The origin is chosen at the center of the sphere and the polar axis used in the calculation points to the right.

$$C_1 = -\frac{3E_o}{\epsilon_r + 2}, \tag{4-170}$$

$$B_n = C_n = 0 \qquad (n > 1). \tag{4-171}$$

Thus

$$V_o(r, \theta) = -\left[1 - \left(\frac{\epsilon_r - 1}{\epsilon_r + 2}\right)\frac{a^3}{r^3}\right] E_o r \cos\theta, \tag{4-172}$$

and

$$V_i(r, \theta) = -\left(\frac{3}{\epsilon_r + 2}\right) E_o r \cos\theta = -\left(\frac{3}{\epsilon_r + 2}\right) E_o z. \tag{4-173}$$

We may calculate the electric field intensity inside and outside the sphere by calculating $-\boldsymbol{\nabla} V$. It will be observed that the electric field inside the sphere is uniform, is along z, and that

$$E_i = \left(\frac{3}{\epsilon_r + 2}\right) E_o. \tag{4-174}$$

Lines of $\boldsymbol{D}$ and equipotentials are shown in Figure 4-24.

4.6 SOLUTION OF POISSON'S EQUATION FOR V

We have as yet dealt only with solutions of Laplace's equation, since we have concerned ourselves only with regions where the electric charge density $\rho_t = 0$. When $\rho_t \neq 0$ we must find a solution of Poisson's equation $\nabla^2 V = -\rho_t/\epsilon_0$ (Eq. 3-42) which is consistent with the given boundary conditions. We have already seen in Section 4.2 that such a solution is unique for a given ρ_t.

Example

p-n JUNCTION DIODE IN SILICON*

We shall use Poisson's equation to study the electrical properties of a silicon *p-n* junction diode in open circuit.

Semiconductors contain two types of fixed charges and two types of mobile charges. There are (a) fixed positive donor atoms, which lack an electron; (b) fixed negative acceptor atoms, which have an extra electron; (c) mobile conduction electrons; (d) mobile positive holes.

A hole is a vacancy left by an electron liberated from the valence bond structure in the material. A hole behaves as a free particle carrying a positive charge numerically equal to that of the electron, and it moves through the semiconductor much as an air bubble rises through water.

Donor and acceptor atoms are atoms of foreign substances that are added in minute amounts (approximately one part per million) to the basic semiconductor material, which is usually germanium or silicon. Since they are fixed in the crystal lattice, they constitute fixed charges. The addition of donor atoms adds conduction electrons without adding corresponding holes and yields *n*-type material, which conducts mostly by negatively charged electrons. Similarly, the addition of acceptor atoms gives *p*-type material, which conducts mostly by positively charged holes.

Let us call P and N the numbers of positive and negative fixed charges per cubic meter, and p and n the numbers of positive and negative mobile charges per cubic meter. Then the electric charge density is

$$\rho_t = e(P - N + p - n), \tag{4-175}$$

and

$$\nabla \cdot \mathbf{E} = -\nabla^2 V = \frac{e}{\epsilon}(P - N + p - n), \tag{4-176}$$

* See R. B. Adler, A. C. Smith, and R. L. Longini, *Introduction to Semiconductor Physics*, and Paul E. Gray, David DeWitt, A. R. Boothroyd, and James F. Gibbons, *Physical Electronics and Circuit Models of Transistors*, SEEC Books, Volumes 1 and 2 (John Wiley, New York, 1964) for a more detailed discussion of semiconductors.

where $\epsilon = \epsilon_r\epsilon_0$. The relative permittivity ϵ_r of germanium is 16, while that of silicon is 12.

Let us assume that the numbers P and N are graded uniformly inside a plane infinite junction as in Figures 4-25a, b, c. This one-dimensional approximation is valid for most solid-state diodes. Then the semiconductor is n-type to the left of the origin and p-type to the right.

The conduction electrons and the holes diffuse inside the material like the molecules of a gas; under thermal equilibrium, p and n satisfy the Boltzmann equation of statistical mechanics:

$$p = n_0 \exp(-eV/kT), \tag{4-177}$$

$$n = n_0 \exp(eV/kT), \tag{4-178}$$

where n_0 is the value of p or of n at the point where the potential V is chosen to be zero—let us say at the origin. It turns out that n_0 is the density of either conduction electrons or holes in the basic semiconductor material. The density p decreases exponentially with increasing V, while n increases exponentially with V (or decreases exponentially with $-V$).

Combining the last three equations and utilizing the fact that all quantities are independent of y and z by hypothesis, we find the differential equation for V:

$$\frac{d^2V}{dx^2} = -\frac{e}{\epsilon}[(P - N) - 2n_0 \sinh(eV/kT)], \tag{4-179}$$

$$= 2n_0\frac{e}{\epsilon}\sinh(eV/kT) - \left(\frac{2eP_0}{\epsilon t}\right)x, \tag{4-180}$$

where t is the thickness of the junction and P_0 is as in Figure 4-25a.

This equation does not appear to have an analytical solution; however, it has been solved numerically by using the relaxation method (see Problem 2-16).* Figure 4-25 shows the curves obtained for a linearly graded p-n junction in silicon at equilibrium and at room temperature.

Such p-n junctions are used as diodes: relatively large currents can flow when the p-type material is made positive with respect to the n-type, and negligible currents flow in the opposite direction. When the applied voltage makes the p-type material positive, positive holes from the p side and negative electrons from the n side are driven into the junction, where they combine. On the other hand, when the p-type material is made negative, the holes and the electrons are removed from the junction, and the current is smaller by a few orders of magnitude.

---

* D. P. Kennedy and R. R. O'Brien, On the Mathematical Theory of the Linearly-Graded *P-N* Junction, *IBM Journal of Research and Development* 11, no. 3 (May 1967), 252.

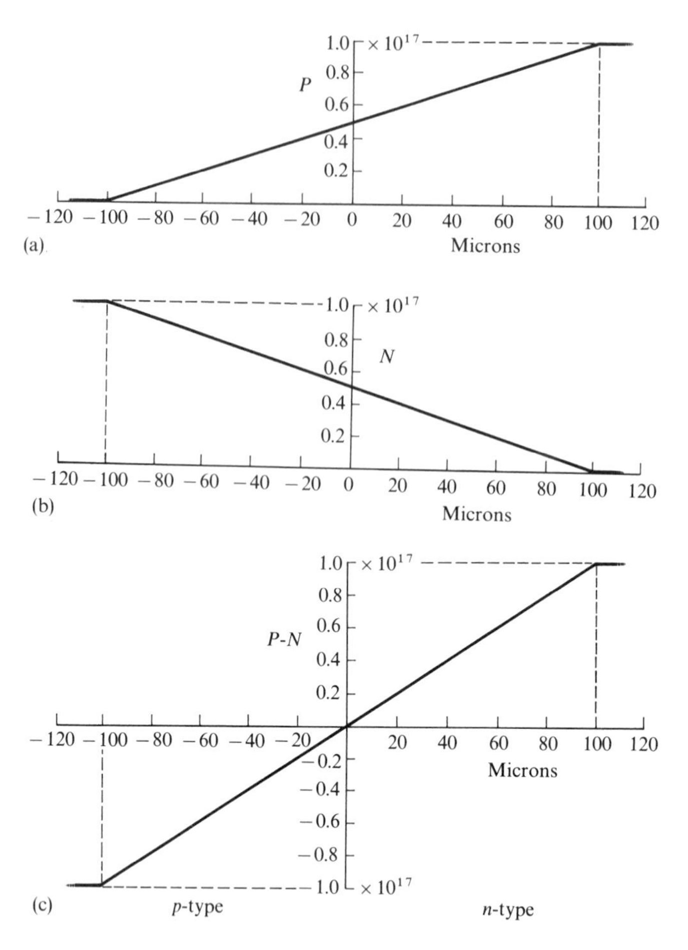

Figure 4-25. Curves for a plane *p-n* junction in silicon at 300 Kelvins. Density of donor and acceptor atoms on either side $N_0 = P_0 = 10^{17}$ meter^{-3}, thickness $t = 200$ microns, or 2×10^{-4} meter. Curves (a) and (b) show respectively the densities of donor and acceptor atoms as functions of x. The x-axis is perpendicular to the junction and the origin is at the center of the junction. Curve (c) shows the net density of *fixed* charges. Remember that donor atoms give fixed positive charges and mobile electrons, while acceptor atoms give fixed negative charges and mobile positive holes. Curve (d) shows the potential V as a function of x. There exists a permanent *contact potential* between the two sides of the junction. Note that V varies roughly like $P - N$. The potential is positive on the side that has a positive net fixed charge (n-type material) and negative on the other side. This is because thermal agitation forces the free electrons from the n side to diffuse into the p side, while it forces holes from the p side to diffuse in the other direction. The electric field intensity $E = -dV/dx$ is shown in the next curve, (e). Since it is negative, it points away from the n-type material, which contains a majority of fixed positive charges. This electric field limits the diffusion of electrons and holes across the junction. The net charge density ρ, curve (f), is proportional to minus the second derivative of V. It has the same sign as V and as the fixed charge density $P - N$. Finally, curves (g) and (h) show the numbers n and p of mobile negative and positive charges. Curve (f) for the charge density ρ is related to curves (c), (g), (h):

$$\rho = e(P - N + p - n).$$

The product np is everywhere equal to n_0^2.

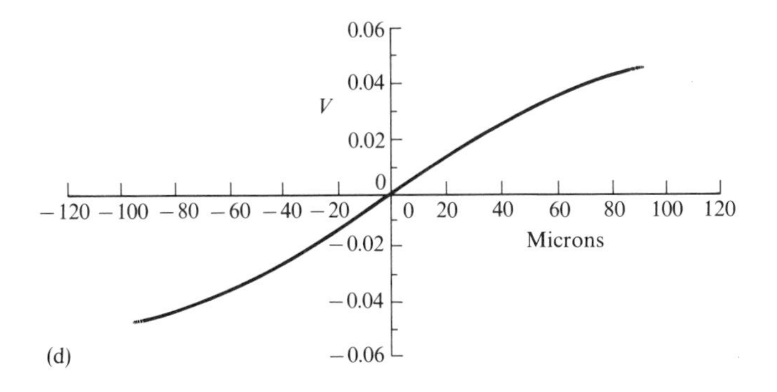
V
0.06
0.04
0.02
0
−0.02
−0.04
−0.06
−120 −100 −80 −60 −40 −20 0 20 40 60 80 100 120
Microns
(d)

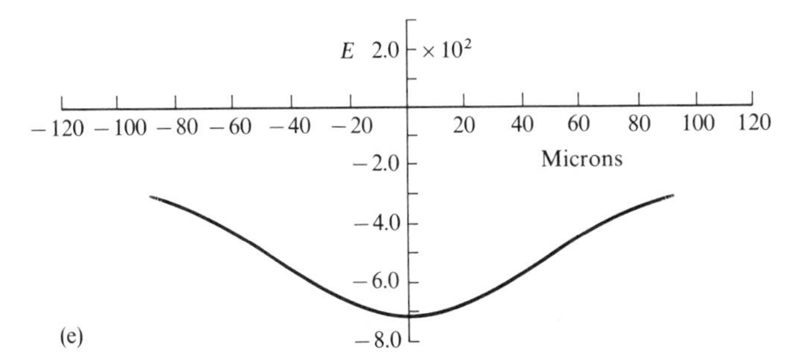
E 2.0 × 10²
−2.0
−4.0
−6.0
−8.0
−120 −100 −80 −60 −40 −20 20 40 60 80 100 120
Microns
(e)

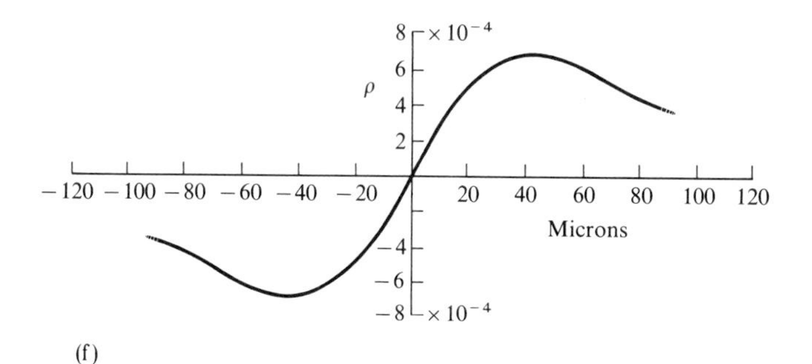
ρ
8 × 10⁻⁴
6
4
2
−4
−6
−8 × 10⁻⁴
−120 −100 −80 −60 −40 −20 20 40 60 80 100 120
Microns
(f)

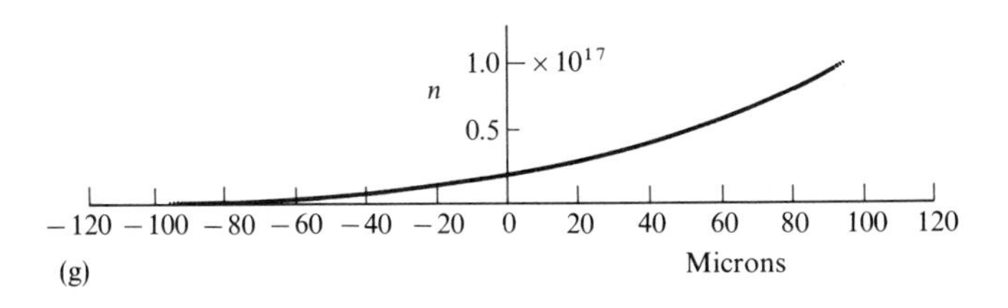
n
1.0 × 10¹⁷
0.5
−120 −100 −80 −60 −40 −20 0 20 40 60 80 100 120
Microns
(g)

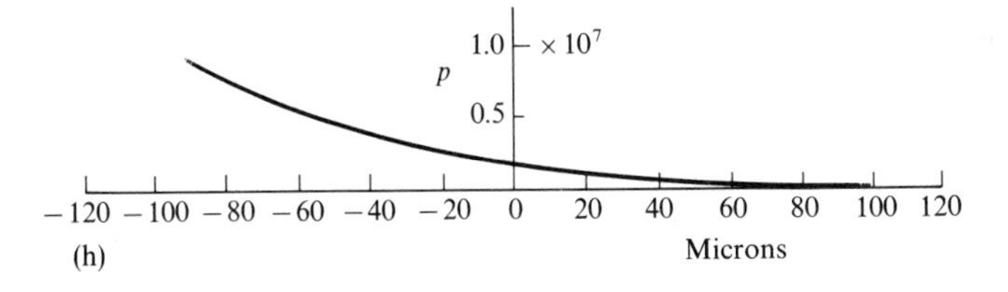
p
1.0 × 10⁷
0.5
−120 −100 −80 −60 −40 −20 0 20 40 60 80 100 120
Microns
(h)

4.7 SOLUTION OF POISSON'S EQUATION FOR E

We can find the Poisson equation for $\boldsymbol{E}$ as follows. In Section 2.3 we found that the curl of $\boldsymbol{E}$ is zero. Now, from Eq. 1-143,

$$\boldsymbol{\nabla} \times \boldsymbol{\nabla} \times \boldsymbol{E} = -\nabla^2 \boldsymbol{E} + \boldsymbol{\nabla}(\boldsymbol{\nabla} \cdot \boldsymbol{E}) = 0, \tag{4-181}$$

and thus

$$\nabla^2 \boldsymbol{E} = \boldsymbol{\nabla}(\boldsymbol{\nabla} \cdot \boldsymbol{E}), \tag{4-182}$$

$$= \frac{\boldsymbol{\nabla}(\rho_f + \rho_b)}{\epsilon_0}, \tag{4-183}$$

from Eq. 3-41.

The solution of this equation (Appendix E) is

$$\boldsymbol{E} = -\frac{1}{4\pi\epsilon_0} \int_{\tau'} \frac{\boldsymbol{\nabla}'(\rho_f + \rho_b)}{r} \, d\tau'. \tag{4-184}$$

We have added a prime to the del to stress the fact that it contains derivatives with respect to the source point (x', y', z').

This equation relates the electric field intensity to the *gradient* of the total charge density. Note the *negative sign* and the *first* power of r in the denominator. It is valid whatever the nature of the media that are present in the field, as long as the gradient is definable.

The more usual expression for the electrostatic field intensity of a volume distribution of charge is a consequence of Coulomb's law and was found in Eq. 3-30:

$$\boldsymbol{E} = \frac{1}{4\pi\epsilon_0} \int_{\tau'} \frac{(\rho_f + \rho_b)\boldsymbol{r}_1}{r^2} \, d\tau'. \tag{4-185}$$

Although the two integrals for $\boldsymbol{E}$ are equal, the integrands are obviously unequal, since there exists no general relationship between ρ_t and $\boldsymbol{\nabla}'\rho_t$ at a point in space. Indeed, the two integrals are equal only if they extend over *all* the charge distribution.

It will be shown in Problem 4-30 that, as a result of the equality of these two integrals,

$$\int_{\tau'} \boldsymbol{\nabla}' \left(\frac{\rho_t}{r} \right) d\tau' = 0 \tag{4-186}$$

for any volume distribution of charge, again, if the gradients exist.

Example Imagine a sphere of charge that has a uniform density throughout its volume, except near the periphery, where the density gradually decreases to zero. Then, if $\boldsymbol{E}$ is calculated by means of the integral of Eq. 4-184, only the region near the surface contributes to the integral, since $\boldsymbol{\nabla}'\rho_t$ is zero everywhere else (see Problem 4-29).

4.8 SUMMARY

In this chapter we have dealt with electric fields that cannot be calculated easily by the methods of Chapter 2.

We first established the conditions that must be satisfied by the potential V, by the normal component D_n of the electric displacement, and by the tangential component E_t of the electric field intensity at the boundary between two media: all three must be continuous across any boundary that does not carry a surface charge. If there is a surface charge, then D_n is discontinuous but V and E_t are still continuous.

We then demonstrated the *uniqueness theorem* according to which, for a given set of boundary conditions, there is only one possible electric field. This theorem is of great practical importance: if we can somehow find a potential $V(x, y, z)$ that satisfies both the boundary conditions and Poisson's equation, then we know that we have found the correct potential.

The method of *images* can sometimes simplify the calculation of electric fields that involve conducting surfaces and dielectrics. For example, the field of a point charge Q near an infinite conducting plane is the same as if the plane were replaced by a charge $-Q$ at the position of the image of Q. The force on Q is of course the same in both cases. The method of images gives the correct field only *outside* the region where the image is situated. In some cases there is an infinite set of images.

We then solved *Laplace's equation* for a number of different boundary conditions requiring either Cartesian or spherical coordinates. Solutions of Laplace's equation are called *harmonic functions*. We found such solutions by separating the variables: in Cartesian coordinates we set

$$V = X(x)Y(y)Z(z), \tag{4-59}$$

where X is a function of x only, Y is a function of y only, Z is a function of z only, while in spherical coordinates we had

$$V = R(r)\Theta(\theta). \tag{4-96}$$

There was no Φ (φ) function, since we considered only fields that were independent of the azimuthal angle φ.

For both of the fields that we calculated in Cartesian coordinates we were able to fit the boundary conditions only by adding an infinite number of solutions involving sine functions. Such infinite series of the form

$$\sum_{n=1}^{\infty} C_n \sin \frac{n\pi y}{b}$$

are called *Fourier series;* an arbitrary boundary condition $V(y)$ can be satisfied by such a series. The value of the C_n coefficients is found by multiplying both the boundary condition $V(y)$ and the Fourier series by $\sin(p\pi y/b)$, integrating from 0 to b, and utilizing the orthogonality property of the sequence of $\sin(n\pi y/b)$ functions:

$$\int_0^b C_n \sin\frac{n\pi y}{b}\sin\frac{p\pi y}{b}\,dy = \begin{cases} 0 & \text{if } p \neq n, \\ C_n \dfrac{b}{2} & \text{if } p = n. \end{cases} \tag{4-81}$$

The separation of variables in spherical coordinates transforms Laplace's equation into two ordinary differential equations:

$$r^2\frac{d^2R}{dr^2} + 2r\frac{dR}{dr} - n(n+1)R = 0, \qquad (4\text{-}101, 4\text{-}103)$$

and *Legendre's equation,*

$$\frac{d}{d\mu}\left[(1-\mu^2)\frac{d\Theta}{d\theta}\right] + n(n+1)\Theta = 0, \tag{4-107}$$

where $\mu = \cos\theta$.

The first equation is readily solved by functions of the type

$$R(r) = Ar^n + \frac{B}{r^{n+1}}. \tag{4-102}$$

The solutions of Legendre's equation are called *Legendre polynomials:*

$$P_n(\cos\theta) = \frac{1}{2^n n!}\frac{\partial^n}{\partial(\cos\theta)^n}(\cos^2\theta - 1)^n. \tag{4-124}$$

The general solution of Laplace's equation in spherical coordinates, for axial symmetry, is then

$$V = \sum_{n=0}^{\infty} A_n r^n P_n(\cos\theta) + \sum_{n=0}^{\infty} B_n r^{-(n+1)} P_n(\cos\theta). \tag{4-125}$$

The individual terms of this equation constitute a complete set of functions; any arbitrary boundary value of the potential having axial symmetry can be satisfied with such a series. The coefficients in the series can be determined by using the specified potentials on the boundaries and by using the orthogonality property of the Legendre functions:

$$\int_{-1}^{+1} P_m(\cos\theta)\,P_n(\cos\theta)\,d(\cos\theta) = \begin{cases} 0 & \text{if } m \neq n, \\ \dfrac{2}{2n+1} & \text{if } m = n. \end{cases} \tag{4-126}$$

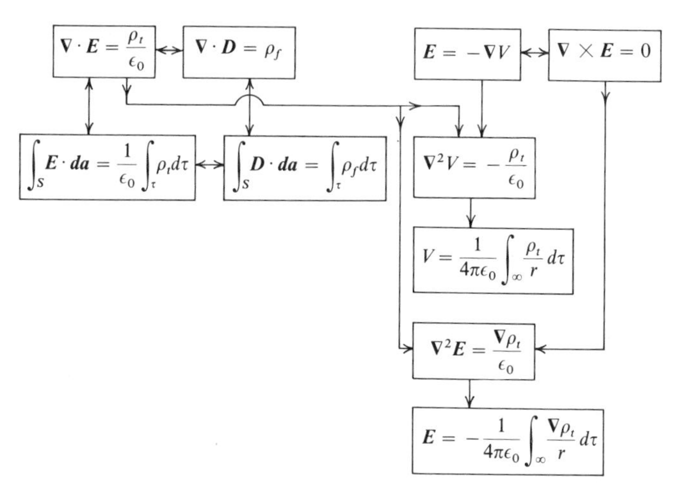

Figure 4-26.

Poisson's equation for V is

$$\nabla^2 V = -\frac{\rho_f + \rho_b}{\epsilon_0} = -\frac{\rho_t}{\epsilon_0}. \tag{3-42}$$

Poisson's equation for $\boldsymbol{E}$ is

$$\nabla^2 \boldsymbol{E} = \frac{\nabla(\rho_f + \rho_b)}{\epsilon_0} = \frac{\nabla \rho_t}{\epsilon_0}, \tag{4-183}$$

and its solution leads to an alternate integral for $\boldsymbol{E}$:

$$\boldsymbol{E} = -\frac{1}{4\pi\epsilon_0} \int_{\tau'} \frac{\nabla'(\rho_f + \rho_b)}{r} d\tau' = -\frac{1}{4\pi\epsilon_0} \int_{\tau'} \frac{\nabla' \rho_t}{r} d\tau'. \tag{4-184}$$

It follows that, for any volume distribution of charge,

$$\int_{\tau'} \nabla' \left(\frac{\rho_t}{r}\right) d\tau' = 0, \tag{4-186}$$

as long, of course, as the gradients exist.

The reasoning that led us to Eqs. 4-183 and 4-184 is shown schematically in Figure 4-26.

PROBLEMS

4-1. It is found experimentally that the electric field intensity in the atmosphere near the surface of the Earth is about 100 volts/meter and that it points downward: the potential increases as we go up. This field is maintained by thunderstorms, which, on the average, deposit negative charges on the earth. Calculate the electric charge carried by the Earth.

4-2. Figure 3-10 shows the **E** and **D** fields of a bar electret.

(a) Show that the lines of **E** are not bent at the cylindrical surface and that they are bent at the end faces.

(b) Show that the inverse is true for the lines of **D**.

4-3. It is interesting to speculate about the electrostatic potential at the surface of the Sun ($M = 2.0 \times 10^{30}$ kilograms, $R = 7.0 \times 10^{8}$ meters).

The electrons in the hot plasma at the surface easily attain the escape velocity $(2\,GM/R)^{1/2}$ for an uncharged star, where $G = 6.67 \times 10^{-11}$ is the gravitational constant. Their *mean* kinetic energy is $(3/2)\,kT$, where $k = 1.38 \times 10^{-23}$ joule/kelvin is Boltzmann's constant and the temperature T is about 6×10^{3} kelvins. The protons have much lower thermal velocities because of their larger mass.

Electrons cannot escape indefinitely, however, because the Sun acquires a positive charge that increases their escape velocity.

(a) Calculate the fraction of the electrons that possess enough energy to escape when the charge is zero. This fraction is

$$\exp\left[-(\text{escape kinetic energy})/kT\right].$$

(b) Calculate the charge that is required to neutralize the gravitational force on a proton. Note that this charge is independent of the radius.

(c) Show that it is only necessary to remove 160 electrons/meter^2 of the surface of the Sun to achieve this charge.

(d) Assuming that equilibrium is achieved when the net attractive force on a proton is zero, show that the average voltage at the surface of the Sun is about 2×10^{3} volts.

4-4. Various devices, such as Van de Graaff particle accelerators for example, have high voltage electrodes maintained under pressure in a metal tank. Let us assume that the electrode is spherical and that it has a radius r_1. The electrode must operate at a voltage V with respect to the tank, which has a radius r_2 and which is grounded.

We focus our attention on the electric field intensity at the surface of the electrode, for that is where it is highest, and we try to optimize r_1 and r_2.

(a) If we disregard the cost of the tank, then its optimum radius will be that for which the electric field intensity in the insulating gas will be minimum, for this will permit using a minimum gas pressure. What is this optimum value of r_2?

(b) In actual practice, cost, weight, and space requirements limit r_2. One must therefore optimize the ratio r_2/r_1 by varying r_1. Show that the E at the surface of the high-voltage electrode ($r = r_1$) has a minimum value of $2V/r_1$ when $r_2 = 2r_1$.

(c) Can you explain qualitatively why there should be an optimum condition?

(d) To determine whether the optimum condition is critical or not, plot the

value of E at $r = r_1$ for an actual case where r_2 was 0.483 meter and for values of r_1 ranging from 0.1 to 0.4 meter.
(e) What range of values of r_1 can be tolerated if one can tolerate an electric field intensity 10% higher than $2V/r_1$?
(f) Calculate $2V/r_1$ for $V = 5 \times 10^5$ volts?

4-5. Perform the same calculations as above for a cylindrical geometry. You should find that, when the radius r_2 of the outer cylinder is fixed, the electric field intensity at the surface of the inner cylinder has a minimum value of V/r_1 when $r_1 = r_2/e$, where e is the base of the natural logarithms.

4-6. According to the uniqueness theorem, the Poisson equation $\nabla^2 V = -\rho_t/\epsilon_0$ can have only one solution if the potential V is determined at the boundaries of the field.

Show that two solutions can differ at most by a constant if the normal component of ∇V is determined everywhere at the boundaries.

4-7. When the space between the plates of a parallel-plate capacitor is filled with a dielectric ϵ, it has a capacitance of C farads. If the dielectric is replaced by a material whose resistivity ρ is much smaller than that of the electrodes, the resistance between the electrodes is R ohms.

(a) Show that $RC = \rho\epsilon$, neglecting edge effects.‡
(b) Show that this result also applies to cylindrical and spherical capacitors.
(c) Show that it applies to *any* pair of electrodes submerged in a medium whose resistivity ρ is much smaller than that of the electrodes.

You should be able to show that the field is unaffected by the conductivity, with the above restriction.

One important application of this fact is the *electrolytic plotting tank*, which is used for plotting electric fields in two, and in some cases three, dimensions.

4-8. Electrified dust particles are ejected into the atmosphere and form an elongated cloud of approximately cylindrical form at an altitude of 40 meters in a 3 meter/second wind. The current feeding the cloud is 80 microamperes.

Calculate the resulting electric field intensity at the surface of the Earth directly under the cloud. Assume the Earth to be flat and conducting.

4-9. A point charge Q is situated between two horizontal parallel conducting plates separated by a distance s and at a distance x above the lower plate.

(a) Calculate the force due to the image charges in the form of an infinite series.
(b) Find an approximate value for the force when Q is situated (i) near one of the plates and (ii) near the position $x = s/2$.
(c) In the Millikan oil-drop experiment a small oil droplet carrying a few excess electrons is situated in the electric field between two charged parallel plates separated by a distance s. The force on the droplet is calculated from the electric field V/s, where V is the difference in potential between the plates. The image force is neglected. Is this serious?

4-10. (a) Show that the force of attraction or repulsion between a point charge q and a conducting sphere of radius R carrying a charge Q is

$$\frac{q}{4\pi\epsilon_0}\left\{\frac{Q + (R/D)q}{D^2} - \frac{Rq}{D[D - (R^2/D)]^2}\right\},$$

where D is the distance from q to the center of the sphere.

(b) Show that the force is attractive when q and Q are of the same sign, for

$$\frac{Q}{q} < \frac{RD^3}{(D^2 - R^2)^2} - \frac{R}{D}.$$

4-11. In the last example in Section 4.3 we calculated the surface charge density induced on the surface of a block of dielectric by a point charge Q at a distance D in front of it.

Show that this surface charge density and Q give the correct electric potential V (a) at the foot of the perpendicular drawn from Q to the dielectric surface, and (b) at a distance D behind the boundary.

4-12. Show that there exist solutions of Laplace's equation that are of the form $X(x) + Y(y) + Z(z)$.

4-13. Calculate the potential at $x = 0$ for the electrodes of Figure 4-16, using the first five terms of the Fourier series. Perform the calculation for $y = 0.1\,b$, $0.2\,b$, and so on.

4-14. Use the fact that $1/r$ is a solution of Laplace's equation to show that

$$f = a\frac{\partial}{\partial x}\left(\frac{1}{r}\right), \qquad g = b\frac{\partial^2}{\partial x^2}\left(\frac{1}{r}\right), \qquad h = c\frac{\partial^2}{\partial x\,\partial z}\left(\frac{1}{r}\right)$$

are also solutions when a, b, c are constants.

4-15. A dielectric sphere of radius R has a uniform polarization $\boldsymbol{P}$.

(a) Show that the electric field intensity inside is uniform and equal to $-\boldsymbol{P}/3\epsilon_0$.

(b) Show that the electric field intensity outside is that of a dipole situated at the center of the sphere and having a dipole moment equal to $\boldsymbol{P}V$, where V is the volume of the sphere.‡

4-16. A grounded, infinite, circular, cylindrical conductor is introduced into a previously uniform electric field with its axis perpendicular to $\boldsymbol{E}_0$.

Show that

$$V = E_0[1 - (a^2/\rho^2)]\rho\cos\varphi.‡$$

4-17. Two point charges $+Q$ and $-Q$ are situated on a diameter of a conducting sphere of radius a at distances $D > a$ to the right and to the left, respectively.

(a) Show that the image charges constitute a dipole of moment $2a^3Q/D^2$ at the center of the sphere.

(b) Now let D and Q approach infinity in such a way that Q/D^2 remains constant. Superpose the fields of $\pm Q$ and of the dipole to find the field outside the sphere.

This is another way of calculating the field around an uncharged conducting sphere that is introduced into a previously uniform field.
(c) Show that the surface charge density is $3\epsilon_0 E_o \cos\theta$, where E_o is the field due to the charges $\pm Q$.

4-18. Show that a sum of terms of the form $r^n \sin n\theta$, where n is any positive or negative integer, is a solution of Laplace's equation in two dimensions.

4-19. A linear distribution of charge of λ coulombs/meter extends along the z-axis from $z = -a$ to $z = +a$. Show that, at any point for which $r > a$,

$$4\pi\epsilon_0 V = \frac{2\lambda a}{r} P_0(\cos\theta) + \frac{2\lambda a^3}{3r^3} P_2(\cos\theta) + \frac{2\lambda a^5}{5r^5} P_4(\cos\theta) + \cdots.$$

4-20. The axial electric field intensity E_z on the axis of the accelerating tube in a particular type of ion accelerator is given approximately by

$$E_z = E_{z0} + kz^2,$$

where z is measured from the center of the tube along its axis. The azimuthal component E_φ is zero.
(a) Show that the radial electric field intensity in the neighborhood of the axis is $-kz\rho$, assuming that the charge density is zero.
(b) Draw a rough sketch of the lines of force in a plane that contains the axis.
(c) What is the maximum charge density that can be tolerated if the value calculated above for the radial field is to be accurate within 5% at the ends of the tube? The accelerating tube is 1.00 meter long, E_{z0} is 7.5×10^5 volts/meter, and k is 1.00×10^6 volts/meter^3.

4-21. It is in some cases necessary to ionize a low pressure gas by bombarding it with electrons. One example is ion pumping. In ion pumps the gas is removed from the system to be evacuated by driving the positive ions into a suitable cathode where they are absorbed.

Now such pumps are especially desirable at very low pressures, at which the mean free path of electrons between collisions with molecules of the residual gas can be of the order of thousands of kilometers. Methods have therefore been devised for increasing the electron path length within the pump.

The *orbitron* is one of several types of ion pumps. It utilizes a cylindrical cathode and an axial rod as anode. Electrons from one or more filaments are injected near one end of the cylinder with sufficient angular momentum to make them orbit around the anode. They are also given a slight axial velocity, and they describe helices along the length of the pump. End plates connected to the cathode reflect electrons at both ends. One end of the cylinder is provided with a flange for connection to the vessel to be evacuated.

One particular orbitron has a diameter of 10 centimeters and a length of 40 centimeters. The anode has a diameter of 5 millimeters and is maintained at +5 kilovolts with respect to the cylinder, which is grounded.

Assume first that there is zero space charge, and neglect both the axial velocity and end effects. It is shown in mechanics that a particle in a central force field will describe a stable circular orbit if

$$\frac{r}{f}\frac{\partial f}{\partial r} > -3,$$

where $f(r)$ is the magnitude of the radial force of attraction exerted on the particle.

(a) Are the orbits stable when there is zero space charge?

(b) Now assume that the gas pressure is zero and that the space charge arises solely from the orbiting electrons. Qualitatively, how will this space charge affect the electric field intensity? Will it eventually lead to unstable orbits?

(c) Assume that, in the presence of gas, the radial electric field intensity is constant. Are the electron orbits again stable?

It is possible to use essentially the same device for transporting charged particles over considerable distances. In this case the particles are captured into spiral orbits at one end of the tube and travel to the other end where they are detected. The distance of travel is limited only by the quality of the vacuum. One particular guide that was used for extracting alpha particles from a target inside a reactor had a length of 6.3 meters and a diameter of 7.5 centimeters. The central wire was maintained at −30 kilovolts.

4-22. In 1959 Lyttleton and Bondi* suggested that the expansion of the Universe could be explained on the basis of Newtonian mechanics if matter carried a net electric charge.

Imagine a spherical volume of astronomical size containing un-ionized atomic hydrogen of uniform density η, and assume that the proton charge $e_p = -(1 + y)e$, where e is the electron charge.

(a) Show that, for $y > 10^{-18}$, the electrostatic repulsion becomes larger than the gravitational attraction and the gas expands.

(b) Show that the force of repulsion on an atom is then proportional to its distance R from the center and that, as a consequence, the radial velocity of an atom at R is proportional to R. Assume that the density is maintained constant by the continuous creation of matter in space.

(c) Show that the velocity $v = R/T$, where T is the time required for the radial distance R of a given atom to increase by a factor of e. This time T can be taken to be the age of the Universe.

(d) In the Millikan oil-drop experiment an electrically charged droplet of oil is suspended in the electric field between two plane horizontal electrodes. It is observed that the charge carried by the droplet changes by integral amounts within an accuracy of about 1 part in 10^5.

Imagine that the proton charge e_p is equal to $-(1 + y)e$, where e is the charge of the electron. Can you show that the Millikan oil-drop experiment leads us to believe that y is less than about 10^{-17}?

4-23. Electrostatic precipitation is used extensively for the elimination of dust particles from industrial gases, for example, for eliminating fly-ash from the smoke of coalfired electric power plants. This method was developed by Cottrell at the beginning of the century. In this process the dust particles are charged by the ions formed in a corona discharge, and they then drift in the electric field to the electrodes, where they are deposited.

---

* *Proc. Roy. Soc.* (London) A *252*, p. 313.

In one type of precipitator the anode is a grounded cylinder having a radius R of 15 centimeters, and the cathode is a wire supported along the axis of the cylinder and maintained at a potential V of -50 kilovolts. The gas is ionized and ions of both signs are formed in the corona discharge near the wire. The positive ions quickly reach the center wire, while the negative ions move out radially to the cylinder. The space charge is thus negative over most of the volume of the cylinder.

Under these conditions it is found experimentally that E is approximately equal to V/R for all values of r. If the dust particles are at least slightly conducting, they acquire a negative charge Q of $12\pi\epsilon_0 Ea^2$, where a is their radius. It turns out that if they are nonconducting Q is somewhat smaller.

(a) Set i to be the electric current per meter and γ to be the mobility (velocity/E) of the negative ions. Show that, for any r, the current due to the ions is

$$i = 2\pi r\rho\gamma E,$$

and that the space charge density

$$\rho = \epsilon_0 E/r.$$

(b) The drift velocity of the *dust* particles is given by Stokes's law: it is the force EQ divided by $6\pi\eta a$, where η is the viscosity of the gas. Show that their drift velocity v is $2\epsilon_0 E^2 a/\eta$.

(c) Calculate i, ρ, v, and the time required for a dust particle to drift from the cathode to the anode, when $\gamma = 2 \times 10^{-4}$ meter^2/volt second, $a = 5$ microns, and $\eta = 2 \times 10^{-5}$ kilogram/meter second.

This simplified theory neglects turbulence, which turns out to be important in practice.

4-24. Would it not be possible to build a perpetual-motion machine that would utilize the contact potential in a junction diode?

4-25. If the number of electron-hole pairs generated per unit volume and per unit time in a semiconductor is G, and if the number of electron-hole recombinations is R, show that

$$\nabla \cdot \boldsymbol{J}_n - e\frac{\partial n}{\partial t} = -e(G - R),$$

$$\nabla \cdot \boldsymbol{J}_p + e\frac{\partial p}{\partial t} = e(G - R),$$

where n and p are, respectively, the number of electrons and the number of holes per unit volume, and where $\boldsymbol{J}_n$ and $\boldsymbol{J}_p$ are, respectively, the electron current density and the hole current density.

The number of recombinations R per unit time and per unit volume is a function of the carrier concentrations n and p, while the number G depends on the temperature, on the illumination, etc.

4-26. In the vacuum diode, electrons are emitted from a hot filament and collected at the anode. Show that, for an idealized diode having two plane parallel electrodes, the electron current density is

$$J = 2.34 \times 10^{-6} V_0^{3/2}/s^2 \text{ amperes/meter}^2,$$

where V_0 is the potential difference and s is the distance between cathode and anode.

This is the *Child-Langmuir* law. It is valid only for the plane parallel diode and for electrons emitted with zero velocity. More generally,

$$J = kV_0^{3/2},$$

where K is a constant, for both ions and electrons and for any geometry, as long as the current is space-charge limited, and as long as u is negligible at the source.‡

4-27. Let us see how current flows through an ionized gas. Two large parallel electrodes of area S are separated by a distance s which is small enough to render edge effects negligible. Pairs of ions are created throughout the gas between the electrodes at the constant rate of n_0 pairs/meter³ second. Let us set

n^+ and n^- to be the numbers of positive and negative ions per cubic meter, respectively;
v^+ and v^-, the velocities of the positive and negative ions;
J^+ and J^- the currents due to the positive and negative ions;
e, the absolute value of the electric charge of an ion;
the x-axis to be perpendicular to the plates, in the direction of the electric field, with the origin at the positive plate. With these conventions v^+ is positive and v^- is negative.

Show that the total current density

$$J = en^+v^+ - en^-v^- = en_0x + en_0(D - x) = en_0D,$$

and that the charge density

$$\rho = en_0\left[\frac{x}{v^+} - \frac{x - D}{v^-}\right].$$

The velocities are both proportional to the electric field intensity E:

$$v^+ = \mu^+E, \qquad v^- = -\mu^-E,$$

where μ^+ and μ^- are the mobilities and where E is itself a function of x. If the negative particles are electrons, $\mu^- \gg \mu^+$.

We have neglected both thermal diffusions and recombination.

4-28. The thrust produced by a rocket motor is equal to $m'v$, where m' is the mass of propellant ejected per unit time and v is the exhaust velocity. Rocket engineers therefore strive to make m' as small as possible by increasing v.

One way of achieving large values of v is to eject a beam of charged particles, as in Figure 4-27, which shows a schematic diagram of an ion motor. The propellant is ionized in the ion source and is ejected as a positive ion beam at a velocity corresponding to the accelerating voltage V. Electrons are injected into the beam to prevent the rocket from charging up.

(a) The current I of positive ions in the beam is carried by particles of mass m and charge ne, where e is the electronic charge.

Show that the thrust is

$$F = I(2Vm/ne)^{1/2}.$$

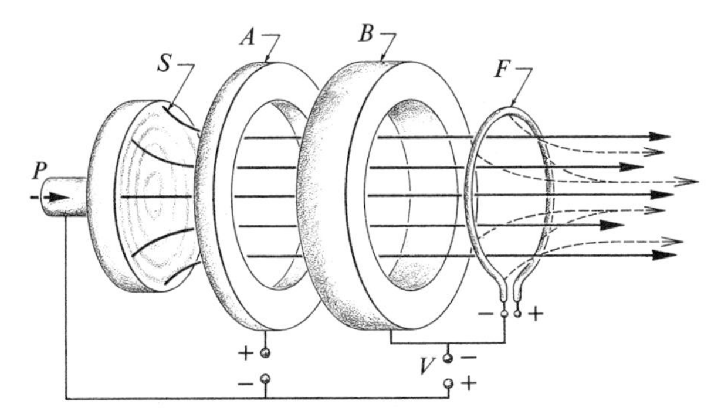

Figure 4-27. Schematic diagram of an ion motor. The propellant is admitted at P and ionized in S; A is a beam-shaping electrode and B is the accelerating electrode, maintained at a voltage V with respect to S; the filament F injects electrons into the ion beam to make it neutral.

(b) What is the value of F for a 0.1 ampere beam of protons accelerated to 50 kilovolts? Ion motors are used in outer space where small thrusts are required to correct either the attitude or the trajectory of satellites.
(c) Show qualitatively that the thrust is independent of the shape of the field or of the presence of space charge in the acceleration region and that it is always given by the above formula.
(d) If we call W the power IV spent in accelerating the beam, show that

$$F = (2Wm')^{1/2} = \frac{2W}{v} = W\left(\frac{2m}{neV}\right)^{1/2}.$$

Thus, for given values of W and m', the thrust is independent of the charge-to-mass ratio of the ions. Or, for a given W, F is *inversely* proportional to v. The last expression shows that, *for a given power expenditure* W, it is preferable to use heavy ions carrying a single charge ($n = 1$) and to use as *low* an accelerating voltage V as possible.
(e) Sketch a graph of the voltage along the axis, inside and outside the motor, (*i*) when the electron source is on, and (*ii*) some time after it has been turned off. Assume that the body of the rocket is connected to electrode B.
(f) If the electron source is turned off, and if the beam current I is one ampere, how long will it take the body of the rocket to attain a voltage equal to the accelerating voltage if V is 50 kilovolts? Assume that the rocket is spherical and that it has a radius of 1 meter. At that point the motor ceases to operate because the ions follow the rocket.

4-29. A sphere of electric charge has a density ρ which is a function of the radius as in Figure 4-28.

(a) Use Gauss's law to show that, at a distance $r > \beta$ from the center of the sphere,

$$E = \frac{\rho_0}{12\epsilon_0 r^2}(\beta^2 + \alpha^2)(\beta + \alpha).$$

(b) Show that one arrives at the same result by using Eq. 4-184. Note that the result is unaffected as $(\beta - \alpha) \to 0$.
(c) Does this value of E make sense when $\alpha = \beta$?

4-30. We have shown in Section 4-7 that, for static fields,

$$E = \frac{1}{4\pi\epsilon_0} \int_{\tau'} \frac{\rho_t r_1}{r^2}\, d\tau' = -\frac{1}{4\pi\epsilon_0} \int_{\tau'} \frac{\nabla' \rho_t}{r}\, d\tau'.$$

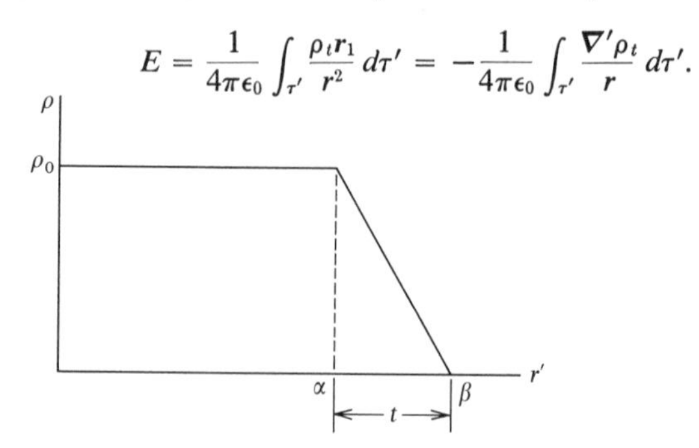

Figure 4-28.

Show that, as a consequence,

$$\int_{\tau'} \nabla' \left(\frac{\rho_t}{r}\right) d\tau' \equiv 0.‡$$

4-31. Show that

$$\int_{\tau'} \nabla' \times \left(\frac{\mathbf{E}}{r}\right) d\tau' \equiv 0$$

for any finite charge distribution.‡

Hints

4-7. The charge density is zero in a uniform conducting medium.

4-15. Choose the z-axis along $\mathbf{P}$ and use Legendre polynomials to calculate V. You can determine the values of all the coefficients from the properties of the field without performing any integration.

4-16. Show that this V obeys Laplace's equation in cylindrical coordinates and that it gives the correct boundary conditions.

4-26. Solve Poisson's equation, setting the charge density ρ equal to J/u, where u is the velocity of the electrons and $(1/2)mu^2 = eV$.

If your calculation involves $(-J)^{1/2}$, disregard the minus sign and consider simply the magnitude of the current.

Assume that the electrons are emitted with $u = 0$.

Set $dV/dx = 0$ at the cathode. This means that there is an unlimited supply of electrons available at the cathode and that the current is limited by the electron space charge, and not by the nature of the cathode.

4-30. Expand this last integral.

4-31. Use the identity of Problem 1-30.

CHAPTER 5

RELATIVITY I

The Basic Concepts

We have now studied at quite some length the fields of stationary electric charges. At this point we can either go on directly to the fields of moving charges, or we can leave the subject of electromagnetism aside, for the moment, and study the basic concepts of relativity. The reason for this digression is that magnetic fields result from relativistic transformations of electric fields.

The longer path is more interesting, as always, but it may not be the better one. There is no doubt that relativity throws much light on the nature of magnetic fields. It also leads directly to the fundamental equations of electromagnetism. But Chapters 5 and 6 do not replace the more conventional approach that begins with Chapter 7. Selecting one path or the other is a matter of time and personal taste.

Therefore, *if you wish to go on with electromagnetism without delay, you can omit Chapters 5 and 6, and go directly to Chapter 7 without losing continuity.*

The present chapter is devoted to the basic concepts of relativity, with little reference to electrical phenomena except near the end. In the next chapter we shall utilize these concepts, first to reveal the origin of magnetic fields, and then to establish several fundamental relations that we shall later rediscover without using relativity.

We shall keep our discussion of relativity as simple as possible. In fact, the only mathematical requirement for this chapter is elementary differential calculus and the vector analysis of Chapter 1.*

---

* For a more detailed introduction to the basic concepts of relativity, see Edwin F. Taylor and John A. Wheeler, *Spacetime Physics* (W. H. Freeman and Company, San Francisco, 1966).

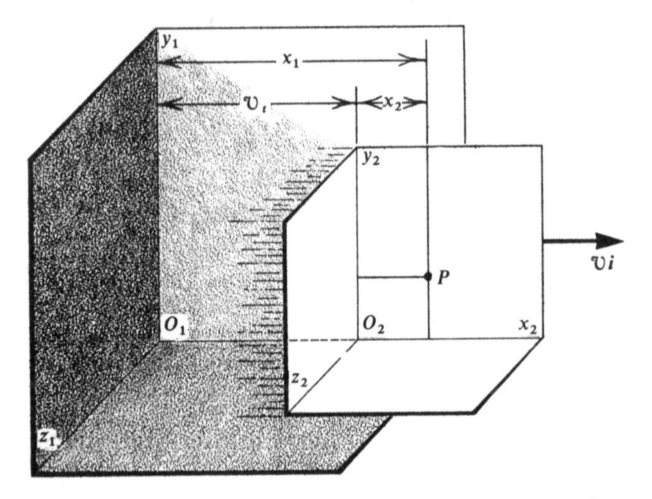

Figure 5-1. Two Cartesian coordinate systems, one moving at a velocity $\mathcal{V}\boldsymbol{i}$ with respect to the other in the positive direction of the common x-axis. The two systems overlap when the origins O_1 and O_2 coincide. *We shall always refer to these two coordinate systems whenever we discuss relativistic effects.*

We could start directly with the Lorentz transformation, which forms the basis of special relativity, but this transformation is so contrary to everyday experience that we shall first demonstrate the inadequacy of the more obvious Galilean transformation.

5.1 THE GALILEAN TRANSFORMATION

The *special theory of relativity* is concerned with the observations made by two different observers, one of whom has a constant velocity with respect to the other. The *general theory of relativity* has to do with gravitation, and we shall have no occasion to use it.

We therefore consider two Cartesian coordinate systems, as depicted in Figure 5-1, where system 2 has a *constant* velocity $\mathcal{V}\boldsymbol{i}$ with respect to system 1 in the direction of the common x-axis. Corresponding axes are parallel and in the same direction. Neither system is accelerated.

We shall constantly refer to these two particular coordinate systems throughout this book whenever we discuss relativistic effects.

According to prerelativistic physics, the two systems are related by the following intuitively obvious equations:

$$x_1 = x_2 + \mathcal{V}t, \qquad y_1 = y_2, \qquad z_1 = z_2. \tag{5-1}$$

We have assumed that the two coordinate systems coincide at the time $t = 0$.

This set of equations constitutes a *Galilean transformation*.

Now, although the Galilean transformation is self-evident, and although it is completely compatible with classical mechanics, it is not generally valid, as we shall see, either for mechanical or for electromagnetic phenomena. We shall have to use a more complex transformation that will reduce to Eqs. 5-1 for everyday mechanical phenomena.

5.2 BREAKDOWN OF THE GALILEAN TRANSFORMATION AND OF CLASSICAL MECHANICS AT HIGH VELOCITIES

Both classical mechanics and the Galilean transformation break down when velocities approach the velocity of light. Many examples can be given.

Example

PARTICLE VELOCITIES NEVER EXCEED c

It is found experimentally that the velocity of a particle does *not* increase indefinitely as its energy increases.* Instead, the velocity approaches asymptotically the velocity of light $c = 3 \times 10^8$ meters/second. The usual expression $(1/2)mv^2$ of classical mechanics for the kinetic energy of a mass m moving with a velocity v cannot therefore be correct.

Example

THE ADDITION OF VELOCITIES

According to the Galilean transformation, vector addition applies to velocities. For example, if a passenger walks at a velocity v toward the front of a train that itself moves forward at a velocity $\mathcal{V}$, then his velocity with respect to the ground is simply the sum of the two velocities $v + \mathcal{V}$. But this simple addition of velocities is found to be incorrect in nuclear reactions where the velocities approach the velocity of light. The resulting velocity is in fact always smaller than $v + \mathcal{V}$, and never exceeds c.

Example

TIME DILATION

The Galilean transformation assumes that the time t, as measured on the train, is the same as that measured on the ground, if the two

* The film "The Ultimate Speed; an Exploration with High Energy Electrons" by W. Bertozzi demonstrates this phenomenon clearly. It was produced by Educational Development Center, Newton, Mass.

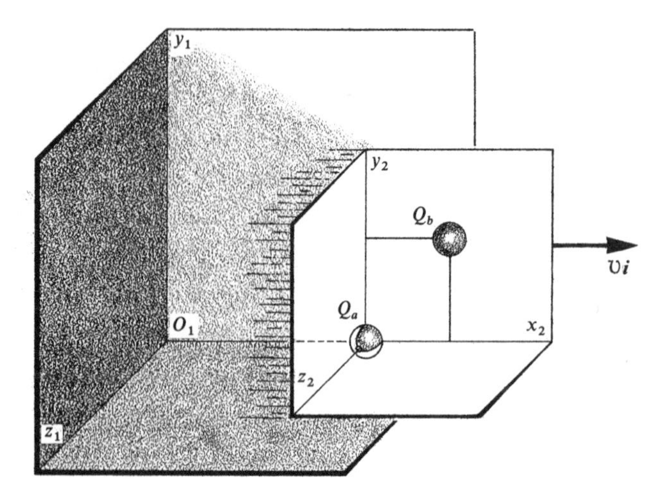

Figure 5-2. The electric charge Q_a is situated at the origin O_2 and Q_b is situated at (x_2, y_2, O), also in 2.

observers have identical synchronized clocks. This also proves to be incorrect at high velocities. In particular, it has been observed that, for the high-energy mesons of the cosmic radiation, time flows about nine times more slowly than the laboratory time. In other words, a period of one microsecond as measured by the meson is equivalent to about nine microseconds measured in the laboratory. This phenomenon is called *time dilation.**

5.3 INADEQUACY OF THE GALILEAN TRANSFORMATION FOR ELECTROMAGNETIC PHENOMENA

The Galilean transformation is totally inadequate for electromagnetic phenomena.

Example THE TROUTON AND NOBLE EXPERIMENT

Consider two electric charges, as in Figure 5-2, Q_a being situated at the origin of reference frame 2, and Q_b at $(x_2, y_2, 0)$, also on frame 2. What is the force exerted *by* Q_a on Q_b, as observed in frame 2.

Trouton and Noble attempted to observe this force in 1903.

---

* See American Journal of Physics, *31*, 342 (1963). See also the film "Time Dilation; an Experiment on Mu-mesons" by F. Friedman, D. Frisch, and J. Smith, produced by Educational Development Center, Newton, Mass.

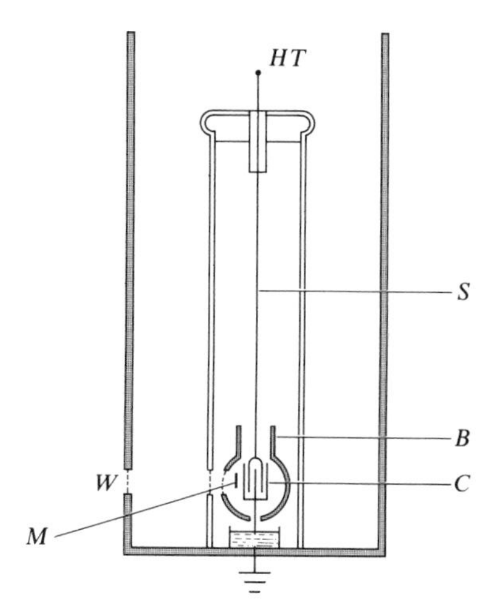

Figure 5-3. In the Trouton and Noble experiment a mica capacitor C (0.0037 microfarad) was suspended from a light phosphor bronze strip S. One side of the capacitor was grounded by a platinum wire that dipped in dilute sulfuric acid, while the other side was connected through the suspension to a Wimshurst machine that operated at 2000 volts. The light celluloid bulb B was covered with gilt paint and grounded to eliminate electrostatic forces on the capacitor C. The mirror M served to reflect a ray of light through the windows W onto a scale situated one meter from C. The deflection of C was predicted taking into account both the Earth's orbital motion and the Sun's proper motion. While the predicted deflection varied from 0.0 to 6.8 centimeters, according to the time at which the reading was taken, the observed deflection did not exceed 0.36 centimeter and was unrelated to the calculated value.

They suspended a charged parallel-plate capacitor in such a way that its plates were vertical and it could rotate around a vertical axis, as in Figure 5-3. Since the Earth rotates around the Sun at an orbital velocity of 3×10^4 meters/second, the force measured in the laboratory is F_2.

The Galilean transformation assumes implicitly the existence of an absolute, fixed frame of reference. If this assumption is correct, we expect from elementary electricity that the charge Q_b will be submitted not only to an electrostatic force but also to a magnetic force, because the moving charges should produce magnetic fields. This magnetic force should tend to turn the capacitor, bringing its plates parallel to its velocity. Trouton and Noble were unable to observe such a torque.

Example

MAGNETIC FIELDS

Another example of the inadequacy of the Galilean transformation for electromagnetic phenomena is simply the existence of a magnetic field in the neighborhood of a conducting wire.

We shall be able to show in the next chapter that the magnetic fields of conduction currents are due to an exceedingly small change, approximately one part in 10^{25}, in the field of the moving electrons. Magnetic fields can in no way be explained solely by Coulomb's law and the Galilean transformation.

Example

THE JASEVA-JAVAN-MURRAY-TOWNES EXPERIMENT*

Although the full potential of this experiment has yet to be achieved, it already provides a highly sensitive method for detecting the effects, if any, of the earth's orbital velocity on the velocity of propagation of light waves.

If there does exist an absolute fixed reference frame, then light waves can be expected to propagate at the fixed velocity c with respect to it, and it should be possible to measure the orbital velocity of the earth. The hypothetical fixed medium of propagation is called the *ether*.

The Jaseva-Javan-Murray-Townes experiment is a modern version of the Michelson-Morley experiment, which was repeated many times between the 1880's and the 1930's and which is described in most books on relativity.

The modern version utilizes two lasers (Figure 5-4) set at right angles on a rotating platform as in Figure 5-5. The two light beams are mixed by means of a half-silvered mirror, and the beat frequency

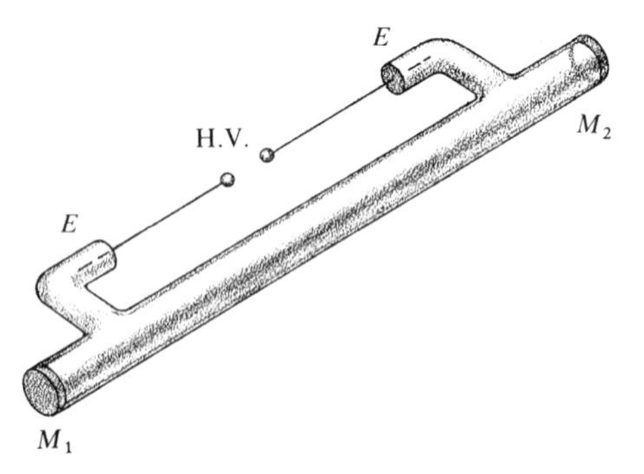

Figure 5-4. Schematic diagram of a gas laser. The glass tube is terminated by the two mirrors M_1 and M_2, and contains an appropriate gas mixture, for example, a mixture of helium and neon. Energy is supplied to the gas by a dc discharge between electrodes E. The space between the two mirrors constitutes a resonant cavity for the light emitted by the gas, and a light beam comes out through mirror M_2, which is half-silvered.

---

* T. S. Jaseva, A. Javan, J. Murray, and C. H. Townes, *Physical Review* 133, A 1221 (1964).

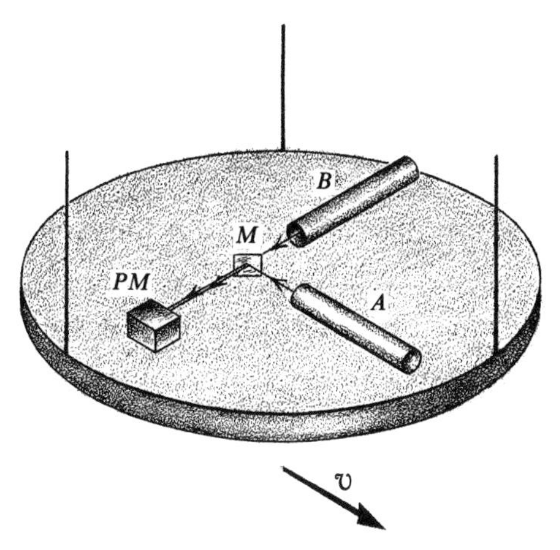

Figure 5-5. The Jaseva-Javan-Murray-Townes experiment. The half-silvered mirror M mixes the light beams of lasers A and B, and the photomultiplier and associated electronic equipment PM measure the beat frequency. The support rotates so that A and B can be alternately parallel and perpendicular to the orbital velocity of the Earth. The object is to observe effects of this velocity on the velocity of propagation of light with respect to the Earth. No effect has been observed so far.

is measured as in Figure 5-5.

Since a laser is a resonant cavity, its frequency

$$\nu = \frac{c}{\lambda} = \frac{c}{2L/n} = \frac{nc}{2L}, \tag{5-2}$$

where $L = n(\lambda/2)$, is the distance between the two mirrors; n is the number of half-wavelengths in the distance L; and c is the velocity of light, as usual.

Let us assume that laser A is parallel to its velocity $\mathcal{V}$ with respect to the ether. Then, in laser A, the time required for light to go from one mirror to the other and back is

$$t_A = \frac{L_A}{c - \mathcal{V}} + \frac{L_A}{c + \mathcal{V}} = \frac{2L_A}{c} \frac{1}{[1 - (\mathcal{V}/c)^2]} = \frac{2\gamma^2 L_A}{c}, \tag{5-3}$$

where

$$\gamma = \frac{1}{[1 - (\mathcal{V}/c)^2]^{1/2}}, \tag{5-4}$$

and laser A should oscillate as if its length were $\gamma^2 L_A$, instead of L_A.

With $\mathcal{V}$ equal to the orbital velocity of the earth, 3×10^4 meters/second, $(\mathcal{V}/c)^2$ is equal to 10^{-8}, and ν_A is smaller than if $\mathcal{V}$ were zero by one part in 10^8.

Laser B is perpendicular to the direction of motion and, from Figure 5-6,

$$t_B = \frac{2L_B'}{c} = \frac{2L_B}{c} \frac{c}{(c^2 - \mathcal{V}^2)^{1/2}} = \frac{2\gamma L_B}{c}. \tag{5-5}$$

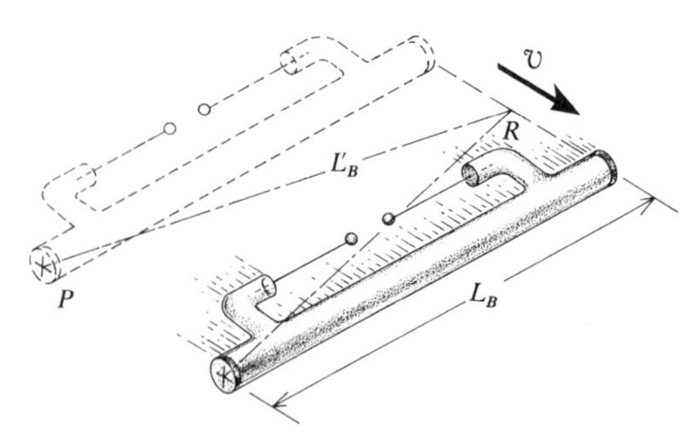

Figure 5-6. When laser B is perpendicular to the orbital velocity of the Earth, light must travel through a distance $2L'_B$ to make a round trip through the tube. It is assumed that light goes from P to R at the velocity c while the laser moves from Q to R at the velocity $\mathcal{V}$. Then $L'_B/L_B = c/(c^2 - \mathcal{V}^2)^{1/2}$.

Then laser B should oscillate as if its length were γL_B, and the beat frequency should be

$$\nu_b = \nu_B - \nu_A = \frac{nc}{2}\left(\frac{1}{L_B\gamma} - \frac{1}{L_A\gamma^2}\right). \tag{5-6}$$

If the complete set-up is now rotated through 90°, the beat frequency should change by

$$\Delta\nu_b = \frac{nc}{2}\left\{\left(\frac{1}{L_B\gamma} - \frac{1}{L_A\gamma^2}\right) - \left(\frac{1}{L_B\gamma^2} - \frac{1}{L_A\gamma}\right)\right\}, \tag{5-7}$$

$$= \frac{nc}{2}\left(\frac{1}{L_b} + \frac{1}{L_A}\right)\left(\frac{1}{\gamma} - \frac{1}{\gamma^2}\right) \approx \frac{nc}{2L}\left(\frac{\mathcal{V}}{c}\right)^2, \tag{5-8}$$

where L is the average length of the lasers, and where the term $(\mathcal{V}/c)^2$ is obtained by expanding $1/\gamma$ using the binomial theorem.
Thus

$$\frac{\Delta\nu_b}{\nu} \approx \left(\frac{\mathcal{V}}{c}\right)^2. \tag{5-9}$$

With $\nu = 3 \times 10^{14}$ hertz, $2\nu_b \approx 3 \times 10^6$ hertz. Since the frequency stability of lasers can be of the order of 20 hertz, it should be possible to observe a frequency shift even 10^5 times smaller than the above $2\nu_b$.

It turns out to be exceedingly difficult to achieve the theoretical accuracy, but it has been demonstrated that the frequency shift is less than one thousandth of the effect predicted on the assumption that light has a fixed velocity with respect to the ether.

5.4 THE FUNDAMENTAL POSTULATE OF RELATIVITY

The experiments of Trouton and Noble, and of Townes *et al.* had two points in common. Both were designed to detect effects of the velocity of the Earth through the ether, and both gave negative results. Many other experiments that had the same objective also gave negative results.*

It is on the basis of such evidence that Einstein proposed in 1905 *the fundamental postulate of relativity*, which can be stated as follows: *it is physically impossible to detect the uniform motion of a frame of reference from observations made entirely within that frame.*

This postulate is quite clear in itself, but it is so fundamental that we shall state it in another way to emphasize its meaning. It means that *any* experiment gives precisely the same result, whether it is performed in reference frame 1 or in reference frame 2, or whether it is performed in a standing or in a moving vehicle, as long as it is not accelerated.

Then, if we are given some physical law like $F_1 = m_1a_1$ which is valid in 1, there exists an identical law $F_2 = m_2a_2$ which is valid in frame 2. In fact, there must exist a transformation, different from the Galilean transformation, that renders *all* the laws of nature identical in frames 1 and 2.

We shall develop this concept of the invariance of a physical law in the next section, using classical mechanics and the Galilean transformation as an example.

5.5 INVARIANCE OF A PHYSICAL LAW AS ILLUSTRATED BY CLASSICAL MECHANICS

The Galilean transformation leaves the laws, and hence the phenomena, of *classical* mechanics unaltered in going from frame 1 to frame 2, and inversely. In other words, the laws of classical mechanics are invariant under a Galilean transformation. Classical, or prerelativistic, mechanics disregards the effects that we shall discuss later, which become prominent when velocities become comparable to the velocity of light.

* For a summary of these experiments see W. K. H. Panofsky and M. Phillips, *Classical Electricity and Magnetism* (Addison-Wesley, Reading, Mass., 1955).

Example

CLASSICAL MECHANICS INSIDE A MOVING TRAIN

If reference frame 1 is fixed with respect to the ground, and 2 is fixed with respect to a railway train moving at a uniform velocity $\mathcal{V}$, a ball thrown vertically upward by a passenger sitting in the train behaves, *for that observer inside the train*, precisely as it would if the experiment were performed on the ground. The reason, of course, is that the ball thrown vertically upward inside the train maintains its horizontal velocity $\mathcal{V}$ throughout its trajectory. In fact, an observer on the ground would see the ball in the moving train describe a parabola.

Many other such experiments can be devised. For example, the passenger can observe the period of a pendulum, or the acceleration of a mass pushed by a spring, or the collision of billiard balls, etc. In all cases the phenomenon *as observed by the passenger in the train* is precisely the same whether the train is stopped or moving at a *constant* velocity.

It is therefore impossible to design a mechanical device that would permit the passenger to measure, or even to detect, the uniform motion of the train from observations made entirely inside the train.

Of course, if the train accelerates, either by changing its speed or its direction, the passenger observes inertial forces, which permit him to measure his acceleration.

Example

THE LAW $\boldsymbol{F} = ma$

Let us write down the equations for a simple experiment that the passenger performs, first when the train is stopped, and then when the train is moving at a constant velocity $\mathcal{V}$. We shall see that the equations of motion in reference frame 1 are of the same form as those in frame 2.

The passenger is given a mass m on which he exerts a known force F in the direction of the track by means of a calibrated spring. As previously, reference frame 1 is fixed with respect to the ground and 2 is fixed with respect to the train. We assume that the law $F = ma$ applies in frame 1, and we deduce the corresponding law in frame 2.

When the train is stopped we can assume that the two frames overlap; then $x_1 = x_2$ and

$$F = m \frac{d^2}{dt^2} x_1 = m \frac{d^2}{dt^2} x_2^2. \tag{5-10}$$

Then $F = ma$ applies in both systems.

When the train has a uniform velocity $\mathcal{V}$, $x_1 = x_2 + \mathcal{V}t$ and

$$F = m\frac{d^2}{dt^2}x_1 = m\frac{d^2}{dt^2}(x_2 + \mathcal{V}t) = m\frac{d^2}{dt^2}x_2, \qquad (5\text{-}11)$$

and the law $F = ma$ again applies in both systems.

What if the force F is exerted in a direction perpendicular to the direction of motion? We again have two similar equations:

$$F = m\frac{d^2}{dt^2}y_1 \qquad (5\text{-}12)$$

with respect to the ground, and

$$F = m\frac{d^2}{dt^2}y_2 \qquad (5\text{-}13)$$

with respect to the train, since $y_1 = y_2$ according to the Galilean transformation.

We therefore say that the law $\mathbf{F} = m\mathbf{a}$ of *classical* mechanics applies in both reference frames, or that it is invariant under a Galilean transformation.

5.6 THE LORENTZ TRANSFORMATION

Toward the turn of the century, several persons discovered independently that invariance of all physical laws could be achieved by using the following equations of transformation:

$$\begin{aligned}
x_1 &= \frac{x_2 + \mathcal{V}t_2}{[1 - (\mathcal{V}/c)^2]^{1/2}}, & x_2 &= \frac{x_1 - \mathcal{V}t_1}{[1 - (\mathcal{V}/c)^2]^{1/2}}, \\
y_1 &= y_2, & y_2 &= y_1, \\
z_1 &= z_2, & z_2 &= z_1, \\
t_1 &= \frac{t_2 + (\mathcal{V}/c^2)x_2}{[1 - (\mathcal{V}/c)^2]^{1/2}}, & t_2 &= \frac{t_1 - (\mathcal{V}/c^2)x_1}{[1 - (\mathcal{V}/c)^2]^{1/2}}.
\end{aligned} \qquad (5\text{-}14)$$

This is the *Lorentz transformation*. It applies to the coordinate systems of Figure 5-1, which are repeated in Figure 5-7.

Since we shall be using these equations frequently, we have rewritten them in a more concise form in Table 5-1.

The Lorentz transformation forms the basis of special relativity, and it has been confirmed in innumerable experiments. We shall spend the rest of this chapter discussing some of its strange consequences. For the moment, we can immediately note a few of its more or less obvious features.

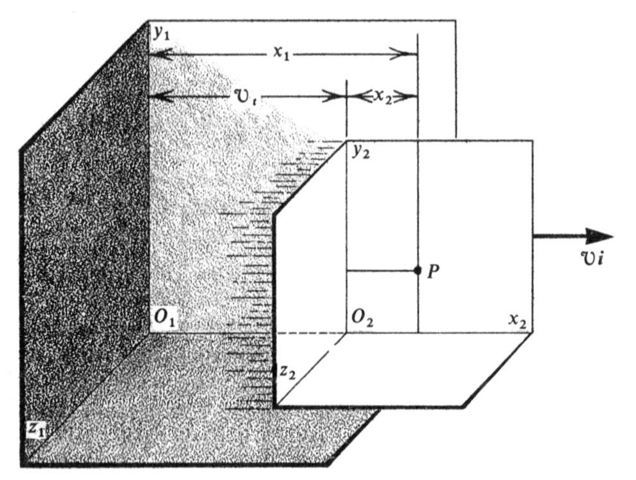

Figure 5-7. The Lorentz transformation refers to these two coordinate systems. System 2 moves at a velocity $\mathcal{V}\boldsymbol{i}$ with respect to 1 in the positive direction of the common x-axis. The two systems overlap when O_1 and O_2 coincide. This figure is identical to Figure 5-1.

(a) The Lorentz transformation reduces to the Galilean transformation if we set the velocity of light c equal to infinity.

(b) The relative velocity $\mathcal{V}$ of the two systems cannot be larger than c, for otherwise x, y, z, t become imaginary in one system or the other.

(c) There are really only four independent equations, since the right-hand column can be deduced from the left-hand one, and vice versa. You should check this immediately.

This is a general rule: the relation between quantities in one frame and the corresponding quantities in the other frame can always be expressed by either one of two equations which are equivalent.

Table 5-1. *The Lorentz Transformation*

$x_1 = \gamma[x_2 + \mathcal{V}t_2]$	$x_2 = \gamma[x_1 - \mathcal{V}t_1]$
$y_1 = y_2$	$y_2 = y_1$
$z_1 = z_2$	$z_2 = z_1$
$t_1 = \gamma[t_2 + (\mathcal{V}/c^2)x_2]$	$t_2 = \gamma[t_1 - (\mathcal{V}/c^2)x_1]$

$$\gamma = \frac{1}{[1 - (\mathcal{V}/c)^2]^{1/2}}$$

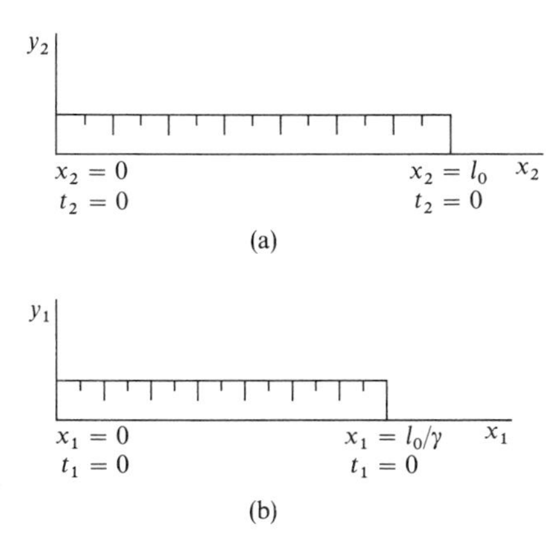

Figure 5-8. (a) A ruler is fixed parallel to the x-axis in reference frame 2 and, for an observer in that frame, it extends from $x_2 = 0$ to $x_2 = l_0$, and hence it has a length l_0. (b) At the time $t_1 = 0$ observer o_1 in reference frame 1 notes the positions of the *two* ends of the same ruler. He finds that it extends from $x_1 = 0$ to $x_1 = l_0/\gamma$, and hence that it is *shorter* than l_0. This is the *Lorentz contraction.* Lengths parallel to the y and z axes are unaffected.

(d) The right-hand column is identical to the left-hand column, except that the subscripts 1 and 2 are interchanged and that $-\mathcal{V}$ is substituted for $+\mathcal{V}$.

This is also a general rule: if a quantity in one frame is known in terms of quantities in the other frame, the inverse relation is obtained by interchanging the subscripts 1 *and* 2, *and changing the sign of* $\mathcal{V}$.

The reason for this rule is simply that we can either consider that frame 2 moves at a velocity $+\mathcal{V}$ with respect to 1, or that frame 1 moves at a velocity $-\mathcal{V}$ with respect to 2.

The Lorentz transformation provides us with transformation equations for the space coordinates and for the time. We shall now deduce from them transformation equations for many other quantities.

5.7 TRANSFORMATION OF A LENGTH

Let us imagine that an observer o_2 in reference frame 2 fixes a ruler of length l_0 on his x-axis, as in Figure 5-8a, so that its extremities are at $x_2 = 0$ and at $x_2 = l_0$. The length l_0 is the length of the ruler as measured in its own reference frame and is called its *proper length.* What will be the length of this *same* ruler for an observer o_1 on 1?

By *observer* we mean either a human being equipped with proper instruments, or some device that can take readings either automatically or by remote control. Observer o_1 is stationary in reference frame 1 and observer o_2 is stationary in 2.

Observer o_1 determines the length of the ruler by noting the positions of both ends *at the same time* t_1. Imagine that he makes his observations at the time $t_1 = 0$ when the two origins coincide. Then o_1 notes that the left-hand end is at $x_1 = 0$, since $x_2 = 0$, $t_1 = t_2 = 0$ at the left-hand end. He also notes that the right-hand end is at

$$l_1 = \gamma(l_0 + \mathcal{V} t_2), \tag{5-15}$$

according to the Lorentz transformation. We must find t_2.

Now we know that $t_1 = 0$ and $x_2 = l_0$ for o_2's reading at the right-hand end. Then

$$t_2 = -(\mathcal{V}/c^2) l_0, \tag{5-16}$$

and

$$l_1 = \gamma l_0[1 - (\mathcal{V}/c)^2] = l_0[1 - (\mathcal{V}/c)^2]^{1/2} = l_0/\gamma. \tag{5-17}$$

Thus the ruler that is fixed in reference frame 2 and that has a proper length l_0 according to observer o_2 appears *shorter* by a factor of $1/\gamma$ to observer o_1, as in Figure 5-8b. In other words, a ruler moving in the direction of its length at a velocity $\mathcal{V}$ relative to the observer appears to be shortened by the factor $1/\gamma = [1 - (\mathcal{V}/c)^2]^{1/2}$. This is the *Lorentz contraction*. It is independent of the sign of $\mathcal{V}$. *Both observers agree on the correctness of this figure; their disagreement bears on the validity of the measurements.* Observer o_1 maintains that his measurement is valid because he has observed the positions of the two ends simultaneously, namely, at $t_1 = 0$. But observer o_2 maintains that, on the contrary, o_1's measurements were *not* simultaneous—that o_1 first noted the position of the right-hand end at the time $t_2 = -(\mathcal{V}/c^2)l_0$ and *later* noted the position of the left-hand end.

Of course the Lorentz contraction would be precisely the same if the ruler were fixed anywhere else on the x_2 axis.

If the ruler is fixed along the x-axis in reference frame 1, observer o_2 finds it shortened by the same factor $1/\gamma$, as in Figure 5-9.

Thus o_1 tells o_2 that o_2's meters are too short, and o_2 tells o_1 that o_1's meters are also too short! This is not really contradictory, because the two comparisons are not really the same. They involve two *different* pairs of measurements.

This is a general rule: *the proper length of an object, measured in a certain direction, is always LONGER than the same length measured in a frame of reference moving in that particular direction.*

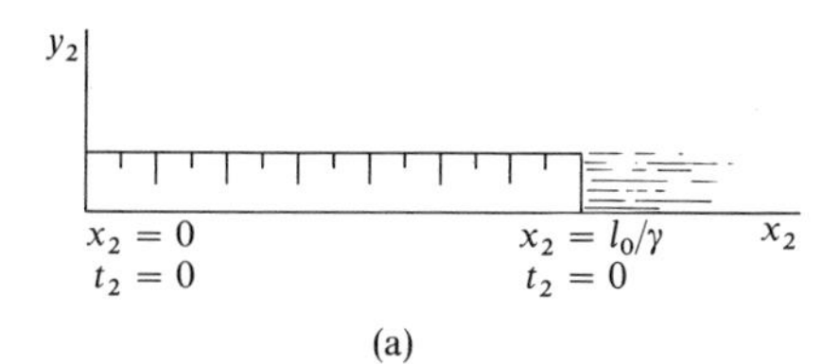

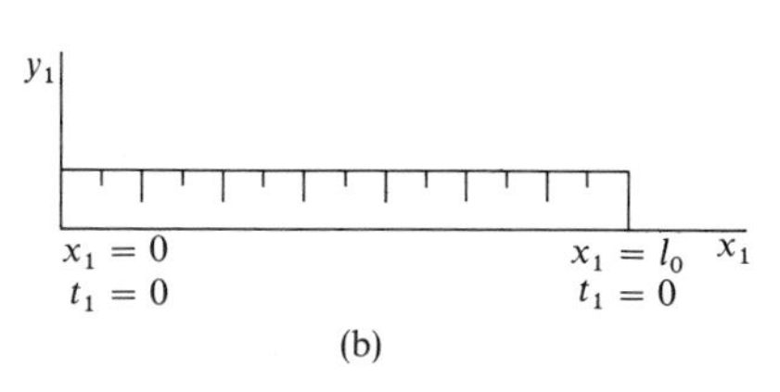

Figure 5-9. The ruler is now fixed in reference frame 1.

What if the ruler moves relative to the observer in the direction perpendicular to its length? It then appears to have its proper length l_0.

Example

THE APPARENT SHAPE OF A RAPIDLY MOVING OBJECT

We have just seen that the Lorentz contraction has the effect of contracting an object in the direction of its motion by the factor $[1 - (\mathcal{V}/c)^2]^{1/2}$ when it moves at a velocity $\mathcal{V}$ with respect to the observer. The other dimensions remain unchanged.

One would therefore expect that a photograph of an object moving rapidly in a direction perpendicular to the line of sight would simply show it to be contracted in the direction of its motion. In fact, it is quite easy to show that the object appears rather to be rotated by an angle $\theta = \arcsin(\mathcal{V}/c)$, as long as the angle subtended by the object at the camera is small. If the object moves in another direction, or if the angle it subtends at the camera is not small, the apparent distortion becomes quite complex.

Figure 5-10 shows a camera taking a photograph of a cube of side l moving at a velocity $\boldsymbol{\mathcal{V}}$. The photograph registers photons that arrive at a given instant. If the cube is quite far away, photons from the face $CDFE$ all arrive at approximately the same time. Edges CE and DF appear shortened by the factor $1/\gamma$, while CD and EF appear to be of the same length as if the cube were not moving with respect to the camera. Thus the square face $CDFE$ appears on the photograph as a rectangle, as in Figure 5-11.

Photons from the edge AB in Figure 5-10 leave the cube some time previously at $A'B'$, where $A'A = B'B$ is the distance traveled by the cube during the time that the light takes to travel the distance l, or $\mathcal{V}(l/c)$. The picture therefore has the appearance shown in Figure 5-11.

Figure 5-12 shows that this is the photograph one would obtain

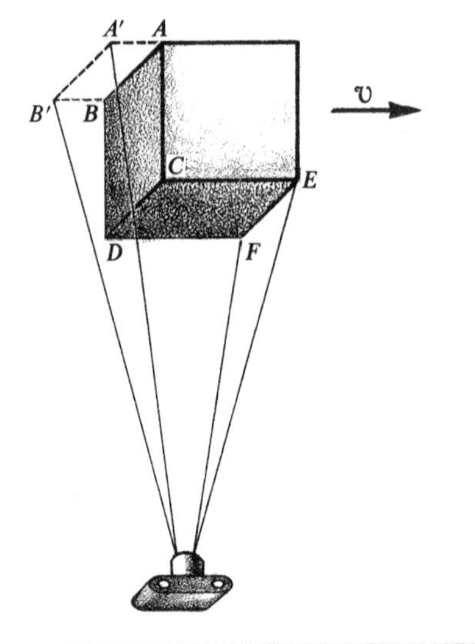

Figure 5-10. A camera photographs a cube moving at a velocity $\mathcal{V}$ perpendicular to the line of sight. Photons from the edge AB leave the cube at $A'B'$, so that the camera sees the left-hand face as in Figure 5-11.

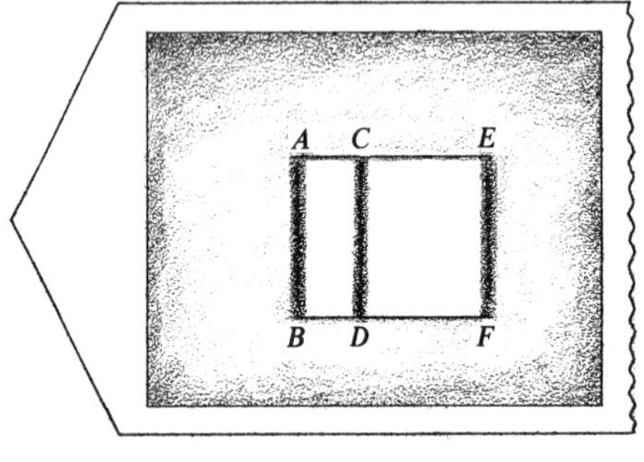

Figure 5-11. The photograph shows the edge AB of the cube displaced to the left and the edges CE and DF contracted by the factor $1/\gamma$.

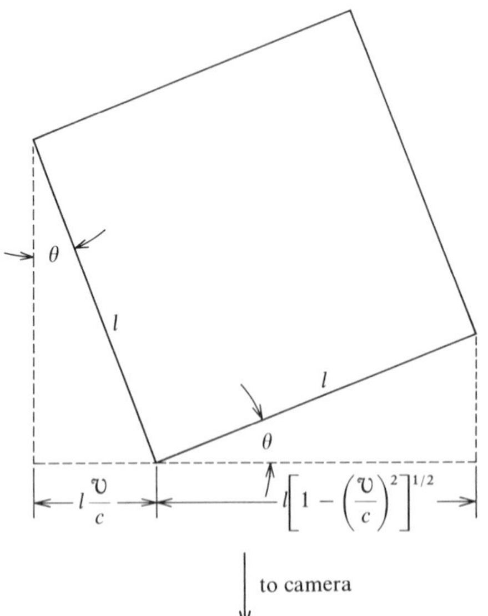

Figure 5-12. If the cube of Figure 5-10 were rotated through an angle $\theta = \text{arc sin}\ (\mathcal{V}/c)$, it would appear on the photograph approximately as in Figure 5-11. The solid angle subtended by the cube at the camera is assumed to be small.

if the cube were rotated through the angle $\theta = \text{arc sin}(\mathcal{V}/c)$. (If the velocity $\mathcal{V}$ of the cube were equal to the velocity of light c, the photograph would only show the left-hand face!)

Of course, the object would have the same appearance to an observer as it does on the photograph, because the eye, like the camera, registers photons that *arrive* together at a given instant.

It will be noticed that the appearance of the object is not exactly the same as if it were rotated, because the face $CDFE$ remains normal to the line of sight.

5.8 TRANSFORMATION OF A TIME INTERVAL

Now imagine that observer o_2 measures the duration of a certain phenomenon that is fixed with respect to his frame and that starts at $t_2 = 0$ and ends at $t_2 = T_0$. The time T_0 is called the *proper time*. What will be the duration of this *same* phenomenon for observer o_1?

To perform his measurement, o_2 has a *single* clock situated at the origin O_2. Observer o_1 uses two identical and synchronized clocks, one at O_1 and one at x_1 as in Figure 5-13. He chooses x_1 so that his second clock is next to that of o_2 at the end of the time interval.

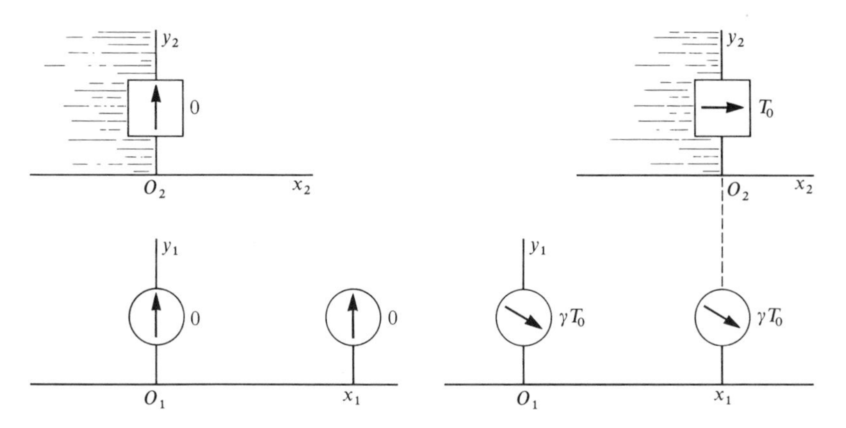

Figure 5-13. Observer o_2 measures the duration of a phenomenon which occurs at the origin O_2. He uses a single clock at O_2 and finds T_0 seconds. To measure the duration of the *same* phenomenon, observer o_1 uses two synchronized clocks, one at $x_1 = 0$ where the time interval begins, and one at $x_1 = \mathcal{V}(\gamma T_0)$ where it ends. Upon comparing the readings of his two clocks, observer o_1 concludes that the time interval is γT_0 seconds, which is *longer* than T_0. Figure (a) shows the clocks at the beginning of the time interval: all three show the same time. Figure (b) shows the clocks at the end of the time interval. Observer o_1 says that o_2's clock runs slow.

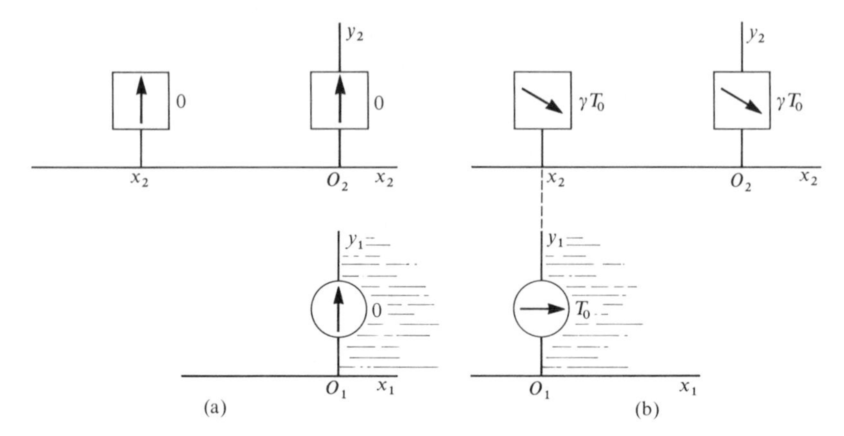

Figure 5-14. The situation here is the inverse of that illustrated in Figure 5-13. The phenomenon now occurs at the origin O_1 instead of O_2, observer o_1 measures its duration with a single clock, and o_2 uses two clocks. The time interval measured by the moving observer is again longer than T_0 by the factor γ.

Thus, for o_1, the time interval begins at $t_1 = 0$ and ends at

$$T_1 = \gamma[T_0 + (\mathcal{V}/c^2)x_2]. \tag{5-18}$$

But $x_2 = 0$; thus

$$T_1 = \gamma\, T_0 = \frac{T_0}{[1 - (\mathcal{V}/c)^2]^{1/2}}. \tag{5-19}$$

Observer o_1 therefore measures a time interval that is *longer* than o_2's and concludes that o_2's clock runs slow.

What if it is o_1 who measures a time interval with a single clock as in Figure 5-14? Then o_2 finds that o_1's clock runs slow.

In other words, a moving clock appears to run slow by a factor of γ whenever its time is checked as above. This is the phenomenon called *time dilation* (see footnote, Section 5.2).

This is also a general rule: *the proper time interval between two phenomena is always SHORTER than the same time interval measured by a moving observer.*

Example

THE TIME READ ON A RAPIDLY MOVING CLOCK

We have just seen that, if o_1 uses two identical clocks at two different points on 1 to measure a time interval that o_2 measures at a fixed point on 2 with a single clock, then o_1 finds that his own time intervals are longer than those of o_2 by the factor $\gamma = 1/[1 - (\mathcal{V}/c)^2]^{1/2}$.

Now, would o_1 arrive at the same conclusion if he used a single clock and *looked* at o_2's moving clock, as in Figure 5-15? We shall assume that o_2's clock, which is situated at the origin O_2 of reference frame 2, is moving *away* from o_1 who stays at O_1. This means that O_2 is to the right of O_1, or that t_1 is positive.

Let us first imagine that observer o_1 has a set of identical synchronized clocks along his x-axis. As o_2's clock goes by each one of these, the relation $t_1 = \gamma t_2$ holds.

But o_1 is at the origin O_1 of his system, and the light from o_2's clock takes some time to reach o_1. Thus o_1 takes his readings later than t_1.

Suppose o_1 reads a time t_2 on o_2's clock. What time is it on o_1's clock? Let us call this time t_1'. Then t_1' is the above t_1 plus the time required for light to travel the distance $\mathcal{V}t_1$ between o_2's clock and o_1's:

$$t_1' = t_1 + \frac{\mathcal{V}t_1}{c} = [1 + (\mathcal{V}/c)]t_1, \tag{5-20}$$

$$= \frac{1 + (\mathcal{V}/c)}{[1 - (\mathcal{V}/c)^2]^{1/2}}\, t_2, \tag{5-21}$$

$$= \left[\frac{1 + (\mathcal{V}/c)}{1 - (\mathcal{V}/c)}\right]^{1/2} t_2 > t_2. \tag{5-22}$$

Thus, if observer o_1 *looks* at o_2's clock when it is moving *away* from him, he arrives at the above result and concludes that o_2's clock is even slower than with the previous method of measurement. If o_1

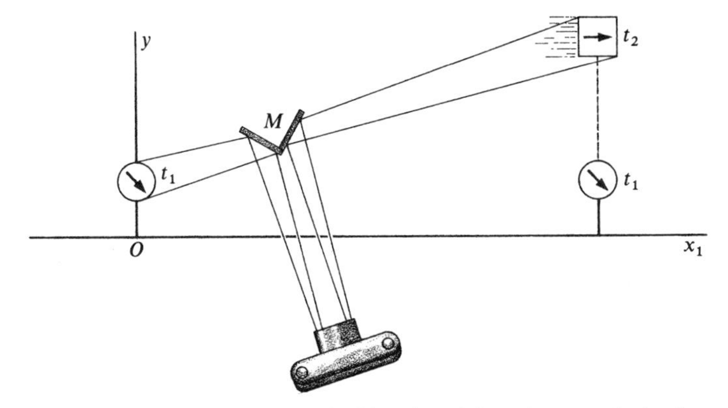

Figure 5-15. Observer o_1 uses a double mirror M to photograph simultaneously his own clock and the moving clock. The relation $t_1 = \gamma t_2$ holds again, but the light from the moving clock does not reach the camera until a later time t_1'.

takes into account the time taken by the signal to arrive to him, then he finds the previous result $t_1 = \gamma t_2$.

What if o_2's clock is moving toward o_1? Then the origin O_2 is to the left of O_1, and both t_1 and t_2 are negative. Therefore the time t_1' at O_1 when o_1 reads t_2 on o_2's clock is

$$t_1' = t_1 + \frac{|\mathcal{V} t_1|}{c} = t_1 - \frac{\mathcal{V} t_1}{c}, \tag{5-23}$$

and t_1' is also negative, since $\mathcal{V} \leq c$. Thus

$$t_1' = \left[\frac{1 - (\mathcal{V}/c)}{1 + (\mathcal{V}/c)}\right]^{1/2} t_2 \tag{5-24}$$

Now o_2's clock appears to run *fast* (t_1' and t_2 are both negative) by the above factor, but, again, if o_1 takes into account the time that the light takes to reach him, he finds that $t_1 = \gamma t_2$.

Example

THE RELATIVISTIC DOPPLER EFFECT FOR ELECTROMAGNETIC WAVES

The shift in frequency that is observed at a detector when it moves with respect to a source of waves is called the *Doppler effect*. This phenomenon is well known in the field of acoustics.

Imagine that we have a source of periodic electromagnetic waves of frequency f_s at O_2 and a detector at O_1. What will be the frequency f_d measured at O_1?

This problem is really identical to the clock problem we have just discussed because the source can be considered as a clock that beats periods $T_s = 1/f_s$, instead of seconds. Therefore, when the source recedes from the observer at a velocity $\mathcal{V}$, as in Figure 5-16a, the apparent period at the detector is

$$T_d = \left[\frac{1 + (\mathcal{V}/c)}{1 - (\mathcal{V}/c)}\right]^{1/2} T_s > T_s, \tag{5-25}$$

and the apparent frequency at the detector is

$$f_d = 1/T_d = \left[\frac{1 - (\mathcal{V}/c)}{1 + (\mathcal{V}/c)}\right]^{1/2} f_s < f_s. \tag{5-26}$$

If the source moves toward the detector as in Figure 5-16b, the apparent frequency is

$$f_d = \left[\frac{1 + (\mathcal{V}/c)}{1 - (\mathcal{V}/c)}\right]^{1/2} f_s > f_s. \tag{5-27}$$

An interesting feature of these equations is that they are always valid, whether it is the source or the detector that moves. This is because *there is no way of knowing which one it is that moves*, according to the fundamental postulate of relativity. The relative velocity $\mathcal{V}$ is a positive quantity in all three equations.

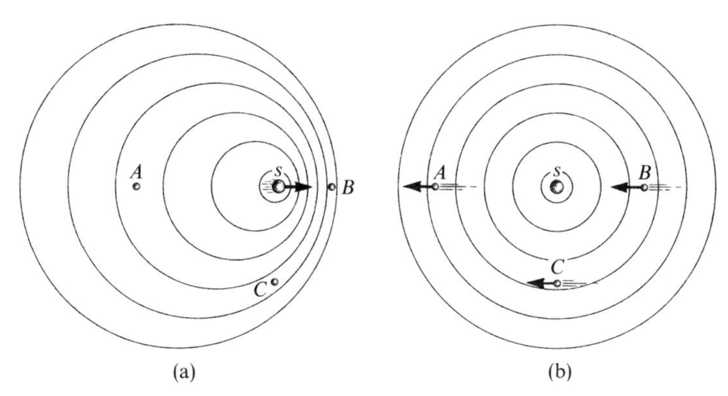

Figure 5-16. The Doppler effect for electromagnetic waves. S is a source and A, B, C are three detectors. At A the source recedes from the detector and $f_d < f_s$. At B the source moves toward the detector and $f_d > f_s$. At C the source moves at right angles to the line joining it to the detector and f_d is either larger or smaller than f_s, as in Eq. 5-28. The frequency shifts are the same in (a) and (b): it is only the *relative* velocity that matters.

If the relative velocity forms a right angle with the line joining the source to the detector, we have either the first or the second equation below, depending on the reference frame in which the angle is measured:

$$f_d = \gamma f_s \text{ (source)} \qquad \text{or} \qquad f_d = \frac{f_s}{\gamma} \text{ (detector)}. \tag{5-28}$$

5.9 SIMULTANEITY

Imagine that observer o_1 sees two events A and B that, to him, occur at the same x-coordinate $x_{A1} = x_{B1}$, and, at the same time, $t_{A1} = t_{B1}$. He maintains that A and B are simultaneous. Are they also simultaneous for o_2?

From the Lorentz transformation,

$$t_{A2} = \gamma[t_{A1} - (\mathcal{V}/c^2)x_{A1}] = \gamma[t_{B1} - (\mathcal{V}/c^2)x_{B1}] = t_{B2}. \tag{5-29}$$

Then two events that occur at the same value of x and that are simultaneous for one observer are also simultaneous for another observer. Note that the two events need *not* occur at the same place, because the y's and the z's can be different.

What if the events do not occur at the same value of x? If they are simultaneous for o_1, then $t_{A1} = t_{B1}$, but since $x_{A1} \neq x_{B1}$, they are *not* simultaneous for the other observer o_2.

In fact, the time interval between the two events, as seen from reference frame 2 is

$$t_{B2} - t_{A2} = \gamma(\mathcal{V}/c^2)(x_{A1} - x_{B1}), \tag{5-30}$$

and it can be either positive or negative, depending on the sign of $x_{A1} - x_{B1}$.

5.10 CAUSALITY AND MAXIMUM SIGNAL VELOCITY

We have just seen that two events that are simultaneous in frame 1 are not simultaneous in frame 2, unless they occur at the same x. We can go even further: the *order* in which two events occur can be different in different frames because

$$t_{B2} - t_{A2} = \gamma[(t_{B1} - t_{A1}) - (\mathcal{V}/c^2)(x_{B1} - x_{A1})], \tag{5-31}$$

and the signs of $t_{B2} - t_{A2}$ and $t_{B1} - t_{A1}$ can be different.

This is a disturbing result indeed: it appears to violate the principle of causality, according to which a cause necessarily occurs before its effect.

For example, imagine that observer o_1 throws a ball in the direction of the x-axis, and the ball, after a flight of a few seconds, breaks a windowpane. The Lorentz transformation surely cannot imply that the series of events, starting with the throwing of the ball and ending with the strewn broken glass, would occur *backwards* in time for certain observers.

Let us imagine two events A and B. Event A occurs at the origins O_1 and O_2 at the moment they coincide. Event A is the cause of B, which occurs at x_{B1} at a later time t_{B1}, and in frame 2, at x_{B2}. Event B cannot occur before A in frame 2. Then t_{B2} must not be negative. Event A could be the throwing of the ball, and event B the breaking of the glass. Event A causes event B through some device, in this case it is the ball, which propagates in some way the signal from A at a velocity v_1 in frame 1. Then

$$t_{B2} = \gamma[t_{B1} - (\mathcal{V}/c^2)x_{B1}] = \gamma[t_{B1} - (\mathcal{V}/c^2)(v_1 t_{B1})], \tag{5-32}$$

$$= \gamma\, t_{B1}[1 - (v_1 \mathcal{V}/c^2)]. \tag{5-33}$$

The factor γ is positive; by definition, t_{B1} is also positive. If v_1 and $\mathcal{V}$ have different signs, then t_{B2} is always positive and event B always comes after A. However, if v_1 and $\mathcal{V}$ have the same sign, we must *always* have

$$v_1\mathcal{V} \leq c^2. \tag{5-34}$$

Let us assume that v_1 and $\mathcal{V}$ are both positive. We have already seen in Section 5-6 that $\mathcal{V}$ can be as large as, but never larger than c. So the above inequality must be satisfied even when $\mathcal{V} = c$. Then

$$v_1 \leq c. \tag{5-35}$$

If v_1 and $\mathcal{V}$ are both negative, we arrive at the same result, by symmetry.

In other words, a signal can never be propagated at a velocity larger than the velocity of light. Otherwise, the principle of causality would be violated, and certain observers would perceive some effects before their causes.

5.11 TRANSFORMATION OF A VELOCITY

If o_2 observes that an object has some velocity v_{2x} in the direction of the x-axis, what is the velocity of this same object according to observer o_1? (The velocity v_{2x} need not be constant.)

The velocity v_{1x}, by definition, is dx_1/dt_1, and

$$v_{1x} = \frac{dx_1}{dt_1} = \frac{dx_1}{dt_2}\frac{dt_2}{dt_1}, \tag{5-36}$$

where

$$\frac{dx_1}{dt_2} = \gamma \frac{d}{dt_2}(x_2 + \mathcal{V}t_2) = \gamma\,(v_{2x} + \mathcal{V}). \tag{5-37}$$

To calculate dt_2/dt_1 we remember that t_2 is a function of both t_1 and x_1, and that x_1 is itself a function of t_1. In more familiar terms, we wish to calculate dz/dx on a curve $z = ax + by$, where y is a specified function of x:

$$\text{if } z = ax + by(x), \qquad \text{then } \frac{dz}{dx} = a + b\frac{dy}{dx}. \tag{5-38}$$

Then

$$\frac{dt_2}{dt_1} = \gamma \frac{d}{dt_1}[t_1 - (\mathcal{V}/c^2)x_1] = \gamma\,[1 - (\mathcal{V}/c^2)v_{1x}], \tag{5-39}$$

and

$$v_{1x} = \gamma^2(v_{2x} + \mathcal{V})[1 - (\mathcal{V}/c^2)v_{1x}]. \tag{5-40}$$

Putting both v_{1x} terms on the left-hand side and simplifying,

$$v_{1x} = \frac{v_{2x} + \mathcal{V}}{1 + (v_{2x}\,\mathcal{V}/c^2)}. \tag{5-41}$$

The velocity v_{1x} is measured at the time t_1, and v_{2x} is measured at the corresponding time t_2.

Therefore the velocity in reference frame 1 is smaller than $v_{2x} + \mathcal{V}$ by the factor $1 + (v_{2x}\,\mathcal{V}/c^2)$.

The inverse transformation is

$$v_{2x} = \frac{v_{1x} - \mathcal{V}}{1 - (v_{1x}\,\mathcal{V}/c^2)}. \tag{5-42}$$

Example This result makes the velocity of light equal to c in both systems. For example, if the moving object is a photon in frame 2, v_{2x} is equal to c, and v_{1x} is also equal to c, for any value of $\mathcal{V}$! Or, if $\mathcal{V} = c$, then $v_{1x} = c$ for any value of v_{2x}.

What if the velocity is not along the x-axis? Let us find a relation between

$$v_{1y} = \frac{dy_1}{dt_1} \quad \text{and} \quad v_{2y} = \frac{dy_2}{dt_2}. \tag{5-43}$$

The relation between v_{1z} and v_{2z} will be of the same form, by symmetry.

We proceed as above:

$$v_{1y} = \frac{dy_1}{dt_1} = \frac{dy_1}{dt_2}\frac{dt_2}{dt_1}, \tag{5-44}$$

and, since $y_1 = y_2$,

$$v_{1y} = \frac{dy_2}{dt_2}\frac{dt_2}{dt_1} = v_{2y}\frac{dt_2}{dt_1} = v_{2y}\,\gamma[1 - (\mathcal{V}/c^2)v_{1x}]. \tag{5-45}$$

Then

$$v_{2y} = \frac{v_{1y}}{\gamma[1 - (v_{1x}\mathcal{V}/c^2)]}. \tag{5-46}$$

The inverse transformation is

$$v_{1y} = \frac{v_{2y}}{\gamma[1 + (v_{2x}\mathcal{V}/c^2)]}. \tag{5-47}$$

Similarly,

$$v_{1z} = \frac{v_{2z}}{\gamma[1 + (v_{2x}\mathcal{V}/c^2)]}, \tag{5-48}$$

$$v_{2z} = \frac{v_{1z}}{\gamma[1 - (v_{1x}\mathcal{V}/c^2)]}. \tag{5-49}$$

Table 5-2 shows all six equations.

Table 5-2. *Transformation of a Velocity*

$v_{1x} = \dfrac{v_{2x} + \mathcal{V}}{1 + (v_{2x}\mathcal{V}/c^2)}$	$v_{2x} = \dfrac{v_{1x} - \mathcal{V}}{1 - (v_{1x}\mathcal{V}/c^2)}$
$v_{1y} = \dfrac{v_{2y}}{\gamma[1 + (v_{2x}\mathcal{V}/c^2)]}$	$v_{2y} = \dfrac{v_{1y}}{\gamma[1 - (v_{1x}\mathcal{V}/c^2)]}$
$v_{1z} = \dfrac{v_{2z}}{\gamma[1 + (v_{2x}\mathcal{V}/c^2)]}$	$v_{2z} = \dfrac{v_{1z}}{\gamma[1 - (v_{1x}\mathcal{V}/c^2)]}$

Example An object whose velocity in reference frame 2 is purely along the y axis has a velocity $\mathcal{V}\boldsymbol{i} + (v_{2y}/\gamma)\boldsymbol{j}$ in frame 1: v_{1x} is simply the relative velocity $\mathcal{V}\boldsymbol{i}$ of the two frames, and v_{1y} is smaller than v_{2y} by the factor $1/\gamma$.

5.12 TRANSFORMATION OF AN ACCELERATION

We can transform an acceleration by proceeding again in the same manner. For example,

$$a_{1x} = \frac{dv_{1x}}{dt_1} = \frac{dv_{1x}}{dt_2}\frac{dt_2}{dt_1}, \tag{5-50}$$

$$= \left\{\frac{d}{dt_2}\frac{(v_{2x} + \mathcal{V})}{[1 + (v_{2x}\mathcal{V}/c^2)]}\right\}\gamma[1 - (\mathcal{V}/c^2)v_{1x}], \tag{5-51}$$

$$= \frac{a_{2x}}{\gamma^3[1 + (v_{2x}\mathcal{V}/c^2)]^3}. \tag{5-52}$$

Table 5-3 gives the six transformation equations.

Table 5-3. *Transformation of an Acceleration*

$$a_{1x} = \frac{a_{2x}}{\gamma^3[1 + (v_{2x}\mathcal{V}/c^2)]^3}$$

$$a_{1y} = \frac{1}{\gamma^2[1 + (v_{2x}\mathcal{V}/c^2)]^2}\left\{a_{2y} - \frac{v_{2y}\mathcal{V}}{c^2 + v_{2x}\mathcal{V}}\,a_{2x}\right\}$$

$$a_{1z} = \frac{1}{\gamma^2[1 + (v_{2x}\mathcal{V}/c^2)]^2}\left\{a_{2z} - \frac{v_{2z}\mathcal{V}}{c^2 + v_{2x}\mathcal{V}}\,a_{2x}\right\}$$

$$a_{2x} = \frac{a_{1x}}{\gamma^3[1 - (v_{1x}\mathcal{V}/c^2)]^3}$$

$$a_{2y} = \frac{1}{\gamma^2[1 - (v_{1x}\mathcal{V}/c^2)]^2}\left\{a_{1y} + \frac{v_{1y}\mathcal{V}}{c^2 - v_{1x}\mathcal{V}}\,a_{1x}\right\}$$

$$a_{2z} = \frac{1}{\gamma^2[1 - (v_{1x}\mathcal{V}/c^2)]^2}\left\{a_{1z} + \frac{v_{1z}\mathcal{V}}{c^2 - v_{1x}\mathcal{V}}\,a_{1x}\right\}$$

5.13 RELATIVISTIC MASS

We have seen that relativity requires that lengths, times, velocities, and accelerations be transformed according to rules that are different from those

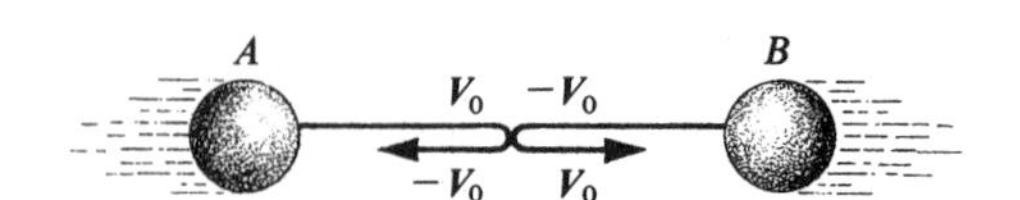

Figure 5-17. This idealized experiment permits us to find a relation between mass and velocity. Observer o_1 slides two equal masses A and B with equal and opposite velocities along the x-axis. They collide and rebound elastically. We conclude from observer o_2's measurements that $m = m_0/[1 - (v/c)^2]^{1/2}$, where v is the velocity of the mass with respect to the observer.

of classical mechanics. We can therefore expect that many other quantities will be affected. Let us find out how mass should be transformed.

Imagine the following experiment, which is illustrated in Figure 5-17. Observer o_1 has two identical masses A and B which slide in opposite directions along the x-axis with zero friction. He observes that their velocities are equal and opposite. The two masses collide and rebound with velocities that are again equal and opposite. The fact that they do rebound with equal and opposite velocities demonstrates that the two masses are equal. We assume that the collision is elastic.

Observer o_2 sees o_1 perform his experiment and measures the velocities of the masses A and B. The velocities with respect to the two frames of reference are shown in Table 5-4.

Let us call m_A and m_B the masses of A and of B in reference frame 2 before the collision. Both are unknown for the moment. Let us set M to be the total mass in frame 2 at the time of the collision.

We assume first that the total mass of A and B remains constant and equal to M in frame 2:

$$m_A + m_B = M. \tag{5-53}$$

During the collision, part of the kinetic energy is transformed into potential energy because the masses are elastically distorted, but the total energy, and hence the total mass, remains constant.

We also assume that there is conservation of momentum during the collision, as seen from 2:

$$m_A v_A + m_B v_B = -M\mathcal{V}. \tag{5-54}$$

On the left-hand side we have written the sum of the momenta of A and of B before the collision, v_A and v_B having the values shown in Table 5-4. The term on the right is the momentum during the collision; it is the total mass M, times the common velocity $-\mathcal{V}$ with respect to frame 2. Remember that during the collision the masses have zero velocity with respect to reference frame 1, and thus a velocity $-\mathcal{V}$ with respect to 2.

Table 5-4. *Velocities of Masses A and B*

Observer	Before Impact		During Impact		After Impact	
	A	B	A	B	A	B
o_1	v_o	$-v_o$	0	0	$-v_o$	v_o
o_2	$\frac{v_o - \mathcal{V}}{1 - (v_o\mathcal{V}/c^2)}$	$\frac{-v_o - \mathcal{V}}{1 + (v_o\mathcal{V}/c^2)}$	$-\mathcal{V}$	$-\mathcal{V}$	$\frac{-v_o - \mathcal{V}}{1 + (v_o\mathcal{V}/c^2)}$	$\frac{v_o - \mathcal{V}}{1 - (v_o\mathcal{V}/c^2)}$

If we multiply the first of these two equations by $\mathcal{V}$ and then add it to the second, we find that

$$\frac{m_A}{m_B} = -\frac{\mathcal{V} + v_B}{\mathcal{V} + v_A}. \tag{5-55}$$

This ratio is always positive. Also,

$$\frac{m_A}{m_B} = \left[\frac{1 - (v_B/c)^2}{1 - (v_A/c)^2}\right]^{1/2}. \tag{5-56}$$

You can show that these two values of m_A/m_B are equal by squaring and equating the right-hand sides. (Hint: do not substitute the values of v_A and v_B until you have simplified the equality.) Note that v_A and v_B cannot both be equal to zero simultaneously.

Then we can write that

$$m_A = \frac{m_0}{[1 - (v_A/c)^2]^{1/2}}, \tag{5-57}$$

$$m_B = \frac{m_0}{[1 - (v_B/c)^2]^{1/2}}, \tag{5-58}$$

where m_0 is called the *rest mass*. These are the masses of A and B as measured by observer o_2, with reference to whom they have velocities v_A and v_B, respectively.

More generally, the mass of an object moving with a velocity v with respect to the observer is

$$m = \frac{m_0}{[1 - (v/c)^2]^{1/2}}. \tag{5-59}$$

The quantity m is the *relativistic mass*.

Note that, if the observer follows the object, $v = 0$ and $m = m_0$, whatever be the velocity of the object with respect to some other observer.

5.14 TRANSFORMATION OF A MASS

Observer o_1 on reference frame 1 observes a moving object and finds that its mass is m_1. Observer o_2 in reference frame 2 observes the *same* object, and he finds a mass m_2. What are the two relations between m_1 and m_2?

For observer o_1 the mass is

$$m_1 = \frac{m_0}{[1 - (v_1/c)^2]^{1/2}}, \tag{5-60}$$

where m_0 is the mass of the object when it is at rest with respect to the observer, and v_1 is the velocity of the object with respect to reference frame 1. Thus

$$m_1 = \frac{m_0}{[1 - (v_{1x}^2 + v_{1y}^2 + v_{1z}^2)/c^2]^{1/2}}. \tag{5-61}$$

Now, from Eqs. 5-41, 5-47, and 5-48,

$$\frac{v_{1x}^2 + v_{1y}^2 + v_{1z}^2}{c^2} = \frac{\gamma^2(v_{2x} + \mathcal{V})^2 + v_{2y}^2 + v_{2z}^2}{c^2\gamma^2[1 + (v_{2x}\mathcal{V}/c^2)]^2}. \tag{5-62}$$

Upon simplifying we find that

$$m_1 = \gamma[1 + (v_{2x}\mathcal{V}/c^2)]m_2. \tag{5-63}$$

The inverse relation is

$$m_2 = \gamma[1 - (v_{1x}\mathcal{V}/c^2)]m_1. \tag{5-64}$$

As usual, these two equations are equivalent: they both give the same value for the ratio m_1/m_2.

Note that the y and z components of velocity do not affect the ratio m_1/m_2. If $v_{2x} = 0$, then $m_1 = \gamma m_2$.

5.15 RELATIVISTIC ENERGY $\mathcal{E}$

What is the physical interpretation for the product mc^2? Dimensionally, it is an energy. From our definition of m,

$$mc^2 = m_0c^2\,[1 - (v/c)^2]^{-1/2}, \tag{5-65}$$

$$= m_0c^2\left[1 + \frac{1}{2}\frac{v^2}{c^2} + \frac{3}{8}\frac{v^4}{c^4} + \cdots\right], \tag{5-66}$$

$$= m_0c^2 + \frac{1}{2}m_0v^2 + \frac{3}{8}m_0\frac{v^4}{c^2} + \cdots \tag{5-67}$$

The term $(1/2)m_0v^2$ is the kinetic energy of classical mechanics. The term

m_0c^2 is an energy that is associated with the rest mass and that is called the *rest energy*.

The quantity mc^2 is the *relativistic energy* $\mathcal{E}$, and the difference $mc^2 - m_0c^2$ is called the *relativistic kinetic energy*. The relativistic kinetic energy is equal to $(1/2)mv^2$ for $v^2 \ll c^2$.

Example An excellent demonstration of the existence of this rest energy is the annihilation of electrons. This phenomenon is well known to nuclear physicists: positive electrons combine with negative electrons as in Figure 5-18 to give two gamma rays, each of which has an energy equal to the rest energy of one electron, or 0.511 MeV (1 million electron volts $= 1.6 \times 10^{-13}$ joule).

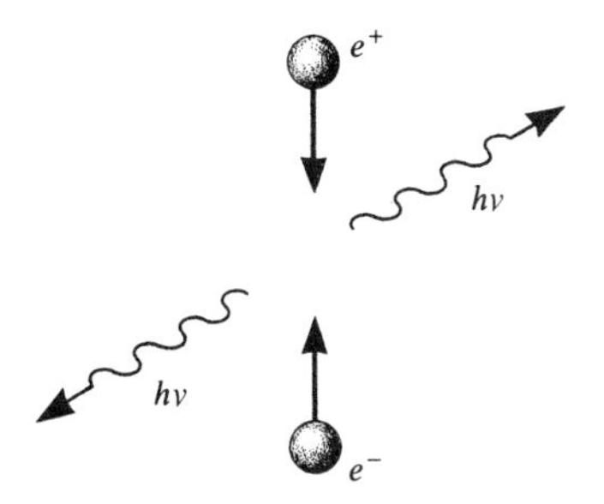

Figure 5-18. A positive electron e^+ and a negative electron e^- are annihilated to form two gamma rays $h\nu$. If the kinetic energy of the electrons is small, the energy $2h\nu$ of the two gamma rays is equal to the rest energy $2m_0c^2$ of the two electrons.

5.16 THE FOUR-VECTOR r

An event such as the emission of a flash of light is characterized by its position x, y, z and by the time t at which it occurs. Let us examine these four variables more closely. We shall find they can be grouped into a single expression that is invariant under a Lorentz transformation and that proves to be useful for discovering other invariant quantities in mechanical and electromagnetic phenomena.

Under a *Galilean* transformation the distance between two points (x_a, y_a, z_a) and (x_b, y_b, z_b) is an invariant:

$$r_{ab}^2 = (x_{a1} - x_{b1})^2 + (y_{a1} - y_{b1})^2 + (z_{a1} - z_{b1})^2, \tag{5-68}$$

$$= (x_{a2} - x_{b2})^2 + (y_{a2} - y_{b2})^2 + (z_{a2} - z_{b2})^2. \tag{5-69}$$

Note that the word invariant is *not* synonymous with the word constant. In fact, if the point a were fixed in frame 1 and b were fixed in 2, r_{ab} would be a function of the time. The distance r_{ab} is said to be invariant because it has the same numerical value in both frames.

Now you can easily show that, with the *Lorentz* transformation, r_{ab} is not an invariant. However, there does exist a corresponding quantity that is invariant under a Lorentz transformation. Imagine that a flash of light is emitted at O_1 at the moment when the two origins coincide. The light propagates in all directions at the velocity c in *both* systems, and

$$\frac{x_1^2 + y_1^2 + z_1^2}{t_1^2} = \frac{x_2^2 + y_2^2 + z_2^2}{t_2^2} = c^2. \tag{5-70}$$

The light arrives at (x_1, y_1, z_1) at the time t_1 and at the point (x_2, y_2, z_2) at the time t_2. Thus, in this case,

$$x_1^2 + y_1^2 + z_1^2 - c^2t_1^2 = x_2^2 + y_2^2 + z_2^2 - c^2t_2^2 = 0. \tag{5-71}$$

More generally, the coordinates x_1, y_1, z_1, t_1 and x_2, y_2, z_2, t_2 for any single event are related by the equation

$$x_1^2 + y_1^2 + z_1^2 - c^2t_1^2 = x_2^2 + y_2^2 + z_2^2 - c^2t_2^2, \tag{5-72}$$

or

$$r_1^2 - c^2t_1^2 = r_2^2 - c^2t_2^2. \tag{5-73}$$

You can check this quite easily. The quantity $r^2 - c^2t^2$ is therefore invariant under a Lorentz transformation.

Now this property of x, y, z, t follows directly from the Lorentz transformation. Then, for *any set of four quantities that transform like x, y, z, t, we have a corresponding invariant quantity.* We shall use this result on several occasions.

Such sets of four quantities are called *four-vectors.*

By analogy with three-dimensional geometry, we specify the coordinates of the event as being

$$(x, y, z, jct),$$

where $j = (-1)^{1/2}$, and the magnitude of the four-dimensional distance between the event and the origin is the square root of the sum of the squares of the components. Thus

$$\mathbf{r} = (r, jct). \tag{5-74}$$

The quantity $(x^2 + y^2 + z^2 - c^2t^2)^{1/2}$ is called the *magnitude* of the four-vector $\mathbf{r}$ and can be considered to be the square root of the scalar product of $\mathbf{r}$ by itself;

$$\mathbf{r}^2 = \mathbf{r} \cdot \mathbf{r} = x^2 + y^2 + z^2 - c^2t^2 = r^2 - c^2t^2. \tag{5-75}$$

This quantity is invariant, as we have shown above.

5.17 TRANSFORMATION OF A MOMENTUM AND OF A RELATIVISTIC ENERGY. THE FOUR-MOMENTUM

The momentum of a mass m moving at a velocity $\boldsymbol{v}$ is

$$\boldsymbol{p} = m\boldsymbol{v} = \frac{m_0}{[1 - (v/c)^2]^{1/2}}\boldsymbol{v}. \tag{5-76}$$

It is to be understood that v is the velocity of m with respect to the observer.

In reference frame 1,

$$p_1 = m_1 v_1, \tag{5-77}$$

and it has the three components

$$p_{1x} = m_1 v_{1x}, \qquad p_{1y} = m_1 v_{1y}, \qquad p_{1z} = m_1 v_{1z}. \tag{5-78}$$

It is a simple matter to transform p_{1x}:

$$p_{1x} = m_1 v_{1x} = \gamma[1 + (v_{2x}\mathcal{V}/c^2)]m_2\left[\frac{v_{2x} + \mathcal{V}}{1 + (v_{2x}\mathcal{V}/c^2)}\right], \tag{5-79}$$

$$= \gamma m_2(v_{2x} + \mathcal{V}), \tag{5-80}$$

$$= \gamma[p_{2x} + \mathcal{V}(\mathcal{E}_2/c^2)], \tag{5-81}$$

where $\mathcal{E}_2 = m_2c^2$ is the relativistic energy of the mass m in reference frame 2. The inverse transformation is

$$p_{2x} = \gamma[p_{1x} - \mathcal{V}(\mathcal{E}_1/c^2)], \tag{5-82}$$

and

$$p_{1y} = p_{2y}, \tag{5-83}$$

$$p_{1z} = p_{2z}. \tag{5-84}$$

It is even simpler to transform a relativistic energy:

$$\mathcal{E}_1 = m_1c^2 = \gamma[1 + (v_{2x}\mathcal{V}/c^2)]m_2c^2, \tag{5-85}$$

$$= \gamma[\mathcal{E}_2 + \mathcal{V}p_{2x}], \tag{5-86}$$

and

$$\mathcal{E}_2 = \gamma[\mathcal{E}_1 - \mathcal{V}p_{1x}]. \tag{5-87}$$

It will be observed in Table 5-5 that $p_x, p_y, p_z, \mathcal{E}/c^2$ transform like x, y, z, t.

Table 5-5. *Transformation of a Four-Momentum*

$p_{1x} = \gamma[p_{2x} + \mathcal{V}(\mathcal{E}_2/c^2)]$	$p_{2x} = \gamma[p_{1x} - \mathcal{V}(\mathcal{E}_1/c^2)]$
$p_{1y} = p_{2y}$	$p_{2y} = p_{1y}$
$p_{1z} = p_{2z}$	$p_{2z} = p_{1z}$
$\mathcal{E}_1/c^2 = \gamma[(\mathcal{E}_2/c^2) + (\mathcal{V}/c^2)p_{2x}]$	$\mathcal{E}_2/c^2 = \gamma[(\mathcal{E}_1/c^2) - (\mathcal{V}/c^2)p_{1x}]$

Table 5-6. *The Four-Momentum* $\mathbf{p}$

Coordinates	x	y	z	t
Corresponding variable	p_x	p_y	p_z	$\mathcal{E}/c^2$
Components of $\mathbf{r}$	x	y	z	jct
Components of $\mathbf{p}$	p_x	p_y	p_z	$j\mathcal{E}/c$
Squared magnitude of $\mathbf{r}$	$x^2 + y^2 + z^2 - c^2t^2 = r^2 - c^2t^2$			
Squared magnitude of $\mathbf{p}$	$p_x^2 + p_y^2 + p_z^2 - (\mathcal{E}^2/c^2) = p^2 - (\mathcal{E}^2/c^2)$			

These four quantities are therefore the components of a four-vector

$$\mathbf{p} = (\boldsymbol{p}, j\mathcal{E}/c), \tag{5-88}$$

which is called the *four-momentum*. See Table 5-6.

The squared magnitude of the four-momentum,

$$p^2 - \frac{\mathcal{E}^2}{c^2} \quad \text{or} \quad p^2 - m^2c^2$$

must be an invariant.

Example

THE RELATION $\mathcal{E}^2 = m_0^2c^4 + p^2c^2$

If one reference frame moves with an object whose rest mass is m_0, and if in some other frame the object has a momentum p and a mass m, then

$$p^2 - m^2c^2 = -m_0^2c^2, \tag{5-89}$$

and

$$\mathcal{E}^2 = m^2c^4 = m_0^2c^4 + p^2c^2. \tag{5-90}$$

This general relation is illustrated in Figure 5-19.

The second term p^2c^2 is negligible when $\gamma^2\beta^2 \equiv \beta^2/(1 - \beta^2) \ll 1$, or $2\beta^2 \ll 1$, where β is the ratio v/c. As expected, at low velocities,

$$\mathcal{E} \approx m_0c^2. \qquad (\beta^2 \ll 1) \tag{5-91}$$

On the other hand, if $\gamma^2\beta^2 \gg 1$, or $\beta^2 \approx 1$,

$$\mathcal{E} \approx pc = mvc \approx mc^2. \qquad (\beta^2 \approx 1) \tag{5-92}$$

For a photon, $m_0 = 0$ and

$$\mathcal{E} = pc. \qquad \text{(Photon)} \tag{5-93}$$

The momentum of a photon is therefore

$$p = \frac{\mathcal{E}}{c} = \frac{h\nu}{c} = \frac{\hbar\omega}{c} = \frac{\hbar}{\lambda\!\!\!^{-}}, \tag{5-94}$$

where h is Planck's constant 6.626×10^{-34} joule second, ν is the frequency, c is as usual the velocity of light in a vacuum 3×10^8 meters/second, $\hbar = h/2\pi = 1.05 \times 10^{-34}$ joule-second, $\omega = 2\pi\nu$, and $\lambda\!\!\!^{-}$ is the wavelength divided by 2π.

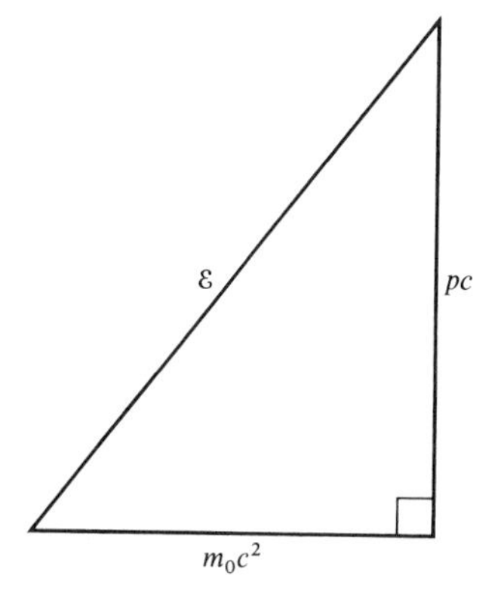

Figure 5-19. The relation between the relativistic energy ℰ, the rest energy m_0c^2, and the momentum p of a mass m is illustrated by this right-angled triangle: $\mathcal{E}^2 = m_0^2c^4 + p^2c^2$. If $p = 0$, $\mathcal{E} = m_0c^2$; if $m_0 = 0$ as for a photon, $\mathcal{E} = pc$.

5.18 TRANSFORMATION OF A FORCE

At the very beginning of the next chapter we shall need to transform a force from one reference frame to another. This is fairly simple now that we know how to transform a momentum. We use the relation

$$\boldsymbol{F} = \frac{d\boldsymbol{p}}{dt}. \tag{5-95}$$

Let the force be $\boldsymbol{F}_1$ in frame 1 and $\boldsymbol{F}_2$ in frame 2:

$$\boldsymbol{F}_1 = X_1\boldsymbol{i} + Y_1\boldsymbol{j} + Z_1\boldsymbol{k} \qquad \text{and} \qquad \boldsymbol{F}_2 = X_2\boldsymbol{i} + Y_2\boldsymbol{j} + Z_2\boldsymbol{k}. \tag{5-96}$$

Then

$$X_1 = \frac{d}{dt_1}\,p_{1x} = \frac{d}{dt_2}\left\{\gamma[p_{2x} + \mathcal{V}(\mathcal{E}_2/c^2)]\right\}\frac{dt_2}{dt_1}, \tag{5-97}$$

$$= \left\{\gamma\,\frac{dp_{2x}}{dt_2} + \gamma\,\frac{\mathcal{V}}{c^2}\,\frac{d}{dt_2}\,\mathcal{E}_2\right\}\frac{1}{\gamma[1 + (v_{2x}\mathcal{V}/c^2)]}, \tag{5-98}$$

$$= \frac{X_2 + (\mathcal{V}/c^2)\,\dfrac{d\mathcal{E}_2}{dt_2}}{1 + (v_{2x}\mathcal{V}/c^2)}. \tag{5-99}$$

Now, from Problem 5-30,

$$\frac{d\mathcal{E}_2}{dt_2} = X_2v_{2x} + Y_2v_{2y} + Z_2v_{2z}, \tag{5-100}$$

since $d\mathcal{E}_2/dt_2$ is the rate at which the energy $\mathcal{E}_2 = m_2c^2$ builds up under the action of the force $\boldsymbol{F}_2$. Finally,

$$X_1 = X_2 + \frac{\mathcal{V}}{c^2 + v_{2x}\mathcal{V}}\,(v_{2y}Y_2 + v_{2z}Z_2). \tag{5-101}$$

Table 5-7. *Transformation of a Force*

$$X_1 = X_2 + \frac{\mathcal{V}}{c^2 + v_{2x}\mathcal{V}}(v_{2y}Y_2 + v_{2z}Z_2)$$

$$Y_1 = \frac{Y_2}{\gamma[1 + (v_{2x}\mathcal{V}/c^2)]},$$

$$Z_1 = \frac{Z_2}{\gamma[1 + (v_{2x}\mathcal{V}/c^2)]},$$

$$X_2 = X_1 - \frac{\mathcal{V}}{c^2 - v_{1x}\mathcal{V}}(v_{1y}Y_1 + v_{1z}Z_1)$$

$$Y_2 = \frac{Y_1}{\gamma[1 - (v_{1x}\mathcal{V}/c^2)]},$$

$$Z_2 = \frac{Z_1}{\gamma[1 - (v_{1x}\mathcal{V}/c^2)]}.$$

See Table 5-7 for the complete set of transformation equations. It is understood that the velocity of the point of application of the force is $\boldsymbol{v}_1$ in frame 1 and $\boldsymbol{v}_2$ in frame 2.

Note that the transformations do *not* involve the coordinates of the point of application. They involve only the velocities $\boldsymbol{v}_1$, $\boldsymbol{v}_2$, $\mathcal{V}\boldsymbol{i}$.

We shall use these relations repeatedly in the next chapter.

Examples

(a) Surprisingly enough, a force directed along the x-axis ($Y = Z = 0$), in the direction of the relative velocity $\mathcal{V}\boldsymbol{i}$, is the same in both systems.

(b) Forces that are equal and opposite in one frame are not necessarily so in another frame. They remain equal and opposite only if their points of application have equal *velocities*.

(c) Whenever the force has a component perpendicular to the $x =$ axis, the transformation changes both its magnitude and its direction. Then two forces that are collinear in one frame are not necessarily collinear in another frame.

5.19 TRANSFORMATION OF AN ELEMENT OF VOLUME

We now wish to transform an electric charge density and an electric current. As a first step we shall transform an element of volume, after which we shall

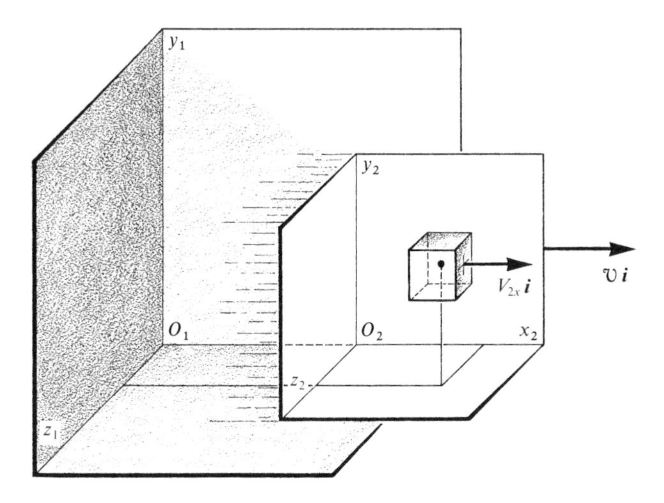

Figure 5-20. An element of volume having the form of a small cube of volume l^3_0 has a velocity v_{2x} with respect to frame 2. It appears to have a smaller volume $d\tau_2 < l^3_0$ for observer o_2 and a different volume $d\tau_1$ for observer o_1.

discuss charge invariance. After that, it will be a simple matter to deduce the required transformations.

An element of volume at rest in reference frame 2 has a volume $d\tau_0$ for observer o_2. What is its volume for o_1? Let us assume that the element of volume is a small cube of volume l_0^3 in 2.

The element of volume is affected by the Lorentz contraction only in the direction of the x-axis. Then, for o_1,

$$d\tau_1 = l_0^3[1 - (\mathcal{V}/c)^2]^{1/2}, \tag{5-102}$$

$$= \frac{d\tau_0}{\gamma}. \tag{5-103}$$

Remember that we have assumed that the element of volume is at rest with respect to reference frame 2. With this assumption the element of volume transforms like a length parallel to the x-axis.

We shall have to deal with elements of volume that have a velocity with respect to frame 2. Let us first set this velocity to be along the x-axis, and let us call it v_{2x} as in Figure 5-20.

Then, for o_2,

$$d\tau_2 = l_0^3\,[1 - (v_{2x}/c)^2]^{1/2} \tag{5-104}$$

and similarly, for o_1,

$$d\tau_1 = l_0^3\,[1 - (v_{1x}/c)^2]^{1/2}, \tag{5-105}$$

where v_{1x} is the velocity of the element of volume with respect to frame 1.

Now we have already calculated v_{1x} in Section 5-11:

$$v_{1x} = \frac{v_{2x} + \mathcal{V}}{1 + (v_{2x}\mathcal{V}/c^2)}. \tag{5-106}$$

Thus substituting this value for v_{1x} and simplifying,

$$d\tau_1 = l_0^3 \frac{\{[1 - (\mathcal{V}/c)^2][1 - (v_{2x}/c)^2]\}^{1/2}}{1 + (v_{2x}\mathcal{V}/c^2)}, \tag{5-107}$$

$$= \frac{d\tau_2}{\gamma[1 + (v_{2x}\mathcal{V}/c^2)]}. \tag{5-108}$$

This transformation is really the same as for a length parallel to the x-axis and moving at a velocity v_{2x} with respect to reference frame 2.

The inverse transformation is

$$d\tau_2 = \frac{d\tau_1}{\gamma[1 - (v_{1x}\mathcal{V}/c^2)]}, \tag{5-109}$$

where the element of volume now has a velocity v_{1x} with respect to frame 1.

We have assumed that the velocity in reference frame 2 is along the x-axis. What happens if it has components along the y- and z-axes? According to the Lorentz transformation, $y_1 = y_2$ and $z_1 = z_2$. Therefore distances perpendicular to the x-axis are the same in both frames, and the y and z components of velocity have no effect on the ratio $d\tau_1/d\tau_2$. Equations 5-108 and 5-109 are thus valid for any value of $\mathbf{v}_1$ and of $\mathbf{v}_2$.

5.20 INVARIANCE OF ELECTRIC CHARGE

From the experience that we have gained to date with relativistic transformations, we might venture to guess that an electric charge of Q_0 coulombs that is stationary with respect to reference frame 2 would appear to carry either $Q_0\gamma$ or Q_0/γ coulombs for an observer in frame 1.

This is in fact *wrong*. *Electric charge is invariant*, and a charged body carries the same electrical charge for all observers.

Possibly the most direct demonstration of the invariance of charge is the fact that the charge-to-mass ratio e/m for a charged particle moving at a velocity v is found experimentally to agree with the law

$$\frac{e}{m} = \frac{e}{m_0}[1 - (v/c)^2]^{1/2}. \tag{5-110}$$

The elementary charge e therefore remains constant and equal to 1.6×10^{-19} coulomb, irrespective of the velocity of the particle, while the mass m

varies with the velocity, as in Section 5-13. This relation is found to apply in particle accelerators up to the highest energies attained to date.

Another convincing demonstration of the invariance of charge is the fact that a metal does not acquire an electric charge when it is heated or cooled, despite the fact that the average kinetic energy of its conduction electrons is much *less* affected than that of its atoms.* One might expect, at first sight, that the extra charge would be negligible, since the change in velocity is small. However, as we saw in Chapter 2, the electric charges in matter are enormous, and it is only because their fields cancel perfectly at all temperatures that ordinary matter remains macroscopically neutral.

Example Ten kilograms of copper contain about 10^{26} atoms and 1.5×10^{7} coulombs of conduction electrons. If the charges on the atoms could be modified by only one part in 10^{15} by heating or cooling, the total charge in ten kilograms of copper would be 1.5×10^{-8} coulomb. If the copper were spherical, it would have a radius of about 6.5 centimeters and, assuming that the extra charge migrated to the surface, the copper would have a potential of about 2 kilovolts. Such electrostatic effects have never been observed.

5.21 TRANSFORMATION OF AN ELECTRIC CHARGE DENSITY AND OF AN ELECTRIC CURRENT. THE FOUR-CURRENT DENSITY

Consider an element of volume that contains dn electric charges of Q coulombs each and that has a velocity v_2 with respect to reference frame 2.

We can show that the total charge Qdn in the element of volume is the same for both observers o_1 and o_2. First, the individual charges Q are the same for both observers, since Q is an invariant, as we have just seen. Second, dn must also be an invariant, because it is simply a number of objects that both observers can count. Then Qdn is invariant.

Then the two densities are

$$\rho_1 = \frac{Qdn}{d\tau_1}, \qquad \rho_2 = \frac{Qdn}{d\tau_2}, \tag{5-111}$$

where $d\tau_1$ is the volume of the element of volume as seen by o_1, and $d\tau_2$ is the volume of the same element as seen by o_2. Thus

* See for example, Charles Kittel, *Introduction to Solid State Physics*, Third Edition (John Wiley, New York, 1966) p. 209 ff.

$$\rho_1 = \rho_2 \frac{d\tau_2}{d\tau_1} = \rho_2\gamma[1 + (v_{2x}\mathcal{V}/c^2)]. \tag{5-112}$$

The inverse relation is

$$\rho_2 = \rho_1 \frac{d\tau_1}{d\tau_2} = \rho_1\gamma[1 - (v_{1x}\mathcal{V}/c^2)]. \tag{5-113}$$

The quantity $\rho_2 v_{2x}$ is the product of the electric charge density and of the velocity of these charges in the direction parallel to the x-axis, as seen by observer o_2. This is analogous to the product ρv, which we used in Section 1.5 for the mass flux in water. The product $\rho_2 v_{2x}$ is the electric charge flowing per second and per square meter at the point considered, in the direction of the x-axis, or the electric current density J_{2x} in amperes/meter^2 as observed by o_2:

$$J_{2x} = \rho_2 v_{2x}. \tag{5-114}$$

Similarly,

$$J_{1x} = \rho_1 v_{1x}. \tag{5-115}$$

Then our two equations for ρ_1 and ρ_2 can be rewritten as follows:

$$\rho_1 = \gamma[\rho_2 + (\mathcal{V}/c^2)J_{2x}], \tag{5-116}$$

$$\rho_2 = \gamma[\rho_1 - (\mathcal{V}/c^2)J_{1x}]. \tag{5-117}$$

These are the equations of transformation for the electric charge density.

We can now deduce the equations of transformation for electric current densities:

$$J_{1x} = \rho_1 v_{1x}, \tag{5-118}$$

$$= \rho_2\gamma[1 + (v_{2x}\mathcal{V}/c^2)] \frac{v_{2x} + \mathcal{V}}{[1 + (v_{2x}\mathcal{V}/c^2)]}, \tag{5-119}$$

$$= \gamma(J_{2x} + \mathcal{V}\rho_2), \tag{5-120}$$

and

$$J_{2x} = \gamma(J_{1x} - \mathcal{V}\rho_1). \tag{5-121}$$

If the charges were fixed with respect to frame 2, J_{2x} would be zero and J_{1x} would be $\mathcal{V}\rho_1$.

We have therefore found transformation equations for the electric charge density and for the electric current density parallel to the x-axis. To be complete, we require corresponding equations for J_{1y}, J_{2y}, J_{1z}, J_{2z}. We need only find a relation between J_{1y} and J_{2y}:

$$J_{1y} = \rho_1 v_{1y}, \tag{5-122}$$

$$= \rho_2\gamma[1 + (v_{2x}\mathcal{V}/c^2)] \frac{v_{2y}}{\gamma[1 + (v_{2x}\mathcal{V}/c^2)]} = J_{2y}, \tag{5-123}$$

and, by symmetry,

$$J_{1z} = J_{2z}. \tag{5-124}$$

See Table 5-8.

Table 5-8. *Transformation of a Four-Current Density*

$J_{1x} = \gamma(J_{2x} + \mathcal{V}\rho_2)$	$J_{2x} = \gamma(J_{1x} - \mathcal{V}\rho_1)$
$J_{1y} = J_{2y}$	$J_{2y} = J_{1y}$
$J_{1z} = J_{2z}$	$J_{2z} = J_{1z}$
$\rho_1 = \gamma[\rho_2 + (\mathcal{V}/c^2)J_{2x}]$	$\rho_2 = \gamma[\rho_1 - (\mathcal{V}/c^2)J_{1x}]$

Examples

If we have *fixed* charges with a density ρ_2 in frame 2, an observer in frame 1 observes a charge density $\gamma\rho_2$ and a corresponding current density $\mathcal{V}\rho_1$, or $\mathcal{V}\gamma\rho_2$.

If we have a conducting wire in frame 2, ρ_2 is zero, and the current density J_{2x} is associated with zero net charge density. If the wire is parallel to the x-axis, and if the current flows in the positive direction of the x-axis,

$$\rho_1 = (\mathcal{V}/c^2)(\gamma J_{2x}), \qquad J_{1x} = \gamma J_{2x}. \tag{5-125}$$

Thus, contrary to what one would expect, J_{1x} is *not* equal to $\mathcal{V}\rho_1$. You will be able to solve this paradox if you solve Problem 5-35.

It will be observed that the equations of transformation for J and ρ are of the same form as those of the Lorentz transformation, with the components of the current playing the role of the coordinates x, y, z, and with the charge density playing the role of the time.

The corresponding four-vector

$$\mathbf{J} = (J, jc\rho) \tag{5-126}$$

is called the four-current density, and its squared magnitude

$$J^2 - c^2\rho^2$$

is an invariant (see Table 5-9).

We shall use this four-vector shortly to state one of the fundamental postulates of electromagnetism, but we must first be able to calculate a divergence in four dimensions.

Table 5-9. *The Four-Current Density* **J**

Coordinates	x	y	z	t
Corresponding variable	J_x	J_y	J_z	ρ
Components of **r**	x	y	z	jct
Components of **J**	J_x	J_y	J_z	$jc\rho$
Squared magnitude of **r**	$x^2 + y^2 + z^2 - c^2t^2 = r^2 - c^2t^2$			
Squared magnitude of **J**	$J_x^2 + J_y^2 + J_z^2 - c^2\rho^2 = J^2 - c^2\rho^2$			

Table 5-10. *Transformation of the Partial Derivatives*

$\frac{\partial}{\partial x_1} = \gamma\left[\frac{\partial}{\partial x_2} - (\mathcal{V}/c^2)\frac{\partial}{\partial t_2}\right]$	$\frac{\partial}{\partial x_2} = \gamma\left[\frac{\partial}{\partial x_1} + (\mathcal{V}/c^2)\frac{\partial}{\partial t_1}\right]$
$\frac{\partial}{\partial y_1} = \frac{\partial}{\partial y_2}$	$\frac{\partial}{\partial y_2} = \frac{\partial}{\partial y_1}$
$\frac{\partial}{\partial z_1} = \frac{\partial}{\partial z_2}$	$\frac{\partial}{\partial z_2} = \frac{\partial}{\partial z_1}$
$\frac{\partial}{\partial t_1} = \gamma\left[\frac{\partial}{\partial t_2} - \mathcal{V}\frac{\partial}{\partial x_2}\right]$	$\frac{\partial}{\partial t_2} = \gamma\left[\frac{\partial}{\partial t_1} + \mathcal{V}\frac{\partial}{\partial x_1}\right]$

5.22 THE FOUR-DIMENSIONAL OPERATOR $\Box$

By analogy with the four-vector (x, y, z, jct), we expect to find that the components of the *four-dimensional operator* $\Box$ (occasionally called "*quad*"), which is the equivalent of the ∇ operator, will be $\partial/\partial x, \partial/\partial y, \partial/\partial z, (1/jc)(\partial/\partial t)$. This is in fact correct. The transformation equations for the partial derivatives will indicate to us which are the proper components:

$$\frac{\partial}{\partial x_1} = \frac{\partial x_2}{\partial x_1}\frac{\partial}{\partial x_2} + \frac{\partial t_2}{\partial x_1}\frac{\partial}{\partial t_2} = \gamma\left[\frac{\partial}{\partial x_2} - (\mathcal{V}/c^2)\frac{\partial}{\partial t_2}\right], \tag{5-127}$$

$$\frac{\partial}{\partial y_1} = \frac{\partial}{\partial y_2}, \tag{5-128}$$

$$\frac{\partial}{\partial z_1} = \frac{\partial}{\partial z_2}, \tag{5-129}$$

$$\frac{\partial}{\partial t_1} = \frac{\partial t_2}{\partial t_1}\frac{\partial}{\partial t_2} + \frac{\partial x_2}{\partial t_1}\frac{\partial}{\partial x_2} = \gamma\left[\frac{\partial}{\partial t_2} - \mathcal{V}\frac{\partial}{\partial x_2}\right]. \tag{5-130}$$

These equations are tabulated in Table 5-10.

These transformation equations are of the same form as the Lorentz equations of Table 5-1 if the operators corresponding to the four components x, y, z, t are selected just as we expected (Table 5-11).

Then

$$\Box = \left(\nabla, \frac{1}{jc}\frac{\partial}{\partial t}\right). \tag{5-131}$$

Formally, the operator $\Box$ is used much like the ∇ of Chapter 1. It can be used for calculating a gradient

$$\Box F = \left(\frac{\partial}{\partial x}, \frac{\partial}{\partial y}, \frac{\partial}{\partial z}, \frac{1}{jc}\frac{\partial}{\partial t}\right) F, \tag{5-132}$$

Table 5-11. *The Four-Dimensional Operator* $\Box$

Coordinates	x	y	z	t
Corresponding operators	$\partial/\partial x$	$\partial/\partial y$	$\partial/\partial z$	$-(1/c^2)(\partial/\partial t)$
Components of $\mathbf{r}$	x	y	z	jct
Components of $\Box$	$\partial/\partial x$	$\partial/\partial y$	$\partial/\partial z$	$(1/jc)(\partial/\partial t)$
Squared magnitude of $\mathbf{r}$	$x^2 + y^2 + z^2 - c^2t^2 = r^2 - c^2t^2$			
$\Box^2$	$(\partial/\partial x)^2 + (\partial/\partial y)^2 + (\partial/\partial z)^2 - (1/c^2)(\partial/\partial t)^2 = \nabla^2 - (1/c^2)(\partial/\partial t)^2$			

or a divergence

$$\Box\cdot\mathbf{a} = \frac{\partial a_x}{\partial x} + \frac{\partial a_y}{\partial y} + \frac{\partial a_z}{\partial z} + \frac{1}{jc}\frac{\partial a_t}{\partial t}. \tag{5-133}$$

The scalar product of $\Box$ by itself is called the *d'Alembertian* and is written $\Box^2$:

$$\Box^2 \equiv \frac{\partial^2}{\partial x^2} + \frac{\partial^2}{\partial y^2} + \frac{\partial^2}{\partial z^2} - \frac{1}{c^2}\frac{\partial^2}{\partial t^2}, \tag{5-134}$$

$$= \nabla^2 - \frac{1}{c^2}\frac{\partial^2}{\partial t^2}. \tag{5-135}$$

Also

$$\Box^2 F = \left(\nabla^2 - \frac{1}{c^2}\frac{\partial^2}{\partial t^2}\right) F. \tag{5-136}$$

Setting the d'Alembertian equal to zero gives the wave equation corresponding to a phase velocity equal to the velocity of light c.

5.23 THE CONSERVATION OF CHARGE

Let us calculate the divergence of the four-current density:

$$\Box\cdot\mathbf{J} = \frac{\partial}{\partial x}J_x + \frac{\partial}{\partial y}J_y + \frac{\partial}{\partial z}J_z + \frac{\partial\rho}{\partial t}, \tag{5-137}$$

$$= \nabla\cdot\boldsymbol{J} + \frac{\partial\rho}{\partial t}. \tag{5-138}$$

This divergence is important. Let us see what it means, physically. Consider a small volume τ. The rate at which the charge $\tau\rho$ enclosed in the volume τ *in*creases with time is $\tau(\partial\rho/\partial t)$. The rate at which the enclosed charge *de*creases with time is $\tau(\nabla\cdot\boldsymbol{J})$, since the current density $\boldsymbol{J}$ is the charge flowing out per unit time and per unit area as in Section 1.5.

Now, in all experiments so far, charge has always been found to be conserved. Then

$$\nabla \cdot \boldsymbol{J} + \frac{\partial \rho}{\partial t} = 0, \tag{5-139}$$

or

$$\nabla \cdot \boldsymbol{J} = -\frac{\partial \rho}{\partial t}. \tag{5-140}$$

This is the *law of conservation of charge*. It says that, *in any given frame of reference*, electric charge is neither created nor destroyed. The law of conservation of charge can also be written as

$$\Box \cdot \mathbf{J} = 0. \tag{5-141}$$

This law must not be confused with charge invariance. Charge invariance means that the electric charge carried by an object is independent of the velocity of the object with respect to the observer. In other words, that the charge is the same in all frames of reference.

5.24 SUMMARY

We constantly refer to the two reference frames of Figure 5-1 (or 5-7), where $\mathcal{V}$ is the velocity of reference frame 2 with respect to reference frame 1—in the positive direction of the common x-axis. It is also the velocity of reference frame 1 with respect to 2, in the negative direction of the common x-axis.

According to classical mechanics, it is possible to transform one set of coordinates into the other by means of the obvious relations

$$x_1 = x_2 + \mathcal{V}t, \qquad y_1 = y_2, \qquad z_1 = z_2. \tag{5-1}$$

This set of equations constitutes a *Galilean transformation*. The Galilean transformation is valid for mechanical phenomena when the velocities are much smaller than the velocity of light. It is *not* valid for electrical phenomena, even at low velocities.

On the assumption that the Galilean transformation was correct, many experiments were performed with the object of measuring the velocity of the Earth with respect to the "ether." The failure of all these experiments led to the fundamental postulate of relativity, which can be stated as follows: *It is physically impossible to detect the uniform velocity of a frame of reference from observations made entirely within that frame.*

This fundamental postulate means that all the laws of nature must be invariant when transformed from reference frame 1 to reference frame 2.

This can be achieved if the Galilean transformation is replaced by the *Lorentz transformation* (Table 5-1):

$$x_1 = \gamma[x_2 + \mathcal{V}t_2], \qquad x_2 = \gamma[x_1 - \mathcal{V}t_1],$$
$$y_1 = y_2, \qquad y_2 = y_1,$$
$$z_1 = z_2, \qquad z_2 = z_1,$$
$$t_1 = \gamma[t_2 + (\mathcal{V}/c^2)x_2], \qquad t_2 = \gamma[t_1 - (\mathcal{V}/c^2)x_1],$$
$$\gamma = 1/[1 - (\mathcal{V}/c)^2]^{1/2}, \qquad c = 3 \times 10^8 \text{ meters/second.}$$

TRANSFORMATION OF A LENGTH PARALLEL TO THE X-AXIS. THE LORENTZ CONTRACTION.

Object of proper length l_0 at rest in 2:

$$l_1 = \frac{l_0}{\gamma} < l_0.$$

Object of proper length l_0 at rest in 1:

$$l_2 = \frac{l_0}{\gamma} < l_0.$$

TRANSFORMATION OF A LENGTH PERPENDICULAR TO THE X-AXIS. No change.

TRANSFORMATION OF A TIME INTERVAL. TIME DILATION.

Proper time interval T_0 measured at position of a single clock at rest in 2, and T_1 measured with two clocks at rest in 1:

$$T_1 = \gamma T_0 > T_0.$$

Proper time interval T_0 measured at position of single clock at rest in 1, and T_2 measured with two clocks at rest in 2:

$$T_2 = \gamma T_0 > T_0.$$

RELATIVISTIC DOPPLER EFFECT FOR ELECTROMAGNETIC WAVES. The velocity $\mathcal{V}$ is the velocity of the source relative to the detector.

(a) The velocity is along the line joining the source to the detector

$$f_d = \left[\frac{1 - (\mathcal{V}/c)}{1 + (\mathcal{V}/c)}\right]^{1/2} f_s, \qquad (5\text{-}25)\ (5\text{-}26)$$

where f_d is the frequency measured at the detector, and f_s is the frequency measured at the source. The velocity $\mathcal{V}$ is taken to be positive when the source recedes from the observer.

(b) The velocity is perpendicular to the line joining the source to the detector

$$f_d = \gamma f_s \text{ (source)} \quad \text{or} \quad f_d = \frac{f_s}{\gamma} \text{ (detector).} \qquad (5\text{-}28)$$

SIMULTANEITY. Two events that are simultaneous in one frame of reference are also simultaneous in other frames of reference if and only if their x coordinates are the same.

MAXIMUM SIGNAL VELOCITY. It is impossible to transmit a signal at a velocity that is larger than the velocity of light.

TRANSFORMATION OF A VELOCITY (SECTION 5.11).

$$v_{1x} = \frac{v_{2x} + \mathcal{V}}{1 + (v_{2x}\mathcal{V}/c^2)}, \qquad v_{2x} = \frac{v_{1x} - \mathcal{V}}{1 - (v_{1x}\mathcal{V}/c^2)},$$

$$v_{1y} = \frac{v_{2y}}{\gamma[1 + (v_{2x}\mathcal{V}/c^2)]}, \qquad v_{2y} = \frac{v_{1y}}{\gamma[1 - (v_{1x}\mathcal{V}/c^2)]},$$

$$v_{1z} = \frac{v_{2z}}{\gamma[1 + (v_{2x}\mathcal{V}/c^2)]}, \qquad v_{2z} = \frac{v_{1z}}{\gamma[1 - (v_{1x}\mathcal{V}/c^2)]}.$$

TRANSFORMATION OF AN ACCELERATION (SECTION 5.12).

$$a_{1x} = \frac{a_{2x}}{\gamma^3[1 + (v_{2x}\mathcal{V}/c^2)]^3},$$

$$a_{1y} = \frac{1}{\gamma^2[1 + (v_{2x}\mathcal{V}/c^2)]^2}\left\{a_{2y} - \frac{v_{2y}\mathcal{V}}{c^2 + v_{2x}\mathcal{V}}\, a_{2x}\right\},$$

$$a_{1z} = \frac{1}{\gamma^2[1 + (v_{2x}\mathcal{V}/c^2)]^2}\left\{a_{2z} - \frac{v_{2z}\mathcal{V}}{c^2 + v_{2x}\mathcal{V}}\, a_{2x}\right\},$$

$$a_{2x} = \frac{a_{1x}}{\gamma^3[1 - (v_{1x}\mathcal{V}/c^2)]^3},$$

$$a_{2y} = \frac{1}{\gamma^2[1 - (v_{1x}\mathcal{V}/c^2)]^2}\left\{a_{1y} + \frac{v_{1y}\mathcal{V}}{c^2 - v_{1x}\mathcal{V}}\, a_{1x}\right\},$$

$$a_{2z} = \frac{1}{\gamma^2[1 - (v_{1x}\mathcal{V}/c^2)]^2}\left\{a_{1z} + \frac{v_{1z}\mathcal{V}}{c^2 - v_{1x}\mathcal{V}}\, a_{1x}\right\}.$$

RELATIVISTIC MASS. The mass m of an object moving with a velocity v with respect to the observer is

$$m = \frac{m_0}{[1 - (v/c)^2]^{1/2}}, \tag{5-59}$$

where m_0 is the *rest mass*.

TRANSFORMATION OF A MASS.

$$m_1 = \gamma[1 + (v_{2x}\mathcal{V}/c^2)]m_2, \tag{5-63}$$

$$m_2 = \gamma[1 - (v_{1x}\mathcal{V}/c^2)]m_1. \tag{5-64}$$

RELATIVISTIC ENERGY.

$$\mathcal{E} = mc^2 = \frac{m_0}{[1 - (v/c)^2]^{1/2}} c^2, \tag{5-65}$$

$$= m_0c^2 + \tfrac{1}{2} m_0v^2 + \tfrac{3}{8} m_0v^4/c^2 + \cdots, \tag{5-67}$$

where m_0c^2 is the *rest energy* and $mc^2 - m_0c^2$, or $(1/2)\, m_0v^2 + \cdots$, is the *relativistic kinetic energy*.

THE FOUR-VECTOR $\mathbf{r} = (\boldsymbol{r}, jct)$.

Components: x, y, z, jct
Invariant squared magnitude: $x^2 + y^2 + z^2 - c^2t^2$, or $r^2 - c^2t^2$

RELATIVISTIC MOMENTUM.

$$\boldsymbol{p} = m\boldsymbol{v} = \frac{m_0}{[1 - (v/c)^2]^{1/2}} \boldsymbol{v} \tag{5-76}$$

TRANSFORMATION OF A MOMENTUM AND OF A RELATIVISTIC ENERGY (SECTION 5.17).

$$p_{1x} = \gamma[p_{2x} + \mathcal{V}(\mathcal{E}_2/c^2)] \qquad p_{2x} = \gamma[p_{1x} - \mathcal{V}(\mathcal{E}_1/c^2)],$$

$$p_{1y} = p_{2y} \qquad p_{2y} = p_{1y}$$

$$p_{1z} = p_{2z} \qquad p_{2z} = p_{1z}$$

$$\mathcal{E}_1/c^2 = \gamma[(\mathcal{E}_2/c^2) + (\mathcal{V}/c^2)p_{2x}] \qquad \mathcal{E}_2/c^2 = \gamma[(\mathcal{E}_1/c^2) - (\mathcal{V}/c^2)p_{1x}]$$

THE FOUR-MOMENTUM $\mathbf{p} = (\boldsymbol{p}, j\mathcal{E}/c)$.

Components: $p_x, p_y, p_z, j\mathcal{E}/c$
Invariant squared magnitude: $p^2 - \mathcal{E}^2/c^2 = p^2 - m^2c^2 = -\, m_0^2c^2$ (5-90)
For a mass m,

$$\mathcal{E}^2 = m_0^2c^4 + p^2c^2. \tag{5-90}$$

For a photon,

$$p = \frac{\mathcal{E}}{c} = \frac{h\nu}{c} = \frac{\hbar\omega}{c} = \frac{\hbar}{\lambda\!\!\!^{-}}, \tag{5-94}$$

where h is Planck's constant 6.626×10^{-34} joule second, ν is the frequency, $c = 3 \times 10^8$ meters/second, $\hbar = h/2\pi = 1.05 \times 10^{-34}$ joule second, $\omega = 2\pi\nu$, and $\lambda\!\!\!^{-}$ is the wavelength divided by 2π.

TRANSFORMATION OF A FORCE (SECTION 5.18).

$$X_1 = X_2 + \frac{\mathcal{V}}{c^2 + v_{2x}\mathcal{V}}(v_{2y}Y_2 + v_{2z}Z_2), \quad X_2 = X_1 - \frac{\mathcal{V}}{c^2 - v_{1x}\mathcal{V}}(v_{1y}Y_1 + v_{1z}Z_1),$$

$$Y_1 = \frac{Y_2}{\gamma[1 + (v_{2x}\mathcal{V}/c^2)]}, \qquad Y_2 = \frac{Y_1}{\gamma[1 - (v_{1x}\mathcal{V}/c^2)]},$$

$$Z_1 = \frac{Z_2}{\gamma[1 + (v_{2x}\mathcal{V}/c^2)]}, \qquad Z_2 = \frac{Z_1}{\gamma[1 - (v_{1x}\mathcal{V}/c^2)]}.$$

TRANSFORMATION OF AN ELEMENT OF VOLUME.

$$d\tau_1 = \frac{d\tau_2}{\gamma[1 + (v_{2x}\mathcal{V}/c^2)]}, \tag{5-108}$$

$$d\tau_2 = \frac{d\tau_1}{\gamma[1 - (v_{1x}\mathcal{V}/c^2)]}. \tag{5-109}$$

INVARIANCE OF ELECTRIC CHARGE. It is found experimentally that the electric charge carried by an object is the same for all observers.

TRANSFORMATION OF AN ELECTRIC CHARGE DENSITY AND OF AN ELECTRIC CURRENT. (SECTION 5.21).

$$J_{1x} = \gamma[J_{2x} + \mathcal{V}\rho_2], \qquad J_{2x} = \gamma[J_{1x} - \mathcal{V}\rho_1],$$

$$J_{1y} = J_{2y}, \qquad J_{2y} = J_{1y},$$

$$J_{1z} = J_{2z}, \qquad J_{2z} = J_{1z},$$

$$\rho_1 = \gamma[\rho_2 + (\mathcal{V}/c^2)J_{2x}], \qquad \rho_2 = \gamma[\rho_1 - (\mathcal{V}/c^2)J_{1x}].$$

THE FOUR-CURRENT DENSITY $\mathbf{J} = (J, jc\rho)$ (SECTION 5.21).

Components: $J_{1x}, J_{1y}, J_{1z}, jc\rho$

Invariant squared magnitude: $J^2 - c^2\rho^2$

TRANSFORMATION OF THE PARTIAL DERIVATIVES (SECTION 5.22).

$$\frac{\partial}{\partial x_1} = \gamma\left[\frac{\partial}{\partial x_2} - (\mathcal{V}/c^2)\frac{\partial}{\partial t_2}\right], \qquad \frac{\partial}{\partial x_2} = \gamma\left[\frac{\partial}{\partial x_1} + (\mathcal{V}/c^2)\frac{\partial}{\partial t_1}\right],$$

$$\frac{\partial}{\partial y_1} = \frac{\partial}{\partial y_2}, \qquad \frac{\partial}{\partial y_2} = \frac{\partial}{\partial y_1},$$

$$\frac{\partial}{\partial z_1} = \frac{\partial}{\partial z_2}, \qquad \frac{\partial}{\partial z_2} = \frac{\partial}{\partial z_1},$$

$$\frac{\partial}{\partial t_1} = \gamma\left[\frac{\partial}{\partial t_2} - \mathcal{V}\frac{\partial}{\partial x_2}\right], \qquad \frac{\partial}{\partial t_2} = \gamma\left[\frac{\partial}{\partial t_1} + \mathcal{V}\frac{\partial}{\partial x_1}\right].$$

THE FOUR-DIMENSIONAL OPERATOR $\Box = \left(\nabla, \frac{1}{jc}\frac{\partial}{\partial t}\right)$.

Components: $\frac{\partial}{\partial x}, \frac{\partial}{\partial y}, \frac{\partial}{\partial z}, \frac{1}{jc}\frac{\partial}{\partial t}$

Four-dimensional gradient:

$$\Box F = \left(\frac{\partial}{\partial x}, \frac{\partial}{\partial y}, \frac{\partial}{\partial z}, \frac{1}{jc}\frac{\partial}{\partial t}\right)F \qquad (5\text{-}132)$$

Four-dimensional divergence:

$$\Box \cdot \mathbf{a} = \frac{\partial a_x}{\partial x} + \frac{\partial a_y}{\partial y} + \frac{\partial a_z}{\partial z} + \frac{1}{jc}\frac{\partial a_t}{\partial t} \qquad (5\text{-}133)$$

Four-dimensional Laplacian, or *d'Alembertian:*

$$\Box^2 F = \left(\nabla^2 - \frac{1}{c^2}\frac{\partial^2}{\partial t^2}\right)F \qquad (5\text{-}136)$$

LAW OF CONSERVATION OF CHARGE. It is found experimentally that, in any given frame of reference, electric charge is always conserved:

$$\nabla \cdot \boldsymbol{J} = -\frac{\partial \rho}{\partial t}, \qquad (5\text{-}140)$$

or

$$\Box \cdot \mathbf{J} = 0. \qquad (5\text{-}141)$$

PROBLEMS

Several of the following problems are adapted, with permission, from a book by Edwin F. Taylor and John A. Wheeler.*

5-1. For what value of $\mathcal{V}$ does the value of γ differ from unity by 1%?

5-2. Draw a graph of γ as a function of $\mathcal{V}$. Try to find "the" best way of conveying the manner in which γ depends on $\mathcal{V}$.‡

5-3. Two events occur at the same place in the laboratory and are separated in time by 3 seconds.

(a) What is the spatial distance between these two events in a moving frame with respect to which the events are separated in time by 5 seconds?
(b)What is the relative speed of the moving and laboratory frames?

5-4. Three men, A, O, and B, ride on a train moving at a velocity $\mathcal{V}$. A is in front, O is in the middle, and B is at the rear. A fourth man, O', stands beside the

* *Spacetime Physics* (W. H. Freeman and Company, San Francisco, 1966).

rails. At the moment O passes O', light signals from A and B reach O and O'. Both O and O' are asked who emitted his light signal first. What do they answer?

5-5. Observers o_1 and o_2 repeat the measurement of the length of the ruler (Section 5.7), but in a different way. Observer o_2 fixes to the ends of his ruler a pair of flash bulbs that can project the shadow of the edge of the ruler on a photographic plate in reference frame 1. Midway along the ruler, he sets up an electronic circuit that can send pulses in both directions to flash the bulbs simultaneously (according to him).

(a) What will be the distance between the edges of the ruler on the photographic plate?
(b) How can o_1 account for this result?

5-6. A straight line passing through the origin O_2 of reference frame 2 forms an angle α_2 with the x-axis.

(a) Calculate the value of α_1 as measured by an observer o_1 on reference frame 1.
(b) What is the value of α_1 when the velocity $\mathcal{V}$ of reference frame 2 with respect to 1 approaches c?

5-7. A physicist is arrested for going through a red light. In court he pleads that he approached at such a speed that the red light appeared green to him. The judge, a graduate of a physics class, changes the charge to speeding and fines the defendant one dollar for every kilometer per hour he exceeded the speed limit of 50 kilometers per hour. What is the fine? ($\lambda_{green} \approx 5.3 \times 10^{-7}$ meter, $\lambda_{red} \approx 6.5 \times 10^{-7}$ meter).

5-8. The radio galaxy 3C295 has a red shift of 46%. The astronomers mean by this that the observed wavelength is 1.46 times the wavelength of the same radiation produced in the laboratory.

(a) Calculate the radial velocity of this galaxy.
(b) Some quasars have red shifts of 200%. What is their radial velocity?

5-9. The twin, or clock, paradox can be illustrated as follows. On their twenty-first birthday, Peter leaves his twin Paul behind on the Earth and goes off in a straight line for seven years of his time at a speed of $0.96c$. Peter then reverses direction and returns at the same speed.

(a) What are the ages of Peter and of Paul at the moment of reunion?
(b) Peter and Paul, expecting this strange result, performed the following experiment during Peter's trip. They both observed a distant variable star whose light alternates from dim to bright at a frequency f when observed from the Earth. The variable star is in a direction perpendicular to Peter's trajectory. They of course both counted the same number of pulsations during the trip. Use the expression for the Doppler shift to verify the difference in age between Peter and Paul at the end of the trip.

5-10. It has been observed that some quasars exhibit light fluctuations with a period of about one day.

Can you infer an upper limit for their size?

5-11. The Lorentz transformation implies that the relative velocity $\mathcal{V}$ of two frames of reference cannot exceed the velocity of light c. We have also shown that a mass and a signal cannot exceed the velocity of light. Discuss the following cases.

(a) A very long straight rod, which is inclined at a small angle θ in relation to a horizontal axis, moves downward at a velocity v.

What is the speed of the point of intersection of the lower edge of the rod with the axis?

Can this speed be greater than c?

Can it be used to transmit a signal?

(b) The same rod is initially at rest with the point of intersection at the origin. The rod is struck a downward blow at the origin with a hammer.

Can the motion of the point of intersection be used to transmit a signal at a velocity greater than the velocity of light?

(c) A powerful laser is rotated rapidly about a vertical axis.

Can the azimuthal velocity of the beam exceed the velocity of light?

Can the beam be used to transmit a signal between two points at a velocity greater than c?

(d) The manufacturers of some oscilloscopes claim writing speeds in excess of the speed of light.

Is this possible?

5-12. We have found six formulas for calculating the velocity components in one frame when the velocity components in the other frame are known. Show that they can be written in the following vector form:

$$\mathbf{v}_1 = [\mathbf{v}_{2\|} + \boldsymbol{\mathcal{V}} + (\mathbf{v}_{2\perp}/\gamma)]/[1 + (\mathbf{v}_2 \cdot \boldsymbol{\mathcal{V}}/c^2)].$$

5-13. Light moves more slowly through a material medium than through a vacuum, its phase velocity v being c/n, where n is the index of refraction of the medium. If now the medium itself moves at a velocity $\mathcal{V} \ll c$ with respect to the laboratory, show that the phase velocity of the light with respect to the laboratory is approximately

$$\frac{c}{n} + \mathcal{V}\left(1 - \frac{1}{n^2}\right).$$

5-14. A ring of electrons rotates about its axis of symmetry and also moves parallel to itself in the direction of this axis.

If v_z is the axial velocity of the ring with respect to the laboratory, and if v_θ is the azimuthal velocity of the electrons in the reference frame of the ring, show that, with respect to the laboratory,

$$\text{(a)}\ \beta^2 = \beta_z^2 + \beta_\theta^2 - \beta_z^2\beta_\theta^2,$$

$$\text{(b)}\ \gamma = \gamma_z\gamma_\theta.$$

5-15. A flash of light is emitted at an angle α_2 with respect to the x-axis.

(a) Show that

$$\tan \alpha_1 = \frac{\sin \alpha_2}{\gamma[\cos \alpha_2 + (\mathcal{V}/c)]}.$$

(b) Show that, as $\mathcal{V}$ approaches c, the angle α_1 is small, except near $\alpha_2 = \pi$. In frame 1 the light is then concentrated in a narrow forward cone. This is called the *headlight effect*.
(c) Show that, for small angles and for $\mathcal{V} \ll c$,

$$\frac{\alpha_2 - \alpha_1}{\alpha_2} \approx \frac{\mathcal{V}}{c}.$$

5-16. Show that the ratio m_A/m_B of Eq. 5-55 is always positive.

5-17. Show that the two values of the ratio m_A/m_B given in Eqs. 5-55 and 5-56 are equal.

5-18. A nuclear bomb is exploded in an underground cavity and the products of the explosion are allowed to cool.
Sketch graphs of $\sum mc^2$ and $\sum m_0c^2$ as functions of the time.

5-19. An atom of hydrogen is accelerated up to a very high velocity.
Show that the ratio of the kinetic energies of the proton and of the electron is equal to the ratio of their rest masses.

5-20. A proton has a kinetic energy of 500 million electron-volts.
(a) Calculate its mass.
(b) Calculate its velocity.

5-21. Draw a table of relativistic kinetic energies, expressed in electron-volts, for electrons and protons that have velocities of $0.1c$, $0.3c$, $0.9c$.
Since both the electron and the proton carry one electronic charge, the energy in electron-volts is equal to the accelerating voltage.

5-22. A cosmic-ray particle, which may be assumed to be a proton, has been observed to have an energy of 16 joules.
(a) Calculate its mass in micrograms.
(b) How long would this particle take to cross our galaxy (diameter 10^5 light years), as measured by a clock moving with the proton? Express your answer in seconds (1 year $\approx \pi \times 10^7$ seconds).

5-23. The Stanford Linear Accelerator is used to accelerate electrons up to energies of 40 GeV (40×10^9 electron-volts).
(a) Calculate the mass of an electron that has the full energy. How does this mass compare with that of a proton at rest (mass 1.7×10^{-27} kilogram). An electron at rest has a mass of 9×10^{-31} kilogram.
(b) What is the length of the accelerator in the reference frame of an electron that has the full energy? The length of the accelerator, as measured on the ground, is 3000 meters.
(c) What time would be required for such an electron to go from one end of the accelerator to the other, (*i*) in the laboratory frame, (*ii*) in the electron's frame of reference?
(d) What is the velocity of this electron, according to its own measurements?

5-24. It is shown by the example in Section 8.9 that the net outward force on an ion that is at the periphery of an ion beam is smaller than the electrostatic force of repulsion by a factor of $1/\gamma^2$, v being the velocity of the ion.

Show that, if the kinetic energy of the ion is equal to its rest energy, the electrostatic force of repulsion is reduced by a factor of 4.

5-25. Visible light can be transformed into high-energy gamma radiation in the following way.

Head-on electron-photon collisions are produced by reflecting a ray of light from a laser backwards on a high-energy electron beam. Let us assume that the photons have an energy $h\nu$ of 2 electron-volts and that the electrons have a kinetic energy of 6×10^9 electron-volts.

Let us disregard the electron recoil, for the moment. In the reference frame of the electrons, the incident photons are Doppler-shifted to $h\nu' \gg h\nu$, and then reflected forward at the same energy $h\nu'$. With respect to the laboratory, there is a further shift to $h\nu'' \gg h\nu'$.

Calculate the final photon energy, using conservation of both momentum and energy in the laboratory frame.

5-26. Since the thrust produced by a rocket motor is equal to the product $m'v$, where m' is the mass of propellant ejected per second and v is the exhaust velocity, the ultimate rocket would transform its propellant into radiation and eject photons backward at the velocity of light. The weight of the propellant would then be minimum.

(a) Show that the power-to-thrust ratio W/F for a photon motor is c.

(b) Show that the thrust F is $c(dM/dt)$, where M is the mass of the propellant remaining at the time t.

Note that these two relations are independent of the frequency: the source of radiation need not be monochromatic.

According to this last equation a photon rocket burning one gram of matter per second would have a thrust of 30 tons. The difficulty is to transform an appreciable fraction of the propellant mass into radiation, as the following example will show.

(c) An ordinary flashlight has a capacity of about 2 ampere hours at about 2 volts. If a flashlight is switched on in outer space, show that its terminal velocity will be of the order of 10^{-4} meter/second.

(d) Show that, in the process, the flashlight loses about one part in 10^{13} of its original mass.

5-27. Let us investigate some of the conditions under which interstellar travel would be possible.

(a) First, time should be contracted by, say, a factor of 10. Then $\gamma = 10$. Show that v/c must then be 0.995.

(b) Imagine a space ship equipped with a photon motor. The motor annihilates the fuel and produces a beam of light directed backwards. This type of motor consumes the least amount of fuel; see the preceding problem.

You can find the fraction f of the initial mass that remains after the ship has attained a velocity of 99.5% c by writing out the equations for the conservation of mass-energy and for the conservation of momentum. Take into account the mass-energy and the momentum of both the rocket and the radiation.

You should find that $f = 0.05$.

The space ship must later be braked to a stop, and this requires 99.5% of the remaining mass. At the end of the return trip we are left with a fraction f^4, or 6.25×10^{-6}, of the initial mass.

(c) It has been suggested that, in principle, the rocket could collect and burn interstellar matter which has a density of about 1 atom of hydrogen per cubic centimeter.

Calculate the mass of gas collected during one year if the rocket sweeps out a volume 1000 square meters in cross-section at the velocity of light.

Is the interstellar gas a useful source of fuel?

5-28. A photon of energy $h\nu$ is emitted at the surface of a star of mass M and radius R.

(a) Show that the fractional frequency change $\Delta\nu/\nu$ is GM/Rc^2 after it has escaped to infinity, where G is the gravitational constant.

This is the *gravitational red shift.*

(b) Calculate $\Delta\nu/\nu$ for the Sun ($G = 6.67 \times 10^{-11}$, $R = 7.0 \times 10^8$ meters, $M = 2.0 \times 10^{30}$ kilograms) and for the Earth ($R = 6.4 \times 10^6$ meters, $M = 6.0 \times 10^{24}$ kilograms).

(c) The star Sirius and a smaller star revolve about one another. By analyzing this rotation using Newtonian mechanics, astronomers have been able to establish that the mass of the smaller star is about equal to that of the Sun. However, light from this star has a $\Delta\nu/\nu$ of 7×10^{-4}.

What is its average density?

(d) The Sun rotates once in about 24.7 days.

What Doppler shift should we observe for light of $\lambda = 5 \times 10^{-7}$ meter from the edge of the Sun's disk at is equator?

(e) Compare this Doppler shift with the gravitational red shift.

5-29. An excited nucleus of Fe^{57} formed by the radioactive decay of Co^{57} emits a gamma ray of energy 14.4×10^3 electron-volts.

(a) By what fraction is the energy of the emitted ray shifted because of the recoil of the nucleus, if the iron atom is completely free?

Assume that the mass of the nucleus is equal to that of 57 protons ($57 \times 1.7 \times 10^{-27}$ kilogram).‡

It was discovered by R. L. Mössbauer in 1958 that, when the iron is in solid form, a significant fraction of the atoms recoil as if they were locked rigidly to the rest of the solid. This is *Mössbauer effect.*

(b) If the solid has a mass of 1 gram, by what fraction is the frequency of the emitted ray shifted in the "recoilless" process?

(c) The natural line width $\Delta\nu/\nu$ of the Fe^{57} gamma ray is 3×10^{-13}. What is the value of $\Delta\nu$?

(d) How does this compare with the $\Delta\nu$ for the decay of a free iron atom and for a recoilless process?

A sample of normal Fe^{57} absorbs gamma rays of 14.4 KeV by the inverse recoilless process much more strongly than it absorbs gamma rays of any nearby energy. The excited nuclei thus formed re-emit the 14.4 KeV radiation in random directions some time later. This process is called *resonant scattering.*

(e) If now the source is moved toward the absorber at a velocity v, what must be the value of v if the absorber is to see gamma rays shifted in frequency by 3 parts in 10^{13}, corresponding to one resonance line width?

(f) What happens to the counting rate of a counter placed behind the absorber when the source is moved (*i*) toward the absorber, (*ii*) away from the absorber at the same speed?

(g) If a 14.4 KeV gamma ray, emitted without recoil by an Fe^{57} nucleus, travels 22.5 meters vertically upward, by what fraction will its energy be reduced?

(h) An Fe^{57} absorber located at this height must move with what speed and in what direction in order to scatter these gamma rays by recoilless processes?

R. V. Pound and G. A. Rebka found that $\Delta\nu/\nu = (2.56 \pm 0.26) \times 10^{-15}$ in this case.

5-30. Show that, if a force F is exerted on a mass m,

$$\boldsymbol{F} \cdot \boldsymbol{v} = \frac{d\mathcal{E}}{dt}.$$

5-31. Use the equation

$$\boldsymbol{F} = \frac{d}{dt}(m\boldsymbol{v})$$

to show that the three vectors $\boldsymbol{F}$, $\boldsymbol{v}$, $\boldsymbol{a}$ are coplanar, where $\boldsymbol{a}$ is the acceleration.

5-32. Show that a mass density transforms as follows:

$$\rho_1 = \gamma^2 \left[1 + \frac{v_{2x}\mathcal{V}}{c^2}\right]^2 \rho_2,$$

$$\rho_2 = \gamma^2 \left[1 - \frac{v_{1x}\mathcal{V}}{c^2}\right]^2 \rho_1.$$

5-33. Imagine that electric charge is not invariant and that $Q = Q_0[1 - (\mathcal{V}/c)^2]^{1/2}$. (Remember that charge is in fact invariant, according to all experiments performed to date.) The charge Q_0 is that measured by an observer moving with the charge, and Q is the charge for an observer moving at a velocity $\mathcal{V}$ with respect to it.

(a) If the electrons in a given sample have an average energy of 100 electron-volts, what percentage increase in their charge must we expect if their velocity increases by 1%?

(b) Would this be enough to produce an observable effect?

5-34. Calculate v, β, γ for electrons in matter, assuming that they have kinetic energies of 10 electon-volts.

5-35. We have seen in Section 5-21 that the charge and current densities in reference frame 1 that correspond to a conduction current parallel to the x-axis in frame 2 are

$$\rho_1 = (\mathcal{V}/c^2)(\gamma J_{2x}) \qquad \text{and} \qquad J_{1x} = \gamma J_{2x},$$

so that $J_{1x} \neq \mathcal{V}\rho_1$.

How can this apply to a conducting wire?‡

Hints

5-2. You might think of using either a semilog or a log-log plot. You might also think of plotting several graphs, or of plotting $\gamma - 1$, etc.

5-29. (a) Call the initial rest mass of the excited nucleus m_0 and its final rest mass m_0'. Then, immediately after the emission of the gamma ray, the nucleus has a mass m' and a velocity u. Eliminate *both* m' and u. If there were no recoil, the gamma-ray energy would be $\Delta E = (m_0 - m_0')c^2$. Show that

$$h\nu = \Delta E[1 - (\Delta E/2m_0c^2)].$$

5-35. Remember that, in a conductor, we have *fixed* positive charges and *mobile* negative electrons. Treat both sets of charges separately, and then calculate the net charge density and the net current density in 1.

CHAPTER 6

RELATIVITY II

*The Electric and Magnetic Fields of Moving Electric Charges**

Our knowledge of relativity is now amply sufficient to permit us to calculate the electric and magnetic fields of moving electric charges, as long as we can neglect accelerations. The fields of .ccelerated charges are more complicated, and we shall not discuss them from the point of view of relativity.

We shall first calculate $\boldsymbol{E}$ and $\boldsymbol{B}$ for an individual charge moving at a constant velocity $\mathcal{V}$ along the x-axis. Then we shall make the usual assumption, which was first made by Lorentz, that all electric and magnetic fields are due to elementary electric charges. Under this assumption, the basic properties of our $\boldsymbol{E}$ and $\boldsymbol{B}$ fields for a single charge must apply to the $\boldsymbol{E}$ and $\boldsymbol{B}$ fields of any set of charges moving at arbitrary constant velocities.

We shall be able to deduce Maxwell's four fundamental equations of electromagnetism, in their general form, but our reasoning will really be valid only for charges moving at constant velocities.

We shall use Coulomb's law and the concepts of relativity that we developed in Chapter 5. You will remember that the latter are all direct consequences of the Lorentz transformation (Section 5.6).

We have already found one of Maxwell's equations (Eq. 3-41), and, in Chapters 7 to 10, we shall rediscover the other three without using relativity. We therefore intend to deduce Maxwell's equations twice, with and without relativity. The two approaches are complementary, and there will be a negligible amount of duplication.

---

* This chapter is based on Chapter 5. If you have not studied Chapter 5 or its equivalent, proceed to Chapter 7.

6.1 FORCE EXERTED ON A MOVING CHARGE BY ANOTHER CHARGE MOVING AT THE SAME CONSTANT VELOCITY

Let us return to the experiment of Trouton and Noble, which we discussed briefly in Section 5.3. You will recall that they attempted to detect a torque on a charged parallel-plate capacitor that was suspended so that its plates were vertical. A torque was expected to arise from a magnetic interaction between the two charges moving along with the Earth around the Sun. No such torque was observed. This result is consistent with the fundamental postulate of relativity, according to which the force between two electric charges must be precisely the same, *for an observer at rest with respect to the charges*, whether the charges move at a constant velocity, or whether they remain fixed with respect to some reference frame.

We shall calculate the force on one of the charges in a reference frame with respect to which the charges move along two parallel paths at the same constant velocity $\mathcal{V}$. For example, if the charges are two electrons emerging from an electron accelerator, the force that we shall calculate is the one that changes the momentum of one of the electrons with respect to the laboratory. The magnitude of $\mathcal{V}$ can have any value less than or equal to c.

After we have studied this relatively simple case we shall go on to more general concepts.

Let us call the two charges Q_a and Q_b as in Figure 6-1, Q_a being situated at O_2, and Q_b at $(x_2, y_2, 0)$. Both charges are fixed in reference frame 2 and move at the constant velocity $\mathcal{V}\boldsymbol{i}$ with respect to reference frame 1.

In frame 2 the force on Q_b is $\boldsymbol{F}_{b2}$, which has the components

$$X_{b2} = \frac{Q_a Q_b x_2}{4\pi\epsilon_0 r_2^3}, \qquad Y_{b2} = \frac{Q_a Q_b y_2}{4\pi\epsilon_0 r_2^3}, \qquad Z_{b2} = 0. \tag{6-1}$$

Let us calculate the force on Q_b, as measured by an observer o_1 in reference frame 1, at the time $t_1 = 0$. The calculation is as follows. Observer o_1 sees the two charges moving at the same velocity $\mathcal{V}\boldsymbol{i}$. We simply use the transformation equations of Table 5-7 to find $\boldsymbol{F}_{b1}$, remembering that $v_{2x} = v_{2y} = v_{2z} = 0$ in this case and that Q_a and Q_b are invariant. This gives

$$X_{b1} = X_{b2} = \frac{Q_a Q_b x_2}{4\pi\epsilon_0 (x_2^2 + y_2^2)^{3/2}}, \tag{6-2}$$

$$Y_{b1} = \frac{Y_{b2}}{\gamma} = \frac{Q_a Q_b y_2}{4\pi\epsilon_0 \gamma (x_2^2 + y_2^2)^{3/2}}, \tag{6-3}$$

$$Z_{b1} = 0. \tag{6-4}$$

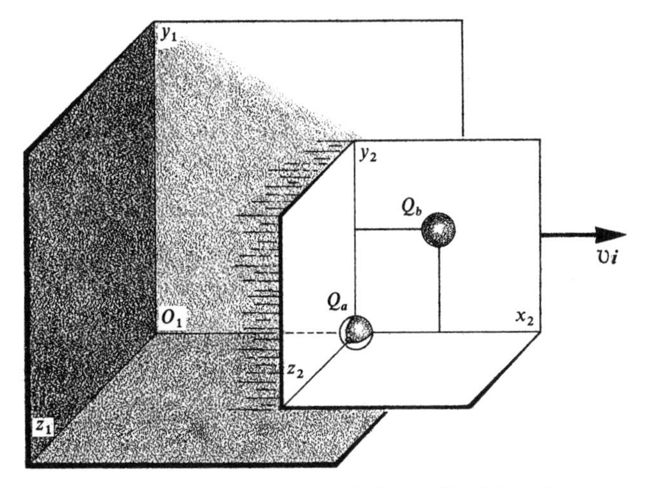

Figure 6-1. Two charges Q_a and Q_b are fixed in reference frame 2 and hence have the same velocity $\mathcal{V}\boldsymbol{i}$ with respect to frame 1. The charge Q_a is at O_2 and Q_b is at (x_2, y_2, O). We calculate the force exerted *by* Q_a *on* Q_b.

Substituting the values of x_2 and y_2 at $t_1 = 0$ from the Lorentz transformation, and omitting the subscript 1 for simplicity,

$$X_b = \frac{\gamma Q_a Q_b x}{4\pi\epsilon_0(\gamma^2 x^2 + y^2)^{3/2}}, \tag{6-5}$$

$$Y_b = \frac{Q_a Q_b y}{4\pi\epsilon_0\gamma(\gamma^2 x^2 + y^2)^{3/2}} = \frac{\gamma Q_a Q_b y}{4\pi\epsilon_0(\gamma^2 x^2 + y^2)^{3/2}}[1 - (\mathcal{V}/c)^2], \tag{6-6}$$

$$Z_b = 0. \tag{6-7}$$

Reference frame 1 can be taken to be fixed with respect to the laboratory; then, the above force and the above coordinates are those measured by an observer in the laboratory.

This is the force exerted *by* Q_a *on* Q_b at the time $t = 0$ when the charge Q_a is at $x = 0$.

We can rewrite $\boldsymbol{F}_b$ as follows:

$$\boldsymbol{F}_b = \frac{\gamma Q_a Q_b \boldsymbol{r}}{4\pi\epsilon_0(\gamma^2 x^2 + y^2)^{3/2}} - \frac{\gamma Q_a Q_b \mathcal{V}^2 y \boldsymbol{j}}{4\pi\epsilon_0 c^2(\gamma^2 x^2 + y^2)^{3/2}}, \tag{6-8}$$

$$= \frac{\gamma Q_a Q_b}{4\pi\epsilon_0(\gamma^2 x^2 + y^2)^{3/2}}\left\{\boldsymbol{r} - \frac{\mathcal{V}^2}{c^2} y\boldsymbol{j}\right\}, \tag{6-9}$$

$$= Q_b\left\{\frac{\gamma Q_a \boldsymbol{r}}{4\pi\epsilon_0(\gamma^2 x^2 + y^2)^{3/2}} + \left[\boldsymbol{\mathcal{V}} \times \frac{\gamma Q_a \mathcal{V} y \boldsymbol{k}}{4\pi\epsilon_0 c^2(\gamma^2 x^2 + y^2)^{3/2}}\right]\right\}. \tag{6-10}$$

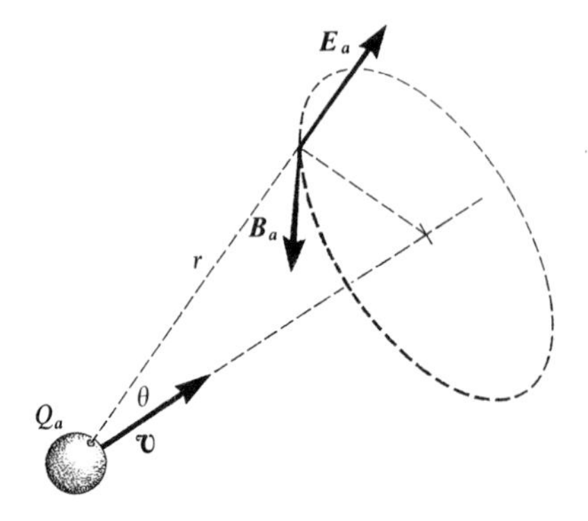

Figure 6-2. The electric field intensity E_a and the magnetic induction B_a due to the charge Q_a moving at the uniform velocity $\mathcal{V}$.

The product $4\pi\epsilon_0 c^2$, which appears in the last term, occurs in many calculations and is *defined* to be 10^7:

$$4\pi\epsilon_0 c^2 \equiv 10^7. \tag{6-11}$$

This, in effect, defines the *coulomb* as the unit of charge, because the units for the other variables F, $\mathcal{V}$, x, y are already defined from elementary mechanics. In fact, the constant ϵ_0 which appears in Coulomb's law is best determined by first measuring the velocity of light c and then using the above equation.

It is customary to write that

$$\frac{1}{\epsilon_0 c^2} \equiv \mu_0 \equiv 4\pi \times 10^{-7} \text{ henry/meter}, \tag{6-12}$$

where μ_0 is the *permeability of free space.*

The first term between the braces in Eq. 6-10 is the electric field intensity E_a of the charge Q_a at the point x, y, z and at the time $t = 0$, when the origins of the two systems coincide, as you can easily verify for the limiting case where $\mathcal{V} = 0$ and $\gamma = 1$.

In fact,

$$F_b = Q_b(E_a + \mathcal{V} \times B_a), \tag{6-13}$$

where

$$E_a = \frac{\gamma Q_a \boldsymbol{r}}{4\pi\epsilon_0(\gamma^2 x^2 + y^2)^{3/2}} \tag{6-14}$$

and

$$B_a = \frac{\mu_0 \gamma Q_a \mathcal{V} y \boldsymbol{k}}{4\pi(\gamma^2 x^2 + y^2)^{3/2}}, \tag{6-15}$$

is the *magnetic induction* due to Q_a at r. The magnetic induction $\boldsymbol{B}$ is expressed in volt-seconds/meter^2, or in webers/meter^2, or in *teslas*. The vectors E_a and B_a, and the corresponding forces are shown in Figures 6-2 and 6-3.

The force F_b is called the *Lorentz force.*

Note that the magnetic field of reference frame 1 has appeared as a result

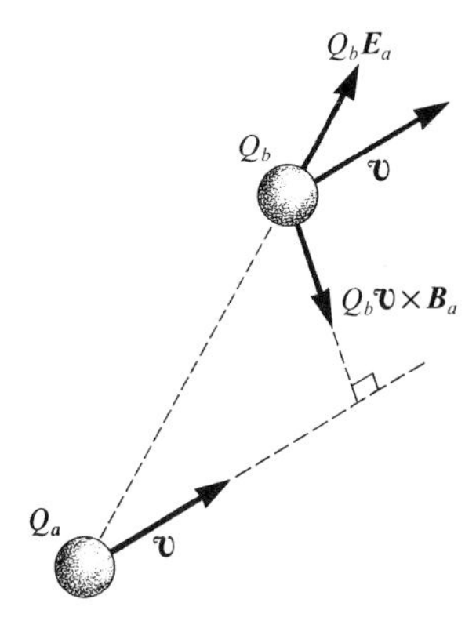

Figure 6-3. The electric force $Q_b\boldsymbol{E}_a$ and the magnetic force $Q_b\boldsymbol{\mathcal{V}} \times \boldsymbol{B}_a$ exerted on Q_b. The charges Q_a and Q_b move at the same uniform velocity $\boldsymbol{\mathcal{V}}$.

of the application of a relativistic transformation to the electric force in reference frame 2.

The electric field intensity $\boldsymbol{E}_a$ is radial, as well as the electric force $Q_b\boldsymbol{E}_a$. Figure 6-4 shows two limiting cases: in position (a), $\boldsymbol{E}_a$ is γ times larger in frame 1 than if $\mathcal{V}$ were zero, but in position (b) it is smaller by the factor $1/\gamma^2 = 1 - (\mathcal{V}/c)^2$.

The magnetic induction $\boldsymbol{B}_a$ is oriented in the positive direction of the z-axis and is therefore related to $\boldsymbol{\mathcal{V}}$ as in Figure 6-2. It exerts a force that is perpendicular to the trajectory of Q_b and that tends to bring Q_a and Q_b closer together if Q_a and Q_b are of the same sign, as in Figure 6-4. In position (a) the magnetic force cancels part of the electric force, and the total force tends to zero when the velocity $\mathcal{V}$ of the charges approaches c. In position (b) the magnetic force is zero.

6.2 FIELD OF A CHARGE MOVING AT A CONSTANT VELOCITY

We now wish to obtain a more general expression for the field of a charge moving at a constant velocity. To do this we shall use two charges as in Figure 6-5: the force exerted *on* Q_b will give us the field *of* Q_a at the position of Q_b. The charge Q_a is situated at O_2 and thus moves at the constant velocity $\mathcal{V}\boldsymbol{i}$ with respect to the laboratory (reference frame 1). The charge Q_b has some unspecified velocity $\boldsymbol{v}_b$, also with respect to the laboratory. The velocity $\boldsymbol{v}_b$ need not be constant.

We calculate again the force exerted *by* Q_a *on* Q_b, first in reference frame 2, and then in reference frame 1.

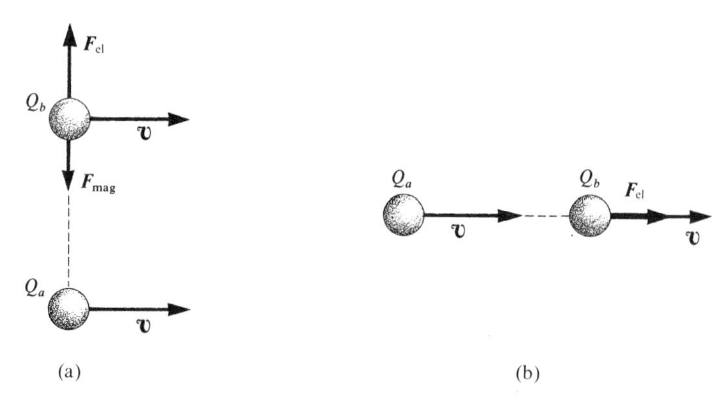

Figure 6-4. The two charges Q_a and Q_b again move at the same uniform velocity $\mathcal{V}$. In (a) the electric and magnetic forces are in opposite directions. In (b) there is only an electric force.

In reference frame 2 we *assume* that the only field acting on Q_b is that of Q_a.

In frame 2, Q_a is stationary and creates a field $\boldsymbol{E}_{a2}$ as in Section 2.2. We also *assume* that the force acting on Q_b in frame 2 is independent of the velocity of Q_b and is given simply by Coulomb's law, or by the product $\boldsymbol{E}_{a2}Q_b$.

Then the force on Q_b in frame 2 has the three components

$$X_{b2} = \frac{Q_aQ_bx_2}{4\pi\epsilon_0 r_2^3}, \qquad Y_{b2} = \frac{Q_aQ_by_2}{4\pi\epsilon_0 r_2^3}, \qquad Z_{b2} = \frac{Q_aQ_bz_2}{4\pi\epsilon_0 r_2^3}. \tag{6-16}$$

We again use the transformation of Table 5-7, except that now $\mathcal{V}_{b2} \neq 0$:

$$X_{b1} = \frac{Q_aQ_b}{4\pi\epsilon_0 r_2^3}\left[x_2 + \frac{v_{b2y}\mathcal{V}}{c^2 + v_{b2x}\mathcal{V}}\,y_2 + \frac{v_{b2z}\mathcal{V}}{c^2 + v_{b2x}\mathcal{V}}\,z_2\right], \tag{6-17}$$

$$Y_{b1} = \frac{Q_aQ_b}{4\pi\epsilon_0 r_2^3\gamma}\left[\frac{y_2}{1 + (v_{b2x}\mathcal{V}/c^2)}\right], \tag{6-18}$$

$$Z_{b1} = \frac{Q_aQ_b}{4\pi\epsilon_0 r_2^3\gamma}\left[\frac{z_2}{1 + (v_{b2x}\mathcal{V}/c^2)}\right]. \tag{6-19}$$

Now we must transform the coordinates and the velocities on the right-hand side. Let us start with the brackets:

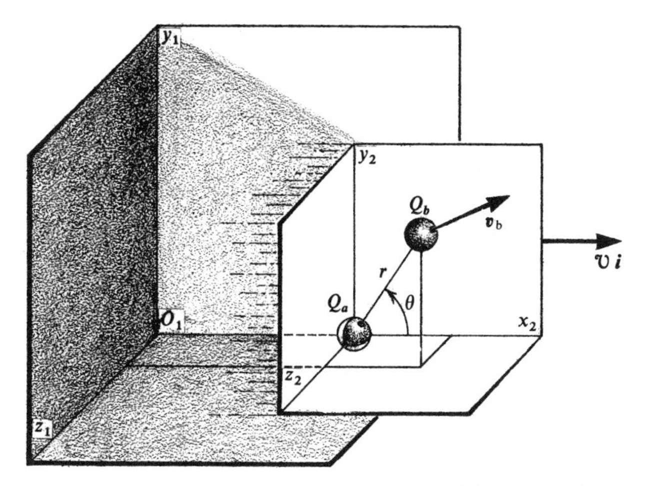

Figure 6-5. Charges Q_a and Q_b have velocities $\mathcal{V}\boldsymbol{i}$ and $\boldsymbol{v}_b$ respectively with respect to reference frame 1.

$$x_2 + \frac{v_{b2y}\mathcal{V}}{c^2 + v_{b2x}\mathcal{V}} y_2 + \frac{v_{b2z}\mathcal{V}}{c^2 + v_{b2x}\mathcal{V}} z_2$$

$$= \gamma(x_1 - \mathcal{V}t_1) + \gamma(v_{b1y}\mathcal{V}/c^2)y_1 + \gamma(v_{b1z}\mathcal{V}/c^2)z_1, \quad (6\text{-}20)$$

$$\frac{y_2}{1 + (v_{b2x}\mathcal{V}/c^2)} = \gamma^2[1 - (v_{b1x}\mathcal{V}/c^2)]y_1, \quad (6\text{-}21)$$

$$\frac{z_2}{1 + (v_{b2x}\mathcal{V}/c^2)} = \gamma^2[1 - (v_{b1x}\mathcal{V}/c^2)]z_1. \quad (6\text{-}22)$$

The vector drawn from Q_a to Q_b is

$$\boldsymbol{r} = (x - \mathcal{V}t)\boldsymbol{i} + y\boldsymbol{j} + z\boldsymbol{k}. \quad (6\text{-}23)$$

Then the force exerted *by* Q_a *on* Q_b in reference frame 1, with respect to which Q_a has a velocity $\mathcal{V}\boldsymbol{i}$ and Q_b has a velocity $\boldsymbol{v}_b$, is

$$\boldsymbol{F}_b = \frac{\gamma Q_a Q_b}{4\pi\epsilon_0 r_2^3} \{\boldsymbol{r} + (\mathcal{V}/c^2)[(v_{by}y + v_{bz}z)\boldsymbol{i} - v_{bx}y\boldsymbol{j} - v_{bx}z\boldsymbol{k}]\}, \quad (6\text{-}24)$$

$$= \frac{\gamma Q_a Q_b}{4\pi\epsilon_0 r_2^3} \{\boldsymbol{r} + (1/c^2)\boldsymbol{v}_b \times (\boldsymbol{\mathcal{V}} \times \boldsymbol{r})\}. \quad (6\text{-}25)$$

We have again omitted the subscript 1 for simplicity. We have made only one assumption: we have assumed that the velocity $\mathcal{V}$ of Q_a is constant. Now

$$r_2^2 = \gamma^2(x - \mathcal{V}t)^2 + y^2 + z^2, \quad (6\text{-}26)$$

$$= \gamma^2[(x - \mathcal{V}t)^2 + (y^2 + z^2)(1 - \beta^2)], \quad (6\text{-}27)$$

$$= \gamma^2 r^2(1 - \beta^2 \sin^2\theta), \quad (6\text{-}28)$$

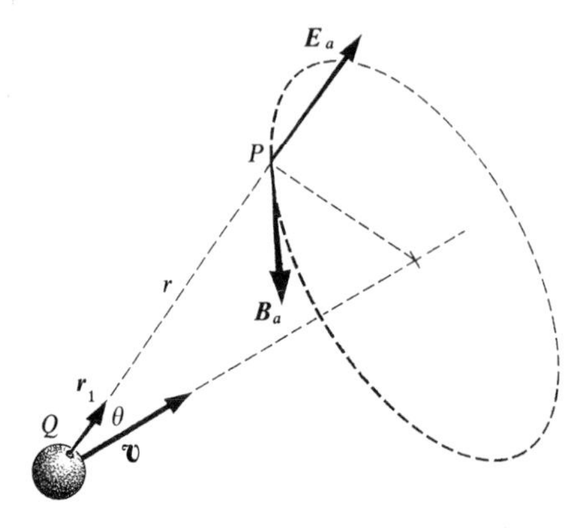

Figure 6-6. The $\boldsymbol{E}$ and $\boldsymbol{B}$ vectors at a point P due to a charge Q moving at a velocity $\mathcal{V}$.

where r and θ are as in Figure 6-5, and $\beta = \mathcal{V}/c$. Then

$$\boldsymbol{F}_b = \frac{Q_a Q_b}{4\pi\epsilon_0 \gamma^2 r^2 (1 - \beta^2 \sin^2\theta)^{3/2}} [\boldsymbol{r}_1 + (1/c^2)\boldsymbol{v}_b \times (\boldsymbol{\mathcal{V}} \times \boldsymbol{r}_1)] \qquad (6\text{-}29)$$

where $\boldsymbol{r}_1 = \boldsymbol{r}/r$ is the *unit vector* pointing from Q_a to Q_b.

The quantity between the braces shows that the force has again two components, the electric force represented by the first term $\boldsymbol{r}_1$, and the magnetic force represented by the double cross-product term:

$$\boldsymbol{F}_b = Q_b(\boldsymbol{E}_a + \boldsymbol{v}_b \times \boldsymbol{B}_a), \qquad (6\text{-}30)$$

where

$$\boldsymbol{E} = \frac{Q}{4\pi\epsilon_0 \gamma^2 r^2 (1 - \beta^2 \sin^2\theta)^{3/2}} \boldsymbol{r}_1, \qquad (6\text{-}31)$$

$$\boldsymbol{B} = \frac{\mu_0 Q \mathcal{V} \sin\theta}{4\pi \gamma^2 r^2 (1 - \beta^2 \sin^2\theta)^{3/2}} \boldsymbol{\varphi}_1, \qquad (6\text{-}32)$$

as in Figure 6-6. We have now omitted the subscript a since we shall not need it from now on. You can verify that these equations reduce to Eqs. 6-9 and 6-10 for the special case where $t = 0$, $z = 0$.

In Cartesian coordinates,

$$\boldsymbol{E} = \frac{\gamma Q[(x - \mathcal{V}t)\boldsymbol{i} + y\boldsymbol{j} + z\boldsymbol{k}]}{4\pi\epsilon_0[\gamma^2(x - \mathcal{V}t)^2 + y^2 + z^2]^{3/2}}, \qquad (6\text{-}33)$$

$$\boldsymbol{B} = \frac{\mu_0}{4\pi} \frac{\gamma Q \mathcal{V}[-z\boldsymbol{j} + y\boldsymbol{k}]}{[\gamma^2(x - \mathcal{V}t)^2 + y^2 + z^2]^{3/2}}. \qquad (6\text{-}34)$$

This is the electric field intensity $\boldsymbol{E}$ and the magnetic induction $\boldsymbol{B}$ at the point (x, y, z) and at the time t due to a charge Q moving at a *constant* velocity $\mathcal{V}\boldsymbol{i}$ and passing through the origin at the time $t = 0$.

We have arrived at this result by calculating the force on a second moving

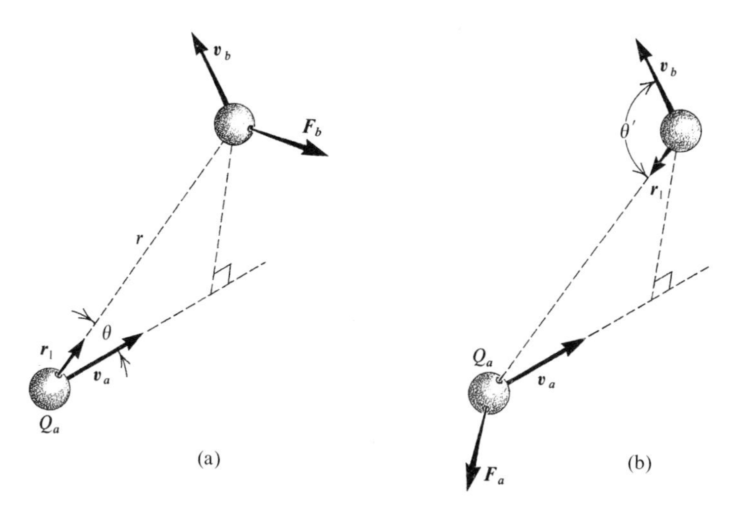

Figure 6-7. Figure (a) shows the force $\boldsymbol{F}_b$ exerted *by* Q_a *on* Q_b. Figure (b) shows the new $\boldsymbol{r}_1$ and the new θ which we use to calculate $\boldsymbol{F}_a$. It is found that $\boldsymbol{F}_a \neq -\boldsymbol{F}_b$, contrary to what one would expect from elementary mechanics.

charge, first in a reference frame moving with Q, and then in the reference frame of the laboratory.

What is the force exerted *by* Q_b *on* Q_a? According to elementary mechanics, $\boldsymbol{F}_a$ should be equal to $-\boldsymbol{F}_b$. However, our experience warns us to be cautious, so let us calculate $\boldsymbol{F}_a$.

We can find $\boldsymbol{F}_a$ from the value of $\boldsymbol{F}_b$ given in Eq. 6-29. In this equation $\mathcal{V}$ is the velocity $\boldsymbol{v}_a$ of Q_a and

$$\boldsymbol{F}_b = \frac{Q_a Q_b}{4\pi\epsilon_0 \gamma^2 r^2 (1 - \beta^2 \sin^2 \theta)^{3/2}} \{\boldsymbol{r}_1 + (1/c^2)\boldsymbol{v}_b \times (\boldsymbol{v}_a \times \boldsymbol{r}_1)\}. \quad (6\text{-}35)$$

This is the force exerted *by* Q_a *on* Q_b if the various terms are defined as in Figure 6-7a, and if

$$\beta = v_a/c, \qquad \gamma^2 = \frac{1}{1 - \beta^2}. \quad (6\text{-}36)$$

To find $\boldsymbol{F}_a$ we interchange the roles of Q_a and Q_b in the above formula. This means

(a) replacing $\boldsymbol{r}_1$ by $-\boldsymbol{r}_1$ as in Figure 6-7b,
(b) interchanging $\boldsymbol{v}_a$ and $\boldsymbol{v}_b$,
(c) replacing θ by θ',
(d) replacing $\beta = v_a/c$ by $\beta' = v_b/c$.

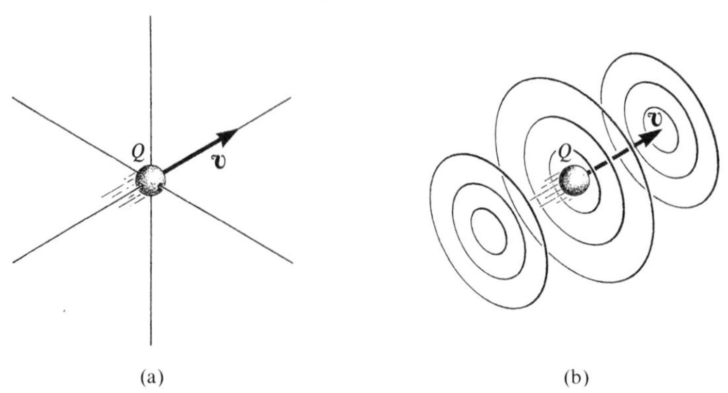

Figure 6-8. (a) Typical line of $\boldsymbol{E}$, and (b) typical lines of $\boldsymbol{B}$ for a charge Q moving at a constant velocity $\mathcal{V}$, as seen by a stationary observer. The electric field is radial. The lines of $\boldsymbol{B}$ are circles centered on the trajectory and perpendicular to it.

Thus

$$\boldsymbol{F}_a = -\frac{Q_a Q_b}{4\pi\epsilon_0 \gamma'^2 r^2 (1 - \beta'^2 \sin^2 \theta')^{3/2}} \{\boldsymbol{r}_1 + (1/c^2)\boldsymbol{v}_a \times (\boldsymbol{v}_b \times \boldsymbol{r}_1)\}. \qquad (6\text{-}37)$$

The first and second terms within the braces give, respectively, the electric and magnetic components of $\boldsymbol{F}_a$.

It is obvious that $\boldsymbol{F}_a \neq -\boldsymbol{F}_b$. The electric components are exactly in opposite directions, but their magnitudes are not the same because of the θ', β', γ' terms, and the magnetic components are completely different.

There is a difference between $\boldsymbol{F}_a$ and $\boldsymbol{F}_b$ that is worth noting. The expression for $\boldsymbol{F}_b$ is valid if $\mathcal{V}$, or $\boldsymbol{v}_a$, is constant. That for $\boldsymbol{F}_a$, on the contrary, is valid if $\boldsymbol{v}_b$ is constant.

The fact that $\boldsymbol{F}_a \neq -\boldsymbol{F}_b$ is *not* peculiar to electric phenomena. It is a purely relativistic effect that would be observed with any type of force.

6.2.1 The Electric Field

Note that the electric field of Q is radial: the lines of force of $\boldsymbol{E}$ are straight lines that converge on Q as in Figure 6-8. The fact that the electric field is radial comes as a surprise. Imagine that you have fixed a light source to Q_a and that you move with Q_b. You will *not* see Q_a where it is at the moment when you look at it, but where it *was* when the light that reaches your eye left Q_a. At the time t you will see Q_a where it was at the previous time $t - (r/c)$, r being the distance traveled by the light that arrives at the time t. You must

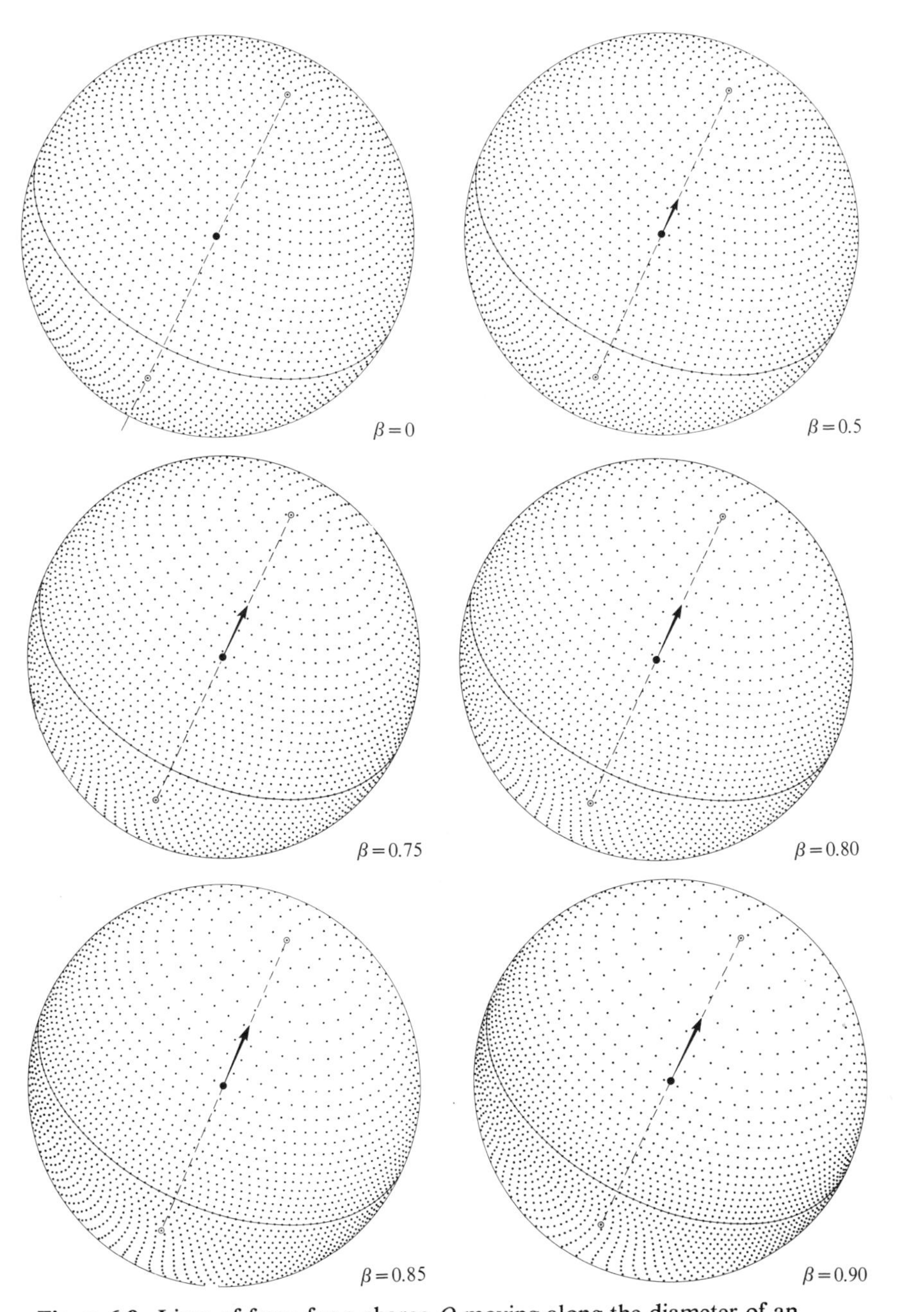

Figure 6-9. Lines of force for a charge Q moving along the diameter of an imaginary sphere. The dots show where lines emerge from the sphere at the instant when the charge is at its center. The density of the dots is a measure of the electric field intensity. The total number of dots is the same in all six figures, so as to satisfy Gauss's law (Section 6.8). Note how the field shifts to the region of $\theta = 90°$ as the velocity increases. For $\mathcal{V} = c$ the field is all concentrated at $\theta = 90°$.

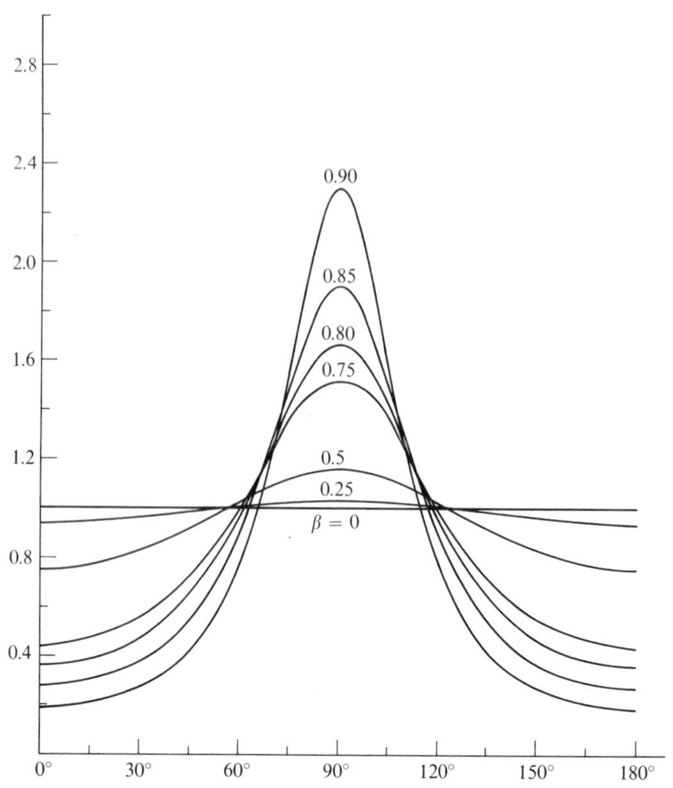

Figure 6-10. The electric field intensity E of a moving point charge as a function of the polar angle θ of Figure 6-6, for seven values of $\beta = \mathcal{V}/c$. The observer is stationary and sees the charge moving at the uniform velocity $\mathcal{V}$. For $\beta = 0$ the field is isotropic. It is hardly disturbed at $\beta = 0.25$. As the velocity increases the field increases near $\theta = 90°$ and decreases both ahead of the charge (near $\theta = 0$) and behind it (near $\theta = 180°$). At extremely high velocities most of the electric field is concentrated near $\theta = 90°$. These curves explain *qualitatively* the validity of Gauss's law for moving charges: as the velocity increases, the flux of $\boldsymbol{E}$ shifts from the regions where $\theta \approx 0$ and $\approx 180°$ to $\theta \approx 90°$ and the total flux of $\boldsymbol{E}$ remains constant. (Then why are the areas under the curves not equal?) Note that the electric field is always symmetrical about 90°. This means that there is no way of telling, from the shape of the field, whether the charge is moving to the right or to the left. The vertical scale gives E divided by $Q/4\pi\epsilon_0 r^2$.

therefore look, not in the direction of the charge, but some distance back; at a given moment, the light rays originating from Q_a are *not* radial lines converging on Q_a.

But the electric field is radial—as if the information concerning the position of the charge traveled at an infinite velocity! Actually, it is only when the velocity of the charge is constant that its electric field is radial; if the charge is accelerated there is a retardation effect, and the field is not radial. We have assumed implicitly that the velocity of the charge has been equal to

$\mathcal{V}\boldsymbol{i}$ for an infinite time; otherwise, any disturbance in the velocity would appear as a distortion of the field.

Figure 6-9 illustrates how the lines of $\boldsymbol{E}$ migrate toward the region $\theta = 90°$ as the velocity increases, and Figure 6-10 shows the magnitude of $\boldsymbol{E}$ as a function of θ for several values of $\beta = \mathcal{V}/c$.

6.2.2 The Magnetic Field

The lines of $\boldsymbol{B}$ are circles centered on the trajectory and perpendicular to it, as in Figure 6-8. At any moment, the magnetic field is symmetrical about the instantaneous position of the charge, as is shown in Figure 6-11. Here again, the field is the same as if the information concerning the position of the charge traveled at an infinite velocity.

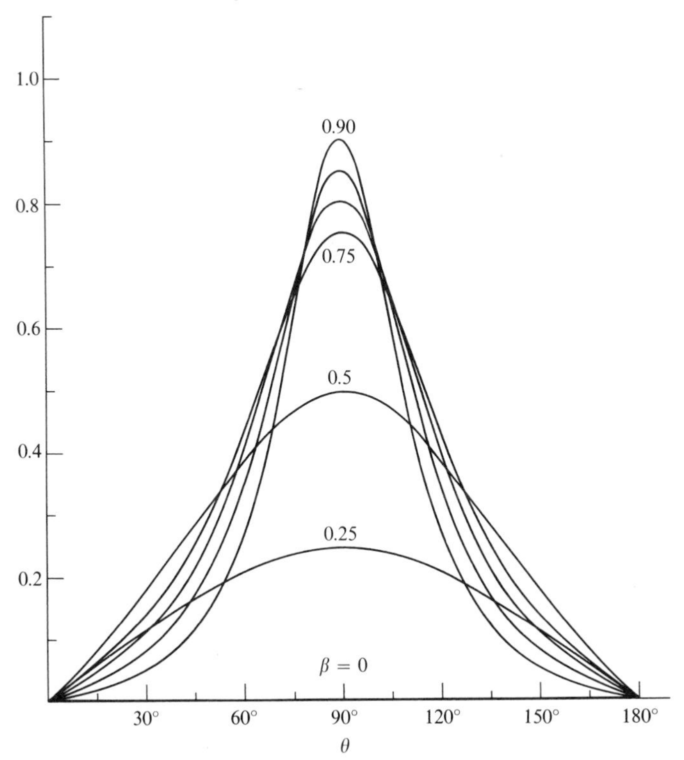

Figure 6-11. The magnetic induction B of a moving point charge as a function of the polar angle θ for seven values of β. For $\beta = 0$ there is no magnetic field. As β increases, B first increases at all angles. Then B continues to increase near $\theta = 90°$, while *de*creasing both ahead of the charge and behind it. At extremely high velocities, most of the magnetic field is concentrated near the plane $\theta = 90°$. Note that the magnetic field is symmetrical with respect to $\theta = 90°$. The vertical scale gives B divided by $\mu Q\gamma c/4\pi r^2$, and thus the maximum ordinate on any curve is β.

The second term on the right in Eq. 6-30 gives the magnetic force on Q_b. The fact that this force is of the form $Q_b \boldsymbol{v}_b \times \boldsymbol{B}_a$ has three important consequences.

(a) The magnetic force can exist, in a given frame of reference, only if the charge Q_b has a velocity with respect to that particular frame.

(b) The magnetic force on Q_b is independent of the component of $\boldsymbol{v}_b$ that is parallel to the $\boldsymbol{B}$ of Q_a.

(c) The magnetic force is always perpendicular to the velocity $\boldsymbol{v}_b$, and it does no work.

Then, if $\mathcal{E}$ is the relativistic energy mc^2 of a particle of mass m which carries the charge Q,

$$\frac{d\mathcal{E}}{dt} = \boldsymbol{F} \cdot \boldsymbol{v} = Q\boldsymbol{E} \cdot \boldsymbol{v}. \tag{6-38}$$

It is only with an *electric* field that one can increase or decrease the kinetic energy of a charged particle; a magnetic field can deflect a charged particle, but cannot change its speed.

Example THE FIELD OF A 10-GeV ELECTRON AT $\theta = 90°$

Imagine that we have an E-meter and a B-meter with infinitely short response times. What will be the maximum values of E and of B at a distance of 1 centimeter from the path of a single 10 GeV electron? One GeV is 10^9 electron-volts.

From Eqs. 6-31 and 6-32, as well as from Figures 6-10 and 6-11, both E and B are maximum at $\theta = 90°$, or when the line joining the particle to the point of observation is perpendicular to the path of the particle. Then $\sin\theta = 1$ and

$$E_{\max} = \frac{\gamma Q}{4\pi\epsilon_0 r^2}, \tag{6-39}$$

in the radial direction,

$$B_{\max} = \frac{\mu_0 \gamma Q v}{4\pi r^2} \tag{6-40}$$

in the azimuthal direction. The relativistic kinetic energy gives us γ:

$$(m - m_0)c^2 = m_0(\gamma - 1)c^2 = 10^{10} \times 1.6 \times 10^{-19} \text{ joule}, \tag{6-41}$$

$$\gamma = \frac{1.6 \times 10^{-9}}{9.1 \times 10^{-31} \times 9 \times 10^{16}} + 1 = 2.0 \times 10^4. \tag{6-42}$$

Then

$$E_{\max} = 9 \times 10^9 \frac{2.0 \times 10^4 \times 1.6 \times 10^{-19}}{10^{-4}} = 0.29 \text{ volt/meter}, \tag{6-43}$$

$$B_{\max} = \frac{10^{-7} \times 2 \times 10^4 \times 1.6 \times 10^{-19} \times 3 \times 10^8}{10^{-4}}$$

$$= 9.6 \times 10^{-10} \text{ tesla.} \quad (6\text{-}44)$$

We have set $\mathcal{V} = c$ because, at such a large value of γ, $v \approx c$.

Although E is 2×10^4 times larger than if the electron were stationary in the laboratory frame, it is still quite small. As for B, it is about 10^{-5} times that of the Earth.

6.3 TRANSFORMATION OF ELECTRIC AND MAGNETIC FIELDS

We have just seen that an electric field in one reference frame becomes both an electric and a magnetic field in another reference frame. There exists a general rule for transforming electric and magnetic fields, and we shall be able to deduce it by again transforming the force on a moving charge.

A charge Q has a velocity $\boldsymbol{v}_1$ in a region where there is an electric field intensity $\boldsymbol{E}_1$ and a magnetic induction $\boldsymbol{B}_1$, all with respect to reference frame 1. Then

$$\boldsymbol{F}_1 = Q(\boldsymbol{E}_1 + \boldsymbol{v}_1 \times \boldsymbol{B}_1). \quad (6\text{-}45)$$

To find $\boldsymbol{E}_2$ and $\boldsymbol{B}_2$ in a reference frame moving at the velocity $\mathcal{V}\boldsymbol{i}$ with respect to the first, we shall transform $\boldsymbol{F}_1$ into $\boldsymbol{F}_2$. We should find that

$$\boldsymbol{F}_2 = Q(\boldsymbol{E}_2 + \boldsymbol{v}_2 \times \boldsymbol{B}_2), \quad (6\text{-}46)$$

and this should give us $\boldsymbol{E}_2$ and $\boldsymbol{B}_2$ in terms of $\boldsymbol{E}_1$, $\boldsymbol{B}_1$, $\mathcal{V}$.

We first write out the components of $\boldsymbol{F}_1$:

$$X_1 = Q[E_{1x} + (v_{1y}B_{1z} - v_{1z}B_{1y})], \quad (6\text{-}47)$$

$$Y_1 = Q[E_{1y} + (v_{1z}B_{1x} - v_{1x}B_{1z})], \quad (6\text{-}48)$$

$$Z_1 = Q[E_{1z} + (v_{1x}B_{1y} - v_{1y}B_{1x})]. \quad (6\text{-}49)$$

Now, from Table 5-7,

$$X_2 = X_1 - \frac{\mathcal{V}}{c^2 - v_{1x}\mathcal{V}}(v_{1y}Y_1 + v_{1z}Z_1), \quad (6\text{-}50)$$

$$Y_2 = \frac{Y_1}{\gamma[1 - (v_{1x}\mathcal{V}/c^2)]}, \quad (6\text{-}51)$$

$$Z_2 = \frac{Z_1}{\gamma[1 - (v_{1x}\mathcal{V}/c^2)]}. \quad (6\text{-}52)$$

We must substitute into these equations the above values of X_1, Y_1, Z_1, and then set

Table 6-1. *Transformation of Electric and Magnetic Fields*

$E_{1x} = E_{2x},$	$E_{2x} = E_{1x},$
$E_{1y} = \gamma[E_{2y} + \mathcal{V}B_{2z}],$	$E_{2y} = \gamma[E_{1y} - \mathcal{V}B_{1z}],$
$E_{1z} = \gamma[E_{2z} - \mathcal{V}B_{2y}],$	$E_{2z} = \gamma[E_{1z} + \mathcal{V}B_{1y}],$
$B_{1x} = B_{2x},$	$B_{2x} = B_{1x},$
$B_{1y} = \gamma[B_{2y} - (\mathcal{V}/c^2)E_{2z}],$	$B_{2y} = \gamma[B_{1y} + (\mathcal{V}/c^2)E_{1z}],$
$B_{1z} = \gamma[B_{2z} + (\mathcal{V}/c^2)E_{2y}],$	$B_{2z} = \gamma[B_{1z} - (\mathcal{V}/c^2)E_{1y}].$

$$v_{1x} = \frac{v_{2x} + \mathcal{V}}{1 + (v_{2x}\mathcal{V}/c^2)}, \tag{6-53}$$

$$v_{1y} = \frac{v_{2y}}{\gamma[1 + (v_{2x}\mathcal{V}/c^2)]}, \tag{6-54}$$

$$v_{1z} = \frac{v_{2z}}{\gamma[1 + (v_{2x}\mathcal{V}/c^2)]}, \tag{6-55}$$

as in Table 5-2.

We start with Y_2 because it is relatively simple to calculate:

$$Y_2 = Q\{\gamma[E_{1y} - \mathcal{V}B_{1z}] + v_{2z}B_{1x} - v_{2x}\gamma[B_{1z} - (\mathcal{V}/c^2)E_{1y}]\}. \tag{6-56}$$

This is of the form

$$Y_2 = Q[E_{2y} + v_{2z}B_{2x} - v_{2x}B_{2z}], \tag{6-57}$$

with

$$E_{2y} = \gamma[E_{1y} - \mathcal{V}B_{1z}], \tag{6-58}$$

$$B_{2x} = B_{1x}, \tag{6-59}$$

$$B_{2z} = \gamma[B_{1z} - (\mathcal{V}/c^2)E_{1y}]. \tag{6-60}$$

We have now found three of the six components of $\boldsymbol{E}$ and $\boldsymbol{B}$ in frame 2. Repeating the calculation with Z_2 gives us two more components:

$$E_{2z} = \gamma[E_{1z} + \mathcal{V}B_{1y}], \tag{6-61}$$

$$B_{2y} = \gamma[B_{1y} + (\mathcal{V}/c^2)E_{1z}]. \tag{6-62}$$

To find E_{2x} we calculate X_2 for the special case where $v_{1y} = v_{1z} = 0$. This immediately gives us

$$E_{2x} = E_{1x}. \tag{6-63}$$

Grouping our results and adding the inverse transformations, we have the equations of Table 6-1.

Now we can demonstrate quite easily that

$$B_1^2 - \frac{E_1^2}{c^2} = B_2^2 - \frac{E_2^2}{c^2}, \tag{6-64}$$

and that

$$\boldsymbol{E}_1 \cdot \boldsymbol{B}_1 = \boldsymbol{E}_2 \cdot \boldsymbol{B}_2. \tag{6-65}$$

Therefore both

$$B^2 - \frac{E^2}{c^2} \qquad \text{and} \qquad \boldsymbol{E} \cdot \boldsymbol{B}$$

are invariants.

Example

THE PARALLEL-PLATE CAPACITOR

As an example, we shall transform the electric field of a moving air-insulated parallel-plate capacitor.

We choose the y-axis perpendicular to the plates and in the direction of the electric field, as in Figure 6-12a. Then

$$E_{2x} = 0, \qquad E_{2y} = E_2, \qquad E_{2z} = 0, \tag{6-66}$$

$$B_{2x} = B_{2y} = B_{2z} = 0. \tag{6-67}$$

If now the capacitor moves at a velocity $\mathcal{V}$ in the direction of the x-axis, a stationary observer sees the field of Figure 6-12b:

$$E_{1x} = 0, \qquad E_{1y} = \gamma E_2, \qquad E_{1z} = 0, \tag{6-68}$$

$$B_{1x} = 0, \qquad B_{1y} = 0, \qquad B_{1z} = \gamma(\mathcal{V}/c^2)E_2. \tag{6-69}$$

This is not too difficult to explain. Let us first consider the electric field. In the reference frame of the capacitor, we have charge densities $\pm\sigma_2$ on the plates, where $\sigma_2 = \epsilon_0 E_2$. Now in frame 1 the plates are shorter by the factor $1/\gamma$, but they carry the same charge. Then σ_1 is larger than σ_2 by the factor γ, and then $\boldsymbol{E}_1$ must also be larger than $\boldsymbol{E}_2$ by the factor γ. (We shall see in Section 6.8 that Gauss's law applies to moving charges.)

What about the potentials on the plates? Since the distance s between the plates is measured in a direction perpendicular to the velocity, it is the same in both frames. But

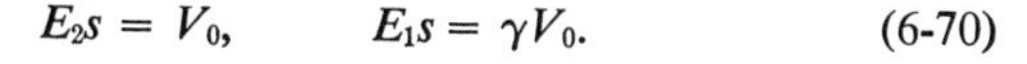

$$E_2 s = V_0, \qquad E_1 s = \gamma V_0. \tag{6-70}$$

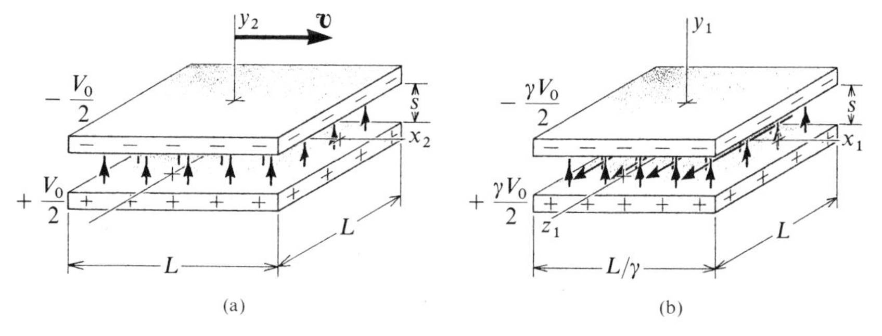

Figure 6-12. Figure (a) shows a parallel-plate capacitor as seen in its own reference frame. The electric field intensity inside it is E_2. For a stationary observer who sees the capacitor moving to the right at a velocity $\mathcal{V}$, the field is that of Figure (b): the electric field intensity is γ times stronger and there is a magnetic induction $\gamma(\mathcal{V}/c^2)E_2$ in the direction of the z-axis.

Then the voltage difference between the plates is γ times larger in frame 1 than in frame 2. We shall return to this subject in the example on page 270.

But why should there be a magnetic field? The answer is that observer o_1 in reference frame 1 sees a current flowing to the right in the lower plate and a current flowing to the left in the upper plate. You will be able to show in Problem 7-27 that

$$B_{1z} = \mu_0\sigma_1\mathcal{V} = \mu_0(\gamma\epsilon_0 E_2)\mathcal{V} = \gamma(\mathcal{V}/c^2)E_2. \tag{6-71}$$

Finally, let us check whether $E^2 - c^2B^2$ and $\boldsymbol{E}\cdot\boldsymbol{B}$ are invariant. In frame 1,

$$E_1^2 - c^2B_1^2 = \gamma^2E_2^2 - c^2\gamma^2(\mathcal{V}/c^2)^2E_2^2, \tag{6-72}$$

$$= \gamma^2[1 - (\mathcal{V}^2/c^2)]E_2^2 = E_2^2, \tag{6-73}$$

which is the value of $E_2^2 - c^2B_2^2$. Similarly,

$$\boldsymbol{E}_1\cdot\boldsymbol{B}_1 = \boldsymbol{E}_2\cdot\boldsymbol{B}_2 = 0. \tag{6-74}$$

6.4 THE VECTOR POTENTIAL *A*

In Section 6-2 we found that the field at x, y, z, t due to a charge Q traveling at a constant velocity $\mathcal{V}$ along the x-axis and passing through the origin at $t = 0$ is given by

$$\boldsymbol{E} = \frac{\gamma Q[(x - \mathcal{V}t)\boldsymbol{i} + y\boldsymbol{j} + z\boldsymbol{k}]}{4\pi\epsilon_0[\gamma^2(x - \mathcal{V}t)^2 + y^2 + z^2]^{3/2}}, \tag{6-75}$$

$$\boldsymbol{B} = \frac{\mu_0}{4\pi}\frac{\gamma Q\mathcal{V}(-z\boldsymbol{j} + y\boldsymbol{k})}{[\gamma^2(x - \mathcal{V}t)^2 + y^2 + z^2]^{3/2}}. \tag{6-76}$$

We are now ready to deduce from these two equations several fundamental relations involving $\boldsymbol{E}$ and $\boldsymbol{B}$.

We start with the vector potential $\boldsymbol{A}$. You will remember that the electrostatic field intensity $\boldsymbol{E}$ can be expressed as minus the gradient of a certain quantity V, which we called the electric potential: $\boldsymbol{E} = -\boldsymbol{\nabla}V$. The magnetic induction $\boldsymbol{B}$ can be expressed in a similar fashion as

$$\boldsymbol{B} = \boldsymbol{\nabla}\times\boldsymbol{A}, \tag{6-77}$$

where the quantity $\boldsymbol{A}$ is the *vector potential* at x, y, z, t. The vector potential is expressed in webers/meter. We shall show that, when the field is due to a single charge Q moving at a constant velocity $\mathcal{V}$ along the x-axis,

$$\boldsymbol{A} = \frac{\mu_0}{4\pi}\frac{\gamma Q\mathcal{V}\boldsymbol{i}}{[\gamma^2(x - \mathcal{V}t)^2 + y^2 + z^2]^{1/2}}. \tag{6-78}$$

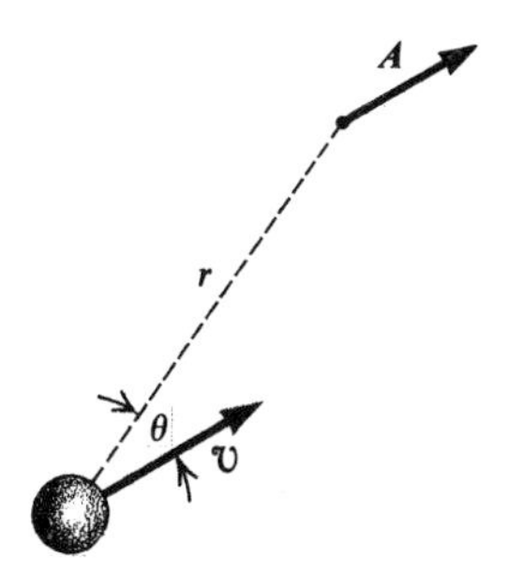

Figure 6-13. The vector potential A is parallel to the velocity of the moving charge when the velocity is constant. It is inversely proportional to r and inversely proportional to $(1 - \beta^2 \sin^2 \theta)^{1/2}$, where $\beta = \mathcal{V}/c$.

It is in some cases more convenient to write A in the following form:

$$A = \frac{\mu_0}{4\pi} \frac{Q\mathcal{V}\boldsymbol{i}}{r(1 - \beta^2 \sin^2 \theta)^{1/2}}, \tag{6-79}$$

where

$$r = [(x - \mathcal{V}t)^2 + y^2 + z^2]^{1/2} \tag{6-80}$$

is the distance between the charge and the point of observation (x, y, z). At the time t, the charge is situated at $(\mathcal{V}t, 0, 0)$.

Note that the vector potential is in the direction of the velocity of the charge as in Figure 6-13.

We first write down the four partial derivatives of the term between brackets in Eq. 6-78, as we shall require them quite often:

$$\frac{\partial}{\partial x}[\gamma^2(x - \mathcal{V}t)^2 + y^2 + z^2] = 2\gamma^2(x - \mathcal{V}t), \tag{6-81}$$

$$\frac{\partial}{\partial y}[\gamma^2(x - \mathcal{V}t)^2 + y^2 + z^2] = 2y, \tag{6-82}$$

$$\frac{\partial}{\partial z}[\gamma^2(x - \mathcal{V}t)^2 + y^2 + z^2] = 2z, \tag{6-83}$$

$$\frac{\partial}{\partial t}[\gamma^2(x - \mathcal{V}t)^2 + y^2 + z^2] = -2\gamma^2\mathcal{V}(x - \mathcal{V}t). \tag{6-84}$$

Then the curl of the above A is

$$\nabla \times A = \begin{vmatrix} \boldsymbol{i} & \boldsymbol{j} & \boldsymbol{k} \\ \frac{\partial}{\partial x} & \frac{\partial}{\partial y} & \frac{\partial}{\partial z} \\ A & 0 & 0 \end{vmatrix} = \frac{\partial A}{\partial z}\boldsymbol{j} - \frac{\partial A}{\partial y}\boldsymbol{k}, \tag{6-85}$$

$$= \frac{\mu_0}{4\pi} \frac{\gamma Q\mathcal{V}(-z\boldsymbol{j} + y\boldsymbol{k})}{[\gamma^2(x - \mathcal{V}t)^2 + y^2 + z^2]^{3/2}} = \boldsymbol{B}, \tag{6-86}$$

as required.

Example

THE VECTOR POTENTIAL FOR A 10-GeV ELECTRON

Let us calculate the maximum value of A for the 10-GeV electron of the example on page 260. From Eq. 6-79, A is maximum at $\theta = 90°$ and

$$A_{max} = \frac{\mu_0 \gamma Q \mathcal{V}}{4\pi r}. \tag{6-87}$$

We have already found that γ is 2.0×10^4 and that $\mathcal{V} \approx c$. Thus, at 1 centimeter from the path of the particle,

$$A_{max} = \frac{4\pi \times 10^{-7} \times 2 \times 10^4 \times 1.6 \times 10^{-19} \times 3 \times 10^8}{4\pi \times 10^{-2}}, \tag{6-88}$$

$$= 9.6 \times 10^{-12} \text{ weber/meter.} \tag{6-89}$$

The vector potential is everywhere parallel to the trajectory, and it points in the same direction as the velocity of the particle.

6.5 THE SCALAR POTENTIAL V. THE ELECTRIC FIELD INTENSITY $\boldsymbol{E}$ EXPRESSED IN TERMS OF V AND $\boldsymbol{A}$

We have seen in Chapter 2 that accumulations of electric charge produce electric fields for which $\boldsymbol{E} = -\nabla V$. Now, if some of the charges are in motion, there is also a magnetic field and, if the magnetic field is not constant, there appears a second electric field that adds to the first one. We shall show that the electric field intensity associated with a changing magnetic field is $-\partial \boldsymbol{A}/\partial t$, and that the general expression for $\boldsymbol{E}$ is

$$\boldsymbol{E} = -\nabla V - \frac{\partial \boldsymbol{A}}{\partial t}. \tag{6-90}$$

as in Figure 6-14.

We shall demonstrate this general result for the special case of a single charge moving at a constant velocity.

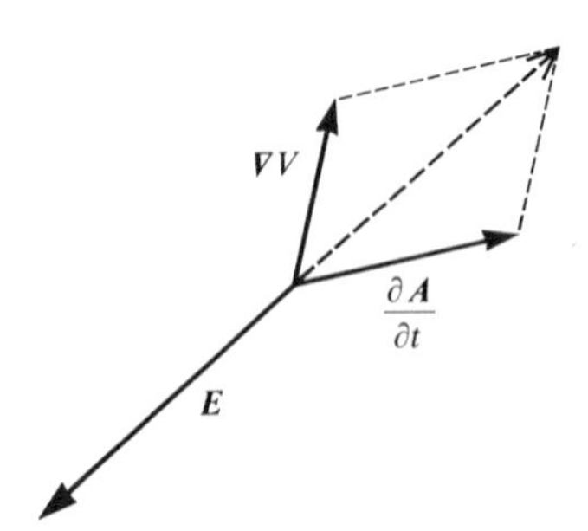

Figure 6-14. The electric field intensity is the vector sum of $-\nabla V$ and $-\partial \boldsymbol{A}/\partial t$.

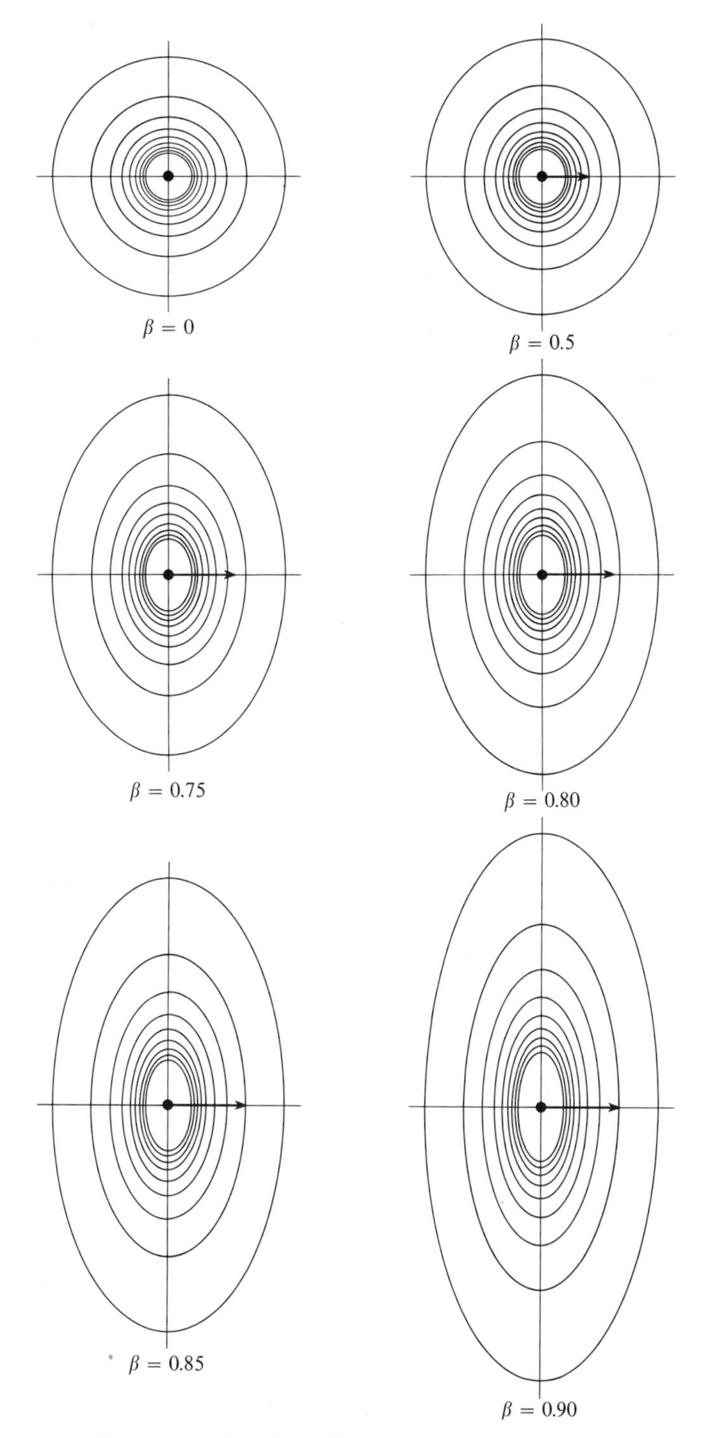

Figure 6-15. Equipotential surfaces V = Constant for a point charge Q moving either to the right or to the left. The equipotentials near Q are not shown because they are too close together.

The electric potential at the point (x, y, z) and at the time t due to our charge Q is

$$V = \frac{1}{4\pi\epsilon_0} \frac{\gamma Q}{[\gamma^2(x - \mathcal{V}t)^2 + y^2 + z^2]^{1/2}}. \tag{6-91}$$

It is again sometimes more convenient to write V in the following form:

$$V = \frac{1}{4\pi\epsilon_0} \frac{Q}{r(1 - \beta^2 \sin^2 \theta)^{1/2}}. \tag{6-92}$$

Note the analogy between the formulas for V and A. Figure 6-15 shows equipotential surfaces for six values of β.

We can prove the validity of Eq. 6-90 by substituting our expressions for V and $\boldsymbol{A}$:

$$\boldsymbol{E} = -\frac{\partial V}{\partial x}\boldsymbol{i} - \frac{\partial V}{\partial y}\boldsymbol{j} - \frac{\partial V}{\partial z}\boldsymbol{k} - \frac{\partial \boldsymbol{A}}{\partial t}, \tag{6-93}$$

$$= \frac{\gamma Q}{[\gamma^2(x - \mathcal{V}t)^2 + y^2 + z^2]^{3/2}} \left\{ \frac{1}{4\pi\epsilon_0} [\gamma^2(x - \mathcal{V}t)\boldsymbol{i} + y\boldsymbol{j} + z\boldsymbol{k}] - \frac{\mu_0}{4\pi} \mathcal{V}^2\gamma^2(x - \mathcal{V}t)\boldsymbol{i} \right\}, \tag{6-94}$$

$$= \frac{\gamma Q[(x - \mathcal{V}t)\boldsymbol{i} + y\boldsymbol{j} + z\boldsymbol{k}]}{4\pi\epsilon_0[\gamma^2(x - \mathcal{V}t)^2 + y^2 + z^2]^{3/2}}, \tag{6-95}$$

which is the electric field intensity $\boldsymbol{E}$ as given in Equation 6-75.

Example V, ∇V, AND $\partial A/\partial t$ FOR A 10-GeV ELECTRON

We return again to the 10-GeV electron and calculate V, ∇V, and $\partial \boldsymbol{A}/\partial t$. The sum of the last two should equal the $\boldsymbol{E}$ we calculated in the example on page 260.

The scalar potential is maximum at $\theta = \pi/2$ and

$$V_{\max} = \frac{\gamma Q}{4\pi\epsilon_0 r}. \tag{6-96}$$

At 1 centimeter from the path of the particle,

$$V_{\max} = \frac{2 \times 10^4 \times 1.6 \times 10^{-19}}{4\pi \times 8.85 \times 10^{-12} \times 10^{-2}}, \tag{6-97}$$

$$= 2.88 \times 10^{-3} \text{ volt}. \tag{6-98}$$

To calculate ∇V at this point, let us use Eq. 6-92 for V, and spherical coordinates. Then

$$\nabla V = \frac{\partial V}{\partial r} \boldsymbol{r}_1 + \frac{1}{r}\frac{\partial V}{\partial \theta} \boldsymbol{\theta}_1, \tag{6-99}$$

$$= \frac{Q}{4\pi\epsilon_0} \left\{ \frac{-1}{r^2(1 - \beta^2 \sin^2 \theta)^{1/2}} \boldsymbol{r}_1 + \frac{\beta^2 \sin\theta \cos\theta}{r^2(1 - \beta^2 \sin^2 \theta)^{3/2}} \boldsymbol{\theta}_1 \right\}, \tag{6-100}$$

$$\nabla V = -\frac{\gamma Q}{4\pi\epsilon_0 r^2}\, r_1 \text{ at } \theta = 90°. \tag{6-101}$$

We can find the value of $\partial A/\partial t$ at $\theta = 90°$ from Eq. 6-79 without performing any calculation. As the particle sweeps by the point 1 centimeter away from its trajectory, r is minimum at $\theta = 90°$. The angle θ changes from 0 to π, but $\sin^2\theta$ has the same value at corresponding angles on either side of $\theta = \pi/2$. Then, as a function of the *time*, A is maximum at the instant considered when $\theta = \pi/2$, and our $\partial A/\partial t$ is zero. Then, for this particular case, $\boldsymbol{E}$ is simply equal to ∇V, and this agrees with the result we obtained in the foregoing example (page 266).

6.6 TRANSFORMATION OF THE ELECTROMAGNETIC POTENTIALS V AND A. THE FOUR-POTENTIAL $\mathbf{A}$

The electric potential V and the components of the vector potential A transform somewhat like x, y, z, t, as is apparent from Table 6-2. We shall show that these transformation equations are correct by showing that they transform $\boldsymbol{E}$ and $\boldsymbol{B}$ correctly.

For the moment, we can see that $A_{1x}, A_{1y}, A_{1z}, V/c^2$ transform like x, y, z, t, and are therefore the components of a four-vector

$$\mathbf{A} = (A, jV/c). \tag{6-102}$$

Then the squared magnitude of $\mathbf{A}$,

$$A^2 - \frac{V^2}{c^2},$$

is an invariant.

Let us first check whether the transformation equations of Table 6-2 for A and V are consistent with the fact that $E_{1x} = E_{2x}$ as in Eq. 6-63. To transform

$$E_{1x} = -\frac{\partial V_1}{\partial x_1} - \frac{\partial A_{1x}}{\partial t_1}, \tag{6-103}$$

we substitute the values of $\partial/\partial x_1$ and $\partial/\partial t_1$ from Eqs. 5-127 and 5-130, and we substitute the values of V_1 and A_{1x} from Table 6-2:

$$-\frac{\partial V_1}{\partial x_1} = -\gamma\left[\frac{\partial}{\partial x_2} - (\mathcal{V}/c^2)\frac{\partial}{\partial t_2}\right] c^2\gamma[(V_2/c^2) + (\mathcal{V}/c^2)A_{2x}], \tag{6-104}$$

$$= \gamma^2\left[-\frac{\partial V_2}{\partial x_2} - \mathcal{V}\frac{\partial A_{2x}}{\partial x_2} + (\mathcal{V}/c^2)\frac{\partial V_2}{\partial t_2} + (\mathcal{V}^2/c^2)\frac{\partial A_{2x}}{\partial t_2}\right]. \tag{6-105}$$

Table 6-2. *Transformation of the Four-Potential*

$A_{1x} = \gamma[A_{2x} + \mathcal{V}(V_2/c^2)],$	$A_{2x} = \gamma[A_{1x} - \mathcal{V}(V_1/c^2)],$
$A_{1y} = A_{2y},$	$A_{2y} = A_{1y},$
$A_{1z} = A_{2z},$	$A_{2z} = A_{1z},$
$V_1/c^2 = \gamma[(V_2/c^2) + (\mathcal{V}/c^2)A_{2x}],$	$V_2/c^2 = \gamma[(V_1/c^2) - (\mathcal{V}/c^2)A_{1x}].$

Similarly,

$$-\frac{\partial A_{1x}}{\partial t_1} = \gamma^2\left\{\mathcal{V}\frac{\partial A_{2x}}{\partial x_2} + (\mathcal{V}^2/c^2)\frac{\partial V_2}{\partial x_2} - \frac{\partial A_{2x}}{\partial t_2} - (\mathcal{V}/c^2)\frac{\partial V_2}{\partial t_2}\right\}. \tag{6-106}$$

Upon simplifying, we do find that

$$E_{1x} = -\frac{\partial V_2}{\partial x_2} - \frac{\partial A_{2x}}{\partial t_2} = E_{2x}, \tag{6-107}$$

as required.

We shall not show that the transformation equations for the other components of $\boldsymbol{E}$ and $\boldsymbol{B}$ are satisfied, but you might wish to try one or two more.

Example

THE PARALLEL-PLATE CAPACITOR

Let us return once more to the parallel-plate capacitor. We have already transformed its field in the example on page 263; we should arrive at the same results by transforming the potentials.

In reference frame 2, with respect to which the capacitor is stationary,

$$A_{2x} = A_{2y} = A_{2z} = 0, \tag{6-108}$$

while the bottom plate is at $+V_0/2$ and the top plate is at $-V_0/2$, as in Figure 6-12a.

Then, from Section 6.6, the bottom plate in frame 1 is at $\gamma V_0/2$ while the top plate is at $-\gamma V_0/2$, so that

$$E_1 = \gamma E_2, \tag{6-109}$$

as previously.

To calculate the vector potential $\boldsymbol{A}$ at any point inside the capacitor, we need the value of V_2 as a function of y:

$$V_2 = -\frac{y}{s}V_{02}. \tag{6-110}$$

Then

$$A_{1x} = -\gamma(\mathcal{V}/c^2)\frac{y}{s}V_{02}, \qquad A_{1y} = 0, \qquad A_{1z} = 0. \tag{6-111}$$

To find $\boldsymbol{B}$, we compute the curl of $\boldsymbol{A}$:

$$B = -\gamma(\mathcal{V}/c^2)\frac{1}{s}V_{02}\begin{vmatrix} i & j & k \\ \frac{\partial}{\partial x} & \frac{\partial}{\partial y} & \frac{\partial}{\partial z} \\ y & 0 & 0 \end{vmatrix}, \tag{6-112}$$

$$= \gamma(\mathcal{V}/c^2)E_2 k, \tag{6-113}$$

again as in the example on page 263.

You can easily check that $A^2 - (V^2/c^2)$ is the same in both frames of reference, and that it is therefore invariant.

6.7 THE LORENTZ CONDITION

The integrals for the vector potential A and for the electric potential V are so similar that one suspects the existence of some general relation linking them together. Such a relation does exist and it is called the *Lorentz condition:*

$$\nabla \cdot A + \epsilon_0\mu_0 \frac{\partial V}{\partial t} = 0, \tag{6-114}$$

or, in four-dimensional notation,

$$\Box \cdot \mathbf{A} = 0, \tag{6-115}$$

where $\Box$ is the four-dimensional quad operator defined in Eq. 5-131, and $\mathbf{A}$ is the four-potential of Eq. 6-102.

You can easily verify the Lorentz condition by substituting the values of A and of V from Eqs. 6-78 and 6-91.

We shall return to the Lorentz condition on several occasions, particularly in Chapters 10 and 14.

We now turn to the four fundamental equations of electromagnetism, which are known as Maxwell's equations.

6.8 GAUSS'S LAW

In discussing electrostatic fields in Chapters 2 and 3 we discovered a simple relation between the divergence of E and the total charge density ρ_t:

$$\nabla \cdot E = \frac{\rho_t}{\epsilon_0}. \tag{6-116}$$

This was Eq. 3-41 and is called Gauss's law. It is one of Maxwell's equations. We shall now show that it applies to the E of Eq. 6-31. Gauss's law applies in fact to all electric fields.

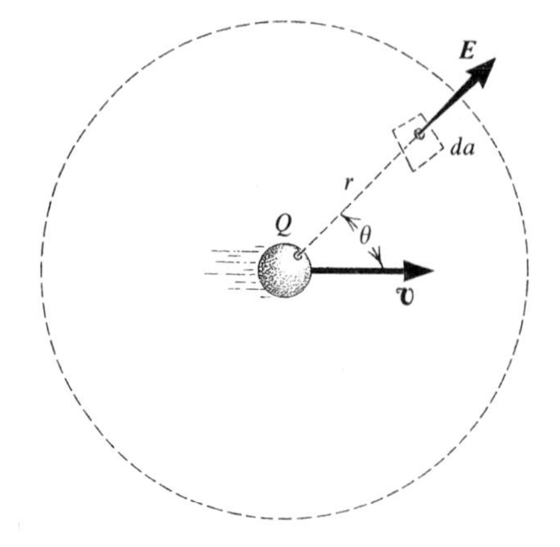

Figure 6-16. Gauss's law applies to a moving charge: the surface integral of $\boldsymbol{E}$ evaluated over the fixed sphere gives Q/ϵ_0, as if the charge were stationary. The excess E near $\theta = \pi/2$ is just compensated by the weak field near $\theta = 0$ and $\theta = \pi$. See Figures 6-9 and 6-10.

Integrating over a volume τ bounded by the surface S, and using the divergence theorem, we obtain the integral form of Gauss's law:

$$\int_S \boldsymbol{E} \cdot d\boldsymbol{a} = \frac{Q_t}{\epsilon_0}, \tag{6-117}$$

where Q_t is the total charge enclosed within the surface S.

We shall check the validity of Gauss's law by integrating the flux of $\boldsymbol{E}$ over a spherical surface, of radius R, which encloses the charge Q at a given moment and which is fixed in reference frame 1. The charge passes through the sphere at the velocity $\mathcal{V}$ as in Figure 6-16.

We can simplify the calculation by choosing t equal to zero. Then

$$\int_S \boldsymbol{E} \cdot d\boldsymbol{a} = \frac{Q}{4\pi\epsilon_0\gamma^2} \int_0^\pi \frac{2\pi r^2 \sin\theta \, d\theta}{r^2(1 - \beta^2 \sin^2\theta)^{3/2}}, \tag{6-118}$$

$$= \frac{Q}{2\epsilon_0\gamma^2} \int_0^\pi \frac{\sin\theta \, d\theta}{[\beta^2 \cos^2\theta + (1/\gamma^2)]^{3/2}}, \tag{6-119}$$

$$= \frac{Q}{2\epsilon_0\gamma^2} \int_{-1}^{+1} \frac{dx}{[\beta^2 x^2 + (1/\gamma^2)]^{3/2}}, \tag{6-120}$$

$$= \frac{Q}{\epsilon_0}. \tag{6-121}$$

Gauss's law therefore applies to the $\boldsymbol{E}$ of point charges moving at constant velocities.

Example

In a previous example (page 263), when we checked our transformation of the field of a parallel-plate capacitor, we used Gauss's law in *both* frames to find the ratio of E_1 to E_2. We were therefore assuming that Gauss's law applies equally well to stationary and mobile charges.

6.9 THE DIVERGENCE OF $\boldsymbol{B}$

The divergence of $\boldsymbol{B}$ is quite easy to calculate from Eq. 6-34:

$$\nabla \cdot \boldsymbol{B} = \frac{\mu_0 \gamma Q \mathcal{V}}{4\pi} \left\{ \frac{\partial}{\partial y} \left(\frac{-z}{[\gamma^2(x - \mathcal{V}t)^2 + y^2 + z^2]^{3/2}} \right) + \frac{\partial}{\partial z} \left(\frac{y}{[\gamma^2(x - \mathcal{V}t)^2 + y^2 + z^2]^{3/2}} \right) \right\} \quad (6\text{-}122)$$

$$= \frac{\mu_0 \gamma Q \mathcal{V}}{4\pi} \left\{ \frac{3yz}{[\ \ ]^{5/2}} - \frac{3yz}{[\ \ ]^{5/2}} \right\} = 0, \quad (6\text{-}123)$$

where we have set

$$[\ \ ] \equiv [\gamma^2(x - \mathcal{V}t)^2 + y^2 + z^2]. \quad (6\text{-}124)$$

This is the second of Maxwell's equations:

$$\boxed{\nabla \cdot \boldsymbol{B} = 0} \quad (6\text{-}125)$$

or, using the divergence theorem,

$$\boxed{\int_S \boldsymbol{B} \cdot d\boldsymbol{a} = 0.} \quad (6\text{-}126)$$

Intuitively, it is quite obvious that the divergence of $\boldsymbol{B}$ should be zero: since the lines of $\boldsymbol{B}$ are circles centered on the path of the moving charge, which is the x-axis in Figures 6-9 and 6-10, the total flux of $\boldsymbol{B}$ flowing out of any imaginary volume must be zero. This is a general result: the divergence of $\boldsymbol{B}$ is always zero.

6.10 THE CURL OF $\boldsymbol{E}$

There is a third Maxwell equation, which is stated as follows:

$$\boxed{\nabla \times \boldsymbol{E} = -\frac{\partial \boldsymbol{B}}{\partial t},} \quad (6\text{-}127)$$

or, using Stokes's theorem,

$$\boxed{\oint_C \boldsymbol{E} \cdot d\boldsymbol{l} = -\frac{\partial}{\partial t} \int_S \boldsymbol{B} \cdot d\boldsymbol{a} = -\frac{\partial \Phi}{\partial t},} \quad (6\text{-}128)$$

where S is any surface bounded by the curve C, and where Φ is the magnetic flux linking the curve C.

Let us verify Eq. 6-127 by substituting the values of $\boldsymbol{E}$ and $\boldsymbol{B}$ from Eqs. 6-33 and 6-34:

$$\boldsymbol{\nabla} \times \boldsymbol{E} = \begin{vmatrix} \boldsymbol{i} & \boldsymbol{j} & \boldsymbol{k} \\ \frac{\partial}{\partial x} & \frac{\partial}{\partial y} & \frac{\partial}{\partial z} \\ E_x & E_y & E_z \end{vmatrix}, \tag{6-129}$$

$$(\boldsymbol{\nabla} \times \boldsymbol{E})_x = \frac{\gamma Q}{4\pi\epsilon_0}\left\{\frac{\partial}{\partial y}\frac{z}{[\ \ ]^{3/2}} - \frac{\partial}{\partial z}\frac{y}{[\ \ ]^{3/2}}\right\} = 0, \tag{6-130}$$

$$(\boldsymbol{\nabla} \times \boldsymbol{E})_y = \frac{\gamma Q}{4\pi\epsilon_0}\left\{\frac{\partial}{\partial z}\frac{x - \mathcal{V}t}{[\ \ ]^{3/2}} - \frac{\partial}{\partial x}\frac{z}{[\ \ ]^{3/2}}\right\}, \tag{6-131}$$

$$= \frac{\gamma Q}{4\pi\epsilon_0}\left\{-\frac{3(x - \mathcal{V}t)z}{[\ \ ]^{5/2}} + \frac{3\gamma^2(x - \mathcal{V}t)z}{[\ \ ]^{5/2}}\right\}, \tag{6-132}$$

$$= \frac{3\gamma Q}{4\pi\epsilon_0}\left\{\frac{(x - \mathcal{V}t)}{[\ \ ]^{5/2}}(\gamma^2 - 1)z\right\}, \tag{6-133}$$

$$(\boldsymbol{\nabla} \times \boldsymbol{E})_z = \frac{\gamma Q}{4\pi\epsilon_0}\left\{\frac{\partial}{\partial x}\frac{y}{[\ \ ]^{3/2}} - \frac{\partial}{\partial y}\frac{x - \mathcal{V}t}{[\ \ ]^{3/2}}\right\}, \tag{6-134}$$

$$= \frac{\gamma Q}{4\pi\epsilon_0}\left\{-\frac{3\gamma^2(x - \mathcal{V}t)y}{[\ \ ]^{5/2}} + \frac{3(x - \mathcal{V}t)y}{[\ \ ]^{5/2}}\right\}, \tag{6-135}$$

$$= \frac{3\gamma Q}{4\pi\epsilon_0}\left\{\frac{(x - \mathcal{V}t)}{[\ \ ]^{5/2}}(1 - \gamma^2)y\right\}, \tag{6-136}$$

while

$$\frac{\partial \boldsymbol{B}}{\partial t} = \frac{\mu_0\gamma Q\mathcal{V}}{4\pi}\frac{3\gamma^2\mathcal{V}(x - \mathcal{V}t)}{[\ \ ]^{5/2}}(-z\boldsymbol{j} + y\boldsymbol{k}). \tag{6-137}$$

We can see that Eq. 6-127 is satisfied since

$$1 - \gamma^2 = -\beta^2\gamma^2 = -(\mathcal{V}/c)^2\gamma^2. \tag{6-138}$$

This is again a general result: the curl of $\boldsymbol{E}$ is always equal to minus the time derivative of $\boldsymbol{B}$. You should be able to show that this is in agreement with the expression for $\boldsymbol{E}$ in terms of V and $\boldsymbol{A}$, which we found in Section 6.5.

6.11 THE CURL OF B

The fourth and last of Maxwell's equations is the following:

$$\boxed{\nabla \times \boldsymbol{B} = \mu_0 \left(\boldsymbol{J}_m + \epsilon_0 \frac{\partial \boldsymbol{E}}{\partial t} \right),} \tag{6-139}$$

where $\boldsymbol{J}_m$ is the current density at the point P, where the electric field intensity is $\boldsymbol{E}$ and the magnetic induction is $\boldsymbol{B}$, or

$$\boxed{\oint_C \boldsymbol{B} \cdot d\boldsymbol{l} = \mu_0 \int_S \left(\boldsymbol{J}_m + \epsilon_0 \frac{\partial \boldsymbol{E}}{\partial t} \right) \cdot d\boldsymbol{a} = \mu_0 I_t.} \tag{6-140}$$

The meaning of the index m, for matter, will become apparent later in this section. Let us disregard it for the moment. The current I_t is the total current linking the curve C.

We shall be able to deduce this equation from the fact that Gauss's law is invariant under a Lorentz transformation.

Charges that are stationary at P in reference frame 1, in which $\boldsymbol{E}$ and $\boldsymbol{B}$ are measured, contribute nothing to $\nabla \times \boldsymbol{B}$. Let us disregard them. Imagine that the density of moving charges at P is ρ and that these charges move at a velocity $\mathcal{V}\boldsymbol{i}$. Then

$$\boldsymbol{J} = \rho \mathcal{V} \boldsymbol{i}. \tag{6-141}$$

We have chosen the x-axis in the direction of $\boldsymbol{J}$ at P.

In reference frame 2, which follows these charges, $J_{2x} = 0$, and we have a charge density

$$\rho_2 = \frac{\rho}{\gamma} \tag{6-142}$$

from Eq. 5-116. In frame 2, $\boldsymbol{B} = 0$, and the only information we have about $\boldsymbol{E}_2$ is Gauss's law:

$$\nabla \cdot \boldsymbol{E}_2 = \frac{\rho_2}{\epsilon_0} = \frac{\rho}{\gamma \epsilon_0}, \tag{6-143}$$

or

$$\frac{\partial}{\partial x_2} E_{2x} + \frac{\partial}{\partial y_2} E_{2y} + \frac{\partial}{\partial z_2} E_{2z} = \frac{\rho}{\gamma \epsilon_0}. \tag{6-144}$$

We can deduce an equation for the field in frame 1 by using the equations of transformation of Table 5-10 for the partial derivatives, and those of Table 6-1 for the components of $\boldsymbol{E}$:

$$\gamma \left(\frac{\partial}{\partial x} + \frac{\mathcal{V}}{c^2} \frac{\partial}{\partial t} \right) E_x + \gamma \frac{\partial}{\partial y} (E_y - \mathcal{V} B_z) + \gamma \frac{\partial}{\partial z} (E_z + \mathcal{V} B_y) = \rho / \gamma \epsilon_0, \tag{6-145}$$

$$\nabla \cdot \boldsymbol{E} + \frac{\mathcal{V}}{c^2} \frac{\partial E_x}{\partial t} - \mathcal{V} \left(\frac{\partial B_z}{\partial y} - \frac{\partial B_y}{\partial z} \right) = \rho \frac{(1 - \beta^2)}{\epsilon_0}. \tag{6-146}$$

Table 6-3. *Maxwell's Equations*

Differential Form	Integral Form
$\nabla \cdot \boldsymbol{E} = \frac{\rho_t}{\epsilon_0}$	$\int_S \boldsymbol{E} \cdot d\boldsymbol{a} = \frac{Q_t}{\epsilon_0}$
$\nabla \cdot \boldsymbol{B} = 0$	$\int_S \boldsymbol{B} \cdot d\boldsymbol{a} = 0$
$\nabla \times \boldsymbol{E} = -\frac{\partial \boldsymbol{B}}{\partial t}$	$\oint_C \boldsymbol{E} \cdot d\boldsymbol{l} = -\frac{\partial}{\partial t} \int_S \boldsymbol{B} \cdot d\boldsymbol{a} = -\frac{\partial \Phi}{\partial t}$
$\nabla \times \boldsymbol{B} - \frac{1}{c^2} \frac{\partial \boldsymbol{E}}{\partial t} = \mu_0 \boldsymbol{J}_m$	$\oint_C \boldsymbol{B} \cdot d\boldsymbol{l} = \mu_0 \int_S \left(\boldsymbol{J}_m + \epsilon_0 \frac{\partial \boldsymbol{E}}{\partial t} \right) \cdot d\boldsymbol{a} = \mu_0 I_t$

We have again omitted the subscripts 1. But Gauss's law also applies in frame 1, and $\nabla \cdot \boldsymbol{E} = \rho/\epsilon_0$. Then, dividing by $\mathcal{V}$,

$$\frac{1}{c^2} \frac{\partial E_x}{\partial t} - \left(\frac{\partial B_z}{\partial y} - \frac{\partial B_y}{\partial z} \right) = -\frac{\rho \mathcal{V}}{\epsilon_0 c^2} = -\frac{J}{\epsilon_0 c^2} = -\mu_0 J. \tag{6-147}$$

We have arrived at this equation by postulating that the current density vector $\boldsymbol{J}$ was directed along the x-axis. In the more general case where $\boldsymbol{J}$ has three components, we have two other equations that can be deduced from this one by rotating the indices. Combining the three,

$$\nabla \times \boldsymbol{B} - \frac{1}{c^2} \frac{\partial \boldsymbol{E}}{\partial t} = \mu_0 \boldsymbol{J}_m. \tag{6-148}$$

Since the exact nature of the current is immaterial, we have added a subscript m on $\boldsymbol{J}$ to indicate that it represents any type of current in matter. For example, in a lossy dielectric, $\boldsymbol{J}_m$ is the sum of the conduction current density $\boldsymbol{J}_f$ plus the polarization current density $\partial \boldsymbol{P}/\partial t$.

6.12 MAXWELL'S EQUATIONS

Maxwell's equations are grouped together in Table 6-3. You will be able to show (Problem 6-18) that they are invariant under a Lorentz transformation.

These are the four fundamental equations of electromagnetism. We shall have many occasions to discuss them, and especially to use them, throughout the remaining chapters. We shall not therefore say more about them for the moment.

Example

THE MAGNETIC FIELD NEAR A STRAIGHT WIRE CARRYING A STEADY ELECTRIC CURRENT

Until now we have studied the electric and magnetic fields of a single point charge. In practice, magnetic fields are produced by electric currents flowing in wires. Such currents are more complex than the simple cases we have studied so far: (a) the conduction electrons are distributed over a finite volume, (b) they move in all directions, (c) they have a broad spectrum of velocities, (d) they drift in the direction of the electric field through a lattice of fixed positive charges, and (e) their drift velocity changes in direction as they go around bends in the wire.

The thermal agitation of the conduction electrons gives a zero average magnetic field; thus we can concentrate our attention on the drift velocity.

We first dispose of the acceleration of the drifting cloud of electrons as it goes around bends in the wire. We have not discussed the fields of accelerated charges, and all that we can say about these accelerations is that they have no appreciable effect on the magnetic field. The accelerations produce fields that are of the order of $\mathcal{V}/c$, or 3×10^{-13} times the one we shall calculate below.

Let us calculate the magnetic field of a current-carrying wire in two simple cases. The first case will be that of a long straight wire; we shall rediscover the well-known formula for **B**. The second case will be that of a short length of wire and will be the subject of the next example (page 280). This will give us an expression that can be integrated over any circuit.

A long straight wire is stationary in the laboratory and carries an electric current I as in Figure 6-17. Inside the wire, the net charge density is zero, as we shall see later on in the example on page 424. There are surface charges, but we can forget about them because they simply superpose an electric field over the magnetic field we are interested in. The electric field depends on the resistance of the wire as well as on the current I flowing through it.

We shall use the same procedure as previously. We shall calculate the force on a charge Q moving at some velocity $\boldsymbol{v}$ in the vicinity of the wire. This force will be of the form $Q\boldsymbol{v} \times \boldsymbol{B}$ and will give us $\boldsymbol{B}$.

We must consider separately the conduction electrons and the positive charges in the wire. Let us call the fixed positive linear charge density inside the wire λ_p and the mobile negative linear charge density λ_n, both measured with respect to the laboratory. Linear charge densities are expressed in coulombs/meter. Then, in the frame of the laboratory,

$$\lambda_p + \lambda_n = 0, \tag{6-149}$$

λ_p being positive and λ_n negative. As usual, reference frame 1 is the laboratory frame, and we omit the subscript 1.

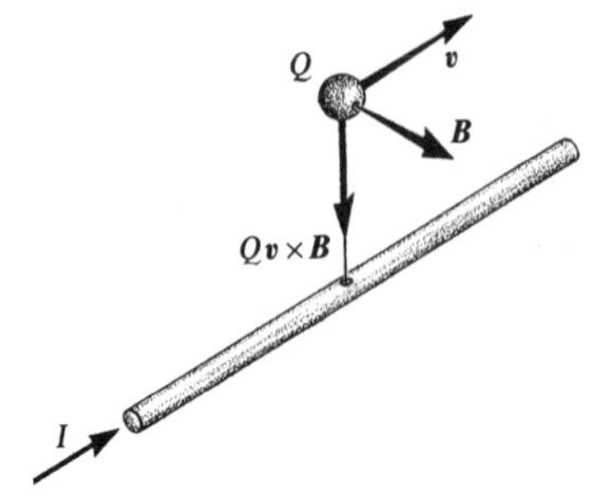

Figure 6-17. A charge Q moves at a velocity $\boldsymbol{v}$ parallel to a straight wire carrying a current I. The magnetic induction $\boldsymbol{B}$ of the wire exerts a force $Q\boldsymbol{v} \times \boldsymbol{B}$ on the moving charge because the wire appears, to the charge Q, to be negatively charged.

Frame 2 moves in the positive direction of the x-axis at the drift velocity $\mathcal{V}\boldsymbol{i}$ of the electrons. The charge Q, on the other hand, has some velocity $\boldsymbol{v}$ with respect to the laboratory as in Figure 6-17. So as to simplify the calculation we choose our y-axes so that Q lies on them at $t_1 = t_2 = 0$, and we calculate the force on Q at that moment in the frame of the laboratory.

The positive charges are fixed with respect to the laboratory and exert on Q a force

$$Y_p = \frac{\lambda_p Q}{2\pi\epsilon_0\rho}, \tag{6-150}$$

where the coefficient of Q is the electric field intensity at a distance ρ, from an infinite line charge of density λ_p coulombs/meter. Note that ρ is the radial distance to Q and *not* a charge density.

The negative charges are stationary in frame 2 and, similarly,

$$Y_{n2} = \frac{\lambda_{n2} Q}{2\pi\epsilon_0\rho}. \tag{6-151}$$

Both the charge Q and the distance ρ, which is perpendicular to the x-axis, are the same in frames 1 and 2. Then, from Table 5-7,

$$Y_n = \frac{\lambda_{n2} Q}{2\pi\epsilon_0\rho}\gamma[1 - (v_x\mathcal{V}/c^2)], \tag{6-152}$$

where v_x is the x component of the velocity of Q.

Now the linear charge density inside the wire is simply the volume charge density multiplied by the cross-sectional area of the wire, the latter quantity being the same in both frames. Thus λ_{n2} must transform like a volume charge density, as in Eq. 5-117:

$$\lambda_{n2} = \gamma[\lambda_n - (\mathcal{V}/c^2)I_n], \tag{6-153}$$

where I_n is the electron current $\lambda_n\mathcal{V}$. Thus

$$\lambda_{n2} = \gamma\lambda_n[1 - (\mathcal{V}/c)^2] = \frac{\lambda_n}{\gamma}, \tag{6-154}$$

and then

$$Y_n = \frac{Q}{2\pi\epsilon_0\rho}[\lambda_n - (v_x/c^2)I_n]. \tag{6-155}$$

The first term is an electric force, while the second is a magnetic force since it depends on the velocity of Q. Then the total force exerted on the charge Q is

$$Y = Y_p + Y_n = -\frac{\mu_0 I_n}{2\pi\rho} Qv_x, \tag{6-156}$$

and *the* electric *fields of the positive and negative charges cancel perfectly*. Remember that $\lambda_n = -\lambda_p$ and that $\epsilon_0 c^2 = 1/\mu_0$ from Eq. 6-15.

Finally, since $I_n = \lambda_n \mathcal{V} = -I$, where I is the electric current flowing in the positive direction of the x-axis,

$$Y = \frac{\mu_0 I}{2\pi\rho} Qv_x. \tag{6-157}$$

If we express this as a magnetic force $Q\boldsymbol{v} \times \boldsymbol{B}$, we find that

$$\boldsymbol{B} = \frac{\mu_0 I}{2\pi\rho} \boldsymbol{\varphi}_1. \tag{6-158}$$

This is the magnetic induction expressed in teslas at a distance of ρ meters from a long straight wire carrying a current of I amperes. The force $Q\boldsymbol{v} \times \boldsymbol{B}$ and the vector $\boldsymbol{B}$ are illustrated in Figure 6-17.

This formula is in principle valid only for an infinitely long wire, because it is based on Eq. 6-150; in practice it is valid within 1% if (a) ρ is less than 7% of the length of the straight portion of the wire and (b) the point considered is near the middle of the wire (Problem 7-9).

Could we have arrived at this same result from Maxwell's equations? Indeed we could have. In the frame of the laboratory there is no electric field, except that which comes from the surface charges and which is irrelevant. Then, from Eq. 6-139, $\boldsymbol{\nabla} \times \boldsymbol{B}$ is equal to $\mu_0 \boldsymbol{J}_f$ and, integrating over a circle centered on the wire and perpendicular to it,

$$\int_S (\boldsymbol{\nabla} \times \boldsymbol{B}) \cdot d\boldsymbol{a} = \mu_0 \int_S \boldsymbol{J}_f \cdot d\boldsymbol{a} = \mu_0 I. \tag{6-159}$$

Or, from Stokes's theorem,

$$\oint_C \boldsymbol{B} \cdot d\boldsymbol{l} = \mu_0 I. \tag{6-160}$$

Now, by symmetry, $\boldsymbol{B}$ must be azimuthal as in Figure 6-17, from what we know about the magnetic field of a single moving charge (Figure 6-8). Then the line integral must be simply $2\pi\rho\, B$ and we have again Eq. 6-158.

This magnetic field results from a relativistic transformation of the electric field of the moving electrons. This is indeed surprising because the drift velocity $\mathcal{V}$ of the electrons in a conductor is only of the order of 10^{-4} meter/second, or *one foot per hour*, or 3×10^{-13} times the velocity of light. Then γ is equal to unity within one part in 10^{25}! Relativistic effects could hardly be expected at such small velocities.

Let us therefore return to Eq. 6-155, and let us calculate the numerical value of the two terms within the bracket for a typical case. Let us say that the wire is made of copper, that it carries a current of one ampere, and that it has a cross-sectional area of one square millimeter. Copper contains 10^{29} atoms/meter^3 and has one conduction electron per atom. Then

$$\lambda_n = -10^{29} \times 10^{-6} \times 1.6 \times 10^{-19} = -1.6 \times 10^4 \text{ coulombs/meter} \quad (6\text{-}161)$$

This is an enormous charge. The force of attraction or repulsion, between two point charges of 1.6×10^4 coulombs each, 10 centimeters apart, would be about 3×10^{14} tons, which is about the force of attraction between the Earth and the Moon.

Let us say that the charge Q is an electron, and that v_x is its drift velocity in a second wire that is parallel to the first one. Then

$$\left(\frac{v_x}{c^2}\right) I_n = \left(\frac{10^{-4}}{9 \times 10^{16}}\right) \times 1 \approx 10^{-21} \text{ coulomb/meter.} \quad (6\text{-}162)$$

Therefore the bracket in Eq. 6-155 is the sum of two terms, the first one of which is about 10^{25} times larger than the second. But the first one is *perfectly* cancelled by λ_p, and we are left with only the second.

Of course the force on the second wire is appreciable for the simple reason that it also contains an enormous number of conduction electrons.

Example

THE FIELD OF A SHORT ELEMENT OF WIRE CARRYING AN ELECTRIC CURRENT

It was interesting to rediscover the well-known formula for the magnetic field of a long straight wire because the calculation showed quite vividly the origin of the magnetic fields of conduction currents. We now wish to be able to calculate $\boldsymbol{B}$ for a wire of arbitrary shape, and, for this purpose, we must know the field of a short element so that we can calculate the total $\boldsymbol{B}$ by integration.

We therefore consider the short length of wire $\boldsymbol{dl}$ at the origin O_1 as in Figure 6-18. It contains a positive charge $\lambda_p\, dl$ which is stationary, and a negative charge $\lambda_n\, dl$ which moves in the direction of the x-axis at the velocity $-\mathcal{V}$. The total charge is zero:

$$\lambda_p\, dl + \lambda_n\, dl = 0. \quad (6\text{-}163)$$

To find the field of this short length of wire we shall add the electrostatic field of the stationary positive charges to the $\boldsymbol{E}$ and $\boldsymbol{B}$ fields of the moving electrons, calculated as in Section 6.2.

As mentioned at the beginning of the last example (page 277), we neglect the fact that the electrons are accelerated in drifting around the curve, and we calculate $\boldsymbol{E}$ and $\boldsymbol{B}$ by considering only their drift velocity $\mathcal{V}\boldsymbol{i}$.

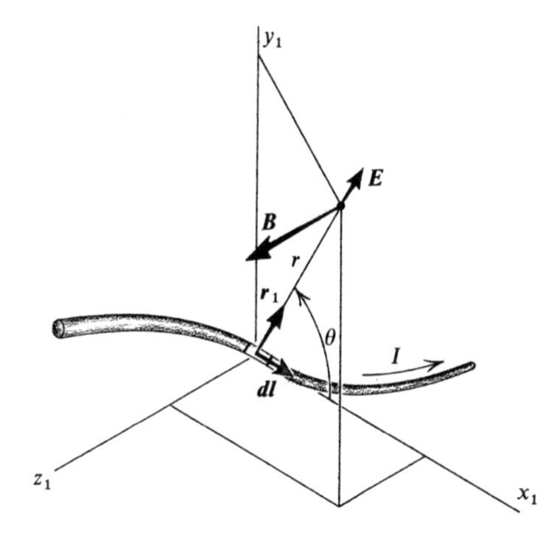

Figure 6-18. Element of wire of length $\boldsymbol{dl}$ carrying a current I. It produces $\boldsymbol{E}$ and $\boldsymbol{B}$ fields at the point $\boldsymbol{r}$. The $\boldsymbol{E}$ field is negligible, in practice.

The positive charges produce at $\boldsymbol{r}(x, y, z)$ an electric field intensity

$$\boldsymbol{E}_p = \frac{\lambda_p \, dl \, \boldsymbol{r}_1}{4\pi\epsilon_0 r^2} \tag{6-164}$$

and no magnetic field.

The negative charges in dl can be treated as a single charge $\lambda_n \, dl$ passing through the origin at a velocity $-\mathcal{V}$. The charge $\lambda_n \, dl$ is negative. Then, from Eqs. 6-31 and 6-32,

$$\boldsymbol{E}_n = \frac{(1 - \beta^2)\lambda_n \, dl}{4\pi\epsilon_0 r^2(1 - \beta^2 \sin^2 \theta)^{3/2}} \boldsymbol{r}_1, \tag{6-165}$$

$$\boldsymbol{B}_n = -\frac{\mu_0}{4\pi} \frac{(1 - \beta^2)\mathcal{V}\lambda_n \, dl \sin \theta}{r^2(1 - \beta^2 \sin^2 \theta)^{3/2}} \boldsymbol{\varphi}_1. \tag{6-166}$$

Therefore, the fields $\boldsymbol{E}$ and $\boldsymbol{B}$ at $\boldsymbol{r}$ must be $\boldsymbol{E}_p + \boldsymbol{E}_n$ and $\boldsymbol{B}_n$:

$$\boldsymbol{E} = \frac{\lambda_p \, dl}{4\pi\epsilon_0 r^2} \left\{1 - \frac{1 - \beta^2}{(1 - \beta^2 \sin^2 \theta)^{3/2}}\right\} \boldsymbol{r}_1, \tag{6-167}$$

$$\boldsymbol{B} = \frac{\mu_0}{4\pi} \frac{(1 - \beta^2)\mathcal{V}\lambda_p \, dl \sin \theta}{r^2(1 - \beta^2 \sin^2 \theta)^{3/2}} \boldsymbol{\varphi}_1. \tag{6-168}$$

The electric field intensity is *not* zero! The electric fields of the moving electrons and of the stationary electric charges do not cancel, although they are of different signs. This result is apparently in contradiction with the previous section. As we shall see below, the $\boldsymbol{E}$ for a short length of wire is positive for a certain range of θ, and negative elsewhere; it turns out that, once integrated over an infinite straight wire, the two contributions cancel perfectly. Integrating the above $\boldsymbol{B}$ over an infinite straight wire also yields the same result as in the previous section.

Let us deduce approximate values for $\boldsymbol{E}$ and $\boldsymbol{B}$ by setting $\beta^2 \ll 1$. The approximations will be extraordinarily good because, as we have seen, $\beta^2 \approx 10^{-25}$. Thus

$$E \approx \frac{\lambda_p\, dl}{4\pi\epsilon_0 r^2}\left\{1 - (1-\beta^2)\left(1 + \frac{3}{2}\beta^2 \sin^2\theta\right)\right\} r_1, \quad (6\text{-}169)$$

$$\approx \frac{\lambda_p\, dl}{4\pi\epsilon_0 r^2}\beta^2\left(1 - \frac{3}{2}\sin^2\theta\right) r_1. \quad (6\text{-}170)$$

Thus E is directed outward for $\sin^2\theta < 2/3$ and inward for larger angles. The explanation is simple: the E of the positive charges points outward and is the same in all directions, while the E of the negative charges points inward and is slightly larger near $\theta = \pi/2$, slightly smaller near $\theta = 0$ and π.

Since $I = \lambda_n(-\mathcal{V}) = \lambda_p \mathcal{V}$, where I is the current in the positive direction of the x-axis, and since $(1/\epsilon_0 c^2) = \mu_0$,

$$E \approx \frac{\mu_0 I\, dl}{4\pi r^2}\mathcal{V}\left(1 - \frac{3}{2}\sin^2\theta\right) r_1. \quad (6\text{-}171)$$

Similarly,

$$B \approx \frac{\mu_0}{4\pi}\frac{I\, dl}{r^2}(1-\beta^2)\left(1 + \frac{3}{2}\beta^2\sin^2\theta\right)\sin\theta\, \varphi_1, \quad (6\text{-}172)$$

$$\approx \frac{\mu_0}{4\pi}\frac{I\, dl}{r^2}\sin\theta\, \varphi_1, \quad (6\text{-}173)$$

$$\approx \frac{\mu_0}{4\pi}\frac{I\, \mathbf{dl} \times r_1}{r^2}, \quad (6\text{-}174)$$

where the vector $\mathbf{dl}$ points in the direction of the current as in Figure 6-17.

For a closed circuit,

$$B = \frac{\mu_0}{4\pi} I \oint \frac{\mathbf{dl} \times r_1}{r^2}. \quad (6\text{-}175)$$

This is the *Biot-Savart law*. We shall return to it in the next chapter.

Example

FORCE ON A SECOND ELEMENT $I\, \mathbf{dl}$

We can use these simplified expressions for E and B to calculate the force on a second wire carrying an electric current.

For two elements of wire as in Figure 6-19, the electric field intensity of $\mathbf{dl}_a$ exerts equal and opposite forces on the positive and negative charges of $\mathbf{dl}_b$, since $\lambda_{nb} = -\lambda_{pb}$. Then the force d^2F_{ab} exerted *by* $\mathbf{dl}_a$ *on* $\mathbf{dl}_b$ is

$$\mathbf{d^2F}_{ab} = (\lambda_{nb}\, dl_b)\mathcal{V}_b \times B_a, \quad (6\text{-}176)$$

$$= (\lambda_{nb}\, dl_b)\mathcal{V}_b \times \frac{\mu_0}{4\pi}\frac{I_a\, \mathbf{dl}_a \times r_1}{r^2}, \quad (6\text{-}177)$$

$$= \frac{\mu_0}{4\pi} I_a I_b \frac{\mathbf{dl}_b \times (\mathbf{dl}_a \times r_1)}{r^2}. \quad (6\text{-}178)$$

We shall use this result at the very beginning of the next chapter.

To find the force $\mathbf{d^2F}_{ba}$ exerted *by* $I_b\mathbf{dl}_b$ *on* $I_a\mathbf{dl}_a$, we exchange

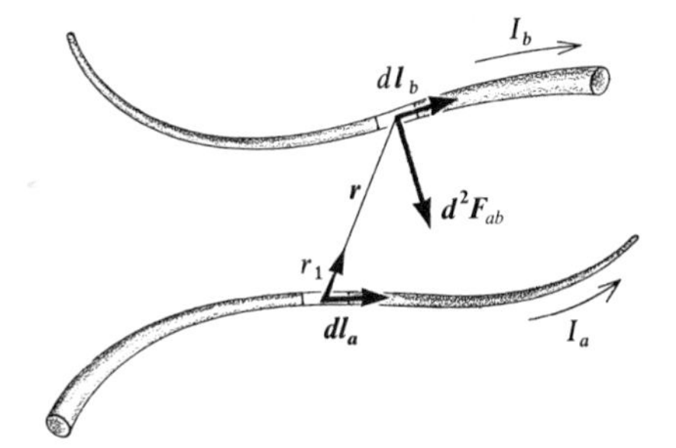

Figure 6-19. The magnetic force $\boldsymbol{d^2F}_{ab}$ exerted *by* the element $\boldsymbol{dl}_a$ on the element $\boldsymbol{dl}_b$.

subscripts a and b, and change the sign of $\boldsymbol{r}_1$. Contrary to what one would expect from elementary mechanics, the result is *not* equal to $-\boldsymbol{d^2F}_{ab}$, unless $\boldsymbol{dl}_a$ and $\boldsymbol{dl}_b$ are parallel. This purely relativistic effect was mentioned earlier in Sections 5.18 and 6.2.

Example

FORCE ON A MOVING CHARGED PARTICLE

If instead we have at the point $\boldsymbol{r}(x, y, z)$ a single charge Q moving at the velocity $\boldsymbol{v}$, then the *electric* force is

$$\boldsymbol{F}_{\text{el}} \approx \frac{\mu_0 I\, dl}{4\pi r^2}\, \mathcal{V}\left(1 - \frac{3}{2}\sin^2\theta\right) Q\, \boldsymbol{r}_1, \tag{6-179}$$

while the magnetic force is

$$\boldsymbol{F}_{\text{mag}} \approx Q\boldsymbol{v} \times \frac{\mu_0}{4\pi}\,\frac{I\, dl \sin\theta}{r^2}\,\boldsymbol{\varphi}_1. \tag{6-180}$$

So as to simplify matters, let us assume that $\boldsymbol{v}$ is along the x-axis. Then

$$\frac{F_{\text{el}}}{F_{\text{mag}}} \approx \frac{\mathcal{V}}{v}\left\{\frac{[1 - (3/2)\sin^2\theta]}{\sin\theta}\right\}. \tag{6-181}$$

The trigonometric term is of the order of unity, except at $\theta = 0$, where the magnetic induction B is zero. The velocity $\mathcal{V}$ of the conduction electrons is so small ($\approx 10^{-4}$ meter/second) that $\boldsymbol{F}_{\text{el}}$ is completely negligible in practice. For example, if we imagine that the particle is a Xe ion (Atomic weight 130) with an energy of only one electron-volt (1.6×10^{-19} joule), its velocity v is of the order of 10^3 meters/second, and even then the electric force is smaller than the magnetic force by a factor of 10^7.

6.13 SUMMARY

We found that the force on a particle carrying a charge Q and moving at a velocity $\boldsymbol{v}$ is of the form

$$\boldsymbol{F} = Q(\boldsymbol{E} + \boldsymbol{v} \times \boldsymbol{B}), \tag{6-13}$$

where $\boldsymbol{E}$ and $\boldsymbol{B}$ are respectively the electric field intensity and the magnetic induction at the point occupied by the particle. This is called the *Lorentz force.*

By investigating the force between two moving particles we were able to show that the *field of a charge Q moving at a constant velocity* $\mathcal{V}$ is given by

$$\boldsymbol{E} = \frac{Q}{4\pi\epsilon_0\gamma^2 r^2(1 - \beta^2 \sin^2\theta)^{3/2}} \boldsymbol{r}_1, \tag{6-31}$$

$$\boldsymbol{B} = \frac{\mu_0 Q \mathcal{V} \sin\theta}{4\pi\gamma^2 r^2(1 - \beta^2 \sin^2\theta)^{3/2}} \boldsymbol{\varphi}_1, \tag{6-32}$$

where r and θ are defined as in in Figure 6-2 and where $\beta = \mathcal{V}/c$, $\gamma = (1 - \beta^2)^{-1/2}$.

The permeability of free space is defined as

$$\mu_0 \equiv 4\pi \times 10^{-7} \equiv \frac{1}{\epsilon_0 c^2} \text{ henrys/meter.} \tag{6-12}$$

Using then the equations of transformation for the components of a force, we found the *equations of transformation for* $\boldsymbol{E}$ *and* $\boldsymbol{B}$ (Section 6.3):

$$\begin{aligned} E_{1x} &= E_{2x}, & E_{2x} &= E_{1x}, \\ E_{1y} &= \gamma[E_{2y} + \mathcal{V}B_{2z}], & E_{2y} &= \gamma[E_{1y} - \mathcal{V}B_{1z}], \\ E_{1z} &= \gamma[E_{2z} - \mathcal{V}B_{2y}], & E_{2z} &= \gamma[E_{1z} + \mathcal{V}B_{1y}], \\ B_{1x} &= B_{2x}, & B_{2x} &= B_{1x}, \\ B_{1y} &= \gamma[B_{2y} - (\mathcal{V}/c^2)E_{2z}], & B_{2y} &= \gamma[B_{1y} + (\mathcal{V}/c^2)E_{1z}], \\ B_{1z} &= \gamma[B_{2z} + (\mathcal{V}/c^2)E_{2y}], & B_{2z} &= \gamma[B_{1z} - (\mathcal{V}/c^2)E_{1y}]. \end{aligned}$$

It follows that

$$E^2 - c^2B^2 \qquad \text{and} \qquad \boldsymbol{E}\cdot\boldsymbol{B}$$

are both invariant.

The *magnetic induction* $\boldsymbol{B}$ is related to the *vector potential* $\boldsymbol{A}$ as follows:

$$\boldsymbol{B} = \nabla \times \boldsymbol{A}, \tag{6-77}$$

where

$$\boldsymbol{A} = \frac{\mu_0}{4\pi} \frac{\gamma Q \mathcal{V} \boldsymbol{i}}{[\gamma^2(x - \mathcal{V}t)^2 + y^2 + z^2]^{1/2}}, \tag{6-78}$$

$$= \frac{\mu_0}{4\pi} \frac{Q \mathcal{V} \boldsymbol{i}}{r(1 - \beta^2 \sin^2\theta)^{1/2}}, \tag{6-79}$$

for a single charge Q moving at the constant velocity $\mathcal{V}\boldsymbol{i}$.

Similarly, the *electric field intensity* $\boldsymbol{E}$ can be related to both the *vector potential* $\boldsymbol{A}$ and the *scalar potential* V:

$$\boldsymbol{E} = -\nabla V - \frac{\partial \boldsymbol{A}}{\partial t}, \tag{6-90}$$

Table 6-4. *Four-Vectors Discussed in Chapters 5 and 6*

Four-Vector	Components		Squared Magnitude
$\mathbf{r}$	(r, jct)	(x, y, z, jct)	$r^2 - c^2t^2$
$\mathbf{p}$	$(p, j\mathcal{E}/c)$	$(p_x, p_y, p_z, j\mathcal{E}/c)$	$p^2 - (\mathcal{E}^2/c^2)$
$\mathbf{J}$	$(J, jc\rho)$	$(J_x, J_y, J_z, jc\rho)$	$J^2 - (c^2\rho^2)$
$\mathbf{A}$	$(A, jV/c)$	$(A_x, A_y, A_z, jV/c)$	$A^2 - (V^2/c^2)$
$\square$	$\left(\nabla, \frac{1}{jc}\frac{\partial}{\partial t}\right)$	$\left(\frac{\partial}{\partial x}, \frac{\partial}{\partial y}, \frac{\partial}{\partial z}, \frac{1}{jc}\frac{\partial}{\partial t}\right)$	$\nabla^2 - \frac{1}{c^2}\frac{\partial^2}{\partial t^2}$

where

$$V = \frac{\gamma Q}{4\pi\epsilon_0[\gamma^2(x - \mathcal{V}t)^2 + y^2 + z^2]^{1/2}}, \tag{6-91}$$

$$= \frac{Q}{4\pi\epsilon_0 r(1 - \beta^2 \sin^2 \theta)^{1/2}}, \tag{6-92}$$

again for a single charge Q having a constant velocity $\mathcal{V}\boldsymbol{i}$.

The three components of A and V are the components of the *four-potential*

$$\mathbf{A} = (A, jV/c), \tag{6-102}$$

and hence A_x, A_y, A_z, V/c^2 transform like x, y, z, t (Section 6.6):

$$A_{1x} = \gamma[A_{2x} + \mathcal{V}(V_2/c^2)], \qquad A_{2x} = \gamma[A_{1x} - \mathcal{V}(V_1/c^2)],$$

$$A_{1y} = A_{2y}, \qquad A_{2y} = A_{1y},$$

$$A_{1z} = A_{2z}, \qquad A_{2z} = A_{1z},$$

$$V_1/c^2 = \gamma[(V_2/c^2) + (\mathcal{V}/c^2)A_{2x}], \qquad V_2/c^2 = \gamma[(V_1/c^2) - (\mathcal{V}/c^2)A_{1x}],$$

and the squared magnitude of $\mathbf{A}$,

$$A^2 - \frac{V^2}{c^2},$$

is an invariant.

The *Lorentz condition* provides us with a relation between V and A:

$$\nabla \cdot A + \epsilon_0\mu_0 \frac{\partial V}{\partial t} = 0, \tag{6-114}$$

or

$$\square \cdot \mathbf{A} = 0. \tag{6-115}$$

We then demonstrated the validity of *Maxwell's equations* for the field of a charge moving at a constant velocity. These four equations are the following:

$$\nabla \cdot \boldsymbol{E} = \frac{\rho_t}{\epsilon_0}, \qquad \nabla \times \boldsymbol{E} = -\frac{\partial \boldsymbol{B}}{\partial t},$$

$$\nabla \cdot \boldsymbol{B} = 0, \qquad \nabla \times \boldsymbol{B} = \mu_0 \left(\boldsymbol{J}_m + \epsilon_0 \frac{\partial \boldsymbol{E}}{\partial t} \right).$$

Finally, we used the transformation equations to find (a) the field of a long straight wire carrying a current I and (b) that of a short element $I\,\boldsymbol{dl}$. For the long wire we found the familiar formula

$$\boldsymbol{B} = \frac{\mu_0 I}{2\pi\rho} \boldsymbol{\varphi}_1. \tag{6-158}$$

This result constituted a striking demonstration of the fact that the magnetic fields associated with conduction currents result from an exceedingly small relativistic correction to the *electric* field of the moving electrons.

For an element of wire $\boldsymbol{dl}$ carrying a current I,

$$\boldsymbol{B} = \frac{\mu_0}{4\pi} \frac{I\,\boldsymbol{dl} \times \boldsymbol{r}_1}{r^2}. \tag{6-174}$$

The magnetic force exerted *by* an element of wire $\boldsymbol{dl}_a$, carrying a current I_a, *on* an element of wire $\boldsymbol{dl}_b$, carrying a current I_b, is

$$d^2\boldsymbol{F}_{ab} = \frac{\mu_0}{4\pi} I_a I_b \frac{\boldsymbol{dl}_b \times (\boldsymbol{dl}_a \times \boldsymbol{r}_1)}{r^2}. \tag{6-178}$$

The element $I\,\boldsymbol{dl}$ also has an electric field, but the resulting electric force is either zero or negligibly small.

PROBLEMS

6-1. Show that the force exerted by Q_a on Q_b, when both charges move at the same velocity $\mathcal{V}$, as in Section 6.1, can also be written

$$X_b = \frac{Q_a Q_b \cos\theta}{4\pi\epsilon_0 \gamma^2 r^2 (1 - \beta^2 \sin^2\theta)^{3/2}},$$

$$Y_b = \frac{Q_a Q_b \sin\theta}{4\pi\epsilon_0 \gamma^4 r^2 (1 - \beta^2 \sin^2\theta)^{3/2}},$$

where r is the distance between the two particles and θ is the angle between r and $\mathcal{V}$, in reference frame 1, as in Figure 6–5.

6-2. Repeat the calculation of Section 6.1 for a pair of equal masses m_a and m_b pulled together by a spring under tension exerting a force $K\boldsymbol{r}$.

(a) Show that, in frame 1,

$$\boldsymbol{F}_b = -\gamma K\left\{\boldsymbol{r} + \boldsymbol{\mathcal{V}} \times \left(\frac{\boldsymbol{\mathcal{V}}}{c^2} y_b \boldsymbol{k}\right)\right\}$$

at $t = 0$ when the two systems coincide.

Note that the "magnetic" force cancels part of the y component of the radial term and that the net force is not along the line joining m_a to m_b.

(b) Show that the force on m_a is equal and opposite to that on m_b. (You should read Section 6.2 before attempting this.)

(c) Show that the torque exerted on the system, in frame 1, is

$$\gamma\beta^2 K(x_{b1} - x_{a1})y_{b1}$$

in the counterclockwise direction.

6-3. Calculate the force, as observed in the laboratory system, between two electrons moving side by side along parallel paths one millimeter apart if they each have a kinetic energy of (*i*) one electron-volt, and (*ii*) one million electron-volts.

6-4. A proton having a kinetic energy of 50 million electron-volts is deflected by a magnetic field of 2.5 teslas, which is perpendicular to the trajectory.

Show that the deflecting electric field "seen" by the proton is about 2.3×10^8 volts/meter.

6-5. Draw a curve of the magnetic induction B as a function of the time at a point 1 centimeter away from the path of a single electron having an energy of 10 GeV (10^{10} electron-volts).

6-6. Show that the equation of motion of a particle of rest mass m_0 and charge Q can be written as

$$\gamma m_0 \frac{d\boldsymbol{v}}{dt} = Q\left(\boldsymbol{E} + \boldsymbol{v} \times \boldsymbol{B} - \frac{\boldsymbol{v}}{c^2}\,\boldsymbol{v}\cdot\boldsymbol{E}\right).$$

For motion in a static magnetic field,

$$\frac{d\boldsymbol{v}}{dt} = \boldsymbol{v} \times \boldsymbol{\omega}$$

where

$$\boldsymbol{\omega} = \frac{Q\boldsymbol{B}}{m}$$

is called the *Larmor frequency*, or *cyclotron frequency*.

6-7. Comment on the following statement. "There is no such thing as a magnetic field; what is called a magnetic field is simply an electric field observed in the wrong frame of reference."

6-8. (a) Show that, if there exists a field $\boldsymbol{E}_1$, $\boldsymbol{B}_1$ with respect to the laboratory, then the field $\boldsymbol{E}_2$, $\boldsymbol{B}_2$ seen by an observer moving at a velocity $\boldsymbol{\mathcal{V}}$ is

$$\boldsymbol{E}_{2\perp} = \gamma(\boldsymbol{E}_{1\perp} + \boldsymbol{\mathcal{V}} \times \boldsymbol{B}_1),$$

$$\boldsymbol{B}_{2\perp} = \gamma\left(\boldsymbol{B}_{1\perp} - \frac{1}{c^2}\boldsymbol{\mathcal{V}} \times \boldsymbol{E}_1\right),$$

$$\boldsymbol{E}_{2\|} = \boldsymbol{E}_{1\|},$$

$$\boldsymbol{B}_{2\|} = \boldsymbol{B}_{1\|},$$

where the symbols $\|$ and $\perp$ identify the components that are, respectively, parallel and perpendicular to $\boldsymbol{\mathcal{V}}$.

Thus, for $\mathcal{V}^2 \ll c^2$,

$$\boldsymbol{E}_2 = \boldsymbol{E}_1 + \boldsymbol{\mathcal{V}} \times \boldsymbol{B}_1,$$

$$\boldsymbol{B}_2 = \boldsymbol{B}_1 - \frac{1}{c^2}\boldsymbol{\mathcal{V}} \times \boldsymbol{E}_1.$$

(b) Show that, if $\boldsymbol{E}$ and $\boldsymbol{B}$ are orthogonal, the field in reference frame 2 is purely magnetic if

$$\boldsymbol{\mathcal{V}} = \frac{\boldsymbol{E}_1 \times \boldsymbol{B}_1}{B_1^2}.$$

(c) Show that, in that case,

$$\boldsymbol{B}_2 = \frac{\boldsymbol{B}_1}{\gamma}.$$

(d) The motion of a charged particle in uniform and mutually perpendicular electric and magnetic fields with $E < cB$ consists of a spiraling motion about the direction of the magnetic field, plus a drifting motion

$$\boldsymbol{v}_D = \frac{\boldsymbol{E}_1 \times \boldsymbol{B}_1}{B_1^2}.$$

Note that $\boldsymbol{v}_D$ is independent of the nature of the particle.

Sketch the trajectories of positive and negative particles in such a field.

Note that, in a plasma, the positive and negative charges all drift together at the same velocity $\boldsymbol{v}_D$; no macroscopic current is generated in this way.

(e) Calculate the drift velocity of a proton at the equator under the combined action of the gravitational attraction and of the magnetic field of the Earth.

In which direction does it drift?

Assume that the magnetic induction of the Earth is horizontal and equal to 4×10^{-5} tesla.

(f) In which direction would an electron drift under the same circumstances?

6-9. Show that $\boldsymbol{B} \cdot \boldsymbol{E}$ is invariant under a Lorentz transformation.

6-10. Show that $B^2 - (E^2/c^2)$ is invariant under a Lorentz transformation.

6-11. Given a field $\boldsymbol{E}$, $\boldsymbol{B}$ in the laboratory frame, under what conditions is it possible to find (a) a reference frame in which either $\boldsymbol{E}$ or $\boldsymbol{B}$ is zero; (b) one in which $\boldsymbol{E}$ and $\boldsymbol{B}$ are parallel, say in the direction of the y-axis; (c) one in which $\boldsymbol{E}$ and $\boldsymbol{B}$ are perpendicular?

6-12. (a) A dielectric-filled parallel-plate capacitor is situated in reference frame 2 with its plates parallel to the xy plane.

Show that $\epsilon_{r1} = \epsilon_{r2}$.

(b) We now turn the capacitor so that its plates are parallel to the yz plane. The capacitor is again situated in frame 2.

Show that ϵ_r is again the same in both reference frames.

6-13. Calculate $\boldsymbol{E}$ and $\boldsymbol{B}$ as functions of the time at a point 1 centimeter away from the path of a 10 million electron-volt proton.

6-14. The capacitor of the examples on pages 263 and 270 moves in a direction perpendicular to its plates.

Calculate the potentials on the plates, and the $\boldsymbol{A}$, $\boldsymbol{E}$, $\boldsymbol{B}$ inside the capacitor in the reference frame of the laboratory.

6-15. The magnetic induction inside a long solenoid of radius R that has N' turns/meter and carries a current I is $\mu_0 N'I$. The magnetic induction outside is zero (example on page 315).

(a) Calculate the values of $\boldsymbol{E}_1$ and $\boldsymbol{B}_1$, both inside and outside such a solenoid, as measured by a stationary observer, when the solenoid moves at a velocity $\mathcal{V}$ in a direction perpendicular to its length.

Assume that the axis of the solenoid is the z_2 axis and that $\boldsymbol{\mathcal{V}} = \mathcal{V}\boldsymbol{i}$.

(b) Now calculate this same field by transforming the potentials.

First show that, in the frame of the solenoid, the vector potential

$$\boldsymbol{A}_2 = -\frac{B_2}{2} y_2 \boldsymbol{i} + \frac{B_2}{2} x_2 \boldsymbol{j}$$

gives a uniform field $\boldsymbol{B}_2$ oriented in the positive direction of the z-axis.

Note that there are an infinite number of possible expressions for $\boldsymbol{A}_2$. For example, we could have set

$$\boldsymbol{A}_2 = B_2 x_2 \boldsymbol{j}.$$

(c) Now calculate $\boldsymbol{A}_1$, V_1, and then $\boldsymbol{E}_1$, $\boldsymbol{B}_1$, inside the solenoid.

You should arrive at the same result as under (a) above.

Both V_1 and $\boldsymbol{A}_1$ depend on the expression chosen arbitrarily for $\boldsymbol{A}_2$, but we always obtain the correct values for $\boldsymbol{E}_1$ and $\boldsymbol{B}_1$ from

$$\boldsymbol{E}_1 = -\nabla V_1 - \frac{\partial \boldsymbol{A}_1}{\partial t_1},$$

$$\boldsymbol{B}_1 = \nabla \times \boldsymbol{A}_1.$$

6-16. The solenoid of the preceding problem moves at a velocity $\mathcal{V}$ in the direction of its length.

Calculate $\boldsymbol{E}_1$ and $\boldsymbol{B}_1$ both inside and outside the solenoid, as measured by a stationary observer.

6-17. Verify that the Lorentz condition applies to the potentials A and V of Eqs. 6-78 and 6-91.

6-18. Show that Maxwell's equations are invariant under a Lorentz transformation.‡

6-19. Calculate the drift velocity of conduction electrons in copper from the following data:

$J = 1$ ampere/millimeter^2;

Atomic weight, 64;

Density, 9.

Express the velocity in meters/hour.

6-20. A charge Q moves at a velocity $V \ll c$ parallel to a wire carrying a current I and zero net charge density. The conduction electrons in the wire move at a velocity $v \ll c$.

Show that, for the charge Q, the wire appears to have a net charge density vV/c^2 times the charge density of the conduction electrons in the wire and that this explains the magnetic force on Q.

6-21. There presumably exists a gravitational equivalent of magnetic forces. Why have such forces not been observed?

6-22. A beam of particles of mass m and velocity v has a radius r. The charge density may be assumed to be constant throughout the cross-section of the beam.

Show that the outward force on an ion situated at the periphery of the beam is

$$\frac{IQ}{2\pi\epsilon_0 rv}\left(1 - \frac{v^2}{c^2}\right),$$

where I is the current in the beam and Q is the magnitude of the charge on an ion.‡

6-23. A certain linear accelerator produces a beam of electrons having a kinetic energy of 100 million electron-volts. The beam is pulsed, each pulse lasting 25 picoseconds (25×10^{-12} second) and containing 3.0×10^9 electrons. At the exit of the accelerator the beam has a diameter of 5.0 millimeters.

You are required to investigate the divergence of the beam and, in particular, to calculate what distance a pulse could travel before blowing up to a diameter of 2.0 centimeters. The beam travels inside a grounded conducting tube that has an inside diameter of 10 centimeters.

One may assume that the charge density inside each pulse is uniform.‡

(a) Calculate the length, diameter, and linear charge density of a pulse at the exit of the accelerator in the reference frame of the electrons.

(b) To calculate the divergence of the beam, we shall first calculate the divergence in the frame of the electrons, and then we shall transform to the laboratory system under (c) below.

You should have found under (a) above that, in the moving frame, the pulses are long cylinders of charge; so let us calculate the radius of a cylinder as a function of the time.

Call the initial radius at the exit of the accelerator r_0 and the radius of the tube r_t, and calculate the beam radius r as a function of the proper time t_2.

To do this you can calculate r as a function of t_2 for an ion at the periphery of the beam. Radial velocities are small compared to c, and the mass can be taken to be the rest mass m_0.

Neglect end effects.

The integral for r is evaluated as follows:

$$\int_{r_0}^{r} \frac{dr}{\left(\ln \frac{r}{r_0}\right)^{1/2}} = \int_{r_0}^{r} \frac{dr}{u} = \int_0^u 2r\, du = 2r_0 \int_0^u \exp u^2\, du.$$

This last integral is tabulated in *Tables of Functions* by E. Jahnke and F. Emde, fourth edition, page 32.

You should find that

$$\int_0^u \exp(u^2)\, du = \frac{1}{2r_0}\left\{\frac{e\lambda_2}{\pi\epsilon_0 m_0}\right\}^{1/2} t_2,$$

where λ_2 is the linear charge density and t_2 is the time, as measured in the moving frame.

(c) Now transform to the laboratory system and draw a curve of r as a function of the distance x_1 traveled by a pulse in the laboratory, for values of r ranging up to 10 millimeters.

You should find that the pulse has to travel about 400 meters before its radius becomes equal to 10 millimeters. (We have neglected scattering by gas molecules.)

(d) We have assumed that the radial velocity dr/dt_2 in the frame of the electrons is small compared to c.

Was this assumption justifiable?

(e) We have also assumed that the charge density was uniform inside each pulse.

Was this assumption really necessary?

(f) We shall see in Problem 8-30 that a continuous beam diverges in the same manner as above.

Why is this?

Hints

6-18. Use Tables 5-8, 5-10, 6-1.

6-22. Calculate the charge per unit length and the force in the reference frame of the ions, and then transform to the reference frame of the laboratory.

6-23. Use the relation Potential Energy + Kinetic Energy = Constant, setting the potential energy equal to zero at r_t, and the radial kinetic energy equal to zero at r_0.

CHAPTER 7

MAGNETIC FIELDS I

Steady Currents and Nonmagnetic Materials

The next three chapters will deal with various aspects of magnetic fields. We shall start by studying in some detail the properties of two vector quantities, namely, the magnetic induction $\boldsymbol{B}$ and the vector potential $\boldsymbol{A}$, which are used to describe magnetic fields. We shall limit ourselves to steady currents and nonmagnetic materials. We shall then discuss varying currents in Chapter 8, and magnetic materials in Chapter 9.

We already know that there are several types of electric current.

(a) There are first the usual *conduction currents through good conductors* such as copper, which are due to the drift of conduction electrons.

(b) There are also the *conduction currents in semiconductors*, which are due to the drift of either or both conduction electrons and holes (see the example on page 176).

(c) There are *electrolytic* currents.

(d) The motion of ions or electrons in a vacuum—for example, in an ion beam or in a vacuum diode—gives *convection currents*.

(e) The *motion of macroscopic charged bodies* also produces electric currents.

All these currents are associated with the motion of free charges, and they are the ones we shall be thinking of in this chapter. We shall use the symbol $\boldsymbol{J}_f$ for the current density associated with free charges, and I or I_f for the corresponding current.

(f) The *polarization current* density $\partial \boldsymbol{P}/\partial t$ in dielectrics (Section 3.2.2) is associated with bound charges, and we shall assume for the moment that it is zero.

(g) There are two other types of electric current, which we shall study later, in Chapters 9 and 10. These are (a) equivalent currents in magnetized substances and (b) displacement currents associated with changing electric fields.

If you have not studied Chapter 6, simply disregard references to it in this chapter and in all following chapters.

7.1 MAGNETIC FORCES

It is common experience that circuits carrying electric currents exert forces on each other. For example, the force between two straight parallel wires carrying currents I_a and I_b is proportional to $I_a I_b/\rho$, where ρ is the distance between the wires. The force is attractive if the currents flow in the same direction, and it is repulsive if they flow in opposite directions.

For the more general case illustrated in Figure 7-1, the force that one current exerts on the other when both are in free space is more complex, but it is again proportional to the product $I_a I_b$. This force is

$$\boldsymbol{F}_{ab} = \frac{\mu_0}{4\pi} I_a I_b \oint_a \oint_b \frac{\boldsymbol{dl}_b \times (\boldsymbol{dl}_a \times \boldsymbol{r}_1)}{r^2}, \tag{7-1}$$

where $\boldsymbol{F}_{ab}$ is the force exerted *by* current I_a *on* current I_b, and where the line integrals are evaluated over the two circuits.

This is the *magnetic force law*. The vectors $\boldsymbol{dl}_a$ and $\boldsymbol{dl}_b$ point in the direction of current flow, $\boldsymbol{r}_1$ is a unit vector pointing *from* $\boldsymbol{dl}_a$ *to* $\boldsymbol{dl}_b$, and r is the distance between the two elements $\boldsymbol{dl}_a$ and $\boldsymbol{dl}_b$. The force is measured in newtons, the currents in amperes, and the lengths in meters. This law is in agreement with Eq. 6-178.

The meaning of the double integral is as follows. We choose a fixed element $\boldsymbol{dl}_b$ on circuit b and add the vectors $\boldsymbol{dl}_b \times (\boldsymbol{dl}_a \times \boldsymbol{r}_1)/r^2$ corresponding to each element $\boldsymbol{dl}_a$ of circuit a. We then repeat the operation for all the other elements $\boldsymbol{dl}_b$ of circuit b and, finally, calculate the over-all sum. In general, this integration cannot be performed analytically. We then divide the circuits into small finite elements and evaluate the sum numerically.

As we saw in Section 6-1, the constant μ_0 is *defined* as follows:

$$\mu_0 \equiv 4\pi \times 10^{-7} \text{ newton/ampere}^2 \tag{7-2}$$

and is called the *permeability of free space*. Since μ_0 is defined in this way, Eq. 7-1 can be considered to be a definition of the unit of current, the ampere or coulomb/second.

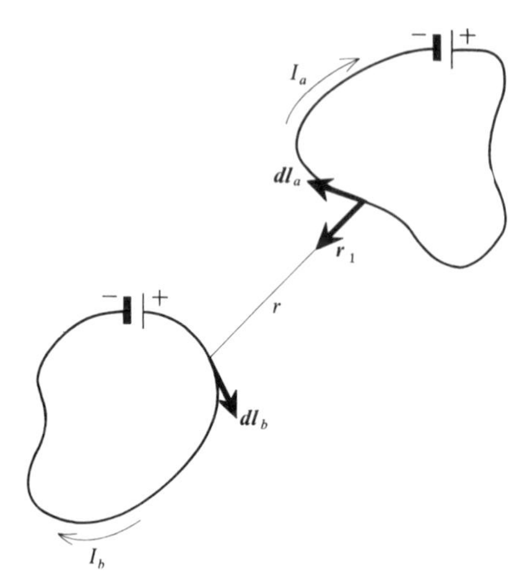

Figure 7-1. Two currents I_a and I_b.

Coulomb's law gave us the force of interaction between stationary electric charges. The magnetic force law now states the force between electric currents. In both laws there are constants of proportionality, ϵ_0 and μ_0, and it is μ_0 that is defined arbitrarily. The coulomb is thus defined, not from Coulomb's law, but from the magnetic force law. It turns out, experimentally, that the value of ϵ_0 in Coulomb's law must be 8.85×10^{-12} farad/meter, as stated previously.

The force $\boldsymbol{F}_{ab}$ is expressed above in such a fashion that $\boldsymbol{dl}_a$ and $\boldsymbol{dl}_b$ do not play symmetrical roles. This is quite disturbing, since Newton's third law surely applies to a pair of circuits carrying steady currents, and $\boldsymbol{F}_{ab}$ must equal $-\boldsymbol{F}_{ba}$.

The force $\boldsymbol{F}_{ab}$ can be expressed in a symmetrical form by expanding the triple vector product under the integral sign:

$$\frac{\boldsymbol{dl}_b \times (\boldsymbol{dl}_a \times \boldsymbol{r}_1)}{r^2} = \frac{\boldsymbol{dl}_a(\boldsymbol{dl}_b \cdot \boldsymbol{r}_1)}{r^2} - \frac{\boldsymbol{r}_1(\boldsymbol{dl}_a \cdot \boldsymbol{dl}_b)}{r^2}. \tag{7-3}$$

We now show that the double integral of the first term on the right is zero. First

$$\oint_a \oint_b \frac{\boldsymbol{dl}_a(\boldsymbol{dl}_b \cdot \boldsymbol{r}_1)}{r^2} = \oint_a \boldsymbol{dl}_a \oint_b \frac{\boldsymbol{dl}_b \cdot \boldsymbol{r}_1}{r^2}. \tag{7-4}$$

Now, on the right-hand side, the second integral is zero for the following reason. It is the integral of dr/r^2 around a *closed* curve, circuit b. It is therefore the integral of $(1/r^2)\, dr$ with identical upper and lower limits of integration, and it is zero.

We are thus left with the double integral of only the second term for the triple vector product $\boldsymbol{dl}_b \times (\boldsymbol{dl}_a \times \boldsymbol{r}_1)$, and

$$\boldsymbol{F}_{ab} = -\frac{\mu_0}{4\pi} I_a I_b \oint_a \oint_b \frac{\boldsymbol{r}_1(\boldsymbol{dl}_a \cdot \boldsymbol{dl}_b)}{r^2}. \tag{7-5}$$

Therefore, $\boldsymbol{F}_{ab} = -\boldsymbol{F}_{ba}$, since the unit vector $\boldsymbol{r}_1$ is directed toward the circuit on which the force is to be calculated, with the result that it is oriented in one direction for $\boldsymbol{F}_{ab}$ and in the opposite direction for $\boldsymbol{F}_{ba}$.

7.2 THE MAGNETIC INDUCTION $\boldsymbol{B}$. THE BIOT-SAVART LAW

Despite the fact that the above integral for $\boldsymbol{F}_{ab}$ is simpler and more symmetrical than that of Eq. 7-1, it is not as useful. The reason is that, with the above integral, the force cannot be expressed as the interaction of current b with the field of current a. We can perform such an operation on Eq. 7-1, however, since

$$\boldsymbol{F}_{ab} = I_b \oint_b \boldsymbol{dl}_b \times \left(\frac{\mu_0}{4\pi} I_a \oint_a \frac{\boldsymbol{dl}_a \times \boldsymbol{r}_1}{r^2} \right), \tag{7-6}$$

$$= I_b \oint_b \boldsymbol{dl}_b \times \boldsymbol{B}_a, \tag{7-7}$$

where

$$\boldsymbol{B}_a = \frac{\mu_0}{4\pi} I_a \oint_a \frac{\boldsymbol{dl}_a \times \boldsymbol{r}_1}{r^2} \tag{7-8}$$

is the *magnetic induction* due to circuit a at the position of the element $\boldsymbol{dl}_b$ of circuit b. As usual, the unit vector $\boldsymbol{r}_1$ points *from* the source, *to* the point of observation: it points *from* the element $\boldsymbol{dl}_a$, *to* the point where $\boldsymbol{B}$ is calculated. The magnetic induction is expressed in *teslas* or in *webers/meter²:*

$$1 \text{ tesla} = 1 \text{ weber/meter}^2 = 10^4 \text{ gauss}. \tag{7-9}$$

The weber is a volt-second. The gauss is not an *MKSA* unit, but it is frequently used because of its convenient order of magnitude.

The above equation for $\boldsymbol{B}$ is the *Biot-Savart law*, which we have already found in the example on page 280. The integration can be performed analytically only for the simplest geometrical forms. It shows that the element of force $\boldsymbol{dF}$ on an element of wire of length $\boldsymbol{dl}$ carrying a current I in a region where the magnetic induction is $\boldsymbol{B}$ is

$$\boldsymbol{dF} = I\, \boldsymbol{dl} \times \boldsymbol{B}. \tag{7-10}$$

If the current I is distributed in space with a current density $\boldsymbol{J}_f$ amperes/meter², then I becomes $J_f\,da$ and must be put under the integral sign. Then $J_f\,da\,\boldsymbol{dl}$ can be written as $\boldsymbol{J}_f\,d\tau$, where $d\tau$ is an element of volume, and

$$\boldsymbol{B} = \frac{\mu_0}{4\pi}\int_{\tau'} \frac{\boldsymbol{J}_f \times \boldsymbol{r}_1}{r^2}\,d\tau'. \tag{7-11}$$

The integration is carried out over any volume τ' that includes all the currents. The subscript f on the current density $\boldsymbol{J}_f$ refers to the fact that the currents we are considering in this chapter are due to free charges. We assume, as we do throughout this chapter, that $\boldsymbol{J}_f$ is not a function of the time and that there are no magnetic materials in the field.

Can this integral be used to calculate $\boldsymbol{B}$ at a point *inside* a current-carrying conductor? Since r is the distance between (a) the point of observation where $\boldsymbol{B}$ is measured and (b) the point where the current density is $\boldsymbol{J}_f$, we are led to expect that the contribution of the local current density will be infinite because of the $1/r^2$ factor. The integral does not, in fact, diverge; it does apply within current-carrying conductors. This can be seen by analogy with electrostatics, where the same problem arises in calculating the electric field intensity $\boldsymbol{E}$ inside a charge distribution. The components of $\boldsymbol{B}$ and of $\boldsymbol{E}$ both vary as $1/r^2$. Since those of $\boldsymbol{E}$ do remain finite within a charge distribution, those of $\boldsymbol{B}$ must also remain finite within a current distribution. We assume that both the charge density and the current density are finite.

As in electrostatics, where we used lines of force to describe an electric field, we can describe a magnetic field by drawing *lines of* $\boldsymbol{B}$ that are everywhere tangent to the direction of $\boldsymbol{B}$.

Similarly, it is convenient to use the concept of flux, the flux of the magnetic induction $\boldsymbol{B}$ through a surface S being defined as the normal component of $\boldsymbol{B}$ integrated over S:

$$\Phi = \int_S \boldsymbol{B} \cdot \boldsymbol{da}. \tag{7-12}$$

The flux Φ is expressed in *webers*.

Example

THE MAGNETIC INDUCTION DUE TO A CURRENT FLOWING IN A LONG STRAIGHT WIRE

An element $\boldsymbol{dl}$ of a long straight wire carrying a current I, as in Figure 7-2, produces a magnetic induction $\boldsymbol{dB}$ as shown in the figure, with

$$\boldsymbol{dB} = \frac{\mu_0 I}{4\pi}\frac{dl \sin\theta}{r^2}\,\boldsymbol{\varphi}_1. \tag{7-13}$$

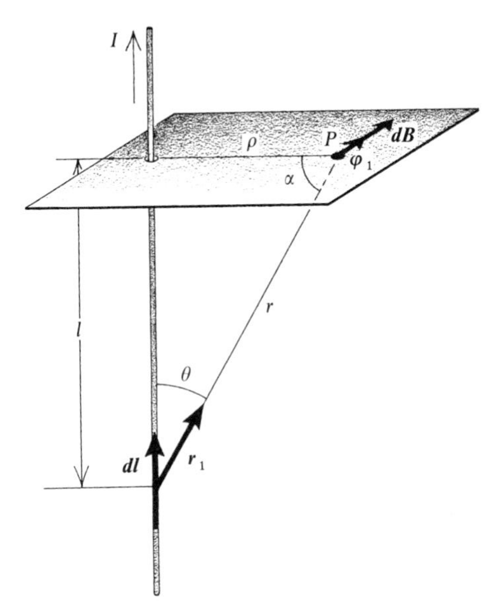

Figure 7-2. The magnetic induction $\boldsymbol{dB}$ produced by an element $I\,\boldsymbol{dl}$ of the current I in an infinitely long straight wire. The vector $\boldsymbol{dB}$ lies in a plane that is perpendicular to the wire and which passes through P.

Expressing dl, $\sin\theta$, and r^2 in terms of α and ρ,

$$\boldsymbol{B} = \frac{\mu_0 I}{4\pi\rho}\int_{-\pi/2}^{+\pi/2}\cos\alpha\, d\alpha\, \boldsymbol{\varphi}_1 = \frac{\mu_0 I}{2\pi\rho}\boldsymbol{\varphi}_1, \tag{7-14}$$

as in Eq. 6-158.

The magnitude of $\boldsymbol{B}$ thus falls off inversely as the first power of the distance from an infinitely long wire and is in the direction perpendicular to a plane containing the wire. The lines of $\boldsymbol{B}$ are circles lying in a plane perpendicular to the wire and are centered on it.

Example

FORCE BETWEEN TWO LONG PARALLEL WIRES

The force between two infinitely long parallel wires carrying currents I_a and I_b, separated by a distance ρ, as in Figure 7-3, follows immediately from the above result. The current I_a produces a magnetic induction $\boldsymbol{B}_a$, as above, at the position of the current I_b. The force acting on an element $I\boldsymbol{dl}_b$ of this current is

$$\boldsymbol{dF} = I_b(\boldsymbol{dl}_b \times \boldsymbol{B}_a), \tag{7-15}$$

$$dF = \frac{I_b dl_b \mu_0 I_a}{2\pi\rho}, \tag{7-16}$$

and the force per unit length is

$$\frac{dF}{dl} = \frac{\mu_0 I_a I_b}{2\pi\rho}. \tag{7-17}$$

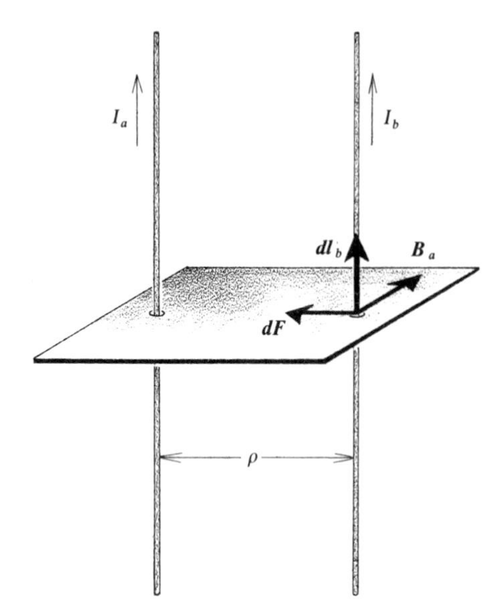

Figure 7-3. Two long parallel wires carrying currents in the same direction. The element of force $\boldsymbol{dF}$ acting on the element $\boldsymbol{dl}_b$ is in the direction shown.

The force is attractive if the currents are in the same direction, and it is repulsive if they are in opposite directions.

This equation is the basis of the international definition of the ampere: two long parallel wires separated by a distance of one meter exert on each other a force of 2×10^{-7} newton per meter of length when the current in each is one ampere. It is assumed that the diameters of the wires are negligible compared to their separation.

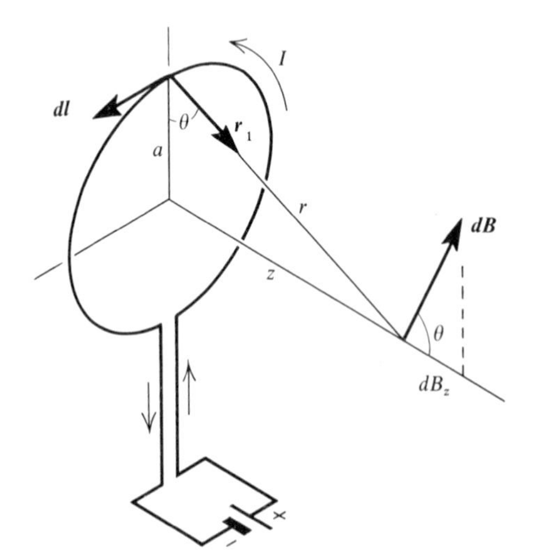

Figure 7-4. The magnetic induction $\boldsymbol{dB}$ produced by an element $I\,\boldsymbol{dl}$ at a point on the axis of circular current loop of radius a. The projection of $\boldsymbol{dB}$ on the axis is dB_z.

Example

THE CIRCULAR LOOP

Let us calculate the magnitude and direction of $\boldsymbol{B}$ on the axis of a circular loop of radius a carrying a current I, as in Figure 7-4.

An element $I\,\boldsymbol{dl}$ of current produces a magnetic induction $\boldsymbol{dB}$, as indicated in the figure. By symmetry, the total magnetic induction will be along the axis, and

$$dB_z = \frac{\mu_0 I\,dl}{4\pi\,r^2}\cos\theta; \tag{7-18}$$

hence

$$B_z = \frac{\mu_0 I}{4\pi}\frac{2\pi a}{r^2}\cos\theta = \frac{\mu_0 I a^2}{2(a^2 + z^2)^{3/2}}. \tag{7-19}$$

The magnetic induction is maximum in the plane of the ring and drops off as z^3 for $z^2 \gg a^2$.

7.3 THE FORCE ON A POINT CHARGE MOVING IN A MAGNETIC FIELD

Equation 7-7 gave us the force on a closed circuit immersed in a magnetic field; it can also give us the force on a single charge Q moving at a velocity $\boldsymbol{v}$ in a magnetic field $\boldsymbol{B}$. We shall see that Eq. 7-7 is consistent with the Lorentz force we have already found in Eq. 6-8.

The force on a current element $I\,dl$ is $I\,\boldsymbol{dl} \times \boldsymbol{B}$. Now, if the cross-sectional area of the wire is da,

$$I = n(da\ v)Q, \tag{7-20}$$

where n is the number of carriers per unit volume, v is their average drift velocity, and Q is the charge on one carrier. The reason for this relation is that the total charge flowing per second is the charge on the carriers that are contained in a length v of the wire.

Then the force on the element $\boldsymbol{dl}$ is

$$n\,da\,dl\,Q\,\boldsymbol{v} \times \boldsymbol{B},$$

and the force on a single charge Q moving at a velocity $\boldsymbol{v}$ in a field $\boldsymbol{B}$ is

$$Q\,\boldsymbol{v} \times \boldsymbol{B}.$$

This force is perpendicular both to the velocity $\boldsymbol{v}$ and to the local magnetic induction $\boldsymbol{B}$.

More generally, if there is also an electric field $\boldsymbol{E}$, the force is

$$Q[\boldsymbol{E} + (\boldsymbol{v} \times \boldsymbol{B})].$$

This is the *Lorentz force.*

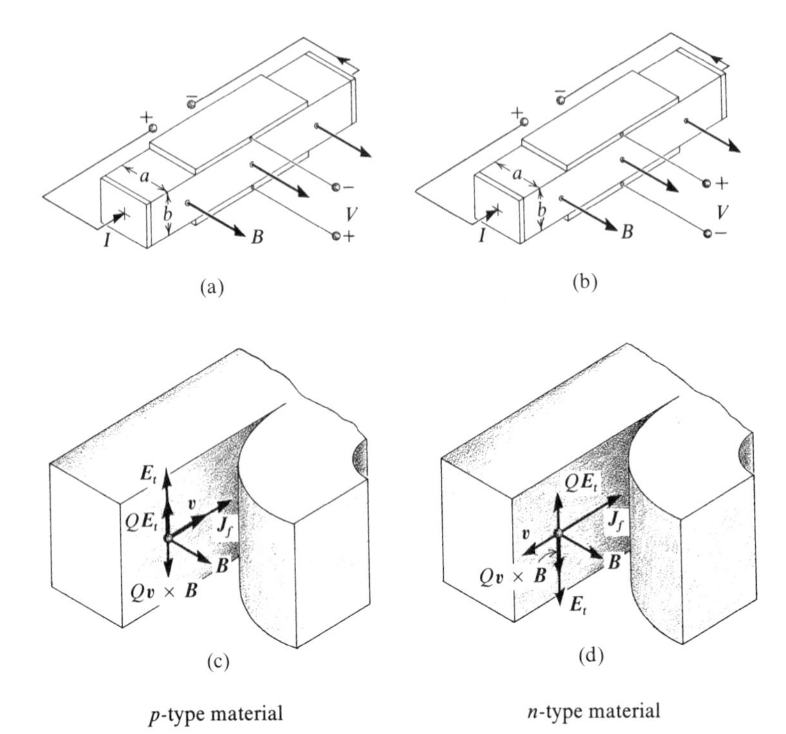

Figure 7-5. Hall effect in semiconductors. In p-type material conduction is due to the drift of positive charges (holes) and the Hall voltage is as shown in (a). In n-type material conduction is due to the conduction electrons and the Hall voltage has the opposite polarity as in (b). Ordinary conductors such as copper behave as in (b). Figure (c) shows the two opposing transverse forces $Q\boldsymbol{E}_t$ and $Q\boldsymbol{v} \times \boldsymbol{B}$ on a positive charge drifting along the axis of the bar at a velocity $\boldsymbol{v}$. Figure (d) shows how these forces are reversed in n-type material.

Example

HALL EFFECT IN SEMICONDUCTORS

We have already seen in the example on page 176, that semiconductors contain either or both of two types of mobile charges, namely conduction electrons and holes. When a current flows through a bar of semiconductor in the presence of a transverse magnetic field, as in Figures 7-5a or 7-5b, the mobile charges drift, not only in the direction of the applied electric field, but also in a direction perpendicular to both the applied electric and magnetic fields. This gives rise to a voltage difference between the upper and lower electrodes.

If the voltmeter connected to these electrodes draws a negligible amount of current, the plates charge up until their field E_t is sufficient to stop the transverse drift. This transverse electric field is called the *Hall field*. The net *transverse force* on the mobile charges inside the bar is then *zero* as in Figures 7-5c and 7-5d.

Let us assume that the conduction is due to holes carrying charges $+e$, e being the magnitude of the electronic charge, and that there are p holes per cubic meter. When the transverse drift has stopped,

$$eE_t = e\frac{V}{b} = evB. \tag{7-21}$$

Now

$$I = Jab = (pev)ab, \tag{7-22}$$

and

$$V = \frac{1}{pea}IB. \tag{7-23}$$

If the conduction is due rather to conduction electrons carrying a charge $-e$, the Hall field has the opposite polarity, as in Figure 7-5b.

Note that in *both* cases the carriers are swept *down* by the magnetic field.

Although the Hall effect is in fact more complex than we have assumed it to be, our value of V is nonetheless approximately correct. The Hall effect is commonly used for measuring magnetic fields and for studying conduction phenomena in semiconductors.

Example

THE HODOSCOPE

The hodoscope is a device that simulates the trajectory of a charged particle in a magnetic field. The principle involved is quite simple: if the charged particle, of mass m, charge Q, and velocity v, is replaced by a light wire fixed at the two ends of the trajectory and carrying a current I, the wire will follow the trajectory if

$$\frac{mv}{Q} = \frac{T}{I}, \tag{7-24}$$

where T is the tension in the wire. This statement is by no means obvious, and we shall have to demonstrate its validity.

The advantage of the so-called *floating wire* lies in the fact that it is much easier to experiment with a wire than with an ion beam; the wire is, in effect, an analog computer.

Let us first consider a region where the ion beam is perpendicular to $\boldsymbol{B}$ as in Figure 7-6a. Then

$$BQv = \frac{mv^2}{R_t}, \tag{7-25}$$

where R_t is the radius of curvature of the trajectory:

$$R_t = \frac{mv}{BQ}. \tag{7-26}$$

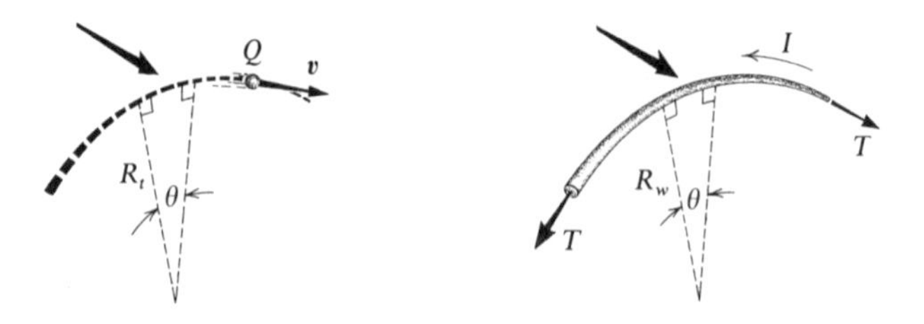

Figure 7-6. (a) Charge Q moving at a velocity $\boldsymbol{v}$ in a magnetic field $\boldsymbol{B}$. The radius of curvature of the trajectory is R_t. (b) Light wire carrying a current I in the *opposite* direction in the same magnetic field. The tension in the wire is T and its radius of curvature is R_w. It is shown that the wire has the same radius of curvature as the trajectory if mv/Q is equal to T/I.

In Figure 7-6b we have replaced the charged particles by a light wire carrying a current I flowing in the *opposite* direction. If the element dl is in equilibrium, the outward force $BI\,dl$ is compensated by the inward component of the tension T:

$$BI\,dl = 2T \sin d(\theta/2) = \frac{T}{R_w}\,dl. \tag{7-27}$$

The radius of curvature of the wire is thus

$$R_w = \frac{T}{BI}, \tag{7-28}$$

and the two radii of curvature will be the same if

$$\frac{mv}{Q} = \frac{T}{I}. \tag{7-29}$$

Note that the particle will be deflected downward if the magnetic force is downward, while the wire will curve downward if the force is *upward*. The magnetic forces must therefore be directed in opposite directions.

If the magnetic field is not uniform, and if the beam is not perpendicular to $\boldsymbol{B}$, then the wire does not always follow the trajectory. For example, magnetic fields are often used both to deflect and focus an ion beam. In such cases the focusing forces on the beam can become defocusing forces on the wire, which is then deflected away from the trajectory.

7.4 THE DIVERGENCE OF THE MAGNETIC INDUCTION $\boldsymbol{B}$

In Section 6.9 we demonstrated that the magnetic fields of moving charges were such that $\boldsymbol{\nabla} \cdot \boldsymbol{B} = 0$. It is also possible to arrive at this same result for steady currents starting from the Biot-Savart law, Eq. 7-8. Since

$$B = \frac{\mu_0}{4\pi} \int_{\tau'} \frac{J_f \times r_1}{r^2}\, d\tau', \tag{7-30}$$

$$\nabla \cdot B = \frac{\mu_0}{4\pi} \int_{\tau'} \nabla \cdot \frac{J_f \times r_1}{r^2}\, d\tau' = \frac{\mu_0}{4\pi} \int_{\tau'} \nabla \cdot \left(J_f \times \frac{r_1}{r^2}\right) d\tau'. \tag{7-31}$$

But, from Problem 1-19

$$\nabla \cdot \left(J_f \times \frac{r_1}{r^2}\right) \equiv \frac{r_1}{r^2} \cdot (\nabla \times J_f) - J_f \cdot \left(\nabla \times \frac{r_1}{r^2}\right), \tag{7-32}$$

where the first term on the right is zero because J_f is a function of the source point x', y', z', while the del operator involves derivatives with respect to the field point x, y, z. The second term on the right is also zero because

$$\nabla \times \frac{r_1}{r^2} = \nabla \times \frac{r}{r^3} = \begin{vmatrix} i & j & k \\ \frac{\partial}{\partial x} & \frac{\partial}{\partial y} & \frac{\partial}{\partial z} \\ (x - x')/r^3 & (y - y')/r^3 & (z - z')/r^3 \end{vmatrix} \equiv 0. \tag{7-33}$$

Then

$$\nabla \cdot B = 0. \tag{7-34}$$

This equation follows from the definition of B given in Eq. 7-30. In Chapter 6 we also showed that it is a consequence of Coulomb's law and of the Lorentz transformation. The fact that $\nabla \cdot B$ is zero means that there cannot exist sources of B.

The net flux of magnetic induction through any closed surface is equal to zero since

$$B \cdot da = \int_{\tau} \nabla \cdot B \, d\tau = 0. \tag{7-35}$$

7.5 THE VECTOR POTENTIAL A

We have shown in Section 6.4 that the magnetic induction B of a single moving charge can be expressed as the curl of a certain quantity A, which we called the *vector potential*. The same relation applies, of course, if we use the Biot-Savart law, as we shall see. The vector potential A is an important quantity; we shall use it repeatedly.

According to the Biot-Savart law, Eq. 7-8, the magnetic induction at the point P of Figure 7-7 is

$$B = \frac{\mu_0}{4\pi} \int_{\tau'} J_f \times \frac{r_1}{r^2}\, d\tau', \tag{7-36}$$

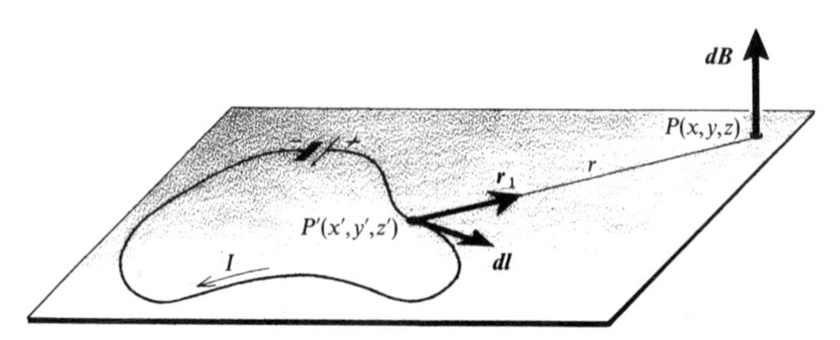

Figure 7-7. A current element $I\,dl$ at a source point P' produces an element of magnetic induction dB at a field point P.

where τ' is the volume of the conductor, or

$$\boldsymbol{B} = \frac{\mu_0}{4\pi}\int_{\tau'} \boldsymbol{\nabla}\left(\frac{1}{r}\right) \times \boldsymbol{J}_f \, d\tau', \tag{7-37}$$

from Problem 1-12. Using the vector identity of Problem 1-18,

$$\boldsymbol{\nabla}\left(\frac{1}{r}\right) \times \boldsymbol{J}_f = \boldsymbol{\nabla} \times \frac{\boldsymbol{J}_f}{r} - \frac{1}{r}\boldsymbol{\nabla} \times \boldsymbol{J}_f, \tag{7-38}$$

where the second term is zero, again because $\boldsymbol{J}_f$ is a function of x', y', z', while $\boldsymbol{\nabla}$ contains derivatives with respect to x, y, z. Then

$$\boldsymbol{B} = \frac{\mu_0}{4\pi}\int_{\tau'} \boldsymbol{\nabla} \times \frac{\boldsymbol{J}_f}{r} \, d\tau'. \tag{7-39}$$

We can change the order of differentiation and integration, and thus

$$\boldsymbol{B} = \boldsymbol{\nabla} \times \left[\frac{\mu_0}{4\pi}\int_{\tau'} \frac{\boldsymbol{J}_f}{r} \, d\tau'\right], \tag{7-40}$$

$$= \boldsymbol{\nabla} \times \boldsymbol{A}, \tag{7-41}$$

where

$$\boldsymbol{A} = \frac{\mu_0}{4\pi}\int_{\tau'} \frac{\boldsymbol{J}_f}{r} \, d\tau' \tag{7-42}$$

is the vector potential of the current distribution.

This integral, like that for $\boldsymbol{B}$, appears to diverge inside a current-carrying conductor because of the $1/r$ factor, but it actually does not. This can be seen from the fact that its components vary as $1/r$, like the electric potential V, which does not diverge within a charge distribution.

The fact that $\boldsymbol{B}$ is equal to the curl of $\boldsymbol{A}$ follows from the equation $\boldsymbol{\nabla} \cdot \boldsymbol{B} = 0$, because the divergence of the curl of a vector is identically equal to zero.

If the current is limited to a conducting wire,

$$A = \frac{\mu_0 I}{4\pi} \oint \frac{\mathbf{dl}}{r}. \tag{7-43}$$

Note that the vector potential $\mathbf{A}$ is *not* uniquely defined by the above integrals. Indeed we can add to these integrals any quantity whose curl is zero without affecting the value of $\mathbf{B}$ in any way. Since we have no reason to add such a term for the moment, we shall use these integrals for defining $\mathbf{A}$, and we shall defer this question until later.

Note also that $\mathbf{B}$ depends on the space derivatives of $\mathbf{A}$, and not on $\mathbf{A}$ itself. The value of $\mathbf{B}$ at a given point can thus be calculated from $\mathbf{A}$ only if $\mathbf{A}$ is known in the *region* around the point considered.

Example

THE LONG STRAIGHT WIRE

We have already found the magnetic induction $\mathbf{B}$ for a long straight wire twice, in Section 6.13 and in Section 7.2.1. Let us calculate this $\mathbf{B}$ once more, using the vector potential A this time.

Elements $I\,\mathbf{dl}$ of the current in Figure 7-8 contribute to the vector potential elements

$$\mathbf{dA} = \frac{\mu_0 I}{4\pi} \frac{\mathbf{dl}}{r}, \tag{7-44}$$

all in the same direction. From the fundamental definition of the curl in terms of a line integral (Eq. 1-88), and from the azimuthal symmetry of the field, $\boldsymbol{\nabla} \times \mathbf{A} = \mathbf{B}$ is in the azimuthal direction around the conductor.

For an infinitely long conductor, $\mathbf{dA}$ is proportional to dl/l for large values of r, since $r \approx l$, and A tends to infinity logarithmically. However, the fact that a function is infinite does not necessarily mean that its derivatives are also infinite; that is, $\mathbf{B}$ can be finite even though A is infinite.

Let us calculate A and $\mathbf{B}$ for a current of finite length $2L$, and then we can let L go to infinity. Referring to Figure 7-8,

$$A_z = 2\frac{\mu_0 I}{4\pi} \int_0^L \frac{dl}{(\rho^2 + l^2)^{1/2}}, \tag{7-45}$$

$$= \frac{\mu_0 I}{2\pi} \ln\left[l + (\rho^2 + l^2)^{1/2}\right]_0^L, \tag{7-46}$$

$$= \frac{\mu_0 I}{2\pi}\left\{\ln L\left[1 + \left(1 + \frac{\rho^2}{L^2}\right)^{1/2}\right] - \ln\rho\right\} \approx \frac{\mu_0 I}{2\pi} \ln\frac{2L}{\rho}, \tag{7-47}$$

for $\rho^2 \ll L^2$.

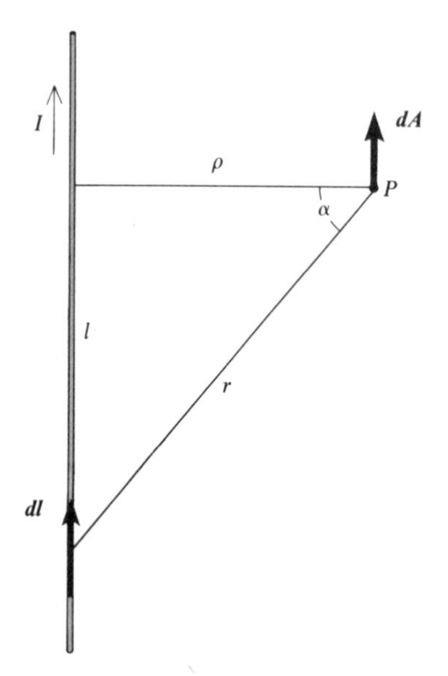

Figure 7-8. An element $I\,dl$ of a current I in a long straight wire produces an element of vector potential dA at the point P.

To calculate $\boldsymbol{B} = \nabla \times \boldsymbol{A}$, we use cylindrical coordinates, keeping in mind that $\boldsymbol{A}$ is parallel to the z-axis and independent of φ:

$$\boldsymbol{B} = \frac{1}{\rho}\begin{vmatrix} \boldsymbol{\rho}_1 & \rho\boldsymbol{\varphi}_1 & \boldsymbol{z}_1 \\ \dfrac{\partial}{\partial\rho} & 0 & \dfrac{\partial}{\partial z} \\ 0 & 0 & A_z \end{vmatrix}, \tag{7-48}$$

$$B_\rho = 0, \tag{7-49}$$

$$B_z = 0, \tag{7-50}$$

$$B_\varphi = -\frac{\partial A_z}{\partial \rho} \tag{7-51}$$

$$\rightarrow \frac{\mu_0 I}{2\pi\rho} \qquad (\rho^2 \ll L^2). \tag{7-52}$$

as in Eqs. 6-158 and 7-14.

Example

PAIR OF LONG PARALLEL WIRES

Figure 7-9 shows two long parallel wires that are separated by a distance R and that carry currents I of equal magnitude but in opposite directions. To calculate $\boldsymbol{A}$ and $\boldsymbol{B}$ we begin with wires of finite length $2L$, use Eq. 7-47 for a single wire, and add the two vector potentials together:

$$A_z = \frac{\mu_0 I}{2\pi}\left(\ln\frac{2L}{\rho_a} - \ln\frac{2L}{\rho_b}\right) \tag{7-53}$$

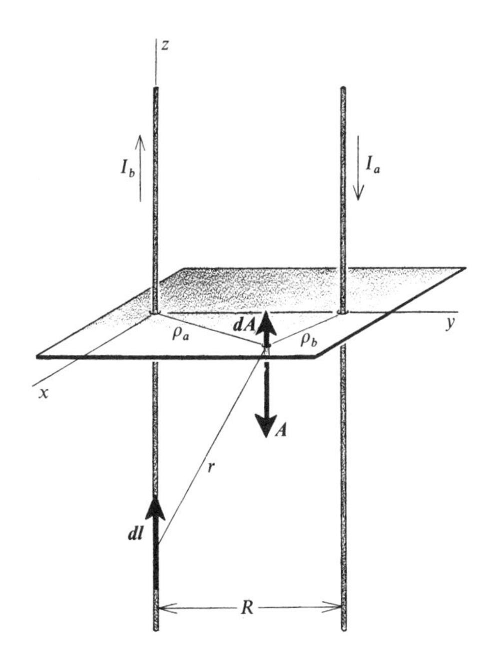

Figure 7-9. Pair of long parallel wires carrying currents of equal magnitude in opposite directions. The element of current $I\, \boldsymbol{dl}$ produces the element of vector potential $\boldsymbol{dA}$ shown. The vector potential $\boldsymbol{A}$ due to both wires is directed in the opposite direction, since the point considered is closer to I_b than to I_a.

In this case we can let L tend to infinity before computing the curl, and

$$A_z = \frac{\mu_0 I}{4\pi} \ln \frac{\rho_b^2}{\rho_a^2} = \frac{\mu_0 I}{2\pi} \ln \frac{\rho_b}{\rho_a}. \tag{7-54}$$

With the Cartesian coordinates shown in Figure 7-9,

$$A_x = 0, \qquad A_y = 0, \qquad A_z = \frac{\mu_0 I}{4\pi} \ln \left[\frac{x^2 + (R-y)^2}{x^2 + y^2} \right], \tag{7-55}$$

$$B_x = \frac{\partial A_z}{\partial y} = -\frac{\mu_0 I}{2\pi} \left[\frac{(R-y)}{\rho_b^2} + \frac{y}{\rho_a^2} \right], \tag{7-56}$$

$$B_y = -\frac{\partial A_z}{\partial x} = -\frac{\mu_0 I}{2\pi} \left(\frac{x}{\rho_b^2} - \frac{x}{\rho_a^2} \right), \tag{7-57}$$

$$B_z = 0. \tag{7-58}$$

At the midpoint between the two conductors, $x = 0$, $y = R/2$, hence

$$B_x = -2 \frac{\mu_0 I}{2\pi \left(\frac{R}{2} \right)} = -\frac{2\mu_0 I}{\pi R}, \qquad B_y = 0, \qquad B_z = 0, \tag{7-59}$$

as expected.

7.5.1 The Line Integral of the Vector Potential A Over a Closed Curve

The line integral of $\boldsymbol{A}\cdot d\boldsymbol{l}$ around a closed curve C is equal to the magnetic flux Φ linking C:

$$\oint_C \boldsymbol{A}\cdot d\boldsymbol{l} = \int_S (\boldsymbol{\nabla}\times\boldsymbol{A})\cdot d\boldsymbol{a} = \int_S \boldsymbol{B}\cdot d\boldsymbol{a} = \Phi, \tag{7-60}$$

where S is any surface bounded by the curve C.

7.6 THE CURL OF THE MAGNETIC INDUCTION B

We have shown that the magnetic induction is always equal to the curl of the vector potential: $\boldsymbol{B} = \boldsymbol{\nabla}\times\boldsymbol{A}$. We shall now show that

$$\boldsymbol{\nabla}\times\boldsymbol{B} = \mu_0 \boldsymbol{J}_f, \tag{7-61}$$

again assuming a steady state and the absence of magnetic materials.

In terms of $\boldsymbol{A}$,

$$\boldsymbol{\nabla}\times\boldsymbol{B} = \boldsymbol{\nabla}\times\boldsymbol{\nabla}\times\boldsymbol{A} = \boldsymbol{\nabla}(\boldsymbol{\nabla}\cdot\boldsymbol{A}) - \boldsymbol{\nabla}^2\boldsymbol{A} \tag{7-62}$$

from Section 1.9.6.

We have already shown in Section 6.7 that $\boldsymbol{\nabla}\cdot\boldsymbol{A}$ is proportional to the time derivative of the electric potential V, and we shall also show this again in Section 10.3. Then, with the above assumptions,

$$\boldsymbol{\nabla}\cdot\boldsymbol{A} = 0. \tag{7-63}$$

For the second term, we have from the definition of $\boldsymbol{A}$ that

$$\boldsymbol{\nabla}^2\boldsymbol{A} = \frac{\mu_0}{4\pi}\int_{\tau'} \boldsymbol{\nabla}^2 \frac{\boldsymbol{J}_f}{r}\, d\tau', \tag{7-64}$$

where we have interchanged the order of differentiation and integration.

The meaning of this equation can be understood by referring to Figure 7-10. At the field point $P(x, y, z)$, where we wish to compute $\boldsymbol{\nabla}^2\boldsymbol{A}$, we form the vector $\boldsymbol{J}_f\, d\tau'/r$, where $\boldsymbol{J}_f$ and $d\tau'$ are respectively the current density and the volume element at the source point P', and where r is the distance from P' to P. We compute the Laplacian of this vector at P by taking the appropriate derivatives with respect to the coordinates x, y, z of P. We then sum

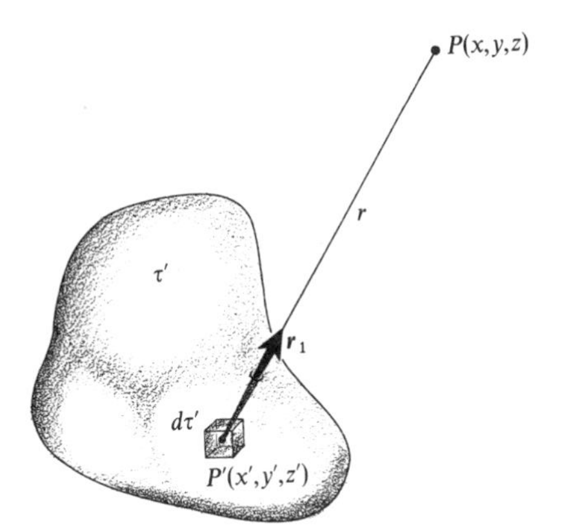

Figure 7-10. Source point P' and field point P for the calculation of $\nabla^2 \boldsymbol{A}$.

the contributions from all such sources in the volume τ', which includes all points at which $\boldsymbol{J}_f$ exists. The volume τ' may include the field point P, where $r = 0$, as will be shown below.

Since $\boldsymbol{J}_f$ is not a function of the coordinates of P, we can write the integral as

$$\nabla^2 \boldsymbol{A} = \frac{\mu_0}{4\pi} \int_{\tau'} \boldsymbol{J}_f \nabla^2 \left(\frac{1}{r}\right) d\tau'. \tag{7-65}$$

Now, by differentiation of

$$\frac{1}{r} = \frac{1}{[(x - x')^2 + (y - y')^2 + (z - z')^2]^{1/2}}, \tag{7-66}$$

we find that $\nabla^2(1/r) = 0$ if $r \neq 0$. There can thus be no contribution to the integral from any element $d\tau'$, except possibly if P and P' coincide and r is zero.

To investigate the integral at $r = 0$ we consider a small volume enclosing a point P, where we wish to calculate $\nabla^2 \boldsymbol{A}$, situated inside the current distribution.

We take the volume so small that $\boldsymbol{J}_f$ does not change appreciably within it; $\boldsymbol{J}_f$ may then be removed from the integral:

$$\nabla^2 \boldsymbol{A} = \frac{\mu_0 \boldsymbol{J}_f}{4\pi} \int_{\tau' \to 0} \nabla^2 \left(\frac{1}{r}\right) d\tau'. \tag{7-67}$$

The meaning of this integral is as follows. For each element of volume $d\tau'$ centered at the point P' within τ', we calculate

$$\nabla^2 \left(\frac{1}{r}\right) = \left(\frac{\partial^2}{\partial x^2} + \frac{\partial^2}{\partial y^2} + \frac{\partial^2}{dz^2}\right) \frac{1}{[(x - x')^2 + (y - y')^2 + (z - z')^2]^{1/2}}, \tag{7-68}$$

multiply by $d\tau'$, and sum the results. Since $\nabla^2(1/r) = \nabla'^2(1/r)$,

$$\nabla^2 A = \frac{\mu_0 J_f}{4\pi} \int_{\tau' \to 0} \nabla'^2 \left(\frac{1}{r}\right) d\tau', \tag{7-69}$$

$$= \frac{\mu_0 J_f}{4\pi} \int_{\tau' \to 0} \nabla' \cdot \nabla' \left(\frac{1}{r}\right) d\tau', \tag{7-70}$$

$$= \frac{\mu_0 J_f}{4\pi} \int_{S' \to 0} \nabla' \left(\frac{1}{r}\right) \cdot d\boldsymbol{a}, \tag{7-71}$$

from the divergence theorem. Then

$$\nabla^2 A = \frac{\mu_0 J_f}{4\pi} \int_{S' \to 0} \frac{\boldsymbol{r}_1 \cdot d\boldsymbol{a}}{r^2}, \tag{7-72}$$

from Problem 1-12, where $\boldsymbol{r}_1$ is the unit vector from the source point to the field point. In this case $\boldsymbol{r}_1$ points inward toward the point P. Thus

$$\nabla^2 A = -\frac{\mu_0 J_f}{4\pi} \int_{S' \to 0} d\Omega, \tag{7-73}$$

where $d\Omega$ is the element of solid angle subtended at the point P by the element of area da. Since the surface S' completely surrounds P,

$$\nabla^2 \boldsymbol{A} = -\mu_0 \boldsymbol{J}_f, \tag{7-74}$$

and

$$\nabla \times \boldsymbol{B} = \mu_0 \boldsymbol{J}_f. \tag{7-75}$$

This result is again valid only for static fields and in the absence of magnetic materials.

7.7 AMPÈRE'S CIRCUITAL LAW

We can express the above result in integral form by integrating the normal component of $\nabla \times \boldsymbol{B}$ over an arbitrary surface S:

$$\int_S (\nabla \times \boldsymbol{B}) \cdot d\boldsymbol{a} = \mu_0 \int_S \boldsymbol{J}_f \cdot d\boldsymbol{a}. \tag{7-76}$$

Then, using Stokes's theorem, we transform the left side of this equation into a line integral around the closed path C, which bounds the surface S, and

$$\oint_C \boldsymbol{B} \cdot d\boldsymbol{l} = \mu_0 \int_S \boldsymbol{J}_f \cdot d\boldsymbol{a} = \mu_0 I. \tag{7-77}$$

This is *Ampère's circuital law:* the line integral of $\boldsymbol{B}$ around a closed path is equal to μ_0 times the total current I crossing any surface bounded by the line integral path. Again, we are limited to nonmagnetic materials and to static fields.

In many cases the same current crosses the surface bounded by the integration path several times. With a solenoid, for example, the integration path could follow the axis and return outside the solenoid. The total current crossing the surface is then the current in each turn multiplied by the number of turns, or the number of *ampere-turns*.

The circuital law can be used to calculate $\boldsymbol{B}$ when it is constant along the path of integration. This law is thus somewhat similar to Gauss's law, which is used to compute $\boldsymbol{E}$ when $\boldsymbol{E}$ is constant over a surface.

Example

LONG CYLINDRICAL CONDUCTOR

Let us investigate the magnitude and direction of the magnetic induction $\boldsymbol{B}$ and of the vector potential $\boldsymbol{A}$ inside and outside a long, straight cylindrical conductor, as in Figure 7-11, carrying a current I uniformly distributed over its cross section with a density

$$J_f = \frac{I}{\pi R^2}. \tag{7-78}$$

Outside the conductor, $\boldsymbol{B}$ is azimuthal and independent of φ, so that, according to the circuital law,

$$B = \frac{\mu_0 I}{2\pi\rho}. \tag{7-79}$$

Inside the conductor, for a circular path of radius ρ,

$$B = \frac{\mu_0 J_f \pi\rho^2}{2\pi\rho} = \frac{\mu_0 I \rho}{2\pi R^2}. \tag{7-80}$$

The magnetic induction B therefore increases linearly with ρ inside the conductor. Outside the conductor, B decreases as $1/\rho$. The curve of B as a function of ρ is shown in Figure 7-12.

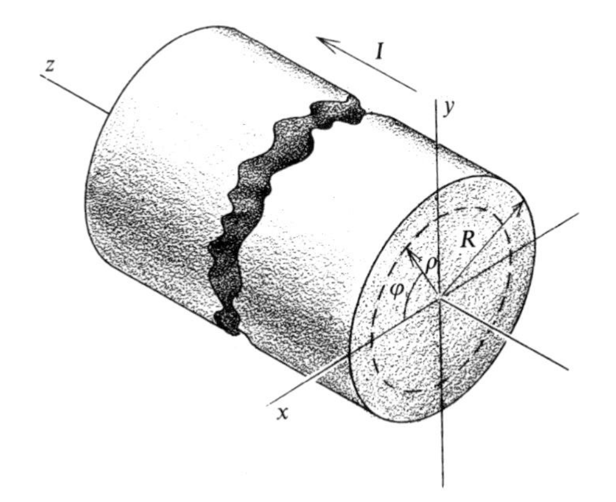

Figure 7-11. Long cylindrical conductor of circular cross section carrying a current I. The circle of radius ρ is a path of integration for calculating $\boldsymbol{B}$ inside.

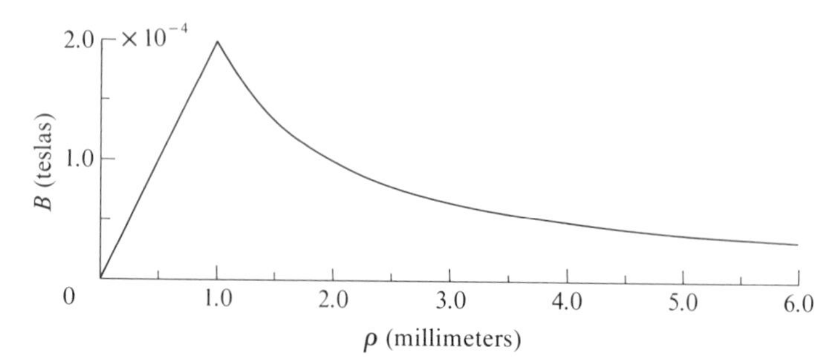

Figure 7-12. The magnetic induction B as a function of radius for a wire of 1 millimeter radius carrying a current of 1 ampere.

To calculate the vector potential $\boldsymbol{A}$ at a point within the conductor, we cannot use Eq. 7-43, which yields an infinite value for $\boldsymbol{A}$, as it did for a single infinite wire; instead, we apply Eq. 7-60 to a path lying in a radial plane, as in Figure 7-13. The path follows the surface of the conductor parallel to the axis and returns at a distance ρ from the axis. Since the vector potential must be parallel to the current density $\boldsymbol{J}_f$, it is everywhere parallel to the axis, and, for a long conductor, it is independent of the axial coordinate z. Hence, from Eq. 7-60,

$$-A(R)l + A(\rho)l = \Phi, \tag{7-81}$$

where $A(R)$ is the magnitude of $\boldsymbol{A}$ at the radius R, $A(\rho)$ is its magnitude at ρ, and Φ is the flux through the radial plane surface enclosed by the path. Now

$$\Phi = \int_\rho^R \frac{\mu_0 I \rho}{2\pi R^2}\, l\, d\rho = \frac{\mu_0 I l}{4\pi R^2}(R^2 - \rho^2). \tag{7-82}$$

Also, the function $A(\rho)$ for the region outside the conductor was found in Eq. 7-47. Setting $\rho = R$ in this equation,

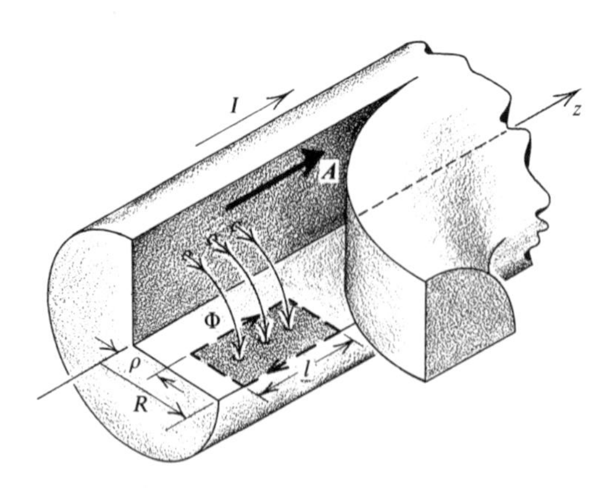

Figure 7-13. Long cylindrical conductor of circular cross section carrying a current I. Part of the wire is removed to show the magnetic flux Φ and the path of integration for calculating $\boldsymbol{A}$ inside.

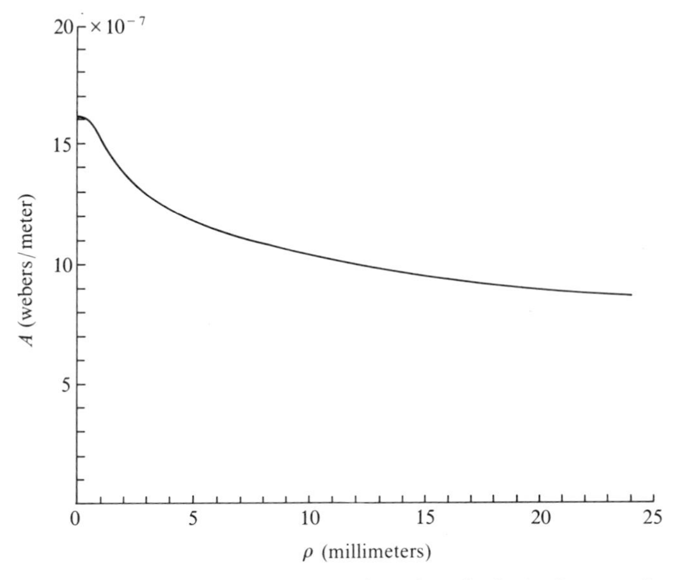

Figure 7-14. The vector potential A for a length of wire 2 meters long and 1 millimeter in radius carrying a current of 1 ampere.

$$A(R) = \frac{\mu_0 I}{2\pi} \ln \frac{2L}{R} \tag{7-83}$$

where L is the half-length of the conductor. Then

$$A(\rho) = \frac{\mu_0 I}{4\pi}\left(\ln \frac{4L^2}{R^2} + 1 - \frac{\rho^2}{R^2}\right) \qquad (\rho \leqslant R). \tag{7-84}$$

The logarithmic term is constant for a given wire and for a given current, whereas the term ρ^2/R^2 increases as the square of the radius ρ inside the conductor. Figure 7-14 shows the value of A both inside and outside a wire that is 1 millimeter in radius and 2 meters long and that carries a current of 1 ampere.

We have assumed that the conductor is nonmagnetic.

The only term of importance inside the parenthesis is ρ^2/R^2, since the others are independent of the coordinates. One may therefore set A equal to

$$-\frac{\mu_0 I}{4\pi}\frac{\rho^2}{R^2}.$$

The curl of this $\boldsymbol{A}$, we find the $\boldsymbol{B}$ of Eq. 7-80, as expected.

Example

THE TOROIDAL COIL

As a second example, consider a close-wound toroidal coil of square cross section, as in Figure 7-15, carrying a current I.

Along path a, the line integral of $\boldsymbol{B}$ is equal to zero, since there is no current linking this path. Then the azimuthal B is zero in this region. The same applies to c and to any similar path outside the toroid. Then the azimuthal B is zero everywhere outside.

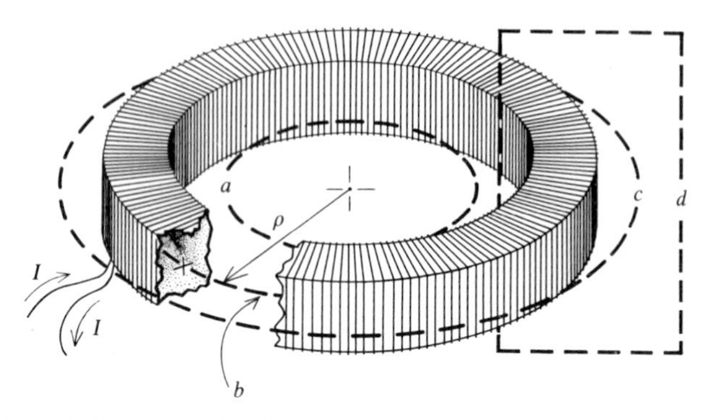

Figure 7-15. Toroidal coil of square cross section carrying a current I. The figures shown by broken lines are paths of integration for $\boldsymbol{B}$.

Inside, along path b, B is fairly constant throughout the section if the toroid is thin in the radial direction. Thus

$$2\pi\rho B = \mu_0 NI, \tag{7-85}$$

where N is the total number of turns, and

$$B = \mu_0 \frac{N}{2\pi\rho} I. \tag{7-86}$$

There exist nonazimuthal components of the magnetic induction outside the toroid. For a path such as d in Figure 7-15, the area bounded by the path is crossed once by the current in the toroidal winding and, at a distance large compared to the outer radius of the toroid, the magnetic induction is that of a single turn along the mean radius.

It is interesting to note that, although the magnetic induction $\boldsymbol{B}$ outside the toroid is essentially zero, the vector potential $\boldsymbol{A}$ is not. This will be evident if one remembers that $\boldsymbol{A}$ is a constant times the integral of $I\,\boldsymbol{dl}/r$, where r is the distance between the element $\boldsymbol{dl}$ and the point where $\boldsymbol{A}$ is calculated. For example, at a point close to the winding, $\boldsymbol{A}$ is due mostly to the nearby turns, it is parallel to the current, and it has approximately the same value inside and outside.

The vector potential $\boldsymbol{A}$ can therefore exist in a region where there is no $\boldsymbol{B}$ field. This simply means that we can have at the same time $\boldsymbol{A} \neq 0$ and $\boldsymbol{\nabla} \times \boldsymbol{A} = 0$, which is entirely plausible. For example, $\boldsymbol{A} = k\boldsymbol{i}$, where k is a constant, satisfies this condition. We are already familiar with a similar situation in electrostatics: the electric potential can have any uniform value in a region where $\boldsymbol{E} = -\boldsymbol{\nabla} V = 0$.*

* See *The Feynman Lectures on Physics* (Addison-Wesley, Reading, Mass., 1964) Vol II, Section 15-5, for a discussion of the quantum-mechanical aspects of this question.

Example

THE LONG SOLENOID

As another example, let us calculate $\boldsymbol{B}$ inside and outside the long solenoid of Figure 7-16. We shall select a region remote from the ends, so that end effects will be negligible; also, we shall assume that the pitch of the winding is small. Let us choose cylindrical coordinates with the z-axis coinciding with the axis of symmetry of the solenoid.

1. We first note that $\boldsymbol{B}$ has the following characteristics *both inside and outside* the solenoid.

(a) By symmetry, $\boldsymbol{B}$ is neither a function of z nor of φ.

(b) Moreover, $B_\rho = 0$ for the following reason. Consider an axial cylinder of length l and radius ρ, either larger than the solenoid radius, or smaller, as in Figure 7-16. The integral of $\boldsymbol{B}\cdot d\boldsymbol{a}$ over its surface is simply $2\pi\rho l B_\rho$ since the integrals over the two end faces cancel. But, according to Eq. 7-35, the integral of $\boldsymbol{B}\cdot d\boldsymbol{a}$ over any closed surface is zero. Then $B_\rho = 0$.

These three characteristics satisfy the condition $\boldsymbol{\nabla}\cdot\boldsymbol{B} = 0$.

(c) Finally, the curl of $\boldsymbol{B}$ must be zero everywhere except inside the wire, and thus

$$\frac{\partial B_z}{\partial \rho} = 0. \qquad \text{(7-87)}$$

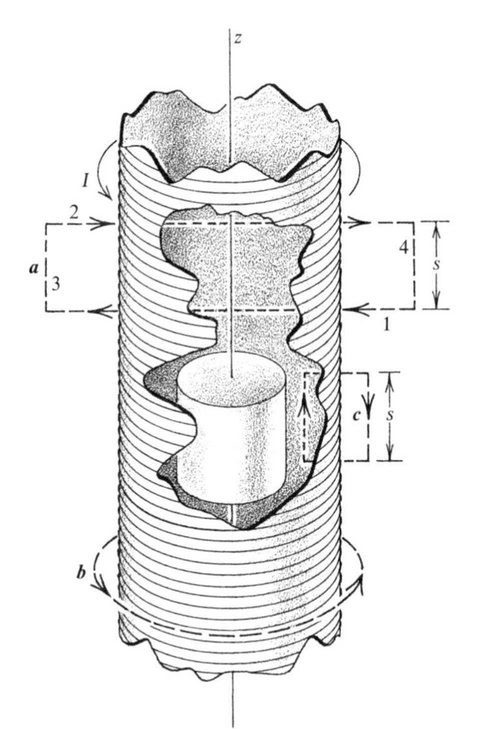

Figure 7-16. Long solenoid carrying a current I. The figures shown by broken lines are paths of integration for $\boldsymbol{B}$.

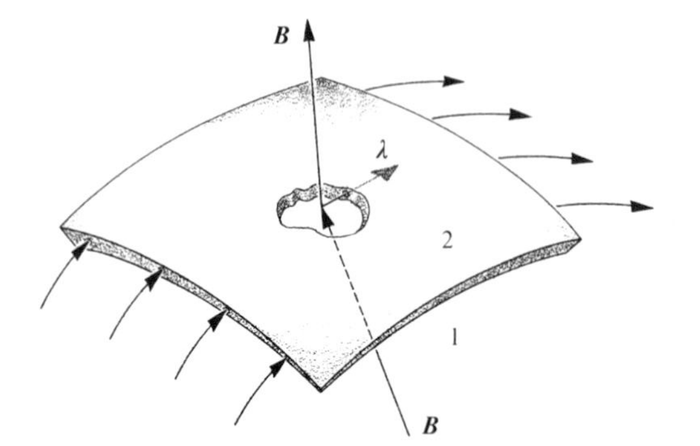

Figure 7-17. Conducting sheet carrying a current density of λ amperes per meter. Since $\nabla \cdot \boldsymbol{B} = 0$, the normal component of $\boldsymbol{B}$ is the same on both sides of the sheet. According to the circuital law, however, the tangential component is not conserved and a line of $\boldsymbol{B}$ is deflected in the direction shown.

2. *Outside* the solenoid.

(a) We can show that $B_z = 0$ by considering path a in Figure 7-16. The net current linking this path is zero, and the line integral of $\boldsymbol{B} \cdot \boldsymbol{dl}$ around it is therefore also zero. Now, since the line integrals along sides 1 and 2 are zero ($B_\rho = 0$), the line integrals along sides 3 and 4 must cancel. But sides 3 and 4 can each be situated at any distance from the solenoid, and B_z must therefore be either zero or constant along them.

Since the lines of $\boldsymbol{B}$ have neither beginning nor end, and since the total return flux in the infinite space outside must equal the finite flux inside the solenoid, B_z must be zero outside.

(b) A path such as b is linked once by the current, and, outside the solenoid,

$$B_\varphi = \frac{\mu_0 I}{2\pi\rho}. \tag{7-88}$$

3. *Inside* the solenoid.

(a) $B_\varphi = 0$ inside because the line integral of $B_\varphi\, dl$ over a circle of radius ρ, say the top edge of the small cylinder shown in Figure 7-16, is $2\pi\rho B_\varphi$, and this must be zero according to Eq. 7-75 because there is no current enclosed by the path.

(b) Considering now path c in Figure 7-16, and remembering that $B_\rho = 0$ both inside and outside, and that $B_z = 0$ outside, we see that $B_z s = \mu_0 N' I s$, and

$$B_z = \mu_0 N' I. \tag{7-89}$$

The magnetic induction inside a long solenoid in the region remote from the ends is therefore uniform and equal to μ_0 times the number of ampere-turns/meter.

Example

REFRACTION OF THE LINES OF *B* AT A CURRENT SHEET

Imagine a thin conducting sheet carrying a current density of λ amperes/meter, as in Figure 7-17. We can find how the lines of $\boldsymbol{B}$

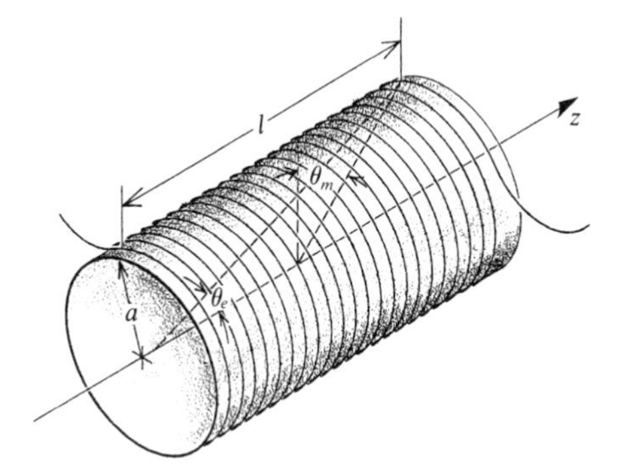

Figure 7-18. A short solenoid.

are refracted in passing through the sheet by proceeding as in Section 4.1.2. Since the divergence of $\boldsymbol{B}$ is zero, the normal component of $\boldsymbol{B}$ is conserved:

$$B_{n1} = B_{n2}. \tag{7-90}$$

Also, if we apply Ampère's circuital law to a path of length L that is perpendicular to the sheet as in Figure 4-2,

$$B_{t_1}L - B_{t_2}L = \mu_0 \lambda L, \tag{7-91}$$

$$B_{t_2} = B_{t_1} - \mu_0 \lambda. \tag{7-92}$$

The line of force is therefore rotated in the clockwise direction for an observer looking in the direction of the vector $\boldsymbol{\lambda}$.

We could have arrived at this result in another way. The magnetic induction $\boldsymbol{B}$ is due to the current sheet itself and the other currents flowing elsewhere in the system. According to Ampère's circuital law, the current sheet produces just below itself in Figure 7-17a $\boldsymbol{B}$ that is directed to the left and whose magnitude is $\mu_0\lambda/2$. Similarly, the $\boldsymbol{B}$ just above the sheet is directed to the right and has the same magnitude. If we add this field to that of the other currents, we see that the tangential components of $\boldsymbol{B}$ must differ as above.

Example

THE SHORT SOLENOID

We can calculate B on the axis of a short solenoid by summing the contributions of the individual turns, using Eq. 7-19. If the length of the solenoid is l and if its radius is a, the magnetic induction at the center is

$$B = \frac{\mu_0}{2} \int_{-l/2}^{+l/2} \frac{a^2 N' I \, dz}{(a^2 + z^2)^{3/2}}, \tag{7-93}$$

$$= \mu_0 N' I \sin \theta_m, \tag{7-94}$$

as in Figure 7-18. We have assumed that the solenoid is close wound. For a long solenoid $\theta_m \rightarrow \pi/2$, and $B \rightarrow \mu_0 N' I$, as in the example on page 315.

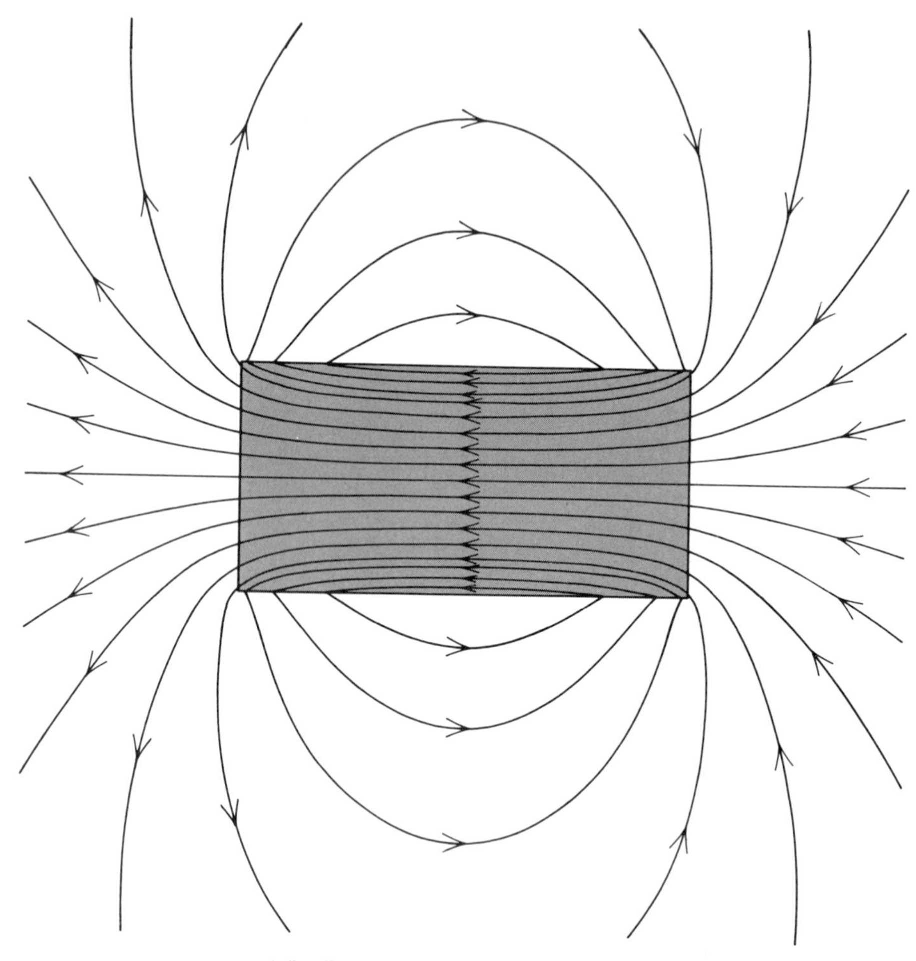

Figure 7-19. Lines of $\boldsymbol{B}$ for a solenoid whose length is equal to twice its diameter.

At one end, again on the axis,

$$B = \mu_0 N' I \frac{\sin \theta_e}{2}. \qquad (7\text{-}95)$$

The magnetic induction thus decreases at both ends of the solenoid, and this is of course due to the fact that the lines of $\boldsymbol{B}$ flare out as in Figure 7-19.

Upon crossing the current sheet, the radial component of $\boldsymbol{B}$ remains unchanged, but the axial component changes both its magnitude and its sign. For example, the axial component in the upper left-hand side of the solenoid in Figure 7-19 changes from, say, $-0.9\ \mu_0 N' I$ to $+0.1\ \mu_0 N' I$, since the surface current density λ is $N' I$.

7.8 THE MAGNETIC DIPOLE

We shall now show that a small loop of wire of area S, situated at the origin in a plane perpendicular to the z-axis and carrying a current I, produces a magnetic field

$$B_r = \frac{\mu_0}{4\pi}\frac{2m}{r^3}\cos\theta, \tag{7-96}$$

$$B_\theta = \frac{\mu_0}{4\pi}\frac{m}{r^3}\sin\theta, \tag{7-97}$$

$$B_\varphi = 0, \tag{7-98}$$

where $m = IS$ is called the *magnetic dipole moment* of the loop. Note that this $\boldsymbol{B}$ field is similar to the $\boldsymbol{E}$ field of an electric dipole (Section 2.9 and Figure 2-10).

We shall start by calculating

$$\boldsymbol{A} = \frac{\mu_0 I}{4\pi}\oint\frac{\boldsymbol{dl}}{r'} \tag{7-99}$$

for a circular loop, as in Figure 7-20, and then we can generalize to any shape of loop.

Since the $\boldsymbol{dl}$ vectors have no z component, $\boldsymbol{A}$ can only have x and y components. A little thought will show that, for any given value of r', we have two symmetrical $\boldsymbol{dl}$ vectors whose y components add and whose x components cancel. Then we need only calculate the y, or azimuthal, component of $\boldsymbol{A}$. If a is the radius of the loop, then

$$\boldsymbol{A} = \frac{\mu_0 I}{4\pi}\int_0^{2\pi}\frac{(a\,d\varphi)\cos\varphi}{r'}\boldsymbol{\varphi}_1, \tag{7-100}$$

where $\boldsymbol{\varphi}_1$ is the azimuthal unit vector in spherical polar coordinates.

We must now express r' in terms of r and of ψ. First

$$r'^2 = r^2 + a^2 - 2ar\cos\psi, \tag{7-101}$$

$$\frac{r}{r'} = \left(1 - \frac{a^2}{r'^2} + \frac{2ar}{r'^2}\cos\psi\right)^{1/2}, \tag{7-102}$$

$$\approx 1 - \frac{a^2}{2r^2} + \frac{a}{r}\cos\psi, \tag{7-103}$$

if we make the assumption that $a \ll r$. Since $r/r' \approx 1$, we have substituted r for r' in the two correction terms on the right-hand side.

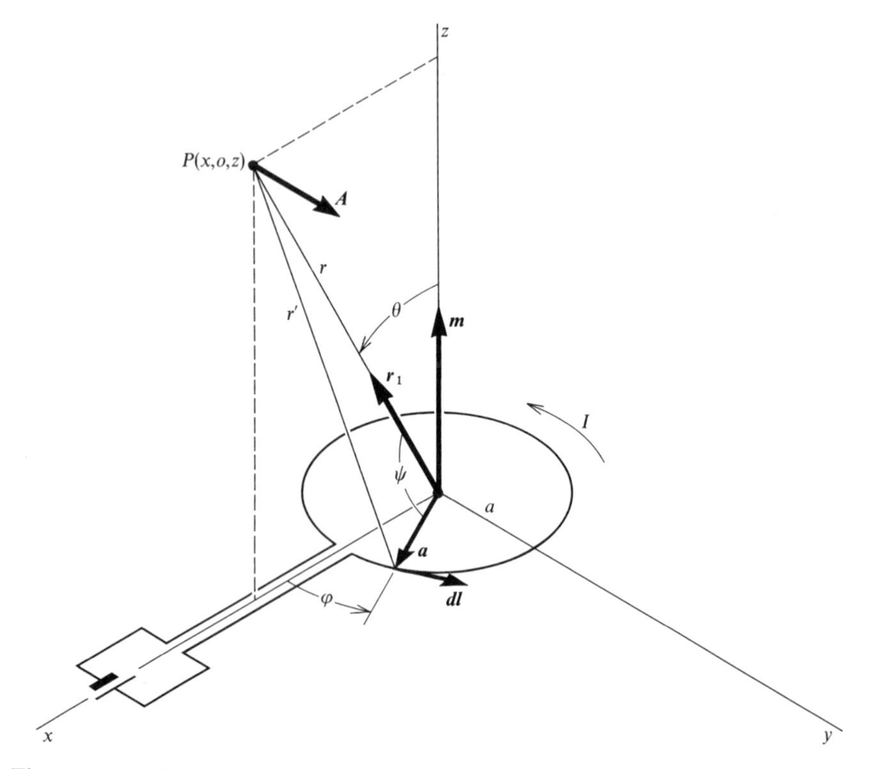

Figure 7-20. Circular loop of wire carrying a current I. The vector potential $\boldsymbol{A}$ at the point P is directed in the azimuthal direction.

To replace the $\cos\psi$ term by a function of φ, we use the fact that the scalar product $\boldsymbol{r}\cdot\boldsymbol{a}$ can be expressed either as

$$(x\boldsymbol{i} + z\boldsymbol{k})\cdot(a\cos\varphi\,\boldsymbol{i} + a\sin\varphi\,\boldsymbol{j}) = xa\cos\varphi \tag{7-104}$$

or as $ra\cos\psi$. Then

$$\cos\psi = \frac{x}{r}\cos\varphi, \tag{7-105}$$

$$\frac{r}{r'} \approx 1 - \frac{a^2}{2r^2} + \frac{ax}{r^2}\cos\varphi, \tag{7-106}$$

and

$$\boldsymbol{A} = \frac{\mu_0 I}{4\pi r}\int_0^{2\pi} a\cos\varphi\left(1 - \frac{a^2}{2r^2} + \frac{ax}{r^2}\cos\varphi\right)d\varphi\,\boldsymbol{\varphi}_1, \tag{7-107}$$

$$= \frac{\mu_0}{4\pi}\frac{I\pi a^2}{r^3}x\,\boldsymbol{\varphi}_1 = \frac{\mu_0}{4\pi}\frac{I\pi a^2}{r^2}\sin\theta\,\boldsymbol{\varphi}_1, \tag{7-108}$$

$$= \frac{\mu_0}{4\pi}\frac{m}{r^2}\sin\theta\,\boldsymbol{\varphi}_1, \tag{7-109}$$

$$= \frac{\mu_0}{4\pi}\frac{\boldsymbol{m}\times\boldsymbol{r}_1}{r^2}, \tag{7-110}$$

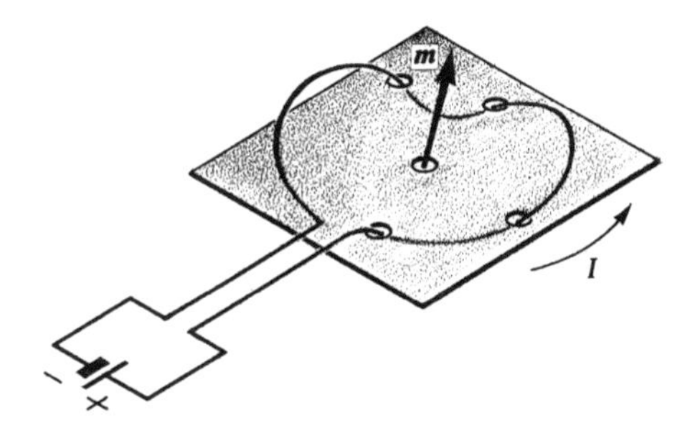

Figure 7-21. Current loop and its magnetic moment $\boldsymbol{m}$.

if the dipole moment is written in the vector form

$$\boldsymbol{m} = IS\,\boldsymbol{k} = I\,\boldsymbol{S}. \tag{7-111}$$

We selected a circular loop so as to make the integral for $\boldsymbol{A}$ tractable. If the loop is plane but not circular, it can be considered to be formed of closely packed circles of various sizes, and the vector potential is then given by the same expression as above, except that πa^2 must be replaced by the area S of the loop.

To find $\boldsymbol{B}$, we simply compute $\boldsymbol{\nabla} \times \boldsymbol{A}$, and we do find that

$$B_r = \frac{\mu_0}{4\pi}\frac{2m}{r^3}\cos\theta, \tag{7-112}$$

$$B_\theta = \frac{\mu_0}{4\pi}\frac{m}{r^3}\sin\theta, \tag{7-113}$$

$$B_\varphi = 0. \tag{7-114}$$

This result is valid for $a \ll r$. The analogy with the field of the electric dipole (Section 2.9) is obvious.

It is possible to generalize the concept of magnetic dipole moment in the following way. For any closed circuit, whether it is planar or not, we can write that

$$\boldsymbol{m} = \tfrac{1}{2}\oint \boldsymbol{r} \times I\,\boldsymbol{dl}, \tag{7-115}$$

where $\boldsymbol{r}$ is the vector from the origin to the current element $I\,\boldsymbol{dl}$, and where the integral is evaluated over the circuit. Figure 7-21 shows the vector $\boldsymbol{m}$ for a simple loop.

The concept of magnetic dipole moment can be further generalized to a current distribution $\boldsymbol{J}_f$ in a volume τ:

$$\boldsymbol{m} = \frac{1}{2}\int_\tau \boldsymbol{r} \times \boldsymbol{J}_f\,d\tau. \tag{7-116}$$

Example THE LONG SOLENOID

To find the magnetic dipole moment of a long solenoid we must calculate the line integral of Eq. 7-115 over the wire of the solenoid. If the pitch of the winding is small, each turn has a dipole moment $\pi R^2 I$, and, for the complete solenoid,

$$\boldsymbol{m} = \pi R^2 N I\, \boldsymbol{k}, \tag{7-117}$$

where N is the total number of turns.

At distances that are large compared to the length of the solenoid, the $\boldsymbol{B}$ field is given by Eqs. 7-112 to 7-114.

7.9 SUMMARY

We are concerned in this chapter with the magnetic fields that are due to the motion of free charges. As a rule, we think in terms of conduction currents in wires.

The *magnetic force* exerted *by* a circuit a carrying a current I_a *on* a circuit b carrying a current I_b is

$$\boldsymbol{F}_{ab} = \frac{\mu_0}{4\pi} I_a I_b \oint_a \oint_b \frac{d\boldsymbol{l}_b \times (d\boldsymbol{l}_a \times \boldsymbol{r}_1)}{r^2}, \tag{7-1}$$

where $\boldsymbol{r}_1$ is the unit vector pointing *from* the element $d\boldsymbol{l}_a$ *to* the element $d\boldsymbol{l}_b$. The constant μ_0 is arbitrarily chosen to be exactly $4\pi \times 10^{-7}$ newton/ampere^2.

The magnetic interaction is best thought of as the interaction between the current I_b and the magnetic field of I_a, or inversely. The *magnetic induction* of current I_a is

$$\boldsymbol{B}_a = \frac{\mu_0}{4\pi} I_a \oint_a \frac{d\boldsymbol{l}_a \times \boldsymbol{r}_1}{r^2} \tag{7-8}$$

and is expressed in *teslas* or in *webers/meter*^{2}. For continuous distributions of current,

$$\boldsymbol{B} = \frac{\mu_0}{4\pi} \int_{\tau'} \frac{\boldsymbol{J}_f \times \boldsymbol{r}_1}{r^2}\, d\tau'. \tag{7-11}$$

The *magnetic flux* is defined as

$$\Phi = \int_S \boldsymbol{B} \cdot d\boldsymbol{a}. \tag{7-12}$$

The *magnetic force* acting on a charge Q moving with a velocity $\boldsymbol{v}$ in a magnetic field $\boldsymbol{B}$ is

$$Q\boldsymbol{v} \times \boldsymbol{B}.$$

If there is also an electric field $\boldsymbol{E}$, then the total force is

$$Q(\boldsymbol{E} + \boldsymbol{v} \times \boldsymbol{B}).$$

This is the *Lorentz force.*

Starting from our definition of $\boldsymbol{B}$, we then showed that

$$\boldsymbol{\nabla} \cdot \boldsymbol{B} = 0. \tag{7-34}$$

Then

$$\int_S \boldsymbol{B} \cdot d\boldsymbol{a} = 0, \tag{7-35}$$

where S is any closed surface, and

$$\boldsymbol{B} = \boldsymbol{\nabla} \times \boldsymbol{A}, \tag{7-41}$$

where

$$\boldsymbol{A} = \frac{\mu_0 I}{4\pi} \oint \frac{d\boldsymbol{l}}{r} \tag{7-42}$$

is the *vector potential.* The line integral is evaluated over the circuit carrying the current I.

If there are only conduction currents,

$$\boldsymbol{\nabla} \times \boldsymbol{B} = \mu_0 \boldsymbol{J}_f, \tag{7-61}$$

and then

$$\oint_C \boldsymbol{B} \cdot d\boldsymbol{l} = \mu_0 \int_S \boldsymbol{J}_f \cdot d\boldsymbol{a}. \tag{7-77}$$

This is *Ampère's circuital law.*

At distances large compared to the size of a current loop of area S and carrying a current I, the magnetic induction $\boldsymbol{B}$ has the same form as the $\boldsymbol{E}$ of an electric dipole if we call

$$\boldsymbol{m} = I\, \boldsymbol{S} \tag{7-111}$$

the *magnetic dipole moment* of the loop.

For a continuous current distribution, the magnetic dipole moment is defined as

$$\boldsymbol{m} = \frac{1}{2} \int_\tau \boldsymbol{r} \times \boldsymbol{J}_f \, d\tau. \tag{7-116}$$

PROBLEMS

Note: An asterisk indicates that the problem requires a knowledge of relativity.

7-1. We have seen in Chapter 2 that ϵ_0 is expressed in farads/meter and, in this chapter, that μ_0 is expressed in newtons/ampere^2. Now $\epsilon_0 \mu_0 = 1/c^2$, where c is the velocity of light. Show that this equation is dimensionally correct.

*7-2. In Section 6.1 we calculated the force exerted by a charge Q_a moving at a velocity V on another charge Q_b moving at the same velocity.

Let us set $x_2 = 0$. Then, from Eqs. 6-2 to 6-4, the force exerted on Q_b is

$$\frac{Q_a Q_b}{4\pi\epsilon_0 y_2^2}(1 - \beta^2)^{1/2}\boldsymbol{j}.$$

Now calculate this same force using Coulomb's law and the magnetic force law of Section 7.1, substituting $Q_a\mathcal{V}$ for I_a and $Q_b\mathcal{V}$ for I_b.

Does the magnetic force law give the correct result in this case?

The magnetic force law should never be used to calculate forces between single particles.

7-3. An electric arc is 5.00 centimeters long and carries a current of 400 amperes in a direction perpendicular to a magnetic field of 50 gauss.

Calculate the magnetic force exerted on the arc.

7-4. Calculate the force due to the Earth's magnetic field on a horizontal wire 100 meters long carrying a current of 50 amperes due north (magnetic). Take the magnetic field to be 0.5×10^{-4} tesla at an angle of 70° with the horizontal.

7-5. (a) Find the current density necessary to float a copper wire in the Earth's magnetic field.

Assume the experiment to be done at the Earth's magnetic equator in a field of 10^{-4} tesla. The density of copper is 8.9 grams/centimeter³.

(b) Will the wire become hot?

The resistivity of copper is 1.7×10^{-8} ohm-meter.

7-6. Show that the total force is zero on a closed circuit carrying a current I in a uniform magnetic field.

7-7. Show that the magnetic force on a coil is zero in regions where $\boldsymbol{J}_f$ is parallel to $\boldsymbol{B}$.

7-8. Show that the magnetic induction $\boldsymbol{B}$ can also be written in the form

$$\boldsymbol{B} = \frac{\mu_0}{4\pi}\int_{\tau'}\frac{\boldsymbol{\nabla}' \times \boldsymbol{J}_f}{r}\,d\tau',$$

where τ' is any volume that encloses all the currents $\boldsymbol{J}_f$, and where the operator $\boldsymbol{\nabla}'$ is evaluated at the source point x', y', z', where the current density is $\boldsymbol{J}_f$.‡

7-9. It was stated in the example on page 277 that the magnetic induction due to the current flowing in a straight wire is given by $\mu_0 I/2\pi\rho$ within 1%, if ρ is less than 7% of the length of the wire, and for points near the middle of the wire. Is this correct? Neglect the field of the rest of the circuit.

7-10. Figure 7-22 shows part of a coil whose cross section has the shape of a pair of intersecting circles.

(a) In what general direction is the field oriented in the region between the conductors?

(b) One author states that the magnetic induction is uniform in the hollow. Is this true?

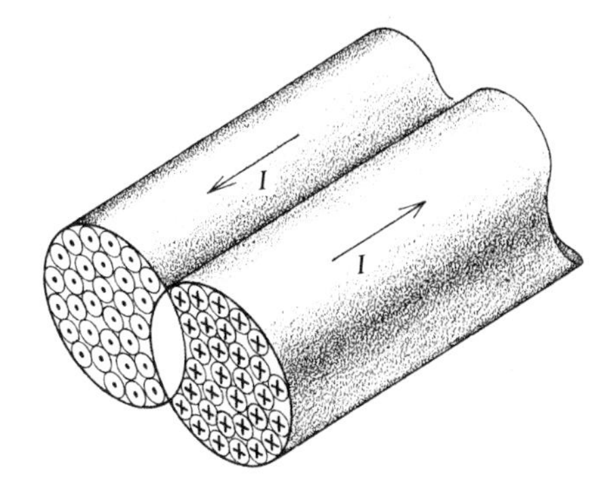

Figure 7-22.

Assume that the length of the coil is infinite.

You can solve this one by straight integration if you wish, but there is a much easier way. . .

7-11. (a) A straight, flat conductor of width $2a$ carries a current I. Show that

$$B_x = -\frac{\mu_0 I}{4\pi a}\alpha,$$

$$B_y = \frac{\mu_0 I}{4\pi a}\ln\frac{r_2}{r_1}$$

in the first quadrant, when the coordinate axes are chosen so that the edges of the conductor are situated at $x = \pm a$, and when the current flows in the direction of the positive z-axis.

The distance from the point where B is measured to the edge at $x = a$ is r_1. The angle α is that between r_1 and r_2 and is positive in the direction from r_1 to r_2.

(b) Calculate B_x, B_y, B at a distance of 26.0 centimeters from the axis of the conductor at an angle of 72.0° from the x-axis in the first quadrant for a strip 10 centimeters wide carrying a current of 5.76 amperes.

(c) Find the magnetic induction $\boldsymbol{B}$ (*i*) at an external point in the plane of the strip at a distance D from its axis and (*ii*) at an external point in the plane perpendicular to the strip and at the same distance D from its axis.

(d) How do the results of (*i*) and (*ii*) compare when $D \gg a$? Explain.

7-12. A long straight conductor has a circular cross section of radius R and carries a current I. Inside the conductor, there is a cylindrical hole of radius a whose axis is parallel to the axis of the conductor and at a distance b from it.

Show that the magnetic induction inside the hole is uniform and is equal to

$$\frac{\mu_0 b I}{2\pi(R^2 - a^2)}.\ddagger$$

7-13. Use the Biot-Savart law to compute the magnetic induction $\boldsymbol{B}$ at the center of a square current loop carrying a current I.

7-14. A short solenoid carries a current I and has N' turns/meter.

(a) Show that, at any point on the axis,

$$B = \frac{1}{2}\mu_0 IN'(\cos\alpha_1 + \cos\alpha_2),$$

where α_1 and α_2 are the angles subtended at the point by a radius R at either end of the solenoid.

For example, if the coil has a length $2L$ and if the point is situated at a distance x from the center,

$$\cos\alpha_1 = \frac{L - x}{[R^2 + (L - x)^2]^{1/2}}, \qquad \cos\alpha_2 = \frac{L + x}{[R^2 + (L + x)^2]^{1/2}}.$$

(b) Plot a curve of $2B/\mu_0 IN'$ as a function of x/R for $L = 10R$ and for values of x/R ranging from 0 to 15.

Note how quickly the field drops off outside the solenoid.

7-15. A solenoid has an inner radius R_1, an outer radius R_2, and a length L.

(a) Show that the magnetic induction at the center, when the solenoid carries a current I, is

$$B = \frac{\mu_0 nIL}{2}\ln\left\{\frac{\alpha + (\alpha^2 + \beta^2)^{1/2}}{1 + (1 + \beta^2)^{1/2}}\right\},$$

where

$$\alpha = \frac{R_2}{R_1}, \qquad \beta = \frac{L}{2R_1},$$

and n is the number of turns per square meter of the cross-sections.

(b) Show that, if V is the volume of the winding, the length of the wire is

$$l = Vn = 2\pi n(\alpha^2 - 1)\beta R_1^3.$$

(c) Fabry has shown that, at the center of *any* solenoid,

$$B = G\left[\frac{P\lambda}{\rho R_1}\right]^{1/2},$$

where G is a factor which depends on the geometry of the solenoid, P is the power dissipated in the solenoid, $\lambda = n\pi r^2$ (where r is the radius of the conductor) is the filling factor (the fraction of the coil cross section that is occupied by the conductor), and ρ is the resistivity of the conductor.

Check the Fabry equation for the above type of solenoid.

7-16. The Zeeman effect observed in the spectra of sunspots reveals the existence of magnetic fields as large as 0.4 tesla.

Let us assume that the magnetic field is due to a disk of electrons 10^7 meters in radius rotating at an angular velocity of 3×10^{-2} radian/second. The thickness of the disk is small compared to its radius.

(a) Show that the density of electrons required to achieve a B of 0.4 tesla is about 10^{19}/meter^2.

(b) Show that the current is about 3×10^{12} amperes.

(c) In view of the enormous size of the Coulomb forces, such charge densities are clearly impossible. Then how could such currents exist?

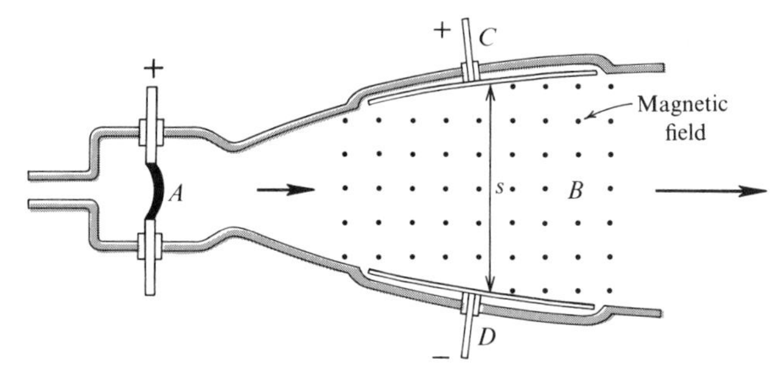

Figure 7-23.

7-17. (a) Compute the period and radius of curvature of the path of an electron moving in a plane perpendicular to the Earth's magnetic field. Assume

$B = 10^{-4}$ tesla,
$e = 1.6 \times 10^{-19}$ coulomb,
$m = 9.1 \times 10^{-31}$ kilogram,
electron energy = 3000 electron-volts,
one electron volt = 1.6×10^{-19} joule.

*(b) Repeat the calculation for an electron having a kinetic energy of 3.0 million electron-volts.

7-18. In the parallel-plate magnetron, the cathode and the anode are flat parallel plates, and a magnetic field is applied in a direction parallel to the plates. Electrons are emitted from the cathode at essentially zero velocity.

If the anode is at a distance s from the cathode, and if it is held at a potential V with respect to the cathode, show that for

$$V \leq \frac{eB^2s^2}{2m}$$

no current will flow to the anode. The magnetic induction is B, e is the magnitude of the electronic charge, and m is the mass of an electron.‡

The parallel-plate magnetron is apparently not a useful device, but certain types of crossed-field amplifiers use a fairly similar geometry.*

7-19. A magnetron consists of an electron-emitting filament at the center of a cylindrical anode situated in a uniform axial magnetic field. Electrons of charge e and mass m are emitted with negligible velocity from the filament.

Find the minimum potential difference between filament and anode for electrons to reach the anode.‡

7-20. Figure 7-23 shows one type of rocket motor that utilizes the magnetic force. An arc A ionizes the gas, which enters from the left, and the ions are blown into the crossed electric and magnetic fields. If the current flowing between

* See, for example, A. F. Harvey, *Microwave Engineering* (Academic Press, New York, 1963), Chapter 12.

the electrodes C and D is I, then the thrust F developed by the motor is BIs, if the values of B and s are assumed to be constant throughout the thrust chamber, and if end effects are neglected.

If m' is the mass of gas flowing through the motor per unit time, and if v is the exhaust velocity, then F is $m'v$, and the kinetic power communicated to the gas is

$$W_G = \tfrac{1}{2}m'v^2 = \tfrac{1}{2}Fv = \tfrac{1}{2}BIsv.$$

If the efficiency of the motor is defined as

$$\eta = \frac{W_G}{W_G + W_D},$$

where W_D is the power dissipated as heat between C and D, show that

$$\eta = \frac{1}{1 + \dfrac{2m'}{\sigma B^2\tau}} = \frac{1}{1 + \dfrac{2J}{\sigma Bv}} = \frac{1}{1 + \dfrac{2E}{Bv}},$$

where σ is the electrical conductivity, τ is the volume of the plasma in the crossed fields, J is the current density, and E is the transverse electric field produced by the electrodes C and D.

7-21. When a direct current flows in a long straight conductor, the current density may be assumed to be the same throughout the cross section, unless the current is subjected to a magnetic field originating elsewhere. Let us see how this comes about.

(a) Calculate the magnetic force on a conduction electron drifting at the surface of a wire 1 millimeter in radius and carrying a current of 1 ampere.

Set the electron velocity equal to 10^{-4} meter/second.

(b) Calculate the radial electric field intensity required to cancel the magnetic force at a radius r inside the wire, and find the corresponding volume charge density.

(c) How many extra electrons are required per meter to supply this field, and how does this compare with the number of conduction electrons per meter?

(d) Is it reasonable to assume that the current density is uniform throughout the cross section of the wire?

7-22. A solenoid has a uniform turn density, except near the ends, where extra turns are added to obtain a higher magnetic induction than near the center.

Show qualitatively that, if the axial velocity is not too large, a charged particle that spirals along the axis in a vacuum inside the solenoid will be reflected back when it reaches the higher magnetic field.

The regions of higher magnetic field are called *magnetic mirrors*.

*7-23. In 1949 Fermi proposed the following mechanism, now called *Fermi acceleration*, to explain the existence of very high energy particles in space.

Imagine a clump of plasma traveling in space at some velocity $v_c\mathbf{i}$. The plasma carries a current, and thus has a magnetic field. Imagine now a particle traveling in the opposite direction at a velocity $-v_a\mathbf{i}$, both v_c and v_a being positive quantities. The particle is deflected in the magnetic field and acquires a velocity $v_b\mathbf{i}$, where v_b is being also a positive quantity.

Set $v_c = v_a = c/2$. Show that, after reflection, the γ of the particle has increased from 1.15 to 2.69.

See also Problem 5-25.

7-24. A sheet of dielectric situated in the *x-y* plane has a velocity $v\mathbf{i}$ in a region where the magnetic induction is $B\mathbf{k}$. Find the polarization $\mathbf{P}$ and the surface charge densities σ_b.

7-25. A current I flows in a wire bent around a square form measuring $2a$ meters on a side.

(a) Calculate the vector potential $\mathbf{A}$ and the magnetic induction $\mathbf{B}$ along an axis passing through the center of the square and parallel to one of the sides.
(b) Draw curves of A and B as functions of distance from the center of the square when I is 1.00 ampere and a is 10.0 centimeters.

7-26. (a) Calculate the axial component of the vector potential $\mathbf{A}$ at the center of a helix of $2N$ turns, of radius R, and of length $2H$, carrying a current I.
(b) Show that the result is the same as that for a single wire of length $2H$ along the side of the helix that carries a current I.
Why is this so?
(c) Is it possible to use this result to calculate the axial component of $\mathbf{B}$ at the center?
(d) Show that the axial component of $\mathbf{B}$ at the center of the helix is

$$\frac{\mu_0 IN}{(R^2 + H^2)^{1/2}}.$$

7-27. A conducting sheet carries a current density of $\boldsymbol{\lambda}$ amperes/meter.

(a) Show that, very close to the sheet, the magnetic induction due to the current in the sheet is $\mu_0\lambda/2$ in a direction perpendicular to $\boldsymbol{\lambda}$.
*(b) Show that the $\mathbf{B}$ given in the example on page 263 in the case of the moving capacitor agrees with the above result.

7-28. In a Van de Graaff high-voltage generator, a charged insulating belt is used to transport electric charges to the high voltage electrode.

(a) Calculate the current that can be carried by a belt 50.0 centimeters wide that is driven by a pulley 10.0 centimeters in diameter and that rotates at 60 revolutions per second, if the electric field intensity at the surface of the belt is 20 kilovolts/centimeter.
(b) Calculate the magnetic induction close to the surface of the belt, neglecting edge effects.

7-29. A circular loop whose axis is coincident with the *z*-axis carries a current I. (a) Calculate $\partial B_\rho/\partial\rho$ at a point on the axis, (b) B_ρ in the neighborhood of the axis, (c) $\partial B_z/\partial\rho$ at a point on the axis, and (d) B_z in the neighborhood of the axis.‡

7-30. Show that the permeability of free space μ_0 can be defined as follows: if an infinitely long solenoid carries a current density of 1.0 ampere/meter, then the magnetic induction in teslas inside the solenoid is numerically equal to μ_0.

*7-31. According to Ampère's circuital law, the magnetic induction at a distance ρ from the axis of a beam of electrons carrying a current I is $\mu_0 I/2\pi\rho$, and is independent of the manner in which the current density J varies with ρ inside the beam. (It is assumed that J is not a function of the azimuthal angle θ, and that ρ is larger than the radius of the beam.)

Show that this is a consequence of Gauss's law.

7-32. A *magnetic bottle* is a field which has the property of containing a plasma.

One type of magnetic bottle operates as follows. An evacuated tube several meters long and about half a meter in diameter is enclosed in a long solenoid that produces a uniform magnetic field. An accelerator injects electrons, which spiral inside the tube, forming in effect a second winding whose field opposes that of the solenoid. Magnetic mirrors (regions of stronger magnetic field; see Problem 7-22) at both ends of the tube reflect the electrons, causing them to spiral back and forth along the length of the tube, thus increasing the magnetic field of the electrons which becomes larger than that of the solenoid.

(a) Sketch a graph of B as a function of radius.
(b) Sketch lines of $\mathbf{B}$ in a plane containing the axis of the solenoid.
(c) Show, qualitatively, that the magnetic field prevents the electrons from straying away from the current sheet.
(d) The electrons ionize the residual gas in the tube. Sketch the path of a low-energy proton formed by the collision of one of the high-energy electrons with a molecule or an atom of hydrogen.

7-33. A conducting sphere of radius R is charged to a potential V and spun about a diameter at an angular velocity ω.

(a) Show that the surface current density is

$$\lambda = \epsilon_0 \omega V \sin\theta = M \sin\theta,$$

where M is $\epsilon_0 \omega V$.
(b) Show that the magnetic induction at the center is

$$B_0 = \frac{2}{3}\frac{V\omega}{c^2} = \frac{2}{3}\mu_0 M.$$

(c) What is the numerical value of this magnetic induction for a sphere 10.0 centimeters in radius, charged to 10.0 kilovolts, and spinning at 1.00×10^4 turns/minute.
(d) Show that the dipole moment is

$$\tfrac{4}{3}\pi R^3 M\, \mathbf{k},$$

where $\mathbf{k}$ is a unit vector along the axis and is related to the direction of rotation by the right-hand screw rule.
(e) What is the dipole moment of a sphere as in (c) above?
(f) What current flowing through a loop 10.0 centimeters in diameter would give the same dipole moment?

7-34. An electron revolves in a circular orbit with an angular momentum $2^{1/2}\hbar$ about a fixed proton. The constant $\hbar = 1.05 \times 10^{-34}$ joule-second is Planck's constant divided by 2π.

(a) Calculate the magnetic moment in terms of $\hbar$.
(b) Calculate B at the position of the proton.

Hints

7-8. Use the results of Problems 1-12 and 1-30. The current density $\boldsymbol{J}_f$ must be zero everywhere on the surface of τ'.

7-12. Imagine that the hole is filled with a conductor carrying current of the same density. Use the circuital law to find $\boldsymbol{B}$. Then imagine that another current of the same density but of opposite direction is superposed in the space occupied by the hole. Superpose the two magnetic fields.

7-18. Use the Lorentz force and remember that its magnetic component does no work. Thus the kinetic energy of an electron is equal to the potential energy it has lost in the electric field. If the y-axis is normal to the plates and if the origin is at the surface of the cathode,

$$\frac{1}{2} mu^2 = e \frac{V}{s} y,$$

and, for the above critical value of v, $u_y = 0$ at $y = s$.

7-19. Use the conservation of energy and Newton's second law in the form Torque = Rate of Change of Angular Momentum. Use polar coordinates.

7-29. Use the divergence and curl properties of $\boldsymbol{B}$.

CHAPTER 8

MAGNETIC FIELDS II

Induced Electromotance and Magnetic Energy

In the last chapter we discussed the magnetic fields of steady currents; we shall now study time-dependent magnetic fields and the nonconservative electric fields which accompany them. We shall then be able to calculate the energy stored in a magnetic field, as well as magnetic forces, torques, and pressures.

We shall continue to disregard magnetic materials. *As usual, you may neglect references to Chapter 6 if you have not worked through it.*

8.1 THE FARADAY INDUCTION LAW

We have seen in Section 2.3 that an electrostatic field is conservative, or that

$$\oint \boldsymbol{E} \cdot d\boldsymbol{l} = 0. \tag{8-1}$$

Thus the work performed by electro*static* forces is zero when a charge moves around a closed path.

Equation 8-1 is not applicable if the path is linked by a changing magnetic flux. Let us consider a simple path having the shape of a loop. Then, if Φ is the magnetic flux linking the loop,

$$\oint \boldsymbol{E} \cdot d\boldsymbol{l} = -\frac{d\Phi}{dt}, \tag{8-2}$$

if the direction in which the flux is taken to be positive is related to the direc-

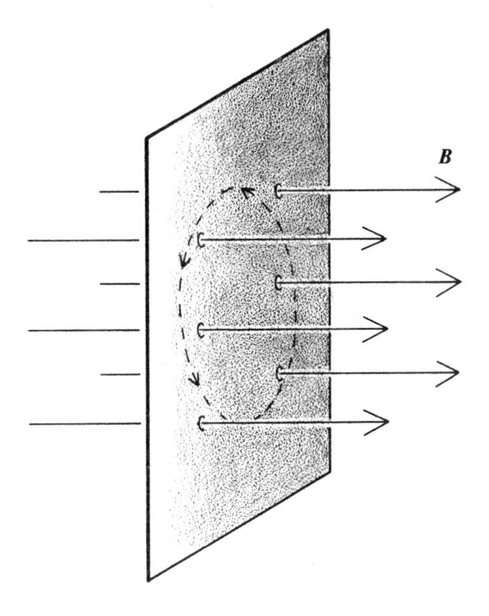

Figure 8-1. The positive direction around an integration path in a magnetic field $\boldsymbol{B}$. The positive direction around the path and the positive direction for $\boldsymbol{B}$ are related according to the right-hand screw rule.

tion in which the line integral is evaluated according to the right-hand screw rule, as in Figure 8-1. This is the *Faraday induction law;* the line integral of $\boldsymbol{E} \cdot \boldsymbol{dl}$ around the path is called the *induced electromotance.*

If there are no sources in a circuit, the current is equal to the induced electromotance divided by the resistance of the circuit, exactly as if the electromotance were replaced by a battery of the same voltage and polarity. The induced electromotance is expressed in volts, and it adds algebraically to the voltages of the other sources that may be present in the circuit.

The above equation can be rewritten as

$$\oint \boldsymbol{E} \cdot \boldsymbol{dl} = -\frac{d}{dt} \int_S \boldsymbol{B} \cdot \boldsymbol{da}, \tag{8-3}$$

where S is any surface bounded by the path of integration chosen for the line integral. This path can be chosen at will, and it need not lie in conducting material.

If the path of integration is not just a simple loop, the surface over which the magnetic induction must be integrated to obtain the flux linkage can be complicated. The procedure for a solenoid is illustrated in Figure 8-2.

If the lines of $\boldsymbol{B}$ are parallel to the axis of the solenoid, they cross the spiral ramp once for each turn of the solenoid, and the total flux crossing the spiral surface, of the *flux linkage*, is N times the magnetic flux crossing the surface corresponding to a single turn. The lines of $\boldsymbol{B}$ are then said to link all N turns. If the lines of $\boldsymbol{B}$ are not parallel, the flux linkage varies from turn to

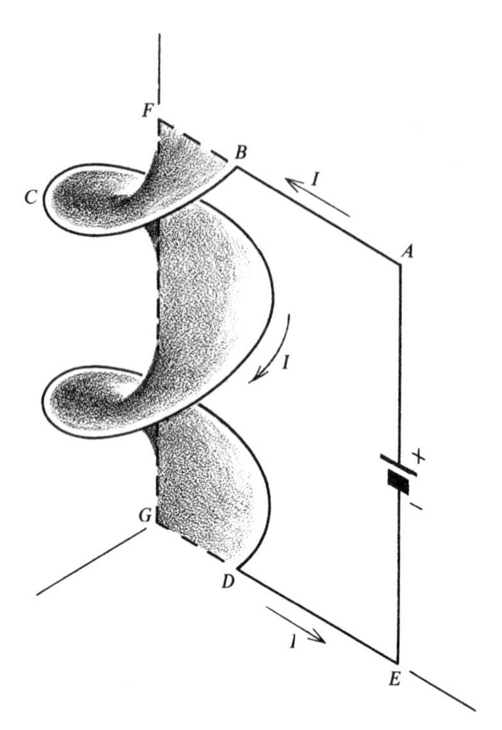

Figure 8-2. Solenoid *BCD* connected to a source. To calculate the flux linkage, the circuit is replaced by *ABFGDE*, in which the current flows downward along the dotted part, plus *BCDGFB*, in which the current flows upward in the dotted part. The flux linkage for the real circuit is then the sum of the flux linkages through the spiral ramp and through the surface *AFGE*.

turn and the electromotance induced in the solenoid is equal to minus the rate of change of the total flux linkage.

In general, therefore, the quantity Φ in Eq. 8-2 must be interpreted as the flux linkage. Flux linkage is measured in weber-turns.

Equation 8-2 is correct, no matter what the origin of the changing flux may be: (1) the current producing the flux may change with time, (2) it may move relative to the closed path around which the electromotance is calculated, or (3) the path, or parts of it, may move relative to the flux-producing circuit.

Induced electromotance due to the motion of all or part of a circuit in a fixed magnetic field is known as *motional-induced electromotance*. It is a direct consequence of the magnetic force (Sections 6.2 and 7.1), which itself follows from the magnetic force law discussed at the beginning of the last chapter.

Transformer-induced electromotance is observed in a rigid, fixed circuit when it is linked by a variable magnetic flux. The distinction between motional and transformer electromotance is artificial in some cases, in that different observers would identify them differently. They would always agree, however, on the electromotance. This will be shown in Appendix C.

If the flux linkage Φ *in*creases, $d\Phi/dt$ is positive, and the electromotance is negative, that is, the induced electromotance is in the negative direction. On the other hand, if Φ *de*creases, $d\Phi/dt$ is negative, and the electromotance is in the positive direction.

The direction of the induced current is always such that it produces a magnetic field that opposes, to a greater or lesser extent, the *change* in flux, depending on the resistance in the circuit. Thus, if Φ increases, the induced current produces an opposing flux. If Φ decreases, the induced current produces an aiding flux. This is *Lenz's law*.

Example THE EXPANDING LOOP

Figure 8-3 shows a closed rectangular circuit, one side of which can slide parallel to itself with a velocity $\boldsymbol{u}$ in a region of uniform mag-

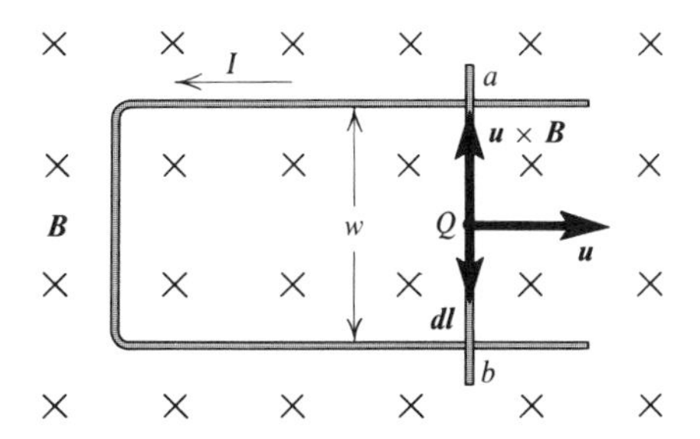

Figure 8-3. A conducting wire ab slides with a velocity $\boldsymbol{u}$ along conducting rails in a region of uniform magnetic induction $\boldsymbol{B}$. The magnetic force on the electrons in the wire produces a current I in the circuit. The electronic charge is taken to be Q.

netic induction $\boldsymbol{B}$. A free charge Q in the moving wire then experiences a magnetic force $Q(\boldsymbol{u} \times \boldsymbol{B})$ (Section 7.3).

The vector $\boldsymbol{u} \times \boldsymbol{B}$ is called the *induced electric field intensity*. It is the magnetic force exerted on a unit charge.

The line integral of the induced electric field intensity around the circuit is not zero. Taking the line integral in the clockwise direction, and $\boldsymbol{B}$ in the direction shown in the figure,

$$\oint \boldsymbol{E}\cdot d\boldsymbol{l} = \oint (\boldsymbol{u} \times \boldsymbol{B})\cdot d\boldsymbol{l} = \int_a^b (\boldsymbol{u} \times \boldsymbol{B})\cdot d\boldsymbol{l} \tag{8-4}$$

$$= -wuB. \tag{8-5}$$

We have assumed that the resistance of the loop is so large that the current flowing through it has a negligible effect on the value of $\boldsymbol{B}$.

The right-hand side is the area swept by the wire per unit time multiplied by the magnetic induction B. It is thus the magnetic flux swept per unit time as in Eq. 8-2.

If we had assumed that two or more sides of the circuit moved in the magnetic field, the end result would have been the same; the induced electromotance is always $-d\Phi/dt$ in such cases.

8.1.1 The Faraday Induction Law in Differential Form

The Faraday law as stated in Eq. 8-2 says nothing about $\boldsymbol{E}$ at a given point; it only gives the electromotance in a complete circuit.

Let us state this law in differential form. Using Stokes's theorem to transform the line integral into a surface integral,

$$\int_S (\nabla \times \boldsymbol{E})\cdot d\boldsymbol{a} = -\frac{d}{dt}\int_S \boldsymbol{B}\cdot d\boldsymbol{a}, \tag{8-6}$$

where S is any surface bounded by the closed integration path. If this path is fixed in space, we may interchange the order of differentiation and integration on the right-hand side and

$$\int_S (\nabla \times \boldsymbol{E})\cdot d\boldsymbol{a} = -\int_S \frac{\partial \boldsymbol{B}}{\partial t}\cdot d\boldsymbol{a}. \tag{8-7}$$

We have used the partial derivative of $\boldsymbol{B}$ because we now require the rate of change of $\boldsymbol{B}$ with time at a fixed point. In a later section we shall examine the more general case in which the path moves.

Since the above equation is valid for arbitrary surfaces, the integrands must be equal at every point, and as in Eq. 6-127,

$$\boxed{\nabla \times \boldsymbol{E} = -\frac{\partial \boldsymbol{B}}{\partial t}.} \tag{8-8}$$

This is one of the four Maxwell equations. The only other one that we found without using relativity, was Eq. 3-41. All four equations were discussed in Chapter 6.

Equation 8-8 is a differential equation that relates the space derivatives of $\boldsymbol{E}$ at a particular point to the time rate of change of $\boldsymbol{B}$ at the same point. The equation does *not* give the value of $\boldsymbol{E}$, unless it can be integrated.

8.2 THE INDUCED ELECTRIC FIELD INTENSITY IN TERMS OF THE VECTOR POTENTIAL *A*

The electric field intensity $\boldsymbol{E}$ induced by a changing magnetic field is related to the vector potential $\boldsymbol{A}$ of Sections 6.4 and 7.5. Since

$$\boldsymbol{B} = \boldsymbol{\nabla} \times \boldsymbol{A}, \tag{8-9}$$

then, from Eq. 8-8,

$$\boldsymbol{\nabla} \times \boldsymbol{E} = -\frac{\partial}{\partial t}(\boldsymbol{\nabla} \times \boldsymbol{A}), \tag{8-10}$$

$$= -\boldsymbol{\nabla} \times \frac{\partial \boldsymbol{A}}{\partial t} \tag{8-11}$$

or

$$\boldsymbol{\nabla} \times \left(\boldsymbol{E} + \frac{\partial \boldsymbol{A}}{\partial t}\right) = 0. \tag{8-12}$$

The term between parentheses must equal a quantity whose curl is zero, namely a gradient. Then

$$\boldsymbol{E} = -\frac{\partial \boldsymbol{A}}{\partial t} - \boldsymbol{\nabla} V. \tag{8-13}$$

For steady currents, $\boldsymbol{A}$ is a constant and Eq. 8-13 reduces to Eq. 2-13. The quantity V is therefore the electric potential of Section 2.3, which is also known as the *scalar potential.*

Equation 8-13 is the general expression for $\boldsymbol{E}$, which we have already found in Section 6.5. This equation states that an electric field intensity can arise both from accumulations of charge, through the $-\boldsymbol{\nabla} V$ term, or from changing magnetic fields, through the $-\partial \boldsymbol{A}/\partial t$ term.

We could also have found Equation 8-12 in the following way. According to the Faraday induction law, the induced electromotance for any given path is

$$\oint \boldsymbol{E} \cdot \boldsymbol{dl} = -\frac{d\Phi}{dt} = -\frac{d}{dt} \oint \boldsymbol{A} \cdot \boldsymbol{dl}, \tag{8-14}$$

from Section 7.5.1. Interchanging the order of differentiation and of integration on the right, which we may do since we are considering fixed paths,

$$\oint \left(\boldsymbol{E} + \frac{\partial \boldsymbol{A}}{\partial t} \right) \cdot \boldsymbol{dl} = 0. \tag{8-15}$$

We have used a partial derivative under the integral sign because we require the time derivative of $\boldsymbol{A}$ at a given point on the path. Thus

$$\oint \left(\boldsymbol{E} + \frac{\partial \boldsymbol{A}}{\partial t} \right) \cdot \boldsymbol{dl} = \int_S \boldsymbol{\nabla} \times \left(\boldsymbol{E} + \frac{\partial \boldsymbol{A}}{\partial t} \right) \cdot \boldsymbol{da} = 0 \tag{8-16}$$

over any fixed closed curve bounding the surface S. The integrand on the right is therefore zero.

Example

THE ELECTROMOTANCE INDUCED IN A LOOP BY A PAIR OF LONG PARALLEL WIRES CARRYING A VARIABLE CURRENT

A pair of parallel wires, as in Figure 8-4, carries equal currents I in opposite directions, and I increases at the rate dI/dt. We shall first calculate the induced electromotance from Eq. 8-2 and then from Eq. 8-13.

The current I in wire a produces a magnetic induction

$$B_a = \frac{\mu_0 I}{2\pi \rho_a}, \tag{8-17}$$

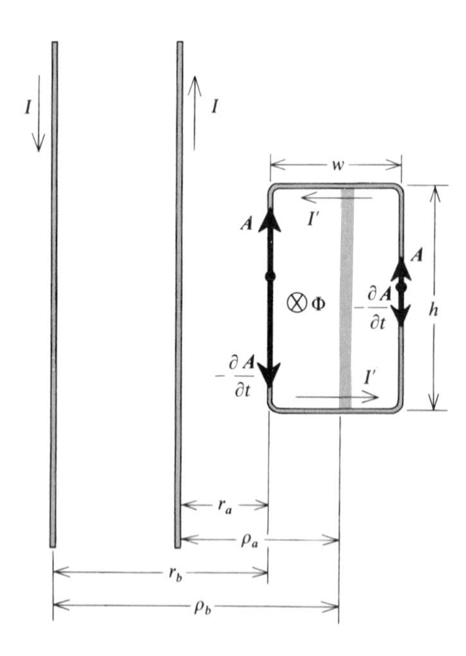

Figure 8-4. Pair of parallel wires carrying equal currents I in opposite directions in the plane of a closed rectangular loop of wire. When I increases, the induced electromotance gives rise to a current I' in the direction shown. The vector potentials $\boldsymbol{A}$ and the induced electric fields $-\partial \boldsymbol{A}/\partial t$ are shown on the vertical wires. The induced current I' flows in the counterclockwise direction because $-\partial \boldsymbol{A}/\partial t$ is larger on the left than on the right.

and a similar relation exists for wire b. The flux through the loop is thus

$$\Phi = \frac{\mu_0 I}{2\pi}\left(\int_{r_a}^{r_a+w} \frac{h\,d\rho_a}{\rho_a} - \int_{r_b}^{r_b+w} \frac{h\,d\rho_b}{\rho_b}\right), \tag{8-18}$$

$$= \frac{\mu_0 h I}{2\pi} \ln\left[\frac{r_b(r_a + w)}{r_a(r_b + w)}\right], \tag{8-19}$$

and it points into the paper as in Figure 8-4. From Eq. 8-2, the induced electromotance is $-\partial\Phi/\partial t$:

$$\oint \boldsymbol{E}\cdot d\boldsymbol{l} = -\frac{\mu_0 h}{2\pi}\frac{dI}{dt} \ln\left[\frac{r_b(r_a + w)}{r_a(r_b + w)}\right]. \tag{8-20}$$

The fact that the line integral is negative for a positive dI/dt indicates that the induced current I' is in the negative direction with respect to $\boldsymbol{B}$. Since $\boldsymbol{B}$ is directed into the paper in Figure 8-4, I' must flow in the counterclockwise direction. This is in agreement with Lenz's law: the current I' produces a magnetic field that opposes the increase in Φ.

Let us now use Eq. 8-13 to calculate this same electromotance from the time derivative of the vector potential $\boldsymbol{A}$, V being equal to zero. From Section 7.5, $\boldsymbol{A}$ is parallel to the wires. Choosing the upward direction as positive,

$$A_L = \frac{\mu_0 I}{2\pi} \ln \frac{r_b}{r_a}, \tag{8-21}$$

$$A_R = \frac{\mu_0 I}{2\pi} \ln\left(\frac{r_b + w}{r_a + w}\right) \tag{8-22}$$

along the left- and right-hand sides of the loop, respectively. Thus

$$E_L = -\frac{\mu_0}{2\pi}\frac{dI}{dt} \ln \frac{r_b}{r_a}, \tag{8-23}$$

$$E_R = -\frac{\mu_0}{2\pi}\frac{dI}{dt} \ln\left(\frac{r_b + w}{r_a + w}\right), \tag{8-24}$$

and

$$\oint \boldsymbol{E}\cdot d\boldsymbol{l} = \frac{\mu_0 h}{2\pi}\frac{dI}{dt} \ln\left[\frac{r_a(r_b + w)}{r_b(r_a + w)}\right]. \tag{8-25}$$

To find the electromotance induced in the loop by a changing current in a single conductor, we set $r_b \to \infty$, and then

$$\oint \boldsymbol{E}\cdot d\boldsymbol{l} = \frac{\mu_0 h}{2\pi}\frac{dI}{dt} \ln\left(\frac{r_a}{r_a + w}\right). \tag{8-26}$$

8.3 INDUCED ELECTROMOTANCE IN A MOVING SYSTEM

Our differential form of the Faraday law, Eq. 8-8, was limited to systems at rest. We shall now consider moving systems.

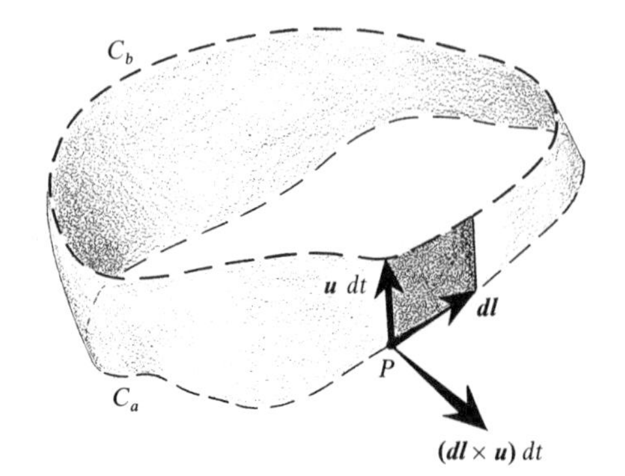

Figure 8-5. A path of integration moves from C_a to C_b in the time dt. The displacement is general and involves a translation, a rotation, and a distortion. The point P is assumed to move with a velocity $\boldsymbol{u}$ in a region where the magnetic induction is $\boldsymbol{B}$.

We return to the integral form of the Faraday law, Eq. 8-2, and consider the rate of change of magnetic flux through a surface bounded by a path C that moves in some arbitrary manner from C_a to C_b in the time dt, as in Figure 8-5. A given point P on the path C moves with a velocity $\boldsymbol{u}$ through a region where the magnetic induction depends both on the coordinates and on time.

The rate of change of flux through the circuit is

$$\frac{d\Phi}{dt} = \frac{\int_{S_b} \boldsymbol{B}_b(t + dt)\cdot d\boldsymbol{a}_b - \int_{S_a} \boldsymbol{B}_a(t)\cdot d\boldsymbol{a}_a}{dt}, \tag{8-27}$$

where $\boldsymbol{B}_b(t + dt)$ is the magnetic induction on some surface S_b bounded by C_b at the time $t + dt$. Similarly, $\boldsymbol{B}_a(t)$ is the magnetic induction on S_a bounded by C_a at the time t.

The magnetic flux emerging *at the time* $t + dt$ from the volume τ swept out during the interval t to $t + dt$ is

$$\int_{S_b} \boldsymbol{B}_b(t + dt)\cdot d\boldsymbol{a}_b - \int_{S_a} \boldsymbol{B}_a(t + dt)\cdot d\boldsymbol{a}_a + dt \int_{C_a} \boldsymbol{B}(t + dt)\cdot(d\boldsymbol{l} \times \boldsymbol{u})$$

$$= \int_\tau \boldsymbol{\nabla}\cdot\boldsymbol{B}(t + dt)\, d\tau = 0, \tag{8-28}$$

since the divergence of $\boldsymbol{B}$ is always zero. We have used $\boldsymbol{B}$ without a subscript for the magnetic induction along the path of integration C_a.

Now, on the surface S_a,

$$\boldsymbol{B}_a(t + dt)\cdot d\boldsymbol{a}_a = \boldsymbol{B}_a(t)\cdot d\boldsymbol{a}_a + \frac{\partial}{\partial t}(\boldsymbol{B}_a\cdot d\boldsymbol{a}_a)\, dt, \tag{8-29}$$

where the time derivative is evaluated at the time t.

Also, from Stokes's theorem,

$$\oint_C \boldsymbol{B}(t+dt)\cdot(\boldsymbol{dl}\times\boldsymbol{u}) = \oint_C [\boldsymbol{u}\times\boldsymbol{B}(t+dt)]\cdot\boldsymbol{dl}, \tag{8-30}$$

$$= \int_{S_a} \boldsymbol{\nabla}\times(\boldsymbol{u}\times\boldsymbol{B})\cdot\boldsymbol{da}_a. \tag{8-31}$$

Substituting 8-29 and 8-31 into 8-28,

$$\int_{S_b} \boldsymbol{B}_b(t+dt)\cdot\boldsymbol{da}_b - \int_{S_a}\boldsymbol{B}_a(t)\cdot\boldsymbol{da}_a - dt\int_{S_a}\frac{\partial \boldsymbol{B}_a}{\partial t}\cdot\boldsymbol{da}_a + dt\int_{S_a}\boldsymbol{\nabla}\times(\boldsymbol{u}\times\boldsymbol{B})\cdot\boldsymbol{da}_a = 0, \tag{8-32}$$

and, finally,

$$\frac{d\Phi}{dt} = -\int_S \boldsymbol{\nabla}\times(\boldsymbol{u}\times\boldsymbol{B})\cdot\boldsymbol{da} + \int_S \frac{\partial \boldsymbol{B}}{\partial t}\cdot\boldsymbol{da}. \tag{8-33}$$

The subscript on S is now unnecessary, since we have really calculated $\Delta\Phi/\Delta t$ in the limit $\Delta t \to 0$. The two terms represent respectively: (a) the flux gained through the sides of the volume traced out by the moving path, and (b) the increase of flux by virtue of the change of $\boldsymbol{B}$ with time.

Thus, from the Faraday induction law,

$$\oint_C \boldsymbol{E}\cdot\boldsymbol{dl} = \int_S\left[\boldsymbol{\nabla}\times(\boldsymbol{u}\times\boldsymbol{B}) - \frac{\partial \boldsymbol{B}}{\partial t}\right]\cdot\boldsymbol{da}, \tag{8-34}$$

or, using Stokes's theorem,

$$\int_S \boldsymbol{\nabla}\times\boldsymbol{E}\cdot\boldsymbol{da} = \int_S\left[\boldsymbol{\nabla}\times(\boldsymbol{u}\times\boldsymbol{B}) - \frac{\partial \boldsymbol{B}}{\partial t}\right]\cdot\boldsymbol{da}. \tag{8-35}$$

Since this equation must be valid for any surface S bounded by any curve C, the integrands must be equal at every point, and

$$\boldsymbol{\nabla}\times\boldsymbol{E} = -\frac{\partial \boldsymbol{B}}{\partial t} + \boldsymbol{\nabla}\times(\boldsymbol{u}\times\boldsymbol{B}). \tag{8-36}$$

This $\boldsymbol{E}$ is the induced electric field intensity as measured in a system moving with a velocity $\boldsymbol{u}$ relative to that in which the magnetic induction is measured as $\boldsymbol{B}$. For example, if $\boldsymbol{E}$ is induced in a moving conductor, $\boldsymbol{u}$ is the velocity of the conductor relative to the laboratory, and $\boldsymbol{B}$ is measured with an instrument that is fixed with respect to the laboratory.*

---

* See Appendix C for a detailed discussion of Eq. 8-36.

Example

THE ELECTROMOTANCE INDUCED IN A FIXED LOOP IN A TIME-DEPENDENT MAGNETIC FIELD

We consider the square loop of Figure 8-6. Let us first suppose that the loop is at rest. There is a uniform magnetic field $\boldsymbol{B}$, which is parallel to the z-axis and which varies with time:

$$B = B_0 \sin \omega t. \tag{8-37}$$

Since the loop is at rest, $\boldsymbol{u} = 0$,

$$\boldsymbol{\nabla} \times \boldsymbol{E} = -\frac{\partial \boldsymbol{B}}{\partial t}, \tag{8-38}$$

$$\oint \boldsymbol{E} \cdot d\boldsymbol{l} = \int_S (\boldsymbol{\nabla} \times \boldsymbol{E}) \cdot d\boldsymbol{a} = -\int_S \frac{\partial \boldsymbol{B}}{\partial t} \cdot d\boldsymbol{a}, \tag{8-39}$$

$$= -B_0 \omega S \cos \theta \cos \omega t, \tag{8-40}$$

where $S = Wh$ is the area of the loop.

For a square loop 10×10 centimeters² normal to a magnetic field varying at a frequency of 60 hertz with a maximum value of 0.01 tesla (100 gauss), the induced electromotance has a frequency of 60 hertz and a maximum value of about 38 millivolts/turn. At any moment it tends to oppose the *change* in flux.

Example

THE ELECTROMOTANCE INDUCED IN A LOOP ROTATING IN FIXED MAGNETIC FIELD

Consider now the same loop of Figure 8-6 rotating with an angular velocity ω about the x-axis in a uniform, time-independent magnetic field $\boldsymbol{B}$ parallel to the z-axis. Then, from Eq. 8-36,

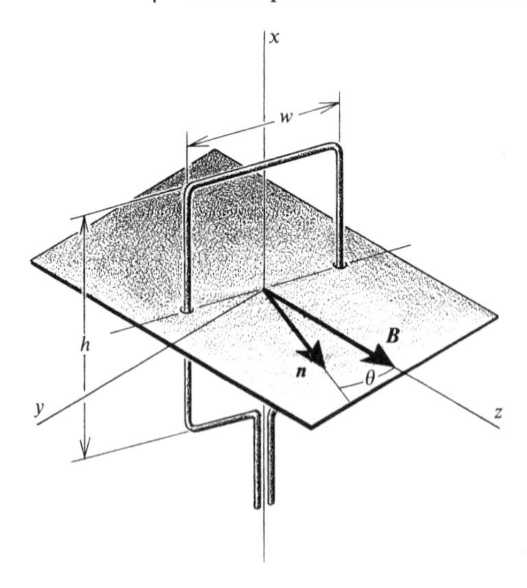

Figure 8-6. Fixed loop in a time-dependent magnetic field $\boldsymbol{B}$. The vector $\boldsymbol{n}$ is normal to the loop.

$$\nabla \times \boldsymbol{E} = \nabla \times (\boldsymbol{u} \times \boldsymbol{B}), \tag{8-41}$$

and

$$\boldsymbol{E} = \boldsymbol{u} \times \boldsymbol{B}, \tag{8-42}$$

as in Section 8.1. Thus, setting $\theta = \omega t$,

$$\oint \boldsymbol{E} \cdot \boldsymbol{dl} = \oint (\boldsymbol{u} \times \boldsymbol{B}) \cdot \boldsymbol{dl}, \tag{8-43}$$

$$= 2\omega \frac{W}{2} hB \sin \omega t. \tag{8-44}$$

The only contributions to the integral are on the vertical sides, since $(\boldsymbol{u} \times \boldsymbol{B})$ is perpendicular to $\boldsymbol{dl}$ along the top and bottom parts of the integration path. Hence

$$\oint \boldsymbol{E} \cdot \boldsymbol{dl} = B\omega S \sin \omega t, \tag{8-45}$$

where S is again the area of the loop. The electromotance is zero when the normal to the loop is parallel to $\boldsymbol{B}$, since the free charges inside the wire are then moving parallel to $\boldsymbol{B}$ and the induced electric field intensity is zero.

The case of a loop rotating in a time-dependent magnetic field is the subject of Problem 8-6.

8.4 INDUCTANCE AND INDUCED ELECTROMOTANCE

To calculate the electromotance induced in one circuit when the current changes in another circuit, it is convenient to express the magnetic flux linking the first one in terms of the current in the second and of a geometrical factor involving both circuits. This factor is known as the *mutual inductance* between the two circuits.

The same procedure can be used to relate the linking flux and the current for a single circuit, in which case the geometrical factor is known as the *self-inductance* of the circuit.

8.4.1 Mutual Inductance

Let us seek an expression for the magnetic flux linking one circuit, but due to the current in another circuit. The current I_a in circuit a produces a magnetic flux Φ_{ab} linking circuit b as in Figure 8-7:

$$\Phi_{ab} = \int_{S_b} \boldsymbol{B}_a \cdot \boldsymbol{da}_b, \tag{8-46}$$

where da_b is an element of area of an arbitrary surface S_b bounded by circuit b, and where $\boldsymbol{B}_a$ is the magnetic induction due to current I_a at a point on S_b.

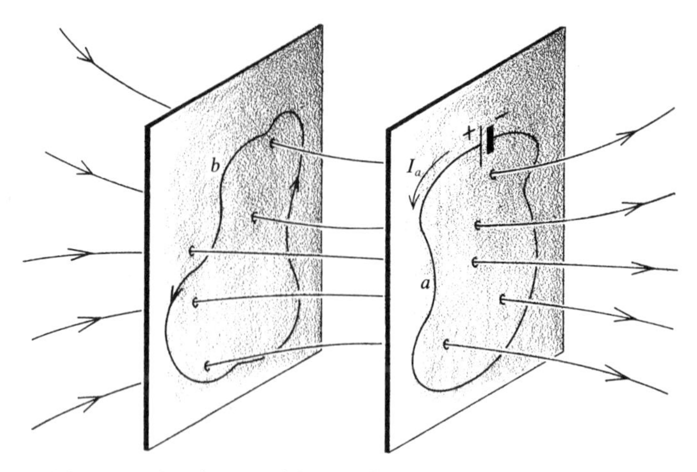

Figure 8-7. Two circuits a and b. The flux Φ_{ab} shown linking b and originating in a is positive. This is because its direction is related by the right-hand screw rule to the direction chosen to be positive around b.

We can calculate Φ_{ab} from the vector potential $\mathbf{A}_a$ produced by I_a:

$$\Phi_{ab} = \int_{S_b} (\nabla \times \mathbf{A}_a) \cdot d\mathbf{a}_b, \tag{8-47}$$

$$= \oint_b \mathbf{A}_a \cdot d\mathbf{l}_b, \tag{8-48}$$

$$= \oint_b \left(\frac{\mu_0 I_a}{4\pi} \oint_a \frac{d\mathbf{l}_a}{r} \right) \cdot d\mathbf{l}_b, \tag{8-49}$$

$$= \frac{\mu_0 I_a}{4\pi} \oint_a \oint_b \frac{d\mathbf{l}_a \cdot d\mathbf{l}_b}{r}, \tag{8-50}$$

$$= M_{ab} I_a, \tag{8-51}$$

where

$$M_{ab} = \frac{\mu_0}{4\pi} \oint_a \oint_b \frac{d\mathbf{l}_a \cdot d\mathbf{l}_b}{r} \tag{8-52}$$

is the mutual inductance between the two circuits. This is the *Neumann equation.*

Similarly, the flux Φ_{ba} linking circuit a is

$$\Phi_{ba} = M_{ba} I_b. \tag{8-53}$$

Since the Neumann equation is symmetrical with respect to the subscripts a and b,

$$M_{ab} = M_{ba}. \tag{8-54}$$

It will be noticed that mutual inductance is a quantity that depends solely on the geometry of the two circuits and that, when multiplied by the current in one circuit, gives the magnetic flux linking the other. A similar situation exists in the case of a capacitance: the charge induced on a grounded conductor is the product of the potential of a neighboring conductor, multiplied by a geometrical factor called the capacitance between the two conductors.

The sign of the mutual inductance is arbitrary. It is positive if a current in the direction arbitrarily chosen as positive in circuit a produces a flux in the direction arbitrarily chosen as positive through b.

Since the mutual inductance is the magnetic flux linking one circuit per unit of current in the other, inductance is measured in webers/ampere, or in *henrys*.

The mutual inductance between two circuits is one henry when a current of one ampere in one of the circuits produces a flux linkage of one weber-turn in the other.

The electromotance induced in circuit b by a change in the current I_a is

$$\oint_b \mathbf{E}\cdot d\mathbf{l} = -\frac{d\Phi_{ab}}{dt} = -M_{ab}\frac{dI_a}{dt}. \tag{8-55}$$

Similarly, the electromotance induced in circuit a by a change in the current I_b is

$$\oint_a \mathbf{E}\cdot d\mathbf{l} = -\frac{d\Phi_{ba}}{dt} = -M_{ab}\frac{dI_b}{dt}. \tag{8-56}$$

This equation is convenient for computing the induced electromotance since it involves only the mutual inductance and dI/dt, both of which can be measured.

This equation gives us a second definition of the henry: the mutual inductance between two circuits is one henry if a current changing at the rate of one ampere/second in one circuit induces an electromotance of one volt in the other.

We shall have an example of mutual inductance after we have studied self-inductance.

8.4.2 Self-inductance

A single circuit is of course linked by its own flux as in Figure 8-8, and

$$\Phi = LI, \tag{8-57}$$

where L is called the self-inductance of the circuit and depends solely on the

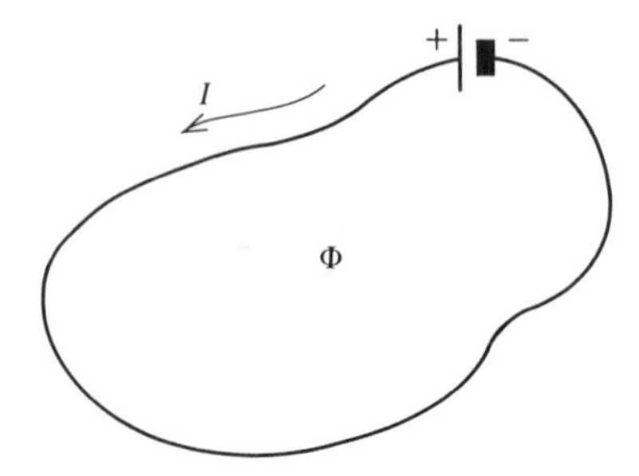

Figure 8-8. Single circuit carrying a current I.

geometry of the circuit. Self-inductance, like mutual inductance, is measured in henrys. It is always positive.

A circuit has a self-inductance of one henry if a current of one ampere produces a flux linkage of one weber-turn.

A change in the current flowing through the circuit produces within the circuit itself an induced electromotance

$$\oint \boldsymbol{E} \cdot \boldsymbol{dl} = -\frac{d\Phi}{dt} = -L\frac{dI}{dt}. \tag{8-58}$$

The induced electromotance tends to oppose the change in current, according to Lenz's law, and adds to whatever other voltages are present.

A circuit therefore has a self-inductance of one henry if a current changing at the rate of one ampere/second induces in it an electromotance of one volt.

If the circuit is truly filamentary, that is, if the cross-sectional area of the conductor is infinitely small, the flux Φ and the inductance L become infinite. This can be explained as follows. As the radius of the conductor tends to zero, $\boldsymbol{B}$ tends to infinity in the immediate neighborhood of the wire. The region where $\boldsymbol{B}$ is infinitely large is itself infinitely small, but the flux tends to infinity logarithmically. This flux clings infinitely close to the wire. In practice, currents are distributed over a finite cross section, and the flux linkage, and therefore the inductance, does not in fact diverge.

We shall calculate self-inductance from the ratio of flux linkage to current for two cases in which the currents are effectively distributed over a surface. Then both the flux and the inductance are finite.

Example

SELF-INDUCTANCE OF A LONG SOLENOID

It was shown in the example on page 315 that the magnetic induction inside a long solenoid, neglecting end effects, is constant, and that

$$B = \mu_0 N' I, \tag{8-59}$$

Table 8-1. *Representative Values of K.*

Radius/Length	K
0	1.00
0.2	0.85
0.4	0.74
0.6	0.65
0.8	0.58
1.0	0.53
1.5	0.43
2.0	0.37
4.0	0.24
10.0	0.12

where N' is the number of turns/meter. Thus

$$\Phi = \frac{\mu_0 NI}{l}\,\pi R^2, \tag{8-60}$$

where N is the total number of turns, l is the length of the solenoid, and R is its radius. Then

$$L = \frac{N\Phi}{I} = \frac{\mu_0 N^2}{l}\,\pi R^2, \tag{8-61}$$

$$= \mu_0 N'^2\, l\, \pi R^2. \tag{8-62}$$

The inductance of a short solenoid is smaller by a factor K, which is a function of R/l. Representative values of K are shown in Table 8-1.

Example

SELF-INDUCTANCE OF A TOROIDAL COIL

A toroidal coil of N turns is wound on a form of nonmagnetic material having a square cross section as in Figure 8-9.

According to the circuital law, or from Eq. 7-86, the magnetic induction in the azimuthal direction, at a radius ρ inside the toroid, is

$$B = \frac{\mu_0 NI}{2\pi\rho}. \tag{8-63}$$

Thus

$$\Phi = \frac{\mu_0 NI}{2\pi}\int_{R-w/2}^{R+w/2} \frac{w\,d\rho}{\rho}, \tag{8-64}$$

$$= \frac{\mu_0 NI}{2\pi}\, w \ln\left[\frac{2R+w}{2R-w}\right], \tag{8-65}$$

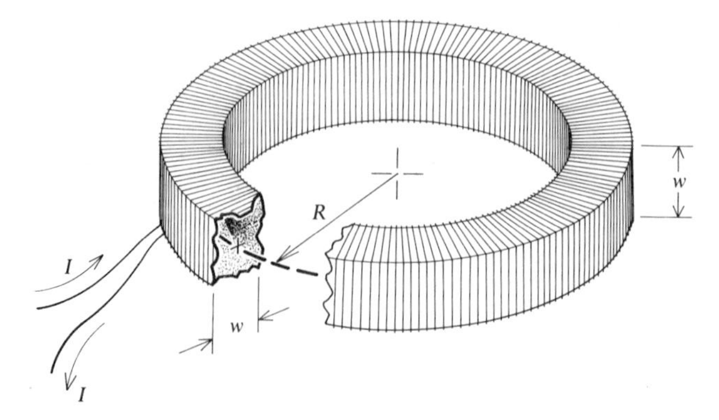

Figure 8-9. Toroidal coil of square cross section and mean radius R.

and

$$L = \frac{\mu_0 N^2 w}{2\pi} \ln\left[\frac{2R + w}{2R - w}\right]. \tag{8-66}$$

For $R \gg w$,

$$\frac{2R + w}{2R - w} \approx \left(1 + \frac{w}{2R}\right)\left(1 + \frac{w}{2R}\right), \tag{8-67}$$

$$\approx 1 + \frac{w}{R}, \tag{8-68}$$

$$\ln\left[\frac{2R + w}{2R - w}\right] \approx \frac{w}{R}, \tag{8-69}$$

and

$$L \approx \frac{\mu_0 N^2}{2\pi R} w^2 \qquad (R \gg w). \tag{8-70}$$

The self-inductance of a toroidal coil is proportional to the square of the number of turns and to the cross-sectional area enclosed by the winding, and inversely proportional to the length of the winding. The reason we obtained a similar expression for the long solenoid is that we neglected end effects.

Example

MUTUAL INDUCTANCE BETWEEN TWO COAXIAL SOLENOIDS

Let us now add a second winding over the solenoid, as in Figure 8-10. We assume for simplicity that both windings are long with respect to the common diameter, in order that end effects may be neglected.

To calculate the mutual inductance between the two coils we shall avoid evaluating the double integrals of the Neumann equation, 8-52, by using again the ratio of flux linkage to current. Let us

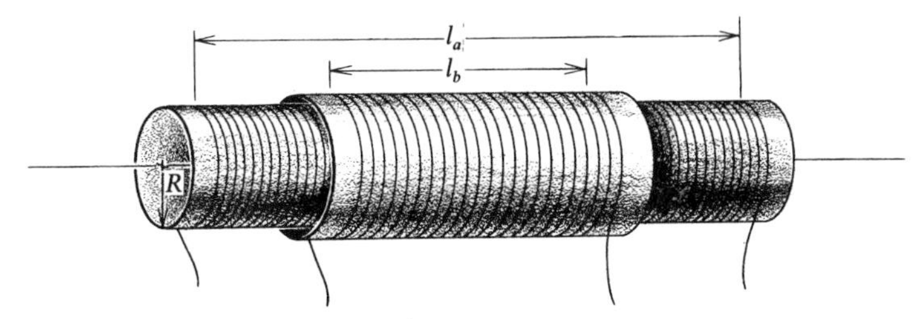

Figure 8-10. Coaxial solenoids. The two radii are taken to be approximately equal.

assume a current I_a in coil a. Then the flux of coil a linking coil b is

$$\Phi_{ab} = \frac{\mu_0 \pi R^2 N_a I_a}{l_a}, \tag{8-71}$$

and the mutual inductance is

$$M_{ab} = \frac{N_b \Phi_{ab}}{I_a} = \frac{\mu_0 \pi R^2 N_a N_b}{l_a}. \tag{8-72}$$

We can also calculate the mutual inductance by assuming a current I_b in coil b. Then

$$\Phi_{ba} = \frac{\mu_0 \pi R^2 N_b I_b}{l_b}. \tag{8-73}$$

This flux links only $(l_b/l_a)N_a$ turns of coil a since B falls rapidly to zero beyond the end of a long solenoid, as we saw in Problem 7-14. Then the mutual inductance is

$$M_{ba} = \frac{l_b N_a \Phi_{ba}}{l_a I_b} = \frac{\mu_0 \pi R^2 N_a N_b}{l_a}, \tag{8-74}$$

as previously.

It is paradoxical that a varying current in the inner solenoid should induce an electromotance in the outer one, since we have shown (example, page 315) that the magnetic induction outside a long solenoid is zero. The explanation is that the induced electric-field intensity at any given point is equal to the negative time derivative of the vector potential at that point and that the vector potential $\boldsymbol{A}$ does not vanish outside an infinite solenoid, despite the fact that $\boldsymbol{B} = \nabla \times \boldsymbol{A}$ does.

We can actually calculate the vector potential just outside the inner solenoid from the mutual inductance. If we consider the direction of the current I_a in the primary as positive, then the electromotance induced in the secondary is $-M dI_a/dt$, or

$$-\frac{\mu_0 \pi R^2 N_a N_b}{l_a}\frac{dI_a}{dt},$$

and the induced electric field intensity $-\partial A/\partial t$ is $2\pi R N_b$ times smaller:

$$-\frac{\partial A}{\partial t} = -\frac{\mu_0 R N_a}{2l_a}\frac{dI_a}{dt}. \tag{8-75}$$

Integrating, we have the vector potential at any point close to the solenoid:

$$A = \frac{\mu_0 R N_a}{2l_a} I_a \tag{8-76}$$

in the azimuthal direction. This can be shown to be correct by calculating the line integral of $\boldsymbol{A}\cdot d\boldsymbol{l}$ as in Section 7.5.1.

8.4.3 Coefficient of Coupling

Let us consider a single-turn coil a, through which a current I_a produces a flux Φ_{aa}, and another single-turn coil b, arranged so that only a fraction k_a of Φ_{aa} passes through it:

$$\Phi_{ab} = k_a \Phi_{aa}. \tag{8-77}$$

The self-inductance of coil a is

$$L_a = \frac{\Phi_{aa}}{I_a}, \tag{8-78}$$

and the mutual inductance between the two coils is

$$M_{ab} = \frac{k_a \Phi_{aa}}{I_a} = k_a L_a. \tag{8-79}$$

Likewise,

$$M_{ba} = k_b L_b. \tag{8-80}$$

Then, since $M_{ba} = M_{ab} = M$,

$$M^2 = k_a k_b L_a L_b, \tag{8-81}$$

$$M = \pm k (L_a L_b)^{1/2}, \tag{8-82}$$

where

$$k = \pm (k_a k_b)^{1/2}, \tag{8-83}$$

called the coefficient of coupling between the two coils, can have values ranging from -1 to $+1$. If the two single-turn coils coincide, the absolute value of k is unity and its sign is the same as that of M.

Although this simple reasoning, which is based on single-turn coils, does not apply to the coils which one uses in practice, Eq. 8-82 remains true and the maximum possible value of k is unity.

We have seen in Section 8.4.2 that the self-inductance of a circuit tends to infinity as the wire radius r tends to zero. The mutual inductance, however, does not tend to infinity if one or both of the circuits is filamentary. The reason is that the intense flux very close to a thin wire does not link another circuit some distance away. That is, as the radius of the wire tends to zero, $L \to \infty$, $k \to 0$, and the mutual inductance remains about the same.

Example The coefficient of coupling for the two coaxial solenoids is

$$k = \frac{M}{(L_a L_b)^{1/2}} = \frac{(\mu_0 \pi R^2 N_a N_b)/l_a}{(\mu_0 N_a^2 \pi R^2/l_a)^{1/2} (\mu_0 N_b^2 \pi R^2/l_b)^{1/2}}, \tag{8-84}$$

$$= \left(\frac{l_b}{l_a}\right)^{1/2}. \tag{8-85}$$

It was assumed in calculating M that l_b was shorter than, or equal to, l_a.

8.5 ENERGY STORED IN A MAGNETIC FIELD

To find the energy stored in a magnetic field, we shall calculate the energy supplied by a source to an isolated circuit, as the current density increases from zero to some value $\boldsymbol{J}_f$.

We shall assume that Ohm's law applies to the conducting medium. Then, at any given point outside a source,

$$\boldsymbol{J}_f = \sigma \boldsymbol{E}, \tag{8-86}$$

where σ is the *conductivity* of the medium, expressed in mhos/meter, at that point. This is *Ohm's Law.*

This electric field intensity is the sum of two terms: $-\boldsymbol{\nabla} V$, produced by accumulations of charge on the terminals of the source and on the surfaces of the conductor, and $-\partial \boldsymbol{A}/\partial t$, due to the changing magnetic field. In a wire, the surface charges adjust themselves so that the total $\boldsymbol{E}$ is along the axis as in Figure 8-11.

Inside a source there is also a third electric field $\boldsymbol{E}_s$, which comes from the local generation of energy and

$$\boldsymbol{J}_f = \sigma \boldsymbol{E} = \sigma \left(-\boldsymbol{\nabla} V - \frac{\partial \boldsymbol{A}}{\partial t} + \boldsymbol{E}_s \right). \tag{8-87}$$

The work done per unit time and per unit volume on the moving charges at any point *outside* the source can be calculated as follows. Consider an element of volume having the form of a rectangular parallelepiped oriented so that one set of sides is parallel to the current density $\boldsymbol{J}_f$, as in Figure 8-11. In one second, a charge $J_f\, da$ goes in through the left-hand face and an equal charge comes out at the other end. The source maintains a difference in potential of $-\boldsymbol{\nabla} V \cdot \boldsymbol{dl}$ across these faces, and the integral of $\boldsymbol{\nabla} V \cdot \boldsymbol{dl}$ around

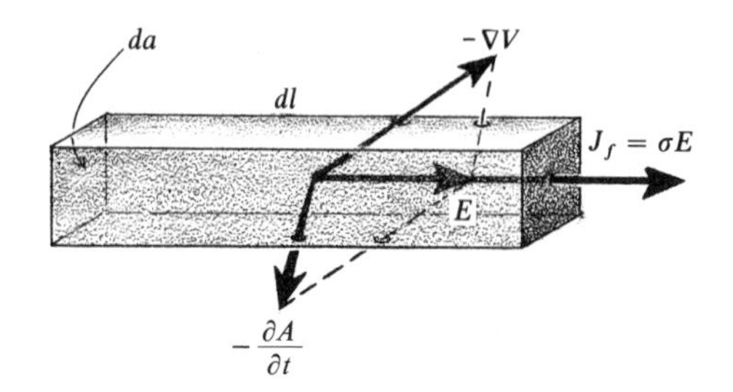

Figure 8-11. Rectangular parallelepiped parallel to the current density $\boldsymbol{J}_f$ in a conductor. The electric field intensity $\boldsymbol{E}$ is the sum of $-\boldsymbol{\nabla}V$ and $-\partial\boldsymbol{A}/\partial t$.

the circuit is equal to the voltage supplied by the source. The source supplies to the element of volume $d\tau$ a power

$$-\boldsymbol{\nabla}V\cdot\boldsymbol{dl}\, J_f\, da = -\boldsymbol{\nabla}V\cdot\boldsymbol{J}_f\, d\tau = \left(\boldsymbol{E} + \frac{\partial \boldsymbol{A}}{\partial t}\right)\cdot \boldsymbol{J}_f\, d\tau, \tag{8-88}$$

and the total power spent by the source is

$$\frac{dW}{dt} = \int_\tau \left(\boldsymbol{E}\cdot\boldsymbol{J}_f + \frac{\partial \boldsymbol{A}}{\partial t}\cdot\boldsymbol{J}_f\right) d\tau, \tag{8-89}$$

$$= \int_\tau \frac{J_f^2}{\sigma}\, d\tau + \int_\tau \frac{\partial \boldsymbol{A}}{\partial t}\cdot\boldsymbol{J}_f\, d\tau. \tag{8-90}$$

The first term on the right gives the power spent in Ohmic or Joule losses (Problem 8-19), while the second gives the rate at which work is done by the source against the induced electromotance. This latter work is that which must be done to establish the magnetic field, and it is the one that concerns us here.

Writing W_m for the energy stored in the magnetic field,

$$\frac{dW_m}{dt} = \int_\tau \frac{\partial \boldsymbol{A}}{\partial t}\cdot\boldsymbol{J}_f\, d\tau, \tag{8-91}$$

where τ is again the volume occupied by the currents.

Example For a long solenoid, $\boldsymbol{A}$ is given by Eq. 8-76 and, replacing $\boldsymbol{J}_f\, d\tau$ by $\boldsymbol{I\, dl}$,

$$\frac{dW_m}{dt} = \frac{\mu_0 RN}{2l}\frac{dI}{dt} I \int dl, \tag{8-92}$$

where the integral is evaluated over the length of the wire, $2\pi RN$. Then

$$\frac{dW_m}{dt} = \frac{\pi\mu_0 R^2 N^2}{l} I \frac{dI}{dt} \tag{8-93}$$

$$W_m = \frac{\pi}{2}\mu_0 I^2 N'^2 l R^2. \tag{8-94}$$

8.5.1 Magnetic Energy in Terms of the Magnetic Induction *B*

It is possible to express the magnetic energy W_m in terms of B, just as the electric energy was expressed in terms of E in Section 2.15.

We assume again that there are no magnetic materials in the field. We also assume that conduction takes place in good conductors so that Eq. 7-75 can be applicable. Then

$$\boldsymbol{J}_f = \frac{1}{\mu_0} \nabla \times \boldsymbol{B}, \tag{8-95}$$

and

$$\frac{dW_m}{dt} = \frac{1}{\mu_0} \int_\tau \frac{\partial \boldsymbol{A}}{\partial t} \cdot (\nabla \times \boldsymbol{B})\, d\tau, \tag{8-96}$$

$$= \frac{1}{\mu_0} \int_\tau \left[\boldsymbol{B} \cdot \left(\nabla \times \frac{\partial \boldsymbol{A}}{\partial t} \right) - \nabla \cdot \left(\frac{\partial \boldsymbol{A}}{\partial t} \times \boldsymbol{B} \right) \right] d\tau, \tag{8-97}$$

$$= \frac{1}{\mu_0} \int_\tau \boldsymbol{B} \cdot \frac{\partial \boldsymbol{B}}{\partial t}\, d\tau + \frac{1}{\mu_0} \int \boldsymbol{B} \times \frac{\partial \boldsymbol{A}}{\partial t} \cdot d\boldsymbol{a}, \tag{8-98}$$

the volume τ being any volume which includes all points where the current density $\boldsymbol{J}_f$ is not zero, and S being the corresponding surface.

As in the corresponding electrostatic case, the above equation becomes simple if we choose τ to include all space, in which case the surface S is at infinity. The magnetic induction B falls off as $1/r^3$ at large distances. This was shown for the case of a current loop in Eqs. 7-112 and 7-113. The induced electric field intensity $-\partial A/\partial t$ falls off as $1/r^2$, because the vector potential A itself falls off as $1/r^2$ at large distances from a current loop (Section 7.8). Since the surface area S increases only as r^2, the surface integral decreases as $1/r^3$ and vanishes as the surface of integration becomes infinite Thus

$$\frac{dW_m}{dt} = \frac{1}{\mu_0} \int_\infty \boldsymbol{B} \cdot \frac{\partial \boldsymbol{B}}{\partial t}\, d\tau, \tag{8-99}$$

On the right we have a partial derivative because $\boldsymbol{B}$ is a function of both the coordinates and of the time. If we remove this derivative from under the integral sign it becomes a total derivative that applies to all the field:

$$\frac{dW_m}{dt} = \frac{1}{2\mu_0} \frac{d}{dt} \int_\infty B^2\, d\tau. \tag{8-100}$$

Setting $W_m = 0$ when $B = 0$,

$$W_m = \frac{1}{2\mu_0} \int_\infty B^2\, d\tau. \tag{8-101}$$

The quantity W_m is the work that must be done to establish a magnetic field in terms of the magnetic induction B, either in free space or in nonmagnetic matter.

It should be noted that the magnetic energy varies as the square of the magnetic induction B. If several fields are superposed, the total energy is therefore *not* just the sum of the energies calculated for each separate field.

Just as in electrostatics, Section 2.15, we may define an energy density

$$\frac{dW_m}{d\tau} = \frac{B^2}{2\mu_0} \tag{8-102}$$

associated with each point in space. We shall arrive at this same result in Section 8.10 from the value of the magnetic pressure.

Example For a long solenoid of length l and radius R, B is $\mu_0 N'I$ throughout the interior, and zero outside. Then

$$W_m = \frac{1}{2\mu_0}\int_\infty B^2\,d\tau = \frac{\pi}{2}\mu_0 I^2 N'^2 l R^2. \tag{8-103}$$

8.5.2 Magnetic Energy in Terms of the Current Density $\boldsymbol{J}_f$ and of the Vector Potential A

It will be recalled from Section 2.15 that we expressed the energy density in an electric field both as $\epsilon_0 E^2/2$ and as $\rho V/2$. We have already expressed the magnetic energy density as $B^2/2\mu_0$; we shall now express it in terms of the free current density $\boldsymbol{J}_f$ and of the vector potential $\boldsymbol{A}$.

We rewrite Eq. 8-101 as follows:

$$W_m = \frac{1}{2\mu_0}\int_\infty (\boldsymbol{B}\cdot\boldsymbol{\nabla}\times\boldsymbol{A})\,d\tau. \tag{8-104}$$

Using the vector identity of Problem 1-19 and the divergence theorem,

$$W_m = \frac{1}{2\mu_0}\int_\infty \boldsymbol{A}\cdot(\boldsymbol{\nabla}\times\boldsymbol{B})\,d\tau + \frac{1}{2\mu_0}\int_\infty (\boldsymbol{A}\times\boldsymbol{B})\cdot d\boldsymbol{a}. \tag{8-105}$$

The surface integral vanishes again as in Eq. 8-98, and when $\boldsymbol{\nabla}\times\boldsymbol{B} = \mu_0\boldsymbol{J}_f$ (Section 7.6),

$$W_m = \frac{1}{2}\int_\tau (\boldsymbol{J}_f\cdot\boldsymbol{A})\,d\tau, \tag{8-106}$$

where τ is any volume that includes all regions where $\boldsymbol{J}_f$ is not zero.

It is convenient to assign an energy density

$$\frac{dW_m}{d\tau} = \frac{1}{2}(\boldsymbol{J}_f\cdot\boldsymbol{A}) \tag{8-107}$$

to conductors carrying a free current density $\boldsymbol{J}_f$.

Example In the case of the long solenoid, the current is distributed, not over a volume, but over the surface of the solenoid with a density $\lambda = N'I$. The vector potential A is parallel to I and is given by Eq. 8-76. Then

$$W_m = \frac{1}{2}\int \lambda A \, da = \frac{\pi}{2}\mu_0 I^2 N'^2 l R^2. \tag{8-108}$$

8.5.3 Magnetic Energy in Terms of the Current I and of the Magnetic Flux Φ

With filamentary circuits we may express the energy in terms of the current I and of the flux Φ linking the circuit. Replacing $\boldsymbol{J}_f\, d\tau$ in Eq. 8-106 by $I\, \boldsymbol{dl}$, $\boldsymbol{dl}$ being an element of the circuit that carries the current I,

$$W_m = \frac{1}{2} I \oint \boldsymbol{A} \cdot \boldsymbol{dl} = \frac{1}{2} I\Phi, \tag{8-109}$$

from Eq. 7-60. The positive directions for I and Φ are related according to the right-hand screw rule.

Example For the long solenoid,

$$W_m = \frac{1}{2} I\Phi = \frac{1}{2} INB(\pi R^2) = \frac{\pi}{2}\mu_0 I^2 N'^2 l R^2. \tag{8-110}$$

8.5.4 Magnetic Energy in Terms of the Currents and of the Inductances

It is possible to express the energy stored in a magnetic field in still another way, in terms of the currents and of the inductances. Since the magnetic flux Φ can be replaced by the product of the self-inductance L and the current I, from Section 8.42,

$$W_m = \frac{1}{2} L I^2. \tag{8-111}$$

For two circuits carrying currents I_a and I_b,

$$W_m = \frac{1}{2} I_a \Phi_a + \frac{1}{2} I_b \Phi_b, \tag{8-112}$$

where Φ_a and Φ_b are the total fluxes linking circuits a and b respectively and consist of contributions from both circuits:

$$\Phi_a = \Phi_{aa} + \Phi_{ba}, \tag{8-113}$$

$$\Phi_b = \Phi_{ab} + \Phi_{bb}. \tag{8-114}$$

Thus

$$W_m = \frac{1}{2} I_a \Phi_{aa} + \frac{1}{2} I_b \Phi_{bb} + \frac{1}{2} I_a \Phi_{ba} + \frac{1}{2} I_b \Phi_{ab}. \tag{8-115}$$

But

$$\Phi_{ba} = MI_b, \qquad \Phi_{ab} = MI_a, \tag{8-116}$$

$$\Phi_{aa} = L_a I_a, \qquad \Phi_{bb} = L_b I_b, \tag{8-117}$$

and therefore

$$W_m = \frac{1}{2} L_a I_a^2 + \frac{1}{2} L_b I_b^2 + MI_a I_b. \tag{8-118}$$

The first two terms on the right are the self-energies due to the interaction of each current with its own field, whereas the third term is the interaction energy due to the mutual inductance.

Example Returning once more to the long solenoid, its stored energy is

$$W_m = \frac{1}{2} LI^2, \tag{8-119}$$

and, using the value of L given in Eq. 8-62, we again find that

$$W_m = \frac{\pi}{2} \mu_0 I^2 N'^2 l R^2. \tag{8-120}$$

8.6 SELF-INDUCTANCE FOR A VOLUME DISTRIBUTION OF CURRENT

The self-inductance of a circuit comprising currents distributed over a finite volume is defined from the energy stored in the system:

$$W_m = \frac{1}{2\mu_0} \int_\infty B^2 \, d\tau = \frac{1}{2} LI^2. \tag{8-121}$$

Then the self-inductance is

$$L = \frac{1}{\mu_0 I^2} \int_\infty B^2 \, d\tau, \tag{8-122}$$

where the integral is evaluated over all space.

Example SELF-INDUCTANCE OF A COAXIAL LINE

We assume that the frequency is low enough to ensure that the currents are distributed uniformly throughout the cross sections of the conductors, and we neglect end effects for simplicity.

We shall calculate successively the magnetic energies per unit

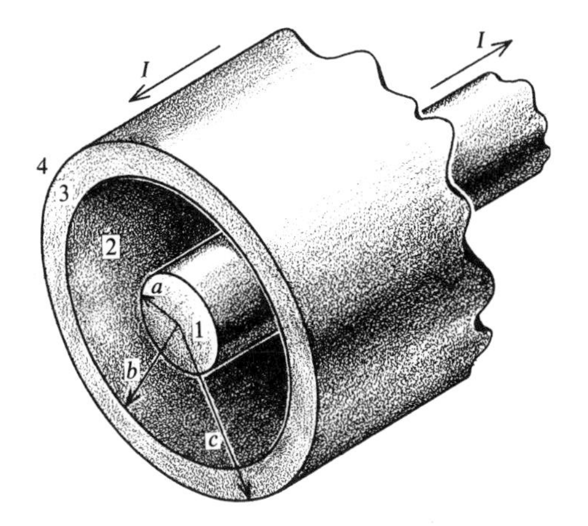

Figure 8-12. Coaxial line of radii a, b, c carrying currents I in opposite directions in the inner and outer conductors.

length of the line in regions 1, 2, 3, 4 as in Figure 8-12, and then set the sum of these energies equal to $(1/2)L'I^2$. This will give us the inductance L' per unit length.

(a) In region 1, from the circuital law (Section 7.7) applied to a path of radius ρ,

$$2\pi\rho B = \mu_0 I \frac{\pi\rho^2}{\pi a^2}, \tag{8-123}$$

and

$$W_{m1} = \frac{1}{2\mu_0}\int_0^a \left(\frac{\mu_0 I\rho}{2\pi a^2}\right)^2 2\pi\rho\, d\rho = \frac{\mu_0 I^2}{16\pi}. \tag{8-124}$$

(b) In region 2,

$$B = \frac{\mu_0 I}{2\pi\rho}, \tag{8-125}$$

$$W_{m2} = \frac{\mu_0 I^2}{4\pi}\ln\frac{b}{a}. \tag{8-126}$$

(c) In region 3, the current within a circular path of radius ρ is that in the center conductor, namely I, less that part of the current that lies in the outer conductor between the radii b and ρ. Thus

$$B = \frac{\mu_0}{2\pi\rho}\left[I - I\left(\frac{\rho^2 - b^2}{c^2 - b^2}\right)\right], \tag{8-127}$$

$$= \frac{\mu_0 I}{2\pi\rho}\left(\frac{c^2 - \rho^2}{c^2 - b^2}\right), \tag{8-128}$$

and

$$W_{m3} = \frac{\mu_0 I^2}{4\pi}\left[\frac{c^4}{(c^2 - b^2)^2}\ln\frac{c}{b} - \frac{3c^2 - b^2}{4(c^2 - b^2)}\right]. \tag{8-129}$$

(d) In region 4, $B = 0$ and $W_{m4} = 0$.

Finally, the inductance per unit length is

$$L' = \frac{2(W_{m1} + W_{m2} + W_{m3} + W_{m4})}{I^2}, \tag{8-130}$$

$$= \mu_0\left\{\frac{1}{8\pi} + \frac{1}{2\pi}\ln\frac{b}{a} + \frac{1}{2\pi}\left[\frac{c^4}{(c^2 - b^2)^2}\ln\frac{c}{b} - \frac{3c^2 - b^2}{4(c^2 - b^2)}\right]\right\}. \tag{8-131}$$

The second term within the braces is normally the most important; it gives the inductance associated with the magnetic energy in the annular region between the conductors.

8.7 MAGNETIC FORCE BETWEEN TWO CIRCUITS

Although the magnetic force law stated in Eqs. 7-1 and 7-5 was the starting point for Chapter 7, we have not really used it to calculate the magnetic force between two circuits, because of the difficulty of evaluating the line integrals.

We are now in a position to find other expressions that are more convenient to use because they are based on mutual inductance. Although mutual inductance is just as difficult to calculate, because it too involves a line integral (Eq. 8-52), it is easily measurable.

We shall use the same method as in the example on page 78, where we calculated forces on conductors. We assume a small virtual translation of one coil, without any rotation, and then apply the principle of the conservation of energy:

$$\begin{aligned}\text{Work done by the sources} &= \text{Increase in magnetic energy}\\ &\quad + \text{Mechanical work done.}\end{aligned}$$

For simplicity we disregard Joule losses. We also assume that the displacements are made infinitely slowly so as to avoid taking kinetic energy into account.

We consider two loops carrying currents I_a and I_b in the same direction as in Figure 8-13. Since the forces shown in the figure indicate that the loops tend to move toward each other, the coils can be kept fixed in space only if the magnetic forces are balanced by equal and opposite mechanical forces.

The virtual displacement can be made in any convenient way: we may, for example, assume that either the currents or the flux linkages are kept constant. Whatever the assumption, the result must be the same, since the force acting between two fixed circuits obviously has some single definite value. We had a similar situation in the example on page 78 and in Problem 2-37, where we found the force on a capacitor plate from the conservation of energy, assuming first that the plates were insulated, and then assuming that they were connected to a battery. The forces were found to be the same.

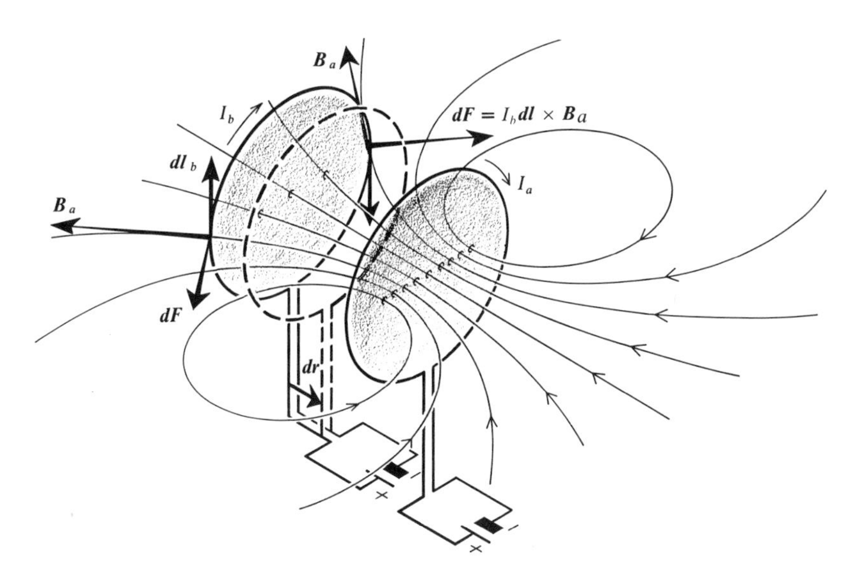

Figure 8-13. The two loops carry currents I_a and I_b. Typical lines of $\boldsymbol{B}$ originating in a are shown linking b. Observe that the elementary forces $\boldsymbol{dF}$ have a component in the direction of coil a, with the result that the total force $\boldsymbol{F}$ is attractive. To calculate $\boldsymbol{F}$, coil b is assumed to be displaced parallel to itself by a distance $\boldsymbol{dr}$ into the position shown by the broken line, and the mechanical work done, $\boldsymbol{F} \cdot \boldsymbol{dr}$, is found from the principle of conservation of energy.

8.7.1 The Magnetic Force When the Currents Are Kept Constant

Let us calculate the magnetic force $\boldsymbol{F}_{ab}$ on loop b with the above method, assuming that the currents are kept constant. We imagine that loop b is allowed to move a distance dr toward loop a. Since the currents are kept constant, only the interaction energy changes; the self-energies remain constant. Hence, from Eq. 8-118,

$$dW_m = I_a I_b \, dM, \tag{8-132}$$

$$= I_a \, d\Phi_{ba} = I_b \, d\Phi_{ab}, \tag{8-133}$$

where Φ_{ba} is the flux originating in b and linking a, and similarly for Φ_{ab}. The mutual inductance M and the fluxes Φ_{ab} and Φ_{ba} all increase in this case, so that dW_m is a positive quantity, and the stored magnetic energy increases. The mutual inductance is positive here.

Consider next the work done by the sources producing the currents in the loops. In loop b, Φ_{ab} increases, and the induced electromotance is in such a direction that it would, by itself, produce a magnetic field opposing $\boldsymbol{B}_a$.

Therefore the induced electromotance in loop b must tend to oppose the current I_b. If I_b is to be kept constant, the source voltage must be increased at each instant by $d\Phi_{ab}/dt$. The source in loop b therefore supplies an amount of work

$$dW_{sb} = I_b \frac{d\Phi_{ab}}{dt} dt = I_b \, d\Phi_{ab}, \tag{8-134}$$

$$= I_b I_a \, dM. \tag{8-135}$$

By symmetry, the work supplied by the source in loop a is the same, and the total work supplied by the sources is

$$dW_s = 2 I_a I_b \, dM, \tag{8-136}$$

which is exactly twice the increase in magnetic energy. The remainder has gone into mechanical work.

In other words, the mechanical work is accompanied by an *equal* increase in magnetic energy, both being supplied by the sources. See Problem 2-37.

The mechanical work done is

$$\boldsymbol{F}_{ab} \cdot d\boldsymbol{r} = I_a I_b \, dM, \tag{8-137}$$

$$= (dW_m)_I, \tag{8-138}$$

where $\boldsymbol{F}_{ab}$ is the force exerted *by* circuit a *on* circuit b. The subscript I indicates that the currents are kept constant.

Since the quantity on the right is positive in the present case, the scalar product $\boldsymbol{F}_{ab} \cdot d\boldsymbol{r}$ must be positive, and the force $\boldsymbol{F}_{ab}$ must point toward coil a, like $d\boldsymbol{r}$. This is correct, since we found at the beginning that the coils tend to move closer to each other.

Note that we would have obtained the wrong sign if we had neglected the work done by the sources.

In the general case, the x-component of $\boldsymbol{F}_{ab}$ is

$$F_{abx} = I_a I_b \frac{\partial M}{\partial x}, \tag{8-139}$$

where the increment of x is the x-component of $d\boldsymbol{r}$. Also,

$$F_{abx} = I_a \left(\frac{\partial \Phi_{ba}}{\partial x} \right)_I = \left(\frac{\partial W_m}{\partial x} \right)_I. \tag{8-140}$$

The partial derivatives are evaluated for constant currents.

Although we have considered a system comprising only two circuits, the procedure remains the same for any number. The only difference is that the expression for the force F_{abx} must then be replaced by a sum of such terms. Stated in another way, the component of force in a given direction on the ith circuit is the product of its current by the space rate of change of flux linking it when it is displaced in the specified direction, all the currents in the system being held constant.

We now have two expressions for the force between two current-carrying circuits: Eq. 7-1, or 7-5, and Eq. 8-137. Let us see whether we can transform the latter equation into the former. We use the Neumann formula, Eq. 8-52. Then

$$\boldsymbol{F}_{ab}\cdot\boldsymbol{dr} = \frac{\mu_0}{4\pi} I_a I_b\, d\left(\oint_a\oint_b \frac{\boldsymbol{dl}_a\cdot\boldsymbol{dl}_b}{r}\right). \tag{8-141}$$

It is to be understood that the differential of the term within the parentheses on the right-hand side must correspond to the displacement $\boldsymbol{dr}$ of the coil b, as in Figure 8-13. Interchanging the order of the d operator and of the line integrals,

$$\boldsymbol{F}_{ab}\cdot\boldsymbol{dr} = \frac{\mu_0}{4\pi} I_a I_b \oint_a\oint_b d\,\frac{\boldsymbol{dl}_a\cdot\boldsymbol{dl}_b}{r}. \tag{8-142}$$

In the process of moving coil b by a distance $\boldsymbol{dr}$, both $\boldsymbol{dl}_a$ and $\boldsymbol{dl}_b$ remain unaffected, and the first d under the integral signs therefore operates only on the $1/r$ factor. If we now define $\boldsymbol{r}_1$ as a unit vector pointing from a to b in the direction of $\boldsymbol{dr}$,

$$d\left(\frac{1}{r}\right) = -\frac{dr}{r^2} = -\frac{\boldsymbol{r}_1\cdot\boldsymbol{dr}}{r^2}, \tag{8-143}$$

and

$$\boldsymbol{F}_{ab}\cdot\boldsymbol{dr} = -\frac{\mu_0}{4\pi} I_a I_b \oint_a\oint_b \frac{\boldsymbol{dl}_a\cdot\boldsymbol{dl}_b}{r^2}\,\boldsymbol{r}_1\cdot\boldsymbol{dr}. \tag{8-144}$$

Since the term $\boldsymbol{dr}$ on the right is independent of both $\boldsymbol{dl}_a$ and $\boldsymbol{dl}_b$, it can be removed from under the integral sign, and, finally,

$$\boldsymbol{F}_{ab} = -\frac{\mu_0}{4\pi} I_a I_b \oint_a\oint_b \frac{\boldsymbol{r}_1(\boldsymbol{dl}_a\cdot\boldsymbol{dl}_b)}{r^2}. \tag{8-145}$$

This is exactly Eq. 7-5, which is an alternative form of the magnetic force law, Eq. 7-1. We have therefore rediscovered the law that was the starting point for our discussion of magnetic fields in Chapter 7.

Example FORCE BETWEEN TWO COAXIAL SOLENOIDS

The case of two long solenoids, one of which extends a distance l within the other as in Figure 8-14, will serve as an example. We must again neglect end effects, for otherwise the calculation would be

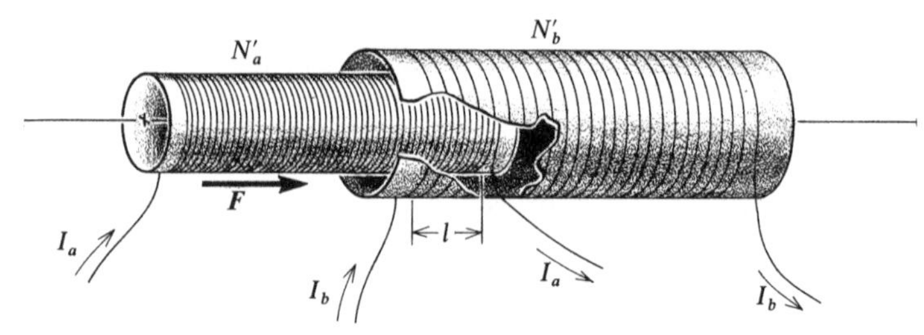

Figure 8-14. Two coaxial solenoids of different lengths and approximately equal diameters, the smaller one penetrating a distance l inside the other. There is an axial attractive force $\boldsymbol{F}$ when the two currents I_a and I_b are in the same direction. The solenoids have N'_a and N'_b turns/meter, respectively.

much more complex. We therefore assume that the solenoids are long and thin. We also assume that their diameters are approximately equal and that the currents flow in the same direction, as in the figure.

The mutual inductance is

$$M = \frac{(N'_a l)\Phi_{ba}}{I_b} = \frac{(N'_b l)\Phi_{ab}}{I_a}, \tag{8-146}$$

where N'_a and N'_b are the numbers of turns per meter on the two solenoids. Or, setting S to be their cross-sectional area,

$$M = \frac{\mu_0 N'_b N'_a I_b S l}{I_b} = \frac{\mu_0 N'_a N'_b I_a S l}{I_a}, \tag{8-147}$$

$$= \mu_0 N'_a N'_b S l, \tag{8-148}$$

$$\frac{\partial M}{\partial l} = \mu_0 N'_a N'_b S, \tag{8-149}$$

$$F = \mu_0 N'_a N'_b S I_a I_b. \tag{8-150}$$

The force is axial and it is attractive, since M increases with l.

Let us repeat the calculation, starting from Eq. 8-101:

$$W_m = \frac{1}{2\mu_0}\int_\infty B^2\,d\tau, \tag{8-151}$$

$$= \frac{1}{2\mu_0}[B_a^2(\tau_a - Sl) + B_b^2(\tau_b - Sl) + (B_a + B_b)^2 Sl], \tag{8-152}$$

$$= \frac{1}{2\mu_0}(B_a^2\tau_a + B_b^2\tau_b + 2B_aB_bSl), \tag{8-153}$$

where $B_a = \mu_0 N'_a I_a$ is the magnetic induction originating in solenoid a and, similarly, B_b is the magnetic induction originating in solenoid b. These are constants, since the currents I_a and I_b are again assumed to be constant. Then

$$dW_m = F\,dl = \frac{B_a B_b S}{\mu_0}\,dl, \tag{8-154}$$

$$F = \mu_0 N_a' N_b' S I_a I_b. \tag{8-155}$$

Note that the magnetic energy W_m is a function of l because the magnetic energy density depends on the square of B. If it depended on the first power of B, W_m would not be a function of l, and the force would be zero.

8.7.2 Magnetic Force When the Fluxes Are Kept Constant

Imagine that during the virtual displacement we adjust the currents so as to keep the flux linkages constant. There is then zero induced electromotance, and the sources supply zero energy to the system, except for the Joule losses, which we are neglecting. Then

$$\mathbf{F}_{ab}\cdot d\mathbf{r} = -(dW_m)_\Phi, \tag{8-156}$$

and

$$F_{abx} = -\left(\frac{\partial W}{\partial x}\right)_\Phi. \tag{8-157}$$

In this case the mechanical work is accompanied by a *de*crease in magnetic energy.

Example FORCE BETWEEN TWO COAXIAL SOLENOIDS

To calculate the force with the above method we can start with Eq. 8-151 for W_m. Thus

$$dW_m = \frac{1}{2\mu_0}$$

$$[2B_a(dB_a)\tau_a + 2B_b(dB_b)\tau_b + 2(dB_a)B_bSl + 2B_a(dB_b)Sl + 2B_aB_bS\,dl]. \tag{8-158}$$

Since the total flux linkage for solenoid a is constant,

$$B_a N_a S + B_b N_a \frac{l}{l_a} S = \text{Constant}, \tag{8-159}$$

or

$$B_a + B_b \frac{l}{l_a} = \text{Constant}, \tag{8-160}$$

and

$$dB_a + dB_b \frac{l}{l_a} + B_b \frac{1}{l_a}\,dl = 0. \tag{8-161}$$

The symmetrical equation for solenoid b is obtained by permuting the indices. Solving these two equations yields

$$dB_a = \frac{lB_a - l_bB_b}{l_al_b - l^2}\,dl, \tag{8-162}$$

and a symmetrical expression for dB_b. Substituting then these two equations into Eq. 8-158, and remembering that τ_a is l_aS while τ_b is l_bS, we find again an axial force

$$F = \frac{1}{\mu_0}B_aB_bS. \tag{8-163}$$

8.8 MAGNETIC TORQUE

To calculate a magnetic torque we again use the method of virtual work, and, by analogy with Eqs. 8-139, 8-140, and 8-157,

$$\Gamma = I_aI_b\frac{\partial M}{\partial\theta}, \tag{8-164}$$

$$= I\left(\frac{\partial\ \Phi}{\partial\theta}\right)_I, \tag{8-165}$$

$$= \left(\frac{\partial W_m}{\partial\theta}\right)_I, \tag{8-166}$$

$$= -\left(\frac{\partial W_m}{\partial\theta}\right)_\Phi, \tag{8-167}$$

where Γ is the torque and Φ is the flux linking the circuit.

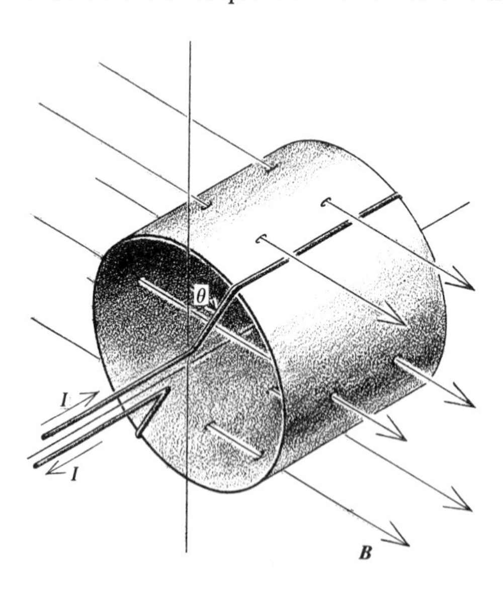

Figure 8-15. Loop carrying a current I in a uniform magnetic field $\boldsymbol{B}$. With the current in the direction shown, the loop is subject to a torque tending to increase the angle θ.

Example

MAGNETIC TORQUE ON A CURRENT LOOP

Figure 8-15 shows a rectangular loop carrying a current I and set at an angle θ in a uniform magnetic field B. The torque is

$$\Gamma = I\frac{\partial}{\partial\theta}(\Phi_0 - BS\cos\theta), \tag{8-168}$$

where Φ_0 is the flux produced by the current I in the loop, and S is the area of the loop. The flux $BS\cos\theta$ is negative because it links the loop in the negative direction with respect to the current I. Thus

$$\Gamma = IBS\sin\theta. \tag{8-169}$$

The torque is positive (in the direction of increasing θ), for $0 < \theta < \pi$. This result can be easily verified from the direction of the elementary forces $I(\boldsymbol{dl} \times \boldsymbol{B})$.

8.9 MAGNETIC FORCES WITHIN AN ISOLATED CIRCUIT

In an isolated circuit the current flows in its own magnetic field and is therefore subjected to a force. This is illustrated in Figure 8-16: the interaction of the current I with its own magnetic induction $\boldsymbol{B}$ produces a force that has the effect of extending the circuit to the right.

Magnetic forces within an isolated circuit are calculated by using either the magnetic force law or the principle of conservation of energy. In the case of Figure 8-16, it is not practical to use the former method because it would be too difficult to calculate the values of $\boldsymbol{B}$ and $\boldsymbol{J}$ *inside* the moving wire. The energy method is therefore indicated. See Problem 8-28.

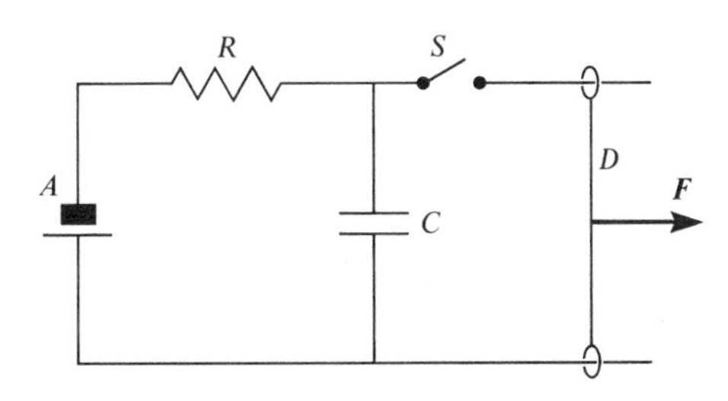

Figure 8-16. Schematic diagram of a rail gun. The battery A charges, through a resistance R, the capacitor C, which can be made to discharge through the line by closing switch S. The role of the capacitor is to store electric charge and to supply a very large current to the loop for a very short time. If side D is allowed to move, it moves to the right under the action of the magnetic force $\boldsymbol{F}$. Compare this figure with Figure 8-3.

Example

THE PINCH EFFECT

An individual ion moving at the periphery of an ion beam, as in Figure 8-17, is subjected to two forces: the outward electrostatic repulsion $\boldsymbol{F}_e$ and the inward magnetic force $\boldsymbol{F}_m$. The magnetic force thus tends to "pinch," or to concentrate the beam along its axis.

At *low* velocities the magnetic force is negligible and the beam diverges. If the axial velocity is sufficiently small, the radial velocity can be of the same order of magnitude or even larger, in which case the beam disappears.

We shall calculate $\boldsymbol{F}_e - \boldsymbol{F}_m$ on the assumption that the charge and current densities have cylindrical symmetry.

If the total current is I and if the radius of the beam is R, then

$$F_m = \frac{\mu_0 I}{2\pi R} Qv, \tag{8-170}$$

since the magnetic induction $\mu_0 I/2\pi R$ is perpendicular to the velocity v.

Also, if Q' is the electric charge per meter in the ion beam,

$$F_e = \frac{Q'}{2\pi\epsilon_0 R} Q \tag{8-171}$$

according to Gauss's law. Since $I = Q'v$,

$$F_e = \frac{IQ}{2\pi\epsilon_0 Rv}, \tag{8-172}$$

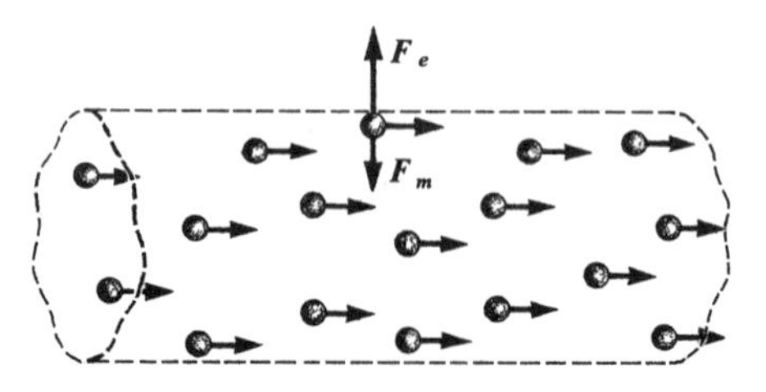

Figure 8-17. The pinch effect. In an ion beam the individual ions are subject to an outward electrostatic force $\boldsymbol{F}_e$ and to an inward magnetic force $\boldsymbol{F}_m$. At low velocities $F_m \ll F_e$ but, as the velocity v approaches the velocity of light c, F_m approaches F_e. If the electrostatic repulsion is cancelled by mixing ions of the opposite charge in the ion beam, $\boldsymbol{F}_m$ acts alone and the beam contracts. This phenomenon is often observed in positive ion accelerators. Residual gas in the path of the ion beam is ionized by impact, and the resulting low-energy electrons are trapped in the positive beam while the low-energy positive ions drift away. Thus, if the vacuum is not too good, the positive space charge in the beam can be reduced and the focusing improved. This phenomenon is known as the *pinch effect*, or as *gas focusing*.

and the net outward force is

$$F_e - F_m = \frac{IQ}{2\pi\epsilon_0 Rv}(1 - \epsilon_0\mu_0 v^2), \tag{8-173}$$

$$= \frac{IQ}{2\pi\epsilon_0 Rv}\left(1 - \frac{v^2}{c^2}\right). \tag{8-174}$$

The net force tends to zero for $v \to c$. This result is valid even if $v \approx c$.

If the kinetic energy of the particle is equal to its rest energy (5×10^5 electron-volts for an electron and 10^9 electron-volts for a proton), then the net force $F_e - F_m$ is smaller than F_e by a factor of four (Problem 5-24).

In practice, a vacuum is never perfect. Let us say the ions are positive. If their energy is of the order of tens of electron-volts or more, they ionize the residual gas, and the resulting positive ions drift away from the positive beam. The low-energy electrons, however, remain trapped in the beam and neutralize part of its space charge, thereby reducing F_e. This type of pinch effect is often called *gas focusing*.

8.10 MAGNETIC PRESSURE

Except for the ion beam of the last example, we have considered as yet only the magnetic forces acting on conducting wires. If, instead, we have a current sheet, then it is appropriate to think in terms of magnetic pressure.

Let us imagine a flat current sheet carrying $\boldsymbol{\lambda}$ amperes/meter and situated in a uniform tangential magnetic field $\boldsymbol{B}/2$ due to currents flowing elsewhere, as in Figure 8-18a, with $\boldsymbol{\lambda}$ normal to $\boldsymbol{B}$. If λ is very small, the magnetic field is not disturbed appreciably by the current sheet and the force per unit area is $\boldsymbol{\lambda}B/2$.

We now increase λ until it produces an aiding field $\boldsymbol{B}/2$ on one side and an opposing field $\boldsymbol{B}/2$ on the other side, as in Figure 8-18b. Then, from Problem 7-27, $\lambda = B/\mu_0$, and the force per unit area, or the pressure, is $B^2/2\mu_0$. We have arrived at this result for a flat current sheet, but the same applies to any current sheet when $B = 0$ on one side.

This value of the magnetic pressure permits us to calculate the energy density (Sections 8.5 to 8.5.4) in still another way. Imagine that the current sheet moves back infinitely slowly by a small distance x. Then the work performed by the magnetic pressure on a small area a of the current sheet is $ax(B^2/2\mu_0)$. This work is supplied by the sources that maintain B constant, and it is equal to the increase in the energy stored in the field (Section 8.7.1),

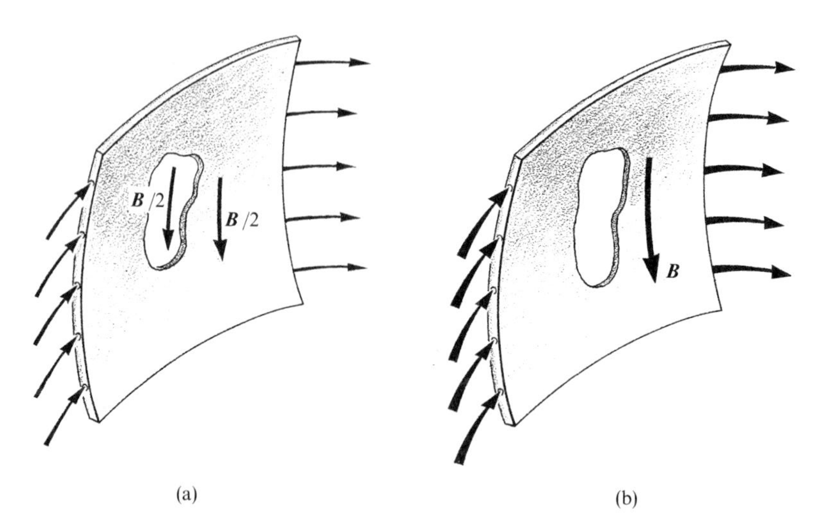

Figure 8-18. (a) A current sheet carries a very weak current density of λ ampere per meter of its width and is situated in a uniform magnetic field $B/2$. (b) The total magnetic field is that of figure (a), plus that of the current sheet. Selecting λ so that it just cancels the magnetic field on the left-hand side of the sheet, we have a total field B on the right.

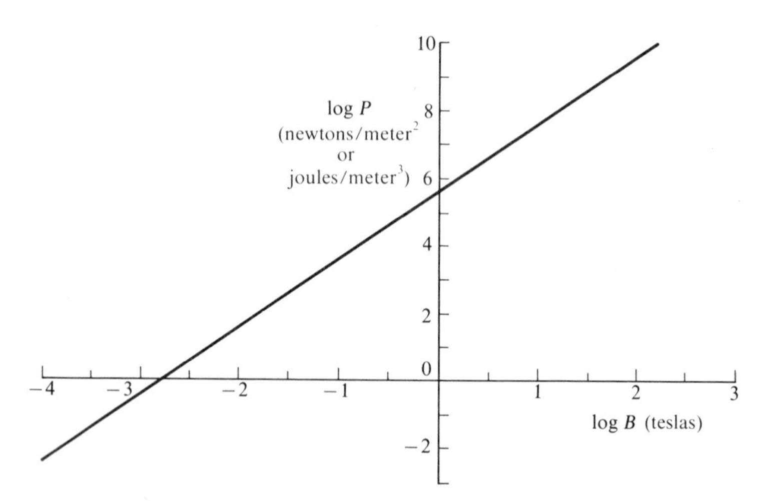

Figure 8-19. Magnetic pressure $B^2/2\mu_0$ as a function of B. The magnetic pressure is equal to the magnetic energy density.

or ax times the energy density. Then the energy density is $B^2/2\mu_0$, as i Section 8.5.1.

We arrived at a similar result in Section 2.16 when we deduced the energy density in an electric field from the electric force exerted on the surface of a conductor. The remarks we made at that time concerning this approach to the energy density in a field are equally valid here.

Figure 8-19 shows the magnetic pressure, or the magnetic energy density $B^2/2\mu_0$ as a function of B.

Example

MAGNETIC PRESSURE INSIDE A LONG SOLENOID

Figure 8-20 shows an end view of a solenoid. We shall use the method of virtual work to show that the magnetic pressure is $B^2/2\mu_0$.

Imagine that the current is maintained constant while the magnetic pressure increases the radius of the solenoid from R to $R + dR$. Then the magnetic energy W_m, which we calculated in Section 8.5, *in*creases by

$$dW_m = \pi\mu_0 I^2 N'^2 l R\, dR, \tag{8-175}$$

and the work performed by the magnetic pressure P is $2\pi Rl\,(dR)P$.

During the expansion, a voltage $Nd\Phi/dt$ is induced in the solenoid; according to Lenz's law, this induced voltage opposes the increase in flux, and hence opposes the applied voltage. So as to maintain the current I constant, the source must supply an extra voltage $Nd\Phi/dt$ and an extra *power* $NI\, d\Phi/dt$. Then the extra *energy* supplied by the source during the expansion is

$$NI\, d\Phi = N'lId(\pi R^2 \mu_0 N'I) = 2\pi\mu_0 I^2 N'^2 l R\, dR. \tag{8-176}$$

One half of this energy is transformed into the magnetic energy dW_m, and the other half into the mechanical work performed by P. Thus

$$2\pi Rl(dR)P = \pi\mu_0 I^2 N'^2 l R\, dR, \tag{8-177}$$

$$P = \frac{\mu_0 I^2 N'^2}{2} = \frac{B^2}{2\mu_0}, \tag{8-178}$$

since B is $\mu_0 IN'$.

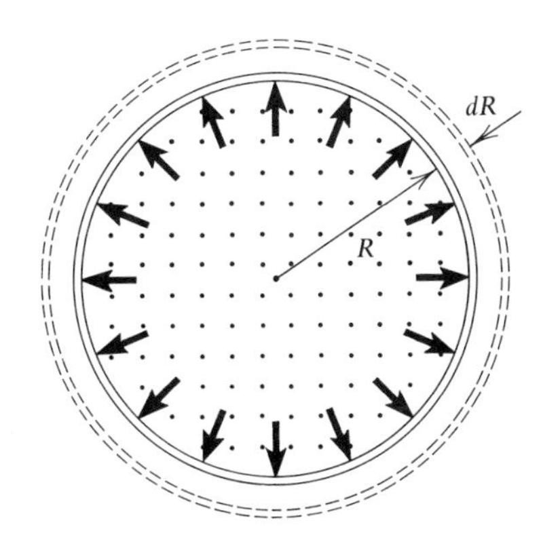

Figure 8-20. End view of a solenoid. The dots represent the lines of ***B***, which are normal to the paper, and the arrows show the direction in which the magnetic pressure is exerted.

8.11 SUMMARY

This chapter is concerned with (a) the nonconservative electric fields associated with time-dependent magnetic fields, (b) the energy stored in magnetic fields, and (c) the forces exerted on current-carrying circuits.

For a closed circuit,

$$\oint \boldsymbol{E}\cdot d\boldsymbol{l} = -\frac{d\Phi}{dt}, \tag{8-2}$$

$$= -\frac{d}{dt}\int_S \boldsymbol{B}\cdot d\boldsymbol{a}, \tag{8-3}$$

where $\boldsymbol{E}$ is the total electric field intensity, including the induced part. The directions in which the line integral is evaluated and the flux Φ is taken to be positive are related according to the right-hand screw rule.

This is the *Faraday induction law*, and the line integral is the *induced electromotance.*

The negative sign indicates that the induced electromotance tends to oppose the change in flux. This is *Lenz's law.*

Induced electromotance is observed whenever the magnetic flux linking a circuit changes with time, whether the circuit moves relative to the magnetic field, or whether a fixed circuit is linked by a variable flux, as in a transformer.

In differential form, the Faraday induction law becomes

$$\boxed{\boldsymbol{\nabla}\times\boldsymbol{E} = -\frac{\partial \boldsymbol{B}}{\partial t}.} \tag{8-8}$$

This is one of Maxwell's four fundamental equations of electromagnetism. It leads to the *general expression for the electric field intensity*

$$\boldsymbol{E} = -\frac{\partial \boldsymbol{A}}{\partial t} - \boldsymbol{\nabla} V, \tag{8-13}$$

where the first term is that part of $\boldsymbol{E}$ that is induced by changing magnetic fields, and the second is the part that is produced by accumulations of charge.

For a conductor moving with a velocity $\boldsymbol{u}$ in a changing magnetic field, Eq. 8-8 becomes

$$\boldsymbol{\nabla}\times\boldsymbol{E} = -\frac{\partial \boldsymbol{B}}{\partial t} + \boldsymbol{\nabla}\times(\boldsymbol{u}\times\boldsymbol{B}) \tag{8-36}$$

when $\boldsymbol{E}$ is measured in a coordinate system moving at a velocity $\boldsymbol{u}$ relative to that in which the magnetic induction is measured as $\boldsymbol{B}$.

The *mutual inductance* M between two circuits is equal to the magnetic flux linking one circuit per unit current flowing in the other:

$$\Phi_{ab} = M_{ab} I_a, \tag{8-51}$$

where Φ_{ab} is the flux originating in circuit a and linking circuit b. It is found that

$$M_{ab} = \frac{\mu_0}{4\pi} \oint_a \oint_b \frac{\boldsymbol{dl}_a \cdot \boldsymbol{dl}_b}{r}. \tag{8-52}$$

This is the *Neumann equation*. It shows that M depends solely on the geometry of the two circuits, if there are no magnetic materials present. By symmetry,

$$M_{ab} = M_{ba}. \tag{8-54}$$

Self-inductance L is a similar quantity, which applies to a single circuit:

$$\Phi = LI. \tag{8-57}$$

In terms of self-inductance, the induced electromotance is

$$\oint \boldsymbol{E} \cdot \boldsymbol{dl} = -L \frac{dI}{dt}; \tag{8-58}$$

a similar relation applies to mutual inductance.

The *coefficient of coupling* k between two circuits is defined by

$$M = \pm k (L_a L_b)^{1/2}, \tag{8-82}$$

and can have values ranging from -1 to $+1$. It is zero if none of the flux of one circuit links the other.

The *energy stored in a magnetic field* is equal to that required to establish the field, and the energy supplied to an element of volume $d\tau$ of conductor per unit time is

$$\frac{dW_m}{dt} = \int_\tau \frac{\partial \boldsymbol{A}}{\partial t} \cdot \boldsymbol{J}_f \, d\tau. \tag{8-91}$$

If there are no magnetic materials present, and if $\boldsymbol{\nabla} \times \boldsymbol{B} = \mu_0 \boldsymbol{J}_f$ is applicable,

$$W_m = \frac{1}{2\mu_0} \int_\infty B^2 \, d\tau, \tag{8-101}$$

$$= \frac{1}{2} \int_\tau (\boldsymbol{J}_f \cdot \boldsymbol{A}) \, d\tau, \tag{8-106}$$

where τ is any volume which includes all regions in which $\boldsymbol{J}_f$ is not zero. We also find that

$$W_m = \frac{1}{2} I\Phi, \tag{8-109}$$

where Φ is the flux linking the current I. As usual, the positive directions for I and Φ are related according to the right-hand screw rule. In terms of the inductances L_a, L_b, M for two circuits a and b,

$$W_m = \frac{1}{2} L_a I_a^2 + \frac{1}{2} L_b I_b^2 + M I_a I_b. \tag{8-118}$$

The *self-inductance of a volume distribution of current* is defined from its stored magnetic energy:

$$L = \frac{1}{\mu_0 I^2} \int_\infty B^2 \, d\tau. \tag{8-122}$$

The *magnetic force* between two current-carrying circuits can be calculated from the principle of conservation of energy applied to a virtual displacement of one of the circuits. The x-component of the force $\boldsymbol{F}_{ab}$ exerted *by a on b* is found to be

$$F_{abx} = I_a I_b \frac{\partial M}{\partial x}, \tag{8-139}$$

$$= I_a \left(\frac{\partial \Phi_{ba}}{\partial x} \right)_I, \tag{8-140}$$

$$= \left(\frac{\partial W_m}{\partial x} \right)_I, \tag{8-140}$$

$$= -\left(\frac{\partial W_m}{\partial x} \right)_\Phi, \tag{8-157}$$

the subscripts indicating which parameter is kept constant.

Equation 8-145 is an alternative of the magnetic force law stated in Ed. 7-1, which was the starting point for our entire discussion of magnetic fields.

Magnetic torque is given by

$$\Gamma_\theta = \left(\frac{dW_m}{d\theta} \right)_I, \tag{8-166}$$

as well as by other expressions similar to the ones above for F_{abx}, with x replaced by θ.

Magnetic forces also occur within an isolated circuit because a current flows in its own magnetic field.

Magnetic pressure on a conductor is observed whenever the current density is not parallel to the local magnetic induction.

If the magnetic induction is B on one side of a current sheet and zero on the other, then the magnetic pressure is equal to the magnetic energy density $B^2/2\mu_0$.

PROBLEMS

Note: An asterisk indicates that the problem requires a knowledge of relativity.

8-1. A carriage runs on two rails on either side of a long tank of water equipped for testing boat models. The rails are 3.0 meters apart and the carriage has a maximum speed of 20 meters/second.

Calculate the maximum voltage between the two rails if the vertical component of the earth's magnetic field is 2.0×10^{-5} tesla.

8-2. A conducting bar slides at a constant velocity u along conducting rails in a region of uniform magnetic induction, as in Figure 8-3. The resistance in the circuit is R and the inductance is negligible.

(a) Calculate the current I flowing in the circuit.
(b) How much power is required to move the bar?
(c) How does this power compare with the power loss in the resistance R?

8-3. If a conductor is given an acceleration $\boldsymbol{a}$, the conduction electrons are subjected to inertia forces $-m\boldsymbol{a}$.

Show that, if $\boldsymbol{E}'$ is the equivalent total electric field intensity in the conductor, then

$$\boldsymbol{\nabla} \times \boldsymbol{E}' = -\frac{\partial \boldsymbol{B}}{\partial t} + \frac{m}{e} \boldsymbol{\nabla} \times \boldsymbol{a}.$$

This effect was predicted by Maxwell, and was observed for the first time in 1916 by Tolman and collaborators.

The inverse effect, namely the acceleration of a body carrying a variable current was also predicted by Maxwell, and was first observed by Barnett and others in 1930.

8-4. In the *betatron*, electrons are held in a circular orbit in a vacuum chamber by a magnetic field $\boldsymbol{B}$. The electrons are accelerated by increasing the magnetic flux linking the orbit.

Show that the average magnetic induction over the plane of the orbit must be twice the magnetic induction at the orbit if the orbit radius is to remain fixed as the electron's energy is increased.‡

8-5. A toroidal coil is fed a voltage $V(t)$ by a variable power supply. The length of the wire is l.

Discuss the electric field intensity E inside the wire and show that E is *not* equal to V/l.

In Problem 8-18, we shall see that the magnitude of the extra term is $\partial A/\partial t$.

8-6. Show that the electromotance induced in a loop rotating at an angular velocity ω in a magnetic field $B_0 \sin \omega t$, as in Figure 8-6, is

$$-B_0 S\omega \cos 2\omega t.$$

8-7. A thin conducting disk of thickness h, diameter D, and conductivity σ is placed in a uniform alternating magnetic field $B = B_0 \sin \omega t$ parallel to the axis of the disk.

(a) Find the induced current density as a function of distance from the axis of the disk.
(b) What is the direction of this current?

8-8. *Electromagnetic flowmeters* operate as follows. A fluid, which must be at least slightly conducting (the blood of an animal, for example), flows in a nonconducting tube between the poles of a magnet. The ions in the fluid are then subjected to magnetic forces $Q\boldsymbol{u} \times \boldsymbol{B}$ in the direction perpendicular to the velocity $\boldsymbol{u}$ of the fluid and to the magnetic induction $\boldsymbol{B}$. Electrodes placed on either side of the tube and in contact with the fluid thus acquire charges of opposite signs and the resulting voltage difference is a measure of u.

For example, if the tube has a rectangular cross section ab with $\boldsymbol{B}$ parallel to side b, and if the velocity were the same throughout the cross section, then the voltage between the electrodes would be uBa.

Note that, in the electromagnetic flowmeter, ions of both signs have the same velocity $\boldsymbol{u}$ and that the polarity of the electrodes is always the same for a given direction of flow and for a given direction of $\boldsymbol{B}$. Compare with the Hall effect discussed in the example on page 300.

Faraday attempted to measure the velocity of the Thames river in this way in 1832. The magnetic field was of course that of the Earth.

(a) In the absence of turbulence, the fluid velocity u in a tube of radius a is of the form

$$u = u_0\left(1 - \frac{\rho^2}{a^2}\right),$$

where ρ is the radial coordinate.

Show that the difference in potential between two points situated at the ends of a diameter perpendicular to $\boldsymbol{B}$ is then

$$\frac{8}{3\pi}\frac{QB}{a},$$

where Q is the volume of fluid flowing through the tube in one second.

This result is in fact *incorrect*, as we shall see below.

(b) Sketch a cross section of the tube, showing by means of arrows of various lengths the magnitude of $\boldsymbol{u} \times \boldsymbol{B}$ at various points inside the tube.

(c) The fact that $\boldsymbol{u} \times \boldsymbol{B}$ is not the same everywhere in the fluid will cause currents to flow.

Sketch another cross section of the tube, showing the lines of current flow.

The current drawn by the electrodes is negligible.

It turns out that this current causes an Ohmic drop that reduces the voltage calculated under (a) above by one quarter, and that the correct voltage difference is

$$\frac{2QB}{\pi a}.$$

(d) Show that this is the voltage one would expect if the velocity u were uniform and equal to the average velocity, and if there were zero Ohmic drop as above.

(e) We have neglected edge effects in the regions where the conducting fluid enters into and emerges from the magnetic field.

Sketch lines of current flow in these two regions.

These currents further increase the Ohmic drop.

8-9. A particle of charge Q, mass m, and kinetic energy T describes a circle of radius r in a uniform magnetic field B.

The magnetic field increases at such a rate that $dr/r \ll 1$ during one revolution.

(a) Show that, during one revolution,

$$\Delta T = \pi r^2 Q \frac{dB}{dt} = T \frac{\Delta B}{B},$$

and that, therefore,

$$\frac{T}{B} = \text{Constant}.$$

Neglect relativistic effects.‡

(b) If the charge of the particle is imagined to be spread out around the circumference of the circle, it corresponds to a current $Q\omega/2\pi$.

Show that the corresponding magnetic moment is T/B and is therefore approximately constant.

(c) Show that the flux linked by the trajectory is constant.

*8-10. Repeat part (a) of the preceding problem without restricting your calculations to nonrelativistic velocities.

You should find that

$$\Delta T = \frac{1}{2} m v^2 \frac{\Delta B}{B}.‡$$

*8-11. In one type of *ion drag accelerator* a ring of high-energy electrons rotating in a magnetic field and containing a small percentage of protons is accelerated as a unit along its axis of symmetry.

A useful analogy for explaining the accelerating force on the protons is that of a marble in a saucer in a railway train. If the acceleration is sufficiently gradual, the marble stays in the saucer and acquires the velocity of the train. the force on the protons is of course due to the electric attraction of the electrons.

The main advantage of this type of accelerator will be demonstrated under (g) below.

Let us assume that the ring is initially at rest and that the electrons have a kinetic energy T_1 of 1.50 million electron-volts. Protons are injected into the ring by admitting a small amount of hydrogen gas, which becomes partly ionized. The kinetic energies of the protons are negligible, initially.

The magnetic induction B_1 is 2.00×10^{-2} tesla.

(a) Show that the initial radius r_1 of the ring is 32.4 centimeters.‡

(b) The magnetic field is then increased by a factor of 50 to $B_2 = 1.00$ tesla.

Show that the radius of the ring shrinks to $r_2 = 4.6$ centimeters, and that the kinetic energy of the electrons increases to 13.4 million electron-volts.‡

(c) Let us call the axis of symmetry of the ring the z-axis.

Show qualitatively that, if B is independent of the time, but decreases in the positive direction of the z-axis, the ring is accelerated in that direction.‡

(d) The ring therefore acquires an axial velocity. But the static magnetic field cannot increase the kinetic energy of the particles, because the magnetic force $Q\boldsymbol{u} \times \boldsymbol{B}$ is perpendicular to the velocity $\boldsymbol{u}$.

Then how can the ring acquire an axial velocity?

(e) Show that, in a plane where the axial component of $\boldsymbol{B}$ is B_{3z}, the ring has a radius

$$r_3 = r_2\left(\frac{B_{2z}}{B_{3z}}\right)^{1/2}.$$

Thus, at a plane where B_{3z} is 0.2 tesla, r_3 is $4.6 \times 5^{1/2} = 10.3$ centimeters.

This process is called *expansion acceleration*.‡

(f) Show that γ_z corresponding to the axial velocity obeys the following equation:

$$\frac{\gamma_{3z}}{\gamma_{2z}} \approx \left(\frac{B_{2z}}{B_{3z}}\right)^{1/2}.$$

(g) The protons have the same γ_z as the electrons, and their transverse velocity is negligible.

(h) Show that the protons have a kinetic energy of 1.2×10^9 electron-volts in the plane where B_{2z} is 0.2 tesla.

The protons therefore acquire energies of the order of 40 times that of the electrons. This is the major advantage of this type of accelerator.

The ring could also be accelerated in an electric field (*electric acceleration*). For example, if, the ring were accelerated through a difference of potential of 10 million volts, the protons would acquire a kinetic energy of 400 million electron-volts. Heavy ions would gain even higher energies.

8-12. A wire bent in the form of a circle of radius R is placed so that its center is at a distance $D = 2R$ from a long straight wire, the two being in the same plane.

Show that the mutual inductance is $0.268\ \mu_0 R$.

8-13. Two long parallel rectangular loops lying in the same plane have lengths l_1 and l_2, and widths w_1 and w_2, respectively. The loops do not overlap, and the distance between the near sides is s.

Show that the mutual inductance between the loops is

$$M = \frac{\mu_0 l_2}{2\pi} \ln \frac{s + w_2}{s\left(1 + \dfrac{w_2}{s + w_1}\right)},$$

if $l_2 < l_1$, and if the loops have but a single turn. Neglect end effects.

8-14. The coefficient of coupling k between two single-turn coils was defined in Section 8.4.3. In general, $k_a \neq k_b$.

Show that

$$\frac{k_a}{k_b} = \frac{L_b}{L_a}.$$

8-15. Show that the self-inductance of a close-wound toroidal coil of n turns, and of major radius R and minor radius r is

$$L = \mu_0 n^2 \{R - (R^2 - r^2)^{1/2}\}.$$

8-16. It is shown in Section 8.7.1 that, if one active circuit is allowed to move with respect to another, the mechanical work performed by the sources is equal to the increase in magnetic energy. The currents are assumed to be maintained constant. Hence the force between two active circuits is given by the rate of change of magnetic energy.

Show that, similarly, if the geometry of an isolated active circuit is altered, the mechanical work performed by the source is equal to the increase in magnetic energy. Assume again that the current is kept constant. It follows that, on this assumption, the force on an element of an active circuit is given by the rate of change of magnetic energy.‡

See also Problem 8-28.

8-17. It is known that high-frequency currents do not penetrate into a conductor as do low-frequency currents. This is called the skin effect.

Would you expect the self-inductance of a coaxial line to increase or to decrease with increasing frequency?

8-18. (a) Show that the magnitude of the vector potential at the surface of a close-wound toroidal coil of circular cross section is $\Phi/2\pi r$, where Φ is the magnetic flux inside the coil and r is the minor radius of the toroid.

(b) Show that the magnitude of the extra term found in Problem 8-5 is $\partial A/\partial t$.

8-19. Show that the power lost by Joule heating per cubic meter in a conductor is EJ_f, or J_f^2/σ, where J_f is the current density and σ is the conductivity.

8-20. Compare the energies per unit volume in (a) a magnetic field of 1.0 tesla and (b) an electrostatic field of 10^6 volts/meter.

8-21. Calculate the magnetic energy in a toroidal coil in at least two different ways.

8-22. Consider n rigid fixed circuits, the ith circuit carrying a current I_i.

Show that the magnetic energy associated with the system can be written as a sum of self-energy terms of the form $(\frac{1}{2}) L_i I_i^2$, plus a sum of mutual energy terms of the form $M_{ij} I_i I_j$, each pair of currents appearing once.

*8-23. In the examples on pages 263 and 270 we transformed the field inside a parallel-plate capacitor.

(a) Show that the field energy in frame 1 is $\gamma(1 + \beta^2)$ times that in frame 2.
(b) How do the energies $\mathcal{E}_1$ and $\mathcal{E}_2$ compare in the case of the solenoid of Problem 6-15?

8-24. A coil of resistance R and inductance L is fed by a dc power supply.

(a) Show that the power expended by the source when the voltage changes is

$$I^2R + \frac{d}{dt}\left(\frac{1}{2}LI^2\right) = I^2R + I\frac{d\Phi}{dt}.$$

(b) Show that this is in agreement with Eq. 8-89.

8-25. *Electromagnetic levitation* forces are sometimes used to support objects in space. For example, if a metal object is situated near a coil carrying an alternating current I, eddy currents will flow in the object and there will result a repulsive force.

Show that the force in the direction of x is

$$F_x = \frac{1}{2} I^2 \frac{\partial L}{\partial x}$$

if the object is allowed to move in the x direction, and if the effective inductance of the coil is L.‡

8-26. (a) Show that a current-carrying coil tends to orient itself in a magnetic field in such a way that the total magnetic flux linking the coil is *maximum.*
(b) Show that the torque exerted on the coil is $\boldsymbol{m} \times \boldsymbol{B}$, where $\boldsymbol{m}$ is the magnetic moment of the coil and $\boldsymbol{B}$ is the magnetic induction when the current in the coil is zero.

8-27. Many satellites require attitude control to keep their instrumentation properly oriented. For example, communication satellites must keep thèir antenna systems on target. Attitude control requires a mechanism for exerting appropriate torques as they are required.

Attitude control can be achieved by means of coils whose magnetic fields interact with that of the Earth.

(a) Show that the torque exerted by such a coil is

$$NIBA \sin \theta,$$

where N is the number of turns, I is the current, B is the magnetic induction due to the earth, A is the area of the coil, and θ is the angle between the earth's magnetic field and the normal to the coil.
(b) Calculate the number of ampere-turns required for a coil wound around the outside surface of a satellite whose diameter is 1.14 meters. The torque required at $\theta = 5$ degrees is 10^{-3} newton-meter, and the magnetic induction at an altitude of 700 kilometers over the equator, where the orbit of the satellite will be situated, is 4.0×10^{-5} tesla.

8-28. Let us calculate the force on the movable link D in the circuit of Figure 8-16.

Instead of being a metallic rod, the link can also be an electric arc. The device then accelerates blobs of plasma and is called a *plasma gun*. Such plasma guns can be used as rocket motors.

We cannot solve this problem by integrating the magnetic force $\boldsymbol{J} \times \boldsymbol{B}$ over the volume of the link, since neither $\boldsymbol{J}$ nor $\boldsymbol{B}$ are known, or even easily calculable. Instead, we shall find the force by investigating the magnetic and mechanical energies involved.

If we set $x = 0$ at the initial position of the link, then the inductance L in the circuit is $L_0 + L'x$, where $L'x$ is the inductance associated with the magnetic flux linking the rails.

During the discharge of the capacitor C we can neglect the battery and its resistance R, since the current flowing in that part of the circuit is negligible.

(a) Show that the force F on the link D is $(i^2/2)L'$, where i is the instantaneous current flowing in the right-hand side of the circuit.

Call the resistance on that side R'.‡

(b) Show that, during the discharge,

$$L\frac{d^2Q}{dt^2} + \frac{dQ}{dt}\frac{dL}{dt} + R'\frac{dQ}{dt} + \frac{Q}{c} = 0,$$

where Q is the charge on the capacitor at the time t.

8-29. A superconducting power transmission line has been proposed that would have the following characteristics. It would carry 10^{11} watts at 200 kilovolts dc over a distance of 10^3 kilometers. The conductors would have a circular cross section of 5 centimeters^2 and would be held 5 centimeters apart, center to center.

(a) Find an approximate value for the magnetic force per meter.‡

(b) Calculate $B^2/2\mu_0$ at the surface of a conductor, on the inner side.‡

(c) Calculate the stored energy and its cost at 0.5 cent per kilowatt-hour.

The inductance of such a line is 0.66 microhenry/meter.

8-30. We have seen in the example on page 366 that an ion situated at the periphery of an ion beam of radius r is subjected to a force

$$\frac{IQ}{2\pi\epsilon_0 rv}\left(1 - \frac{v^2}{c^2}\right).$$

(a) Show that the divergence of the beam is given by the following equation:

$$\frac{dr}{dt} = \left(C\ln\frac{r}{r_0}\right)^{1/2},$$

where r_0 is the radius of the beam at a point where the radial velocity is zero, and where

$$C = \frac{IQ}{\pi\epsilon_0 vm}\left(1 - \frac{v^2}{c^2}\right).$$

You can solve this differential equation by referring to Problem 6-23(b).

*(b) Show that this differential equation is equivalent to the equation

$$\frac{dr}{dt_2} = \left[\frac{e\lambda_2}{\pi\epsilon_0 m_0}\ln\frac{r}{r_0}\right]^{1/2}$$

which we found in Problem 6-23 for a pulsed beam of electrons of charge e.

*8-31. Show that F_m is equal to F_e (Figure 8-17) within 10% when the beam energy (the kinetic energy of a particle) is at least equal to twice the rest energy of a particle, approximately.

For electrons, this beam energy is 1 MeV, and for protons it is 2 GeV.

8-32. Check the graph of Figure 8-19 and show that magnetic pressure is equal to one atmosphere when B is only about 0.5 tesla, or 5000 gauss.

8-33. (a) Show that magnetic pressure is about $4B^2$ atmospheres, if B is expressed in teslas.

(b) Draw a log-log plot of the electric force per unit area $\frac{1}{2}\epsilon_0 E^2$ as a function of E.

(c) Discuss the similarities and differences between the magnetic pressure and the electric force per unit area.

8-34. A thin tube of wall thickness t and inside radius R carries a current I.

(a) Show that the magnetic pressure is directed inward and is $(\mu_0/8\pi^2)(I/R)^2$.

(b) Calculate the pressure in atmospheres for a current of 1000 amperes and a tube of 1 centimeter radius.

8-35. Magnetic fields are used for performing various mechanical tasks that require a high power level for a very short time.

For example, the magnetic pressure inside a coil can be used to crush a light aluminum tube that acts as a shutter to turn off a beam of light or of soft X-rays. When the coil is suddenly connected to a charged capacitor, the induced electromotance $d\Phi/dt$ produces a large current in the aluminum tube, which collapses under the magnetic pressure.

Let us calculate the pressure exerted on a conducting tube placed inside a long solenoid. If the current I in the solenoid is increased gradually from zero to some arbitrarily large value, the current induced in the tube is small and the magnetic pressure is negligible. Let us assume that dI/dt is so large that the induced current in the tube maintains zero magnetic field inside it. Then there is a magnetic field B only in the annular region between the solenoid and the conducting tube.

(a) Calculate the pressure on the tube for $B = 1$ tesla.

(b) What would be the pressure if the conducting tube were parallel to the axis, but off center?

8-36. Extremely high magnetic fields can be obtained by discharging a capacitor through a low-inductance coil. The capacitor itself must of course have a low inductance. Such fast capacitors cost approximately one dollar per joule of stored-energy capacity.

(a) Estimate the cost of a capacitor that could store an energy equal to that of a 100 tesla (one megagauss) magnetic field occupying a volume of one liter.

(b) Estimate the cost of the electricity required to charge the capacitors.
(c) Calculate the magnetic pressure in atmospheres, at 100 teslas.

8-37. Flux compression is one method of obtaining large values of B. For example, a magnetic field B_0 is established inside a conducting tube of radius R_0, and the tube is then imploded by means of an explosive placed around the tube. Currents flow in the tube, and the internal magnetic pressure builds up until it is equal to the external gas pressure.

(a) Show that, if the radius of the tube shrinks very rapidly,

$$B \approx B_0(R_0/R)^2,$$

where B is the magnetic induction when the radius has been reduced to R.
(b) Show that the surface current density in the tube must be 10^9 amperes/meter to achieve a field of 10^3 teslas.
(c) If the initial B is 20 teslas, and if the tube has initially a diameter of 20 centimeters, what should be the value of B when the tube is compressed to a diameter of about 5 millimeters?
(d) Calculate the resulting increase in magnetic energy, and hence the explosive energy, required to compress the field. Assume that the cylinder is 20 centimeters long, and neglect end effects.
(e) Flux compression can also be achieved by means of a conducting piston shot axially into a solenoid.

If the radius of the solenoid is R_0, and if the radius of the piston is R, show that the magnetic induction in the annular region between the piston and the solenoid reaches a value of

$$B \approx \frac{B_0}{1 - \left(\frac{R}{R_0}\right)^2}.$$

Hints

8-4. Relate the centripetal acceleration to the magnetic force acting on the electron, and then find the condition that lets the linear momentum of the electron increase with fixed orbit radius. Use Newton's second law with the tangential force on the electron given by the Faraday induction law.

8-9. (a) If you find a ΔT that is too large by a factor of 2, you have probably made an approximation that is not legitimate. The increase in kinetic energy ΔT is the work done by the force $Q(\partial A/\partial t)$.

8-10. When the magnetic field is constant, the mass of an electron remains constant and the magnetic force BeV is equal to the centripetal force mv^2/r, as for the nonrelativistic case, except that m is the relativistic mass γm_0.

8-11. (a) See the hint for the preceding problem.
(b) Use the result of the preceding problem.
(c) Draw a set of lines of $\boldsymbol{B}$ in a plane containing the axis. See also Problem 7-22.

(e) Remember that the particles are submitted to a central force. What can you say about their angular momentum?

(f) Use the fact that $\gamma_{2z}\gamma_{2\theta} = \gamma_{3z}\gamma_{3\theta} = \gamma$, as demonstrated in Problem 5-14, and remember that $v_{2\theta} \approx v_{3\theta} \approx c$ in the reference frame of the ring.

8-16. Consider the general case of a single-turn circuit.

8-25. Consider the object as a secondary winding whose impedance is $j\omega L_2$. (The resistance R_2 is made negligible by selecting a high enough frequency.) The effective inductance L of the primary is then $L_1(1 - k^2)$, and displacing the object changes k. Then calculate the work done by the source in the primary for a small Δk which takes place during the time $t = 0$ to $t = T$, where T is the period $2\pi/\omega$.*

8-28. (a) Equate the power supplied by the capacitor to the rate at which the energy increases in the right-hand side of the circuit. The voltage across a variable inductance is $d(Li)/dt$.

8-29. Assume that the current is concentrated along the axes of the conductors. Can you show that this gives the correct result?

---

* For a more general proof of this result, see Sir James Jeans, *The Mathematical Theory of Electricity and Magnetism*, Fifth Edition, Cambridge University Press (1951), Paragraph 564.

CHAPTER 9

MAGNETIC FIELDS III

Magnetic Materials

Thus far we have studied only those magnetic fields that are due to the motion of free charges. Now, on the atomic scale, all bodies contain spinning electrons that move around in orbits, and these electrons also produce magnetic fields.

Our purpose in this chapter is to express the magnetic fields of these atomic currents in macroscopic terms.

Magnetic materials are similar to dielectrics in that individual charges or systems of charges can possess magnetic moments, and these moments, when properly oriented, produce a resultant magnetic moment in a macroscopic body. Such a body is then said to be *magnetized.*

In most atoms the magnetic moments due to the orbital and spinning motions of the electrons cancel. If the cancellation is not complete, the material is said to be *paramagnetic*. The so-called transition elements, like manganese, are examples of paramagnetic substances. When such a substance is placed in a magnetic field, its atoms are subjected to a torque that tends to align them with the field, but thermal agitation tends to destroy the alignment. This phenomenon is analogous to the alignment of polar molecules in dielectrics (Section 3.9).

In *diamagnetic* materials, the elementary moments are not permanent but are induced according to the Faraday induction law (Section 8.1). All materials are diamagnetic, but orientational polarization may predominate.

Most magnetic devices use *ferromagnetic* materials, such as iron, in which the magnetization can be orders of magnitude larger than that of either para- or diamagnetic substances. This large magnetization is attributed to electron spin and is associated with group phenomena in which all the elementary moments in a small region, known as a *domain*, are aligned. The magnetization of one domain may be oriented at random with respect to that of a neighboring domain. The large magnetizations that are characteristic of ferromagnetic materials are the result of the collective orientation of whole domains.

Although the behavior of magnetic materials seems at first glance to be analogous to that of dielectric materials, there is one important difference. While most dielectrics are linear, ferromagnetic materials are not only highly nonlinear, but their behavior also depends on their previous magnetic history. The calculation of the fields associated with magnetic materials is therefor largely empirical.

9.1 THE MAGNETIZATION $\boldsymbol{M}$

If $\boldsymbol{m}$ is the *average magnetic dipole moment per atom*, and if N is the number of atoms per unit volume, the *magnetization* is defined as

$$\boldsymbol{M} = N\boldsymbol{m}, \tag{9-1}$$

if the individual dipole moments in the element of volume $d\tau$ considered are all aligned in the same direction. If the moments are not all aligned, then the magnetization is the net magnetic moment per unit volume. The volume $d\tau$ is subject to the limitations discussed in Section 3.1.

The magnetization $\boldsymbol{M}$ is measured in amperes/meter, and it corresponds to the polarization $\boldsymbol{P}$ in dielectrics.

9.2 MAGNETIC INDUCTION $\boldsymbol{B}$ AT AN EXTERIOR POINT

Let us calculate the magnetic induction at a point P outside a magnetized body, as in Figure 9-1. Our discussion will parallel that of Section 3.2.

From Section 7-8, the vector potential of a current loop is

$$\boldsymbol{A} = \frac{\mu_0}{4\pi} \frac{\boldsymbol{m} \times \boldsymbol{r}_1}{r^2}, \tag{9-2}$$

where $\boldsymbol{m}$ is the magnetic dipole moment of the loop, and where the distance r is large compared to the size of the loop. The unit vector $\boldsymbol{r}_1$ points, as usual, from the source, which is the loop, to the point P where the field is calculated.

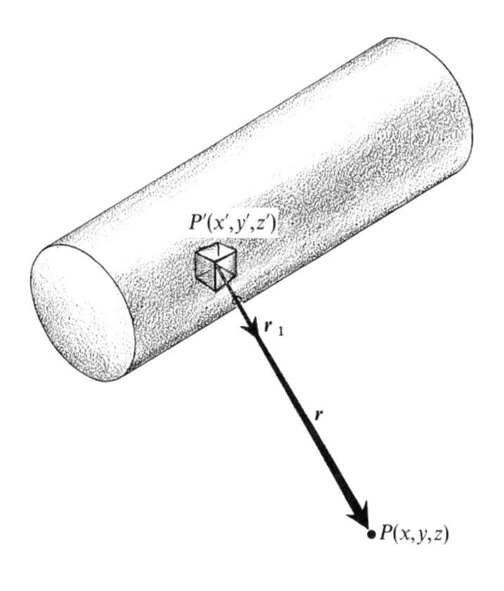

Figure 9-1. Element of volume inside a magnetized body, and an external point P.

Then the element of dipole moment $\boldsymbol{M}\,d\tau'$ situated at $P'(x', y', z')$ produces at $P(x, y, z)$ a vector potential

$$d\boldsymbol{A} = \frac{\mu_0}{4\pi}\frac{\boldsymbol{M} \times \boldsymbol{r}_1}{r^2}\,d\tau', \tag{9-3}$$

and

$$\boldsymbol{A} = \frac{\mu_0}{4\pi}\int_{\tau'}\frac{\boldsymbol{M} \times \boldsymbol{r}_1}{r^2}\,d\tau', \tag{9-4}$$

$$= \frac{\mu_0}{4\pi}\int_{\tau'}\boldsymbol{M} \times \boldsymbol{\nabla}'\left(\frac{1}{r}\right)d\tau', \tag{9-5}$$

where τ' is the volume of the magnetized material, or

$$\boldsymbol{A} = -\frac{\mu_0}{4\pi}\int_{\tau'}\boldsymbol{\nabla}' \times \frac{\boldsymbol{M}}{r}\,d\tau' + \frac{\mu_0}{4\pi}\int_{\tau'}\frac{\boldsymbol{\nabla}' \times \boldsymbol{M}}{r}\,d\tau', \tag{9-6}$$

from Problem 1-18. Then, from Problem 1-30,

$$\boldsymbol{A} = \frac{\mu_0}{4\pi}\int_{S'}\frac{\boldsymbol{M} \times \boldsymbol{n}}{r}\,da' + \frac{\mu_0}{4\pi}\int_{\tau'}\frac{\boldsymbol{\nabla}' \times \boldsymbol{M}}{r}\,d\tau', \tag{9-7}$$

where S' is the surface bounding the volume τ' of the magnetized body. We may omit the prime on the $\boldsymbol{\nabla}$ that operates on $\boldsymbol{M}$ since $\boldsymbol{\nabla}$ clearly involves derivatives with respect to the coordinates x', y', z' of the point where the magnetization is $\boldsymbol{M}$.

Although these expressions for $\boldsymbol{A}$ are all equivalent, the last one lends itself to a simple physical interpretation: both integrands have the form of the vector potential calculations already familiar to us from Section 7.5,

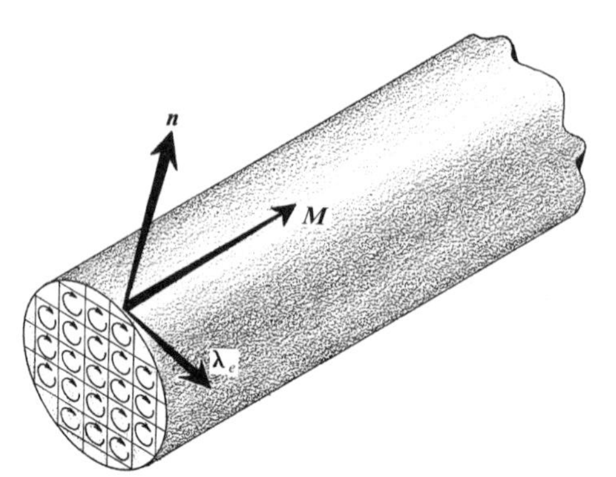

Figure 9-2. Ampère's model for the equivalent currents.

provided the numerators represent a surface current density in the first term and a volume current density in the second term. We therefore define an *equivalent surface current density*

$$\boldsymbol{\lambda}_e = \boldsymbol{M} \times \boldsymbol{n} \tag{9-8}$$

and an *equivalent volume current density*

$$\boldsymbol{J}_e = \boldsymbol{\nabla} \times \boldsymbol{M}. \tag{9-9}$$

Then the vector potential at P is

$$\boldsymbol{A} = \frac{\mu_0}{4\pi} \int_{S'} \frac{\boldsymbol{\lambda}_e}{r}\, da' + \frac{\mu_0}{4\pi} \int_{\tau'} \frac{\boldsymbol{J}_e}{r}\, d\tau'. \tag{9-10}$$

The equivalent surface current density $\boldsymbol{\lambda}_e$ for a uniformly polarized rod is shown in Figure 9-2.

The physical nature of the equivalent or *Ampèrian currents* can be understood from a model due to Ampère. Consider Figure 9-2 and imagine currents λ_e flowing around cells in a piece of material whose length is one meter. The current in one cell is nullified by the currents in the adjoining cells, except at the periphery of the material. The net current density on the surface of the rod is thus λ_e. Equating the magnetic moment of the current λ_e around the periphery to the sum of the magnetic moments of the currents in the cells, we have

$$\lambda_e S = MS, \tag{9-11}$$

where S is the cross-sectional area of the rod and M is the magnetization, or the magnetic moment per unit volume of the material, and

$$\lambda_e = M. \tag{9-12}$$

The origin of the equivalent surface current density $\boldsymbol{\lambda}_e$ can thus be understood in terms of currents on the atomic scale.

There is a volume current density $\boldsymbol{J}_e$ when currents in adjacent cells are unequal. Since $\boldsymbol{J}_e$ is $\boldsymbol{\nabla} \times \boldsymbol{M}$, $\boldsymbol{\nabla} \cdot \boldsymbol{J}_e$ is always zero. Therefore charge cannot accumulate at a point by virtue of $\boldsymbol{J}_e$. Furthermore, the equivalent currents do not produce heating, since they do not involve electron drift and scattering processes like those associated with conduction currents.

Since $\boldsymbol{A}$ can be ascribed to the current distributions $\boldsymbol{\lambda}_e$ and $\boldsymbol{J}_e$, we can immediately write down the value of $\boldsymbol{B}$ using the Biot-Savart law (Section 7.2):

$$\boldsymbol{B} = \frac{\mu_0}{4\pi} \int_{S'} \frac{\boldsymbol{\lambda}_e \times \boldsymbol{r}_1}{r^2} \, da' + \frac{\mu_0}{4\pi} \int_{\tau'} \frac{\boldsymbol{J}_e \times \boldsymbol{r}_1}{r^2} \, d\tau'. \tag{9-13}$$

Thus, if we know the magnetization $\boldsymbol{M}$, we can find the equivalent current densities. Then we can use the equivalent currents to calculate $\boldsymbol{B}$ as if they were in a vacuum.

We have demonstrated the validity of this procedure only for points situated outside the material, but the procedure is equally valid inside, as we shall see in the next section.

9.3 MAGNETIC INDUCTION $\boldsymbol{B}$ AT AN INTERIOR POINT. THE DIVERGENCE OF $\boldsymbol{B}$

We now calculate the macroscopic magnetic induction, or the space and time average of the magnetic induction on the atomic scale, at an interior point in magnetized material.

We proceed as we did with dielectrics and divide the material into two regions, one near the point P and one farther away. The near region is a small sphere of radius R as in Figure 9-3, and the far region is the remaining volume of magnetized material.

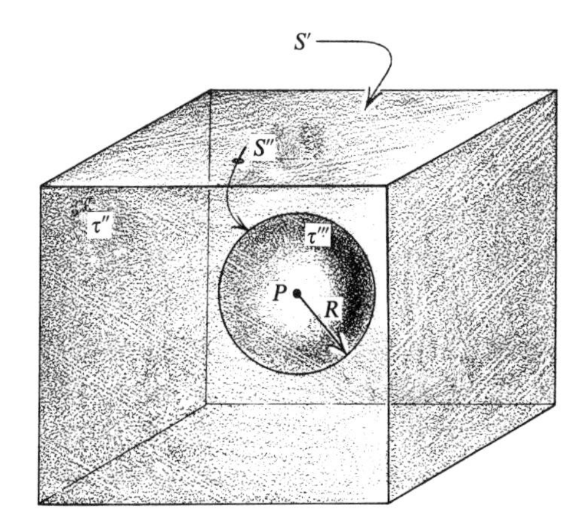

Figure 9-3. Small imaginary sphere of radius R and volume τ''' centered at P within a magnetized body. The remaining volume is τ''.

If the radius R is chosen large enough, the magnetic induction $\boldsymbol{B}''$ at the center of the sphere due to the magnetized material in the outer volume τ'' is adequately given by

$$\boldsymbol{B}'' = \frac{\mu_0}{4\pi}\int_{\tau''}\frac{\boldsymbol{J}_e \times \boldsymbol{r}_1}{r^2}\,d\tau'' + \frac{\mu_0}{4\pi}\int_{S'}\frac{\boldsymbol{\lambda}_e \times \boldsymbol{r}_1}{r^2}\,da' + \frac{\mu_0}{4\pi}\int_{S''}\frac{\boldsymbol{\lambda}_e \times \boldsymbol{r}_1}{r^2}\,da''. \quad (9\text{-}14)$$

We now calculate the macroscopic magnetic induction at P due to the dipoles within the small sphere of volume τ'''. We had a similar task in Section 3.3.2, and the average $\boldsymbol{E}$ turned out to be proportional to the polarization $\boldsymbol{P}$ and in the *opposite* direction. In magnetic materials the same type of calculation gives a different result because the $\boldsymbol{B}$ field in the vicinity of a current loop is different from the $\boldsymbol{E}$ field near an electric dipole, as shown in Figure 9-4, although at large distances the two fields have the same form. In fact, we shall find that the average $\boldsymbol{B}$ due to the near dipoles is in the *same* direction as $\boldsymbol{M}$.

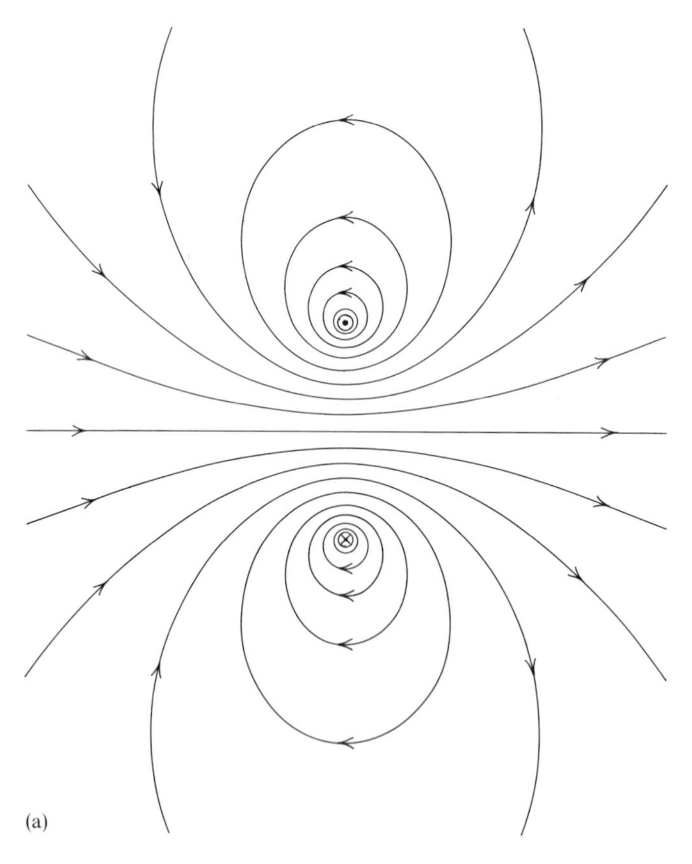

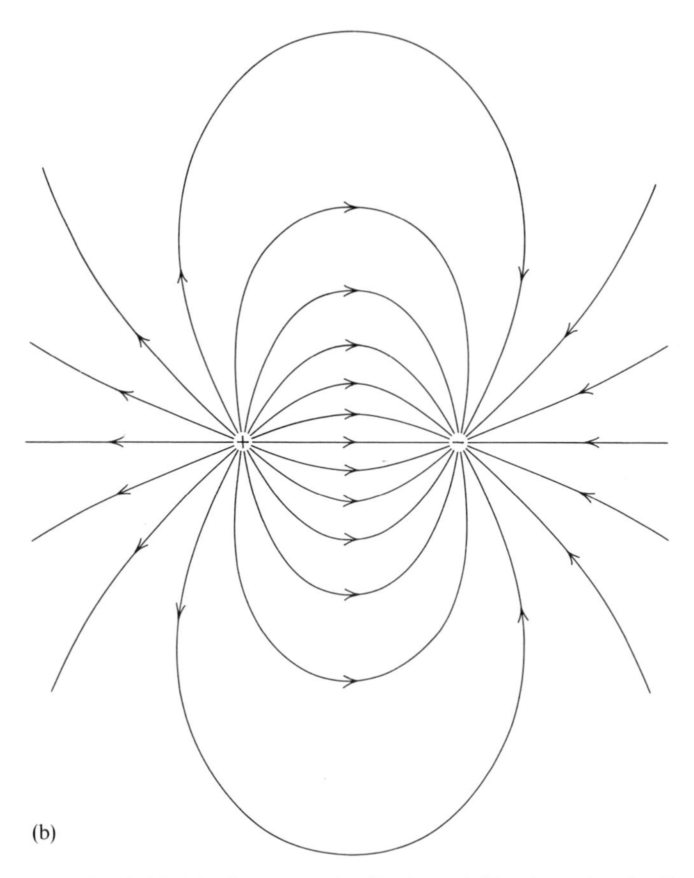

Figure 9-4. The fields (a) of a magnetic dipole and (b) of an electric dipole.

Let us first compute the average magnetic induction produced over the volume of a sphere by a small current loop inside it. For simplicity let us place the loop of magnetic moment $\boldsymbol{m}$ at the center of the sphere, as in Figure 9-5. By definition, the average field in the sphere is

$$\overline{\boldsymbol{B}} = \frac{1}{\tau}\int_\tau \boldsymbol{B}\, d\tau = \frac{1}{\tau}\int_\tau (\boldsymbol{\nabla} \times \boldsymbol{A})\, d\tau, \tag{9-15}$$

where τ is the volume of the sphere. Since we do not know the value of $\boldsymbol{A}$ in the immediate vicinity of an atom, we transform the volume integral into a surface integral, using the result of Problem 1-30:

$$\overline{\boldsymbol{B}} = \frac{1}{\tau}\int_S (\boldsymbol{n} \times \boldsymbol{A})\, da, \tag{9-16}$$

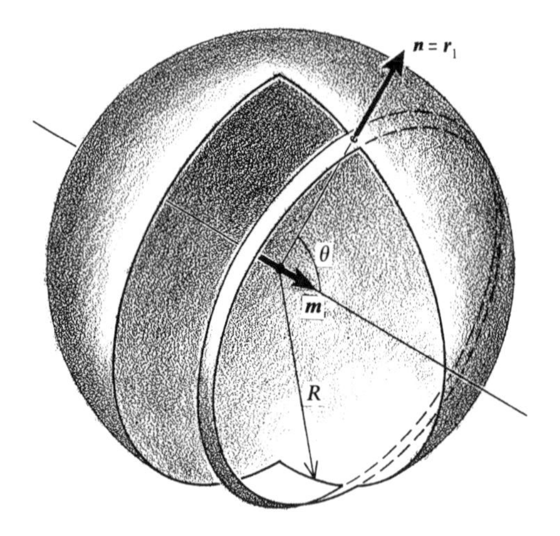

Figure 9-5. Current loop of magnetic moment $\boldsymbol{m}$ at the center of a sphere of radius R large enough for the dipole approximation to be valid on the surface.

where S is the surface bounding τ, and $\boldsymbol{n}$ is the outward drawn normal to S. At the surface the dipole approximation is valid and

$$\boldsymbol{A} = \frac{\mu_0}{4\pi}\frac{\boldsymbol{m} \times \boldsymbol{r}_1}{r^2}. \tag{9-17}$$

We can now evaluate the volume integral of Eq. 9-15 without knowing the value of $\boldsymbol{A}$ inside, provided we know its value at the surface:

$$\int_\tau \boldsymbol{B}\, d\tau = \frac{\mu_0}{4\pi}\int_S \boldsymbol{n} \times \frac{\boldsymbol{m} \times \boldsymbol{r}_1}{R^2}\, da \tag{9-18}$$

or, expanding the double vector product as in Problem 1-8,

$$\int_\tau \boldsymbol{B}\, d\tau = \frac{\mu_0}{4\pi}\int_S \frac{\boldsymbol{m}(\boldsymbol{n}\cdot\boldsymbol{r}_1) - \boldsymbol{r}_1(\boldsymbol{n}\cdot\boldsymbol{m})}{R^2}\, da. \tag{9-19}$$

The first term in the numerator on the right is simply $\boldsymbol{m}$. The second term is $\boldsymbol{r}_1 m \cos\theta$ and, by virtue of the symmetry, we take its component along $\boldsymbol{m}$, which makes it $\boldsymbol{m}\cos^2\theta$. Therefore

$$\int_\tau \boldsymbol{B}\, d\tau = \frac{\mu_0}{4\pi} 4\pi\, \boldsymbol{m} - \frac{\mu_0}{4\pi}\int_0^\pi \boldsymbol{m}\cos^2\theta(2\pi\sin\theta\, d\theta), \tag{9-20}$$

$$= \frac{2}{3}\mu_0 \boldsymbol{m}, \tag{9-21}$$

and the average magnetic induction inside a sphere of radius R due to the single loop of magnetic moment $\boldsymbol{m}$ at its center is

$$\frac{(2/3)\mu_0\boldsymbol{m}}{(4/3)\pi R^3}, \quad \text{or} \quad \frac{\mu_0\boldsymbol{m}}{2\pi R^3}.$$

If the current loop is not situated at the center of the sphere, the average $\boldsymbol{B}$ is unaffected, as you will be able to show in Problem 9-3.

Since there are many atomic current loops inside the sphere, each one produces an average field as above, and the total average field due to the near dipoles is

$$\boldsymbol{B}''' = \frac{4\pi R^3}{3} N \frac{\mu_0 \boldsymbol{m}}{2\pi R^3} = \frac{2}{3}\mu_0 \boldsymbol{M} \tag{9-22}$$

if they are all aligned in the same direction.

This result is open to question, because some of the loops will be too close to the spherical surface for the dipole approximation of Eq. 9-17 to be valid. However, the error can be made arbitrarily small by increasing the size of the sphere. Molecules are so small that it is easy to choose a microscopic sphere that is large enough to make this error negligible. This type of problem was encountered previously in Section 3.3.

The magnetization $\boldsymbol{M}$ may be taken to be uniform inside the small sphere S'', and we may assume that the $\boldsymbol{B}$ at the center is this average $\boldsymbol{B}$ we have just calculated, or $(2/3)\mu_0\boldsymbol{M}$.

Therefore the magnetic induction at a point P inside magnetized material as in Figure 9-3 is the sum of $\boldsymbol{B}''$ due to the remote material as in Eq. 9-14, plus $\boldsymbol{B}'''$ due to the material inside the small sphere of radius R as above:

$$\boldsymbol{B} = \frac{\mu_0}{4\pi}\int_{\tau''}\frac{\boldsymbol{J}_e \times \boldsymbol{r}_1}{r^2}\,d\tau'' + \frac{\mu_0}{4\pi}\int_{S'}\frac{\boldsymbol{\lambda}_e \times \boldsymbol{r}_1}{r^2}\,da' + \frac{\mu_0}{4\pi}\int_{S''}\frac{\boldsymbol{\lambda}_e \times \boldsymbol{r}_1}{r^2}\,da'' + \frac{2}{3}\mu_0\boldsymbol{M}. \tag{9-23}$$

Let us evaluate the third term. This term gives the magnetic induction at the center of the sphere due to the currents flowing on the surface S''. The equivalent current density $\boldsymbol{\lambda}_e$ on S'' is $\boldsymbol{M} \times \boldsymbol{n}$ and is in the direction shown in Figure 9-6. Thus, at the center of the sphere,

$$\frac{\boldsymbol{\lambda}_e \times \boldsymbol{r}_1}{r^2} = \frac{(\boldsymbol{M} \times \boldsymbol{n}) \times \boldsymbol{n}}{R^2}. \tag{9-24}$$

These currents give a magnetic induction that is in the direction *opposite* to $\boldsymbol{M}$, and we must therefore take the component of $(\boldsymbol{M} \times \boldsymbol{n}) \times \boldsymbol{n}$ in the direction of $-\boldsymbol{M}$:

$$\frac{\mu_0}{4\pi}\int_{S''}\frac{\boldsymbol{\lambda}_e \times \boldsymbol{r}_1}{r^2}\,da' = -\frac{\mu_0}{4\pi}\int_0^\pi \frac{\boldsymbol{M}\sin^2\theta}{R^2}\,2\pi R\sin\theta\, R\,d\theta \tag{9-25}$$

$$= -\frac{2}{3}\mu_0\boldsymbol{M}. \tag{9-26}$$

The equivalent currents on the surface S'' therefore cancel exactly the

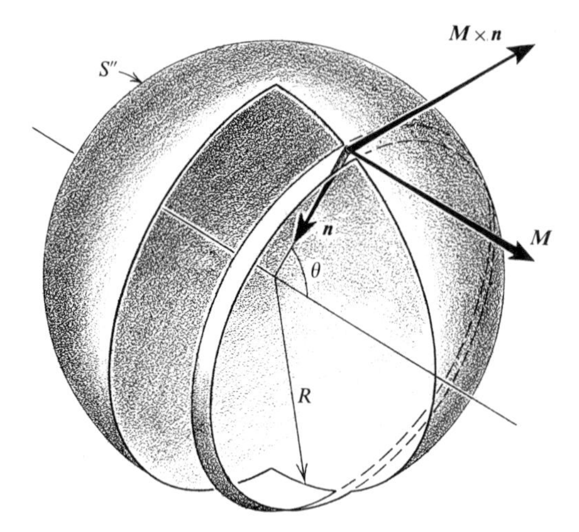

Figure 9-6. Equivalent current density $\boldsymbol{M} \times \boldsymbol{n}$ on the imaginary surface S''.

magnetic induction due to the near dipoles. Thus, at an inside point; we are left with only the first two integrals of Eq. 9-23:

$$\boldsymbol{B} = \frac{\mu_0}{4\pi}\int_{\tau''} \frac{\boldsymbol{J}_e \times \boldsymbol{r}_1}{r^2}\, d\tau'' + \frac{\mu_0}{4\pi}\int_{S'} \frac{\boldsymbol{\lambda}_e \times \boldsymbol{r}_1}{r^2}\, da'. \tag{9-27}$$

Now, if we extended the first integral to the complete volume τ', which is $\tau'' + \tau'''$, we would be adding the integral of $(\boldsymbol{J}_e \times \boldsymbol{r}_1)/r^2$ over the volume τ''' of the small sphere. This latter integral is zero, by symmetry, because $\boldsymbol{J}_e$ is constant throughout the small volume τ'''. Finally,

$$\boldsymbol{B} = \frac{\mu_0}{4\pi}\int_{\tau'} \frac{\boldsymbol{J}_e \times \boldsymbol{r}_1}{r^2}\, d\tau' + \frac{\mu_0}{4\pi}\int_{S'} \frac{\boldsymbol{\lambda}_e \times \boldsymbol{r}_1}{r^2}\, da', \tag{9-28}$$

where $\boldsymbol{J}_e = \boldsymbol{\nabla} \times \boldsymbol{M}$ and $\boldsymbol{\lambda}_e = \boldsymbol{M} \times \boldsymbol{n}$.

The equivalent currents therefore yield the correct value for the magnetic induction $\boldsymbol{B}$ at points inside magnetized material as well as outside. In other words, *we can always calculate* $\boldsymbol{B}$ *by replacing the magnetized material by its equivalent currents*. We obtained a corresponding result for dielectrics in Section 3.3.2.

Therefore the relation

$$\boxed{\boldsymbol{\nabla} \cdot \boldsymbol{B} = 0} \tag{9-29}$$

of Sections 6.9 and 7.4 applies even in the presence of magnetic materials. This is one of Maxwell's equations.

It also follows that the vector potential due to magnetized material is

$$A = \frac{\mu_0}{4\pi}\int_{\tau'} \frac{J_e}{r}\,d\tau' + \frac{\mu_0}{4\pi}\int_{S'} \frac{\lambda_e}{r}\,da' \tag{9-30}$$

at any point in space.

More generally, if we include the free current densities J_f and λ_f,

$$A = \frac{\mu_0}{4\pi}\int_{\tau'} \frac{J_f + \nabla' \times M}{r}\,d\tau' + \frac{\mu_0}{4\pi}\int_{S'} \frac{\lambda_f + M \times n}{r}\,da'. \tag{9-31}$$

Example

THE $\boldsymbol{B}$ FIELD OF A UNIFORMLY MAGNETIZED CYLINDER

We return to the cylinder that is uniformly magnetized in the direction of its axis, as in Figure 9-2. This is an idealized case, as we shall see later on, but it is a good example to use at this point.

We have shown in Section 9.2 that the cylindrical surface carries an equivalent current density λ_e, which is numerically equal to M. Inside the magnet, $\nabla \times M = 0$, and the volume current density J_e is zero. Over the end faces, M is parallel to the normal unit vector n, and λ_e is zero.

The B field of a cylinder uniformly magnetized in the direction of its axis of symmetry is therefore identical to that of a solenoid of the same dimensions that carries a current density $N'I = M$, where N' is the number of turns/meter. Figure 7-19 shows an example of such a field.

Note that B and M are in the *same* general direction inside the magnet. Near the center of a long uniformly magnetized cylinder, B would be equal to $\mu_0 M$.

9.4 THE MAGNETIC FIELD INTENSITY $\boldsymbol{H}$. AMPÈRE'S CIRCUITAL LAW

We have found (in Section 7.6) that, for steady currents and nonmagnetic materials,

$$\nabla \times B = \mu_0 J_f. \tag{9-32}$$

Now we have just seen that magnetized material can always be replaced by its equivalent currents for calculating B. Consequently, if we have magnetized material as well as steady currents,

$$\nabla \times B = \mu_0 (J_f + J_e). \tag{9-33}$$

where J_e is the equivalent volume current density. This equation is of course valid only at points where the derivatives of B with respect to the coordinates

x, y, z exist. It is therefore not applicable at the surface of magnetic materials, and we can substitute $\nabla \times \boldsymbol{M}$ for $\boldsymbol{J}_e$. Thus

$$\nabla \times \boldsymbol{B} = \mu_0(\boldsymbol{J}_f + \nabla \times \boldsymbol{M}), \tag{9-34}$$

$$\nabla \times \left(\frac{\boldsymbol{B}}{\mu_0} - \boldsymbol{M}\right) = \boldsymbol{J}_f. \tag{9-35}$$

The vector between parentheses, whose curl is equal to the *free* current density at the point, is the *magnetic field intensity*

$$\boldsymbol{H} = \frac{\boldsymbol{B}}{\mu_0} - \boldsymbol{M} \tag{9-36}$$

and is expressed in amperes/meter. Note that $\boldsymbol{H}$ and $\boldsymbol{M}$ are expressed in the same units.

Thus

$$\boldsymbol{B} = \mu_0(\boldsymbol{H} + \boldsymbol{M}). \tag{9-37}$$

This equation is to be compared with Eq. 3-45, which applies to dielectrics:

$$\boldsymbol{D} = \epsilon_0 \boldsymbol{E} + \boldsymbol{P}. \tag{9-38}$$

Rewriting now Eq. 9-35, we have that, for steady currents, either inside or outside magnetic material,

$$\nabla \times \boldsymbol{H} = \boldsymbol{J}_f. \tag{9-39}$$

Integrating over a surface S,

$$\int_S (\nabla \times \boldsymbol{H}) \cdot d\boldsymbol{a} = \int_S \boldsymbol{J}_f \cdot d\boldsymbol{a} \tag{9-40}$$

or, using Stokes's theorem on the left-hand side,

$$\int_C \boldsymbol{H} \cdot d\boldsymbol{l} = I_f, \tag{9-41}$$

where C is the curve bounding the surface S, and I_f is the current of free charges linking the curve C. Note that I_f does *not* include the equivalent currents. The term on the left is called the *magnetomotance.*

This is a more general form of *Ampère's circuital law* (Section 7.7) in that it can be used to calculate $\boldsymbol{H}$ even in the presence of magnetic materials. It is rigorously valid, however, only for steady currents; we shall deal with variable currents in the next chapter.

Sections 9.8.1.1 and 9.8.1.2 will illustrate the use of $\boldsymbol{H}$ and of Ampère's circuital law.

9.5 MAGNETIC SUSCEPTIBILITY χ_m AND RELATIVE PERMEABILITY μ_r

As for dielectrics, it is convenient to define a *magnetic susceptibility* χ_m and a *relative permeability* μ_r such that

$$\boldsymbol{M} = \chi_m \boldsymbol{H}. \tag{9-42}$$

Now, since

$$\boldsymbol{B} = \mu_0(\boldsymbol{H} + \boldsymbol{M}), \tag{9-43}$$

then

$$\boldsymbol{B} = \mu_0(1 + \chi_m)\boldsymbol{H} = \mu_0\mu_r\boldsymbol{H} = \mu\boldsymbol{H}, \tag{9-44}$$

where

$$\mu_r = 1 + \chi_m. \tag{9-45}$$

The quantity μ is called the *permeability*. Both χ_m and μ_r are dimensionless quantities.

Equation 9-43 is general, but Eqs. 9-44 and 9-45 are based on the assumption that the material is both isotropic and linear, in other words, that $\boldsymbol{M}$ is proportional to $\boldsymbol{H}$ and in the same direction. This assumption is unfortunately never completely true in *ferromagnetic* materials. In a permanent magnet, for example, $\boldsymbol{B}$ and $\boldsymbol{H}$ point in roughly *opposite* directions (example, page 402). Even in so-called *soft* magnetic materials, such as soft iron, $\boldsymbol{B}$ and $\boldsymbol{H}$ do not always point in the same direction, and when they do, μ_r is by no means constant. It can vary by orders of magnitude, depending on the value of H and on the previous history of the material (Section 9.6). The quantities χ_m and μ_r are therefore not as meaningful as the corresponding quantities χ_e and ϵ_r for dielectrics.

For iron and for $B \approx 0.5$ tesla, $\mu_r \approx \chi_m \approx 10^3$.

In *paramagnetic* materials M is given by the Langevin equation 3-98 with p replaced by m and E_{loc} by B. As for dielectrics, the magnetic energy mB is much less than kT at room temperature. For example, m is of the order of 10^{-23} ampere-meter^2, and, even in a magnetic field of one tesla, $mB/kT \approx 2.5 \times 10^{-3}$. Expanding the exponentials as in Section 3.9.1,

$$M \approx \frac{Nm^2}{3kT} B, \tag{9-46}$$

$$\chi_m = \frac{M}{H} \approx \mu_0 \frac{M}{B} \approx \frac{\mu_0 N m^2}{3kT}. \tag{9-47}$$

The magnetic susceptibility of paramagnetic substances is smaller than unity by several orders of magnitude and is proportional to the inverse of the absolute temperature.

In *diamagnetic* materials the magnetization is in the direction *opposite* to the external field; the relative permeability is *less* than unity and is independent of the temperature.

If orientational magnetization predominates, the resultant permeability is greater than unity.

9.5.1 The Equivalent Current Density $\boldsymbol{\nabla} \times \boldsymbol{M}$ and the Free Current Density $\boldsymbol{J}_f$

In a linear and isotropic magnetic material, $\boldsymbol{B} = \mu \boldsymbol{H}$. If, moreover, μ is not a function of the coordinates, we have from Eq. 9-39 that

$$\mu \boldsymbol{\nabla} \times \boldsymbol{H} = \boldsymbol{\nabla} \times \mu \boldsymbol{H} = \boldsymbol{\nabla} \times \boldsymbol{B} = \mu \boldsymbol{J}_f \tag{9-48}$$

or, using Eq. 9-37,

$$\mu_0 \boldsymbol{\nabla} \times (\boldsymbol{H} + \boldsymbol{M}) = \mu \boldsymbol{J}_f = \mu_0 \mu_r \boldsymbol{J}_f, \tag{9-49}$$

$$\boldsymbol{\nabla} \times \boldsymbol{M} = \mu_r \boldsymbol{J}_f - \boldsymbol{\nabla} \times \boldsymbol{H}, \tag{9-50}$$

$$\boldsymbol{J}_e = (\mu_r - 1)\boldsymbol{J}_f. \tag{9-51}$$

Thus, in linear, isotropic, and homogeneous magnetic materials, the equivalent volume current density $\boldsymbol{\nabla} \times \boldsymbol{M}$ is a multiple of the free current density $\boldsymbol{J}_f$, the factor of proportionality being positive.

This relation reminds one of Eq. 3-58, which is the corresponding relation for dielectrics.

Also,

$$\mu \boldsymbol{J}_f = \mu_0(\boldsymbol{J}_f + \boldsymbol{\nabla} \times \boldsymbol{M}), \tag{9-52}$$

which corresponds to Eq. 3-60. The above equation is valid only for steady fields, as we shall see in Chapter 10.

It follows from Eq. 9-31 that, in a homogeneous medium μ,

$$\boldsymbol{A} = \frac{\mu}{4\pi} \int_{\tau'} \frac{\boldsymbol{J}_f}{r}\, d\tau' + \frac{\mu_0}{4\pi} \int_{S'} \frac{\boldsymbol{\lambda}_f + \boldsymbol{M} \times \boldsymbol{n}}{r}\, da'. \tag{9-53}$$

The surface integral cannot be transformed like the volume integral, because $\boldsymbol{\lambda}_f$ and $\boldsymbol{M} \times \boldsymbol{n}$ are unrelated; for example, a piece of soft iron placed in a magnetic field acquires a magnetization $\boldsymbol{M}$ even though its $\boldsymbol{\lambda}_f$ is zero.

9.6 HYSTERESIS

One can measure B as a function of H with a ring-shaped sample as in Figure 9-7. The function of winding a, which has N turns and carries a

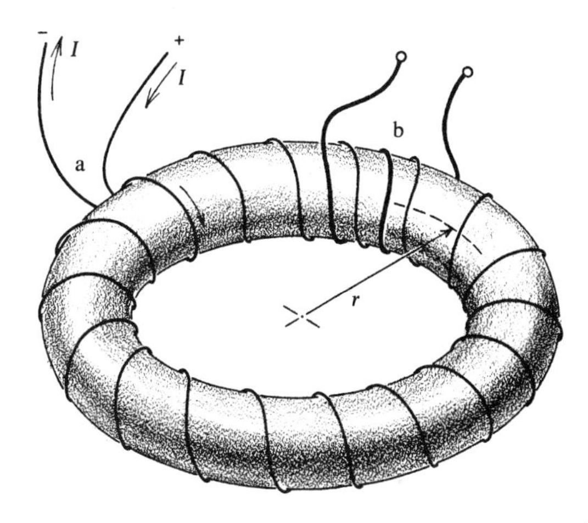

Figure 9-7. Rowland ring for the determination of B as a function of H in a ferromagnetic substance.

current I, is to apply an azimuthal magnetic field intensity

$$H = \frac{NI}{2\pi r}. \tag{9-54}$$

The magnetic induction is

$$B = \frac{\Phi}{S}, \tag{9-55}$$

where S is the cross-sectional area of the core and Φ is the magnetic flux. One can measure changes in Φ, and hence changes in B, by changing I and integrating the electromotance induced in winding b as in Problem 9-11. Both B and H are therefore easily measurable.

If we start with an unmagnetized sample of iron and increase the current in coil a of Figure 9-7 to some specified magnitude, the magnetic induction increases along a curve such as ab in Figure 9-8. This curve is known as the *magnetization curve.* If the current in the winding is reduced to zero, B decreases along bc. The magnitude of the magnetic induction at c is the *remanence* or the *retentivity* for the particular sample of material. If the current is then reversed in direction and increased, B reaches a point d where it is reduced to zero. The magnitude of H at this point is known as the *coercive force.* On further increasing the current in the same direction a point e, symmetrical to point b, is reached. If the current is now reduced, reversed, and increased, the point b is again reached. The closed curve $bcdefgb$ is known as a *hysteresis loop*. If, at any point, the current is varied in a smaller cycle, a small hysteresis loop is described.

The *differential permeability* is defined as the ratio

$$\frac{1}{\mu_0}\frac{dB}{dH}.$$

It is not a constant.

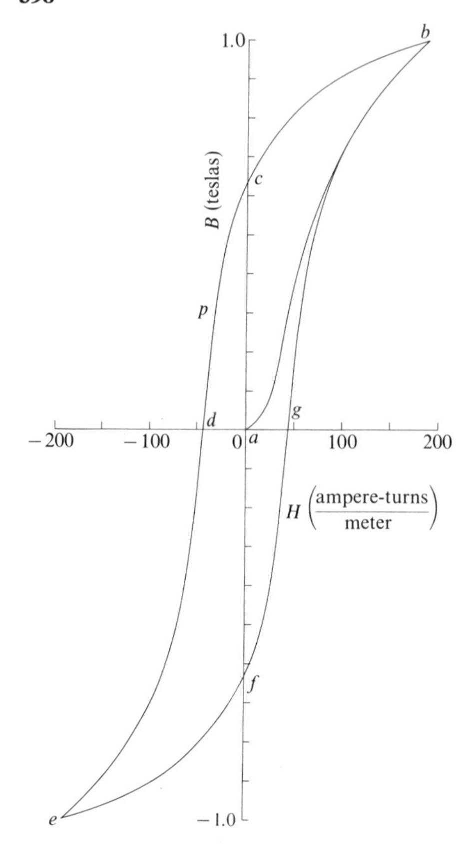

Figure 9-8. Magnetization curve *ab* and hysteresis loop *bcdefg*. It is shown in the example on page 402 that B and H inside a permanent magnet give a point such as p in the second quadrant.

A characteristic feature of the hysteresis cycle is the *saturation* induction beyond which M increases no further. Maximum alignment of the domains is then achieved, and a further increase in H increases B only as the contribution from the conduction current increases. The saturation induction is characteristically in the range from 1 to 2 teslas.

In a permanent magnet, the point p representing B and H on the hysteresis loop of Figure 9-8 is in the second quadrant.

9.6.1 Energy Dissipated in a Hysteresis Cycle

Energy is required to describe a hysteresis cycle. This can be shown by considering Figure 9-7. When the current is increasing, the electromotance induced in the winding opposes the increase in current, according to Lenz's law (Section 8.1), and the extra power spent by the source is

$$\frac{dW}{dt} = I\left(N\frac{d\Phi}{dt}\right) = INS\frac{dB}{dt}, \tag{9-56}$$

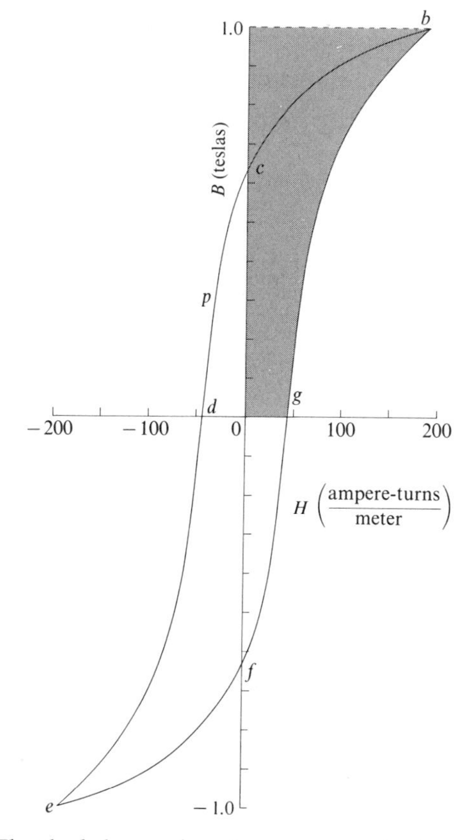

Figure 9-9. The shaded area gives the energy required, per unit volume, in going from g to b on the hysteresis loop. The total energy per unit volume required to describe a complete hysteresis loop is equal to the area enclosed by the loop.

where S is the cross-sectional area of the ring, N is the number of turns, and B is the average magnetic induction in the core. Also,

$$\frac{dW}{dt} = \frac{INSl}{l}\frac{dB}{dt} = H\tau\frac{dB}{dt}, \tag{9-57}$$

where l is the mean circumference of the ring, $\tau = Sl$ is its volume, and $H = NI/l$, as in Eq. 9-54. Thus

$$W_1 = \tau\int_g^b H\,dB \tag{9-58}$$

is the energy supplied by the source in going from the point g to the point b in Figure 9-9. This integral corresponds to the crosshatched area in the figure and is equal to the energy supplied per unit volume of the magnetic core.

When the current is in the same direction but is decreasing, the polarity of the induced electromotance is reversed, according to Lenz's law, with the result that the energy

$$W_2 = \tau \int_b^c H \, dB \tag{9-59}$$

is returned to the source.

Finally, the energy supplied by the source during one cycle is

$$W = \tau \oint H \, dB, \tag{9-60}$$

where the integral is evaluated around the hysteresis loop. The area of the hysteresis loop in tesla-ampere turns/meter or in weber-ampere turns/meter³ is therefore the number of joules dissipated per cubic meter and per cycle in the core.

Note that this energy loss is unrelated to the losses associated with the eddy currents. Eddy-current losses, which are caused by changes in the magnetic field, can be minimized by using laminated materials, as in transformers, by using powders dispersed in an insulator, or by using nonconducting materials. Hysteresis losses, however, can be minimized only by selecting a material with a narrow hysteresis loop.

Example For the case illustrated in Figure 9-9, the area enclosed by the hysteresis loop is roughly 150 weber-ampere turns/meter³, or 150 joules per cubic meter and per cycle, or 600 calories per cubic meter and per cycle. This energy loss could be important if the material were subjected to an alternating field.

9.7 BOUNDARY CONDITIONS

Let us examine the continuity conditions that $\boldsymbol{B}$ and $\boldsymbol{H}$ must obey at the interface between two media. We shall proceed as in Section 4.1.

Figure 9-10a shows a short cylindrical volume whose top and bottom faces are parallel and infinitely close to the interface. Since there is zero flux through the cylindrical surface, the flux through the top face must equal that through the bottom, and

$$B_{n1} = B_{n2}. \tag{9-61}$$

The normal component of $\boldsymbol{B}$ is therefore continuous across the interface.

Consider now Figure 9-10b. The closed path has two sides parallel to

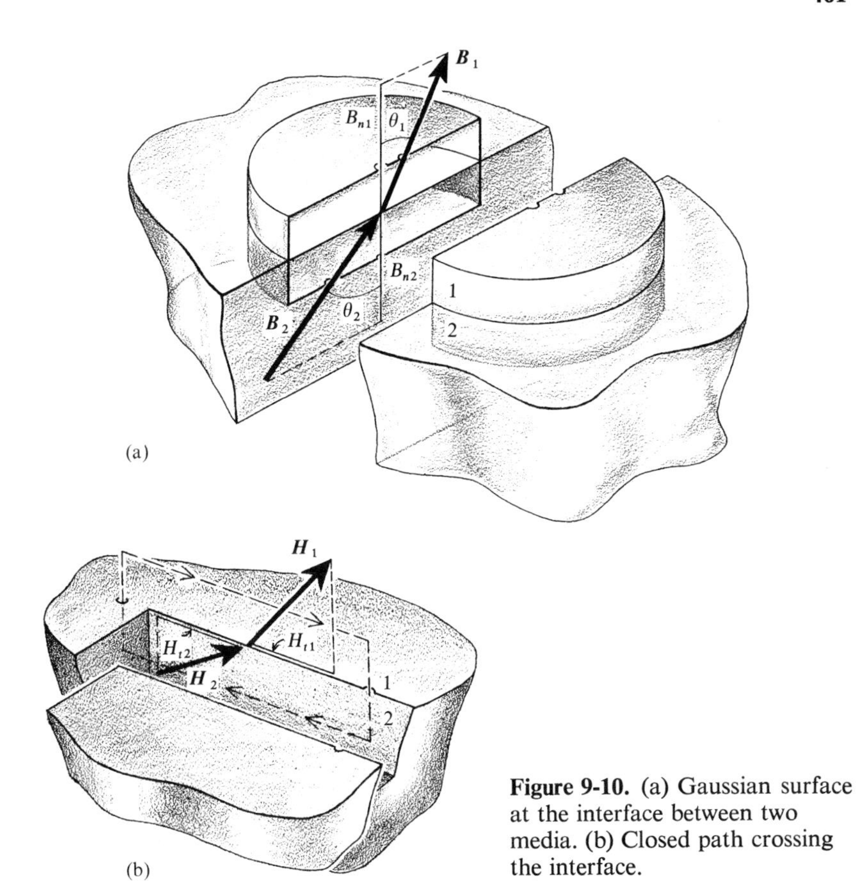

Figure 9-10. (a) Gaussian surface at the interface between two media. (b) Closed path crossing the interface.

the interface and close to it. From the circuital law, Eq. 9-41,

$$\oint \boldsymbol{H} \cdot d\boldsymbol{l} = I, \tag{9-62}$$

where I is the conduction current linking the path.

If the two sides parallel to the interface are *infinitely* close to it, I is equal to zero since there are no true surface currents, except in superconductors. Thus

$$H_{t1} = H_{t2}, \tag{9-63}$$

and the tangential component of $\boldsymbol{H}$ is continuous across an interface.

If we can set $\boldsymbol{B}$ equal to $\mu\boldsymbol{H}$, the relative permeabilities being those which correspond to the actual values of $\boldsymbol{H}$, then

$$\mu_{r1}\mu_0 H_1 \cos\theta_1 = \mu_{r2}\mu_0 H_2 \cos\theta_2, \tag{9-64}$$

from the continuity of the normal component of $\boldsymbol{B}$, and

$$H_1 \sin \theta_1 = H_2 \sin \theta_2, \tag{9-65}$$

from the continuity of the tangential component of $\boldsymbol{H}$. Then

$$\frac{\tan \theta_1}{\tan \theta_2} = \frac{\mu_{r1}}{\mu_{r2}}. \tag{9-66}$$

The lines of $\boldsymbol{B}$, or of $\boldsymbol{H}$, are farthest away from the normal in the medium having the larger permeability.

9.8 MAGNETIC FIELD CALCULATIONS

It should be obvious that magnetic fields involving ferromagnetic materials do not lend themselves to rigorous mathematical analyses. Approximate methods of calculation have therefore been devised, and these are usually satisfactory. If more accuracy is required, one must resort to making measurements on models.

As a first example of magnetic field calculations we shall return to the field of a bar magnet and apply the concepts that we have developed so far. Then we shall discuss a different approach that utilizes the concept of magnetic circuit.

Example

THE $\boldsymbol{B}$ AND $\boldsymbol{H}$ FIELDS OF A BAR MAGNET

We first return to the idealized case of the uniformly magnetized cylinder of Figure 9-2. This will help us understand the field of a real bar magnet.

As we have seen previously in the example on page 393, the $\boldsymbol{B}$ field, both inside and outside the bar magnet, is that of a solenoid of the same size and carrying a current density $N'I = M$, where N' is the number of turns/meter. This field is shown in Figure 9-11. Note that the lines of $\boldsymbol{B}$ are refracted at the cylindrical surface as explained in the example on page 316.

Outside the magnet, $\boldsymbol{H}$ is simply $\boldsymbol{B}/\mu_0$.

Inside the magnet,

$$\boldsymbol{H} = \frac{\boldsymbol{B}}{\mu_0} - \boldsymbol{M}, \tag{9-67}$$

from Eq. 9-36. (Remember that the relation $\boldsymbol{H} = \boldsymbol{B}/\mu$ is valid *only*

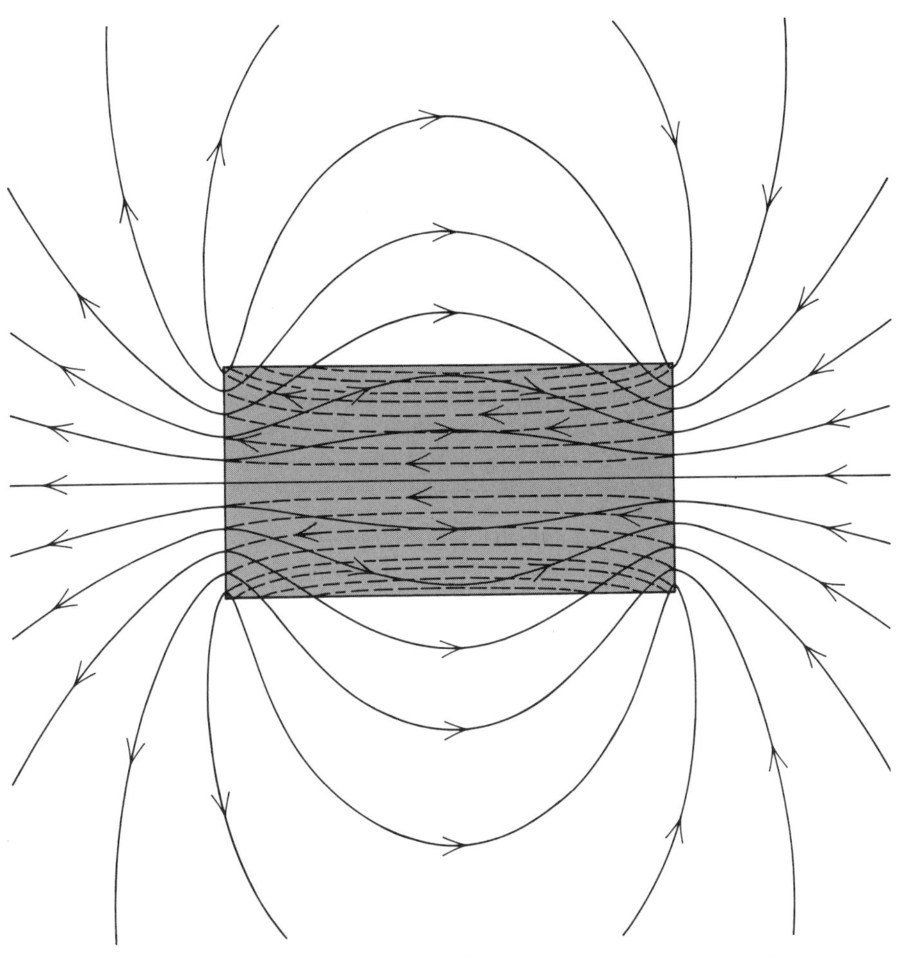

Figure 9-11. Lines of $\boldsymbol{H}$ (solid) for a uniformly magnetized cylinder. Lines of $\boldsymbol{M}$ (not shown) are parallel to the axis and point to the left. Lines of $\boldsymbol{B}$ are shown broken inside the magnet; outside, they follow the lines of $\boldsymbol{H}$.

This figure is identical to Figure 3-10, and the above $\boldsymbol{B}$ field is the same as that of Figure 7-19.

in soft magnetic materials that are both linear and isotropic.) Near the center of a long bar magnet, $\boldsymbol{B} \approx \mu_0 \boldsymbol{M}$ and $\boldsymbol{H} \approx 0$. For a short bar magnet we can use the values of $\boldsymbol{B}$ that we calculated for a short solenoid in the example on page 317, substituting again the magnetization M for the current density IN'. Thus, at the center of the magnet,

$$B = \mu_0 M \sin \theta_m, \qquad H = -M(1 - \sin \theta_m), \tag{9-68}$$

while at the center of one face

$$B = \mu_0 M \frac{\sin \theta_e}{2}, \qquad H = -M\left(1 - \frac{\sin \theta_e}{2}\right). \tag{9-69}$$

The $\boldsymbol{H}$ field is also shown in Figure 9-11.

It will be noticed that, *inside* the magnet, $\boldsymbol{B}$ and $\boldsymbol{H}$ point in approximately *opposite* directions. This can be seen in Figure 9-11, and also in Figure 9-8. Also, taking the curl of Eq. 9-67,

$$\boldsymbol{\nabla} \times \boldsymbol{B} = \mu_0(\boldsymbol{\nabla} \times \boldsymbol{H} + \boldsymbol{\nabla} \times \boldsymbol{M}) = 0, \tag{9-70}$$

$\boldsymbol{\nabla} \times \boldsymbol{H}$ being zero because $\boldsymbol{J}_f$ is zero (Eq. 9-39), and $\boldsymbol{\nabla} \times \boldsymbol{M}$ being also zero since the magnetization $\boldsymbol{M}$ is uniform, by hypothesis.

Note that the lines of $\boldsymbol{H}$ are *not* refracted at the cylindrical surface of the magnet.

The field of a *real* bar magnet is not as simple to analyze as the one we have just described. Since the magnetic moments of the individual atoms tend to align themselves with the $\boldsymbol{B}$ field, the magnetization $\boldsymbol{M}$, and hence also the equivalent current density on the cylindrical surface, are weaker near the ends. Moreover, since $\boldsymbol{M}$ is not uniform, there can be Ampèrian currents inside the magnet. The end faces also carry Ampèrian currents since $\boldsymbol{M} \times \boldsymbol{n}$ is zero only on the axis. The net result is that there are "poles" near the ends of the magnet from which lines of $\boldsymbol{B}$ appear to radiate in all directions. The poles are most conspicuous if the bar magnet is long and thin.

Example

THE BAR MAGNET AND THE BAR ELECTRET

It is instructive to compare the field of our uniformly magnetized cylinder of magnetic material, Figure 9-12a, with that of its electric equivalent, the uniformly polarized cylinder of dielectric of Figure 9-12b. We discussed the field of the bar electret in the example on page 113.

In the case of the magnet, the $\boldsymbol{B}$ field is that of the equivalent solenoid, Figure 9-12c which has a current density M. Then, to find the $\boldsymbol{H}$ field, we use the relation

$$\boldsymbol{B} = \mu_0 \boldsymbol{H} + \mu_0 \boldsymbol{M}. \tag{9-71}$$

Thus $\boldsymbol{H}$ is $\boldsymbol{B}/\mu_0$ outside and $\boldsymbol{B}/\mu_0 - \boldsymbol{M}$ inside.

In the case of the electret, the $\boldsymbol{E}$ field is that of the bound charges on the end faces as in Figure 9-12d, or of a pair of parallel and oppositely charged disks carrying charge densities $\pm P$. The $\boldsymbol{D}$ field is then given by

$$\boldsymbol{D} = \epsilon_0 \boldsymbol{E} + \boldsymbol{P} \tag{9-72}$$

and is therefore $\epsilon_0 \boldsymbol{E}$ outside and $\epsilon_0 \boldsymbol{E} + \boldsymbol{P}$ inside.

Mathematically, the fields of the magnet and of the electret obey similar equations:

Electret	*Magnet*	
$\boldsymbol{D} = \epsilon_0 \boldsymbol{E} + \boldsymbol{P}$	$\boldsymbol{B} = \mu_0 \boldsymbol{H} + \mu_0 \boldsymbol{M}$	
$\boldsymbol{\nabla} \cdot \boldsymbol{D} = 0$	$\boldsymbol{\nabla} \cdot \boldsymbol{B} = 0$	(9-73)
$\boldsymbol{\nabla} \times \boldsymbol{E} = 0$	$\boldsymbol{\nabla} \times \boldsymbol{H} = 0$	

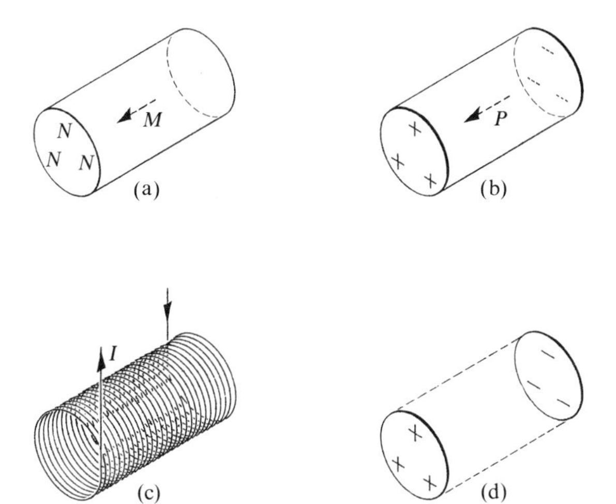

Figure 9-12. (a) Bar magnet. (b) Bar electret. (c) At any point in space the magnetic induction $\boldsymbol{B}$ due to the magnet is the same as that of a solenoid of the same size carrying a current density $N'I$ equal to M. The number of turns per meter on the solenoid is N'. (d) Similarly, the electric field intensity $\boldsymbol{E}$ due to the electret is the same as that of a pair of disks occupying the positions of the end faces and carrying charge densities $\pm P$.

Indeed, the $\boldsymbol{H}$ field of the magnet can be calculated like the $\boldsymbol{E}$ field of the electret if one assumes the existence of a magnetic pole density equal to $\pm M$ on the end faces.

9.8.1 Magnetic Circuits

Figure 9-13 shows a ferromagnetic core around which is wound a coil of N turns carrying a current I. We wish to calculate the magnetic flux Φ through some cross section of the core.

In the absence of ferromagnetic material, the lines of $\boldsymbol{B}$ are as shown in the figure. At first sight, one expects the $\boldsymbol{B}$ inside the core to be much larger close to the winding than on the opposite side. This is not the case, however, and $\boldsymbol{B}$ is of the same order of magnitude at all points within the ferromagnetic material.

This can be understood as follows. The magnetic induction due to the current I magnetizes the core in the region near the coil, and this magnetization produces Ampèrian currents that both increase $\boldsymbol{B}$ and extend it along the core. This further increases and extends the magnetization, and hence $\boldsymbol{B}$, until the lines of $\boldsymbol{B}$ extend all around the core.

Of course some of the lines of $\boldsymbol{B}$ escape into the air and then return to

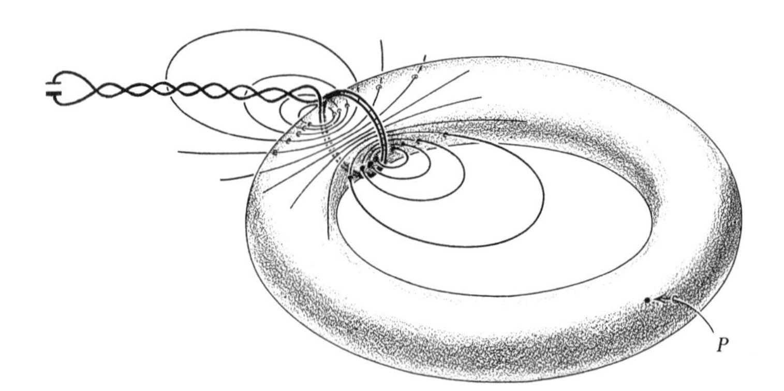

Figure 9-13. Ferromagnetic toroid with concentrated winding. The lines of force shown are in the plane of the toroid and are similar to those of Figure 9-4a. They apply only when there is no iron present.

the core to pass again through the coil. This constitutes the *leakage flux* that may, or may not, be negligible. For example, if the toroid is made up of a long thin iron wire, the flux at P is negligible compared to that near the coil.

Let us assume that the cross section of the toroid is large enough to render the leakage flux negligible. Then, applying Ampère's circuital law, Eq. 9-41, to a circular path of radius r going all around inside the toroid,

$$\oint \boldsymbol{H}\cdot d\boldsymbol{l} = NI, \tag{9-74}$$

$$2\pi r \frac{B}{\mu} = NI, \tag{9-75}$$

$$B = \frac{\mu NI}{2\pi r}. \tag{9-76}$$

Then, taking R_1 to be the radius corresponding to the average value of B, and R_2 to be the minor radius of the toroid,

$$\Phi = \frac{\mu\pi R_2^2}{2\pi R_1} NI. \tag{9-77}$$

The flux through the core is therefore the same as if we had a toroidal coil (example, page 313) of the same size and if the number of ampere-turns were increased by a factor of μ_r. In other words, for each ampere-turn in the coil there are $\mu_r - 1$ ampere-turns in the core. The amplification can be as high as 10^5 or more.

This equation shows that the magnetic flux is given by the magnetomotance NI multiplied by the factor

$$\frac{\mu\pi R_2^2}{2\pi R_1},$$

which is called the *permeance* of the magnetic circuit. The inverse of the permeance is called the *reluctance*. Thus

$$\text{Magnetic Flux} = \text{Permeance} \times \text{Magnetomotance} \tag{9-78}$$

$$= \frac{\text{Magnetomotance}}{\text{Reluctance}}. \tag{9-79}$$

The analogy with Ohm's law is obvious: if an electromotance were induced in the core, the current would be

$$I = \frac{\sigma\pi R_2^2}{2\pi R_1}\,\mathcal{V}. \tag{9-80}$$

Thus the corresponding quantities in electric and magnetic circuits are as follows:

Current I	Magnetic flux Φ
Current density $\boldsymbol{J}$	Magnetic induction $\boldsymbol{B}$
Conductivity σ	Permeability μ
Electromotance	Magnetomotance
Electric field intensity $\boldsymbol{E}$	Magnetic field intensity $\boldsymbol{H}$
Conductance	Permeance
Resistance	Reluctance

There is one important difference between electric and magnetic circuits: the magnetic flux cannot be made to follow a magnetic circuit like an electric current follows a conducting path. Indeed, a magnetic circuit behaves much as an electric circuit would if it were submerged in salt water: part of the current would flow through the components, and the rest would flow through the water.

If a magnetic circuit is not properly designed, the leakage flux can easily be an order of magnitude *larger* than that flowing around the circuit.

9.8.1.1 *Magnetic Circuit With an Air Gap.* Figure 9-14 shows a circuit with an air gap whose cross section is different from that of the soft-iron yoke. Each winding provides $NI/2$ ampere-turn.

We shall see that the magnetic flux in the air gap is equal to the magnetomotance NI divided by the sum of the reluctances of the iron yoke and of the air gap.

This is a general law: *reluctances and permeances in a magnetic circuit add in the same way as resistances and conductances in an electric circuit.*

We assume that the leakage flux is negligible. As we shall see, this will result in quite a large error.

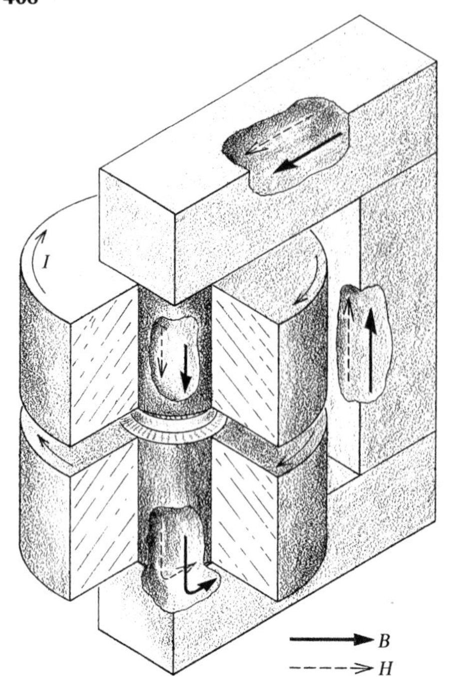

Figure 9-14. Electromagnet. The coils have been cut out to expose the iron yoke.

Applying again Ampère's circuital law to the circuit, we see that

$$NI = H_iL_i + H_gL_g \approx H_gL_g, \tag{9-81}$$

where the subscript i refers to the iron yoke and g to the air gap. As we shall see below, the requirement that the reluctance of the yoke be much smaller than that of the gap makes $H_iL_i \ll H_gL_g$. The path length L_i in the iron can be taken to be the length measured along the center of the cross section of the yoke.

Now, according to Eq. 9-29, or 6-125, the net outward flux of $\boldsymbol{B}$ through *any* closed surface must be equal to zero. This results from the fact that $\boldsymbol{\nabla}\cdot\boldsymbol{B}$ is *always* zero; indeed, this is one of Maxwell's equations. Thus, if we neglect leakage flux, the flux of $\boldsymbol{B}$ must be the same over any cross section of the magnetic circuit and

$$B_iA_i = B_gA_g, \tag{9-82}$$

where A_i and A_g are respectively the cross sections of the iron yoke and of the air gap.

Combining these two equations,

$$B_gA_g\left[\frac{L_i}{\mu A_i} + \frac{L_g}{\mu_0 A_g}\right] = NI, \tag{9-83}$$

and the magnetic flux is

$$\Phi = B_g A_g = \frac{NI}{\dfrac{L_i}{\mu A_i} + \dfrac{L_g}{\mu_0 A_g}} \approx \frac{NI\,\mu_0 A_g}{L_g}. \tag{9-84}$$

Now we have two reluctances in series, $L_i/\mu A_i$ in the yoke, and $L_g/\mu_0 A_g$ in the gap.

You can now show that $H_i L_i \ll H_g L_g$, as in Eq. 9-81, if the reluctance of the yoke is much smaller than that of the gap.

Since we have neglected leakage flux, this equation can only serve to provide an upper limit for Φ.

Example

If there is a total of 10,000 turns in the two windings, and if $I = 1.00$ ampere, $A_i = 100$ centimeters², $A_g = 50.0$ centimeters², $\mu_r = 1{,}000$, $L_i = 90.0$ centimeters, $L_g = 1.00$ centimeter, then

$$\Phi = \frac{10^4}{\dfrac{0.9}{10^3 \times 4\pi \times 10^{-7} \times 10^{-2}} + \dfrac{10^{-2}}{4\pi \times 10^{-7} \times 5 \times 10^{-3}}}, \tag{9-85}$$

$$= 6.3 \times 10^{-3} \text{ weber}, \tag{9-86}$$

$$B = \frac{6.3 \times 10^{-3}}{5 \times 10^{-3}} = 1.3 \text{ teslas} = 1.3 \times 10^4 \text{ gauss}, \tag{9-87}$$

$$L = \frac{6.3 \times 10^{-3} \times 10^4}{1.00} = 63 \text{ henrys} \tag{9-88}$$

and the stored energy is

$$\tfrac{1}{2}LI^2 = \tfrac{1}{2} \times 63 \times 1.00 \approx 32 \text{ joules}. \tag{9-89}$$

In this particular case the leakage flux is 70% of the flux in the gap. In other words, the magnetic induction in the gap is not 1.3 teslas, but only $1.3/1.7 = 0.77$ tesla.*

9.8.1.2 *Magnetic Circuit Energized by a Permanent Magnet.* Figure 9-15 shows a magnetic circuit that is similar to that of Figure 9-14, except that the two windings and part of the yoke have been replaced by permanent magnets. Let us see how we can calculate the magnetic flux. We neglect the leakage flux, as previously.

The $\boldsymbol{B}$ and $\boldsymbol{H}$ vectors are oriented as shown in the figure. In the *soft-iron yoke* and in the *pole pieces*, $\boldsymbol{B}$ and $\boldsymbol{H}$ are in the same direction and the operating point on the hysteresis curve (Section 9.6) is situated somewhere in the first quadrant. In the *air gap* $\boldsymbol{H} = \boldsymbol{B}/\mu_0$ and the two vectors are again oriented

* There exist empirical formulas for calculating leakage fluxes. See, for example, R. K. Tenzer, *Estimating Leakage Factors for Magnetic Circuits by a Simple Method*, Electrical Manufacturing, February 1957.

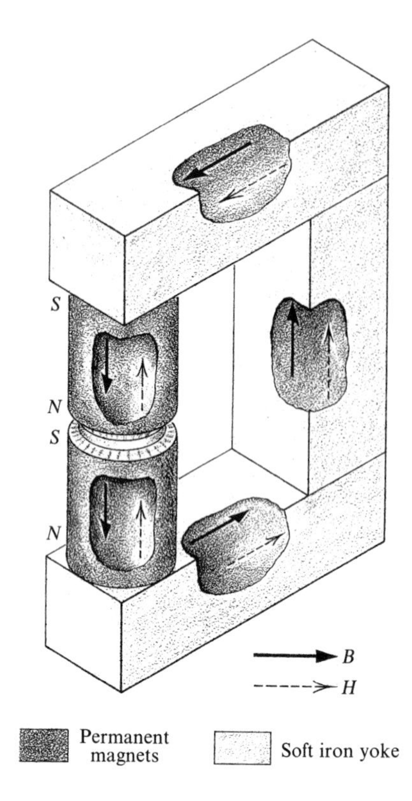

Figure 9-15. Magnetic circuit energized by a pair of permanent magnets.

in the same direction. As we saw in the example on page 402, the operating point for the *permanent magnets* is in the second quadrant on the hysteresis curve and the vectors $\boldsymbol{H}$ and $\boldsymbol{B}$ are oriented in opposite directions.

This is a good illustration of the fact that the flux of $\boldsymbol{H}$ over a closed surface is not necessarily equal to zero (Problem 9.19). In fact, the ends of the permanent magnets act as either sources or sinks of $\boldsymbol{H}$. For example, the lower end of the top magnet acts as a source of $\boldsymbol{H}$ whereas the top end acts as a sink of $\boldsymbol{H}$.

To calculate the magnetic flux we proceed as follows. Since there are neither free currents nor displacement currents, the line integral of $\boldsymbol{H} \cdot \boldsymbol{dl}$ around the magnetic circuit must be zero. Then

$$H_i L_i + H_g L_g - H_m L_m = 0, \tag{9-90}$$

$$H_m L_m = H_i L_i + H_g L_g \approx H_g L_g, \tag{9-91}$$

where the subscripts i, g, m refer, respectively, to the soft iron, the air gap, and the permanent magnets. The magnetomotance $H_i L_i$ is again negligible if the reluctance of the iron is much smaller than that of the air gap.

Also, if the leakage flux is negligible, Φ is the same over any cross section, and

$$A_i B_i = A_g B_g = A_m B_m. \tag{9-92}$$

These two equations are similar to Eqs. 9-81 and 9-82, except that NI is replaced by $H_m L_m$. Thus

$$\Phi = \frac{H_m L_m}{\dfrac{L_i}{\mu A_i} + \dfrac{L_g}{\mu_0 A_g}}. \tag{9-93}$$

Remember that L_i is the length of the path in the soft-iron yoke only, and that the total distance around the circuit is $L_i + L_m + L_g$.

The design of a magnetic circuit energized by a permanent magnet is somewhat complicated by the fact that the position of the operating point on the hysteresis curve (Section 9.6), and hence the value of H_m, depends on the permeance of the magnetic circuit.

As a rule, one tries to use an operating point that makes the *energy product* $H_m B_m$ maximum. The reason for this is the following. The purpose of the magnetic circuit is to produce a magnetic field in an air gap, and the

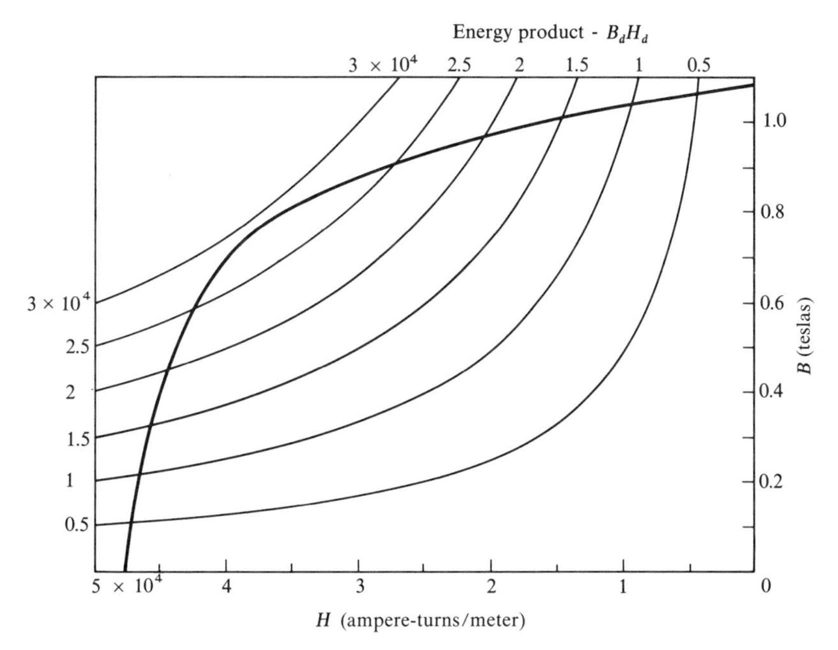

Figure 9-16. Demagnetization curve for Alnico 5. The energy product is maximum for $B \approx 0.8$ tesla.

magnetic energy required is the volume of the air gap, $A_g L_g$, multiplied by the energy density $\frac{1}{2}\mu_0 B_g^2$. Now

$$\frac{1}{2\mu_0} A_g L_g B_g^2 = \frac{1}{2}(H_g L_g)(A_g B_g), \tag{9-94}$$

$$\approx \frac{1}{2}(H_m L_m)(A_m B_m) = \frac{1}{2}(H_m B_m)(L_m A_m), \tag{9-95}$$

from Eqs. 9-91 and 9-92.

The magnetic energy in the air gap is therefore equal to one-half the energy product multiplied by the volume of the permanent magnet. As a rule, one requires that the volume $L_m A_m$ be as small as possible, for reasons of economy, size, and weight. Then the operating point on the hysteresis curve is chosen so that the energy product $H_m B_m$ is maximum.

Figure 9-16 shows the demagnetization curve for the alloy Alnico 5, which is commonly used for permanent magnets.

Example

Let us design a magnetic circuit that will produce a magnetic field of 1.3 teslas in an air gap 1.00 centimeter long and 50.0 centimeters² in area as in the previous example. We again neglect the leakage flux and the percentage error will be the same as in the previous example.

With the alloy Alnico 5, the energy product $H_m B_m$ is maximum at

$$B_m = 8000 \text{ gauss} = 0.80 \text{ tesla}, \tag{9-96}$$

$$H_m = 450 \text{ oersteds} = 3.6 \times 10^4 \text{ ampere-turns/meter}. \tag{9-97}$$

Then

$$A_m = \frac{B_g A_g}{B_m} = \frac{1.3}{0.80} \times 5.0 \times 10^{-3} = 8.1 \times 10^{-3} \text{ meter}^2 \tag{9-98}$$

and the permanent magnets should be about 10 centimeters in diameter.

To find the value of L_m, we use Eq. 9-91. If we use the same geometry as in the previous example, the reluctance of the yoke is even more negligible because L_i is now shorter since it does not include the lengths of the permanent magnets and we can set

$$H_m L_m = H_g L_g. \tag{9-99}$$

Then

$$L_m = \frac{H_g L_g}{H_m} = \frac{B_g L_g}{\mu_0 H_m} \tag{9-100}$$

$$= \frac{1.3 \times 10^{-2}}{4\pi \times 10^{-7} \times 3.6 \times 10^4} = 0.29 \text{ meter}. \tag{9-101}$$

Each magnet should therefore be about 15 centimeters long. If permanent magnets of the proper shape are not available, smaller magnets can be assembled in an appropriate series-parallel combination.

We have again neglected the leakage flux and B_g is in fact only 0.77 tesla.

9.8.2 Solution of Poisson's Equation for $\boldsymbol{B}$

You will recall that there exists a Poisson equation for $\boldsymbol{E}$ (Section 4.7). There exists a similar equation for $\boldsymbol{B}$: taking the curl of Eq. 9-34,

$$\nabla \times \nabla \times \boldsymbol{B} \equiv -\nabla^2 \boldsymbol{B} + \nabla(\nabla \cdot \boldsymbol{B}) = \mu_0 \nabla \times (\boldsymbol{J}_f + \nabla \times \boldsymbol{M}) \quad (9\text{-}102)$$

and, since $\nabla \cdot \boldsymbol{B}$ is zero (Eq. 9-29),

$$\nabla^2 \boldsymbol{B} = -\mu_0 \nabla \times (\boldsymbol{J}_f + \nabla \times \boldsymbol{M}). \quad (9\text{-}103)$$

This is Poisson's equation for $\boldsymbol{B}$.

The solution of this equation will be found in Section E.9 of Appendix E. It is

$$\boldsymbol{B} = \frac{\mu_0}{4\pi} \int_{\tau'} \frac{\nabla' \times (\boldsymbol{J}_f + \nabla' \times \boldsymbol{M})}{r} \, d\tau'. \quad (9\text{-}104)$$

We have used primes on the dels on the right-hand side because they operate at the source point (x', y', z'). Note the analogy with the corresponding equation 4-184 for $\boldsymbol{E}$. The above equation relates $\boldsymbol{B}$, not to the current density $\boldsymbol{J}_f + \nabla \times \boldsymbol{M}$ itself, but to its *curl.* We again have r to the first power in the denominator, but a positive sign. Of course the integral can be calculated only if the curl exists.

You will remember from Section 9.3 that, for a volume distribution of charge,

$$\boldsymbol{B} = \frac{\mu_0}{4\pi} \int_{\tau'} \frac{(\boldsymbol{J}_f + \nabla' \times \boldsymbol{M}) \times \boldsymbol{r}_1}{r^2} \, d\tau'. \quad (9\text{-}105)$$

Since the two integrals for $\boldsymbol{B}$ are equal, it follows (Problem 9-22) that

$$\int_{\tau'} \nabla' \times \frac{\boldsymbol{J}_f + \nabla' \times \boldsymbol{M}}{r} \, d\tau' = 0 \quad (9\text{-}106)$$

for any current distribution, as long as the curls exist.

Example THE MAGNETIC INDUCTION AT THE CENTER OF A ROTATING DISK OF CHARGE

A thin disk of charge of radius R and thickness $2t \ll R$ has a density

$$K(R - \rho)(t - z)^2$$

in cylindrical coordinates, ρ and z being the radial and axial coordinates, respectively. When the disk rotates at an angular velocity ω the current density is

$$J_f = K(R - \rho)(t - z)^2 \omega \rho \, \boldsymbol{\varphi}_1. \tag{9-107}$$

We shall calculate $\boldsymbol{B}$, first with Eq. 9-104, and then with Eq. 9-105.

The curl of $\boldsymbol{J}_f$ is

$$\boldsymbol{\nabla}' \times \boldsymbol{J}_f = -\frac{\partial J_f}{\partial z} \boldsymbol{\rho}_1 + \frac{1}{\rho} \frac{\partial}{\partial \rho} (\rho J_f) \, \boldsymbol{z}_1 \tag{9-108}$$

and $\boldsymbol{\nabla}' \times \boldsymbol{M}$ is zero in this case. By symmetry, we can neglect the first term on the right. Then $\boldsymbol{B}$ is axial and

$$B = \frac{\mu_0}{4\pi} \int_{\tau'} \frac{\frac{1}{\rho} \frac{\partial}{\partial \rho} (\rho J_f)}{r} \, d\tau'. \tag{9-109}$$

Since, by hypothesis, the disk is very thin, we may substitute ρ for r, and then

$$B = \frac{\mu_0}{4\pi} \int_{z=-t}^{z=+t} \int_{\rho=0}^{\rho=R} \frac{\omega K(2R - 3\rho)(t - z)^2}{\rho} 2\pi\rho \, d\rho \, dz, \tag{9-110}$$

$$= \frac{2}{3} \mu_0 K \omega R^2 t^3. \tag{9-111}$$

We can arrive at the same result if we use Eq. 9-105. So as to simplify the calculation, we shall use the formula which we found in the example on page 299 for a current ring, setting $z = 0$. The ring of inner radius ρ and outer radius $\rho + d\rho$ carries a current

$$I = K(R - \rho)\omega\rho \, d\rho \int_{-t}^{+t} (t - z)^2 \, dz, \tag{9-112}$$

$$= \frac{8}{3} K(R - \rho)\omega\rho t^3 \, d\rho, \tag{9-113}$$

and

$$B = \frac{\mu_0}{2} \int_0^R \frac{(8/3)K(R - \rho)\omega\rho t^3 \, d\rho}{\rho}, \tag{9-114}$$

$$= \frac{2}{3} \mu_0 K \omega R^2 t^3, \tag{9-115}$$

as before.

9.9 SUMMARY

The *magnetization* $\boldsymbol{M}$ corresponds to the polarization $\boldsymbol{P}$ in dielectrics and is the net magnetic moment per unit volume.

According to the Ampèrian model, the magnetization produces an *equivalent surface current density* $\boldsymbol{\lambda}_e$ and an *equivalent volume current density* $\boldsymbol{J}_e$, where

$$\boldsymbol{\lambda}_e = \boldsymbol{M} \times \boldsymbol{n}, \tag{9-8}$$

$$\boldsymbol{J}_e = \boldsymbol{\nabla} \times \boldsymbol{M}. \tag{9-9}$$

We have shown that $\boldsymbol{B}$ can be calculated correctly, both inside and outside magnetic material, by treating the equivalent currents as if they were real conduction currents flowing in a vacuum. For example, the $\boldsymbol{B}$ field of a cylinder uniformly magnetized in the direction of its axis of symmetry is identical to that of a solenoid carrying a current density IN' equal to M.

The *magnetic field intensity* $\boldsymbol{H}$ is related to $\boldsymbol{B}$ and to $\boldsymbol{M}$ as follows:

$$\boldsymbol{H} = \frac{\boldsymbol{B}}{\mu_0} - \boldsymbol{M}, \tag{9-36}$$

or

$$\boldsymbol{B} = \mu_0(\boldsymbol{H} + \boldsymbol{M}). \tag{9-37}$$

This equation is to be compared with

$$\boldsymbol{D} = \epsilon_0 \boldsymbol{E} + \boldsymbol{P}, \tag{9-38}$$

which applies to dielectrics.

In terms of $\boldsymbol{H}$, *Ampère's circuital law* is stated as follows:

$$\boldsymbol{\nabla} \times \boldsymbol{H} = \boldsymbol{J}_f \tag{9-39}$$

or

$$\oint_C \boldsymbol{H} \cdot d\boldsymbol{l} = I_f, \tag{9-41}$$

where I is the current of free charges linking the curve C. This is a more general form of Ampère's law (Section 7-7), in that it is valid even in the presence of magnetic materials, but it is still rigorously valid only for steady currents.

It is the custom to write that

$$\boldsymbol{M} = \chi_m \boldsymbol{H}, \tag{9-42}$$

where χ_m is the *magnetic susceptibility*. Then

$$\boldsymbol{B} = \mu_0(1 + \chi_m)\boldsymbol{H} = \mu_0\mu_r\boldsymbol{H} = \mu\boldsymbol{H}, \tag{9-44}$$

where μ_r is the *relative permeability* and μ is the *permeability*.

In linear, isotropic, and homogeneous materials,

$$\boldsymbol{J}_e = (\mu_r - 1)\boldsymbol{J}_f, \tag{9-51}$$

and

$$\mu \boldsymbol{J}_f = \mu_0(\boldsymbol{J}_f + \boldsymbol{\nabla} \times \boldsymbol{M}). \tag{9-52}$$

The *hysteresis loop* of Figure 9-8 is a plot of B as a function of H for one complete cycle. It shows clearly that $\mu_r = B/\mu_0 H$ is definitely *not* a constant for a given ferromagnetic material. The area of the hysteresis loop is equal to the energy dissipated per cubic meter and per cycle in the material.

At the interface between two media 1 and 2, the *boundary conditions* for $\boldsymbol{B}$ and $\boldsymbol{H}$ are as follows:

$$B_{n1} = B_{n2}, \tag{9-61}$$

$$H_{t1} = H_{t2}. \tag{9-63}$$

If the media are linear and isotropic, then the $\boldsymbol{B}$ and $\boldsymbol{H}$ vectors are parallel, both in medium 1 and in medium 2, and

$$\frac{\tan\theta_1}{\tan\theta_2} = \frac{\mu_{r1}}{\mu_{r2}}, \tag{9-66}$$

where the angles are those between the $\boldsymbol{B}$ vectors and the normal as in Figure 9-10a.

The concept of *magnetic circuit* is widely used whenever the magnetic flux is guided mostly through magnetic material. We then have a magnetic equivalent of Ohm's law:

$$\text{Magnetic Flux} = \text{Permeance} \times \text{Magnetomotance} \tag{9-78}$$

$$= \frac{\text{Magnetomotance}}{\text{Reluctance}}. \tag{9-79}$$

The correspondence between electric and magnetic circuits is as follows:

Current I	Magnetic Flux Φ
Current density $\boldsymbol{J}$	Magnetic induction $\boldsymbol{B}$
Conductivity σ	Permeability μ
Electromotance	Magnetomotance
Electric field intensity $\boldsymbol{E}$	Magnetic field intensity $\boldsymbol{H}$
Conductance	Permeance
Resistance	Reluctance

The magnetomotance of a coil is NI, where N is the total number of turns, while that of a cylindrical permanent magnet is $H_m L_m$, where H_m is the value of H inside the magnet and L_m is the length of the magnet.

Poisson's equation for $\boldsymbol{B}$ is

$$\nabla^2 \boldsymbol{B} = -\mu_0 \nabla \times (\boldsymbol{J}_f + \nabla \times \boldsymbol{M}), \tag{9-103}$$

and thus

$$\boldsymbol{B} = \frac{\mu_0}{4\pi} \int_{\tau'} \frac{\nabla' \times (\boldsymbol{J}_f + \nabla \times \boldsymbol{M})}{r} \, d\tau' \tag{9-104}$$

This is an alternate expression for $\boldsymbol{B}$, which should be compared with Eq. 4-184.

PROBLEMS

9-1. A long tube is uniformly magnetized in the direction parallel to its axis.

What is the value of $\boldsymbol{B}$ inside the tube?

9-2. Show that the torque exerted on a small permanent magnet of dipole moment $\boldsymbol{m}$ situated in a magnetic field $\boldsymbol{B}$ is $\boldsymbol{m} \times \boldsymbol{B}$.

9-3. A small magnetic dipole of moment $\boldsymbol{m}$ is arbitrarily oriented at an off-center point within a spherical surface of radius R.

Show that the average magnetic induction inside the sphere is $\mu_0 \boldsymbol{m}/2\pi R^3$ as in Section 9.3.‡

9-4. Draw a parallel between the bound charges of Chapter 3 and the equivalent currents.

9-5. The magnetization of iron can contribute as much as 2 teslas to the magnetic induction in iron. If each electron can contribute a magnetic moment of 9.27×10^{-22} ampere-meter^2 (one Bohr magneton), how many electrons per atom, on the average, contribute to the polarization?

9-6. Show that the equivalent surface current density on the interface between two different magnetic materials with magnetizations $\boldsymbol{M}_1$ and $\boldsymbol{M}_2$ is

$$(\boldsymbol{M}_2 - \boldsymbol{M}_1) \times \boldsymbol{n},$$

where $\boldsymbol{n}$ is a unit vector that is normal to the interface and that points from medium 2 to medium 1.

9-7. Draw a parallel between the relative permeability μ_r and the relative permittivity ϵ_r.

9-8. Imagine a parallel-plate capacitor whose plates are charged and insulated.

(a) How are E and D affected by the introduction of a dielectric between the plates?

(b) Show a polar molecule of the dielectric oriented in the field.

(c) Is its energy maximum or minimum?

(d) Now imagine a long solenoid carrying a fixed current.

How are B and H affected by the introduction of a cylinder of ferromagnetic material inside the solenoid?

(e) Show a small current loop representing the magnetic dipole moment of an atom, and show the direction of the current on the loop.

(f) Is its energy maximum or minimum?

9-9. A conducting wire carrying a current I is embedded in a nonconducting magnetic material.

Find $\boldsymbol{H}$, $\boldsymbol{B}$, $\boldsymbol{M}$ and the equivalent currents.

9-10. A long wire of radius a carries a current I and is surrounded by a long hollow iron cylinder. The inner radius of the cylinder is b and the outer radius c.

(a) Compute the flux of B inside a section of the cylinder l meters long.
(b) Find the equivalent current density on the inner and outer iron surfaces, and find the direction of the equivalent currents relative to the current in the wire.
(c) Find the equivalent current density inside the iron.
(d) Find B at distances $r > c$ from the wire.
How would this value be affected if the iron cylinder were removed?

9-11. It was mentioned in Section 9.6 that the hysteresis loop for a given sample of material can be obtained by using a ring-shaped sample, as in Figure 9-7.

Show that, if the ratio R/L (resistance/inductance) for the circuit of winding b is large, and if the current in winding a is not changed too rapidly,

$$\Delta\Phi = RQ/N,$$

where $\Delta\Phi$ is the change in flux, Q is the charge that flows through the circuit, and N is the number of turns in winding b.

9-12. The *peaking strip* is a device that is used for measuring B. It consists of a fine wire of magnetic material, oriented in the direction of $\mathbf{B}$, and centered on the axis of a solenoid as in Figure 9-17. A small pick-up coil of a few thousand turns is fixed to the center of the wire.

The peaking strip is operated as follows. If the solenoid carries a direct current that just cancels the field one wishes to measure, plus a small alternating current, then the H on the axis of the solenoid is that due to the alternating current alone, and the peaking strip goes through a hysteresis loop at every cycle. The electromotance induced in the pick-up coil is $N\, d\Phi/dt$, where N is the number of turns in the coil and Φ is the magnetic flux in the strip. Now the right- and left-hand sides of the hysteresis loop (Figure 9-8) are nearly vertical in certain materials, such a molybdenum permalloy, and the electromotance then has two sharp peaks, one positive and the other negative, which can be observed on an oscilloscope. If the oscilloscope sweep is synchronized with the alternating current in the solenoid, the two peaks are symmetrical when the time average of H on the axis of the solenoid is zero, or when the dc field of the solenoid exactly cancels the measured field.

The peaking strip has a rather limited range of applications, first because of the large size of the solenoid: for maximum sensitivity the length of the wire should be much larger than its diameter, because end effects reduce the value of B inside it, and the length of the solenoid must be several times that of the wire, again to reduce end effects. In practice, the solenoid is 10 to 20 centimeters long. Also, because, of limitations in the power that can be dissipated in the solenoid, the peaking strip can be used only up to fields of a few hundred gauss. Finally, the field of the solenoid can alter the B in the pole pieces and thus change their permeability locally.

Calculate the peak value of the electromotance induced in the pick-up coil under the following conditions.

Peaking strip diameter, 2.5×10^{-5} meter;
Number of turns in pick-up coil, 1000;
Maximum value of dB/dH, 0.8 tesla-meter/ampere-turn;

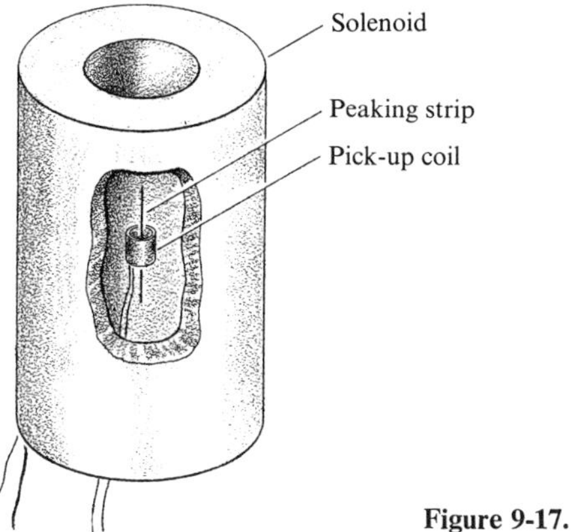

Figure 9-17.

Frequency of applied alternating current, 60 hertz;
Amplitude of the alternating H, 0.25 ampere-turn/meter.

So as to simplify the calculation, replace the sine wave for H by straight lines, one for each half cycle.

9-13. If $\boldsymbol{n}$ is the unit vector normal to an interface, show that $\boldsymbol{n} \cdot \nabla \times (\boldsymbol{A}_1 - \boldsymbol{A}_2) = 0$ and that $\boldsymbol{n} \times (\nabla \times \boldsymbol{A}_1/\mu_1 - \nabla \times \boldsymbol{A}_2/\mu_2) = 0$.

9-14. Show that both the normal and tangential components of the vector potential $\boldsymbol{A}$ must be continuous across the interface between two media if the currents are constant.

9-15. A thin disk of iron of radius a and thickness t is magnetized in the direction parallel to its axis.

Calculate $\boldsymbol{H}$ and $\boldsymbol{B}$ on the axis, both inside and outside the iron.

9-16. A coil of 300 turns is wound on an iron ring ($\mu_r = 500$) of 40 centimeters mean diameter and 10 centimeters^2 in cross section.

(a) Calculate the magnetic flux in the ring when the current in the coil is one ampere.
(b) Calculate the flux when there is a gap of 1.0 millimeter in the ring.

9-17. Show that the energy stored in a magnetic circuit is

$$W = \tfrac{1}{2}\Phi^2 R,$$

where Φ is the magnetic flux and R is the reluctance.

9-18. A magnetic circuit links a one-turn coil.

(a) Show that the inductance L of the coil is equal to the inverse of the reluctance R of the magnetic circuit.
(b) Show that, if the coil has N turns,

$$L = \frac{N^2}{R}.$$

9-19. We have seen in Sections 6.9 and 7.4 that $\nabla \cdot \boldsymbol{B}$ is zero. This equation is valid even in nonlinear, nonhomogeneous and nonisotropic media.

Show that it does *not* follow that $\nabla \cdot \boldsymbol{H}$ is always zero.

9-20. Calculate A_m and L_m for a magnetic circuit, similar to that of Figure 9-15, which will produce a field of one tesla in an air gap 1.00 centimeter² in area and 2.00 millimeters long. Assume a leakage factor of 50%.

9-21. In the process of designing magnetic devices it is often necessary to predict in detail the configuration of a magnetic field. Now magnetic fields can seldom be calculated analytically. One solution is to perform a numerical calculation on a computer. Another solution, which can be more convenient in some cases, is to use an electrolytic tank in which the magnetic field is simulated by an electric field as follows.

Consider a two-dimensional static magnetic field in the xy plane. In regions where $\boldsymbol{J}_f = 0$ and where there is no permanent magnetism,

$$\nabla \cdot \boldsymbol{B} = 0, \qquad \text{and} \qquad \nabla \times \boldsymbol{B} = 0.$$

Note that these equations are analogous to

$$\nabla \cdot \boldsymbol{E} = 0 \qquad \text{and} \qquad \nabla \times \boldsymbol{E} = 0,$$

which apply to electrostatic fields.

(a) Show that

$$B_x = \frac{\partial A}{\partial y}, \qquad B_y = -\frac{\partial A}{\partial x},$$

where A is the z-component of the vector potential.
(b) Show that a line of constant A is a line of $\boldsymbol{B}$.
(c) Show that $\boldsymbol{B}$ is derivable from a potential u such that

$$B_x = -\frac{\partial u}{\partial x}, \qquad B_y = -\frac{\partial u}{\partial y},$$

and that lines of constant u are orthogonal to lines of $\boldsymbol{B}$ and thus to lines of constant A.
(d) Hence show how a magnetic field can be simulated in an electrolytic tank, using conducting electrodes to simulate the pole pieces. The pole pieces are assumed to have infinite permeability.

Note that this method is not applicable if there are current-carrying conductors or permanent magnets in the field.
(e) Show that

$$\nabla^2 u = 0.$$

The relaxation method described in Problem 2-16 can therefore also be used for plotting magnetic fields.

9-22. Show that

$$\int \nabla' \times \left(\frac{\boldsymbol{J}_f + \nabla' \times \boldsymbol{M}}{r} \right) d\tau' = 0.‡$$

Hints

9-3. The dipole $\boldsymbol{m}$ has two components, one that is radial and another that is perpendicular to the diameter passing through $\boldsymbol{m}$.

You should be able to show that, for a radial dipole $\boldsymbol{m}_r$,

$$\frac{4\pi}{\mu_0}\int \boldsymbol{B}\,d\tau = 2\pi \boldsymbol{m}_r \int_0^\pi \frac{\sin^3\theta\, d\theta}{\{1 + (S/R)^2 - 2(S/R)\cos\theta\}^{3/2}},$$

where S is the distance between the dipole and the center of the sphere, and $\theta = 0$ at the position of the dipole. The integral on the right is equal to 4/3 for any value of S.

Similarly, you can show that, for the other component $\boldsymbol{m}_t$,

$$\frac{4\pi}{\mu_0}\int \boldsymbol{B}\,d\tau = \pi \boldsymbol{m}_t \int_0^\pi \frac{1 + \cos^2\theta - 2(S/R)\cos\theta}{\{1 + (S/R)^2 - 2(S/R)\cos\theta\}^{3/2}} \sin\theta\, d\theta.$$

The integral on the right is equal to 8/3 and is also independent of S.

9-22. See Section 9.8.2.

CHAPTER 10

MAXWELL'S EQUATIONS

At this stage we have found all four of Maxwell's equations in Chapter 6, but we have found only three of them without using relativity. These were Eqs. 3-41, 8-8, and 9-29. Our first objective in this chapter is therefore to deduce the fourth one, namely the equation for the curl of $\boldsymbol{B}$, without using the methods of relativity. Then we shall reexamine all four equations as a group and deduce from them two general, and rather surprising, properties of electromagnetic fields.

The reasoning that will lead us to Maxwell's equation for the curl of $\boldsymbol{B}$ is unfortunately rather involved (see Figure 10-11), but we shall learn many things on the way.

We assume, as usual, that ϵ, μ, σ are not functions of the time.

The following definitions will be useful. A medium is *homogeneous* if its properties do not vary from point to point; it is *isotropic* if its properties are the same in all directions from any given point; it is *linear and isotropic* if

$$\boldsymbol{D} = \epsilon \boldsymbol{E}, \qquad \boldsymbol{H} = \boldsymbol{B}/\mu, \qquad \boldsymbol{J}_f = \sigma \boldsymbol{E}, \tag{10-1}$$

where ϵ, μ, σ are constants independent of $\boldsymbol{E}$ and $\boldsymbol{H}$, and independent of direction. Crystalline media are usually *an*isotropic.

10.1 THE CONSERVATION OF ELECTRIC CHARGE

According to all experiments performed to date, it appears that electric charge is never created nor destroyed (Section 5.23). Therefore, if we consider a surface S' enclosing a volume τ', the net free charge escaping through S' during a time interval Δt must be equal to the net free charge lost by τ' during this same time interval. Thus

$$\int_{S'} \boldsymbol{J}_f \cdot d\boldsymbol{a}' = -\int_{\tau'} \frac{\partial \rho_f}{\partial t}\, d\tau'. \tag{10-2}$$

This is the *law of conservation of free charge* stated in integral form.

We can also state this law in differential form if we replace the surface integral by a volume integral. Then

$$\int_{\tau'} \boldsymbol{\nabla} \cdot \boldsymbol{J}_f \, d\tau' = -\int_{\tau'} \frac{\partial \rho_f}{\partial t}\, d\tau', \tag{10-3}$$

and, since this equation is valid for any τ',

$$\boldsymbol{\nabla} \cdot \boldsymbol{J}_f = -\frac{\partial \rho_f}{\partial t}. \tag{10-4}$$

There is also conservation of bound charges, since

$$\boldsymbol{\nabla} \cdot \frac{\partial \boldsymbol{P}}{\partial t} = \frac{\partial}{\partial t}(\boldsymbol{\nabla} \cdot \boldsymbol{P}) = -\frac{\partial \rho_b}{\partial t}, \tag{10-5}$$

where $\partial \boldsymbol{P}/\partial t$ is the polarization current density (Section 3.2.2), and $-\boldsymbol{\nabla} \cdot \boldsymbol{P}$ is the bound charge density ρ_b (Section 3.2.1).

Finally, the charges involved in magnetization currents (Section 9.2) are conserved because $\boldsymbol{\nabla} \cdot \boldsymbol{\nabla} \times \boldsymbol{M} \equiv 0$. The net outward flux of these charges through any closed surface is therefore zero.

Thus, if we set $\boldsymbol{J}_m$ to be the current density in matter,

$$\boldsymbol{J}_m = \boldsymbol{J}_f + \frac{\partial \boldsymbol{P}}{\partial t} + \boldsymbol{\nabla} \times \boldsymbol{M}, \tag{10-6}$$

and

$$\boldsymbol{\nabla} \cdot \boldsymbol{J}_m = -\frac{\partial \rho_f}{\partial t} - \frac{\partial \rho_b}{\partial t} = -\frac{\partial \rho_t}{\partial t}. \tag{10-7}$$

There is also a fourth type of current that can exist even in a vacuum and that we shall use in Section 10.6. The corresponding current density is $\epsilon_0 \partial \boldsymbol{E}/\partial t$ and

$$\boldsymbol{\nabla} \cdot \epsilon_0 \frac{\partial \boldsymbol{E}}{\partial t} = \frac{\partial}{\partial t} \epsilon_0 \boldsymbol{\nabla} \cdot \boldsymbol{E} = \frac{\partial \rho_t}{\partial t}, \tag{10-8}$$

since the divergence of $\boldsymbol{E}$ is equal to ρ_t/ϵ_0, from Section 3.6.

Then the total volume current density is

$$\boldsymbol{J}_t = \boldsymbol{J}_f + \frac{\partial \boldsymbol{P}}{\partial t} + \boldsymbol{\nabla} \times \boldsymbol{M} + \epsilon_0 \frac{\partial \boldsymbol{E}}{\partial t} = \boldsymbol{J}_m + \epsilon_0 \frac{\partial \boldsymbol{E}}{\partial t}, \tag{10-9}$$

and

$$\nabla \cdot \boldsymbol{J}_t = 0. \tag{10-10}$$

The second and fourth terms in $\boldsymbol{J}_t$ are often grouped together, and

$$\epsilon_0 \frac{\partial \boldsymbol{E}}{\partial t} + \frac{\partial \boldsymbol{P}}{\partial t} = \frac{\partial}{\partial t}(\epsilon_0 \boldsymbol{E} + \boldsymbol{P}) = \frac{\partial \boldsymbol{D}}{\partial t} \tag{10-11}$$

is called the *displacement current density*, since $\boldsymbol{D}$ is the electric displacement (Section 3.6).

Thus

$$\boldsymbol{J}_t = \boldsymbol{J}_f + \frac{\partial \boldsymbol{D}}{\partial t} + \nabla \times \boldsymbol{M}. \tag{10-12}$$

Example

CHARGE DENSITY IN A CONDUCTOR

The current density in a linear and isotropic conductor obeys Ohm's law

$$\boldsymbol{J}_f = \sigma \boldsymbol{E}, \tag{10-13}$$

where σ is the electric conductivity. Then, from Eqs. 10-4 and 3-54, and if σ is not a function of the coordinates,

$$\nabla \cdot \boldsymbol{J}_f = -\frac{\partial \rho_f}{\partial t} = \sigma \nabla \cdot \boldsymbol{E} = \sigma \frac{\rho_f}{\epsilon}, \tag{10-14}$$

and

$$\rho_f = \rho_{f0} e^{-(\sigma/\epsilon)t}. \tag{10-15}$$

Therefore, *as long as our assumptions are satisfied*, the free charge density can only decrease! To achieve a nonzero ρ_f in a homogeneous, isotropic, and linear conductor, one must inject charges in some manner that does not satisfy Eq. 10-13. For example, a conductor can be bombarded with a pulse of high-energy electrons that come to rest inside the material.

"The inverse of the coefficient of t in the exponent is the *relaxation time*. Our calculation is not entirely valid because we have disregarded the fact that the conductivity σ is frequency dependent, and thus a function of the relaxation time. The calculation is valid for poor conductors, but it leads to values which are too small by orders of magnitude in the case of good conductors. Relaxation times in good conductors are nevertheless short and $\rho_f = 0$ in practice. For copper at room temperature, for example, the correct value is $\approx 4 \times 10^{-14}$ second." *

10.2 THE POTENTIALS V AND $\boldsymbol{A}$

We have seen in Sections 3.3.2 and 9.3 that, for a finite charge and current distribution as in Figure 10-1, the scalar potential V and the vector potential

* W. M. Saslow and G. Wilkinson, American Journal of Physics, *39*, 1244 (1971).

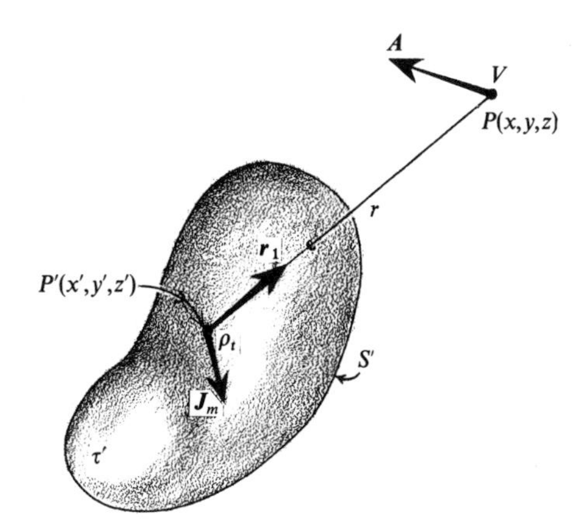

Figure 10-1. The charges and currents within the volume τ' produce a scalar potential V and a vector potential $\boldsymbol{A}$ at the field point $P(x, y, z)$. The current density $\boldsymbol{J}_m$ at the source point $P'(x', y', z')$ is $\boldsymbol{J}_f + \boldsymbol{\nabla} \times \boldsymbol{M}$ and the charge density ρ_t is $\rho_f + \rho_b$.

$\boldsymbol{A}$ are given respectively by the following integrals:

$$V = \frac{1}{4\pi\epsilon_0} \int_{\tau'} \frac{\rho_f - \boldsymbol{\nabla}\cdot\boldsymbol{P}}{r} \, d\tau' + \frac{1}{4\pi\epsilon_0} \int_{S'} \frac{\sigma_f + \boldsymbol{P}\cdot\boldsymbol{n}}{r} \, da', \tag{10-16}$$

$$\boldsymbol{A} = \frac{\mu_0}{4\pi} \int_{\tau'} \frac{\boldsymbol{J}_f + \boldsymbol{\nabla} \times \boldsymbol{M}}{r} \, d\tau' + \frac{\mu_0}{4\pi} \int_{S'} \frac{\boldsymbol{\lambda}_f + \boldsymbol{M} \times \boldsymbol{n}}{r} \, da', \tag{10-17}$$

where

ρ_f is the volume density of free charge,
$\boldsymbol{P}$ is the electric polarization (Section 3.1),
$-\boldsymbol{\nabla}\cdot\boldsymbol{P}$ is the volume density of bound charge (Section 3.2.1), ρ_b,
σ_f is the surface density of free charge,
$\boldsymbol{P}\cdot\boldsymbol{n}$ is the surface density of bound charge (Section 3.2.1), σ_b,
$\boldsymbol{J}_f$ is the volume density of current due to the motion of free charges,
$\boldsymbol{M}$ is the magnetization (Section 9.1),
$\boldsymbol{\nabla} \times \boldsymbol{M}$ is the volume density of equivalent currents (Section 9.2),
$\boldsymbol{\lambda}_f$ is the surface density of current due to free charges,
$\boldsymbol{M} \times \boldsymbol{n}$ is the surface density of equivalent currents (Section 9.2),
r is the distance between the points P and P',
$P(x, y, z)$ is the field point where V and $\boldsymbol{A}$ are calculated,
$P'(x', y', z')$ is the source point where the charge and current densities are ρ_f, ρ_b, σ_f, etc.

The integrals are evaluated over volumes τ' and surfaces S', which enclose all charges and currents. If the charge or current distributions extend to infinity, these integrals do not converge, as a rule.

We have also seen in Sections 3.7 and 9.5.1 that, in a linear, isotropic, and homogeneous medium ϵ, μ,*

$$V = \frac{1}{4\pi}\int_{\tau'}\frac{\rho_f}{\epsilon r}\,d\tau' + \frac{1}{4\pi\epsilon_0}\int_{S'}\frac{\sigma_f + \boldsymbol{P}\cdot\boldsymbol{n}}{r}\,da', \tag{10-18}$$

$$\boldsymbol{A} = \frac{1}{4\pi}\int_{\tau'}\frac{\mu_r\boldsymbol{J}_f}{r}\,d\tau' + \frac{\mu_0}{4\pi}\int_{S'}\frac{\boldsymbol{\lambda}_f + \boldsymbol{M}\times\boldsymbol{n}}{r}\,da'. \tag{10-19}$$

If you have studied Chapter 6, you will recall that we found in Sections 6.4 and 6.5 that, for an isolated charge passing through the origin at a constant velocity $\mathcal{V}$ along the x-axis,

$$V = \frac{\gamma Q}{4\pi\epsilon_0\{\gamma^2(x - \mathcal{V}t)^2 + y^2 + z^2\}^{1/2}}, \tag{10-20}$$

$$\boldsymbol{A} = \frac{\mu_0}{4\pi}\frac{\gamma Q\mathcal{V}\boldsymbol{i}}{\{\gamma^2(x - \mathcal{V}t)^2 + y^2 + z^2\}^{1/2}}. \tag{10-21}$$

The potentials V and $\boldsymbol{A}$ are related to the field vectors $\boldsymbol{E}$ and $\boldsymbol{B}$ by the equations

$$\boldsymbol{E} = -\boldsymbol{\nabla}V - \frac{\partial\boldsymbol{A}}{\partial t}, \tag{10-22}$$

$$\boldsymbol{B} = \boldsymbol{\nabla}\times\boldsymbol{A}. \tag{10-23}$$

These are Eqs. 6-90 and 6-77, or 8-13 and 7-41 respectively. In the first equation, $-\boldsymbol{\nabla}V$ is that part of $\boldsymbol{E}$ that is due to accumulations of electric charge, whereas $-\partial\boldsymbol{A}/\partial t$ is the part that is due to changing magnetic fields. Once V and $\boldsymbol{A}$ are known, it is merely a matter of differentiation to find $\boldsymbol{E}$ and $\boldsymbol{B}$.

Therefore the six components of $\boldsymbol{E}$ and $\boldsymbol{B}$ can be deduced from only four quantities, V and the three components of $\boldsymbol{A}$.

---

* Our reasoning with respect to $\boldsymbol{A}$ will not be quite consistent. In Section 9.5.1 we used the relation

$$\mu_r\boldsymbol{J}_f = \mu_0(\boldsymbol{J}_f + \boldsymbol{\nabla}\times\boldsymbol{M})$$

to transform Eq. 10-17 into 10-19. Now this relation is valid only for steady fields. The more general expression for $\mu_r\boldsymbol{J}_f$ contains a time derivative on the right, and it can be deduced from Eqs. 10-94 and 10-99. It is as follows:

$$\mu_r\boldsymbol{J}_f = \mu_0(\boldsymbol{J}_f + \boldsymbol{\nabla}\times\boldsymbol{M}) - (\mu - \mu_0)\frac{\partial\boldsymbol{D}}{\partial t}.$$

The integrand $\mu_r\boldsymbol{J}_f/r$ in Eq. 10-19 is valid for steady fields; is it also valid for variable fields? It is, but we have not shown that the above time derivative is involved.

Note that $\partial\boldsymbol{D}/\partial t$ is the displacement current. This quantity appears in one of Maxwell's equations (See Sections 6.11 and 10.6).

10.2.1 The Retarded Potentials

The integrals of Eqs. 10-18 and 10-19 have one serious fault: they do not take into account the finite velocity of propagation of electric and magnetic fields. For example, if the charge density changes in one region, the integrals imply that V changes simultaneously throughout all space. This is of course inadmissible (Section 5.10). The potentials V and A at the time t cannot correspond exactly to the charge and current distributions at that particular moment, unless the charges are fixed in space. The analogy with astronomy is obvious: one cannot see a star as it is now, but only as it was thousands or millions of years ago.

If we have charge and current distributions ρ_f and $\mathbf{J}_f$ in a vacuum, then

$$V = \frac{1}{4\pi\epsilon_0}\int_{\tau'} \frac{[\rho_f]_c}{r}\, d\tau', \tag{10-24}$$

$$\mathbf{A} = \frac{\mu_0}{4\pi}\int_{\tau'} \frac{[\mathbf{J}_f]_c}{r}\, d\tau', \tag{10-25}$$

where the brackets indicate that the charge and current distributions are those at the previous time $t - (r/c)$. These are the *retarded potentials.*

From now on, square brackets will be used exclusively to identify retarded values of charge density, current density, position, velocity, etc.

The expressions for V and $\mathbf{A}$ in the presence of matter are simple in one particular case, namely that of an infinite linear, homogeneous, and isotropic medium ϵ, μ. Then the surface integrals are zero, and

$$V = \frac{1}{4\pi\epsilon}\int_{\tau'} \frac{[\rho_f]_v}{r}\, d\tau', \tag{10-26}$$

$$\mathbf{A} = \frac{\mu}{4\pi}\int_{\tau'} \frac{[\mathbf{J}_f]_v}{r}\, d\tau', \tag{10-27}$$

where ρ_f and $\mathbf{J}_f$ are now calculated at the retarded time $t - (r/v)$, v being the velocity of electromagnetic waves in the medium.

The integrals are evaluated over any volume τ' that encloses all the charges and currents.

In material media the velocity of electromagnetic waves depends on the frequency. This phenomenon is known as *dispersion.* The velocity v is therefore strictly definable only if ρ_f and $\mathbf{J}_f$ are pure sinusoidal functions of the time, although, in practice, dispersion can often be neglected.

Fortunately, it is not always necessary to take retardation into account. Retardation can be either important or neglibible, depending on the magnitude of the time delay, on the nature of the phenomena involved, and on the time resolution required. The following two examples illustrate the importance of retardation for extended pulsating sources.

Example

THE RETARDED POTENTIALS FOR AN OSCILLATING ELECTRIC DIPOLE

One can imagine an electric dipole whose moment is a sinusoidal function of the time, as in Figure 10-2, with

$$Q = Q_0 e^{j\omega t}. \tag{10-28}$$

Then

$$\boldsymbol{p} = Q_0 e^{j\omega t} \boldsymbol{s} = \boldsymbol{p}_0 e^{j\omega t}, \tag{10-29}$$

where

$$\boldsymbol{p}_0 = Q_0 \boldsymbol{s}. \tag{10-30}$$

For example, the charges could be situated on a pair of spheres joined by a thin wire of negligible resistance and capacitance.

The upward current I flowing through the connecting wire is

$$I = \frac{dQ}{dt} = j\omega Q_0 e^{j\omega t} = I_0 e^{j\omega t}, \tag{10-31}$$

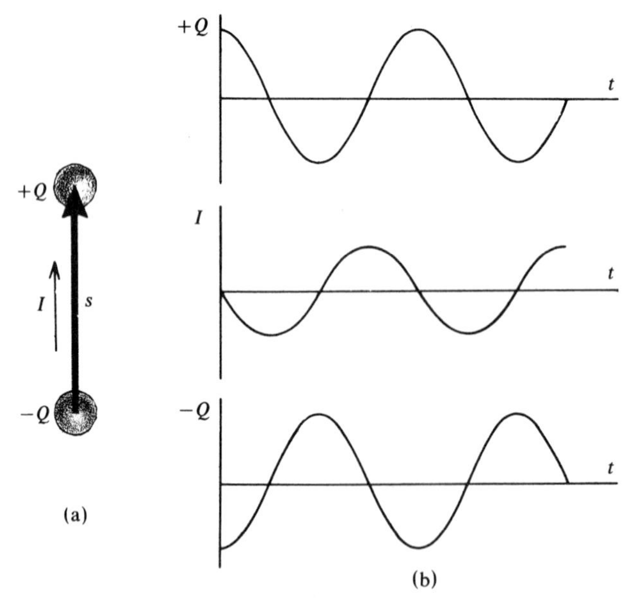

Figure 10-2. (a) An oscillating electric dipole. The total charge is zero and the vector $\boldsymbol{s}$ is oriented from $-Q$ to $+Q$ as shown. (b) The charges and the current are functions of the time.

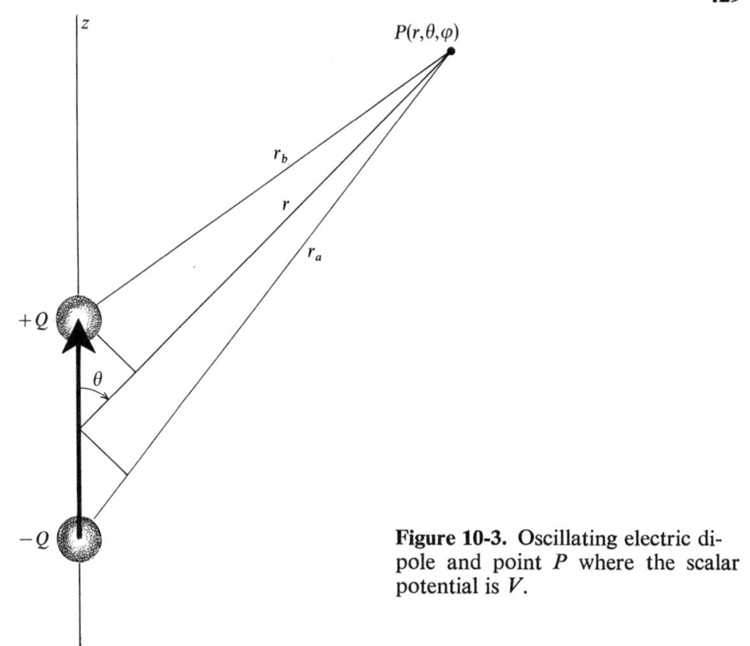

Figure 10-3. Oscillating electric dipole and point P where the scalar potential is V.

and

$$I_0 = j\omega Q_0, \tag{10-32}$$

$$I_0 \mathbf{s} = j\omega \mathbf{p}_0. \tag{10-33}$$

$$I\mathbf{s} = j\omega \mathbf{p}. \tag{10-34}$$

If the wave propagates in a vacuum, the retarded scalar potential is

$$V = \frac{Q_0 \exp j\omega\left(t - \frac{r_b}{c}\right)}{4\pi\epsilon_0 r_b} - \frac{Q_0 \exp j\omega\left(t - \frac{r_a}{c}\right)}{4\pi\epsilon_0 r_a}, \tag{10-35}$$

where the numerators are the charges, as they appear at the point $P(r, \theta, \varphi)$ of Figure 10-3 at the time t.

Note that the charges $-Q$ and $+Q$ give scalar potentials which differ not only in amplitude but also in phase. The amplitudes differ because of the terms r_a and r_b in the denominators, and the phases differ because of the r_a/c and r_b/c terms in the exponentials.

Similarly,

$$\mathbf{A} = \frac{\mu_0}{4\pi r} I_0 \exp j\omega\left(t - \frac{r}{c}\right)\mathbf{s}. \tag{10-36}$$

We can find a more convenient expression for V that is valid for $r \gg s$ as follows. Since

$$r_a \approx r + \frac{s}{2}\cos\theta, \tag{10-37}$$

$$r_b \approx r - \frac{s}{2}\cos\theta, \tag{10-38}$$

then

$$\omega\left(t - \frac{r_b}{c}\right) \approx \omega\left(t - \frac{r}{c} + \frac{s\cos\theta}{2c}\right), \tag{10-39}$$

$$\approx \omega\left(t - \frac{r}{c}\right) + \frac{s}{2ƛ}\cos\theta, \tag{10-40}$$

and

$$V \approx \frac{Q_0 \exp j\omega\left(t - \frac{r}{c}\right)}{4\pi\epsilon_0 r}\left\{\frac{\exp\left(j\frac{s}{2ƛ}\cos\theta\right)}{1 - \frac{s}{2r}\cos\theta} - \frac{\exp\left(-j\frac{s}{2ƛ}\cos\theta\right)}{1 + \frac{s}{2r}\cos\theta}\right\} \tag{10-41}$$

where $ƛ = \lambda/2\pi$ is called the *radian length*. Expanding the exponential functions and the denominators of the two terms within the braces as power series, and neglecting terms of the third order and higher in s/r and $s/ƛ$, we obtain

$$V = \frac{p_0 \exp j\omega\left(t - \frac{r}{c}\right)}{4\pi\epsilon_0 rƛ}\left(\frac{ƛ}{r} + j\right)\cos\theta, \tag{10-42}$$

where we have substituted p_0 for $Q_0 s$. We have assumed the length of the dipole s to be small compared to both r and $ƛ$: $s \ll r$ and $s \ll ƛ$. This is our only approximation. We have made *no* assumption as to the relative orders of magnitude of r and $ƛ$.

It is important to check this result with that for the static dipole (Eq. 2-74). For zero frequency, $\omega = 0$, $ƛ \to \infty$, and the two expressions agree as expected.

Example

THE RETARDED POTENTIAL A FOR AN OSCILLATING MAGNETIC DIPOLE

Figure 10-4 shows a magnetic dipole that is similar to that of Figure 7-20, but which is fed by a source of alternating current. It carries a current $I_0 \exp j\omega t$.

Instead of Eq. 7-100, we now have that

$$\mathbf{A} = \frac{\mu_0}{4\pi}\int_0^{2\pi} \frac{I_0 \exp j\omega\left(t - \frac{r'}{c}\right)}{r'} a\cos\varphi\, d\varphi\, \boldsymbol{\varphi}_1. \tag{10-43}$$

This is the sum of the $d\mathbf{A}$'s due to the elements $I\,d\mathbf{l}$ around the loop, all retarded by the appropriate time intervals r'/c. Rewriting,

$$\mathbf{A} = \frac{\mu_0 a I_0 \exp j\omega\left(t - \frac{r}{c}\right)}{4\pi}\int_0^{2\pi} \frac{\exp j\left(\frac{r - r'}{ƛ}\right)}{r'}\cos\varphi\, d\varphi\, \boldsymbol{\varphi}_1. \tag{10-44}$$

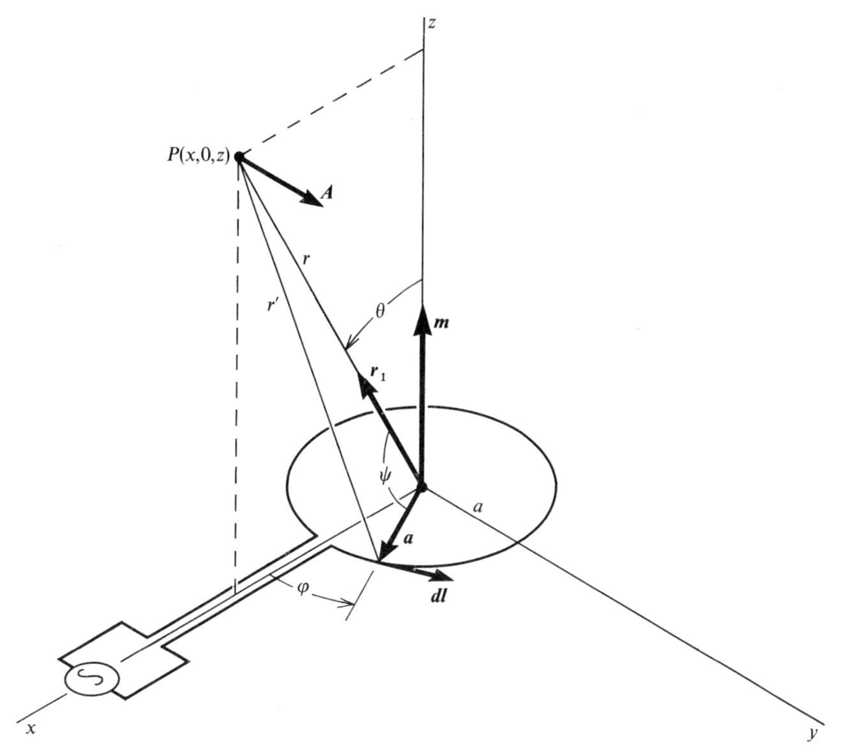

Figure 10-4. Oscillating magnetic dipole.

If the loop is small with respect to $\lambda\!\!\!^{-} = \lambda/2\pi$, then

$$|r - r'| \ll \lambda\!\!\!^{-}, \tag{10-45}$$

$$\exp j\left(\frac{r - r'}{\lambda\!\!\!^{-}}\right) \approx 1 + j\frac{r - r'}{\lambda\!\!\!^{-}} \tag{10-46}$$

and

$$\boldsymbol{A} = \frac{\mu_0 a I_0 \exp j\omega\left(t - \dfrac{r}{c}\right)}{4\pi r} \int_0^{2\pi} \left\{\frac{r}{r'} + j\frac{r}{\lambda\!\!\!^{-}}\left(\frac{r}{r'} - 1\right)\right\} \cos\varphi \, d\varphi \, \boldsymbol{\varphi}_1. \tag{10-47}$$

Also, from Eq. 7-106,

$$\frac{r}{r'} \approx 1 - \frac{a^2}{2r^2} + \frac{ax}{r^2}\cos\varphi, \tag{10-48}$$

for $a \ll r$. If we now integrate and then substitute $r \sin\theta$ for x, we find that

$$\boldsymbol{A} = \frac{\mu_0}{4\pi} \frac{I_0 \pi a^2 \exp j\omega\left(t - \dfrac{r}{c}\right)}{r\lambda\!\!\!^{-}} \left(\frac{\lambda\!\!\!^{-}}{r} + j\right) \sin\theta \, \boldsymbol{\varphi}_1, \tag{10-49}$$

or, setting $\pi a^2 I_0 = m_0$,

$$A = \frac{\mu_0}{4\pi} \frac{m_0 \exp j\omega\left(t - \frac{r}{c}\right)}{r\lambda} \left(\frac{\lambda}{r} + j\right) \sin\theta\, \varphi_1, \tag{10-50}$$

$$= \frac{\mu_0}{4\pi} \frac{\boldsymbol{m}_0 \times \boldsymbol{r}_1}{r\lambda} \left(\frac{\lambda}{r} + j\right) \exp j\omega\left(t - \frac{r}{c}\right), \tag{10-51}$$

where

$$\boldsymbol{m}_0 = I_0 \boldsymbol{S} \tag{10-52}$$

is the magnetic moment of the loop. The magnitude of the vector $\boldsymbol{S}$ is the area πa^2 of the loop, and its direction is related to the direction of current flow by the right-hand screw rule.

For zero frequency, $\omega = 0$, $\lambda \to \infty$, and

$$A = \frac{\mu_0}{4\pi} \frac{\boldsymbol{m} \times \boldsymbol{r}_1}{r^2}, \tag{10-53}$$

as in Section 7.8.

10.3 THE LORENTZ CONDITION

In view of the fact that ρ_f and $\boldsymbol{J}_f$, which appear in the integrals for V and $\boldsymbol{A}$ in Eqs. 10-26 and 10-27, are related together by Equation 10-4 for the conservation of free charge, we may suspect the existence of an equation linking together the potentials V and $\boldsymbol{A}$.

We have already found such a relation for the potentials of a single charge moving at a constant velocity, in Section 6.7.

We shall now show that a similar relation exists between the potentials of Eqs. 10-24 and 10-25. We shall use the *un*retarded potentials V and $\boldsymbol{A}$ in order to simplify the calculation. This means that we shall assume that the time delay r/v is negligible.

We calculate the divergence of $\boldsymbol{A}$:

$$\nabla \cdot \boldsymbol{A} = \frac{\mu}{4\pi} \nabla \cdot \int_{\tau'} \frac{\boldsymbol{J}_f}{r} d\tau'. \tag{10-54}$$

Since the divergence of $\boldsymbol{A}$ is calculated at the field point $P(x, y, z)$, while the integral is evaluated over the source points $P'(x', y', z')$ as in Figure 10-1, the order of the del and of the integral sign can be inverted and

$$\nabla \cdot \boldsymbol{A} = \frac{\mu}{4\pi} \int_{\tau'} \nabla \cdot \frac{\boldsymbol{J}_f}{r} d\tau'. \tag{10-55}$$

Now

$$\nabla \cdot \frac{\boldsymbol{J}_f}{r} = \frac{1}{r} \nabla \cdot \boldsymbol{J}_f + \boldsymbol{J}_f \cdot \nabla \frac{1}{r}. \tag{10-56}$$

The first term on the right is zero since $\boldsymbol{J}_f$ is a function of x', y', z', and not of the coordinates x, y, z of the field point that are involved in the operator $\boldsymbol{\nabla}$. Also, from Problem 1-12, $\boldsymbol{\nabla}(1/r)$ is equal to $-\boldsymbol{\nabla}'(1/r)$, and thus

$$\boldsymbol{\nabla}\cdot\frac{\boldsymbol{J}_f}{r} = -\boldsymbol{J}_f\cdot\boldsymbol{\nabla}'\frac{1}{r}. \tag{10-57}$$

Similarly,

$$\boldsymbol{\nabla}'\cdot\frac{\boldsymbol{J}_f}{r} = \frac{1}{r}\boldsymbol{\nabla}'\cdot\boldsymbol{J}_f + \boldsymbol{J}_f\cdot\boldsymbol{\nabla}'\frac{1}{r}. \tag{10-58}$$

Then, combining these last two equations, we find that

$$\boldsymbol{\nabla}\cdot\frac{\boldsymbol{J}_f}{r} = \frac{1}{r}\boldsymbol{\nabla}'\cdot\boldsymbol{J}_f - \boldsymbol{\nabla}'\cdot\frac{\boldsymbol{J}_f}{r}, \tag{10-59}$$

and

$$\boldsymbol{\nabla}\cdot\boldsymbol{A} = \frac{\mu}{4\pi}\int_{\tau'}\frac{\boldsymbol{\nabla}'\cdot\boldsymbol{J}_f}{r}\,d\tau' - \frac{\mu}{4\pi}\int_{\tau'}\boldsymbol{\nabla}'\cdot\frac{\boldsymbol{J}_f}{r}\,d\tau'. \tag{10-60}$$

The second integral is similar to the one we started from (Eq. 10-55), except that now the divergence of $\boldsymbol{J}_f/r$ is computed at P' instead of at P. The second integral is zero for the following reason. According to the divergence theorem, it is equal to the integral of $(\boldsymbol{J}_f/r)\cdot d\boldsymbol{a}$ over the surface S' bounding the volume τ'. Now because, by definition, all currents are contained within the surface S', everywhere on S' the vector $\boldsymbol{J}_f$ is either zero or tangential, and $(\boldsymbol{J}_f/r)\cdot d\boldsymbol{a}$ is zero. Then

$$\boldsymbol{\nabla}\cdot\boldsymbol{A} = \frac{\mu}{4\pi}\int_{\tau'}\frac{\boldsymbol{\nabla}'\cdot\boldsymbol{J}_f}{r}\,d\tau'. \tag{10-61}$$

Now we have had to use a prime on the del operator that appears under the integral sign because it was essential for us to discriminate between derivatives with respect to x, y, z and derivatives with respect to x', y', z'. But the $\boldsymbol{\nabla}'\cdot\boldsymbol{J}_f$ term is really the same as the $\boldsymbol{\nabla}\cdot\boldsymbol{J}_f$ which appears in Eq. 10-4. In this latter equation we were only concerned with the source point, and a prime on the del would have been superfluous. Then, from Eq. 10-4, $\boldsymbol{\nabla}'\cdot\boldsymbol{J}_f$ is $-\partial\rho_f/\partial t$ and

$$\boldsymbol{\nabla}\cdot\boldsymbol{A} = -\frac{\mu}{4\pi}\int_{\tau'}\frac{(\partial\rho_f/\partial t)}{r}\,d\tau'. \tag{10-62}$$

Now r is not a function of the time, since it is the distance between the two fixed points P and P'. We can therefore remove the $\partial/\partial t$ operator from under the integral sign and

$$\boldsymbol{\nabla}\cdot\boldsymbol{A} = -\epsilon\mu\frac{\partial}{\partial t}\frac{1}{4\pi\epsilon}\int_{\tau'}\frac{\rho_f}{r}\,d\tau', \tag{10-63}$$

$$= -\epsilon\mu\frac{\partial V}{\partial t}, \tag{10-64}$$

from Eq. 10-26.

The relation

$$\nabla \cdot A + \epsilon\mu \frac{\partial V}{\partial t} = 0 \tag{10-65}$$

is called the *Lorentz condition*. It is an *identity* if V and A are defined as in Eqs. 10-26 and 10-27, despite the fact that the above proof is valid only for slowly varying fields.* It is based on the well-founded assumption that there is conservation of charge.

For charges and currents in a vacuum,

$$\nabla \cdot A + \epsilon_0\mu_0 \frac{\partial V}{\partial t} = 0 \tag{10-66}$$

and, for steady fields,

$$\nabla \cdot A = 0. \tag{10-67}$$

We have already noted in Section 10.2 that the six components of $\boldsymbol{E}$ and $\boldsymbol{B}$ can be computed from V and the three components of A; with the Lorentz condition it suffices to know the three components of A to find V, *if* V is a function of the time, and therefore to calculate the six components of the field vectors $\boldsymbol{E}$ and $\boldsymbol{B}$.

Example

THE LEAKY SPHERICAL CAPACITOR

A spherical capacitor as in Figure 10-5 contains a slightly conducting dielectric. We wish to calculate $\boldsymbol{B}$ and A when the capacitor discharges through its own dielectric.

a) By symmetry, A_θ and A_φ are both zero. Then

$$\boldsymbol{B} = \nabla \times A = 0. \tag{10-68}$$

There are other ways of showing that $\boldsymbol{B} = 0$. First, we recall from Section 7.4 that

$$\int_S \boldsymbol{B} \cdot d\boldsymbol{a} = 0 \tag{10-69}$$

over any closed surface S. If we choose S to be a spherical surface concentric with the capacitor, then $\boldsymbol{B} \cdot d\boldsymbol{a}$ is $B_r da$. By symmetry, B_r must be independent of θ and φ, so that $4\pi r^2 B_r$ is zero. Therefore B_r is also zero. But, again by symmetry, B_θ and B_φ must also be zero, and thus $\boldsymbol{B} = 0$.

Also, from Section 8.1.1,

$$\nabla \times \boldsymbol{E} = -\frac{\partial \boldsymbol{B}}{\partial t}. \tag{10-70}$$

In the present case $\boldsymbol{E}$ is radial and independent of θ and φ, so that its curl is zero, from Eq. 1-137. Then $\boldsymbol{B}$ must be constant. But $\boldsymbol{B}$ is zero for $t \to \infty$. Then $\boldsymbol{B}$ is always zero. See also Problem 10-17.

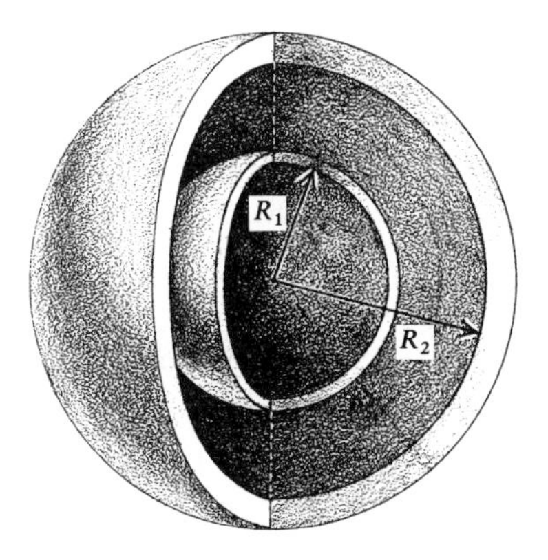

Figure 10-5. Spherical capacitor.

b) To calculate A_r we use the Lorentz condition. We first obtain an approximate value for V from the following equation:

$$-\frac{\partial V}{\partial r} \approx E = \frac{J_f}{\sigma} = \frac{I}{4\pi r^2 \sigma} \tag{10-71}$$

where J_f is the conduction current density and σ is the conductivity. We have neglected the term $-\partial A/\partial t$ of Equation 10-22; as we shall see later, this is well justified. Then

$$I \approx I_0 e^{-t/RC}, \tag{10-72}$$

where R and C are, respectively, the resistance and capacitance of the spherical capacitor.

Let us assume that the outer conductor is grounded and that the inner conductor is at the potential V_1. Then

$$V_1 \approx \int_{R_1}^{R_2} \frac{I_0 e^{-t/RC}}{4\pi r^2 \sigma}\, dr = \frac{I_0 e^{-t/RC}}{4\pi\sigma}\left(\frac{1}{R_1} - \frac{1}{R_2}\right) \tag{10-73}$$

and, at a radius r,

$$V \approx \frac{I_0 e^{-t/RC}}{4\pi\sigma}\left(\frac{1}{r} - \frac{1}{R_2}\right). \tag{10-74}$$

According to the Lorentz condition,

$$\nabla \cdot A = -\epsilon\mu \frac{\partial V}{\partial t} \approx \epsilon\mu \frac{I_0 e^{-t/RC}}{4\pi\sigma RC}\left(\frac{1}{r} - \frac{1}{R_2}\right) \tag{10-75}$$

or, setting the coefficient before the parenthesis on the right-hand side equal to K,

$$\frac{1}{r^2}\frac{\partial}{\partial r}(r^2 A_r) \approx K\left(\frac{1}{r} - \frac{1}{R_2}\right), \tag{10-76}$$

$$\frac{\partial}{\partial r}(r^2 A_r) \approx K\left(r - \frac{r^2}{R_2}\right), \tag{10-77}$$

$$r^2 A_r \approx K\left(\frac{r^2}{2} - \frac{r^3}{3R_2}\right). \tag{10-78}$$

We have disregarded the constant of integration because it is independent of all four variables r, θ, φ, t for the following reasons. It is of course independent of r, and it is independent of θ and φ by symmetry. It is also independent of t because A decreases exponentially with time, like V and I, and the exponential function is already contained in the coefficient K. Thus

$$A_r \approx \epsilon\mu \frac{I_0 e^{-t/RC}}{4\pi\sigma RC}\left(\frac{1}{2} - \frac{r}{3R_2}\right). \tag{10-79}$$

Finally we must justify our assumption that

$$\frac{\partial A_r}{\partial t} \approx -\frac{\epsilon\mu}{R^2C^2}\frac{I}{4\pi\sigma}\left(\frac{1}{2} - \frac{r}{3R_2}\right) \tag{10-80}$$

is negligible compared to $\partial V/\partial r$ in Equation 10-71. First, we note that the above parenthesis is of the order of unity. Now, since $(\epsilon\mu)^{-1/2} \approx$ velocity of light in the dielectric (Section 11.4), $RC/(\epsilon\mu)^{1/2}$ is the distance traveled by light in one time constant RC. This, squared, is vastly larger than the r^2 of Equation 10-71, under our assumption that the dielectric is only slightly conducting.

10.4 THE DIVERGENCE OF $\boldsymbol{E}$ AND THE NONHOMOGENEOUS WAVE EQUATION FOR V

It was in Section 3.6 that we found the first of our four Maxwell's equations, namely

$$\boxed{\boldsymbol{\nabla}\cdot\boldsymbol{E} = \frac{\rho_t}{\epsilon_0},} \tag{10-81}$$

where ρ_t is the total charge density $\rho_f + \rho_b$. Although this result was arrived at in discussing electrostatic fields, it is in fact general, as one can see from the discussion in Section 6.8.

For a linear, homogeneous, and isotropic dielectric, we have Eq. 3-54,

$$\boldsymbol{\nabla}\cdot\boldsymbol{E} = \frac{\rho_f}{\epsilon}, \tag{10-82}$$

and

$$\boldsymbol{\nabla}\cdot\left(-\boldsymbol{\nabla}V - \frac{\partial \boldsymbol{A}}{\partial t}\right) = \frac{\rho_f}{\epsilon}. \tag{10-83}$$

Rewriting,

$$\boldsymbol{\nabla}^2 V + \frac{\partial}{\partial t}\boldsymbol{\nabla}\cdot\boldsymbol{A} = -\frac{\rho_f}{\epsilon}, \tag{10-84}$$

and, using the Lorentz condition,

$$\nabla^2 V - \epsilon\mu \frac{\partial^2 V}{\partial t^2} = -\frac{\rho_f}{\epsilon}. \tag{10-85}$$

This is the nonhomogeneous wave equation for V. At points where $\rho_f = 0$ we have the usual wave equation (Appendix E); the phase velocity is $1/(\epsilon\mu)^{1/2}$.

If ρ_f is not a function of the time, neither is V and, in a homogeneous linear and isotropic dielectric,

$$\nabla^2 V = -\frac{\rho_f}{\epsilon} = -\frac{\rho_t}{\epsilon_0}, \tag{10-86}$$

which is Eq. 3-55.

The solution of the nonhomogeneous wave equation 10-85 for an infinite medium ϵ, μ (*not* functions of x, y, z, t) is the retarded scalar potential of Eq. 10-24 (Appendix E).

10.5 THE NONHOMOGENEOUS WAVE EQUATION FOR $\boldsymbol{A}$

We can find the wave equation for $\boldsymbol{A}$ by analogy with that for V.

The vector equation 10-25 for $\boldsymbol{A}$ is equivalent to three scalar equations for the three Cartesian components of $\boldsymbol{A}$, each one of these equations being similar to the single scalar equation for V. Then the three components of $\boldsymbol{A}$ must satisfy three scalar differential equations similar to Eq. 10-85, and, combining them together,

$$\nabla^2 \boldsymbol{A} - \epsilon\mu \frac{\partial^2 \boldsymbol{A}}{\partial t^2} = -\mu \boldsymbol{J}_f. \tag{10-87}$$

If $\boldsymbol{J}_f$ is constant, $\boldsymbol{A}$ is also constant and

$$\nabla^2 \boldsymbol{A} = -\mu \boldsymbol{J}_f. \tag{10-88}$$

We have therefore found three functions of $\boldsymbol{A}$: $\nabla \times \boldsymbol{A}$, which is the magnetic induction $\boldsymbol{B}$; $\nabla \cdot \boldsymbol{A}$, which is related to $\partial V/\partial t$ by the Lorentz condition; and $\nabla^2 \boldsymbol{A}$ as above.

10.6 THE CURL OF $\boldsymbol{B}$

We have already found the curl of $\boldsymbol{B}$ in Section 9.4, but that equation is strictly valid only for steady currents. We have also found the curl of $\boldsymbol{B}$ in

Section 6.11, but we were not then familiar with the equivalent volume current density $\nabla \times \boldsymbol{M}$.

It will not be difficult to find the correct value of $\nabla \times \boldsymbol{B}$, now that we have the Lorentz condition and the wave equation for $\boldsymbol{A}$.

In terms of the vector potential $\boldsymbol{A}$,

$$\nabla \times \boldsymbol{B} = \nabla \times \nabla \times \boldsymbol{A} = -\nabla^2 \boldsymbol{A} + \nabla(\nabla \cdot \boldsymbol{A}). \tag{10-89}$$

Substituting now the value of $\nabla^2 \boldsymbol{A}$ from the wave equation 10-87 and the value of $\nabla \cdot \boldsymbol{A}$ from the Lorentz condition 10-65,

$$\nabla \times \boldsymbol{B} = \mu \boldsymbol{J}_f - \epsilon\mu \frac{\partial^2 \boldsymbol{A}}{\partial t^2} + \nabla\left(-\epsilon\mu \frac{\partial V}{\partial t}\right) \tag{10-90}$$

$$= \mu \boldsymbol{J}_f - \epsilon\mu \frac{\partial^2 \boldsymbol{A}}{\partial t^2} - \epsilon\mu \frac{\partial}{\partial t} \nabla V \tag{10-91}$$

$$= \mu \boldsymbol{J}_f - \epsilon\mu \frac{\partial^2 \boldsymbol{A}}{\partial t^2} - \epsilon\mu \frac{\partial}{\partial t}\left\{-\boldsymbol{E} - \frac{\partial \boldsymbol{A}}{\partial t}\right\} \tag{10-92}$$

$$= \mu \boldsymbol{J}_f + \epsilon\mu \frac{\partial \boldsymbol{E}}{\partial t}. \tag{10-93}$$

If we remove the ϵ and the μ which restrict the equation to linear, homogeneous, and isotropic media, we obtain the following equation:

$$\boxed{\nabla \times \boldsymbol{H} = \boldsymbol{J}_f + \frac{\partial \boldsymbol{D}}{\partial t}.} \tag{10-94}$$

The term $\partial \boldsymbol{D}/\partial t$ is called the displacement current density (Section 10.1).

This is another of Maxwell's equations. Let us rewrite it in terms of $\boldsymbol{E}$ and $\boldsymbol{B}$, and without the ϵ and the μ:

$$\nabla \times \boldsymbol{B} \equiv \mu_0 \nabla \times (\boldsymbol{H} + \boldsymbol{M}) = \mu_0(\nabla \times \boldsymbol{H} + \nabla \times \boldsymbol{M}), \tag{10-95}$$

$$= \mu_0\left(\boldsymbol{J}_f + \frac{\partial \boldsymbol{D}}{\partial t} + \nabla \times \boldsymbol{M}\right), \tag{10-96}$$

$$= \mu_0\left(\boldsymbol{J}_f + \epsilon_0 \frac{\partial \boldsymbol{E}}{\partial t} + \frac{\partial \boldsymbol{P}}{\partial t} + \nabla \times \boldsymbol{M}\right), \tag{10-97}$$

or

$$\boxed{\nabla \times \boldsymbol{B} = \mu_0 \boldsymbol{J}_t,} \tag{10-98}$$

where the total current density $\boldsymbol{J}_t$ is defined in Eq. 10-9.

One convenient form for this equation is the following:

$$\boxed{\nabla \times \boldsymbol{B} - \frac{1}{c^2}\frac{\partial \boldsymbol{E}}{\partial t} = \mu_0 \boldsymbol{J}_m,} \tag{10-99}$$

where $c = 1/(\epsilon_0\mu_0)^{1/2}$ is the velocity of light in a vacuum, as we shall see in the next chapter, and where $\boldsymbol{J}_m$ is the current density in matter, as defined in Eq. 10-6.

This result is in agreement with the value of $\boldsymbol{\nabla} \times \boldsymbol{B}$ which we found in Section 9.4 when we were dealing with steady currents. We can see now that

$$\boldsymbol{\nabla} \times \boldsymbol{B} \approx \mu_0(\boldsymbol{J}_f + \boldsymbol{\nabla} \times \boldsymbol{M}) \tag{10-100}$$

if the displacement current $(\partial/\partial t)\ (\epsilon_0 \boldsymbol{E} + \boldsymbol{P})$ is negligible compared to $\boldsymbol{J}_f + \boldsymbol{\nabla} \times \boldsymbol{M}$. When this condition is satisfied the current is said to be *quasi-stationary*.

Example

THE DISPLACEMENT CURRENT DENSITY IN A CONDUCTOR

It is instructive to estimate the ratio of displacement current density $\partial \boldsymbol{D}/\partial t$ to conduction current density $\boldsymbol{J}_f$ in a conductor. For an alternating electric field $\boldsymbol{E}_0 \cos \omega t$ within a material of conductivity σ,

$$\frac{\partial \boldsymbol{D}}{\partial t} = -\omega\epsilon \boldsymbol{E}_0 \sin \omega t, \tag{10-101}$$

$$\boldsymbol{J}_f = \sigma \boldsymbol{E}_0 \cos \omega t, \tag{10-102}$$

$$\left| \frac{\dfrac{\partial D}{\partial t}}{J_f} \right| = \frac{\omega\epsilon}{\sigma}. \tag{10-103}$$

The relative permittivity ϵ_r of a conductor is not readily measured, since polarization effects are usually overshadowed by conduction. However, for purposes of estimation, we may set $\epsilon_r \approx 1$ and $\sigma \approx 10^7$ mhos/meter for a good conductor. Then

$$\left| \frac{\dfrac{\partial D}{\partial t}}{J_f} \right| \approx 10^{-17} f, \tag{10-104}$$

where $f = \omega/2\pi$ is the frequency. The displacement current in a good conductor is therefore negligible compared to the conduction current at any frequency lower than optical frequencies, where $f \approx 10^{15}$ hertz.

It is interesting to note that the conduction current is in phase with the electric field intensity, while the displacement current leads the electric field by $\pi/2$ radians.

10.7 MAXWELL'S EQUATIONS

Let us group together the four Maxwell equations, which we found successively as Eqs. 10-81, 7-34, 8-8, and 10-99:

$$\nabla \cdot \boldsymbol{E} = \frac{\rho_t}{\epsilon_0}, \tag{10-105}$$

$$\nabla \cdot \boldsymbol{B} = 0, \tag{10-106}$$

$$\nabla \times \boldsymbol{E} + \frac{\partial \boldsymbol{B}}{\partial t} = 0, \tag{10-107}$$

$$\nabla \times \boldsymbol{B} - \frac{1}{c^2}\frac{\partial \boldsymbol{E}}{\partial t} = \mu_0 \boldsymbol{J}_m. \tag{10-108}$$

These equations are also given in Table 6.3.

As usual,

$\boldsymbol{E}$ is the electric field intensity in volts/meter;
$\rho_t = \rho_f + \rho_b$ is the total electric charge density in coulombs/meter^3;
ρ_f is the free charge density;
ρ_b is the bound charge density $-\nabla \cdot \boldsymbol{P}$, $\boldsymbol{P}$ being the electric polarization in coulombs/meter^2;
$\boldsymbol{B}$ is the magnetic induction in teslas;
$\boldsymbol{J}_m = \boldsymbol{J}_f + \partial \boldsymbol{P}/\partial t + \nabla \times \boldsymbol{M}$ is the current density due to the flow of charges in matter, in amperes/meter^2;
$\boldsymbol{J}_f$ is the current density of free charges;
$\partial \boldsymbol{P}/\partial t$ is the polarization current density;
$\nabla \times \boldsymbol{M}$ is the equivalent current density in magnetized matter;
$\boldsymbol{M}$ is the magnetization in amperes/meter;
c is the velocity of light, 3×10^8 meters/second, and $c^2 = 1/(\epsilon_0\mu_0)$;
ϵ_0 is the permittivity of free space, 8.85×10^{-12} farad/meter;
μ_0 is the permeability of free space, $4\pi \times 10^{-7}$ henry/meter.

In conductors obeying Ohm's law,

$$\boldsymbol{J}_f = \sigma \boldsymbol{E}, \tag{10-109}$$

where σ is the conductivity in mho/meter. Inside a source such as a battery, there is also another electric field intensity $\boldsymbol{E}_s$, due to the local generation of energy, and

$$\boldsymbol{J}_f = \sigma(\boldsymbol{E} + \boldsymbol{E}_s). \tag{10-110}$$

Maxwell's equations are partial differential equations involving space and time derivatives of the field vectors $\boldsymbol{E}$ and $\boldsymbol{B}$, the total charge density ρ_t, and the current density $\boldsymbol{J}_m$. They do *not* yield the values of $\boldsymbol{E}$ and $\boldsymbol{B}$ directly, but only after integration and after taking into account the proper boundary conditions.

These are the four fundamental equations of electromagnetism. They apply to all electromagnetic phenomena in media which are at rest with respect to the coordinate system used. They are valid for *non*homogeneous, *non*linear, and even *non*isotropic media. The remaining chapters will be based on them.

Maxwell's equations are not all independent of each other. If we take the divergence of Eq. 10-107, recalling that the divergence of the curl of any vector is zero, then

$$\nabla \cdot \frac{\partial \boldsymbol{B}}{\partial t} = 0, \tag{10-111}$$

or, inverting the order of the operations,

$$\frac{\partial}{\partial t}(\nabla \cdot \boldsymbol{B}) = 0. \tag{10-112}$$

The quantity $\nabla \cdot \boldsymbol{B}$ is therefore independent of the time at any point in space. We can set the divergence of $\boldsymbol{B}$ equal to zero everywhere if we assume that, for each point of space, it is equal to zero at some time, either in the past or in the future. Under this assumption, Eq. 10-106 can be deduced from Eq. 10-107. Equations 10-106 and 10-107 are sometimes called the *first pair of Maxwell's equations.*

Similarly, taking the divergence of Eq. 10-108,

$$\nabla \cdot \frac{\partial \boldsymbol{E}}{\partial t} = -\frac{1}{\epsilon_0} \nabla \cdot \boldsymbol{J}_m. \tag{10-113}$$

If we assume that there is conservation of charge, as we did in Section 10-1,

$$\nabla \cdot \frac{\partial \boldsymbol{E}}{\partial t} = \frac{1}{\epsilon_0} \frac{\partial \rho_t}{\partial t}, \tag{10-114}$$

$$\frac{\partial}{\partial t}(\nabla \cdot \boldsymbol{E}) = \frac{\partial}{\partial t} \frac{\rho_t}{\epsilon_0}, \tag{10-115}$$

$$\nabla \cdot \boldsymbol{E} = \frac{\rho_t}{\epsilon_0} + C, \tag{10-116}$$

where C is some quantity that can be a function of the coordinates, but that is independent of the time. If we further assume that, at every point of space at some time either in the past or in the future, both $\nabla \cdot \boldsymbol{E}$ and ρ_t are simultaneously equal to zero, then the constant of integration C must be zero, and we are left with Eq. 10-105. Under these two assumptions, Eqs. 10-105 and 10-108 are therefore not independent. They form the *second pair of Maxwell's equations.*

If we write out ρ_t and $\boldsymbol{J}_m$ in full, Maxwell's equations become

$$\nabla\cdot\boldsymbol{E} = \frac{1}{\epsilon_0}(\rho_f - \nabla\cdot\boldsymbol{P}), \tag{10-117}$$

$$\nabla\cdot\boldsymbol{B} = 0, \tag{10-118}$$

$$\nabla\times\boldsymbol{E} + \frac{\partial\boldsymbol{B}}{\partial t} = 0, \tag{10-119}$$

$$\nabla\times\boldsymbol{B} - \frac{1}{c^2}\frac{\partial\boldsymbol{E}}{\partial t} = \mu_0\left(\boldsymbol{J}_f + \frac{\partial\boldsymbol{P}}{\partial t} + \nabla\times\boldsymbol{M}\right), \tag{10-120}$$

and, if we use all four field vectors $\boldsymbol{E}$, $\boldsymbol{D}$, $\boldsymbol{B}$, $\boldsymbol{H}$, remembering that $\boldsymbol{D}$ is $\epsilon_0\boldsymbol{E} + \boldsymbol{P}$, while $\boldsymbol{B}$ is $\mu_0(\boldsymbol{H} + \boldsymbol{M})$,

$$\nabla\cdot\boldsymbol{D} = \rho_f, \tag{10-121}$$

$$\nabla\cdot\boldsymbol{B} = 0, \tag{10-122}$$

$$\nabla\times\boldsymbol{E} + \frac{\partial\boldsymbol{B}}{\partial t} = 0, \tag{10-123}$$

$$\nabla\times\boldsymbol{H} - \frac{\partial\boldsymbol{D}}{\partial t} = \boldsymbol{J}_f. \tag{10-124}$$

All three sets of equations (10-105 to 10-108, 10-117 to 10-120, and 10-121 to 10-124) are valid in nonhomogeneous, nonlinear, and nonisotropic media.

In the remaining chapters we shall be concerned mostly with fields where $\boldsymbol{E}$ and $\boldsymbol{H}$ are sinusoidal functions of the time. In such cases we can replace the time derivatives by factors of $j\omega$ (Appendix D). Setting also $\boldsymbol{D} = \epsilon\boldsymbol{E}$ and $\boldsymbol{B} = \mu\boldsymbol{H}$,

$$\nabla\cdot\epsilon\boldsymbol{E} = \rho_f, \tag{10-125}$$

$$\nabla\cdot\mu\boldsymbol{H} = 0, \tag{10-126}$$

$$\nabla\times\boldsymbol{E} + j\omega\mu\boldsymbol{H} = 0, \tag{10-127}$$

$$\nabla\times\boldsymbol{H} - j\omega\epsilon\boldsymbol{E} = \boldsymbol{J}_f. \tag{10-128}$$

10.7.1 Maxwell's Equations in Integral Form

We have stated Maxwell's equations in differential form; let us now state them in integral form, as we did in Chapter 6, in order that we may arrive at a better understanding of their physical meaning. We shall use Eqs. 10-105 to 10-108.

In Section 3.6 we deduced Eq. 10-105 from

$$\int_S \boldsymbol{E}\cdot d\boldsymbol{a} = \frac{1}{\epsilon_0}\int_\tau \rho_t\, d\tau = \frac{Q_t}{\epsilon_0}, \tag{10-129}$$

where S is the surface bounding the volume τ and where Q_t is the net charge contained within τ. The outward flux of $\boldsymbol{E}$ through any closed surface S is

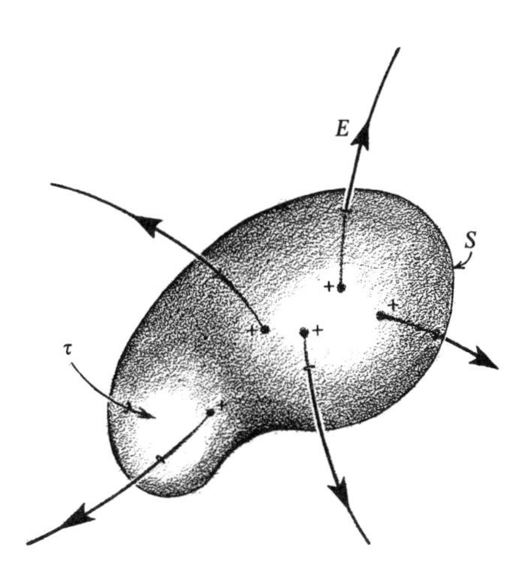

Figure 10-6. Lines of $\boldsymbol{E}$ emerging from a volume τ containing a net charge Q_t. The outward flux of $\boldsymbol{E}$ is equal to Q_t/ϵ_0.

therefore equal to $1/\epsilon_0$ times the net charge inside, as illustrated in Figure 10-6. This is Gauss's law (Sections 3.6 and 6.8).

Integrating Eq. 10-106 in a similar manner, we find that the outward flux of $\boldsymbol{B}$ through any closed surface S is equal to zero (Sections 6.9 and 9.3):

$$\int_S \boldsymbol{B} \cdot d\boldsymbol{a} = 0. \tag{10-130}$$

This is shown in Figure 10-7.

Equation 10-107 can be integrated over a surface S bounded by a curve C:

$$\int_S \boldsymbol{\nabla} \times \boldsymbol{E} \cdot d\boldsymbol{a} = -\int_S \frac{\partial \boldsymbol{B}}{\partial t} \cdot d\boldsymbol{a}, \tag{10-131}$$

or, if we use Stokes's theorem on the left and invert the operations on the right, the surface S being fixed in space,

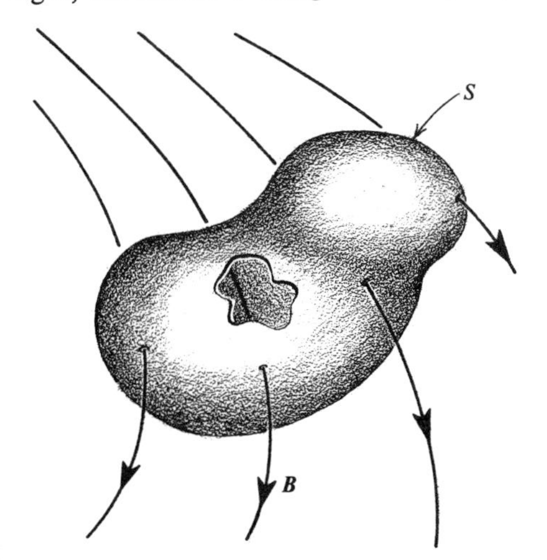

Figure 10-7. Lines of $\boldsymbol{B}$ through a closed surface S. The net outward flux of $\boldsymbol{B}$ is equal to zero.

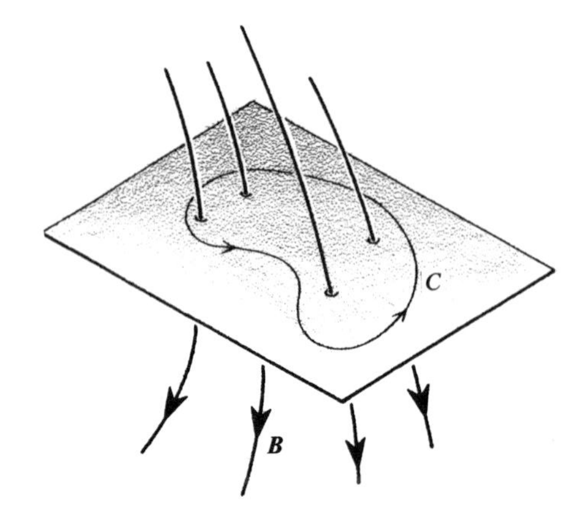

Figure 10-8. The direction of the electromotance induced around C is indicated by an arrow for the case where the magnetic induction $\boldsymbol{B}$ is in the direction shown and *in*creases. The electromotance is in the same direction if $\boldsymbol{B}$ is upward and decreases.

$$\oint_C \boldsymbol{E}\cdot d\boldsymbol{l} = -\frac{\partial}{\partial t}\int_S \boldsymbol{B}\cdot d\boldsymbol{a} = -\frac{d\Phi}{dt}. \qquad (10\text{-}132)$$

Then the electromotance induced around curve C is equal to minus the rate of change of the magnetic flux Φ linking C, as in Figure 10-8. The positive directions for $\boldsymbol{B}$ and around C are related according to the right-hand screw rule (Section 8.1).

Finally, if we also integrate Eq. 10-108 over an area S bounded by a curve C,

$$\oint_C \boldsymbol{B}\cdot d\boldsymbol{l} = \mu_0\int_S\left(\boldsymbol{J}_m + \epsilon_0\frac{\partial \boldsymbol{E}}{\partial t}\right)\cdot d\boldsymbol{a} = \mu_0 I_t. \qquad (10\text{-}133)$$

If we perform the same operation on Eq. 10-124,

$$\oint_C \boldsymbol{H}\cdot d\boldsymbol{l} = \int_S\left(\boldsymbol{J}_f + \frac{\partial \boldsymbol{D}}{\partial t}\right)\cdot d\boldsymbol{a}, \qquad (10\text{-}134)$$

and the magnetomotance around C is equal to the sum of the free current plus the displacement current linking C. This is illustrated in Figure 10-9. The positive directions are again related by the right-hand screw rule.

10.8 DUALITY

If ρ_f and $\boldsymbol{J}_f$ are both zero, Maxwell's equations 10-121 to 10-124 reduce to

$$\boldsymbol{\nabla}\cdot\boldsymbol{D} = 0, \qquad (10\text{-}135)$$

$$\boldsymbol{\nabla}\cdot\boldsymbol{B} = 0, \qquad (10\text{-}136)$$

$$\boldsymbol{\nabla}\times\boldsymbol{E} + \frac{\partial \boldsymbol{B}}{\partial t} = 0, \qquad (10\text{-}137)$$

$$\boldsymbol{\nabla}\times\boldsymbol{H} - \frac{\partial \boldsymbol{D}}{\partial t} = 0. \qquad (10\text{-}138)$$

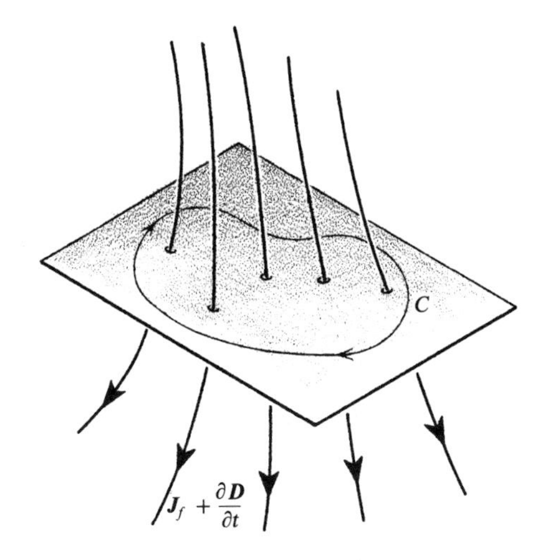

Figure 10-9. The direction of the magnetomotance around C is indicated by an arrow for the case where the current $\boldsymbol{J}_f + (\partial \boldsymbol{D}/\partial t)$ is in the direction shown. The displacement current is downward if $\boldsymbol{D}$ is downward and increases, or if it is upward and decreases.

Let us consider one field where the electric field intensity is $\boldsymbol{E}$ and the magnetic induction is $\boldsymbol{B}$. Then the above equations are necessarily satisfied. Let us now consider a *different* field $\boldsymbol{E}'$, $\boldsymbol{B}'$ such that

$$\boldsymbol{E}' = -K\boldsymbol{B} = -K\mu\boldsymbol{H}, \qquad (10\text{-}139)$$

$$\boldsymbol{H}' = +K\boldsymbol{D} = +K\epsilon\boldsymbol{E}. \qquad (10\text{-}140)$$

Substituting $\boldsymbol{E}'$ and $\boldsymbol{B}'$ into Eqs. 10-121 to 10-124, and assuming that ϵ and μ are independent of x, y, z, t, we find that Maxwell's equations also apply to the primed field:

$$\nabla \cdot \boldsymbol{D}' = 0, \qquad (10\text{-}141)$$

$$\nabla \cdot \boldsymbol{B}' = 0, \qquad (10\text{-}142)$$

$$\nabla \times \boldsymbol{E}' + \frac{\partial \boldsymbol{B}'}{\partial t} = 0, \qquad (10\text{-}143)$$

$$\nabla \times \boldsymbol{H}' - \frac{\partial \boldsymbol{D}'}{\partial t} = 0. \qquad (10\text{-}144)$$

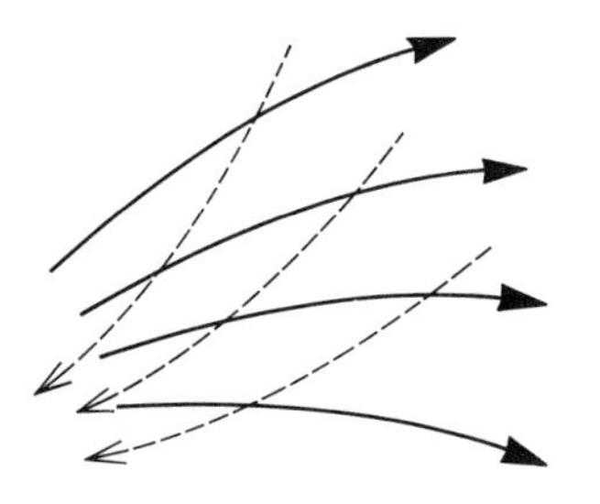

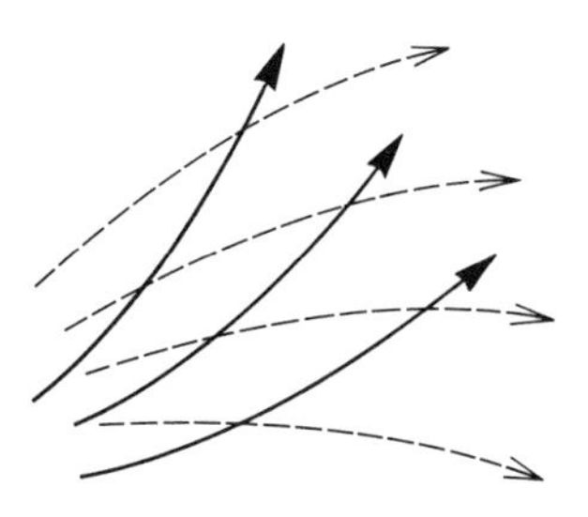

Figure 10-10. Pair of symmetrical fields. Lines of $\boldsymbol{E}$ are solid and lines of $\boldsymbol{B}$ are hatched.

This duality property of electromagnetic fields is illustrated in Figure 10-10.

Example If we have a purely electric field $\{\boldsymbol{E}(x, y, z), \boldsymbol{B} = 0\}$, then it is possible to produce a purely magnetic field $\{\boldsymbol{B}'(x, y, z), \boldsymbol{E}' = 0\}$, where $\boldsymbol{B}'$ is the same function of x, y, z as $\boldsymbol{E}$, within a constant factor. The inverse is also true. Experimentally, of course, one of the two fields can be more difficult to obtain than the other.

Example The simplest electric field is

$$\boldsymbol{E} = \text{Constant} \times \boldsymbol{i}. \tag{10-145}$$

It is the field inside a parallel-plate capacitor when edge effects are negligible.

The corresponding magnetic field

$$\boldsymbol{B}' = \text{Constant} \times \boldsymbol{i} \tag{10-146}$$

is found inside a long solenoid.

Example We have found in Section 2.9 that the $\boldsymbol{E}$ field of an electric dipole is given by

$$E_r = \frac{1}{4\pi\epsilon_0}\frac{2p}{r^3}\cos\theta, \tag{10-147}$$

$$E_\theta = \frac{1}{4\pi\epsilon_0}\frac{p}{r^3}\sin\theta, \tag{10-148}$$

$$E_\varphi = 0, \tag{10-149}$$

as long as r is much larger than the length of the dipole. Then, in Section 7.8, we showed that the $\boldsymbol{B}$ field of a magnetic dipole is given by

$$B_r = \frac{\mu_0}{4\pi}\frac{2m}{r^3}\cos\theta, \tag{10-150}$$

$$B_\theta = \frac{\mu_0}{4\pi}\frac{m}{r^3}\sin\theta, \tag{10-151}$$

$$B_\varphi = 0, \tag{10-152}$$

if r is again much larger than the dipole.

The analogy is obvious: these two fields are mathematically identical, within a constant factor.

10.9 LORENTZ'S LEMMA

We again consider two distinct fields, $\boldsymbol{E}_a$, $\boldsymbol{B}_a$, and $\boldsymbol{E}_b$, $\boldsymbol{B}_b$, which are *not* necessarily related as in Section 10.8. According to the principle of superposition, these two fields can either exist separately or be superimposed without disturbing each other, giving a third field $\boldsymbol{E}_a + \boldsymbol{E}_b$, $\boldsymbol{B}_a + \boldsymbol{B}_b$.

For these two fields we have the vector identity

$$\boldsymbol{\nabla}\cdot(\boldsymbol{E}_a \times \boldsymbol{B}_b - \boldsymbol{E}_b \times \boldsymbol{B}_a) = \boldsymbol{B}_b\cdot\boldsymbol{\nabla} \times \boldsymbol{E}_a - \boldsymbol{E}_a\cdot\boldsymbol{\nabla} \times \boldsymbol{B}_b - \boldsymbol{B}_a\cdot\boldsymbol{\nabla} \times \boldsymbol{E}_b + \boldsymbol{E}_b\cdot\boldsymbol{\nabla} \times \boldsymbol{B}_a, \quad (10\text{-}153)$$

and, using the Maxwell equations 10-107 and 10-108,

$$\boldsymbol{\nabla}\cdot(\boldsymbol{E}_a \times \boldsymbol{B}_b - \boldsymbol{E}_b \times \boldsymbol{B}_a) = -\boldsymbol{B}_b \cdot \frac{\partial \boldsymbol{B}_a}{\partial t} - \boldsymbol{E}_a\cdot\mu_0 \left(\boldsymbol{J}_{mb} + \epsilon_0 \frac{\partial \boldsymbol{E}_b}{\partial t}\right) + \boldsymbol{B}_a \cdot \frac{\partial \boldsymbol{B}_b}{\partial t} + \boldsymbol{E}_b\cdot\mu_0 \left(\boldsymbol{J}_{ma} + \epsilon_0 \frac{\partial \boldsymbol{E}_a}{\partial t}\right). \quad (10\text{-}154)$$

If the two fields are harmonic functions of the time and if they are of the same frequency, the $\partial/\partial t$ operators can be replaced by $j\omega$ and the time derivatives cancel. Then

$$\boldsymbol{\nabla}\cdot(\boldsymbol{E}_a \times \boldsymbol{B}_b - \boldsymbol{E}_b \times \boldsymbol{B}_a) = -\mu_0 \boldsymbol{E}_a \cdot \left(\boldsymbol{J}_{fb} + \frac{\partial \boldsymbol{P}_b}{\partial t} + \boldsymbol{\nabla} \times \boldsymbol{M}_b\right) + \mu_0 \boldsymbol{E}_b \cdot \left(\boldsymbol{J}_{fa} + \frac{\partial \boldsymbol{P}_a}{\partial t} + \boldsymbol{\nabla} \times \boldsymbol{M}_a\right). \quad (10\text{-}155)$$

If, moreover, the medium is linear and isotropic, $\boldsymbol{P}$ is $\epsilon_0 \chi_e \boldsymbol{E}$, and the above time derivatives also cancel. Finally, if the point considered is not inside a source and if Ohm's law applies, $\boldsymbol{J}_f = \sigma \boldsymbol{E}$ and the $\boldsymbol{J}_f$ terms cancel. We are then left with the $\boldsymbol{\nabla} \times \boldsymbol{M}$ terms. If the medium is nonmagnetic, then $\boldsymbol{M}$ is zero and

$$\boldsymbol{\nabla}\cdot(\boldsymbol{E}_a \times \boldsymbol{B}_b - \boldsymbol{E}_b \times \boldsymbol{B}_a) = 0. \quad (10\text{-}156)$$

If the medium is magnetic, we can perform a similar calculation using $\boldsymbol{H}$'s instead of $\boldsymbol{B}$'s and Eq. 10-124. This gives

$$\boldsymbol{\nabla}\cdot(\boldsymbol{E}_a \times \boldsymbol{H}_b - \boldsymbol{E}_b \times \boldsymbol{H}_a) = 0. \quad (10\text{-}157)$$

In summary, therefore, if two fields a and b are sinusoidal and of the same frequency, if also the medium is linear and isotropic, and if the point considered is not inside a source, then Eq. 10-157 applies. If, moreover, the medium is nonmagnetic, then we have Eq. 10-156.

This result is known as *Lorentz's lemma*. It is remarkable in that it establishes a relation between two independent electromagnetic fields.

A more general discussion of this type will lead us to an important theorem in Section 14-8.

Example If the a field is purely electric and the b field is purely magnetic, then

$$\nabla \cdot (\boldsymbol{E}_a \times \boldsymbol{H}_b) = 0, \tag{10-158}$$

subject *only* to the above limitations.

This surprising statement is easily shown to be correct by expanding the divergence:

$$\nabla \cdot (\boldsymbol{E}_a \times \boldsymbol{H}_b) = \boldsymbol{H}_b \cdot (\nabla \times \boldsymbol{E}_a) - \boldsymbol{E}_a \cdot (\nabla \times \boldsymbol{H}_b), \tag{10-159}$$

$$= -\boldsymbol{H}_b \cdot \frac{\partial \boldsymbol{B}_a}{\partial t} - \boldsymbol{E}_a \cdot \left(\boldsymbol{J}_{fb} + \frac{\partial \boldsymbol{D}_b}{\partial t} \right) \equiv 0, \tag{10-160}$$

since, by hypothesis, $\boldsymbol{B}_a$ and $\boldsymbol{E}_b$ are both zero.

Example The field of a small permanent magnet is that of a magnetic dipole and is given by Eqs. 7-112 to 7-114. If the magnet carries an electric charge Q, it also has an electric field

$$E_r = \frac{Q}{4\pi\epsilon_0 r^2} \tag{10-161}$$

and

$$\nabla \cdot (\boldsymbol{E}_a \times \boldsymbol{H}_b) = \nabla \cdot \begin{vmatrix} \boldsymbol{r}_1 & \boldsymbol{\theta}_1 & \boldsymbol{\varphi}_1 \\ E_r & 0 & 0 \\ B_r & B_\theta & 0 \end{vmatrix}, \tag{10-162}$$

$$= \nabla \cdot (E_r B_\theta \, \boldsymbol{\varphi}_1), \tag{10-163}$$

$$= \nabla \cdot \left(\frac{Q}{4\pi\epsilon_0 r^2} \frac{\mu_0 m}{4\pi r^3} \sin\theta \, \boldsymbol{\varphi}_1 \right) \equiv 0. \tag{10-164}$$

10.10 THE NONHOMOGENEOUS WAVE EQUATIONS FOR *E* AND *B*

These wave equations follow immediately from Maxwell's equations.

Taking the curl of Eq. 10-107 and remembering from Section 1.9.6 that

$$\nabla \times \nabla \times \boldsymbol{E} = -\nabla^2 \boldsymbol{E} + \nabla(\nabla \cdot \boldsymbol{E}), \tag{10-165}$$

we find that

$$\nabla^2 \boldsymbol{E} - \nabla(\nabla \cdot \boldsymbol{E}) = \frac{\partial}{\partial t} (\nabla \times \boldsymbol{B}), \tag{10-166}$$

and, if ϵ and μ are independent of the coordinates and of the time,

$$\nabla^2 \boldsymbol{E} - \nabla(\nabla \cdot \boldsymbol{E}) = \frac{\partial}{\partial t} \left(\mu \boldsymbol{J}_f + \epsilon\mu \frac{\partial \boldsymbol{E}}{\partial t} \right), \tag{10-167}$$

$$\nabla^2 \boldsymbol{E} - \epsilon\mu \frac{\partial^2 \boldsymbol{E}}{\partial t^2} = \nabla \left(\frac{\rho_f}{\epsilon} \right) + \mu \frac{\partial \boldsymbol{J}_f}{\partial t}. \tag{10-168}$$

This is the nonhomogeneous wave equation for $\boldsymbol{E}$ in a homogeneous, linear, and isotropic medium ϵ, μ.

In regions where ρ_f and $\boldsymbol{J}_f$ are both zero, we have the homogeneous wave equation

$$\boldsymbol{\nabla}^2\boldsymbol{E} - \epsilon\mu\,\frac{\partial^2\boldsymbol{E}}{\partial t^2} = 0. \tag{10-169}$$

For steady fields we have again Eq. 4-183:

$$\boldsymbol{\nabla}^2\boldsymbol{E} = \boldsymbol{\nabla}\,\frac{\rho_f}{\epsilon}, \tag{10-170}$$

$$= \frac{\boldsymbol{\nabla}(\rho_f + \rho_b)}{\epsilon_0}. \tag{10-171}$$

This latter equation is valid for any medium, since we could also have deduced it from Eqs. 10-107 and 10-105, as we did in Section 4.7.

For an infinite medium ϵ, μ, the solution of the complete nonhomogeneous equation 10-168 is

$$\boldsymbol{E} = -\frac{1}{4\pi\epsilon}\int_{\tau'}\frac{[\boldsymbol{\nabla}'\rho_f]_v}{r}\,d\tau' - \frac{\mu}{4\pi}\int_{\tau'}\frac{[\partial\boldsymbol{J}_f/\partial t]_v}{r}\,d\tau'. \tag{10-172}$$

Note that the velocity v of the $\boldsymbol{E}$ wave is $(\epsilon\mu)^{1/2}$ (see Appendix E), so that the terms between brackets are taken at the time $t - (r/v)$, if $\boldsymbol{E}$ is calculated at the time t. It will be shown in Problem 10-19 that the first term on the right is $-\boldsymbol{\nabla}V$, while the second is $-\partial\boldsymbol{A}/\partial t$, so that $\boldsymbol{E}$ is $-\boldsymbol{\nabla}V - \partial\boldsymbol{A}/\partial t$, as usual.

We can find the nonhomogeneous wave equation for $\boldsymbol{B}$ which corresponds to Eq. 10-168 by multiplying Eq. 10-124 by μ and then taking its curl, assuming that μ is independent of the coordinates. Then

$$\boldsymbol{\nabla}^2\boldsymbol{B} - \boldsymbol{\nabla}(\boldsymbol{\nabla}\cdot\boldsymbol{B}) = -\mu\boldsymbol{\nabla}\times\left(\boldsymbol{J}_f + \epsilon\,\frac{\partial\boldsymbol{E}}{\partial t}\right) \tag{10-173}$$

or, using Eqs. 10-106 and 10-107, and assuming that ϵ is also independent of the coordinates,

$$\boldsymbol{\nabla}^2\boldsymbol{B} - \epsilon\mu\,\frac{\partial^2\boldsymbol{B}}{\partial t^2} = -\mu\boldsymbol{\nabla}\times\boldsymbol{J}_f. \tag{10-174}$$

In regions where $\boldsymbol{J}_f$ is zero we have the homogeneous equation

$$\boldsymbol{\nabla}^2\boldsymbol{B} - \epsilon\mu\,\frac{\partial^2\boldsymbol{B}}{\partial t^2} = 0. \tag{10-175}$$

For steady fields,

$$\boldsymbol{\nabla}^2\boldsymbol{B} = -\mu\boldsymbol{\nabla}\times\boldsymbol{J}_f \tag{10-176}$$

or, since μ is, by hypothesis, independent of the coordinates, and because of Eq. 9-52, we find again Eq. 9-103:

$$\boldsymbol{\nabla}^2\boldsymbol{B} = -\mu_0\boldsymbol{\nabla}\times\mu_r\boldsymbol{J}_f, \tag{10-177}$$

$$= -\mu_0\boldsymbol{\nabla}\times(\boldsymbol{J}_f + \boldsymbol{\nabla}\times\boldsymbol{M}). \tag{10-178}$$

The second equation is valid for any medium.

The solution of the non-homogeneous wave equation 10-174 for $\boldsymbol{B}$ is

$$\boldsymbol{B} = \frac{\mu}{4\pi}\int_{\tau'} \frac{[\nabla' \times \boldsymbol{J}_f]_v}{r}\, d\tau' \tag{10-179}$$

for an infinite medium ϵ, μ. This is the curl of $\boldsymbol{A}$, as will be shown in Problem 10-18.

See also Problem 7-8.

10.11 SUMMARY

There is *conservation of free charge*, and therefore

$$\nabla \cdot \boldsymbol{J}_f = -\frac{\partial \rho_f}{\partial t}. \tag{10-4}$$

There is also *conservation of bound charge*, and, setting

$$\boldsymbol{J}_m = \boldsymbol{J}_f + \frac{\partial \boldsymbol{P}}{\partial t} + \nabla \times \boldsymbol{M} \tag{10-6}$$

to be the current flowing in matter,

$$\nabla \cdot \boldsymbol{J}_m = -\frac{\partial \rho_t}{\partial t}, \tag{10-7}$$

where $\rho_t = \rho_f + \rho_b$ is the total charge density.

The *total volume current density* is

$$\boldsymbol{J}_t = \boldsymbol{J}_f + \epsilon_0 \frac{\partial \boldsymbol{E}}{\partial t} + \frac{\partial \boldsymbol{P}}{\partial t} + \nabla \times \boldsymbol{M}, \tag{10-9}$$

$$= \boldsymbol{J}_f + \frac{\partial \boldsymbol{D}}{\partial t} + \nabla \times \boldsymbol{M}, \tag{10-12}$$

and

$$\nabla \cdot \boldsymbol{J}_t = 0. \tag{10-10}$$

The term $\partial \boldsymbol{D}/\partial t$ is called the *displacement current density*.

In an infinite medium ϵ, μ, the *retarded potentials* are

$$V = \frac{1}{4\pi\epsilon}\int_{\tau'} \frac{[\rho_f]_v}{r}\, d\tau', \tag{10-26}$$

$$\boldsymbol{A} = \frac{\mu}{4\pi}\int_{\tau'} \frac{[\boldsymbol{J}_f]_v}{r}\, d\tau', \tag{10-27}$$

where $[\rho_f]_v$ and $[\boldsymbol{J}_f]_v$ are the free charge and current densities at the source point $P'(x', y', z')$ at the time $t - (r/v)$, r is the distance between P' and

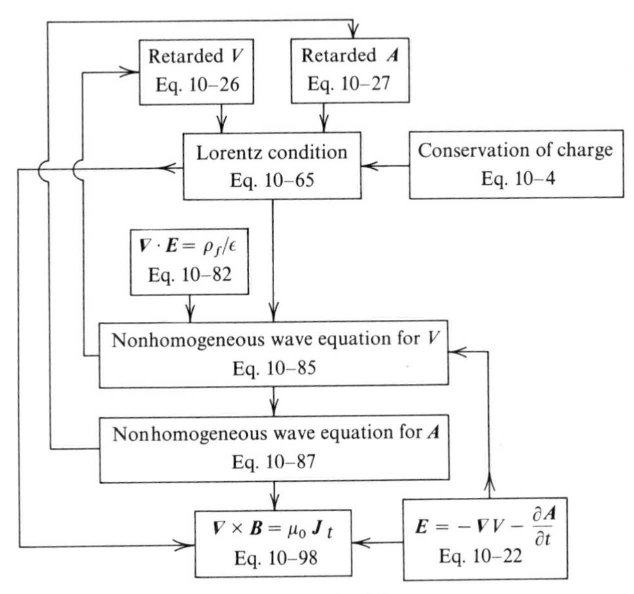

Figure 10-11.

the point $P(x, y, z)$ where V and A are calculated at the time t, and $v = 1/(\epsilon\mu)^{1/2}$ is the phase velocity of electromagnetic waves in the medium.

The *Lorentz condition* relates together the electromagnetic potentials V and $\boldsymbol{A}$:

$$\nabla \cdot \boldsymbol{A} + \epsilon\mu \frac{\partial V}{\partial t} = 0. \tag{10-65}$$

When V is a function of the time, the six components of the field vectors $\boldsymbol{E}$ and $\boldsymbol{B}$ can be obtained from the three components of $\boldsymbol{A}$.

The *nonhomogeneous wave equations for V and $\boldsymbol{A}$* in a homogeneous medium ϵ, μ are as follows:

$$\nabla^2 V - \epsilon\mu \frac{\partial^2 V}{\partial t^2} = -\frac{\rho_f}{\epsilon}, \tag{10-85}$$

$$\nabla^2 \boldsymbol{A} - \epsilon\mu \frac{\partial^2 \boldsymbol{A}}{\partial t^2} = -\mu \boldsymbol{J}_f. \tag{10-87}$$

The reasoning that led us to the equation for the curl of $\boldsymbol{B}$ is shown schematically in Figure 10-11. This is one of Maxwell's equations:

$$\nabla \times \boldsymbol{B} - \frac{1}{c^2}\frac{\partial \boldsymbol{E}}{\partial t} = \mu_0 \boldsymbol{J}_m, \tag{10-99}$$

or

$$\nabla \times \boldsymbol{H} = \boldsymbol{J}_f + \frac{\partial \boldsymbol{D}}{\partial t}. \tag{10-94}$$

Maxwell's equations can be written in various forms:

$$\nabla \cdot \boldsymbol{E} = \frac{\rho_t}{\epsilon_0}, \tag{10-105}$$

$$\nabla \cdot \boldsymbol{B} = 0, \tag{10-106}$$

$$\nabla \times \boldsymbol{E} + \frac{\partial \boldsymbol{B}}{\partial t} = 0, \tag{10-107}$$

$$\nabla \times \boldsymbol{B} - \frac{1}{c^2}\frac{\partial \boldsymbol{E}}{\partial t} = \mu_0 \boldsymbol{J}_m; \tag{10-108}$$

or

$$\nabla \cdot \boldsymbol{E} = \frac{1}{\epsilon_0}(\rho_f - \nabla \cdot \boldsymbol{P}), \tag{10-117}$$

$$\nabla \cdot \boldsymbol{B} = 0, \tag{10-118}$$

$$\nabla \times \boldsymbol{E} + \frac{\partial \boldsymbol{B}}{\partial t} = 0, \tag{10-119}$$

$$\nabla \times \boldsymbol{B} - \frac{1}{c^2}\frac{\partial \boldsymbol{E}}{\partial t} = \mu_0\left(\boldsymbol{J}_f + \frac{\partial \boldsymbol{P}}{\partial t} + \nabla \times \boldsymbol{M}\right); \tag{10-120}$$

or

$$\nabla \cdot \boldsymbol{D} = \rho_f, \tag{10-121}$$

$$\nabla \cdot \boldsymbol{B} = 0, \tag{10-122}$$

$$\nabla \times \boldsymbol{E} + \frac{\partial \boldsymbol{B}}{\partial t} = 0, \tag{10-123}$$

$$\nabla \times \boldsymbol{H} - \frac{\partial \boldsymbol{D}}{\partial t} = \boldsymbol{J}_f; \tag{10-124}$$

or, for fields which are sinusoidal functions of the time in a medium ϵ, μ,

$$\nabla \cdot \epsilon \boldsymbol{E} = \rho_f, \tag{10-125}$$

$$\nabla \cdot \mu \boldsymbol{H} = 0, \tag{10-126}$$

$$\nabla \times \boldsymbol{E} + j\omega\mu \boldsymbol{H} = 0, \tag{10-127}$$

$$\nabla \times \boldsymbol{H} - j\omega\epsilon \boldsymbol{E} = \boldsymbol{J}_f. \tag{10-128}$$

In integral form, Eqs. 10-105 to 10-108 are written as follows:

$$\int_S \boldsymbol{E} \cdot d\boldsymbol{a} = \frac{Q_t}{\epsilon_0}, \qquad (10\text{-}129)$$

$$\int_S \boldsymbol{B} \cdot d\boldsymbol{a} = 0, \qquad (10\text{-}130)$$

$$\oint_C \boldsymbol{E} \cdot d\boldsymbol{l} = -\frac{d\Phi}{dt}, \qquad (10\text{-}132)$$

$$\oint_C \boldsymbol{B} \cdot d\boldsymbol{l} = \mu_0 I_t, \qquad (10\text{-}133)$$

where S is any closed surface, $Q_t = Q_f + Q_b$ is the total charge enclosed by S, C is any closed curve, Φ is the magnetic flux linking C, and I_t is the total current flowing through C:

$$\boldsymbol{I}_t = \oint_S \left(\boldsymbol{J}_m + \epsilon_0 \frac{\partial \boldsymbol{E}}{\partial t} \right) \cdot d\boldsymbol{a}, \qquad (10\text{-}133)$$

this S being any surface bounded by the curve C.

It follows from Maxwell's equations that, to any field $\boldsymbol{E}$, $\boldsymbol{B}$, with $\rho_f = 0$, $J_f = 0$, there corresponds another field

$$\boldsymbol{E}' = -K\boldsymbol{B}, \qquad (10\text{-}139)$$

$$\boldsymbol{H}' = +K\boldsymbol{D}. \qquad (10\text{-}140)$$

If we have two fields $\boldsymbol{E}_a$, $\boldsymbol{H}_a$, and $\boldsymbol{E}_b$, $\boldsymbol{H}_b$, which are sinusoidal functions of the time and of the same frequency, then, in a region outside the sources where the medium is linear and isotropic,

$$\boldsymbol{\nabla} \cdot (\boldsymbol{E}_a \times \boldsymbol{H}_b - \boldsymbol{E}_b \times \boldsymbol{H}_a) = 0. \qquad (10\text{-}157)$$

This is *Lorentz's lemma.*

Finally, we found the nonhomogeneous wave equations for $\boldsymbol{E}$ and $\boldsymbol{B}$ in a homogeneous medium ϵ, μ:

$$\nabla^2 \boldsymbol{E} - \epsilon\mu \frac{\partial^2 \boldsymbol{E}}{\partial t^2} = \boldsymbol{\nabla} \left(\frac{\rho_f}{\epsilon} \right) + \mu \frac{\partial \boldsymbol{J}_f}{\partial t}, \qquad (10\text{-}168)$$

$$\nabla^2 \boldsymbol{B} - \epsilon\mu \frac{\partial^2 \boldsymbol{B}}{\partial t^2} = -\mu \boldsymbol{\nabla} \times \boldsymbol{J}_f. \qquad (10\text{-}174)$$

PROBLEMS

10-1. What is the relaxation time for a semiconductor having a rather low conductivity of about 100 mhos/meter and a rather large relative permittivity of 15?

10-2. Two plane parallel electrodes are separated by a plate of thickness s whose conductivity σ varies linearly from σ_0 near the positive plate to $\sigma_0 + a$ near the negative plate.

(a) Calculate the space charge density ρ_f when the current density is J_f.
(b) Calculate ρ_f near both plates for $\sigma_0 = 1.00 \times 10^7$ mhos/meter, $\sigma_0 + a = 2.00 \times 10^7$ mhos/meter, $J_f = 1.00$ ampere/meter², $\epsilon_r = 3$, $s = 1.00$ centimeter.

10-3. Show that the nonhomogeneous wave equations for V and A are compatible with the conservation of charge.

10-4. Calculate the magnetic induction inside and outside a large flat conducting plate if the current density J decreases linearly with depth z inside the plate, J being equal to $J_0(1 - az)$. The plate thickness is $1/a$.

10-5. Deduce Eq. 2-5 for the electric field intensity due to a point charge Q, starting from Maxwell's equations.

10-6. Write out Maxwell's equations in terms of $\boldsymbol{E}$ and $\boldsymbol{H}$ only for a nonhomogeneous medium in which ϵ_r and μ_r are functions of the coordinates.

10-7. We have seen in Section 10.2 that

$$\boldsymbol{E} = -\boldsymbol{\nabla} V - \frac{\partial \boldsymbol{A}}{\partial t},$$

$$\boldsymbol{B} = \boldsymbol{\nabla} \times \boldsymbol{A},$$

where V and $\boldsymbol{A}$ are the scalar and vector potentials, respectively.

Rewrite Maxwell's equations in terms of these potentials, for linear homogeneous media.

10-8. At very high frequencies currents are limited to the region close to the surface of a conductor and there is essentially zero electric and magnetic field in the interior.

(a) Setting E_t and B_t to be the tangential electric field intensity and magnetic induction in the conductor, show that

$$\frac{\partial E_t}{\partial z} = -\frac{\partial B_t}{\partial t},$$

where the z-axis is normal to the surface of the conductor and points outward, E_t is taken to be positive in the direction of the x-axis, and B_t in the direction of the y-axis.
(b) Show that, outside the conductor, $\boldsymbol{B}$ is tangential to the surface, or very nearly so.
(c) Show that $\boldsymbol{B}$ is related to the surface current density $\boldsymbol{\lambda}$ by the equations

$$\boldsymbol{B} = \mu_0(\boldsymbol{\lambda} \times \boldsymbol{n}) \qquad \text{and} \qquad B = \mu_0 \lambda,$$

where $\boldsymbol{n}$ is a unit vector normal to the surface of the conductor and pointing outward.
(d) Does this result depend on the way in which the current varies with depth inside the conductor?
(e) Can there be a normal component of $\boldsymbol{E}$ outside?

10-9. Inside a superconductor

$$\boldsymbol{E} = 0 \qquad \text{and} \qquad \boldsymbol{B} = 0,$$

under ideal conditions. This is true (except for a thin region, of the order of a tenth of a micron thick, near the surface), even if the material is cooled in a magnetic field.

(a) Show that, just outside, $\boldsymbol{B}$ is tangential to the surface.
(b) Show that $\boldsymbol{J}_f = 0$ inside. Therefore currents must flow very close to the surface.

Assume steady-state conditions and set the magnetization $\boldsymbol{M}$ equal to zero.
(c) Does there exist a relation between the tangential $\boldsymbol{B}$ and the surface current density?
(d) What can you say about the magnetic forces?
(e) Show that the magnetic flux linking a curve C which is entirely situated inside a pure superconducting material must be constant.
(f) What happens if a superconducting ring is moved from a region where the magnetic induction is $\boldsymbol{B}_1$ to another where it is $\boldsymbol{B}_2$?

10-10. The conduction current density in a liquid metal is

$$\boldsymbol{J}_f = \sigma(\boldsymbol{E} + \boldsymbol{u} \times \boldsymbol{B}),$$

where σ is the conductivity, $\boldsymbol{E}$ is the electric field intensity, $\boldsymbol{u}$ is the velocity of the fluid, and $\boldsymbol{B}$ is the magnetic induction, all quantities being measured in the frame of reference of the laboratory.

The displacement current density is negligible.

(a) Show that the Lorentz force per unit volume of fluid is

$$\sigma(\boldsymbol{E} + \boldsymbol{u} \times \boldsymbol{B}) \times \boldsymbol{B}.$$

For example, a conducting fluid is pushed in the direction of $\boldsymbol{E} \times \boldsymbol{B}$ in crossed electric and magnetic fields. This is the principle of operation of *electromagnetic pumps*. See also Problem 8-8. The field $\boldsymbol{u} \times \boldsymbol{B}$ then opposes $\boldsymbol{E}$.

The Lorentz force is equal to

$$\sigma(\boldsymbol{E}_\perp + \boldsymbol{u}_\perp \times \boldsymbol{B}) \times \boldsymbol{B},$$

where the subscript $\perp$ indicates a component perpendicular to $\boldsymbol{B}$.
(b) Show that the Lorentz force can also be written as

$$\sigma B^2(\boldsymbol{v} - \boldsymbol{u}_\perp),$$

where

$$\boldsymbol{v} = \frac{\boldsymbol{E} \times \boldsymbol{B}}{B^2}.$$

is a velocity perpendicular to $\boldsymbol{B}$.

This means that the Lorentz force tends to make the component of the velocity that is perpendicular to the magnetic lines of force equal to the velocity $\boldsymbol{v}$.
(c) Show that

$$\frac{\partial \boldsymbol{B}}{\partial t} = \boldsymbol{\nabla} \times (\boldsymbol{u} \times \boldsymbol{B}) + \frac{1}{\sigma\mu_0} \boldsymbol{\nabla}^2 \boldsymbol{B}.$$

In liquid metals $1/\sigma\mu_0$ is of the order of unity. For example, mercury has a conductivity of about 10^6 mhos/meter, and $1/\sigma\mu_0$ is about $10/4\pi$.

10-11. Imagine an expanding spherically symmetrical universe in which there is continuous creation of charge. Then, by symmetry, the vector potential $\boldsymbol{A}$ must be purely radial.

(a) Show that, under steady-state conditions, the current density $\boldsymbol{J}_m$ must be everywhere zero, according to Maxwell's equations.
Since this did not appear reasonable to them, Lyttleton and Bondi (see Problem 4-22) suggested that, if continuous charge creation does exist, then Maxwell's equations must be modified.

They suggested that one should write

$$\boldsymbol{\nabla} \times \boldsymbol{B} = \mu_0 \boldsymbol{J}_m + \frac{1}{c^2}\frac{\partial \boldsymbol{E}}{\partial t} - \left(\frac{1}{l^2}\boldsymbol{A}\right),$$

$$\boldsymbol{\nabla}\cdot\boldsymbol{E} = \frac{\rho_t}{\epsilon_0} - \left(\frac{1}{l^2}V\right),$$

where the new terms are set between parentheses. The quantities V and $\boldsymbol{A}$ are the usual scalar and vector potentials, which are related to $\boldsymbol{E}$ and $\boldsymbol{B}$ as in Section 10.2:

$$\boldsymbol{E} = -\boldsymbol{\nabla} V - \frac{\partial \boldsymbol{A}}{\partial t},$$

$$\boldsymbol{B} = \boldsymbol{\nabla} \times \boldsymbol{A}.$$

The other two equations of Maxwell for $\boldsymbol{\nabla} \times \boldsymbol{E}$ and $\boldsymbol{\nabla}\cdot\boldsymbol{B}$ would remain unchanged.

Lyttleton and Bondi suggested that the constant l, which has the dimensions of a length, would be of the order of the radius of the Universe. The new terms would therefore be negligible in all but cosmological problems.
(b) If these modified Maxwell equations were correct, would V and $\boldsymbol{A}$ be measurable, in principle?

Remember that, with the above equations for $\boldsymbol{E}$ and $\boldsymbol{B}$, it is only the rates of change of V and $\boldsymbol{A}$ that affect $\boldsymbol{E}$ and $\boldsymbol{B}$.
(c) Write out the equation for the conservation of the total charge (Section 10.1) in its modified form, assuming that charge is everywhere created at the constant rate of q coulombs/meter^3-second.
(d) Would the Lorentz condition (Section 10.3) still be valid?
(e) Now set $\boldsymbol{A} = A'\boldsymbol{r}$, where A' is a constant, and assume V to be a constant. Show that

$$\boldsymbol{B} = 0, \qquad \boldsymbol{E} = 0, \qquad \boldsymbol{J}_m = (q/3)\boldsymbol{r}, \qquad \rho_t = \epsilon_0 V/l^2.$$

Assuming that the velocity of the outward flow of matter is the same as that of the charge, namely $\boldsymbol{J}_m/\rho_t$, it follows that the radial velocity is proportional to r, which is consistent with the linear velocity-distance relation observed by astronomers: $v = r/T$, where $T \approx 3 \times 10^{17}$ seconds is the *Hubble constant*.
(f) Show that $\rho_t = qT/3$.

Now the space-charge density

$$\rho_t = \frac{\eta}{m} ye,$$

where $\eta \approx 10^{-26}$ kilogram/meter^3 is the mass density, m is the proton mass (the Universe is mostly hydrogen), and ye is the excess charge of a hydrogen atom (See Problem 4-22). Similarly, the rate of charge creation q is related to the rate of mass creation Q as follows:

$$q = \left(\frac{Q}{m}\right) ye.$$

(g) Show that, if this theory is correct, then $Q \approx 1$ hydrogen atom per 2×10^{16} meters^3/second.

10-12. (a) Starting from an electromagnetic field characterized by the field vectors $\boldsymbol{E}$ and $\boldsymbol{H}$, utilize the duality property of electromagnetic fields four times to find successively the fields $\boldsymbol{E}^{\text{i}}$, $\boldsymbol{H}^{\text{i}}$; $\boldsymbol{E}^{\text{ii}}$, $\boldsymbol{H}^{\text{ii}}$; $\boldsymbol{E}^{\text{iii}}$, $\boldsymbol{H}^{\text{iii}}$; and $\boldsymbol{E}^{\text{iv}}$, $\boldsymbol{H}^{\text{iv}}$.
(b) Assuming that $\boldsymbol{E}$ and $\boldsymbol{H}$ are respectively parallel to the x- and y-axes, show the 10 vectors on 5 separate diagrams.
(c) Compare the fields when the constant of proportionality K is equal to the velocity of light in the medium, that is, $1/(\epsilon\mu)^{1/2}$.

10-13. (a) Show that Maxwell's equations for free space are invariant under the transformation

$$\boldsymbol{E}' = \boldsymbol{E}\cos\theta + c\boldsymbol{B}\sin\theta,$$
$$\boldsymbol{B}' = -(\boldsymbol{E}/c)\sin\theta + \boldsymbol{B}\cos\theta.$$

The transformation $\boldsymbol{E}' = -K\boldsymbol{B}$, $\boldsymbol{H}' = K\boldsymbol{D}$ of Section 10.8 and the transformation $\boldsymbol{E}' = -\boldsymbol{E}$, $\boldsymbol{B}' = -\boldsymbol{B}$ are special cases corresponding to $\theta = \pi/2$ and to $\theta = \pi$ respectively.
(b) Show that the energy density

$$\frac{1}{2}\epsilon_0 E^2 + \frac{1}{2\mu_0} B^2$$

and the Poynting vector $\boldsymbol{E} \times \boldsymbol{H}$ for a plane electromagnetic wave (Section 11.1.1) are also invariant under this transformation.‡

10-14. A rectangular plate $ABCD$ of resistive material has a uniform thickness s and a conductivity σ. If conducting electrodes are fixed to the two edges AB and CD, the resistance is R_1. When conducting electrodes are fixed to the edges BC and DA, the resistance is R_2.
Show that

$$R_1R_2 = \frac{1}{\sigma^2 s^2}.$$

This equation also applies to any region $ABCD$ bounded by lines of current flow and equipotentials (which are interchanged when the connections are modified as above). We shall use this theorem* in Problem 13-7.

* Proc. IEEE **55,** 1238 (1967). See also D. J. Epstein, Proc. IEEE, **56,** 198 (1968).

10-15. Verify Lorentz's lemma graphically by considering two pairs of orthogonal vectors $\boldsymbol{E}_a$, $\boldsymbol{H}_a$, and $\boldsymbol{E}_b$, $\boldsymbol{H}_b$ lying in the same plane.

10-16. Imagine various pairs of electromagnetic fields in free space, and verify whether Lorentz's lemma applies.‡

10-17. In the example on page 434 we showed that the leakage current in a spherical capacitor does not produce a magnetic field. There are two other ways of demonstrating this.

(a) Show that $\boldsymbol{\nabla} \times \boldsymbol{A} = 0$.

(b) Use the fact that

$$\boldsymbol{B} = \frac{\mu_0}{4\pi} \int_{\tau'} \frac{\boldsymbol{\nabla}' \times \boldsymbol{J}_f}{r}\, d\tau'.$$

(c) Use also Problem 4-7 to show that $\boldsymbol{\nabla} \times \boldsymbol{H}$ is zero.

10-18. We have shown in Section 10.10 that

$$\boldsymbol{B} = \frac{\mu}{4\pi} \int_{\tau'} \frac{\boldsymbol{\nabla}' \times \boldsymbol{J}_f}{r}\, d\tau'$$

for an infinite medium μ and if retardation effects are negligible.

Show that the term on the right is $\boldsymbol{\nabla} \times \boldsymbol{A}$.‡

10-19. Show that the two terms on the right-hand side of Eq. 10-172 are respectively $-\boldsymbol{\nabla} V$ and $-\partial \boldsymbol{A}/\partial t$.

Assume that the retardation is negligible and omit the brackets.‡

Hints

10-13(b). Use Eq. 11-17.

10-16. You can try a capacitor in a magnetic field, the fields of the electric and magnetic dipoles, the electric and magnetic fields in a cylindrical magnetron, and so forth.

10-18. Use Eq. 9-106.

10-19. Use Eq. 4-186.

CHAPTER 11

PROPAGATION OF ELECTROMAGNETIC WAVES I

Plane Waves in Infinite Media

In this chapter we shall study the basic aspects of the propagation of electromagnetic waves in infinite media; reflection and refraction phenomena will be the subject of the next chapter, after which we shall be prepared to study guided waves in Chapter 13. We shall not deal with sources of electromagnetic waves until Chapter 14.

Let us start the present discussion with propagation in free space, and then take up three types of media: dielectrics, conductors, and ionized gases. Dielectrics, which are defined as nonconductors, can be either magnetic or nonmagnetic.

At this stage you would be well advised to work through Appendixes D and E, unless you are familiar with the exponential notation for representing the cosine function, and with wave propagation.

Maxwell's equations impose no limit on the frequency of electromagnetic waves. To date, the spectrum that has been investigated experimentally is shown in Figure 11-1. It extends continuously from the long radio waves to the very high energy gamma rays observed in cosmic radiation. In the former the frequencies are about 10^4 hertz and the wavelengths about 3×10^4 meters; in the latter the frequencies are of the order of 10^{24} hertz (and higher), and the wavelengths of the order of 3×10^{-16} meter (and shorter). The known spectrum thus covers a range of 20 or more orders of magnitude. Radio, light, and heat waves, X-rays and gamma rays—all are electromagnetic, although the sources and the detectors, as well as the modes of interaction with matter, vary widely as the frequency changes by orders of magnitude.

The fundamental identity of all these types of waves is demonstrated by

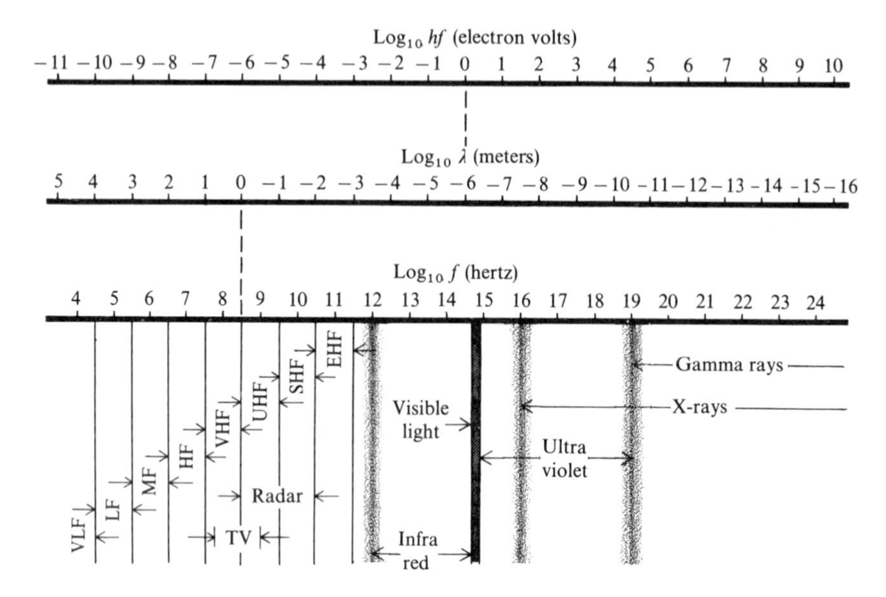

Figure 11-1. The spectrum of electromagnetic waves. The abbreviations *VLF, LF, MF, . . .* mean, respectively, very low frequency, low frequency, medium frequency, high frequency, very high frequency, ultrahigh frequency, super high frequency, and extremely high frequency. The limits indicated by the shaded regions are approximate. The energy hf, where h is Planck's constant (6.63×10^{-34} joule-second) and f is the frequency, is that of a photon or quantum of radiation.

many experiments covering overlapping parts of the spectrum. It is also demonstrated by the fact that in free space they are all transverse waves with a common velocity of propagation c given by Eq. 11-7. For example, simultaneous radio and optical observations on flare stars has shown that the velocity of propagation is the same within experimental error for wavelengths differing by more than six orders of magnitude.

We shall use $\boldsymbol{H}$ rather than $\boldsymbol{B}$ in discussing electromagnetic waves, in spite of the fact that, until now, we have used $\boldsymbol{H}$ only for magnetic materials. There are two reasons for using $\boldsymbol{H}$ instead of $\boldsymbol{B}$ in dealing with electromagnetic waves: one is that $\boldsymbol{E} \times \boldsymbol{H}$ is an energy flux density; the other is that E/H has the dimensions of an impedance. These two concepts prove to be of great practical value.

11.1 PLANE ELECTROMAGNETIC WAVES IN FREE SPACE

Let us start with the relatively simple case of plane sinusoidal waves propagating in a vacuum in a region infinitely remote from matter. Then Maxwell's equations 10-125 to 10-128 reduce to

$$\nabla \cdot \boldsymbol{E} = 0, \tag{11-1}$$

$$\nabla \cdot \boldsymbol{H} = 0, \tag{11-2}$$

$$\nabla \times \boldsymbol{E} + j\omega \, \mu_0 \boldsymbol{H} = 0, \tag{11-3}$$

$$\nabla \times \boldsymbol{H} - j\omega \, \epsilon_0 \boldsymbol{E} = 0. \tag{11-4}$$

We obtain wave equations for $\boldsymbol{E}$ and $\boldsymbol{H}$, as in Section 10-10 by taking the curls of Eqs. 11-3 and 11-4:

$$\nabla^2 \boldsymbol{E} + \epsilon_0 \mu_0 \, \omega^2 \boldsymbol{E} = 0, \tag{11-5}$$

$$\nabla^2 \boldsymbol{H} + \epsilon_0 \mu_0 \, \omega^2 \boldsymbol{H} = 0. \tag{11-6}$$

These differential equations are those of an unattenuated wave traveling at the velocity $1/(\epsilon_0 \mu_0)^{1/2}$. It follows that the field vectors can be propagated as waves in free space at the velocity

$$c = \frac{1}{(\epsilon_0 \mu_0)^{1/2}}. \tag{11-7}$$

These are two remarkable results. We have deduced from our investigation of the basic electromagnetic phenomena: (a) the possibility of the existence of electromagnetic waves and (b) the velocity of such waves in free space.

The above expression for c is in itself remarkable. It links three basic constants of electromagnetism: the velocity of an electromagnetic wave c, the permittivity of free space ϵ_0, which we first met in Section 2.1 while discussing Coulomb's law, and the permeability of free space μ_0, which enters into the magnetic force law of Section 7.1.

It will be remembered from Sections 6.1 and 7.1 that the constant μ_0 was *defined* arbitrarily to be *exactly* $4\pi \times 10^{-7}$ henry/meter. The constant ϵ_0 can thus be deduced from the measured value for the velocity of electromagnetic waves,

$$c = 2.9979 \times 10^8 \text{ meters/second}: \tag{11-8}$$

$$\epsilon_0 = \frac{1}{c^2 \mu_0} = 8.8542 \times 10^{-12} \text{ farad/meter}. \tag{11-9}$$

The permittivity of free space ϵ_0 can also be determined directly from measurements involving electrostatic phenomena. The measurements lead to the above value within experimental error, thereby confirming the theory.

For a plane electromagnetic wave propagating in the positive direction of the z-axis, $\boldsymbol{E}$ is independent of x and y,

$$\nabla \cdot \boldsymbol{E} = \frac{\partial E_z}{\partial z} = 0, \tag{11-10}$$

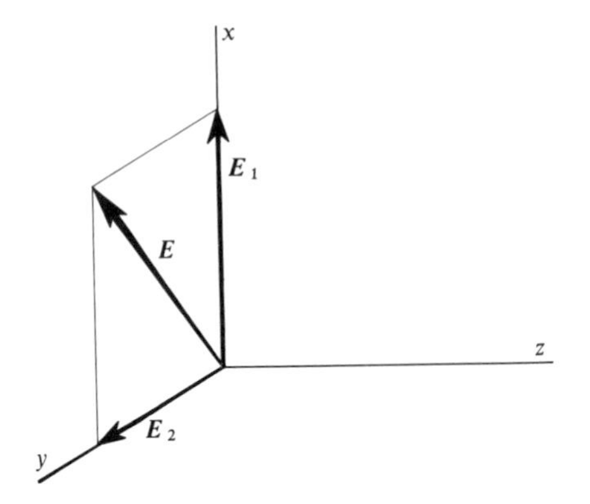

Figure 11-2. Decomposition of the vector $\boldsymbol{E}$ into two vectors $\boldsymbol{E}_1$ and $\boldsymbol{E}_2$.

and the z component of $\boldsymbol{E}$ cannot be a function of z. We shall set

$$E_z = 0, \tag{11-11}$$

since we are interested in waves and not in uniform fields.

The same argument applies to the $\boldsymbol{H}$ vector, and

$$H_z = 0. \tag{11-12}$$

A plane electromagnetic wave propagating in free space is therefore transverse, since it has no longitudinal components.

We now assume that the wave is plane-polarized with its $\boldsymbol{E}$ vector pointing in the direction of the x-axis.* This does not involve any loss of generality since any plane-polarized wave can be considered to be the sum of two waves that are plane-polarized in perpendicular directions and in phase. For example, in Figure 11-2, the vector $\boldsymbol{E}$ can be resolved into two mutually perpendicular vectors $\boldsymbol{E}_1$ and $\boldsymbol{E}_2$.†

For a plane-polarized wave having its $\boldsymbol{E}$ vector in the direction of the x-axis (Figure 11-3),

$$\boldsymbol{E} = E_0 \exp j\omega \left(t - \frac{z}{c}\right) \boldsymbol{i}, \tag{11-13}$$

---

* The *plane of polarization* was originally taken to be the plane containing the direction of propagation and the $\boldsymbol{H}$ vector. It is now common either to avoid the use of the term *plane of polarization* entirely or to use the term *plane of vibration*, which is defined to be the plane containing the direction of propagation and the $\boldsymbol{E}$ vector. Radio engineers say that a wave is polarized in the direction of its $\boldsymbol{E}$ vector. We shall always refer to the orientations of the $\boldsymbol{E}$ vector as above.

† One can also add two plane-polarized waves that differ in phase. Then, at a given point, the maxima of $\boldsymbol{E}_1$ and of $\boldsymbol{E}_2$ do not occur at the same time, and their sum $\boldsymbol{E}$ describes an ellipse about the z-axis. We then have an *elliptically polarized wave*.

If $\boldsymbol{E}_1$ and $\boldsymbol{E}_2$ have equal amplitudes but are $\pi/2$ out of phase, the ellipse becomes a circle and the wave is said to be *circularly polarized*. The polarization is said to be right- or left-handed according to whether the vectors $\boldsymbol{E}$ and $\boldsymbol{H}$ rotate clockwise or counterclockwise for an observer looking toward the *source*.

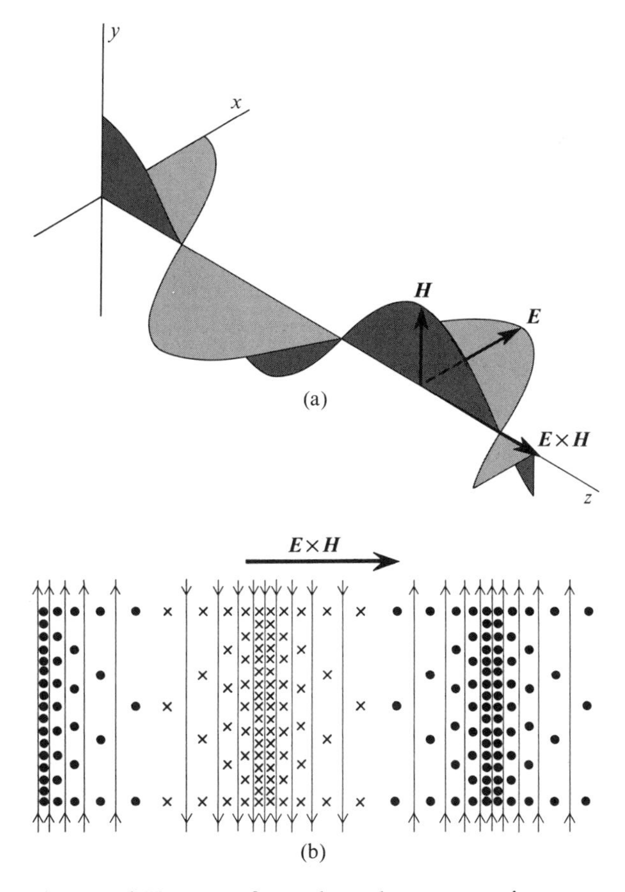

Figure 11-3. The $\boldsymbol{E}$ and $\boldsymbol{H}$ vectors for a plane electromagnetic wave traveling in the positive direction along the z-axis. (a) The variation of $\boldsymbol{E}$ and $\boldsymbol{H}$ with z at a particular moment. The two vectors are in phase, but perpendicular to each other. (b) The corresponding lines of force as seen when looking down on the xz-plane. The lines represent the electric field. The dots represent magnetic lines of force coming out of the paper, and the crosses represent magnetic lines of force going into the paper. The vector $\boldsymbol{E} \times \boldsymbol{H}$ gives the direction of propagation.

and, from either Eqs. 11-3 or 11-4,

$$H_x = 0, \tag{11-14}$$

$$\boldsymbol{H} = \epsilon_0 c\, E_0 \exp j\omega \left(t - \frac{z}{c}\right) \boldsymbol{j}. \tag{11-15}$$

Therefore $\boldsymbol{H}$ is perpendicular to $\boldsymbol{E}$, and

$$\frac{E}{H} = \frac{1}{\epsilon_0 c} = \mu_0 c = \left(\frac{\mu_0}{\epsilon_0}\right)^{1/2} = 377 \text{ ohms}, \tag{11-16}$$

$$\frac{E}{B} = c = 3.00 \times 10^8 \text{ meters/second}. \tag{11-17}$$

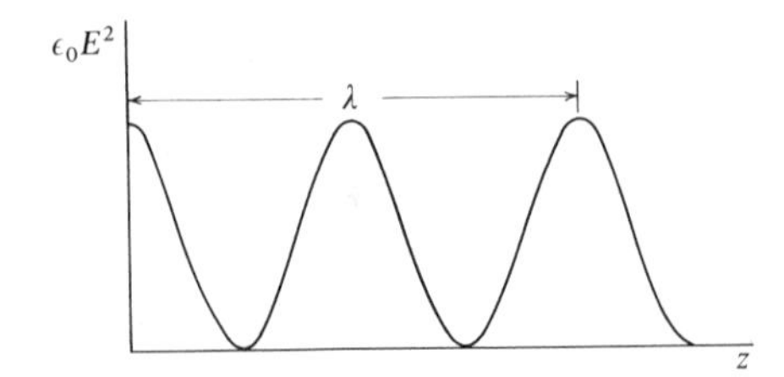

Figure 11-4. The energy density $\epsilon_0 E^2 = \mu_0 H^2$ as a function of z for a plane electromagnetic wave traveling along the z-axis in free space.

The $\boldsymbol{E}$ and $\boldsymbol{H}$ vectors are perpendicular and oriented in such a way that their vector product $\boldsymbol{E} \times \boldsymbol{H}$ points in the direction of propagation, as in Figure 11-3. The $\boldsymbol{E}$ and $\boldsymbol{H}$ vectors are in phase, since E_x/H_y is real, and they have the same relative magnitudes at all points at all times.

The electric and magnetic energy densities are in phase and are equal, since

$$\frac{\frac{1}{2}\epsilon_0 E^2}{\frac{1}{2}\mu_0 H^2} = 1. \tag{11-18}$$

At any instant, the total energy density is therefore distributed as in Figure 11-4.

11.1.1 The Poynting Vector

We have found that a plane electromagnetic wave in free space propagates in the direction of the vector $\boldsymbol{E} \times \boldsymbol{H}$. Let us calculate the divergence of this vector for any electromagnetic field in free space:

$$\boldsymbol{\nabla} \cdot (\boldsymbol{E} \times \boldsymbol{H}) = -\boldsymbol{E} \cdot (\boldsymbol{\nabla} \times \boldsymbol{H}) + \boldsymbol{H} \cdot (\boldsymbol{\nabla} \times \boldsymbol{E}), \tag{11-19}$$

$$= -\left(\boldsymbol{E} \cdot \epsilon_0 \frac{\partial \boldsymbol{E}}{\partial t}\right) - \left(\boldsymbol{H} \cdot \mu_0 \frac{\partial \boldsymbol{H}}{\partial t}\right), \tag{11-20}$$

$$= -\frac{\partial}{\partial t}\left\{\frac{\epsilon_0 E^2}{2} + \frac{\mu_0 H^2}{2}\right\}. \tag{11-21}$$

Integrating over a volume τ bounded by a surface S, and using the divergence theorem,

$$\int_S (\boldsymbol{E} \times \boldsymbol{H}) \cdot \boldsymbol{da} = -\frac{\partial}{\partial t}\int_\tau \left(\frac{\epsilon_0 E^2}{2} + \frac{\mu_0 H^2}{2}\right) d\tau. \tag{11-22}$$

The integral on the right is the sum of the electric and magnetic energies, according to Sections 2.15 and 8.5.1. The right-hand side is thus the energy lost per unit time by the volume τ and the left-hand side must be the total outward flux of energy through the surface S bounding τ.

The quantity

$$\boldsymbol{\mathcal{S}} = \boldsymbol{E} \times \boldsymbol{H} \tag{11-23}$$

is called the *Poynting vector.* When integrated over a closed surface, it gives the total outward flow of energy per unit time. It will be noticed that the vector $\mathbf{S}$ points in the direction of propagation of the wave.

The instantaneous value of the Poynting vector at a given point in space is $\boldsymbol{E} \times \boldsymbol{H}$ or, according to Eq. 11-16,

$$\mathbf{S} = \frac{1}{\mu_0 c} E^2 \boldsymbol{k} = c\epsilon_0 E^2 \boldsymbol{k}. \tag{11-24}$$

For a plane sinusoidal wave, the average value of $\mathbf{S}$ is

$$\mathbf{S}_{\mathrm{av}} = c\epsilon_0 E_{\mathrm{rms}}^2 \boldsymbol{k} = \tfrac{1}{2} c\epsilon_0 E_0^2 \boldsymbol{k}. \tag{11-25}$$

$$= 2.66 \times 10^{-3} E_{\mathrm{rms}}^2 \boldsymbol{k} \text{ watts/meter}^2. \tag{11-26}$$

The energy can therefore be considered to travel with an average density

$$\tfrac{1}{2}\epsilon_0 E_{\mathrm{rms}}^2 + \tfrac{1}{2}\mu_0 H_{\mathrm{rms}}^2 = \epsilon_0 E_{\mathrm{rms}}^2 = \tfrac{1}{2}\epsilon_0 E_0^2 \tag{11-27}$$

at the velocity of propagation $c\boldsymbol{k}$.*

Example

A laser beam has a power of 20 gigawatts and a diameter of 2 millimeters. Let us calculate the peak values of E and of B.

From Eq. 11-26,

$$E_0 = 2^{1/2} E_{rms} = 2^{1/2}\left(\frac{10^3}{2.66} \times \frac{20 \times 10^9}{\pi \times 10^{-6}}\right)^{1/2}, \tag{11-28}$$

$$= 2.2 \times 10^9 \text{ volts/meter.} \tag{11-29}$$

This is an enormous electric field; it corresponds to a voltage difference of about a quarter of a volt over a distance of one angstrom (10^{-10} meter), which is about the diameter of an atom.

We can now find B_0 from Eq. 11-17:

$$B_0 = \frac{E_0}{c} = \frac{2.2 \times 10^9}{3 \times 10^8} = 7.3 \text{ teslas.} \tag{11-30}$$

This is about ten times larger than the magnetic induction between the pole pieces of a powerful permanent magnet.

Example

ENERGY FLOW THROUGH AN IMAGINARY CYLINDER

Figure 11-5 shows wave fronts of a plane electromagnetic wave flowing through an imaginary cylinder. If the amplitude of the wave is constant, the electric and magnetic energy densities averaged over one period are constant, and the net outward energy flow is zero.

---

* See *The Feynman Lectures on Physics* (Addison Wesley, Reading, Mass., 1964), Volume II, Section 27-4 for an interesting discussion on energy density and energy flow in electromagnetic fields.

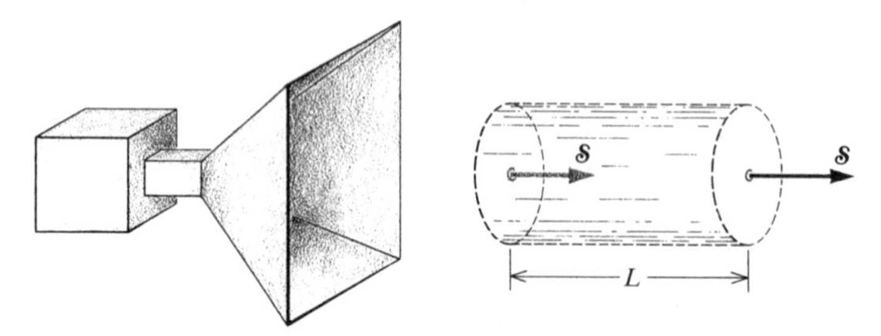

Figure 11-5. The electromagnetic wave emanating from the horn sweeps through the imaginary cylinder. The amplitude gradually decreases at the source, and the intensity, represented by the Poynting vectors, is larger on the right than on the left, as in Figure 11-6.

However, if the electromagnetic wave *de*creases in amplitude, the average energy density *de*creases with time, and there must be a net *out*ward flux of energy.

The instantaneous value of E for a plane wave traveling in the positive direction along the z-axis is the same as that near the source at a previous time $t - (z/c)$, where z is the distance from the source. If the amplitude decreases linearly with time, the electric field intensity at the source is

$$E = E_0(1 - at) \exp j\omega t, \tag{11-31}$$

whereas, at a distance z,

$$E = E_0\left\{1 - a\left(t - \frac{z}{c}\right)\right\} \exp j\omega\left(t - \frac{z}{c}\right). \tag{11-32}$$

At any given time t, the wave then has the general shape shown in Figure 11-6.

We now calculate the rate at which the electromagnetic energy enclosed within the cylindrical volume decreases with time. We assume that E changes only slightly during a period $1/f$, so that averages can be calculated as above. This requires that a/f be much smaller than unity.

The net average outward energy flow is obtained by evaluating the surface integral of $\boldsymbol{E} \times \boldsymbol{H}$ over the two ends of the cylinder, the left-hand end being at z_0 and the right-hand end at $z_0 + L$. Since $H = \epsilon_0 c E$,

$$\int_S (\boldsymbol{E} \times \boldsymbol{H})\cdot d\boldsymbol{a}$$

$$= \frac{1}{2} c\epsilon_0 S E_0^2 \left\{\left(1 - at + \frac{a(z_0 + L)}{c}\right)^2 - \left(1 - at + \frac{az_0}{c}\right)^2\right\}, \tag{11-33}$$

$$= \frac{\epsilon_0 S E_0^2}{2}\left\{2a(1 - at)L + \frac{a^2}{c}(2z_0 + L)L\right\}, \tag{11-34}$$

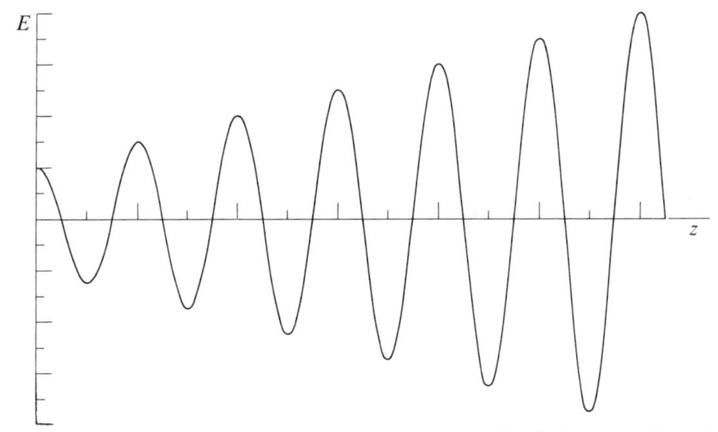

Figure 11-6. Sine wave from a source whose amplitude decreases linearly with time. The wave travels to the right.

whereas the average energy enclosed at the time t is

$$W = \int_{z_0}^{z_0+L} \frac{\epsilon_0}{2} E_0^2 \left(1 - at + \frac{az}{c}\right)^2 S\,dz, \tag{11-35}$$

$$= \frac{\epsilon_0 S E_0^2}{2} \left\{(1 - at)^2 L + \frac{a^2}{3c^2}(L^3 + 3z_0 L^2 + 3z_0^2 L) + \frac{a}{c}(1 - at)(2z_0 + L)L\right\}. \tag{11-36}$$

Then

$$-\frac{dW}{dt} = \frac{\epsilon_0 S E_0^2}{2} \left\{2a(1 - at)L + \frac{a^2}{c}(2z_0 + L)L\right\}. \tag{11-37}$$

The net average outward energy flow per unit time is thus equal to the rate at which the enclosed energy decreases with time, as expected.

11.2 THE *E* AND *H* VECTORS IN HOMOGENEOUS, ISOTROPIC, LINEAR, AND STATIONARY MEDIA

We now go on to the propagation of plane electromagnetic waves in homogeneous, isotropic, linear, and stationary media. We repeat the definitions of these terms that we gave in the introduction to Chapter 10.

A medium is *homogeneous* if its properties do not vary from point to point; it is *isotropic* if its properties are the same in all directions from any given point; it is *linear and isotropic* if

$$\mathbf{D} = \epsilon \mathbf{E}, \qquad \mathbf{H} = \mathbf{B}/\mu, \qquad \mathbf{J}_f = \sigma \mathbf{E}, \tag{11-38}$$

where ϵ, μ, and the conductivity σ are constants independent of $\boldsymbol{E}$ and $\boldsymbol{H}$, and independent of direction. Crystalline media are usually *an*isotropic. We shall call a medium *stationary* if it is at rest with respect to the coordinate system used.

We assume that there are no sources in the region under consideration. We also assume that the medium of propagation is infinite in extent. This will avoid reflection and refraction; these phenomena will be the subject of the next chapter.

The wave equations for $\boldsymbol{E}$ and $\boldsymbol{H}$ in such media follow from Eqs. 10-168 and 10-174 if we simply substitute $\sigma\boldsymbol{E}$ for $\boldsymbol{J}_f$ and $\mu\boldsymbol{H}$ for $\boldsymbol{B}$:

$$\nabla^2\boldsymbol{E} - \epsilon\mu\frac{\partial^2\boldsymbol{E}}{\partial t^2} - \sigma\mu\frac{\partial\boldsymbol{E}}{\partial t} = \nabla\frac{\rho_f}{\epsilon}, \tag{11-39}$$

$$\nabla^2\boldsymbol{H} - \epsilon\mu\frac{\partial^2\boldsymbol{H}}{\partial t^2} - \sigma\mu\frac{\partial\boldsymbol{H}}{\partial t} = 0. \tag{11-40}$$

In both equations the second term on the left comes from the displacement current, while the third comes from the conduction current. The two equations are identical, except for the fact that there is no magnetic equivalent of the electric charge density ρ_f. These differential equations describe an attenuated wave (see Eq. E-55 in Appendix E).

Let us first dispose of the charge density that appears in Eq. 11-39. We consider again a plane wave propagating in the positive direction of the z-axis. Then all derivatives with respect to x and to y are zero, so that

$$\nabla\cdot\boldsymbol{E} = \frac{\partial}{\partial z}E_z = \frac{\rho_f}{\epsilon}. \tag{11-41}$$

Then

$$\nabla\frac{\rho_f}{\epsilon} = \frac{\partial}{\partial z}\left(\frac{\rho_f}{\epsilon}\right)\boldsymbol{k} = \frac{\partial^2}{\partial z^2}E_z\,\boldsymbol{k}, \tag{11-42}$$

and the wave equation for $\boldsymbol{E}$ becomes

$$\frac{\partial^2}{\partial z^2}(E_x\boldsymbol{i} + E_y\boldsymbol{j}) - \mu\left(\epsilon\frac{\partial^2}{\partial t^2} + \sigma\frac{\partial}{\partial t}\right)(E_x\boldsymbol{i} + E_y\boldsymbol{j} + E_z\boldsymbol{k}) = 0. \tag{11-43}$$

The longitudinal component E_z must therefore satisfy the equation

$$\epsilon\frac{\partial^2 E_z}{\partial t^2} + \sigma\frac{\partial E_z}{\partial t} = 0, \tag{11-44}$$

with the result that E_z, if it exists, must be of the form

$$E_z = a + b\exp(-\sigma t/\epsilon), \tag{11-45}$$

where a and b are constants of integration independent of t. Thus E_z must decrease exponentially with time and there is no E_z wave. If $\sigma = 0$, E_z is of the form $a + bt$ and again there is no wave. Hence we may set

$$E_z = 0, \tag{11-46}$$

since we are concerned solely with wave propagation. Also, from Eq. 11-41, we may set

$$\rho_f = 0 \tag{11-47}$$

for a plane wave in a conducting medium. This is in agreement with our discussion of the example on page 424.

In a nonconductor, $\sigma = 0$ and the wave again has no longitudinal component of $\boldsymbol{E}$.

We have therefore found that, for plane electromagnetic waves, (a) we may set $\rho_f = 0$ in Eq. 11-39, and (b) the $\boldsymbol{E}$ vector is transverse.

It is easy to show that the $\boldsymbol{H}$ vector is also transverse: the divergence of $\boldsymbol{H}$ being equal to zero,

$$\frac{\partial H_z}{\partial z} = 0, \tag{11-48}$$

because the derivatives with respect to x and to y are both zero, by hypothesis. Then H_z is not a function of z, and, if we are concerned only with waves, we may set

$$H_z = 0. \tag{11-49}$$

Then the $\boldsymbol{H}$ vector is also transverse.

Plane electromagnetic waves are therefore transverse in any homogeneous, isotropic, linear, and stationary medium.

Now that we have shown that $\boldsymbol{E}$ and $\boldsymbol{H}$ are transverse, let us investigate their relative orientations. We assume again a plane-polarized wave with the $\boldsymbol{E}$ vector parallel to the x-axis:

$$\boldsymbol{E} = E_0 \exp j(\omega t - kz)\, \boldsymbol{i} \tag{11-50}$$

where the *wave number* k (Appendix E) is in general complex. Then, from the Maxwell equation 11-3,

$$\begin{vmatrix} \boldsymbol{i} & \boldsymbol{j} & \boldsymbol{k} \\ 0 & 0 & -jk \\ E & 0 & 0 \end{vmatrix} = -j\omega\,\mu(H_x \boldsymbol{i} + H_y \boldsymbol{j}). \tag{11-51}$$

Then the x component of $\boldsymbol{H}$ is zero and

$$\boldsymbol{H} = \frac{k}{\omega\mu} E_0 \exp j(\omega t - kz)\, \boldsymbol{j}, \tag{11-52}$$

$$\frac{E}{H} = \frac{\omega\mu}{k}. \tag{11-53}$$

The $\boldsymbol{E}$ and $\boldsymbol{H}$ vectors in a plane-polarized wave are therefore (a) mutually perpendicular; (b) oriented so that their vector product $\boldsymbol{E} \times \boldsymbol{H}$ points in the direction of propagation; (c) not necessarily in phase, because the wave number k can be complex.

11.3 PROPAGATION OF PLANE ELECTROMAGNETIC WAVES IN NONCONDUCTORS

In nonconductors, $\sigma = 0$, as a rule $\rho_f = 0$, and the wave equations 11-39 and 11-40 reduce to

$$\nabla^2 \boldsymbol{E} - \epsilon\mu \frac{\partial^2 \boldsymbol{E}}{\partial t^2} = 0, \tag{11-54}$$

$$\nabla^2 \boldsymbol{H} - \epsilon\mu \frac{\partial^2 \boldsymbol{H}}{\partial t^2} = 0. \tag{11-55}$$

Substituting the values of $\boldsymbol{E}$ and $\boldsymbol{H}$ from Eqs. 11-50 and 11-52, we find that

$$-k^2 + \epsilon\mu\,\omega^2 = 0. \tag{11-56}$$

The wave number

$$k = \omega(\epsilon\mu)^{1/2} \tag{11-57}$$

is real and there is no attenuation.

The *phase velocity*

$$u = \frac{\omega}{k} = \frac{1}{(\epsilon\mu)^{1/2}} = \frac{c}{(\epsilon_r\mu_r)^{1/2}}. \tag{11-58}$$

The phase velocity in nonconductors is therefore *less* than in free space and the *index of refraction*, is

$$n \equiv \frac{c}{u} = (\epsilon_r\mu_r)^{1/2}. \tag{11-59}$$

In a nonmagnetic medium $\mu_r = 1$ and

$$n = \epsilon_r^{1/2}. \tag{11-60}$$

Note however that n and ϵ_r are both functions of the frequency. We have discussed briefly the variation of ϵ_r with frequency in Sections 3.10 and 10.2.1. Since tables of n are usually compiled at optical frequencies, whereas ϵ_r is usually measured at much lower frequencies, such pairs of values cannot be expected to correspond.

Also, from Eq. 11-53,

$$H_y = \left(\frac{\epsilon}{\mu}\right)^{1/2} E_x \tag{11-61}$$

and

$$\boldsymbol{H} = H_0 \exp j(\omega t - kz)\,\boldsymbol{j} = \left(\frac{\epsilon}{\mu}\right)^{1/2} E_0 \exp j(\omega t - kz)\,\boldsymbol{j}. \tag{11-62}$$

In nonconductors, the $\boldsymbol{E}$ and $\boldsymbol{H}$ vectors are in phase and the electric and magnetic energy densities are equal:

$$\tfrac{1}{2}\,\epsilon E^2 = \tfrac{1}{2}\,\mu H^2. \tag{11-63}$$

The total instantaneous energy density is thus ϵE^2 or μH^2, and the average total energy density is $\epsilon E^2_{\text{rms}}$, or μH^2_{rms}. The average value of the Poynting vector is

$$\mathbf{S}_{av} = \frac{1}{2}\, E_0 H_0\, \boldsymbol{k} = \frac{1}{2}\left(\frac{\epsilon}{\mu}\right)^{1/2} E_0^2\, \boldsymbol{k}, \tag{11-64}$$

$$= \left(\frac{\epsilon}{\mu}\right)^{1/2} E^2_{\text{rms}}\, \boldsymbol{k} = \frac{1}{(\epsilon\mu)^{1/2}}\, \epsilon E^2_{\text{rms}}\, \boldsymbol{k}, \tag{11-65}$$

$$= u\epsilon E^2_{\text{rms}}\, \boldsymbol{k}, \tag{11-66}$$

$$= 2.66 \times 10^{-3} \left(\frac{\epsilon_r}{\mu_r}\right)^{1/2} E^2_{\text{rms}}\, \boldsymbol{k} \text{ watts/meter}^2. \tag{11-67}$$

The average value of the Poynting vector is equal to the phase velocity u multiplied by the average energy density.

Example

If the 20 gigawatt laser beam of the example given at the end of Section 11.1.1 passes through a glass whose index of refraction is 1.6, then $\epsilon_r^{1/2}$ is 1.6, μ_r is unity, and E_0 is *smaller* than in air by a factor of $1.6^{1/2}$, or by 1.26. Then E_0 in the glass is 1.7×10^9 volts/meter.

To calculate B_0 we can use the fact that the Poynting vector $(\frac{1}{2})E_o H_0$ is the same in the glass as in the air; since E_0 is smaller in the glass by a factor of 1.26, B_0 must be *larger* by the same factor.

You can check that this makes the electric energy density equal to the magnetic energy density (Eq. 11-63).

11.4 PROPAGATION OF PLANE ELECTROMAGNETIC WAVES IN CONDUCTING MEDIA

We must now solve the two wave equations 11-39 and 11-40 for a wave traveling along the z-axis, with $\rho_f = 0$. Substituting again Eqs. 11-50 and 11-52, we find that

$$-k^2 + \omega^2\epsilon\mu - j\omega\sigma\mu = 0, \tag{11-68}$$

$$k^2 = \omega^2\epsilon\mu - j\omega\sigma\mu = \frac{\epsilon_r\mu_r}{\lambda_0^2}\left(1 - \frac{j\sigma}{\omega\epsilon}\right), \tag{11-69}$$

where $\lambda\!\!\!^{-}_0 = \lambda_0/2\pi = c/\omega$ is the radian length for a wave of the same angular frequency ω propagating in free space.

The quantity $j\omega\epsilon/\sigma$ is the ratio of the displacement current density $\partial D/\partial t$ to the conduction current density $J_f = \sigma E$. We shall call the magnitude of this ratio the $\mathcal{Q}$ *of the medium:*

$$\mathcal{Q} = \left|\frac{\dfrac{\partial D}{\partial t}}{J_f}\right| = \left|\frac{\dfrac{\partial \epsilon E}{\partial t}}{\sigma E}\right| = \frac{\omega\epsilon}{\sigma} = \frac{\epsilon_r}{60\sigma\lambda\!\!\!^{-}_0}. \tag{11-70}$$

For nonconductors, $\mathcal{Q} \to \infty$. For common types of conductors, σ is of the order of 10^7 mhos/meter (5.8×10^7 for copper) and we can set $\epsilon_r \approx 1$ (example, page 439). The ratio $\mathcal{Q}$ is thus very small for the usual conductors. At optical frequencies σ is a function of frequency, as explained in the Example on page 424.

Then

$$k^2 = \frac{\epsilon_r\mu_r}{\lambda\!\!\!^{-}_0{}^2}\left(1 - \frac{j}{\mathcal{Q}}\right)^{\dagger} \tag{11-71}$$

and the wave number k is complex:

$$k = k_r - jk_i \tag{11-72}$$

where k_r and k_i are both positive:

$$k_r = \frac{1}{\lambda\!\!\!^{-}_0}\left(\frac{\epsilon_r\mu_r}{2}\right)^{1/2}\left\{\left(1 + \frac{1}{\mathcal{Q}^2}\right)^{1/2} + 1\right\}^{1/2}, \tag{11-73}$$

$$k_i = \frac{1}{\lambda\!\!\!^{-}_0}\left(\frac{\epsilon_r\mu_r}{2}\right)^{1/2}\left\{\left(1 + \frac{1}{\mathcal{Q}^2}\right)^{1/2} - 1\right\}^{1/2}, \tag{11-74}$$

$$k = \frac{(\epsilon_r\mu_r)^{1/2}}{\lambda\!\!\!^{-}_0}\left(1 + \frac{1}{\mathcal{Q}^2}\right)^{1/4} \exp\{-j \arctan(k_i/k_r)\}. \tag{11-75}$$

Our expression for the arc tan function is correct, since k_r is a positive quantity (Appendix E). In a vacuum, $k_r = 1/\lambda\!\!\!^{-}_0$, $k_i = 0$.

The real part k_r of the wave number is $1/\lambda\!\!\!^{-} = 2\pi/\lambda$, where λ is the wavelength in the medium; the imaginary part k_i is the reciprocal of the distance δ over which the amplitude is attenuated by a factor of e. The quantity $\delta = 1/k_i$ is called the *attenuation distance.*

Again from Appendix E, the phase velocity is

$$u = \frac{\omega}{k_r}, \tag{11-76}$$

† Note that k^2 is used here for the square of the complex number k, and *not* for the square of its modulus.

and the index of refraction is complex

$$n = \lambda\!\!\!^{-}_0 k = \lambda\!\!\!^{-}_0 (k_r - jk_i). \tag{11-77}$$

The ratio E/H is again given by Eq. 11-53:

$$\frac{E}{H} = \frac{\omega\mu}{k} = \left(\frac{\mu}{\epsilon}\right)^{1/2} \frac{1}{\left(1 + \frac{1}{Q^2}\right)^{1/4}} \exp j\theta. \tag{11-78}$$

where

$$\theta = \arctan \frac{k_i}{k_r} \tag{11-79}$$

is the phase of $\boldsymbol{E}$ with respect to $\boldsymbol{H}$.

Also,

$$\boldsymbol{E} = E_0 \exp\{j(\omega t - k_r z) - k_i z\}\ \boldsymbol{i}, \tag{11-80}$$

$$\boldsymbol{H} = H_0 \exp\{j(\omega t - k_r z - \theta) - k_i z\}\ \boldsymbol{j}, \tag{11-81}$$

with

$$\frac{E_0}{H_0} = \left(\frac{\mu}{\epsilon}\right)^{1/2} \frac{1}{\left(1 + \frac{1}{Q^2}\right)^{1/4}}. \tag{11-82}$$

The electric and magnetic energy densities are in the ratio

$$\left|\frac{\frac{1}{2}\epsilon E^2}{\frac{1}{2}\mu H^2}\right| = \frac{1}{\left(1 + \frac{1}{Q^2}\right)^{1/2}}, \tag{11-83}$$

and the average total energy density is

$$\frac{1}{2}\left(\frac{1}{2}\epsilon E_0^2 + \frac{1}{2}\mu H_0^2\right) \exp(-2k_i z) = \frac{1}{4}\epsilon E_0^2 \left\{1 + \left(1 + \frac{1}{Q^2}\right)^{1/2}\right\} \exp(-2k_i z). \tag{11-84}$$

11.4.1 The Poynting Vector in Conducting Media

The Poynting vector

$$\mathbf{S} = \boldsymbol{E} \times \boldsymbol{H} \tag{11-85}$$

was discussed in Section 11.1.1 for a wave propagating in free space. We shall now reexamine this vector for the more general case of conducting media, and then we shall find a general expression for calculating the time average of the energy flux in a field.

We again have the vector identity

$$\boldsymbol{\nabla} \cdot (\boldsymbol{E} \times \boldsymbol{H}) = \boldsymbol{H} \cdot (\boldsymbol{\nabla} \times \boldsymbol{E}) - \boldsymbol{E} \cdot (\boldsymbol{\nabla} \times \boldsymbol{H}). \tag{11-86}$$

Now, from Eqs. 10-123 and 10-124 for linear and isotropic media,

$$\boldsymbol{\nabla}\cdot(\boldsymbol{E}\times\boldsymbol{H}) = -\boldsymbol{H}\cdot\mu\frac{\partial\boldsymbol{H}}{\partial t} - \boldsymbol{E}\cdot\left(\epsilon\frac{\partial\boldsymbol{E}}{\partial t} + \boldsymbol{J}_f\right), \tag{11-87}$$

$$= -\frac{\partial}{\partial t}\left(\frac{1}{2}\mu H^2 + \frac{1}{2}\epsilon E^2\right) - \boldsymbol{E}\cdot\boldsymbol{J}_f. \tag{11-88}$$

Integrating over a volume τ and using the divergence theorem on the left-hand side, and then changing signs,

$$-\int_S (\boldsymbol{E}\times\boldsymbol{H})\cdot d\boldsymbol{a} = \frac{\partial}{\partial t}\int_\tau \left(\frac{1}{2}\epsilon E^2 + \frac{1}{2}\mu H^2\right) d\tau + \int_\tau \boldsymbol{E}\cdot\boldsymbol{J}_f d\tau, \tag{11-89}$$

where S is the closed surface bounding the volume τ.

The first term on the right gives the increase in the electric and magnetic energies in the volume τ per unit time; the second integral gives the electromagnetic energy removed from the same volume and transformed into heat energy, also per unit time. Then the term on the left, with its negative sign, must represent the rate at which electromagnetic energy flows *into* the volume τ.

Then

$$\int_S (\boldsymbol{E}\times\boldsymbol{H})\cdot d\boldsymbol{a}$$

again gives the total *out*ward flow of electromagnetic energy per unit time through the surface S, as in Section 11.1.1.

As a rule, both $\boldsymbol{E}$ and $\boldsymbol{H}$ are expressed as exponential functions of the time and of the space coordinates. However, since we require the *product* of two such quantities, we must revert to the cosine functions (Appendix D). For example, if $\boldsymbol{E}$ is of the form

$$\boldsymbol{E}_0 \exp j(\omega t - kz)$$

one must write instead

$$\boldsymbol{E}_0 \exp(-k_i z)\cos(\omega t - k_r z).$$

In practice, it is more convenient to use the following formula, which gives directly the time average of $\mathbf{S}$, with $\boldsymbol{E}$ and $\boldsymbol{H}$ expressed as exponential functions:

$$\mathbf{S}_{av} = \frac{1}{2}\,\mathrm{Re}\,(\boldsymbol{E}\times\boldsymbol{H}^*), \tag{11-90}$$

where $\boldsymbol{H}^*$ is the complex conjugate of $\boldsymbol{H}$†, and where the operator Re means "Real part of."

† The complex conjugate is obtained by substituting $-j$ for j.

We can show this equation to be correct in the following way. Let us write $\boldsymbol{E}$ and $\boldsymbol{H}$ in the form

$$\boldsymbol{E} = (\boldsymbol{E}_{0r} + j\,\boldsymbol{E}_{0i})e^{j\omega t}, \tag{11-91}$$

$$\boldsymbol{H} = (\boldsymbol{H}_{0r} + j\,\boldsymbol{H}_{0i})e^{j\omega t}, \tag{11-92}$$

where $\boldsymbol{E}_{0r}$, $\boldsymbol{E}_{0i}$, $\boldsymbol{H}_{0r}$, $\boldsymbol{H}_{0i}$ are real functions of x, y, z.

Then

$$\mathbf{S}_{av} = (\text{Re}\,\boldsymbol{E} \times \text{Re}\,\boldsymbol{H})_{av} \tag{11-93}$$

$$= \left\{(\boldsymbol{E}_{0r}\cos\omega t - \boldsymbol{E}_{0i}\sin\omega t) \times (\boldsymbol{H}_{0r}\cos\omega t - \boldsymbol{H}_{0i}\sin\omega t)\right\}_{av}. \tag{11-94}$$

Now the average value of $\cos^2 \omega t$ is $\frac{1}{2}$, the average of $\sin^2 \omega t$ is also $\frac{1}{2}$, and the average of $\sin \omega t \cos \omega t$ is zero. Thus

$$\mathbf{S}_{av} = \tfrac{1}{2}\,(\boldsymbol{E}_{0r} \times \boldsymbol{H}_{0r} + \boldsymbol{E}_{0i} \times \boldsymbol{H}_{0i}). \tag{11-95}$$

But

$$\tfrac{1}{2}\,\text{Re}\,(\boldsymbol{E} \times \boldsymbol{H}^*) = \tfrac{1}{2}\,\text{Re}\left\{(\boldsymbol{E}_{0r} + j\,\boldsymbol{E}_{0i}) \times (\boldsymbol{H}_{0r} - j\,\boldsymbol{H}_{0i})\right\}, \tag{11-96}$$

$$= \tfrac{1}{2}\,(\boldsymbol{E}_{0r} \times \boldsymbol{H}_{0r} + \boldsymbol{E}_{0i} \times \boldsymbol{H}_{0i}), \tag{11-97}$$

as above. We have therefore demonstrated the validity of Eq. 11-90 for electromagnetic fields that are sinusoïdal functions of the time. We have made no assumption as to the relative orientations and phases of the vectors $\boldsymbol{E}$ and $\boldsymbol{H}$.

It is not difficult to show that

$$\mathbf{S}_{av} = (\text{Average energy density}) \times (\text{Phase velocity}). \tag{11-98}$$

We can thus consider again, in this more general case, that the average energy density is propagated at the phase velocity.

11.5 PROPAGATION OF PLANE ELECTROMAGNETIC WAVES IN GOOD CONDUCTORS

Let us return to the values of k_r and k_i for conducting media as given in Eqs. 11-73 and 11-74. In good conductors $\mathcal{Q}$ is much smaller than unity and

$$\left\{\left(1 + \frac{1}{\mathcal{Q}^2}\right)^{1/2} \pm 1\right\}^{1/2} \approx \left(\frac{1}{\mathcal{Q}}\right)^{1/2}\left\{1 + \frac{\mathcal{Q}^2}{2} \pm \mathcal{Q}\right\}^{1/2}, \tag{11-99}$$

$$\approx \left(\frac{1}{\mathcal{Q}}\right)^{1/2}\left(1 \pm \frac{\mathcal{Q}}{2}\right), \tag{11-100}$$

$$\approx \left(\frac{1}{\mathcal{Q}}\right)^{1/2}, \tag{11-101}$$

within 1% when

$$\mathcal{Q} = \frac{\omega\epsilon}{\sigma} \leqq \frac{1}{50}, \tag{11-102}$$

or when the conduction current density σE is at least 50 times larger than the displacement current density $\partial D/\partial t$.

Good conductors will be defined as those for which the above condition is satisfied. One must remember here that σ is a function of ω at optical frequencies. According to this definition, copper is a "good" conductor up to frequencies of about 2×10^{16} hertz, or to the ultraviolet.

For good conductors, Eq. 11-68 simplifies to

$$k^2 = -j\omega\sigma\mu, \tag{11-103}$$

and

$$k = (-j\omega\sigma\mu)^{1/2} = \left(\frac{\omega\sigma\mu}{2}\right)^{1/2}(1 - j). \tag{11-104}$$

Then $k_i = k_r$ and

$$\delta = \lambda = \left(\frac{2}{\omega\sigma\mu}\right)^{1/2}. \tag{11-105}$$

From Eq. 11-53, we have that, for plane-polarized waves,

$$\frac{E}{H} = \frac{\omega\mu}{k} = \left(\frac{\omega\mu}{\sigma}\right)^{1/2} e^{j\pi/4}. \tag{11-106}$$

Therefore E leads H by $\pi/4$ radian in good conductors, while E and H are in phase in nonconductors. This difference comes from the fact that the current that is associated with H in conductors is the conduction current, and not the displacement current as in nonconductors.

From Eqs. 11-50 and 11-52,

$$\boldsymbol{E} = E_0 \exp\left\{j\left(\omega t - \frac{z}{\delta}\right) - \frac{z}{\delta}\right\}\boldsymbol{i}, \tag{11-107}$$

$$\boldsymbol{H} = \left(\frac{\sigma}{\omega\mu}\right)^{1/2} E_0 \exp\left\{j\left(\omega t - \frac{z}{\delta} - \frac{\pi}{4}\right) - \frac{z}{\delta}\right\}\boldsymbol{j} \tag{11-108}$$

or, in terms of cosine functions,

$$\boldsymbol{E} = E_0 e^{-z/\delta} \cos\left(\omega t - \frac{z}{\delta}\right)\boldsymbol{i}, \tag{11-109}$$

$$\boldsymbol{H} = \left(\frac{\sigma}{\omega\mu}\right)^{1/2} E_0 e^{-z/\delta} \cos\left(\omega t - \frac{z}{\delta} - \frac{\pi}{4}\right)\boldsymbol{j}, \tag{11-110}$$

$$= H_0 e^{-z/\delta} \cos\left(\omega t - \frac{z}{\delta} - \frac{\pi}{4}\right)\boldsymbol{j}. \tag{11-111}$$

Figure 11-7 shows the curves of E/E_0 and of H/H_0 for $t = 0$.

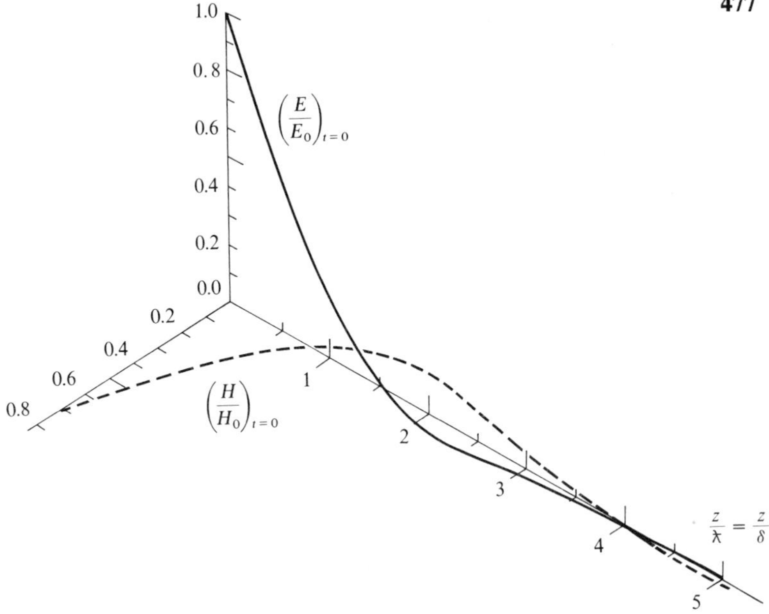

Figure 11-7. The ratios

$$E/E_o = e^{-z/\delta} \cos \{\omega t - (z/\delta)\}$$

and

$$H/H_o = e^{-z/\delta} \cos \{\omega t - (z/\delta) - (\pi/4)\}$$

at $t = 0$ as functions of $z/\lambda\!\!\!^- = z/\delta$ for an electromagnetic wave propagating in the positive direction of the z-axis in a good conductor.

The amplitude of the wave is attenuated by a factor of $1/e = 0.368$ in one radian length $\lambda\!\!\!^-$, and by a factor of $(1/e)^{2\pi} \approx 2 \times 10^{-3}$ in one wavelength λ, whereas the Poynting vector is attenuated by $(1/e)^2 = 0.135$ in $\lambda\!\!\!^-$ and by $(1/e)^{4\pi} \approx 4 \times 10^{-6}$ in λ. The attenuation is so rapid that the wave is barely discernible (Figure 11-7).

The attenuation distance δ in conductors is called the *skin depth*, or the *depth of penetration*. The skin depth *de*creases if either the conductivity σ, the relative permeability μ_r, or the frequency f *in*creases. Good conductors are therefore always opaque to light, except in the form of extremely thin films. It does *not* follow, however, that substances that are nonconducting at low frequencies are necessarily transparent at optical frequencies.

Table 11-1 shows the skin depth δ for various conductors at four typical frequencies. Note that the attenuation in iron is much larger than in silver, despite the fact that iron is a relatively poor conductor.

Table 11-1. *Skin Depths δ for Conductors*

Conductor	Conductivity σ (mho/meter)	Relative Permeability μ_r*	$\delta f^{1/2}$ (meters/second$^{1/2}$)	Skin Depth δ: 60 Hertz (centimeters)	1 Kilohertz (millimeters)	1 Megahertz (millimeters)	3 Gigahertz (microns)
Aluminum	3.54×10^7	1.00	0.085	1.1	2.7	0.085	1.6
Brass (65.8 Cu, 34.2 Zn)	1.59×10^7	1.00	0.126	1.63	3.98	0.126	2.30
Chromium	3.8×10^7	1.00	0.081	1.0	2.6	0.081	1.5
Copper	5.80×10^7	1.00	0.066	0.85	2.1	0.066	1.2
Gold	4.50×10^7	1.00	0.075	0.97	2.38	0.075	1.4
Graphite	1.0×10^5	1.00	1.59	20.5	50.3	1.59	29.0
Magnetic iron	1.0×10^7	2×10^2	0.011	0.14	0.35	0.011	0.20
Mumetal (75 Ni, 2 Cr, 5 Cu, 18 Fe)	0.16×10^7	2×10^4	0.0029	0.037	0.092	0.0029	0.053
Nickel	1.3×10^7	1×10^2	0.014	0.18	4.4	0.014	0.26
Sea water	≈ 5.0	1.00	2×10^2	3×10^3	7×10^3	2×10^2	†
Silver	6.15×10^7	1.00	0.064	0.83	2.03	0.064	1.2
Tin	0.870×10^7	1.00	0.171	2.21	5.41	0.171	3.12
Zinc	1.86×10^7	1.00	0.117	1.51	3.70	0.117	2.14

SOURCE: Adapted from the *American Institute of Physics Handbook* (McGraw-Hill, New York, 1963).

* At $B = 0.002$ tesla.

† At this frequency Q is about 2, and sea water is not a "good" conductor ($\epsilon_r \approx 70$).

The phase velocity

$$u = \frac{\omega}{k_r} = \omega \lambda\!\!\!\bar{} = \left(\frac{2\omega}{\sigma\mu}\right)^{1/2} \tag{11-112}$$

is proportional to the square root of the frequency.

The ratio of the electric to the magnetic energy density is

$$\frac{\frac{1}{2}\epsilon E_0^2}{\frac{1}{2}\mu H_0^2} = \frac{\omega\epsilon}{\sigma} = \mathcal{Q} \leq \frac{1}{50}, \tag{11-113}$$

and the energy is essentially all in the magnetic form. This results from the large conductivity σ, which causes E/J_f to be small. The electric field intensity is weak, but the current density, and hence H, are relatively large.

From Eqs. 11-90 and 11-106, the average value of the Poynting vector is

$$\mathbf{S}_{av} = \frac{1}{2}\,\mathrm{Re}\,(\boldsymbol{E} \times \boldsymbol{H}^*), \tag{11-114}$$

$$= \frac{1}{2}\left(\frac{\sigma}{2\omega\mu}\right)^{1/2} e^{-2z/\delta} E_0^2\, \boldsymbol{k}. \tag{11-115}$$

Example

PROPAGATION IN COPPER AT 1 MEGAHERTZ

Copper has a conductivity σ of 5.80×10^7 mhos/meter. Then, at 1 megahertz.

$$\mathcal{Q} = \frac{2\pi \times 10^6 \times 8.85 \times 10^{-12}}{5.80 \times 10^7} \approx 10^{-12}. \tag{11-116}$$

The requirement that $\mathcal{Q}$ be less than $\frac{1}{50}$ is indeed well met!

The skin depth, which is also the radian length, is

$$\delta = \lambda\!\!\!\bar{} = \left(\frac{2}{2\pi \times 10^6 \times 5.8 \times 10^7 \times 4\pi \times 10^{-7}}\right)^{1/2}$$

$$= 6.6 \times 10^{-5} \text{ meter}, \tag{11-117}$$

$$= 66 \text{ microns}. \tag{11-118}$$

The wavelength $2\pi\lambda\!\!\!\bar{}$ at 1 megahertz is about 0.4 millimeter in copper, while it is 300 meters in air. In fact, the ratio of the wavelengths in copper and in air is

$$\frac{\lambda}{\lambda_0} = 2\pi \left(\frac{2}{\omega\sigma\mu_0}\right)^{1/2} f(\epsilon_0\mu_0)^{1/2} = 1.4 \times 10^{-9} f^{1/2}. \tag{11-119}$$

The phase velocity is correspondingly low:

$$u = \omega\lambda\!\!\!\bar{} = 2\pi \times 10^6 \times 6.6 \times 10^{-5} = 4.1 \times 10^2 \text{ meters/second}. \tag{11-120}$$

This is about ten times *smaller* than the velocity of *sound* in copper, which is 3.6×10^3 meters/second. Also,

$$\left|\frac{E}{H}\right| = \left(\frac{2\pi \times 10^6 \times 4\pi \times 10^{-7}}{5.8 \times 10^7}\right)^{1/2} = 3.7 \times 10^{-4} \text{ ohm}, \quad (11\text{-}121)$$

as against 120π, or 377 in free space (Eq. 11-16).

Example

JOULE LOSSES IN GOOD CONDUCTORS

Let us compare the energy lost by the wave and that gained by the medium through Joule losses. We consider a thin sheet of conductor, perpendicular to the direction of propagation, ab meters² in area, and of thickness Δz, as in Figure 11-8. If the amplitude of E on the left-hand face is E_0 then, on the right-hand face, it is

$$E_0 \, e^{-\Delta z/\delta},$$

and $\mathcal{S}_{av}$ decreases from

$$\frac{1}{2}\left(\frac{\sigma}{2\omega\mu}\right)^{1/2} E_0^2 \quad \text{to} \quad \frac{1}{2}\left(\frac{\sigma}{2\omega\mu}\right)^{1/2} E_0^2 \, e^{-2\Delta z/\delta}$$

within the sheet. Then the average power lost by the wave in the sheet is

$$\frac{ab}{2}\left(\frac{\sigma}{2\omega\mu}\right)^{1/2} E_0^2 \, (1 - e^{-2\Delta z/\delta}) \approx \frac{ab}{2}\left(\frac{\sigma}{2\omega\mu}\right)^{1/2} E_0^2 \, 2\frac{\Delta z}{\delta}, \quad (11\text{-}122)$$

$$\approx \frac{1}{2} ab\sigma \, \Delta z \, E_0^2 \text{ watts} \quad (11\text{-}123)$$

if Δz is small compared to δ. We could also have arrived at this result by calculating $(d\mathcal{S}_{av}/dz)\,\Delta z$.

Now the space average of the peak voltage across the sheet is

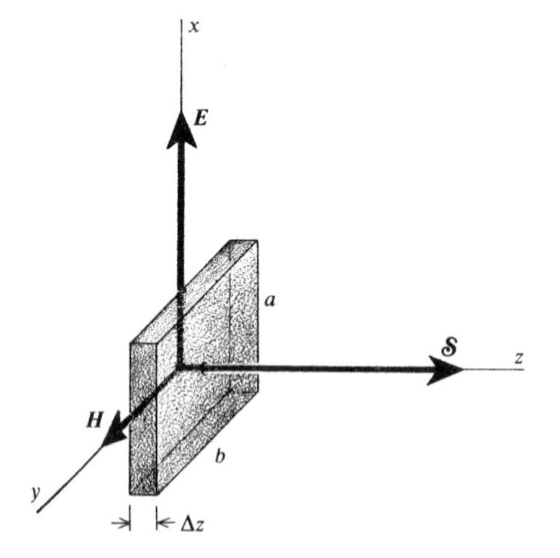

Figure 11-8. Element of volume ab meter² in area and Δz meters thick, normal to the direction of propagation in a good conductor.

$$\frac{1}{2} E_0(1 + e^{-\Delta z/\delta})a \approx E_0\left(1 - \frac{\Delta z}{2\delta}\right) a \approx E_0 a \qquad (11\text{-}124)$$

and the resistance in the direction of current flow is $a/(\sigma b\ \Delta z)$. Then the Joule losses must dissipate

$$\frac{1}{2} \frac{E_0^2 a^2}{a/\sigma b\ \Delta z} = \frac{1}{2} ab\sigma\ \Delta z\ E_0^2 \text{ watts.} \qquad (11\text{-}125)$$

The factor of $\frac{1}{2}$ is required because $E_0 a$ is the peak voltage across the sheet, and not the rms voltage.

The energy lost by the wave is therefore equal to the heat energy generated in the medium, as we could have expected.

11.6 PROPAGATION OF PLANE ELECTROMAGNETIC WAVES IN LOW-PRESSURE IONIZED GASES

We shall use the general results of Section 11.2, but first we must find the conductivity σ of an ionized gas.

We shall assume that the gas pressure is low, so that collisions and energy losses can be neglected. Under those conditions the conduction process in ionized gases is very different from that in metals, because conduction electrons in metals suffer large numbers of collisions with the crystal lattice.

We also neglect thermal agitation. This is equivalent to setting the temperature T equal to zero.

Let us consider a plane electromagnetic wave traveling in the positive direction along the z-axis with the $\boldsymbol{E}$ and $\boldsymbol{H}$ vectors respectively parallel to the x- and the y-axes. An ion of charge Q, mass m, velocity $\boldsymbol{u}$, situated at x, y, z, is subjected to a Lorentz force as in Sections 6.1 and 7.3:

$$\boldsymbol{f} = Q\ \{\boldsymbol{E} + (\boldsymbol{u} \times \boldsymbol{B})\}, \qquad (11\text{-}126)$$

$$= Q\ \left\{E\boldsymbol{i} - B\frac{dz}{dt}\boldsymbol{i} + B\frac{dx}{dt}\boldsymbol{k}\right\}. \qquad (11\text{-}127)$$

Then

$$\frac{d^2x}{dt^2} = \frac{Q}{m}\left(E - B\frac{dz}{dt}\right), \qquad (11\text{-}128)$$

$$= Q\frac{E}{m}\left(1 - \frac{1}{c}\frac{dz}{dt}\right), \qquad (11\text{-}129)$$

where we have used the free-space relationship $E/B = c$ (Eq. 11-17). As we shall see later on in Section 11.6.3, the ratio E/B is even larger in ionized gases so that we are overestimating the magnetic force. Similarly,

$$\frac{d^2y}{dt^2} = 0, \tag{11-130}$$

$$\frac{d^2z}{dt^2} = Q\frac{E}{mc}\frac{dx}{dt}. \tag{11-131}$$

We neglect thermal agitation because random velocities produce zero net current.

Let us assume that the velocity of the ion along the z-axis is much smaller than the velocity of light:

$$\frac{1}{c}\left|\frac{dz}{dt}\right| \ll 1. \tag{11-132}$$

We shall justify this assumption later on. Then, in the direction of the x-axis, the magnetic force is negligible compared to the electric force and

$$\frac{d^2x}{dt^2} = \frac{Q}{m}E = \frac{Q}{m}E_0 \cos \omega t. \tag{11-133}$$

We must *not* use the exponential notation here, because we shall have to solve Eq. 11-131, which involves the product of the two variables E and dx/dt (see Appendix D).

Thus

$$x = -\frac{1}{\omega^2}\frac{Q}{m}E_0 \cos \omega t. \tag{11-134}$$

We have neglected the constants of integration, which are related to the mean position and velocity of the charge during one cycle. Also,

$$\frac{dx}{dt} = \frac{1}{\omega}\frac{Q}{m}E_0 \sin \omega t. \tag{11-135}$$

Substituting in Eq. 11-131,

$$\frac{d^2z}{dt^2} = \frac{Q^2E_0^2}{2\omega m^2c}\sin 2\omega t, \tag{11-136}$$

$$\frac{dz}{dt} = -\frac{Q^2E_0^2}{4\omega^2m^2c}\cos 2\omega t, \tag{11-137}$$

$$z = -\frac{Q^2E_0^2}{8\omega^3m^2c}\sin 2\omega t. \tag{11-138}$$

Before proceeding further, let us find the conditions under which the inequality expressed in Eq. 11-132 is satisfied. We expect that the ion velocity will remain low, either if the field is not intense, or if the frequency is so high that the time available for acceleration is small. Let us calculate

$$\frac{1}{c}\left(\frac{dz}{dt}\right)_{\max} = \frac{Q^2E_0^2}{4\omega^2m^2c^2}. \tag{11-139}$$

For an electron, $Q = 1.60 \times 10^{-19}$ coulomb, $m = 9.11 \times 10^{-31}$ kilogram, and

$$\frac{1}{c}\left(\frac{dz}{dt}\right)_{max} = \frac{2.17 \times 10^{3}}{f^{2}} E_0^2. \qquad (11\text{-}140)$$

Let us assume that the average Poynting vector is calculated as in free space (Eq. 11-26):

$$\mathcal{S}_{av} = 2.66 \times 10^{-3} \frac{E_0^2}{2}. \qquad (11\text{-}141)$$

We shall see later on in Section 11.6.3 that this is correct only at high frequencies. Then

$$\frac{1}{c}\left(\frac{dz}{dt}\right)_{max} = 1.63 \times 10^{6} \frac{\mathcal{S}_{av}}{f^{2}}. \qquad (11\text{-}142)$$

The velocity dz/dt is proportional to the average flux of energy $\mathcal{S}_{av}$, and inversely proportional to the square of the frequency. This seems to indicate that it is possible for $(dz/dt)_{max}$ to exceed the velocity of light c. This is because we have neglected relativistic effects in our calculation. The ratio $(dz/dt)_{max}/c$ is usually much smalller than unity in practice.

Examples

At one kilometer from an antenna radiating 50 kilowatts of power isotropically at a frequency of one megahertz the ratio is about 10^{-8}.

Much higher power densities with $\mathcal{S}_{av} \approx 10^{8}$ watts/meter^2 are common in wave guides, but the frequencies are then of the order of several gigahertz and $(dz/dt)_{max}/c \approx 10^{-5}$.

Laser beams can carry power densities of the order of 10^{16} watts/meter^2 at $f \approx 10^{15}$ hertz, and then the ratio is about 10^{-8}.

Our assumption of Eq. 11-132 is therefore well justified for electrons; it is even better justified for ions, since $(dz/dt)/c$ is proportional to $1/m^2$.

In the case of the antenna, the ratio z_{max}/x_{max}, which is equal to $QE_0/8\omega mc$, is about 2×10^{-5}, x_{max} is about 8 millimeters, z_{max} about 1.6×10^{-7} meter, and dz/dt is only 2 meters/second. This velocity corresponds to an energy of about 10^{-11} electron-volt, while E_0 is 1.7 volts/meter.

The ion motion is thus due almost exclusively to the electric field, and the space charge density in a plane wave remains unaffected by the ion motion, since ions of a given type move as a group in a plane perpendicular to the direction of propagation.

11.6.1 The Conductivity of an Ionized Gas

Since the ion drift velocity can be ascribed almost exclusively to the electric field intensity E, we can consider the ionized gas as having a conductivity σ such that

Now

$$\sigma E = J_f. \tag{11-143}$$

$$J_f = \sum_i N_i Q_i \left(\frac{dx}{dt}\right)_i, \tag{11-144}$$

where N_i is the number of ions or electrons of a given type per cubic meter, Q_i is the charge per ion or electron in coulombs, and $(dx/dt)_i$ is the drift velocity along the x-axis, which we have found above. The quantity J_f is the *convection current density*. Then

$$\sigma E_0 \exp j\omega t = \sum_i \frac{N_i Q_i^2}{\omega m_i} E_0 \exp j\left(\omega t - \frac{\pi}{2}\right) \tag{11-145}$$

and

$$\sigma = -\frac{j}{\omega}\sum_i \frac{N_i Q_i^2}{m_i}. \tag{11-146}$$

Since the masses of the ions are larger than the electron mass by several orders of magnitude, whereas their charges are at most a few times larger than that of the electron, we can retain only the term corresponding to the electrons and

$$\sigma = -j\frac{N_e Q_e^2}{\omega m_e} = -j\,4.47 \times 10^{-9}\frac{N_e}{f} \text{ mho/meter}, \tag{11-147}$$

where N_e is the number of free electrons per cubic meter.

Let us now try to understand the meaning of this imaginary conductivity. Imagine a cube of some resistive material such as carbon, with copper electrodes covering two opposite faces. If it has a volume of one cubic meter, the resistance R' between the electrodes is

$$\frac{\mathcal{V}}{I} = R' = \rho\frac{l}{A} = \rho = \frac{1}{\sigma} \tag{11-148}$$

where $\mathcal{V}$ is the applied voltage, I is the current, ρ is the resistivity, l is the distance between the electrodes (1 meter), A is the cross section (1 meter^2), and σ is the conductivity of the medium. The quantity R' is numerically equal to ρ, but is expressed in ohms.

Thus the conductivity σ of a medium is the conductance $G' = 1/R'$ between two opposite faces of an imaginary cube measuring one meter on the side. More generally, σ is the admittance $Y' = 1/Z'$ between the same electrodes.

In the present case σ is imaginary and negative, so that

$$Z' = \frac{1}{\sigma} = j\frac{\omega m_e}{N_e Q_e^2} = j\omega L' \tag{11-149}$$

where

$$L' = \frac{m_e}{N_e Q_e^2} \tag{11-150}$$

is the equivalent inductance of an imaginary cubic meter of plasma. The quantity L' is expressed in henry-meters.

The electron current *lags* the electric field intensity by $\pi/2$ radians and the electron current is inductive. Since $\boldsymbol{E}$ and $\boldsymbol{J}_f$ are $\pi/2$ radians out of phase, the scalar product $\boldsymbol{E}\cdot\boldsymbol{J}_f$ is also purely imaginary, and there is no energy loss in the medium (Problem 11-22). This means that, on the average, the oscillating electrons do not gain energy from the field once it is established. It will be remembered that we assumed at the beginning that the electrons do not lose energy by collision with the gas molecules.

Since the displacement current $\partial D/\partial t = j\omega\epsilon_0 E$ *leads* the electric field intensity E by $\pi/2$ radians, whereas the electron current *lags* by the same angle, the displacement and electron currents are therefore π radians out phase.

Example

If there is an ionized gas between the plates of a parallel-plate capacitor, the total current density is

$$J_t = \frac{\partial D}{\partial t} + J_f = j\omega\epsilon_0 E - j\frac{N_e Q_e^2}{\omega m_e} E. \tag{11-151}$$

The two components of $\boldsymbol{J}_t$ are shown as functions of the frequency

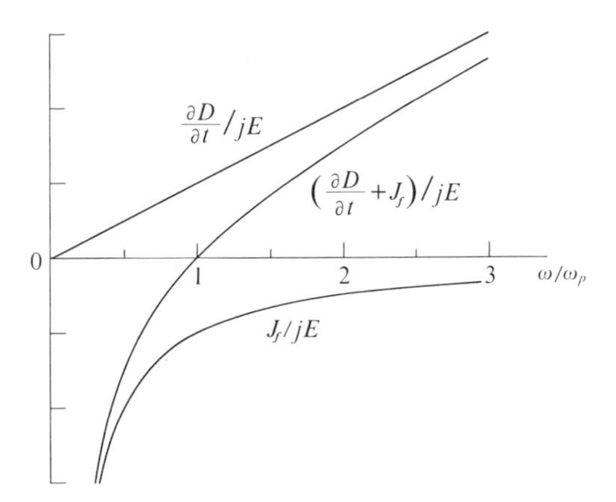

Figure 11-9. The displacement current density $\partial D/\partial t$ and the convection current density J_f in an ionized gas are plotted here as functions of the circular frequency ω. The total current density is inductive below ω_p, zero at ω_p, and capacitive above ω_p.

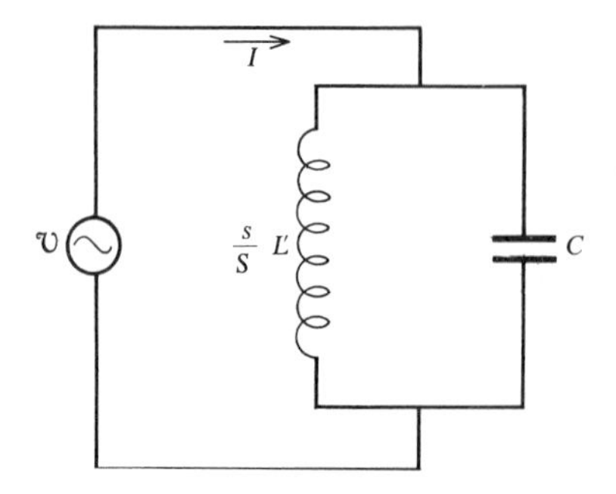

Figure 11-10. If there is an ionized gas between the plates of a parallel-plate capacitor of capacitance C, the impedance is the same as that of a capacitance C in parallel with an inductance $(s/S)L'$, where L' is given by Eq. 11-150.

in Figure 11-9. If S is the area of the plates and s their separation, the total current is

$$I = j\omega\epsilon_0 S(\mathcal{V}/s) + \frac{N_e Q_e^2 S}{j\omega m_e}(\mathcal{V}/s), \tag{11-152}$$

$$= \left(j\omega C + \frac{1}{j\omega L'}\frac{S}{s}\right)\mathcal{V}, \tag{11-153}$$

where $\mathcal{V}$ is the applied voltage. The capacitor therefore behaves as if were connected in parallel with an inductance $L's/S$ as in Figure 11-10.

If an alternating voltage is applied to the circuit, the current through C and that through sL'/S both remain finite but become equal and opposite in sign at the resonant frequency.

11.6.2 The Plasma Angular Frequency ω_p

We have just calculated the total current density in an ionized gas. Let us rewrite J_t in a different form:

$$J_t = j\omega\epsilon_0 E\left(1 - \frac{N_e Q_e^2}{\omega^2 \epsilon_0 m_e}\right), \tag{11-154}$$

$$= j\omega\epsilon_0 E\left(1 - \frac{\omega_p^2}{\omega^2}\right), \tag{11-155}$$

where the second term between the parentheses is the ratio of the convection to the displacement current density and where

$$\omega_p = \left(\frac{N_e Q_e^2}{\epsilon_0 m_e}\right)^{1/2} \tag{11-156}$$

is called the *plasma angular frequency*. This quantity depends solely on the properties of the gas considered. It corresponds to a frequency

$$f_p = 8.98 N_e^{1/2} \text{ hertz}. \tag{11-157}$$

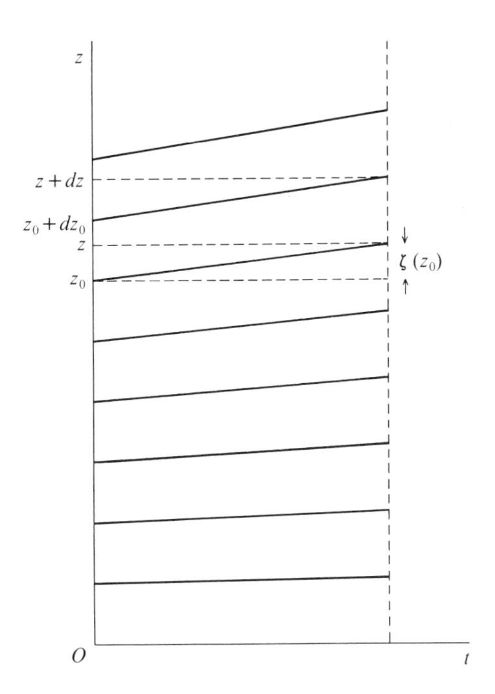

Figure 11-11. The vertical line on the left at $t = 0$ shows schematically the initial uniform distribution of the electrons along the z-axis at positions labeled z_o. The vertical line on the right shows the new distribution for a positive displacement ζ increasing gradually with z_o. The new positions are labeled z. We have assumed for simplicity that ζ increases uniformly with z; however, ζ can be any continuous function of z.

In the plasma of a gas discharge, N_e is typically of the order of 10^{18} electrons/meter^3, and f_p is about 10^4 megahertz, whereas in the ionosphere N_e is of the order of 10^{11} electrons/meter^3 and f_p is about 3 megahertz.

Let us investigate further this frequency, which is characteristic of an ionized gas. The phenomena involved are complex, but it will nevertheless be instructive to use the following simple model. We consider a neutral ionized gas. The ions being much heavier than the electrons, we shall investigate group motions of the electrons in the *absence* of an electromagnetic wave, assuming that the ions remain essentially fixed in position. We neglect thermal agitation. Under these conditions, the charge density due to the ions is uniform throughout the volume considered, and the electron density N_e can vary from point to point.

Let us assume that the electrons move in the direction of the z-axis distances $\zeta(z_0)$ as in Figure 11-11, $\zeta(z_0)$ being a function only of the original value z_0 of the coordinate z. We make no assumption as to the function $\zeta(z_0)$. Then an electron originally at z_0 moves out to

$$z_0 + \zeta(z_0),$$

while an electron at $z_0 + dz_0$ moves out to

$$z_0 + dz_0 + \zeta(z_0 + dz_0),$$

or to

$$z_0 + dz_0 + \zeta(z_0) + \frac{d\zeta}{dz}\,dz_0,$$

where the derivative $d\zeta/dz$ is evaluated at z_0.

In the process, the charge originally occupying an element of volume of thickness dz_0 and area S comes to occupy a volume

$$\left(dz_0 + \frac{d\zeta}{dz}\,dz_0\right) S,$$

and the electron density at z changes from N_e to

$$N_e' = \frac{N_e}{1 + \dfrac{d\zeta}{dz}}. \tag{11-158}$$

Assuming now that the displacement ζ is everywhere small and that it varies smoothly with z, we can set

$$\left|\frac{d\zeta}{dz}\right| \ll 1, \tag{11-159}$$

and then

$$N_e' = N_e\left(1 - \frac{d\zeta}{dz}\right). \tag{11-160}$$

Then the net charge density is that due to the ions, minus that due to the electrons, or

$$\rho_f = \sum_i N_i Q_i - N_e Q_e\left(1 - \frac{d\zeta}{dz}\right) = N_e Q_e \frac{d\zeta}{dz}, \tag{11-161}$$

where Q_e, the magnitude of the electronic charge, is positive, and where N_i is the number of *ions*, either positive or negative, of a given type. The net charge density ρ_f is positive if $d\zeta/dz$ is positive, because the electrons are then spread out.

From the Maxwell equation 10-105, this net charge density gives rise to an electric field in the direction of the z-axis and

$$\frac{dE_z}{dz} = \frac{N_e Q_e}{\epsilon_0}\frac{d\zeta}{dz}. \tag{11-162}$$

Integrating and neglecting uniform fields,

$$E_z = \frac{N_e Q_e}{\epsilon_0}\zeta. \tag{11-163}$$

Then E_z is proportional to ζ, and the equation of motion for an electron of mass m_e is

$$m_e \frac{d^2 z}{dt^2} = -\frac{N_e Q_e^2}{\epsilon_0}\zeta. \tag{11-164}$$

Now $z = z_0 + \zeta$, where z_0 is the initial position and is a constant, so that

$$m_e \frac{d^2\zeta}{dt^2} = -\frac{N_e Q_e^2}{\epsilon_0}\zeta. \tag{11-165}$$

This is the equation for an undamped simple harmonic vibration of angular frequency

$$\omega_p = \left(\frac{N_e Q_e^2}{\epsilon_0 m_e}\right)^{1/2}, \tag{11-166}$$

which is the value of ω_p given in Eq. 11-156. Each electron can therefore execute a simple harmonic vibration about its initial position.

We have therefore calculated the plasma angular frequency ω_p in two different ways. We first considered the motion of the electrons under the action of an applied alternating electric field and found that for one particular value of ω, namely ω_p, the relation between the applied electric field E and the resulting electron motion was such that the total current was equal to zero. We then considered the motion of the electrons under the action of the electric field created by their own nonuniform displacement and found that they could oscillate about their initial position with the same angular frequency ω_p.

We can also calculate ω_p in the following manner. If we assume that the electron density can oscillate at some angular frequency ω_p, the resulting electric field intensity E will oscillate at the same frequency. Then we can use Eq. 11-147 for the conductivity σ of the ionized gas. Now Eq. 10-15 for the free charge density ρ_f assumes only the conservation of charge and Ohm's law $J_f = \sigma E$. Substituting the value of σ and setting $\epsilon_r = 1$,

$$\rho = \rho_0 \exp j \frac{N_e Q_e^2}{\omega_p m_e \epsilon_0} t. \tag{11-167}$$

The imaginary exponent shows that ρ oscillates, as expected, at an angular frequency given by the coefficient of jt. This angular frequency must again be ω_p, and

$$\omega_p = \left(\frac{N_e Q_e^2}{m_e \epsilon_0}\right)^{1/2} \tag{11-168}$$

as previously.

11.6.3 Wave Propagation at High Frequencies Where $\omega > \omega_p$

We use the wave number k for a medium of conductivity σ as given by Eq. 11-69, set $\epsilon_r = 1$, $\mu_r = 1$, and substitute the value of σ from Eq. 11-147. Then

$$k = \pm \frac{\left(1 - \dfrac{N_e Q_e^2}{\omega^2 \epsilon_0 m_e}\right)^{1/2}}{\lambda_0}, \tag{11-169}$$

$$= \pm \frac{\left\{1 - \left(\dfrac{\omega_p}{\omega}\right)^2\right\}^{1/2}}{\lambda_0}. \tag{11-170}$$

For a wave propagating in the positive direction of the z-axis and with its $\boldsymbol{E}$ vector parallel to the x-axis,

$$\boldsymbol{E} = E_0 \exp j(\omega t - kz)\, \boldsymbol{i} \tag{11-171}$$

where k is positive. From the general relationship for E/H, Eq. 11-53,

$$H = \frac{k}{\omega\mu} E = \left(\frac{\epsilon_0}{\mu_0}\right)^{1/2} \left\{1 - \left(\frac{\omega_p}{\omega}\right)^2\right\}^{1/2} E_0 \exp j(\omega t - kz). \tag{11-172}$$

When $\omega > \omega_p$, the wave number k is real and the $\boldsymbol{E}$ and $\boldsymbol{H}$ vectors are in phase, the ratio E/H being larger than in free space by a factor of $1/\{1 - (\omega_p/\omega)^2\}^{1/2}$. As ω approaches ω_p, H tends to zero as expected, since the total current density $J_f + (\partial D/\partial t)$ tends to zero. The Poynting vector $\mathbf{S}$ also tends to zero.

The index of refraction is

$$n = \left\{1 - \left(\frac{\omega_p}{\omega}\right)^2\right\}^{1/2}, \tag{11-173}$$

$$= \left\{1 - 80.5\, \frac{N_e}{f^2}\right\}^{1/2}, \tag{11-174}$$

and the phase velocity is

$$u = \frac{c}{\left\{1 - \left(\dfrac{\omega_p}{\omega}\right)^2\right\}^{1/2}}. \tag{11-175}$$

At high frequencies where $\omega > \omega_p$, the *phase* velocity u is therefore *greater* than the velocity of light, the wave number k is real, and there is no attenuation. Figure 11-12 shows both n and $1/n$ as functions of ω/ω_p. Since the phase velocity increases with increasing ion density, waves tend to bend away from regions of high ion density.

For high frequencies where $\omega^2 \gg \omega_p^2$, the transmission is hardly affected by the presence of ionized gas.

Note that the *phase velocity* is the velocity at which a given *phase* is propagated. This is *not* the velocity at which a *signal* can be transmitted. The reason for this is that a signal can be transmitted only if the wave is modulated in some way, for example by varying its amplitude at an acoustic

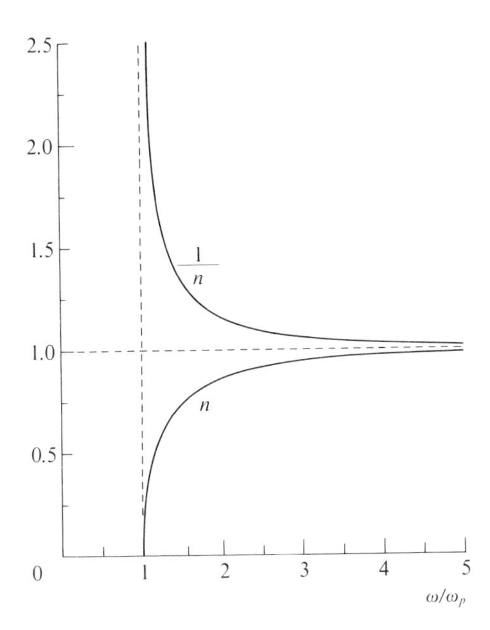

Figure 11-12. The index of refraction n for an ionized gas and its inverse $1/n$ as functions of the ratio ω/ω_p.

frequency, as in amplitude-modulation radio broadcasting. Since any modulation involves frequencies other than the carrier frequency, a signal necessarily involves more than one frequency.

A single frequency corresponds to a pure sine wave extending from $t = -\infty$ to $t = +\infty$. Clearly, such a wave can transmit no information. Since the phase velocity in an ionized gas is frequency dependent, the various frequency components of a signal travel at different velocities. The result is that a signal travels at a velocity that is different from those of its component waves and that is always less than the velocity of light in a vacuum.* It was shown in Section 5.10 that this general law is required to preserve the principle of causality.

Since the component waves travel at different velocities, the shape of the signal, that is, the envelope of the wave, changes with time as the waves progress through the dispersive medium.

11.6.4 Wave Propagation at Low Frequencies Where $\omega < \omega_p$

For $\omega < \omega_p$, the wave number k of Eq. 11-170 is imaginary. The vectors $\boldsymbol{E}$ and $\boldsymbol{H}$ are then out of phase by $\pi/2$ radian.

---

* Léon Brillouin, *Wave Propagation and Group Velocity* (Academic Press, New York, 1960).

The ratio $|E/H|$ is smaller than in free space for $\omega < (\omega_p/2^{1/2})$ and larger at higher frequencies, from Eq. 11-172.

The average Poynting vector $\mathbf{S}_{av} = (1/2)\mathrm{Re}(\boldsymbol{E} \times \boldsymbol{H}^*)$ is zero, and there is no energy transmission.

The index of refraction and the phase velocity are also imaginary, and

$$\boldsymbol{E} = \boldsymbol{E}_0 \exp(j\omega t - k'z), \tag{11-176}$$

$$\boldsymbol{H} = \boldsymbol{H}_0 \exp(j\omega t - k'z), \tag{11-177}$$

where $k' = jk$ is a real number. There is no wave, since the phases of $\boldsymbol{E}$ and of $\boldsymbol{H}$ are independent of z, and the amplitude decreases exponentially with z.

An electromagnetic wave is therefore either transmitted without attenuation, or not transmitted at all through an ionized gas, depending on the ratio ω/ω_p. High frequencies are transmitted, whereas low frequencies are not. Our theory, however, is much simplified. In particular, we have neglected energy losses, which is justifiable only at low gas pressures.

Example

Telemetry measurements during a Saturn I-b launching showed that there was complete radio blackout at a certain moment at all frequencies up to 10 gigahertz (10^{10} hertz). At that moment, the minimum electron density in the plasma surrounding the rocket was such that

$$80.5 \frac{N_e}{f^2} = 80.5 \frac{N_e}{10^{20}} = 1, \tag{11-178}$$

and the electron density was $\approx 10^{18}$ electrons/meter^3.

Example

THE IONOSPHERE

In the region of the upper atmosphere ranging in altitude from approximately 50 to 1000 kilometers the ionization is sufficient to interfere with the propagation of electromagnetic waves. The ionization is attributed mostly to the ultraviolet radiation of the Sun. On the whole, the ion density first increases with altitude and then decreases, but it shows "ledges" where the ion density varies more slowly with altitude. The ledges are commonly called the D, E, F_1, and F_2 *layers*. These layers are due to the fact that both the nature of the solar radiation and the composition of the atmosphere change with altitude. Both the heights and the intensities of ionization of these layers change with the hour of the day, with the season, with the sunspot cycle, and so on.

It appears that the conductivity σ is due almost exclusively to the electron density, except possibly near the lower limit of the ionosphere.

The free electron density is typically 10^{11}/meter^3, varying from about 10^{10} to 10^{12}/meter^3 from the lowest to the highest layer. Over this range of altitudes the number of molecules per cubic meter varies from about 10^{22} to 10^{15}. The percent ionization therefore increases very rapidly with altitude, but it always remains low. For $N_e = 10^{11}$/meter^3, the plasma frequency $f_p \approx 3$ megahertz.

Frequencies lower than f_p are not transmitted. At radio frequencies, waves are bent back toward the Earth, since the electron density increases with increasing height, and since the phase velocity increases with increasing electron density.

The assumption that there are no collisions between the electrons and the gas atoms or molecules is not satisfactory in the lowest regions of the ionosphere, where the pressure is highest, at frequencies of the order of 1 megahertz or lower.

In the presence of the Earth's magnetic field the ionized gas becomes doubly refracting, with the result that there are two distinct phase velocities, depending on whether the $\boldsymbol{E}$ vector of the wave is parallel or perpendicular to the $\boldsymbol{B}$ of the Earth.

11.7 SUMMARY

We began our discussion of electromagnetic waves by considering the simplest case, namely that of *plane waves propagating in a vacuum.* Then $\epsilon_r = 1$, $\mu_r = 1$, $\sigma = 0$, $\rho_t = 0$ and, from Maxwell's equations,

$$\nabla^2 \boldsymbol{E} + \epsilon_0 \mu_0 \omega^2 \boldsymbol{E} = 0, \tag{11-5}$$

$$\nabla^2 \boldsymbol{H} + \epsilon_0 \mu_0 \omega^2 \boldsymbol{H} = 0. \tag{11-6}$$

These are the wave equations for $\boldsymbol{E}$ and $\boldsymbol{H}$ in free space. They show that the phase velocity of such waves is

$$c = \frac{1}{(\epsilon_0 \mu_0)^{1/2}}. \tag{11-7}$$

Plane electromagnetic waves in free space are transverse. Their $\boldsymbol{E}$ and $\boldsymbol{H}$ vectors are orthogonal and oriented so that $\boldsymbol{E} \times \boldsymbol{H}$ points in the direction of propagation. The magnitudes of $\boldsymbol{E}$ and $\boldsymbol{H}$ are related to give equal electric and magnetic energy densities:

$$\frac{E}{H} = \left(\frac{\mu_0}{\epsilon_0}\right)^{1/2} = 377 \text{ ohms}. \tag{11-16}$$

The quantity

$$\mathbf{S} = \boldsymbol{E} \times \boldsymbol{H} \tag{11-23}$$

is called the *Poynting vector.* It is expressed in watts/meter^2 and, when its normal component is integrated over a closed surface, it gives the electromagnetic power flowing out of the surface. In the case of a plane wave, $\mathbf{S}$ is equal to the energy density multiplied by the phase velocity.

In any *homogeneous, isotropic, linear, and stationary medium*, a plane electromagnetic wave has the following characteristics:

(1) $\rho_f = 0$ (Eq. 11-47);
(2) $\boldsymbol{E}$ and $\boldsymbol{H}$ are transverse (Eqs. 11-46 and 11-49);
(3) $\boldsymbol{E}$ and $\boldsymbol{H}$ are mutually perpendicular;
(4) $E/H = \omega\mu/k$ (Eq. 11-53);
(5) $\boldsymbol{E} \times \boldsymbol{H}$ points in the direction of propagation.
(6) $\boldsymbol{E} \times \boldsymbol{H}$ equals the energy density multiplied by the phase velocity.

In *nonconductors*, the phase velocity is

$$u = \frac{c}{(\epsilon_r \mu_r)^{1/2}} \tag{11-57}$$

and, in nonmagnetic media ($\mu_r = 1$), the index of refraction n is related to the relative permittivity by the relation

$$n = \epsilon_r^{1/2}. \tag{11-59}$$

The vectors $\boldsymbol{E}$ and $\boldsymbol{H}$ are in phase, and the electric and magnetic energy densities are equal (Eq. 11-63).

In *conducting media* the ratio

$$\mathcal{Q} = \frac{\omega\epsilon}{\sigma} \tag{11-70}$$

is the magnitude of the ratio of the displacement current density to the conduction current density. The wave number k is complex (Eq. 11-75), and $\boldsymbol{E}$ leads $\boldsymbol{H}$ (Eq. 11-78). The average value of the Poynting vector is then given by

$$\mathbf{S}_{\text{av}} = \tfrac{1}{2}\text{Re}(\boldsymbol{E} \times \boldsymbol{H}^*). \tag{11-90}$$

Good conductors are defined as media for which

$$\mathcal{Q} \leqq \frac{1}{50}. \tag{11-102}$$

Then

$$k = \frac{1 - j}{\delta}, \tag{11-104}$$

where

$$\delta = \lambda\!\!\!^{-} = \left(\frac{2}{\omega\sigma\mu}\right)^{1/2} \tag{11-105}$$

is the *attenuation distance*. The attenuation is so rapid that the wave is hardly discernible (Figure 11-7). The $\boldsymbol{E}$ vector leads the $\boldsymbol{H}$ vector by $\pi/4$ radian, and

$$\frac{E}{H} = \left(\frac{\omega\mu}{\sigma}\right)^{1/2} e^{j\pi/4}. \tag{11-106}$$

Most of the energy is in the magnetic form.

In a *low-pressure ionized gas* the motion of the ions is not appreciably affected by the magnetic field and the conductivity σ, defined by

$$\sigma E = J_f, \tag{11-143}$$

is imaginary:

$$\sigma = -j\frac{N_e Q_e^2}{\omega m_e}, \tag{11-147}$$

where N_e is the number of electrons per cubic meter, Q_e is the charge of the electron, and m_e is its mass. The total current density is less than in free space:

$$J_t = \frac{\partial D}{\partial t} + J_f = j\omega\epsilon_0\left\{1 - \left(\frac{\omega_p}{\omega}\right)^2\right\}E, \tag{11-155}$$

where

$$\omega_p = \left(\frac{N_e Q_e^2}{\epsilon_0 m_e}\right)^{1/2} \tag{11-156}$$

is the *plasma angular frequency*.

At high frequencies where $\omega^2 \gg \omega_p^2$, the wave is unaffected by the presence of the ionized gas. As the frequency decreases, the attenuation remains equal to zero, and the phase velocity increases to values greater than c until it becomes equal to infinity at $\omega = \omega_p$. At angular frequencies smaller than ω_p, the wave number is imaginary, the field is attenuated exponentially with z, and there is no wave.

PROBLEMS

Note: An asterisk indicates that the problem requires a knowledge of relativity.

*11-1. Show that the wave equations for $\boldsymbol{E}$ and $\boldsymbol{B}$ is free space are invariant under a Lorentz transformation, but not under a Galilean transformation.

11-2. A plane electromagnetic wave of circular frequency ω propagates in free space in the direction of the unit vector $\boldsymbol{n}_1$.

Setting the wave number $\boldsymbol{k} = k\boldsymbol{n}_1$, show that

$$\boldsymbol{k}\cdot\boldsymbol{E} = 0, \qquad \boldsymbol{k}\times\boldsymbol{E} - \omega\mu_0\boldsymbol{H} = 0,$$

$$\boldsymbol{k}\cdot\boldsymbol{H} = 0, \qquad \boldsymbol{k}\times\boldsymbol{H} + \omega\epsilon_0\boldsymbol{E} = 0.$$

11-3. A 3.0-megahertz plane electromagnetic wave propagates in free space, and its peak electric field intensity is 100 millivolts/meter.

Calculate the peak voltage induced in a 1.00 meter^2, 10-turn receiving loop oriented so that its plane contains the normal to a wave front and forms an angle of 30 degrees with the electric vector.

11-4. (a) Calculate the electric field intensity due to the radiation at the surface of the Sun from the following data: power radiated by the Sun, 3.8×10^{26} watts; radius of the Sun, 7.0×10^{8} meters.

(b) What is the electric field intensity due to solar radiation at the surface of the Earth? The average distance between the Sun and the Earth is 1.5×10^{11} meters.

(c) Show that the average solar energy incident on the Earth is 2 calories/centimeter^2-minute.

11-5. A circularly polarized wave results from the superposition of two waves that are (a) of the same frequency and amplitude, (b) plane-polarized in perpendicular directions, and (c) 90° out of phase.

Show that the average value of the Poynting vector for such a wave is the sum of the average values of the Poynting vectors for the two plane-polarized waves.

11-6. (a) Show that, for an arbitrary electromagnetic field in a vacuum,

$$\boldsymbol{\nabla}\cdot(\boldsymbol{E} \times \dot{\boldsymbol{E}} + c^2\boldsymbol{B} \times \dot{\boldsymbol{B}}) = -\frac{\partial}{\partial t}(\boldsymbol{E}\cdot\boldsymbol{\nabla} \times \boldsymbol{E} + c^2\boldsymbol{B}\cdot\boldsymbol{\nabla} \times \boldsymbol{B}),$$

where the dots indicate partial derivatives with respect to the time.

Note that this equation has the form of a conservation law and that the spatial density of the conserved quantity is given by the parenthesis on the right-hand side. This spatial density is expressed in volts^2/meter^3.

The flux of the conserved quantity is given by the parenthesis on the left.

(b) Show that this flux vanishes for a linearly polarized wave.

(c) Show that the flux is nonvanishing for a circularly polarized wave, that it is proportional to the frequency, and that it is in the direction of propagation if the $\boldsymbol{E}$ and $\boldsymbol{B}$ vectors rotate in the positive direction, but contrary to the direction of propagation if $\boldsymbol{E}$ and $\boldsymbol{B}$ rotate in the negative direction.†

11-7. A charged particle travels at a velocity $\boldsymbol{u}$ in the field of a plane electromagnetic wave in free space, the velocity of the particle being parallel to the direction of propagation of the wave.

Show that the net transverse force is in the direction of $\boldsymbol{E}$, but that it tends to zero as the velocity of the particle approaches c.

11-8. Several phenomena currently under investigation involve the motion of the particles of a plasma in the presence of both an electromagnetic wave and a constant magnetic field. Examples are the propagation of electromagnetic waves through the ionosphere or through the Earth's magnetosphere, and the heating of plasmas.

In this problem we shall be concerned with the energy transfer from the wave to a single particle of mass m and charge e.

---

† Daniel M. Lipkin, Jour. Math. Phys. **5**, 696 (1964).

(a) The particle is illuminated with a plane-polarized wave in the presence of a constant and uniform magnetic field $\mathcal{B}$ parallel to the direction of propagation.

Show that, if the angular frequency ω of the wave is equal to the rest-mass cyclotron frequency $e\mathcal{B}/m_0$, and if we neglect both the change of mass with energy and the magnetic field of the wave, then the kinetic energy averaged over one cycle increases indefinitely as the square of the time t:

$$\frac{1}{2} m_0 v^2 = \frac{e^2 E_0^2}{8 m_0} t^2,$$

where E_0 is the maximum value of E. We assume that the particle is at rest at $t = 0$ and that $\omega^2 t^2 \gg 1$.‡

(b) We now consider the more complicated case where the wave is circularly polarized and where the $\boldsymbol{B}$ of the wave is not negligible.

Use the following expressions for the $\boldsymbol{E}$ and $\boldsymbol{B}$ of the wave:

$$\boldsymbol{E} = E_0\{\sin(\omega t - kz)\,\boldsymbol{i} - \cos(\omega t - kz)\,\boldsymbol{j}\},$$

$$\boldsymbol{B} = B_0\{\cos(\omega t - kz)\,\boldsymbol{i} + \sin(\omega t - kz)\,\boldsymbol{j}\}.$$

We shall require these particular expressions under (e) below.

Show that

$$\frac{d\mathcal{E}}{dt} = c\frac{dp_z}{dt},$$

where $\mathcal{E}$ is the total energy of the particle, or $m_0 c^2$ plus its kinetic energy.

This result is valid even at relativistic velocities. It shows that the wave cannot increase the energy of the particle without also increasing the z component of its momentum. This is because the wave has a transverse $\boldsymbol{B}$.‡

*(c) Since the particle has a forward velocity, it feels a frequency which is lower than if it were stationary.

Use the results of the example on page 212 and of Section 5.14 to show that the condition for resonance is then

$$\omega = \frac{e\mathcal{B}}{m} + \frac{1}{\lambda}\frac{dz}{dt},$$

or

$$\frac{e\mathcal{B}}{m} = \omega\left(1 - \frac{\dot{z}}{c}\right),$$

where the mass m is the mass measured in the laboratory system, and where dz/dt is the longitudinal velocity of the particle.

*(d) You can now use the result you obtained under (b) above to show that, if there is resonance at $t = 0$, then resonance will be maintained indefinitely afterward. The frequency of the wave, measured at the source, is assumed to be constant.

This result is most interesting. As the velocity of the particle increases, its mass increases and its cyclotron frequency $e\mathcal{B}/m$ decreases. However, the axial velocity also increases, and the resulting Doppler effect gives a frequency shift which exactly compensates for the increase in mass.

(e) It is quite difficult to find the energy $\mathcal{E}$ as a function of z, but you can do part of the calculation.

First, you can show—if you have not already done so under (b)—that

$$\frac{dp_x}{dt} = e\{(E_0 - v_z B_0)\sin(\omega t - kz) + v_y \mathcal{B}\},$$

$$\frac{dp_y}{dt} = e\{-(E_0 - v_z B_0)\cos(\omega t - kz) - v_x \mathcal{B}\},$$

$$\frac{dp_z}{dt} = eB_0\{v_x \sin(\omega t - kz) - v_y \cos(\omega t - kz)\}.$$

C. S. Roberts and S. J. Buchsbaum have shown† that these three equations lead to the following equation for $\mathcal{E}$:

$$\left(\frac{d\mathcal{E}}{dt}\right)^2 + V(\mathcal{E}) = 0,$$

where

$$V(\mathcal{E}) = -2\frac{\omega^2}{c^2}\frac{E_0^2}{\mathcal{B}^2}\mathcal{E}_0^3\frac{\mathcal{E} - \mathcal{E}_0}{\mathcal{E}^2}$$

for a particle which is initially at rest ($\mathcal{E}_0 = m_0c^2$). Since $\mathcal{E} \geq \mathcal{E}_0$, V is always negative, and $\mathcal{E}$ increases indefinitely.

It is useful to sketch a curve of V as a function of $\mathcal{E}$ to understand qualitatively the behavior of $d\mathcal{E}/dt$.

(f) We have found $d\mathcal{E}/dt$ in (e) above, and dz/dt in (c); you can show, finally, that

$$\frac{\mathcal{E} - \mathcal{E}_0}{\mathcal{E}_0} = \frac{1}{2^{1/3}}\left(3\frac{E_0ez}{\mathcal{E}_0}\right)^{2/3}.$$

The left-hand side is the kinetic energy $(m - m_0)c^2$ expressed as a multiple of the rest energy m_0c^2.

The above theory, or rather a similar one that applies to standing waves, has been checked (see second reference in footnote) with electrons at a frequency of one gigahertz. The predicted electron kinetic energy, which was 5.80×10^5 electron-volts, agreed with the experimental value. The electron energy was deduced from the X-ray spectrum and also from the radius of the electron trajectory.

11-9. A plane elliptically polarized wave results from the superposition of two plane-polarized waves whose $\boldsymbol{E}$ vectors are oriented in perpendicular directions and out of phase.

For example, $\boldsymbol{E}$ can have the two components

$$E_x = E_{x0}\exp j(\omega t - kz),$$

$$E_y = E_{y0}\exp j(\omega t - kz + \varphi),$$

where E_{x0} and E_{y0} are real.

---

† C. S. Roberts and S. J. Buchsbaum, Phys. Rev. **135**, A383 (1964). See also H. R. Jory and A. W. Trivelpiece, Jour. Appl. Phys. **39**, 3053 (1968).

Show that the $\boldsymbol{E}$ and $\boldsymbol{H}$ vectors for an elliptically polarized wave are orthogonal if $\varphi = 0$ *or* if the wave impedance E/H is real.

11-10. Show that $\mathcal{S}_{av}$ for a plane wave is equal to the average energy density multiplied by the phase velocity, in any homogeneous, isotropic, linear, and stationary medium.

11-11. In the general case of a uniform plane wave in a nonconductor,

$$\boldsymbol{E} = (E_{0x}\boldsymbol{i} + E_{0y}\boldsymbol{j} + E_{0z}\boldsymbol{k}) \exp j(\omega t - k_x x - k_y y - k_z z),$$

where the coefficients E_{0x}, E_{0y}, E_{0z}, k_x, k_y, k_z can be complex, but are independent of the coordinates x, y, z and of the time t.

The vector $\boldsymbol{H}$ is given by a similar expression with the same values of k_x, k_y, k_z.

(a) Show that

$$k_x^2 + k_y^2 + k_z^2 = k^2$$

where k is the wave number corresponding to the medium and the frequency under consideration.

(b) Discuss the relative orientations of $\boldsymbol{E}$, $\boldsymbol{H}$, and $\boldsymbol{k}$.

11-12. Discuss the propagation of a plane electromagnetic wave in a medium where (a) $\mathcal{Q} = 1$, (b) $\mathcal{Q}^2 \gg 1$.

11-13. The quantity

$$\epsilon - j\frac{\sigma}{\omega}$$

is called the *complex permittivity* of a medium. Justify the use of this term.

11-14. It was shown in the example on page 424 that the charge density ρ in a conductor decreases exponentially with a relaxation time of ϵ/σ:

$$\rho = \rho_0 \exp(-\sigma t/\epsilon)$$

Show that $\mathcal{Q}$ must be at most about $\frac{1}{3}$ if ρ/ρ_0 is to be less than 1% within one fourth of a period.

11-15. Using the values of $\boldsymbol{E}$ and of $\boldsymbol{H}$ that we have found for plane electromagnetic waves in good conductors, verify that all of Maxwell's equations apply.

11-16. A plane electromagnetic wave propagates in a good conductor in the positive direction of the z-axis.

Calculate the total power lost per square meter by Joule heating between $z = 0$ and $z \to \infty$, and show that this is equal to $\mathcal{S}_{av}$ at $z = 0$.

11-17. (a) Show that the real part n_r of the index of refraction n, the attenuation distance δ and the $\mathcal{Q}$ of a good conductor are related by the following equations:

$$\mathcal{Q}\delta^2 = \frac{2\epsilon}{\mu\sigma^2},$$

$$\frac{\mathcal{Q}}{\delta^2} = \frac{\epsilon_r\mu_r}{2\lambda_0^2},$$

$$\mathcal{Q}n_r^2 = \frac{\epsilon_r\mu_r}{2}.$$

(b) Show that the minimum value of n_r is 5 for "good" conductors.

11-18. Compare the ratios $|E/H|$ for an electromagnetic wave in air and in sea water at 20 kilohertz and at 20 megahertz.

11-19. Calculate the attenuation of an electromagnetic wave in sea water and in copper at 20 kilohertz and at 20 megahertz. State your result in decibels/meter. (See Problem 13-25 for a definition of the decibel.)

11-20. A conducting body can be heated by immersing it in an alternating magnetic field.

Consider the simple case of a plane conducting surface exposed to a tangential magnetic field intensity $H_0 \exp j\omega t$.

Show that a plane electromagnetic wave penetrates perpendicularly into the conductor and that the energy dissipated per square meter is

$$\frac{1}{2}\left(\frac{\omega\mu}{2\sigma}\right)^{1/2} H_0^2,$$

if the conductor is much thicker than the skin depth δ.

Note that this power dissipation is proportional to the square root of the frequency and that it occurs mostly within one skin depth of the surface.

11-21. It is interesting to draw a parallel between the flow of heat in a thermally conducting medium and the propagation of an electric or magnetic field in an electrically conducting medium.

Letting $\boldsymbol{\Phi}$ be the heat flux density in watts/meter^2,

$$\boldsymbol{\Phi} = -\lambda \boldsymbol{\nabla} T,$$

where λ is the thermal conductivity in watts/meter-kelvin and T is the temperature.

Then, for conservation of energy,

$$\boldsymbol{\nabla}\cdot\boldsymbol{\Phi} = -\rho c \frac{\partial T}{\partial t} + Q,$$

where ρ is the density in kilograms/meter^3, c is the specific heat in joules/kilogram-kelvin, and Q is the heat produced within the medium in watts/meter^3.

For $Q = 0$,

$$\boldsymbol{\nabla}^2 T - \frac{\rho c}{\lambda}\frac{\partial T}{\partial t} = 0.$$

This equation is identical in form to that for an electromagnetic wave in a good conductor, $\rho c/\lambda$ corresponding to $\mu\sigma$. Its solution for heat flow in one dimension is entirely similar to Eq. 11-107:

$$T = T_0 \exp\left\{j\left(\omega t - \frac{z}{\delta_t}\right) - \frac{z}{\delta_t}\right\},$$

where

$$\delta_t = \left(\frac{2\lambda}{\omega\rho c}\right)^{1/2}.$$

Compare the velocities of propagation of T and of B in copper and in iron.

Property	*Copper*	*Iron*	*Units*
σ	5.8×10^7	1.0×10^7	mhos/meter
c	0.092	0.11	calories/gram-kelvin
ρ	8.9	7.9	grams/centimeter^3
λ	1.0	0.15	calories/second-centimeter-kelvin
μ_r	1.0	200	

11-22. Show that there is zero energy dissipation in a medium when $\boldsymbol{E}$ and $\boldsymbol{J}_f$ are out of phase by $\pi/2$ radians.

11-23. (a) A mass m slides on a frictionless horizontal surface under the action of a spring of stiffness (stretching force/elongation) k fixed to a rigid support, and of a force $F = F_0 \cos \omega t$ in line with the spring.

Show that this mechanical system is mathematically equivalent to a series-resonant electric circuit.

Discuss the amplitudes of the displacement and of the velocity as functions of ω.

(b) A mass m slides on a horizontal frictionless surface under the action of a force $F = F_0 \cos \omega t$, which is applied to it through a spring of stiffness k.

Show that this mechanical system is mathematically equivalent to a parallel-resonant circuit.

Discuss the amplitudes and the velocities of the two ends of the spring as functions of ω.

11-24. Show that the conductivity of an ionized gas is

$$\frac{\epsilon_0 \omega_p^2}{j\omega}.$$

11-25. Two plane electromagnetic waves of equal amplitude propagate in the ionosphere where the electron density is N_e electrons/meter^3. One wave has a circular frequency ω_1 and a corresponding wavelength λ_1; the other has a slightly different circular frequency ω_2 and a wavelength λ_2.

(a) At a given time t there exist values of z for which the two waves are in phase and values of z for which they are opposite in phase. What is the distance between the maxima?

(b) What is their velocity?

This velocity is called the *group velocity* u_g.

(c) Show that, in the limit,

$$u_g = \frac{1}{\dfrac{dk}{d\omega}}.$$

(d) Show that

$$u_p u_g = c^2,$$

where u_p is the phase velocity given in Eq. 11-175.

(e) Calculate the phase velocities and the group velocity for $f_1 = 5.3$ megahertz, $f_2 = 5.4$ megahertz, and $N_e = 5 \times 10^{10}$ electrons/meter^3.

(f) Calculate the distance and the number of waves between two minima.

11-26. *Pulsars*, or pulsating radio stars, emit sharp bursts of radio energy about 5 to 50 milliseconds wide at intervals of about one second. For any given pulsar the repetition frequency is stable within about one part in 10^8. The amplitude and shape of the pulses vary widely, but each pulsar has its own characteristic mean pulse profile.

Within a few months after the discovery of pulsars, distance estimates were obtained in the following manner. It was observed that the arrival time of a pulse depends on the frequency of observation, the arrival time being later at lower frequencies. This delay is attributed to dispersion in the interstellar medium which is ionized hydrogen with an electron density $N_e \approx 10^5$ per cubic meter.

(a) Show that, if $\omega^2 \gg \omega_p$, a plot of the time delay Δt as a function of

$$\frac{1}{f^2} - \frac{1}{(f + \Delta f)^2}$$

is a straight line whose slope is a measure of the distance to the pulsar.
(b) In the case of pulsar CP 0328, arrival times measured at 151, 408, and 610 megahertz gave the results shown below.

f Megahertz	Δt Seconds
151	
	4.18
408	
	0.367
610	

Show that, according to these measurements, the distance to CP 0328 is 268 parsec. (The *parsec* is 3.086×10^{16} meters. It is the distance from which the radius of the Earth's orbit, 1.495×10^{11} meters, would subtend an angle of one second).

The fact that such plots give straight lines passing through the origin indicates that the assumption $\omega^2 \gg \omega_p^2$ is correct. The delay therefore occurs over large distances in a low-density plasma such as that of interstellar space, and not inside the pulsar itself, which is quite small.

11-27. The *solar wind* is a high-conductivity plasma which is emitted radially from the surface of the Sun. Let us calculate the flux of electromagnetic energy in the solar wind at the orbit of the Earth.

In the plane of the Earth's orbit, the magnetic field of the Sun is approximately radial, pointing outwards in certain regions and inwards in others. This field is "frozen" in the high-conductivity plasma. Since the Sun rotates (with a period of 27 days), and the plasma has a radial velocity, the lines of $\boldsymbol{B}$ are in fact Archimedes spirals and, at the Earth, they form an angle of about 45° with the Sun-Earth direction. This is the so-called *garden-hose* effect.

At the orbit of the Earth the solar wind has a density of about 10^7 proton masses/meter^3 and a velocity of about 4×10^5 meters/second, while the magnetic field of the Sun is about 5×10^{-9} tesla.

(a) First show that, in an electrically neutral ($\rho_t = 0$) and nonmagnetic fluid of conductivity σ and velocity $\boldsymbol{v}$, Maxwell's equations become

$$\nabla \cdot \boldsymbol{E} = 0, \qquad \nabla \cdot \boldsymbol{B} = 0,$$

$$\nabla \times \boldsymbol{E} = -\frac{\partial \boldsymbol{B}}{\partial t}, \qquad \nabla \times \boldsymbol{B} = \mu_0 \left\{ \sigma(\boldsymbol{E} + \boldsymbol{v} \times \boldsymbol{B}) + \epsilon_0 \frac{\partial \boldsymbol{E}}{\partial t} \right\},$$

the polarization currents being negligibly small compared to the conduction currents.

In these equations, $\boldsymbol{E}$, $\boldsymbol{B}$, and $\boldsymbol{v}$ are measured with respect to a "fixed" frame of reference, and not with respect to the solar wind. Since ρ_t is zero, the current density $\boldsymbol{J}_f$ is the same in both frames, except for a factor γ, which is very close to unity. See Section 5.21. Also, since the positive and negative charges drift at the same velocity, $\boldsymbol{J}_f$ is zero and

$$\boldsymbol{E} = -\boldsymbol{v} \times \boldsymbol{B}.$$

(b) Show that the component of $\boldsymbol{v}$ which is normal to $\boldsymbol{B}$ is

$$v_n = \frac{1}{B^2} \boldsymbol{B} \times (\boldsymbol{v} \times \boldsymbol{B})$$

(c) Show that the Poynting vector for the solar wind is

$$\mathcal{S} = \frac{B^2}{\mu_0} v_n \approx 6 \text{ microwatts/meter}^2.$$

This is about 4×10^{-9} times the average value of the Poynting vector of the solar radiation, which is about 1.4 kilowatts/meter^2.

The Poynting vector of the solar wind is normal to the local $\boldsymbol{B}$ and it points at an angle 45° away from the Sun-Earth direction.

(d) You can show that the kinetic, magnetic, and electric energy densities are related as follows:

$$U_K \gg U_M \gg U_E.$$

Hints

11-8. (a) Remember that

$$\frac{d\boldsymbol{p}}{dt} = m \frac{d\boldsymbol{v}}{dt} = e(\boldsymbol{E} + \boldsymbol{v} \times \boldsymbol{\mathcal{B}})$$

and solve for V_x and V_y.

(b) Since a magnetic field can only deflect a particle without changing the magnitude of its velocity, the gain in energy is due solely to the electric field and

$$\frac{d\mathcal{E}}{dt} = e\boldsymbol{v} \cdot \boldsymbol{E}.$$

CHAPTER 12

PROPAGATION OF ELECTROMAGNETIC WAVES II

Reflection and Refraction

In Chapter 11 we studied the propagation of electromagnetic waves in infinite, continuous media. We shall now examine the effects of a discontinuity in the medium of propagation, as in Figure 12-1.

We shall investigate again the same three types of media as in Chapter 11: dielectrics, good conductors, and low-pressure ionized gases. It will be recalled from the introduction to Chapter 11 that dielectrics are defined as nonconductors and that they may be either magnetic or nonmagnetic; we shall limit ourselves to nonmagnetic dielectrics.

We assume an ideally thin, infinite, plane interface between two linear, homogeneous, isotropic media. Then an incident wave along $\boldsymbol{n}_i$ gives rise to both a reflected wave along $\boldsymbol{n}_r$ and a transmitted wave along $\boldsymbol{n}_t$. The three waves of course satisfy the conditions of continuity for the tangential components of $\boldsymbol{E}$ and of $\boldsymbol{H}$ at the interface. For the time being, we exclude total reflection; this will be discussed later, in Section 12.4. So as to avoid multiple reflections, we assume that the media extend to infinity on both sides of the interface.

There are many cases where a wave incident on a discontinuity is partly reflected and partly transmitted. For example, a sound wave incident upon a wall gives both a reflected wave, which comes back into the room, and a transmitted wave, which proceeds into the wall. A similar phenomenon occurs at a discontinuity in an electrical transmission line, at the junction between two different types of coaxial line for example. Waves on strings show the same type of behavior, as is shown in Section E-4 of Appendix E.

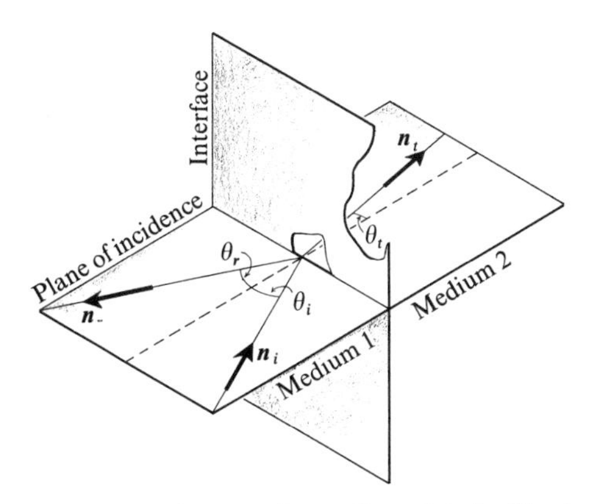

Figure 12-1. An electromagnetic wave in medium 1 is incident on the interface between media 1 and 2 and gives rise to both a reflected and a transmitted wave. The vectors $\boldsymbol{n}_i$, $\boldsymbol{n}_r$, and $\boldsymbol{n}_t$ are unit vectors normal to the respective wave fronts, and point in the direction of propagation. The angles θ_i, θ_r, θ_t are, respectively, the angles of incidence, of reflection, and of refraction. (Radio engineers use a different convention and call the complementary angles $(\pi/2) - \theta_i$, $(\pi/2) - \theta_r$, and $(\pi/2) - \theta_t$ the angles of incidence, reflection, and refraction, respectively.) Total reflection will be treated separately in Section 12.4.

12.1 THE LAWS OF REFLECTION AND SNELL'S LAW OF REFRACTION

If we assume that the electromagnetic wave incident on the interface as in Figure 12-1 is both plane and plane-polarized, then its electric field intensity $\boldsymbol{E}_i$ is of the form

$$\boldsymbol{E}_i = \boldsymbol{E}_{0i} \exp j\omega_i \left(t - \frac{\boldsymbol{n}_i \cdot \boldsymbol{r}}{u_1} \right), \tag{12-1}$$

where u_1 is the phase velocity of the wave in medium 1. (See Appendix E.) The time $t = 0$ and the origin $\boldsymbol{r} = 0$ can be chosen arbitrarily. This equation defines a plane wave for all values of t and for all values of $\boldsymbol{r}$, that is, a wave that extends throughout all time and all space. However, it is used only in medium 1.

Both the reflected and the refracted waves from a plane interface are then also plane and plane-polarized, since the laws of reflection and of refraction for any given incident ray must be the same at all points on the interface. Then the reflected and the transmitted waves are of the form

$$\boldsymbol{E}_r = \boldsymbol{E}_{0r} \exp j\omega_r \left(t - \frac{\boldsymbol{n}_r \cdot \boldsymbol{r}}{u_1} \right), \tag{12-2}$$

$$\boldsymbol{E}_t = \boldsymbol{E}_{0t} \exp j\omega_t \left(t - \frac{\boldsymbol{n}_t \cdot \boldsymbol{r}}{u_2}\right), \tag{12-3}$$

where u_2 is the phase velocity of the wave in medium 2. Note that we have made no assumption whatever as to the amplitudes, phases, frequencies, or directions of the reflected and transmitted waves. The amplitudes $\boldsymbol{E}_{0r}$ and $\boldsymbol{E}_{0t}$ can be complex to take phase differences into account.

We shall choose the origin at some convenient point on the interface.

We shall be able to determine the characteristics of both the reflected and the transmitted waves from the fact that the *tangential* components of both $\boldsymbol{E}$ and $\boldsymbol{H}$ must be continuous across the interface (Sections 4.1.3 and 9.7). In other words, the sum of the tangential components of $\boldsymbol{E}_i$ and $\boldsymbol{E}_r$ just above the interface must be equal to the tangential component of $\boldsymbol{E}_t$ just below the interface; a similar situation holds for $\boldsymbol{H}$. The reflected and the transmitted waves could also be obtained from the continuity of the *normal* components of $\boldsymbol{D}$ and of $\boldsymbol{B}$ across the interface.

To obtain continuity of the tangential components of $\boldsymbol{E}$ and of $\boldsymbol{H}$ at the interface, some valid relation must exist between $\boldsymbol{E}_i$, $\boldsymbol{E}_r$, $\boldsymbol{E}_t$ for all time t and for all points on the interface. Such a relation will be possible if (a) all three vectors $\boldsymbol{E}_i$, $\boldsymbol{E}_r$, $\boldsymbol{E}_t$ are identical functions of the time t, (b) all three vectors are identical functions of position $\boldsymbol{r}_I$ on the interface, and (c) there exist certain relations between $\boldsymbol{E}_{0i}$, $\boldsymbol{E}_{0r}$, $\boldsymbol{E}_{0t}$.

From condition (a),

$$\omega_i = \omega_r = \omega_t. \tag{12-4}$$

All three waves must therefore be of the same frequency. This is intuitively quite obvious, since they are all superpositions of the wave emitted by the source and of those waves emitted by the electrons executing forced vibrations in media 1 and 2. It will be recalled from mechanics that forced vibrations have the same frequency as the applied force.

From condition (b), we must also have, at any point $\boldsymbol{r}_I$ on the interface,

$$\frac{\boldsymbol{n}_i \cdot \boldsymbol{r}_I}{u_1} = \frac{\boldsymbol{n}_r \cdot \boldsymbol{r}_I}{u_1}, \tag{12-5}$$

$$= \frac{\boldsymbol{n}_t \cdot \boldsymbol{r}_I}{u_2}. \tag{12-6}$$

Then from the first of these equations,

$$(\boldsymbol{n}_i - \boldsymbol{n}_r) \cdot \boldsymbol{r}_I = 0. \tag{12-7}$$

Since the vector $\boldsymbol{r}_I$ lies in the interface, the vector $\boldsymbol{n}_i - \boldsymbol{n}_r$ must be normal to the interface and the tangential components of these two vectors must be equal and of the same sign as in Figure 12-2. Then

$$\theta_i = \theta_r, \tag{12-8}$$

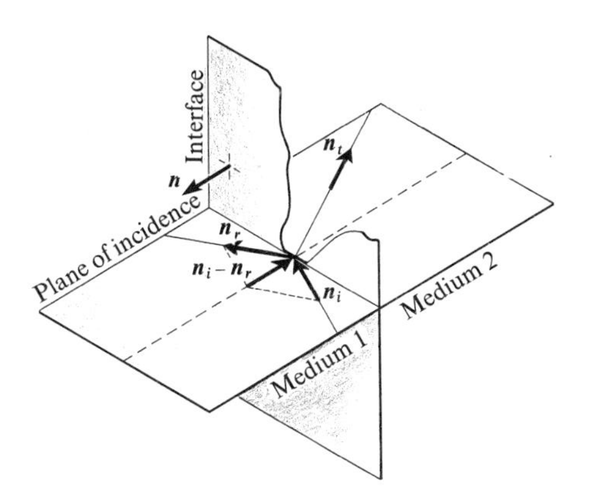

Figure 12-2. The vectors $\boldsymbol{n}_i$, $\boldsymbol{n}_r$, $\boldsymbol{n}_t$, $\boldsymbol{n}$.

and *the angle of reflection is equal to the angle of incidence.* Since $(\boldsymbol{n}_i - \boldsymbol{n}_r)$ is parallel to the normal $\boldsymbol{n}$, *the three vectors* $\boldsymbol{n}_i$, $\boldsymbol{n}_r$, $\boldsymbol{n}$ *are coplanar. These are the laws of reflection.*

The plane of these three vectors is normal to the interface and is called the *plane of incidence.*

Considering now Eq. 12-6,

$$\left(\frac{\boldsymbol{n}_i}{u_1} - \frac{\boldsymbol{n}_t}{u_2}\right) \cdot \boldsymbol{r}_I = 0. \tag{12-9}$$

Hence the vector between parentheses must also be normal to the interface, so that $\boldsymbol{n}_i$, $\boldsymbol{n}_t$, $\boldsymbol{n}$ are coplanar and all four vectors $\boldsymbol{n}_i$, $\boldsymbol{n}_r$, $\boldsymbol{n}_t$, $\boldsymbol{n}$ are in the plane of incidence. Moreover, the tangential components of $\boldsymbol{n}_i/u_1$ and of $\boldsymbol{n}_t/u_2$ must be equal, and

$$\frac{\sin\theta_i}{u_1} = \frac{\sin\theta_t}{u_2} \tag{12-10}$$

or, since the wave number k is ω/u,

$$k_1 \sin\theta_i = k_2 \sin\theta_t. \tag{12-11}$$

The quantity $k \sin\theta$ is therefore conserved in crossing the interface. We can also write that

$$\frac{\sin\theta_t}{\sin\theta_i} = \frac{k_1}{k_2}, \tag{12-12}$$

$$= \frac{n_1}{n_2}, \tag{12-13}$$

since $k = n/\lambda_0$, where n is the index of refraction. This is *Snell's law.*

It is important to note that this law, as well as the laws of reflection, are general. They apply to *any* two media; they even hold true for total reflection, as will be shown later on.

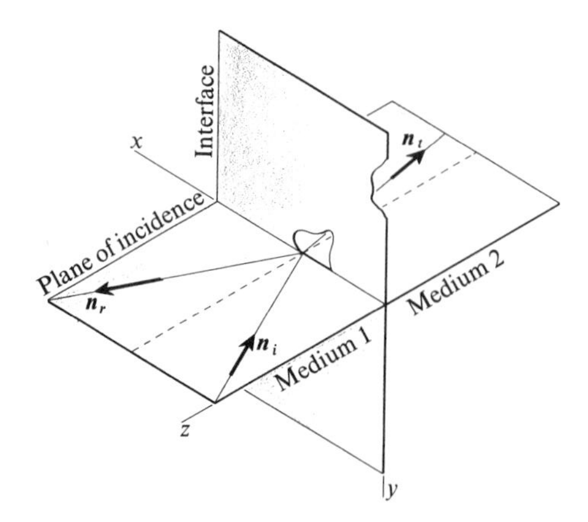

Figure 12-3. Coordinate system used for the study of reflection and refraction.

Thus, choosing the origin in the interface and axes as in Figure 12-3,

$$\boldsymbol{E}_i = \boldsymbol{E}_{0i} \exp j\{\omega t - k_1(x \sin \theta_i - z \cos \theta_i)\}, \tag{12-14}$$

$$\boldsymbol{E}_r = \boldsymbol{E}_{0r} \exp j\{\omega t - k_1(x \sin \theta_i + z \cos \theta_i)\}, \tag{12-15}$$

$$\boldsymbol{E}_t = \boldsymbol{E}_{0t} \exp j\{\omega t - k_2(x \sin \theta_t - z \cos \theta_t)\}. \tag{12-16}$$

12.2 FRESNEL'S EQUATIONS

We shall turn now to the third condition mentioned in the preceding section. We must find the relations between the quantities $\boldsymbol{E}_{0i}$, $\boldsymbol{E}_{0r}$, $\boldsymbol{E}_{0t}$ that will ensure continuity of the tangential components of $\boldsymbol{E}$ and of $\boldsymbol{H}$ at the interface.

We recall from Section 11-2 that the $\boldsymbol{E}$ and $\boldsymbol{H}$ vectors in a plane electromagnetic wave are always perpendicular to the direction of propagation and to each other. The $\boldsymbol{E}$ vector of the incident wave can thus be oriented in any direction perpendicular to the vector $\boldsymbol{n}_i$.

It will be convenient to divide the discussion into two parts. We shall consider successively incident waves polarized with their $\boldsymbol{E}$ vectors normal, and then parallel to the plane of incidence. Any incident wave can be separated into two such components.

12.2.1 Incident Wave Polarized with Its $\boldsymbol{E}$ Vector *Normal* to the Plane of Incidence

The $\boldsymbol{E}$ and the $\boldsymbol{H}$ vectors of the incident wave are oriented as in Figure 12-4. If the media are isotropic, as we assumed at the beginning of this chapter, the

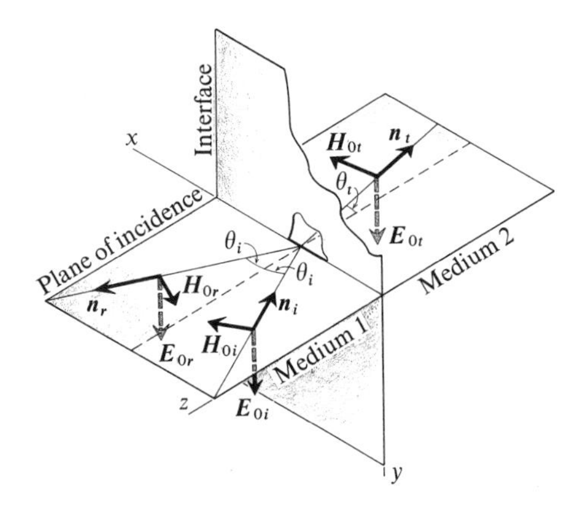

Figure 12-4. The incident, reflected, and transmitted waves when the incident wave is polarized with its **E** vector *normal* to the plane of incidence. The arrows indicate the directions in which the $\mathbf{E}_{0i}$, $\mathbf{H}_{0i}$, $\mathbf{E}_{0r}$, etc., vectors are taken to be positive at the interface. The vectors $\mathbf{E} \times \mathbf{H}$ point everywhere in the direction of propagation.

E vectors of both the reflected and the transmitted waves will also be perpendicular to the plane of incidence, as in the figure.

Considering the electric and magnetic field intensities of the incident wave to be known, we have four unknowns: $\mathbf{E}_{0r}$, $\mathbf{E}_{0t}$, $\mathbf{H}_{0r}$, $\mathbf{H}_{0t}$. We also have four equations that are provided (a) by the continuity of the tangential components of **E** and **H** at the interface and (b) by the relations between **E** and **H** for plane waves in medium 1 and in medium 2 as given in Eq. 11-53. It will therefore suffice to calculate only the **E** vectors.

Instead of using the continuity of the tangential components of **E** and **H**, we could also use the continuity of the normal components of **D** and **B**. This is not desirable, however, because if we did so, and if we wished our results to be applicable to the reflection from the surface of a conductor, we would have to take into account the surface charge density. This would introduce still another unknown.

The continuity of the tangential component of **E** at the interface requires that

$$E_{0i} + E_{0r} = E_{0t} \tag{12-17}$$

at any given time and at any given point on the interface. Similarly, the continuity of the tangential component of **H** requires that

$$H_{0i} \cos \theta_i - H_{0r} \cos \theta_i = H_{0t} \cos \theta_t \tag{12-18}$$

or, from Eq. 11-53,

$$\frac{k_1}{\omega \mu_1} (E_{0i} - E_{0r}) \cos \theta_i = \frac{k_2}{\omega \mu_2} E_{0t} \cos \theta_t . \tag{12-19}$$

Thus, recalling that $k = n/\lambda_0$, where n is the index of refraction and λ_0 is the free-space wavelength divided by 2π,

$$\left(\frac{E_{0r}}{E_{0i}}\right)_N = \frac{\dfrac{n_1}{\mu_{r1}} \cos\theta_i - \dfrac{n_2}{\mu_{r2}} \cos\theta_t}{\dfrac{n_1}{\mu_{r1}} \cos\theta_i + \dfrac{n_2}{\mu_{r2}} \cos\theta_t}, \tag{12-20}$$

$$\left(\frac{E_{0t}}{E_{0i}}\right)_N = \frac{2\dfrac{n_1}{\mu_{r1}} \cos\theta_i}{\dfrac{n_1}{\mu_{r1}} \cos\theta_i + \dfrac{n_2}{\mu_{r2}} \cos\theta_t}, \tag{12-21}$$

where the subscript N indicates that E_{0i} is normal to the plane of incidence.

These are two of *Fresnel's equations;* the other pair will be deduced in the next section. Fresnel's equations give the ratios of the amplitudes of the incident, reflected, and transmitted waves. They apply to *any* two media. We shall find later on that they are valid even for total reflection.

12.2.2 Incident Wave Polarized with Its E Vector *Parallel* to the Plane of Incidence

In this case the $\mathbf{E}$ vectors of all three waves must be in the plane of incidence as in Figure 12-5. We have chosen the orientations of $\mathbf{E}_{0r}$ and of $\mathbf{E}_{0t}$ so that Figures 12-4 and 12-5 become identical at normal incidence, except for a rotation of 90° around the normal to the interface.

We now have

$$H_{0i} - H_{0r} = H_{0t}, \tag{12-22}$$

or

$$\frac{k_1}{\omega\mu_1}(E_{0i} - E_{0r}) = \frac{k_2}{\omega\mu_2} E_{0t}, \tag{12-23}$$

and

$$(E_{0i} + E_{0r}) \cos\theta_i = E_{0t} \cos\theta_t. \tag{12-24}$$

Then, solving for E_{0r} and E_{0t},

$$\left(\frac{E_{0r}}{E_{0i}}\right)_P = \frac{-\dfrac{n_2}{\mu_{r2}} \cos\theta_i + \dfrac{n_1}{\mu_{r1}} \cos\theta_t}{\dfrac{n_2}{\mu_{r2}} \cos\theta_i + \dfrac{n_1}{\mu_{r1}} \cos\theta_t}, \tag{12-25}$$

$$\left(\frac{E_{0t}}{E_{0i}}\right)_P = \frac{2\dfrac{n_1}{\mu_{r1}} \cos\theta_i}{\dfrac{n_2}{\mu_{r2}} \cos\theta_i + \dfrac{n_1}{\mu_{r1}} \cos\theta_t}. \tag{12-26}$$

Equations 12-20, 12-21, 12-25, and 12-26 are *Fresnel's equations.*

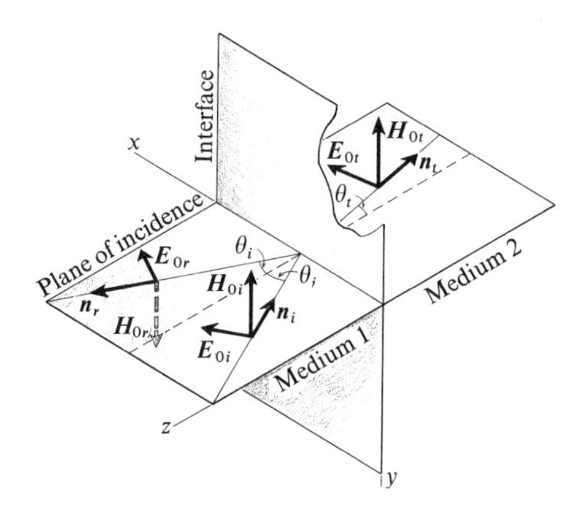

Figure 12-5. The incident, reflected, and transmitted waves when the incident wave is polarized with its $\boldsymbol{E}$ vector *parallel* to the plane of incidence. The arrows for $\boldsymbol{E}$ and for $\boldsymbol{H}$ have the same meaning as in Figure 12-4.

At normal incidence $\theta_i = \theta_t = 0$, the plane of incidence becomes undetermined, and the pair 12-20 and 12-21 is identical to the pair 12-25 and 12-26.

12.3 REFLECTION AND REFRACTION AT THE INTERFACE BETWEEN TWO NONMAGNETIC NONCONDUCTORS

Now that we have established the laws of reflection, Snell's law of refraction, and Fresnel's equations, all of which apply to the interface between any two media, we shall consider the relatively simple case of the interface between two nonmagnetic nonconductors. The indexes of refraction n_1 and n_2 are real numbers equal to or larger than unity, and

$$\frac{\sin\theta_t}{\sin\theta_i} = \frac{n_1}{n_2}, \tag{12-27}$$

as in Eq. 12-13. The larger angle is always in the medium with the lower index of refraction.

From Fresnel's equations,

$$\left(\frac{E_{0r}}{E_{0i}}\right)_N = \frac{\left(\dfrac{n_1}{n_2}\right)\cos\theta_i - \cos\theta_t}{\left(\dfrac{n_1}{n_2}\right)\cos\theta_i + \cos\theta_t}, \tag{12-28}$$

$$\left(\frac{E_{0t}}{E_{0i}}\right)_N = \frac{2\left(\dfrac{n_1}{n_2}\right)\cos\theta_i}{\left(\dfrac{n_1}{n_2}\right)\cos\theta_i + \cos\theta_t}, \tag{12-29}$$

for a wave polarized with its $\boldsymbol{E}$ vector normal to the plane of incidence.

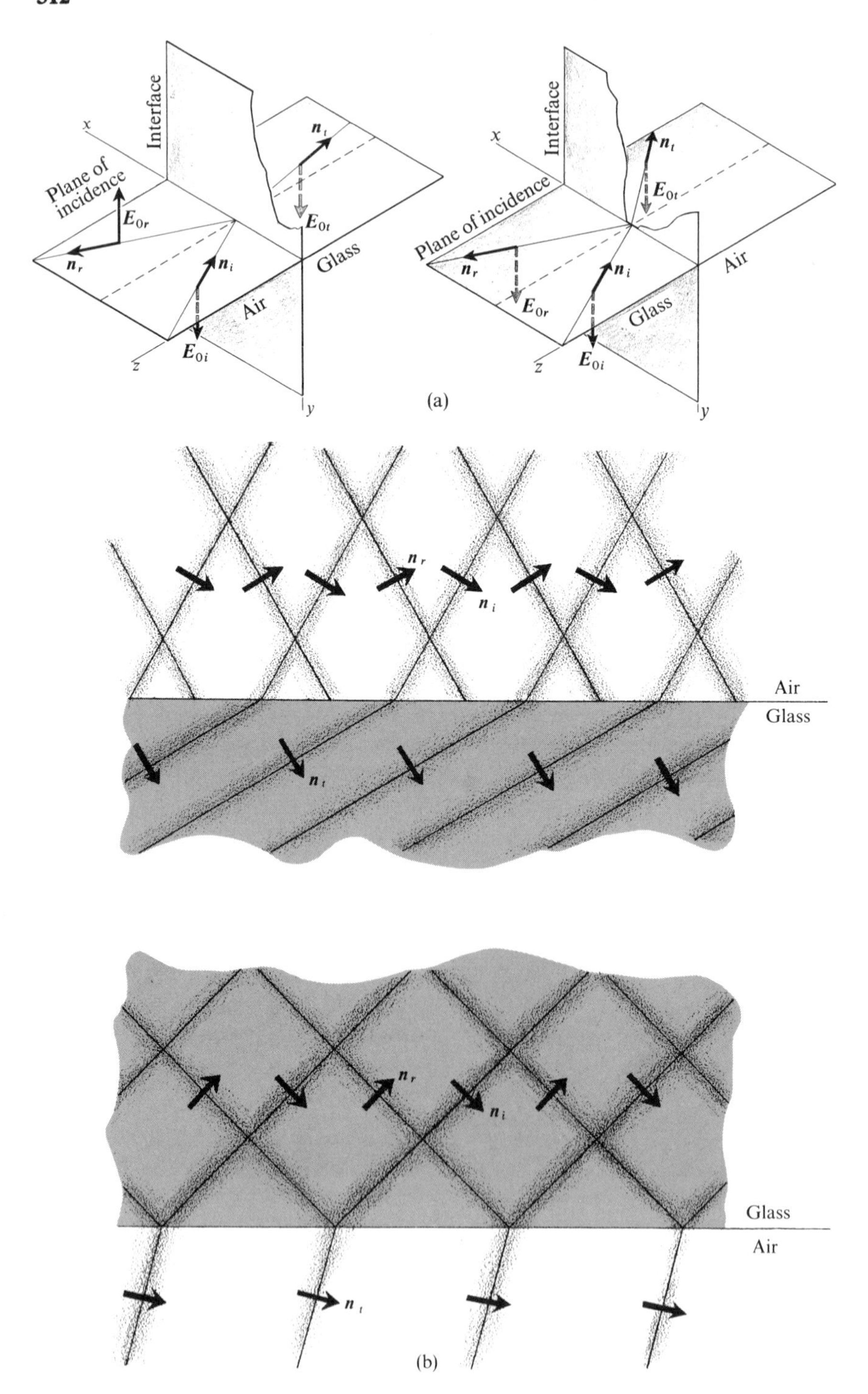

Interface
x
Plane of incidence
E_{0r}
n_t
E_{0t}
n_r
n_i
Glass
Air
E_{0i}
z
y
Interface
x
Plane of incidence
n_t
E_{0t}
n_r
n_i
Air
E_{0r}
Glass
z
E_{0i}
y
(a)
n_r
n_i
Air
Glass
n_t
n_r
n_i
Glass
Air
n_t
(b)

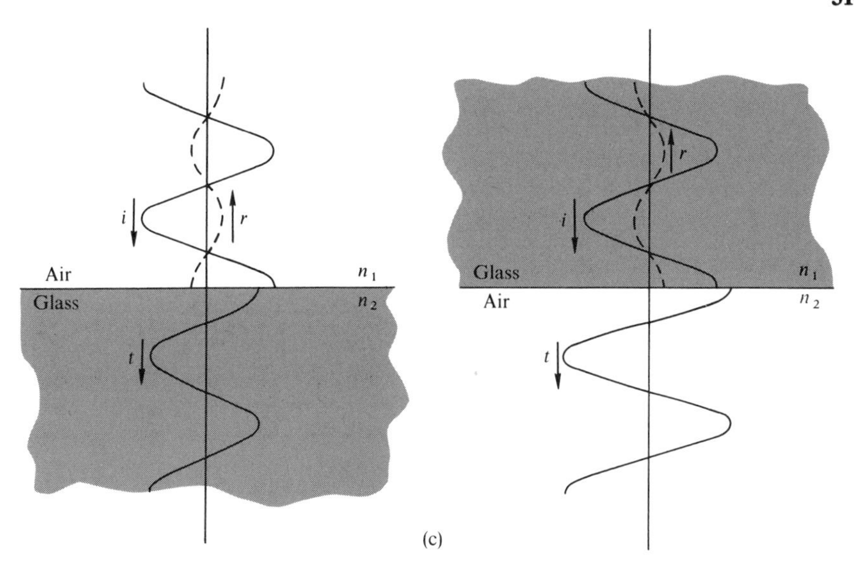

Figure 12-6. (a) The relative phases, at the interface, of the electric field intensities in the reflected and transmitted waves for $n_2 > n_1$ and for $n_2 < n_1$ with $\boldsymbol{E}_{0i}$ *normal* to the plane of incidence. In the first case the reflected wave is π radians out of phase with the incident wave. The transmitted wave is in phase in both cases. (b) "Crests" of $\boldsymbol{E}$ corresponding to (a) at some particular time. They are spaced one wavelength apart and travel in the directions shown. Note the phase shift of π on reflection from a glass surface. Note also the interference pattern resulting from the superposition of the incident and reflected waves. Constructive interference occurs wherever two crests or two troughs meet; destructive interference occurs where a crest meets a trough. (c) Graphs of E_i, E_r, E_t at normal incidence on an air-glass interface. Where the wave emerges from glass into air, as on the right, E_t is larger than E_i. Conservation of energy still applies, however. See Section 12.3.2.

It will be observed that $(E_{0t}/E_{0i})_N$ is always real and positive. This means that at the interface the transmitted wave is always in phase with the incident wave. The ratio $(E_{0r}/E_{0i})_N$ can, however, be either positive or negative, depending on the value of n_1/n_2, for if $n_1/n_2 > 1$, then $\theta_t > \theta_i$ and $\cos\theta_i > \cos\theta_t$; whereas if $n_1/n_2 < 1$, then $\theta_t < \theta_i$ and $\cos\theta_i < \cos\theta_t$. The reflected wave is thus either in phase with the incident wave at the interface if $n_1 > n_2$ or is π radians out of phase if $n_1 < n_2$. Figure 12-6 illustrates the $\boldsymbol{E}$ vectors for both types of reflection; Figure 12-7 shows the ratios of Eqs. 12-28 and 12-29 for $n_1/n_2 = 1/1.5$. This corresponds, for example, to a light wave incident in air on a glass with an index of refraction of 1.5.

For an incident wave polarized with its $\boldsymbol{E}$ vector parallel to the plane of incidence,

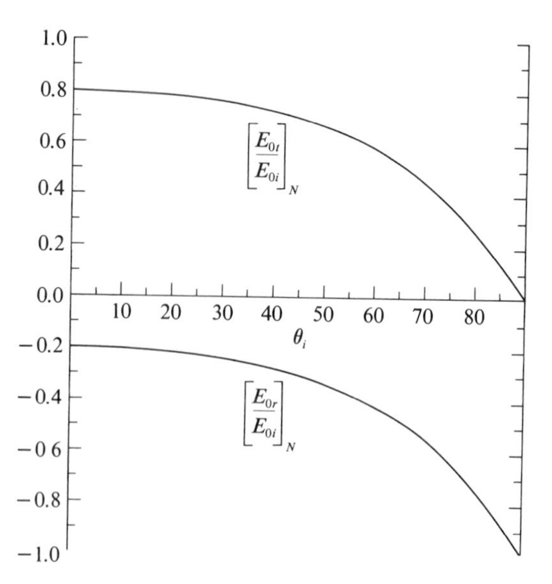

Figure 12-7. The ratios $(E_{0r}/E_{0i})_N$ and $(E_{0t}/E_{0i})_N$ as functions of the angle of incidence θ_i for $n_1/n_2 = 1/1.5$. This corresponds to light incident in air on a glass with $n = 1.5$. The wave is polarized with its **E** vector *normal* to the plane of incidence.

$$\left(\frac{E_{0r}}{E_{0i}}\right)_P = \frac{-\cos\theta_i + \left(\dfrac{n_1}{n_2}\right)\cos\theta_t}{\cos\theta_i + \left(\dfrac{n_1}{n_2}\right)\cos\theta_t}, \tag{12-30}$$

$$\left(\frac{E_{0t}}{E_{0i}}\right)_P = \frac{2\left(\dfrac{n_1}{n_2}\right)\cos\theta_i}{\cos\theta_i + \left(\dfrac{n_1}{n_2}\right)\cos\theta_t}. \tag{12-31}$$

The second ratio is always positive. This indicates that the relative phases of $\boldsymbol{E}_{0t}$ and $\boldsymbol{E}_{0i}$ are as in Figure 12-5, which we used in arriving at this result; that is, the incident and transmitted electric field intensities are in phase at the interface.

On the other hand, the ratio for $\boldsymbol{E}_{0r}$ can be either positive or negative, which indicates that $\boldsymbol{E}_{0r}$ can point either as in Figure 12-5 or in the opposite direction. The tangential components of $\boldsymbol{E}_{0i}$ and of $\boldsymbol{E}_{0r}$ can thus be either in phase or π radians out of phase. The $\boldsymbol{E}_{0r}$ component is in phase with $\boldsymbol{E}_{0i}$ at the interface if

$$\left(\frac{n_1}{n_2}\right)\cos\theta_t - \cos\theta_i > 0 \tag{12-32}$$

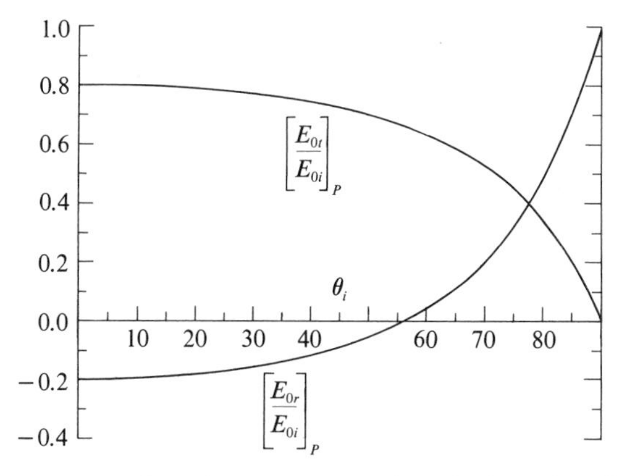

Figure 12-8. The ratios $(E_{0r}/E_{0i})_P$ and $(E_{0t}/E_{0i})_P$ as functions of the angle of incidence θ_i for $n_1/n_2 = 1/1.5$. This corresponds to light incident in air on a glass with $n = 1.5$. The wave is polarized with its $\mathbf{E}$ vector *parallel* to the plane of incidence.

or if

$$\sin\theta_t\cos\theta_t - \sin\theta_i\cos\theta_i > 0, \tag{12-33}$$

$$\sin 2\theta_t - \sin 2\theta_i > 0, \tag{12-34}$$

$$\sin(\theta_t - \theta_i)\cos(\theta_t + \theta_i) > 0. \tag{12-35}$$

This inequality will be satisfied either if

$$\theta_t > \theta_i \qquad \textit{and} \qquad \theta_t + \theta_i < \frac{\pi}{2} \tag{12-36}$$

or if

$$\theta_t < \theta_i \qquad \textit{and} \qquad \theta_t + \theta_i > \frac{\pi}{2}. \tag{12-37}$$

The phase of the reflected wave in this case does *not* therefore depend only on the ratio n_2/n_1; it depends on both θ_i and θ_t. The ratio E_{0r}/E_{0t} can be either positive or negative, both for $n_2 > n_1$ and for $n_2 < n_1$. Figure 12-8 shows the ratios of Eqs. 12-30 and 12-31, again for $n_1/n_2 = 1/1.5$.

12.3.1 The Brewster Angle

We have seen in the foregoing that the electric field intensity E_{0r} of the reflected wave is either in phase or π radians out of phase with the incident wave, depending on whether $\sin(\theta_t - \theta_i)\cos(\theta_t + \theta_i)$ is greater or less than zero. It will be gathered from this that there is *no* reflected wave when this expression is equal to zero, that is, when $\theta_i = \theta_t = 0$ or when $\theta_i + \theta_t = \pi/2$. The first condition is incorrect, however; it arises from the fact that we have multiplied the inequality

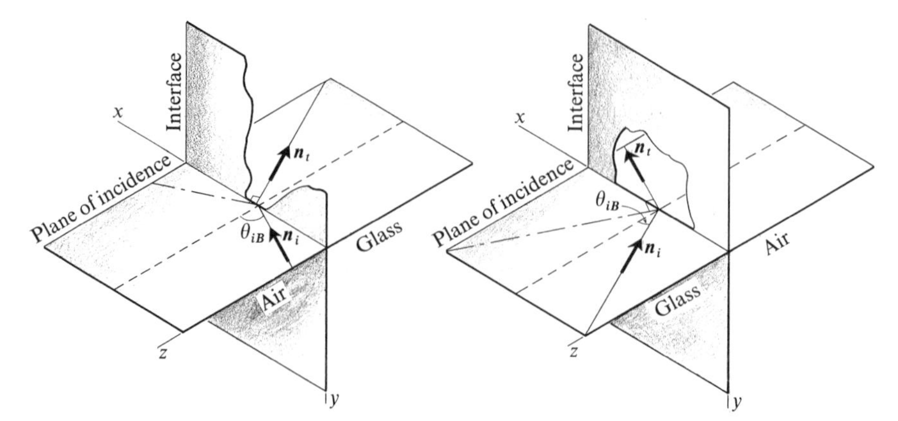

Figure 12-9. When the incident wave is polarized with its $\boldsymbol{E}$ vector *parallel* to the plane of incidence, there is *no* reflected wave at $\theta_i + \theta_t = \pi/2$. The angle of incidence θ_i is then called the Brewster angle. The position of the missing reflected ray is at 90° to the transmitted ray. For any pair of media, the sum of the two angles θ_{iB} is 90°.

$$\left(\frac{n_1}{n_2}\right) \cos \theta_t - \cos \theta_i > 0 \tag{12-38}$$

by $\sin \theta_i$, which is equal to zero at $\theta_i = 0$.

Thus, for

$$\theta_i + \theta_t = \frac{\pi}{2}, \tag{12-39}$$

there is a reflected wave only when the incident wave is polarized with its $\boldsymbol{E}$ vector normal to the plane of incidence. This is rather remarkable because it involves the passage of a wave through a discontinuity in the medium of propagation *without* the production of a reflected wave. The conditions of continuity at the interface are then satisfied by two waves only—the incident and the transmitted waves—instead of the usual three. This is illustrated in Figure 12-9. This angle of incidence is called the *Brewster angle*. It is also called the *polarizing angle*, since an unpolarized wave incident on an interface at this angle is reflected as a polarized wave with its $\boldsymbol{E}$ vector normal to the plane of incidence.*

---

* The Brewster angle is often explained *incorrectly* as follows. For this particular angle of incidence, the missing reflected ray is at 90° to the transmitted ray. It is argued that the electrons excited in medium 2 do not radiate in their direction of oscillation (Section 14.1.4) and hence cannot give rise to a reflected ray in medium 1 in this case. This explanation is incorrect, since the Brewster angle is observed even when medium 2 is a vacuum.

At the Brewster angle,

$$\frac{n_1}{n_2} = \frac{\sin \theta_t}{\sin \theta_{iB}} = \frac{\sin\left(\frac{\pi}{2} - \theta_{iB}\right)}{\sin \theta_{iB}} = \cot \theta_{iB}. \tag{12-40}$$

Example

MEASURING THE RELATIVE PERMITTIVITY OF THE MOON'S SURFACE AT RADIO FREQUENCIES

The nature of the Moon's surface can be inferred, to some extent, from the value of its relative permittivity $\epsilon_r = n^2$ (Section 11.3) at radio frequencies. The Brewster angle has been used to measure this quantity in the following way. When the Moon is illuminated with a radio wave originating from a satellite in lunar orbit, the reflection observed at the Earth can be compared to the reflection of sunlight from the surface of a lake: most of the light comes from the regions that happen to be correctly oriented for specular reflection. The surface of the Moon thus glistens over an area of the order of 100 kilometers in diameter, the area depending on the height of the satellite above the surface of the Moon, and also on the roughness of the surface. (If the detector on the Earth receives both the reflected wave and a direct wave from the satellite, it is possible to discriminate between the two by making use of the fact that the Doppler effect makes the two frequencies slightly different.) A plot of the intensity of the reflected wave as a function of the angle of incidence when the $\boldsymbol{E}$ vector lies in the plane of incidence shows zero reflection at the Brewster angle.

In one such measurement, performed at a frequency of 0.14 gigahertz, the Brewster angle was found to be $60 \pm 1°$ in the mare northwest of Hanstein. This gives an ϵ_r of 3.0 ± 0.2.

It is possible to perform similar measurements at other points on the surface of the Moon because of the relative motions of the three bodies involved, namely the satellite, the Moon, and the Earth.

12.3.2 The Coefficients of Reflection and of Transmission at an Interface Between Two Nonconductors

It is useful to define coefficients of reflection and of transmission that are related to the flow of energy across the interface. The average energy flux per unit area in the incident wave is given by the average value of the Poynting vector, Eq. 11-64. Setting $\mu_r = 1$,

$$\mathbf{S}_{i\,\mathrm{av}} = \frac{1}{2}\left(\frac{\epsilon_1}{\mu_0}\right)^{1/2} E_{0i}^2\, \boldsymbol{n}_i, \tag{12-41}$$

$$\mathbf{S}_{r\,\mathrm{av}} = \frac{1}{2}\left(\frac{\epsilon_1}{\mu_0}\right)^{1/2} E_{0r}^2\, \boldsymbol{n}_r, \tag{12-42}$$

$$\mathbf{S}_{t\,\mathrm{av}} = \frac{1}{2}\left(\frac{\epsilon_2}{\mu_0}\right)^{1/2} E_{0t}^2\, \boldsymbol{n}_t. \tag{12-43}$$

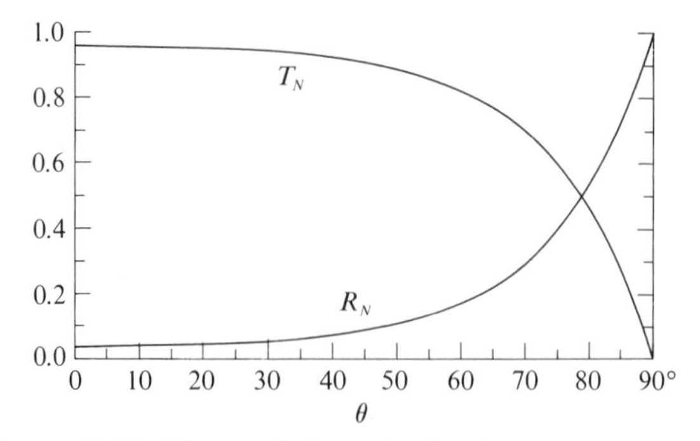

Figure 12-10. The coefficient of reflection R_N and the coefficient of transmission T_N as functions of the angle of incidence θ_i for $n_1/n_2 = 1/1.5$. The $\boldsymbol{E}$ vector of the incident wave is *normal* to the plane of incidence.

The *coefficients of reflection R and of transmission T* are defined as the ratios of the average energy fluxes per unit time and per unit area at the interface:

$$R = \left|\frac{\mathbf{S}_{r\text{ av}}\cdot\boldsymbol{n}}{\mathbf{S}_{i\text{ av}}\cdot\boldsymbol{n}}\right| = \frac{E_{0r}^2}{E_{0i}^2}, \tag{12-44}$$

where $\boldsymbol{n}$ is the unit vector normal to the interface;

$$T = \frac{\mathbf{S}_{t\text{ av}}\cdot\boldsymbol{n}}{\mathbf{S}_{i\text{ av}}\cdot\boldsymbol{n}} = \left(\frac{\epsilon_{r2}}{\epsilon_{r1}}\right)^{1/2}\frac{E_{0t}^2\cos\theta_t}{E_{0i}^2\cos\theta_i}, \tag{12-45}$$

$$= \frac{n_2\,E_{0t}^2\cos\theta_t}{n_1\,E_{0i}^2\cos\theta_i}. \tag{12-46}$$

Then, from Fresnel's equations for nonconductors,

$$R_N = \left\{\frac{\left(\dfrac{n_1}{n_2}\right)\cos\theta_i - \cos\theta_t}{\left(\dfrac{n_1}{n_2}\right)\cos\theta_i + \cos\theta_t}\right\}^2, \tag{12-47}$$

$$T_N = \frac{4\left(\dfrac{n_1}{n_2}\right)\cos\theta_i\cos\theta_t}{\left\{\left(\dfrac{n_1}{n_2}\right)\cos\theta_i + \cos\theta_t\right\}^2}, \tag{12-48}$$

$$R_P = \left\{\frac{-\cos\theta_i + \left(\dfrac{n_1}{n_2}\right)\cos\theta_t}{\cos\theta_i + \left(\dfrac{n_1}{n_2}\right)\cos\theta_t}\right\}^2, \tag{12-49}$$

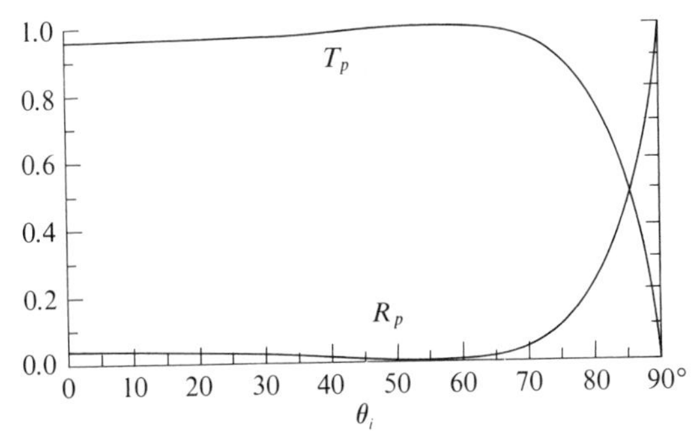

Figure 12-11. The coefficient of reflection R_P and the coefficient of transmission T_P as functions of the angle incidence θ_i for $n_1/n_2 = 1/1.5$. The ***E*** vector of the incident wave is *parallel* to the plane of incidence.

$$T_P = \frac{4\left(\dfrac{n_1}{n_2}\right)\cos\theta_i\cos\theta_t}{\left\{\cos\theta_i + \left(\dfrac{n_1}{n_2}\right)\cos\theta_t\right\}^2}. \tag{12-50}$$

In both cases, $R + T = 1$, as expected, since there must be conservation of energy. At the Brewster angle, defined by Eq. 12-40, $R_P = 0$ and $T_P = 1$, again as expected.

Figures 12-10 and 12-11 show the coefficients of reflection R and of transmission T as functions of the angle of incidence θ_i for $n_1/n_2 = 1/1.5$, whereas Figure 12-12 shows R and T as functions of n_1/n_2 at normal incidence.

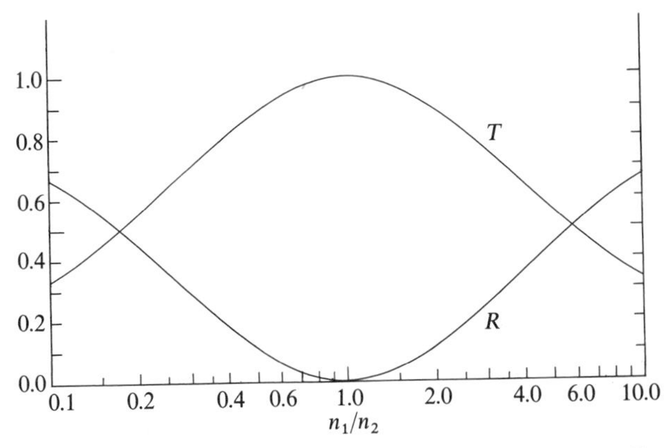

Figure 12-12. The coefficient of reflection R and the coefficient of transmission T at normal incidence as functions of the ratio n_1/n_2.

12.4 TOTAL REFLECTION AT AN INTERFACE BETWEEN TWO NONMAGNETIC NONCONDUCTORS

We shall now consider the phenomenon of total reflection, which we have excluded in Sections 12.3 to 12.3.2.

If $n_1 > n_2$, and if θ_i is sufficiently large, Snell's law

$$\sin \theta_t = \frac{n_1}{n_2} \sin \theta_i \tag{12-51}$$

leads to the apparently absurd result that $\sin \theta_t$ is greater than unity. The critical angle of incidence, for which $\sin \theta_t = 1$ and $\theta_t = 90°$, is

$$\sin \theta_{ic} = \frac{n_2}{n_1}. \tag{12-52}$$

It is observed experimentally that, when $\theta_i \geq \theta_{ic}$, the wave originating in medium 1 and incident on the interface is totally reflected back into medium 1 as in Figure 12-13. This phenomenon, which is called *total reflection*, does not depend on the orientation of the $\boldsymbol{E}$ vector in the incident wave. For light propagating in glass with an index of refraction of 1.6, the critical angle of incidence is 38.7°.

It turns out, as we shall see in Section 12.4.1, that Snell's law (Eq. 12-13), the laws of reflection and refraction (Section 12.1), and Fresnel's equations (12-20, 12-21, and 12-25, 12-26) are all applicable to total reflection if we disregard the fact that $\sin \theta_t > 1$ and if we set

$$\cos \theta_t = -(1 - \sin^2 \theta_t)^{1/2}, \tag{12-53}$$

$$= -\left\{1 - \left(\frac{n_1}{n_2}\right)^2 \sin^2 \theta_i\right\}^{1/2}, \tag{12-54}$$

$$= -j\frac{n_1}{n_2}\left\{\sin^2 \theta_i - \left(\frac{n_2}{n_1}\right)^2\right\}^{1/2}. \tag{12-55}$$

Note the negative sign before the square root.

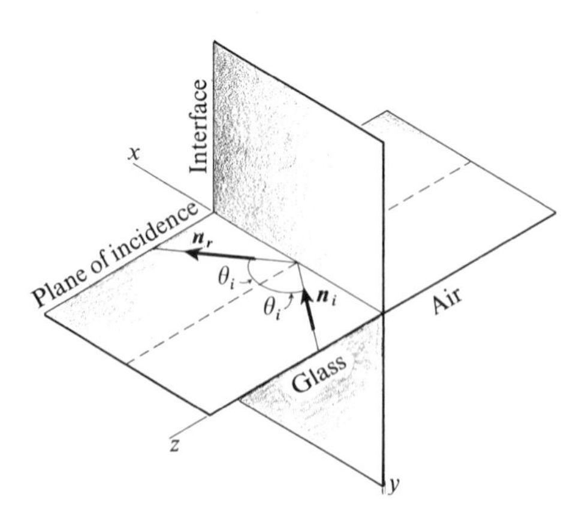

Figure 12-13. For angles of incidence θ_i equal to or greater than the critical angle θ_{ic}, the wave is totally reflected back into medium 1.

Then, from Eqs. 12-15 and 12-16,

$$\boldsymbol{E}_r = \boldsymbol{E}_{0r} \exp j\{\omega t - k_1(x \sin \theta_i + z \cos \theta_i)\}, \tag{12-56}$$

$$\boldsymbol{E}_t = \boldsymbol{E}_{0t} \exp j\{\omega t - k_2(x \sin \theta_t - z \cos \theta_t)\}, \tag{12-57}$$

$$= \boldsymbol{E}_{0t} \exp \left\{ j\left(\omega t - \frac{n_1}{\lambda_0} x \sin \theta_i\right) + \frac{n_2}{\lambda_0}\left((n_1/n_2)^2 \sin^2\theta_i - 1\right)^{1/2} z \right\}. \tag{12-58}$$

INCIDENT WAVE POLARIZED WITH ITS $\boldsymbol{E}$ VECTOR NORMAL TO THE PLANE OF INCIDENCE. From Fresnel's equations 12-20 and 12-21, and for nonmagnetic nonconductors,

$$\left(\frac{E_{0r}}{E_{0i}}\right)_N = \exp 2j \arctan \frac{\left\{\sin^2 \theta_i - \left(\frac{n_2}{n_1}\right)^2\right\}^{1/2}}{\cos \theta_i} = \exp j\alpha, \tag{12-59}$$

$$\left(\frac{E_{0t}}{E_{0i}}\right)_N = \frac{2 \cos \theta_i}{\cos \theta_i - j\left\{\sin^2 \theta_i - \left(\frac{n_2}{n_1}\right)^2\right\}^{1/2}}, \tag{12-60}$$

$$= \frac{2 \cos \theta_i}{\left\{1 - \left(\frac{n_2}{n_1}\right)^2\right\}^{1/2}} \exp j\frac{\alpha}{2} \tag{12-61}$$

We first notice that the amplitude of the *reflected* wave is equal to that of the incident wave, so that the coefficient of reflection R is equal to unity. The energy is totally reflected and there is zero net flux of energy through the interface.

The phase jump upon reflection varies from 0° at the critical angle of incidence ($\sin \theta_i = n_2/n_1$) to 180° at glancing incidence, through positive angles. This is shown in Figure 12-14.

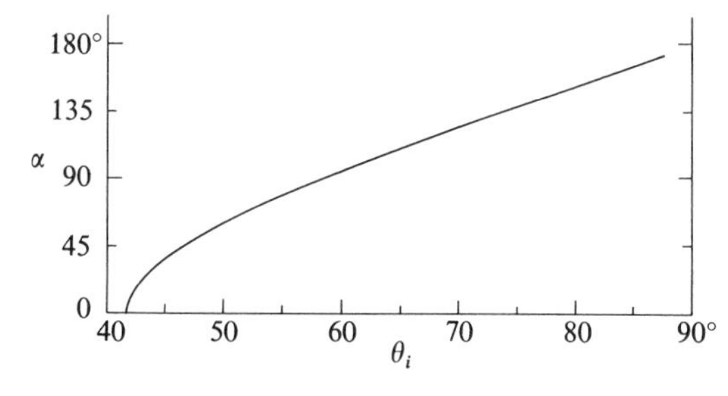

Figure 12-14. The phase α of the reflected wave with respect to that of the incident wave in the case of total reflection. The incident wave is polarized with its $\boldsymbol{E}$ vector *normal* to the plane of incidence; the ratio n_1/n_2 is equal to 1.50.

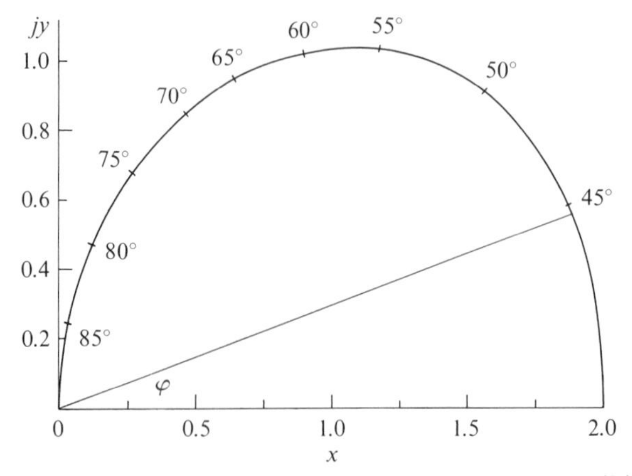

Figure 12-15. The ratio $(E_{0t}/E_{0i})_N$ of Eq. 12-60 is plotted here in the complex plane for various angles of incidence θ_i larger than the critical angle and for $n_1/n_2 = 1.50$. The amplitude of the transmitted wave is represented by the distance from a point on the curve to the origin, and is greatest at the critical angle. The transmitted wave leads the incident wave by an angle equal to the argument of the complex ratio—for example, the angle φ shown.

As regards the *transmitted* wave, it is obvious that $\mathbf{E}_{0t}$ is not zero, despite the fact that the net flux of energy across the interface is zero. Medium 2 can be considered to act like a pure inductance fed by a source of alternating current. The average power input to the inductance is zero, the power flow being alternately one way and then the other, but there is nevertheless a current through the inductance.

Figure 12-15 shows how the electric field intensity of the transmitted wave varies both in amplitude and in phase with the angle of incidence.

The transmitted wave is quite remarkable. According to Eq. 12-58, it travels unattenuated (as in Figure 12-16) parallel to the interface, with a wavelength

$$\lambda_x = \frac{\lambda_0}{n_1 \sin\theta_i} = \frac{\lambda_1}{\sin\theta_i}, \qquad (12\text{-}62)$$

λ_1 being the wavelength in medium 1 above the interface. The wavelength λ_x is exactly the distance along the x-axis between two neighboring equiphase points in the incident wave. This was to be expected, since the continuity conditions must be satisfied at all points on the interface.

This result, namely that the wave travels unattenuated parallel to the interface, is most surprising if we think of an incident wave of finite cross-

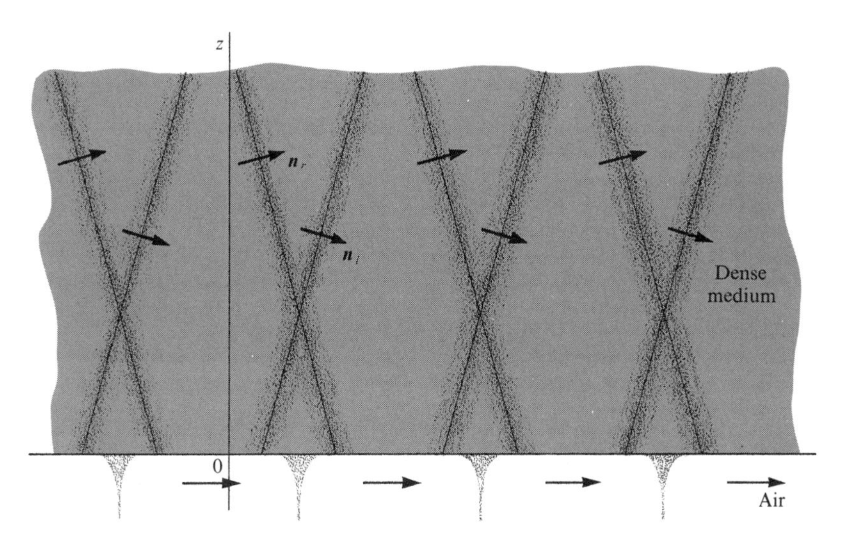

Figure 12-16. "Crests" of E for the incident, reflected, and transmitted waves are represented here schematically for the case of total reflection. They are spaced one wavelength apart. The transmitted wave travels unattenuated below the interface, and its amplitude decreases exponentially with depth in medium 2. The data used for the figure are the following: $n_1 = 3.0$, $n_2 = 1.0$, $\theta_i = 75°$. (See Problem 12-12.)

section. Does the transmitted wave on the other side of the interface extend beyond the illuminated region? Our discussion cannot provide us with an answer, since it is based on the assumption that the incident wave is infinite in extent. Physically, what happens is this: a given incident ray, instead of being reflected abruptly at the interface, penetrates into medium 2, where it is bent back into medium 1. It is this phenomenon which gives rise to the "transmitted" wave.*

The transmitted wave is damped exponentially in the direction perpendicular to the interface in such a way that its amplitude decreases by a factor of e over a distance

$$\delta_z = \frac{\lambda_2}{\left\{\left(\frac{n_1}{n_2}\right)^2 \sin^2\theta_i - 1\right\}^{1/2}}. \tag{12-63}$$

This is illustrated qualitatively in Figure 12-16.

The ratio δ_z/λ_2 is shown in Figure 12-17 as a function of the angle of incidence θ_i for $n_1/n_2 = 1.5$.

---

* See A. von Hippel, *Dielectrics and Waves* (Wiley, New York, 1954), p. 54, for a brief account of work by F. Goos and H. Hanchen on this subject.

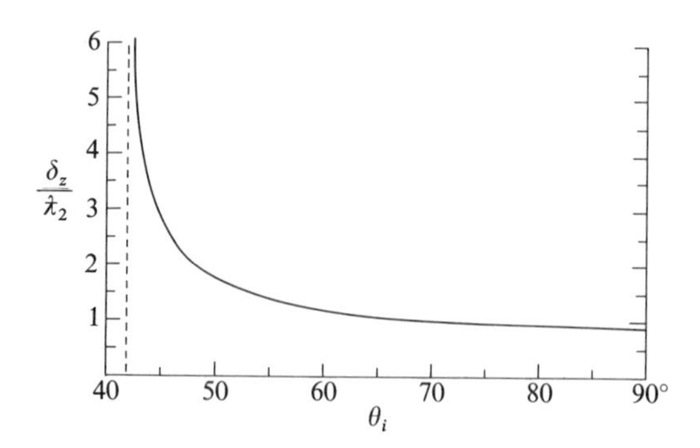

Figure 12-17. The ratio of δ_z (the depth of penetration) to $\lambda\!\!\!^{-}_2$ (the wavelength in medium 2 divided by 2π) for the transmitted wave when there is total reflection. The index of refraction of the first medium is 1.50 times that of the second medium.

We shall find in Section 12.4.1.4 that there is an energy flow parallel to the interface in the positive direction of the x-axis. The flux is a function of the angle of incidence as in Figure 12-18, and it decreases exponentially with z. There is zero average energy flux in the direction normal to the interface.

The transmitted wave, has been observed both with visible light and with microwaves.*

INCIDENT WAVE POLARIZED WITH ITS $\boldsymbol{E}$ VECTOR PARALLEL TO THE PLANE OF INCIDENCE. Equations 12-56 and 12-57 for $\boldsymbol{E}_t$ and $\boldsymbol{E}_r$ are again valid, the

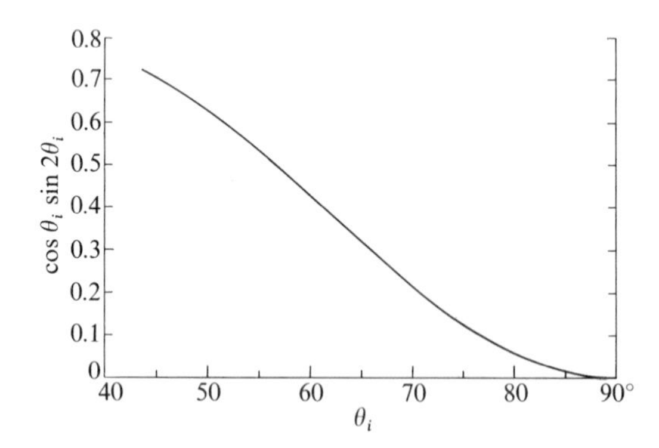

Figure 12-18. The flux of energy $\mathcal{S}_t$, parallel to and just below the interface, is proportional to $\cos\theta_i \sin 2\theta_i$ in the case of total reflection.

* See, for example, J. Strong, *Concepts of Classical Optics* (W. H. Freeman and Company, San Francisco, 1958), Appendix J by G. F. Hull, Jr., p. 516.

coefficient of reflection R is again equal to unity, but the phase jump upon reflection is different. Total reflection of a wave polarized in some arbitrary direction therefore gives an elliptically polarized reflected wave.

Example

LIGHT EMISSION FROM A CATHODE RAY TUBE

In a cathode ray tube, light is generated in a fluorescent screen, which is in contact with the tube face. The light is of course emitted in all directions but, as a rule, the part that is emitted backwards in the direction of the electron gun is reflected by a thin aluminum coating laid over the fluorescent screen as in Figure 12-19. This increases the light output by a factor of two.

The figure shows that most of the light emitted by the screen is trapped inside the tube by total reflection. We can calculate the fraction of the light that comes out through the tube face as follows.

Let us assume that, inside the cone $2\theta_{ic}$, all the light passes through the interface. Then the fraction F is the solid angle corresponding to the critical angle θ_{ic}, divided by 2π. We divide by 2π and not by 4π because, with the reflecting coating, all the light is emitted within a solid angle of 2π. Now the solid angle corresponding to θ_{ic} is the area of the surface represented by a dashed line in the figure, divided by R^2. Thus

$$F = \frac{1}{2\pi}\int_0^{\theta_{ic}} \frac{2\pi R \sin\theta R\, d\theta}{R^2} = (1 - \cos\theta_{ic}). \qquad (12\text{-}64)$$

For a glass with an index of refraction of 1.6, F is only 22%, with our assumption that the coefficient of transmission is equal to unity for $\theta < \theta_{ic}$. The fraction F is, in fact, even smaller than 22%, but not very much so because the coefficient of transmission is approximately equal to unity, except near the critical angle.

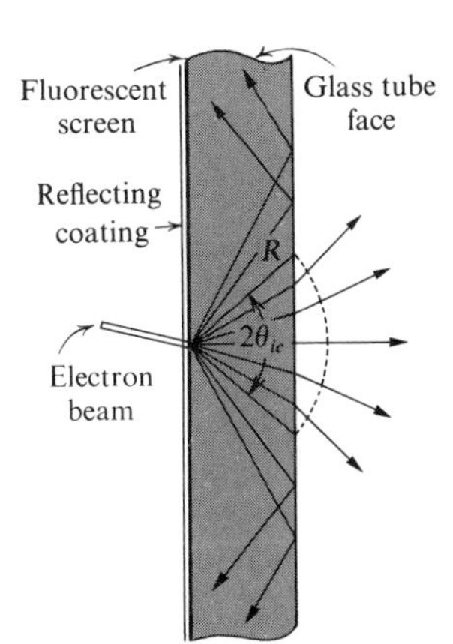

Figure 12-19. Section of the face of a cathode ray tube.

Example

THE CRITICAL ANGLE AND BREWSTER'S ANGLE

The critical angle arc sin (n_2/n_1) (Eq. 12-52) is somewhat larger than the Brewster angle arc tan (n_2/n_1) (Eq. 12-40). For example, again for light propagating inside a glass with an index of refraction of 1.6 (n_1), the wave is totally *transmitted* into the air at a glass-air interface when the angle of incidence is the Brewster angle, 32.0°, and is totally *reflected* into the glass at the critical angle of 38.7°.

Figure 12-20 shows these two angles as functions of the ratio n_1/n_2. For large values of n_1/n_2, that is, for light incident in a relatively "dense" medium, θ_{ic} is nearly equal to θ_{iB}. For media with more similar indexes of refraction, the Brewster angle approaches 45°, whereas the critical angle approaches 90°.

It is interesting to note that, for a wave polarized with its $\boldsymbol{E}$ vector parallel to the plane of incidence, the amplitude of the reflected wave changes rapidly when the angle of incidence lies between the Brewster angle and the critical angle.

This peculiar behavior of the reflected wave could be used for measuring small angular displacements.

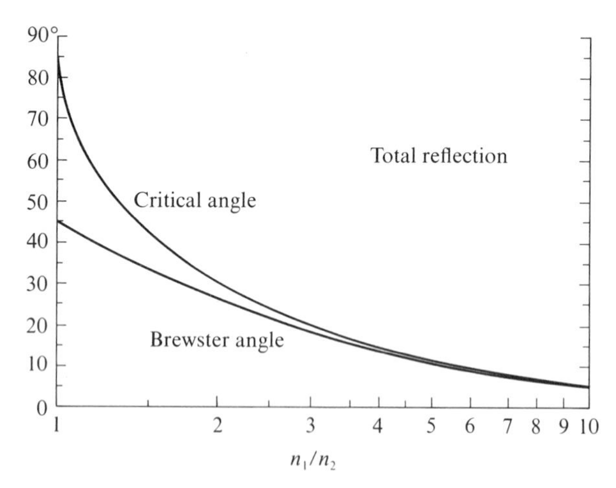

Figure 12-20. The critical angle and the Brewster angle as functions of the ratio n_1/n_2 of the indexes of refraction on either side of the interface. The wave is incident in medium 1. It is polarized with its $\boldsymbol{E}$ vector *parallel* to the plane of incidence for the Brewster angle curve.

12.4.1 Demonstration of the Validity of Snell's Law, of the Laws of Reflection and Refraction, and of Fresnel's Equations in the Case of Total Reflection*

We shall study only the case where the incident wave is polarized with its $\boldsymbol{E}$ vector *normal* to the plane of incidence.

It turns out that it is impossible to satisfy the requirement of continuity of the tangential components of $\boldsymbol{E}$ and $\boldsymbol{H}$, and of the normal components of $\boldsymbol{D}$ and $\boldsymbol{B}$ at the interface, with only the incident and reflected waves. We therefore conclude that there must exist some sort of transmitted wave.

The transmitted wave must, however, be of a rather special nature, since it is not observable under ordinary conditions. It must of course satisfy the general wave equation for nonmagnetic nonconductors (Eq. 11-54),

$$\frac{\partial^2 \boldsymbol{E}_t}{\partial x^2} + \frac{\partial^2 \boldsymbol{E}_t}{\partial z^2} = \epsilon_2 \mu_0 \frac{\partial^2 \boldsymbol{E}_t}{\partial t^2}. \tag{12-65}$$

We have set the derivative with respect to y equal to zero because, by hypothesis, the field does not vary with the y-coordinate (Figure 12-3).

We can use Eq. 12-14 for the incident wave:

$$\boldsymbol{E}_i = \boldsymbol{E}_{0i} \exp j\{\omega t - k_1(x \sin \theta_i - z \cos \theta_i)\}. \tag{12-66}$$

For the reflected wave, we can write

$$\boldsymbol{E}_r = \boldsymbol{E}_{0r} \exp j(\omega t - k_{1x}x - k_{1z}z), \tag{12-67}$$

where $\boldsymbol{E}_{0r}$, k_{1x}, k_{1z} are unknown constants. We have used the same value of ω as for the incident wave since, as in Section 12.1, all three waves have the same frequency. We have also used only x and z terms in the exponent since the derivative with respect to y must, again, be zero.

We set

$$\boldsymbol{E}_t = (E_{0tx}\boldsymbol{i} + E_{0ty}\boldsymbol{j} + E_{0tz}\boldsymbol{k}) \exp j(\omega t - k_{2x}x - k_{2z}z), \tag{12-68}$$

where E_{0tx}, E_{0ty}, E_{0tz}, k_{2x}, k_{2z} are also unknown constants. Again we have the same ω, and we have no y term in the exponent. We let the unknowns be complex in order that the various components of $\boldsymbol{E}_r$ and of $\boldsymbol{E}_t$ can be out of phase with each other and in order that the dependence on x and on z can be more general than with a simpler expression, such as that for the incident wave.

---

* You may wish to omit Sections 12.4.1 to 12.4.1.4 inclusively.

All the unknowns are independent both of the coordinates and of the time, the only dependence on x, z, t being expressed by the exponential functions.

We represent the $\boldsymbol{H}$ vectors by similar expressions, in which the letter E is replaced by the letter H and, for simplicity, we choose the origin of coordinates at the interface, as in Figure 12-3.

We shall determine all these unknowns by using the boundary conditions at the interface, the wave equation, and Maxwell's equations.

12.4.1.1 *The Wave Numbers k_{1x} and k_{1z} for the Reflected Wave.* To satisfy the boundary conditions at the interface, we proceed just as we did in Section 12.1. We first require that the exponents be equal at the interface $z = 0$. Then, from Eqs. 12-66 and 12-67,

$$k_{1x} = k_1 \sin \theta_i = \frac{n_1}{\lambda_0} \sin \theta_i. \tag{12-69}$$

To find k_{1z}, we turn to the wave equation in medium 1.

$$k_{1x}^2 + k_{1z}^2 = \epsilon_1 \mu_0 \omega^2 = \left(\frac{n_1}{\lambda_0}\right)^2, \tag{12-70}$$

and

$$k_{1z}^2 = \left(\frac{n_1}{\lambda_0}\right)^2 (1 - \sin^2 \theta_i), \tag{12-71}$$

$$k_{1z} = \frac{n_1}{\lambda_0} \cos \theta_i. \tag{12-72}$$

We choose the positive sign in order to have a wave propagating in the proper direction.

The reflected wave is thus of the form

$$\boldsymbol{E}_r = \boldsymbol{E}_{0r} \exp j\{\omega t - k_1(x \sin \theta_i + z \cos \theta_i)\}. \tag{12-73}$$

This is a plane wave reflected from the interface at an angle equal to the angle of incidence.

12.4.1.2 *The Wave Numbers k_{2x} and k_{2z} for the Transmitted Wave.* Equating similarly the exponents for $\boldsymbol{E}_i$ and for $\boldsymbol{E}_t$, again at the interface $z = 0$, we find the wave number k_{2x} of the transmitted wave:

$$k_{2x} = k_1 \sin \theta_i = \frac{n_1}{\lambda_0} \sin \theta_i. \tag{12-74}$$

To find k_{2z}, we again use the wave equation. Using the expression for $\boldsymbol{E}_t$,

$$k_{2x}^2 + k_{2z}^2 = \epsilon_2\mu_0\omega^2, \tag{12-75}$$

$$k_{2z} = \pm\,(\epsilon_2\mu_0\omega^2 - k_{2x}^2)^{1/2}, \tag{12-76}$$

$$= \pm\left(\frac{n_2^2}{\lambda_0^2} - \frac{n_1^2}{\lambda_0^2}\sin^2\theta_i\right)^{1/2}, \tag{12-77}$$

$$= \pm j\frac{n_2}{\lambda_0}\left\{\left(\frac{n_1}{n_2}\right)^2\sin^2\theta_i - 1\right\}^{1/2}. \tag{12-78}$$

The exponential function for the transmitted wave is thus

$$\exp\left\{j\left\{\omega t - \frac{n_1}{\lambda_0}x\sin\theta_i\right\} + \frac{n_2}{\lambda_0}\left\{\left(\frac{n_1}{n_2}\right)^2\sin^2\theta_i - 1\right\}^{1/2} z\right\}.$$

We have replaced the $\pm$ sign before the z term by a $+$ sign since the electric field intensity must not become infinite as $z \to -\infty$. Then k_{2z} can be written with a plus sign:

$$k_{2z} = +j\frac{n_2}{\lambda_0}\left\{\left(\frac{n_1}{n_2}\right)^2\sin^2\theta_i - 1\right\}^{1/2}. \tag{12-79}$$

If θ_t were real, we would expect to have

$$k_{2z} = -\frac{n_2}{\lambda_0}\cos\theta_t. \tag{12-80}$$

Comparing these two equations, we set*

$$\cos\theta_t = -\left\{1 - \left(\frac{n_1}{n_2}\right)^2\sin^2\theta_i\right\}^{1/2}, \tag{12-81}$$

$$= -\,(1 - \sin^2\theta_t)^{1/2}. \tag{12-82}$$

We have elected to place the negative sign before the radical so as to preserve the formalism. This is the value of $\cos\theta_t$ that we used earlier (Eqs. 12-53 and 12-54).

12.4.1.3 *The Amplitudes of **E** and **H** in the Reflected and Transmitted Waves.* Now that we have found the wave numbers for the reflected and transmitted waves, we only have to find the amplitudes of their $\boldsymbol{E}$ and $\boldsymbol{H}$ vectors.

---

* It is probably useful to recall here that, if A is some *positive real* number, then

$$(-A)^{1/2} = jA^{1/2},$$

and

$$A^{1/2} = -j(-A)^{1/2}.$$

These equations do *not* apply if A is negative. If you have any doubt concerning such operations you should plot A, $-A$, and their square roots in the complex plane. It is rather easy to be misled; for example,

$$A = (A^2)^{1/2} = \{(-1)(-A^2)\}^{1/2} = j^2A = -A!$$

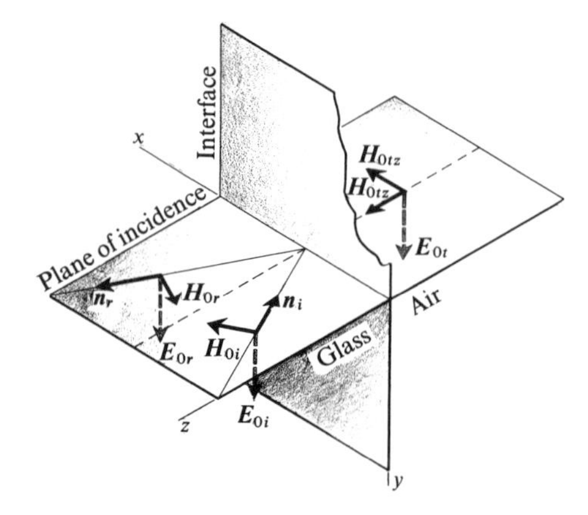

Figure 12-21. The $\boldsymbol{E}$ and $\boldsymbol{H}$ vectors for the incident, reflected, and transmitted waves in the case of total reflection when the incident wave is polarized with its $\boldsymbol{E}$ vector normal to the plane of incidence. The continuity of the tangential component of $\boldsymbol{H}$ across the interface makes $H_{0ty} = 0$.

We can dispose of $\boldsymbol{H}_{0r}$ immediately because the relation between $\boldsymbol{H}_{0r}$ and $\boldsymbol{E}_{0r}$ is that which applies to plane waves in dielectrics (Eq. 11-61).

We can assume, as in Section 12.2.1, that the vectors $\boldsymbol{E}_{0r}$ and $\boldsymbol{E}_{0t}$ for the reflected and transmitted waves are normal to the plane of incidence as in Figure 12-21. Then

$$\boldsymbol{E}_i = E_{0i} \exp j(\omega t - k_1 \boldsymbol{n}_i \cdot \boldsymbol{r})\, \boldsymbol{j}, \tag{12-83}$$

$$\boldsymbol{E}_r = E_{0r} \exp j(\omega t - k_1 \boldsymbol{n}_r \cdot \boldsymbol{r})\, \boldsymbol{j}, \tag{12-84}$$

$$\boldsymbol{E}_t = E_{0t} \exp j(\omega t - k_{2x} x - k_{2z} z)\, \boldsymbol{j}. \tag{12-85}$$

We have written the exponents in the first two equations in a more compact form than previously. The wave numbers k_{2x} and k_{2z} are now known (Eqs. 12-74 and 12-79).

For the incident and reflected waves, $\boldsymbol{H}$ is in the plane of incidence, just as in Figure 12-4, and it has both x- and z-components:

$$\boldsymbol{H}_i = H_{0i}(\cos\theta_i \boldsymbol{i} + \sin\theta_i \boldsymbol{k}) \exp j(\omega t - k_1 \boldsymbol{n}_i \cdot \boldsymbol{r}), \tag{12-86}$$

$$\boldsymbol{H}_r = H_{0r}(-\cos\theta_i \boldsymbol{i} + \sin\theta_i \boldsymbol{k}) \exp j(\omega t - k_1 \boldsymbol{n}_r \cdot \boldsymbol{r}). \tag{12-87}$$

For the transmitted wave, however, we must use a more general expression, since we know nothing as yet concerning its magnetic field intensity. We therefore set

$$\boldsymbol{H}_t = (H_{0tx}\boldsymbol{i} + H_{0ty}\boldsymbol{j} + H_{0tz}\boldsymbol{k}) \exp j(\omega t - k_{2x} x - k_{2z} z), \tag{12-88}$$

and we utilize the fact that the tangential component of $\boldsymbol{E}$ must be continuous across the interface. At the origin, which we have chosen at some arbitrary point in the interface, the exponential functions of $\boldsymbol{E}_i$, $\boldsymbol{E}_r$, $\boldsymbol{E}_t$ all reduce to $\exp j\omega t$; thus

$$E_{0i} + E_{0r} = E_{0t}. \tag{12-89}$$

Similarly, the continuity of the tangential component of $\boldsymbol{H}$ requires that

$$0 = H_{0ty}, \tag{12-90}$$

$$(H_{0i} - H_{0r}) \cos \theta_i = H_{0tx}, \tag{12-91}$$

or, from Eq. 11-61, which applies to plane waves in dielectrics,

$$\left(\frac{\epsilon_0}{\mu_0}\right)^{1/2} n_1(E_{0i} - E_{0r}) \cos \theta_i = H_{0tx}. \tag{12-92}$$

From the continuity of the normal component of $\boldsymbol{B}$ (or of $\boldsymbol{H}$, since the media are nonmagnetic) at the interface,

$$\left(\frac{\epsilon_0}{\mu_0}\right)^{1/2} n_1(E_{0i} + E_{0r}) \sin \theta_i = H_{0tz}. \tag{12-93}$$

We are now left with four unknowns, E_{0r}, E_{0t}, H_{0tx}, H_{0tz}, and only three equations. We therefore turn to Maxwell's equations. We choose one of the simpler ones and apply it to the transmitted wave. Since

$$\boldsymbol{\nabla} \cdot \boldsymbol{H}_t = 0, \tag{12-94}$$

then

$$k_{2x} H_{0tx} + k_{2z} H_{0tz} = 0. \tag{12-95}$$

Solving and substituting the values of k_{2x} and of k_{2z}, we find Eqs. 12-59 and 12-60, and

$$\left(\frac{H_{0tx}}{E_{0i}}\right)_N = -\left(\frac{\epsilon_0}{\mu_0}\right)^{1/2} \frac{2jn_1 \cos \theta_i \left\{\sin^2 \theta_i - \left(\frac{n_2}{n_1}\right)^2\right\}^{1/2}}{\cos \theta_i - j\left\{\sin^2 \theta_i - \left(\frac{n_2}{n_1}\right)^2\right\}^{1/2}}, \tag{12-96}$$

$$\left(\frac{H_{0tz}}{E_{0i}}\right)_N = \left(\frac{\epsilon_0}{\mu_0}\right)^{1/2} \frac{n_1 \sin 2\theta_i}{\cos \theta_i - j\left\{\sin^2 \theta_i - \left(\frac{n_2}{n_1}\right)^2\right\}^{1/2}}. \tag{12-97}$$

To obtain $\boldsymbol{E}_r$, the value of E_{0r} must be substituted into Eq. 12-84. Similarly, to obtain $\boldsymbol{E}_t$, H_{tx}, H_{tz}, the values of $\boldsymbol{E}_{0t}$, H_{0tx}, H_{0tz} must be multiplied by the exponential function of Section 12.4.1.2 for the transmitted wave.

Since both denominators on the right are identical, whereas one numerator is real and the other imaginary, the two components of $\boldsymbol{H}$ are $\pi/2$ radians out of phase, and, at any given point, it has neither a fixed amplitude nor a fixed direction but rotates in the plane of incidence. The transmitted wave is thus rather complex.

12.4.1.4 *The Poynting Vector for the Transmitted Wave.* It is instructive to calculate the Poynting vector for the transmitted wave. From Eq. 11-90,

$$(\mathcal{S}_{av})_N = \frac{1}{2} \operatorname{Re} (\boldsymbol{E}_t \times \boldsymbol{H}_t^*) = \frac{1}{2} \operatorname{Re} \begin{vmatrix} \boldsymbol{i} & \boldsymbol{j} & \boldsymbol{k} \\ 0 & E_t & 0 \\ H_{tx}^* & 0 & H_{tz}^* \end{vmatrix}, \tag{12-98}$$

$$= \frac{1}{2} \operatorname{Re} (E_t H_{tz}^*) \, \boldsymbol{i} - \frac{1}{2} \operatorname{Re} (E_t H_{tx}^*) \, \boldsymbol{k}, \tag{12-99}$$

where

$$E_t = \frac{2 \cos \theta_i}{\cos \theta_i - j\left\{\sin^2 \theta_i - \left(\frac{n_2}{n_1}\right)^2\right\}^{1/2}} E_{0i}$$

$$\times \exp\left\{ j\{\omega t - \frac{n_1}{\lambda_0} x \sin \theta_i\} + \frac{n_2}{\lambda_0} \left\{ \left(\frac{n_1}{n_2}\right)^2 \sin^2 \theta_i - 1 \right\}^{1/2} z \right\} \tag{12-100}$$

and where the stars identify complex conjugates ($-j$ substituted for j).

Then the x-component of $\mathcal{S}_{av}$ is

$$\frac{1}{2} \operatorname{Re} (E_t H_{tz}^*) = \left(\frac{\epsilon_0}{\mu_0}\right)^{1/2} \frac{n_1 \cos \theta_i \sin 2\theta_i}{1 - \left(\frac{n_2}{n_1}\right)^2} E_{0i}^2 \exp 2 \frac{n_2}{\lambda_0} \left\{ \left(\frac{n_1}{n_2}\right)^2 \sin^2 \theta_i - 1 \right\}^{1/2} z, \tag{12-101}$$

and the z-component is zero.

The energy flow in the transmitted wave was discussed near the end of Section 12.4.

12.5 REFLECTION AND REFRACTION AT THE SURFACE OF A GOOD CONDUCTOR

We have found the wave number of a good conductor in Section 11.5:

$$k_2 = \frac{n_2}{\lambda_0} = \left(\frac{\sigma_2 \mu_2 \omega}{2}\right)^{1/2} (1 - j) = \frac{1}{\delta} (1 - j), \tag{12-102}$$

where δ is the skin depth. Since the index of refraction n_2 is complex and very large, the direct application of Snell's law, Eq. 12-13, leads to a very small complex value for $\sin \theta_t$.

As we shall see in Section 12.5.1, the laws of reflection, Snell's law, and Fresnel's equations are again valid if we disregard the fact that n_2 is complex and if we set

$$\sin \theta_t = \frac{n_1}{n_2} \sin \theta_i, \tag{12-103}$$

$$\cos \theta_t = + \left\{ 1 - \left(\frac{n_1}{n_2}\right)^2 \sin^2 \theta_i \right\}^{1/2}. \tag{12-104}$$

Note that we require a *positive* sign before the square root, whereas we had a negative sign in the corresponding equation 12-54 for total reflection.

The second term between the braces is complex, because of n_2, but negligible since $n_2 \gg n_1$ (Section 11.5). Then

$$\cos\theta_t \approx 1, \qquad \theta_t \approx 0, \qquad \sin\theta_t \approx 0 \tag{12-105}$$

to a high degree of approximation. The wave penetrates into the conductor essentially along the normal to the surface, whatever the angle of incidence.

The exponential function for E_t is

$$\exp\left\{j\left(\omega t - \frac{n_1}{\lambda_0} x \sin\theta_i + \frac{z}{\delta}\right) + \frac{z}{\delta}\right\}.$$

This wave travels in a direction that is nearly perpendicular to the interface, and it is attenuated by a factor of e over one skin depth δ.

The concept of skin depth therefore applies to an electromagnetic wave incident at any angle on a good conductor. Whatever the angle of incidence θ_i, the transmitted wave can be considered to be a plane wave propagating along the normal to the surface, with the enormous damping which is characteristic of electromagnetic waves in good conductors.

INCIDENT WAVE POLARIZED WITH ITS $\boldsymbol{E}$ VECTOR NORMAL TO THE PLANE OF INCIDENCE. From Fresnel's equations 12-20 and 12-21,

$$\left(\frac{E_{0r}}{E_{0i}}\right)_N = \frac{n_1\mu_{r2}\cos\theta_i - \dfrac{\lambda_0}{\delta}(1-j)}{n_1\mu_{r2}\cos\theta_i + \dfrac{\lambda_0}{\delta}(1-j)} \approx -1, \tag{12-106}$$

$$\left(\frac{E_{0t}}{E_{0i}}\right)_N = \frac{2n_1\mu_{r2}\cos\theta_i}{n_1\mu_{r2}\cos\theta_i + \dfrac{\lambda_0}{\delta}(1-j)} \ll 1. \tag{12-107}$$

The fact that the first ratio is negative means that the $\boldsymbol{E}$ vector of the reflected wave is in the direction opposite to that shown in Figure 12-4, which was used for the calculation. As to the second ratio, it is necessarily quite small since $E_{0t} = E_{0i} - E_{0r}$ and $E_{0r} \approx E_{0i}$. Also, the E in the conductor can be expected to be small. The $\boldsymbol{E}$ vectors are shown in Figure 12-22.

Reflection from the surface of a dielectric with $n_2 \gg n_1$ would also give $E_{0r}/E_{0i} \approx -1$ and a weak transmitted wave.

It is interesting to note that there is a small loss of intensity on reflection from a good conductor, E_{0r} being somewhat smaller than E_{0i}. You will remember that, with total reflection, there is no loss of intensity and R is equal to unity (Eq. 12-59).

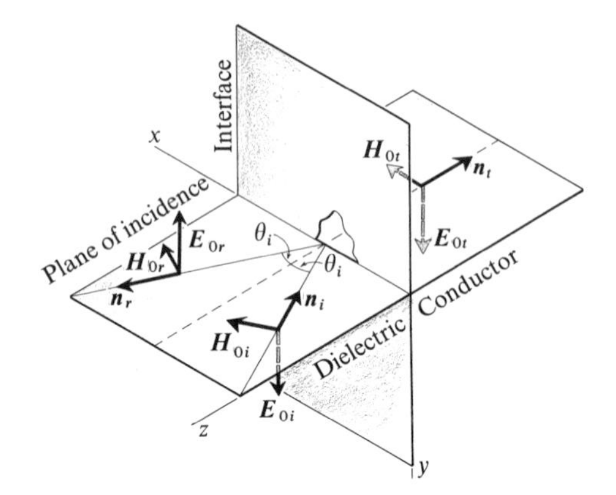

Figure 12-22. The incident, reflected, and transmitted waves at the interface between a dielectric and a good conductor. The incident wave is in the dielectric and is polarized with its $\boldsymbol{E}$ vector normal to the plane of incidence. The $\boldsymbol{H}$ vector of the transmitted wave is parallel to the interface, as shown, but it lags the $\boldsymbol{E}$ vector by $\pi/4$ radian.

The coefficient of reflection R is approximately equal to unity; the electric fields of the incident and reflected waves nearly cancel on the surface, and a weak, highly attenuated wave penetrates perpendicularly into the conductor.

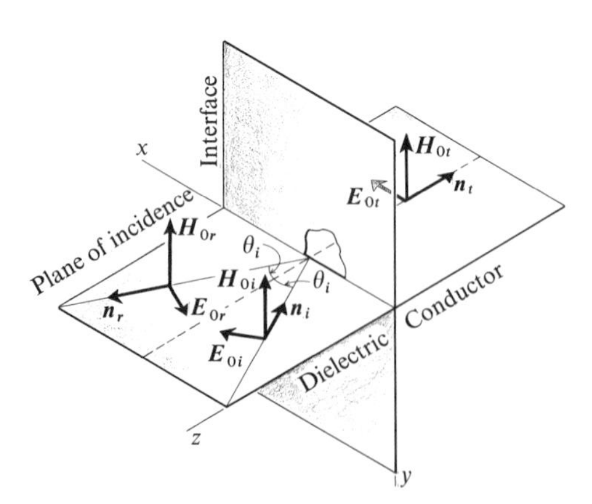

Figure 12-23. Reflection and refraction at the interface between a dielectric and a good conductor. The incident wave is in the dielectric and is polarized with its $\boldsymbol{E}$ vector *parallel* to the plane of incidence. The $\boldsymbol{E}$ vector of the transmitted wave is as shown, but it leads the $\boldsymbol{H}$ vector by $\pi/4$ radian. The coefficient of reflection R is approximately unity, the tangential components of the $\boldsymbol{E}$ vectors of the incident and reflected waves nearly cancel on the interface, and a weak attenuated wave penetrates perpendicularly into the conductor.

INCIDENT WAVE POLARIZED WITH ITS $\boldsymbol{E}$ VECTOR PARALLEL TO THE PLANE OF INCIDENCE. The three waves are as in Figure 12-23, again with

$$\left(\frac{E_{0r}}{E_{0i}}\right)_P \approx -1, \qquad \left(\frac{E_{0t}}{E_{0i}}\right)_P \ll 1. \tag{12-108}$$

The tangential components of $\boldsymbol{E}_{0i}$ and of $\boldsymbol{E}_{0r}$ nearly cancel at the surface of the conductor, as expected, and the transmitted wave is again a weak, highly attenuated plane wave which penetrates perpendicularly into the conductor.

Example

COMMUNICATING WITH SUBMARINES AT SEA

One interesting application of the above discussion is the problem of communicating with submarines at sea. For shore-to-ship communication with the submarine antenna submerged, the efficiency is extremely low, first because most of the incident energy is reflected upwards at the surface of the sea, and second because the weak transmitted wave is highly attenuated. It was shown in Problem 11-19 that the attenuation in sea water is about 53 decibels/foot at 20 megahertz, and 1.7 decibels/foot at 20 kilohertz. Very low frequencies ($\approx$ 15 kilohertz) are used at very high power. The response of a submerged receiving antenna will be calculated in Problem 14-24.

Ship-to-shore communication is presently impossible at these low frequencies since the power required at the transmitter is too large, and since it is impossible to use a sufficiently large antenna on the submarine. Two-way communication is achieved at frequencies of a few megahertz with the submarine antenna projecting from the water.

Example

STANDING WAVES AT NORMAL INCIDENCE

The three waves are shown in Figure 12-24. Since the direction of propagation of the reflected wave is opposite to that of the incident wave, and since $\boldsymbol{E} \times \boldsymbol{H}$ is in the direction of propagation, the $\boldsymbol{H}$ of the reflected wave must be in phase with that of the incident wave as in the figure. At the reflecting surface the electric field intensities nearly cancel, and we have a node of $\boldsymbol{E}$; the magnetic field intensities add, and we have a loop of $\boldsymbol{H}$. This is shown in Figure 12-25. The nodes of $\boldsymbol{E}$ and $\boldsymbol{H}$ are thus not coincident but are spaced a quarter wavelength apart.

A little thought will show that a similar situation exists for reflection from any surface. Either the $\boldsymbol{E}$ or the $\boldsymbol{H}$ vector must change direction on reflection, in order that the Poynting vector $\boldsymbol{E} \times \boldsymbol{H}$ can change direction.

Example

TRANSMISSION OF AN ELECTROMAGNETIC WAVE THROUGH A THIN SHEET OF COPPER AT NORMAL INCIDENCE

We assume that the sheet is in air as in Figure 12-26.

We use the subscript t for the wave transmitted through the first interface, the subscript tr for the wave reflected back at the second interface, and the subscript tt for the wave that emerges on the other side of the sheet, as in the figure. We neglect multiple reflections for the moment.

From Section 12.5 Eq. 12-107, setting $n_1 = 1$, $\mu_{r2} = 1$, and remembering that the t wave progresses in the negative direction of the z-axis,

$$E_t = E_{0t} \exp\left\{j\left(\omega t + \frac{z}{\delta}\right) + \frac{z}{\delta}\right\}, \tag{12-109}$$

with

$$E_{0t} = \frac{2\delta}{\lambda_0(1-j)} E_{0i}, \tag{12-110}$$

and, again from Section 12.5,

$$H_t = H_{0t} \exp\left\{j\left(\omega t + \frac{z}{\delta}\right) + \frac{z}{\delta}\right\}, \tag{12-111}$$

with

$$H_{0t} = 2\left(\frac{\epsilon_0}{\mu_0}\right)^{1/2} E_{0i} \tag{12-112}$$

so as to ensure continuity of H_{0t} at the interface. At the second interface, we again have three waves: the t wave with E_t and H_t as above, the tr wave with

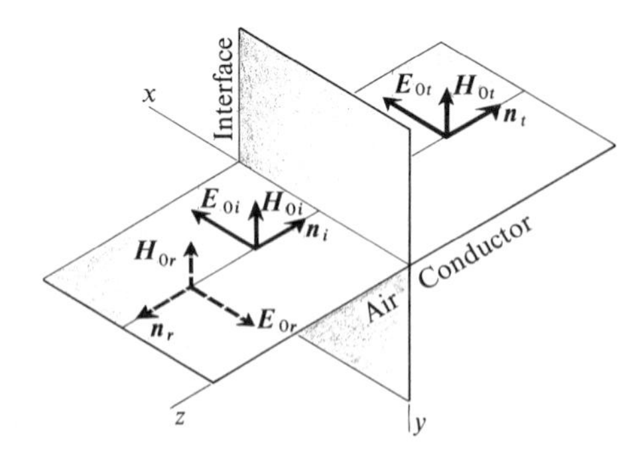

Figure 12-24. Reflection at normal incidence from the surface of a good conductor. The electric fields of the incident and reflected waves cancel at the interface, or nearly so. The magnetic field intensities add, however, with the result that the amplitude of the $\boldsymbol{H}_{0t}$ vector at the interface is $2H_{0i}$.

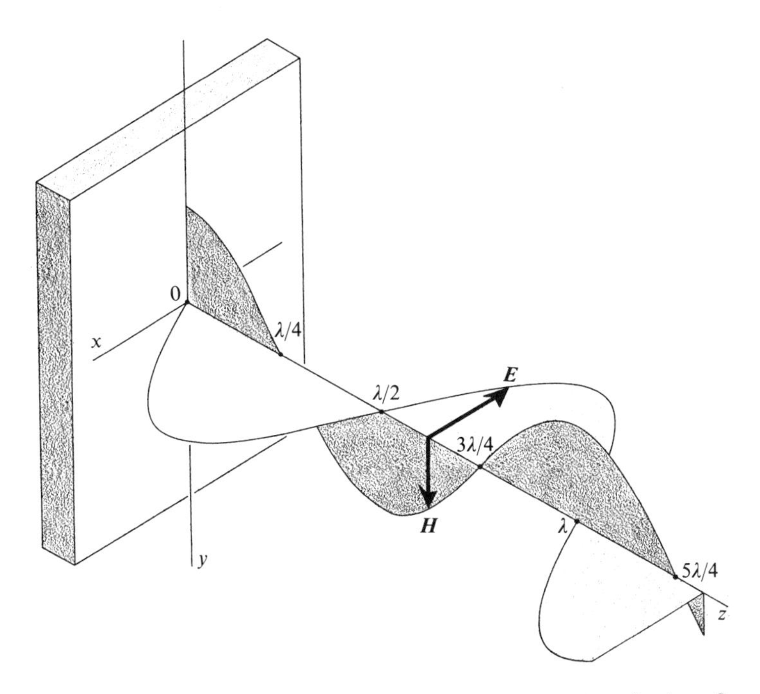

Figure 12-25. The standing-wave pattern resulting from the reflection of an electromagnetic wave at the surface of a good conductor. The curves show the standing waves of $\boldsymbol{E}$ and of $\boldsymbol{H}$ at some particular time. The nodes of $\boldsymbol{E}$ and of $\boldsymbol{H}$ are not coincident but are spaced $\lambda/4$ apart as shown.

$$E_{tr} = E_{0tr} \exp\left\{j\left(\omega t - \frac{z}{\delta}\right) - \frac{z}{\delta}\right\}, \tag{12-113}$$

$$H_{tr} = \left(\frac{\sigma}{2\mu_0\omega}\right)^{1/2}(1-j)E_{0tr} \exp\left\{j\left(\omega t - \frac{z}{\delta}\right) - \frac{z}{\delta}\right\}, \tag{12-114}$$

and the tt wave with

$$E_{tt} = E_{0tt} \exp j\left(\omega t + \frac{z}{\lambda\!\!\!^{-}_0}\right), \tag{12-115}$$

$$H_{tt} = \left(\frac{\epsilon_0}{\mu_0}\right)^{1/2} E_{0tt} \exp j\left(\omega t + \frac{z}{\lambda\!\!\!^{-}_0}\right). \tag{12-116}$$

The continuity equations require that

$$E_t + E_{tr} = E_{tt}, \tag{12-117}$$

and

$$H_t - H_{tr} = H_{tt} \tag{12-118}$$

or

$$(1-j)\frac{\lambda\!\!\!^{-}_0}{\delta}(E_t - E_{tr}) = E_{tt} \tag{12-119}$$

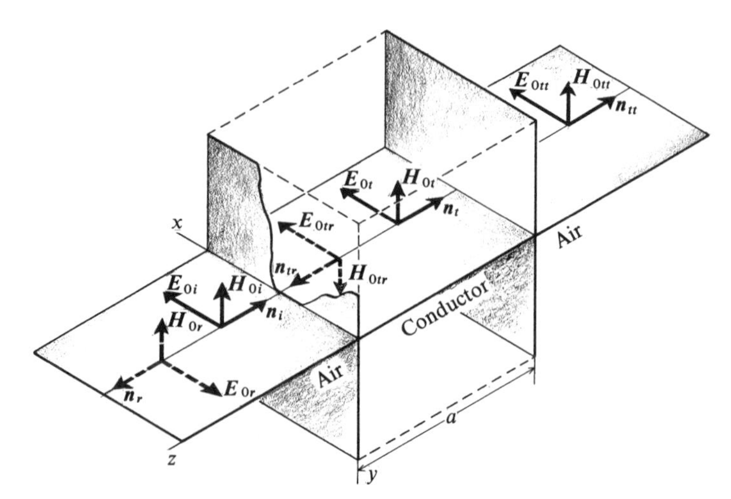

Figure 12-26. An electromagnetic wave incident normally on a conducting sheet of thickness a. The subscripts i, r, t, tr, tt refer, respectively, to the incident wave, to the wave reflected at the first interface, to the wave transmitted at the first interface, and so on. Multiple reflections inside the sheet are neglected.

at $z = -a$. Keeping in mind that $(\sigma/\epsilon_0\omega)^{1/2} \gg 1$ for a good conductor, we find that

$$\frac{E_{0tr}}{E_{0t}} = \exp\left\{-\frac{2a}{\delta}(1+j)\right\}, \tag{12-120}$$

$$\frac{E_{0tt}}{E_{0t}} = 2\exp\left\{j\frac{a}{\lambda_0} - \frac{a}{\delta}(1+j)\right\}, \tag{12-121}$$

$$E_{tr} = E_{0t}\exp\left\{j\left(\omega t - \frac{z+2a}{\delta}\right) - \frac{z+2a}{\delta}\right\}, \tag{12-122}$$

$$E_{tt} = 2E_{0t}\exp\left\{j\left(\omega t + \frac{z+a}{\lambda_0} - \frac{a}{\delta}\right) - \frac{a}{\delta}\right\}. \tag{12-123}$$

It will be noticed that at $z = -a$, with the above approximations, the $\mathbf{E}$ vector of the tr wave has the same phase and amplitude as the t wave, while the $\mathbf{E}$ vector of the tt wave is also in phase but has twice the amplitude.

In the incident wave,

$$\mathbf{S}_{i\,\text{av}} = \frac{1}{2}\operatorname{Re}(\mathbf{E}_i \times \mathbf{H}_i^*), \tag{12-124}$$

$$\mathcal{S}_{i\,\text{av}} = \frac{1}{2}\left(\frac{\epsilon_0}{\mu_0}\right)^{1/2} E_{0i}^2, \tag{12-125}$$

whereas, in the t wave,

$$\mathbf{S}_{t\,\mathrm{av}} = \frac{1}{2}\,\mathrm{Re}\,(\boldsymbol{E}_t \times \boldsymbol{H}_t^*), \tag{12-126}$$

$$\mathcal{S}_{t\,\mathrm{av}} = 2^{3/2}\left(\frac{\omega\epsilon_0}{\sigma}\right)^{1/2} e^{2z/\delta}\mathcal{S}_{i\,\mathrm{av}}. \tag{12-127}$$

Just below the first interface, at $z = 0$, $\mathcal{S}_{t\,\mathrm{av}} \ll \mathcal{S}_{i\,\mathrm{av}}$, since $(\omega\epsilon_0/\sigma)^{1/2} \approx \mathcal{Q}^{1/2}$ and $\mathcal{Q}$ is small (Section 11.5 and the example on page 479). Then, inside the sheet, the Poynting vector is further reduced by a factor of $e^{-2a/\delta}$. On the far side of the sheet, from Eqs. 12-121 and 12-110,

$$\mathcal{S}_{tt\,\mathrm{av}} = \frac{1}{2}\left(\frac{\epsilon_0}{\mu_0}\right)^{1/2} E_{0tt}^2, \tag{12-128}$$

$$= 16\,\frac{\omega\epsilon_0}{\sigma}\,e^{-2a/\delta}\mathcal{S}_{i\,\mathrm{av}}, \tag{12-129}$$

$$= \left\{2^{3/2}\left(\frac{\omega\epsilon_0}{\sigma}\right)^{1/2}\right\}(e^{-2a/\delta})\left\{2^{5/2}\left(\frac{\omega\epsilon_0}{\sigma}\right)^{1/2}\right\}\mathcal{S}_{i\,\mathrm{av}}, \tag{12-130}$$

where the quantities in enclosures show respectively the loss in energy flux that occurs at the first interface, in traversing the sheet of thickness a, and at the second interface.

For a sheet of copper ($\sigma = 5.80 \times 10^7$ mho/meter) 25.4 microns thick and at 1 megahertz, $\delta = 6.6 \times 10^{-5}$ meter (Table 11-1), and

$$\mathcal{S}_{tt\,\mathrm{av}} = (2.8 \times 10^{-6})(0.46)(5.5 \times 10^{-6})\mathcal{S}_{i\,\mathrm{av}}, \tag{12-131}$$

$$= 7.1 \times 10^{-12}\mathcal{S}_{i\,\mathrm{av}}. \tag{12-132}$$

It will be observed that the enormous attenuation arises mostly from reflection at the faces of the conducting sheet, for the thickness we have chosen.

Let us now take into consideration multiple reflections within the copper sheet. We have seen that reflection is nearly perfect at the second interface. This reflected wave proceeds to the left through the conductor and is similarly reflected to the right at the first interface, and so on. To take into account the wave that is reflected back at the second and then at the first interface, we must add to the above E_{tt} the term

$$2E_{0t}\exp\left\{j\left(\omega t + \frac{z+a}{\lambda\!\!\!^{-}_0}\right) - \frac{2a}{\delta}(1+j)\right\}.$$

For the wave which is reflected four times, we add the term

$$2E_{0t}\exp\left\{j\left(\omega t + \frac{z+a}{\lambda\!\!\!^{-}_0}\right) - \frac{5a}{\delta}(1+j)\right\},$$

and so forth. Then, taking into accoumt multiple reflections,

$$E'_{tt} = 2E_{0t}\exp\left\{j\left(\omega t + \frac{z+a}{\lambda\!\!\!^{-}_0}\right) - \frac{a}{\delta}(1+j)\right\}$$

$$\times\left\{1 + \exp\left\{-\frac{2a}{\delta}(1+j)\right\} + \exp\left\{-\frac{4a}{\delta}(1+j)\right\} + \cdots\right\}, \tag{12-133}$$

where the quantity between braces is of the form

$$1 + x^2 + x^4 + \cdots = \frac{1}{1 - x^2}. \tag{12-134}$$

Then

$$E'_{tt} = \frac{E_{0t} \exp\left\{j\left(\omega t + \dfrac{z + a}{\lambda_0}\right)\right\}}{\sinh\left\{\dfrac{a}{\delta}(1 + j)\right\}}, \tag{12-135}$$

$$\frac{E'_{0tt}}{E_{0t}} = \frac{\exp j\dfrac{a}{\lambda_0}}{\sinh\left\{\dfrac{a}{\delta}(1 + j)\right\}}, \tag{12-136}$$

and

$$\mathcal{S}_{tt'\,\mathrm{av}} = \frac{1}{2}\left(\frac{\epsilon_0}{\mu_0}\right)^{1/2} |E'_{0tt}|^2, \tag{12-137}$$

$$= \frac{\dfrac{1}{2}\left(\dfrac{\epsilon_0}{\mu_0}\right)^{1/2}}{\sinh^2\dfrac{a}{\delta}\cos^2\dfrac{a}{\delta} + \cosh^2\dfrac{a}{\delta}\sin^2\dfrac{a}{\delta}} |E_{0t}|^2, \tag{12-138}$$

$$= \frac{4\dfrac{\omega\epsilon_0}{\sigma}}{\sinh^2\dfrac{a}{\delta}\cos^2\dfrac{a}{\delta} + \cosh^2\dfrac{a}{\delta}\sin^2\dfrac{a}{\delta}} \mathcal{S}_{i\,\mathrm{av}}, \tag{12-139}$$

$$= 1.29 \times 10^{-11}\ \mathcal{S}_{i\,\mathrm{av}}, \tag{12-140}$$

which is larger than when multiple reflections are neglected. This is of course true only for the particular case we have chosen.

12.5.1 Demonstration of the Validity of Snell's Law, of the Laws of Reflection and Refraction, and of Fresnel's Equations at the Interface Between a Dielectric and a Good Conductor*

We shall proceed in the same manner as for total reflection and write that

$$\boldsymbol{E}_i = \boldsymbol{E}_{0i} \exp j\{\omega t - k_1(x \sin\theta_i - z\cos\theta_i)\}, \tag{12-141}$$

$$\boldsymbol{E}_r = \boldsymbol{E}_{0r} \exp j\{\omega t - k_1(x \sin\theta_i + z\cos\theta_i)\}, \tag{12-142}$$

$$\boldsymbol{E}_t = \boldsymbol{E}_{0t} \exp j(\omega t - k_{2x}x - k_{2z}z), \tag{12-143}$$

with axes as in Figure 12-27. We have not taken the trouble to show that $\boldsymbol{E}_r$ is of the form shown above; the demonstration is identical to that which led to

* You may wish to omit Sections 12.5.1 and 12.5.1.1.

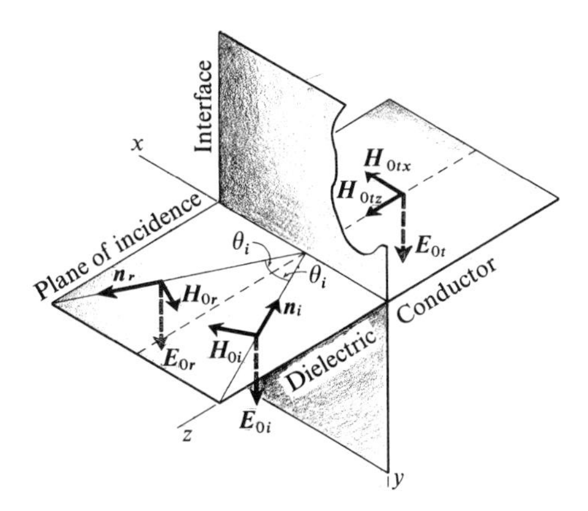

Figure 12-27. An electromagnetic wave incident on a good conductor, with its $\boldsymbol{E}$ vector *normal* to the plane of incidence. The angle of reflection is equal to the angle of incidence. The continuity of the tangential component of $\boldsymbol{H}$ across the interface makes $H_{0ty} = 0$.

Eq. 12-73 for total reflection. We have set all three circular frequencies equal, as previously.

12.5.1.1 *The Wave Numbers k_{2x} and k_{2z} for Refraction into a Good Conductor.* We can again equate the exponential functions for $\boldsymbol{E}_i$ and $\boldsymbol{E}_t$ on the interface, just as we did in Sections 12.1 and 12.4.1.2. Then

$$k_{2x} = k_1 \sin\theta_i = \frac{n_1}{\lambda_0} \sin\theta_i, \tag{12-144}$$

as for total reflection.

We can find k_{2z} from the wave equation for good conductors,

$$\frac{\partial^2 E_t}{\partial x^2} + \frac{\partial^2 E_t}{\partial z^2} = j\omega\sigma_2\mu_2 E_t. \tag{12-145}$$

The derivative with respect to y is zero since the wave is assumed to be independent of the y-coordinate. Then

$$-k_{2x}^2 - k_{2z}^2 = j\omega\sigma_2\mu_2 = -k_2^2, \tag{12-146}$$

where, from Eqs. 11-104 and 11-105,

$$k_2 = \frac{1-j}{\delta} = \frac{n_2}{\lambda_0}. \tag{12-147}$$

Thus

$$k_{2z} = \pm k_2 \left(1 - \frac{k_{2x}^2}{k_2^2}\right)^{1/2}, \tag{12-148}$$

$$= \pm k_2 \left\{1 - \left(\frac{n_1}{n_2}\right)^2 \sin^2\theta_i\right\}^{1/2}. \tag{12-149}$$

Since the "index of refraction" n_2 of the conductor is much larger than that of the dielectric n_1, the second term under the square root can be neglected. Thus

$$k_{2z} \approx - k_2 = \frac{-1 + j}{\delta}, \tag{12-150}$$

and

$$k_{2x} \ll k_{2z}. \tag{12-151}$$

We have selected the negative sign before k_2 so as to make the imaginary part of k_{2z} positive. This is required to prevent the wave from building up to infinite amplitude as $z \to -\infty$.

If both n_1 and n_2 were real, we would have

$$k_{2z} = - k_2 \cos \theta_t. \tag{12-152}$$

Comparing Eqs. 12-149 (with a negative sign before k_2) and 12-152, we see that we were justified in writing $\cos \theta_t$ as in Eq. 12-104.

12.5.1.2 *The Amplitudes of **E** and **H** in the Reflected and Transmitted Waves.* We assume that $\boldsymbol{E}_i$ is *normal* to the plane of incidence. Then the $\boldsymbol{E}$ vectors are all normal to the plane of incidence and parallel to the interface as in Figure 12-22.

At first sight one expects a discontinuity in the tangential component of the magnetic field intensity at the interface, because of the existence of surface currents. This is incorrect, however, and there is continuity for the following reason: currents flow *in*side the conductor (unless it is superconducting), and they are distributed over a finite thickness. There is *zero* surface current. Thus, if we draw a closed path like the one in Figure 9-10*b*, with its two long sides infinitely close to the surface, there is zero current enclosed by the path and the tangential component of $\boldsymbol{H}$ must be the same on both sides of the interface. If we make the path a little wider, keeping one side some distance away from the conductor and pushing the other one into the conductor, then the current linking the path is not zero and the tangential H's are not equal. But then we are not comparing the tangential components of $\boldsymbol{H}$ at the interface. This point was mentioned in Section 9.7.

A similar argument shows that the tangential component of $\boldsymbol{E}$ is the same on both sides of the interface. Thus

$$E_{0i} + E_{0r} = E_{0t} \tag{12-153}$$

$$0 = H_{0ty} \tag{12-154}$$

$$\left(\frac{\epsilon_0}{\mu_0}\right)^{1/2} n_1(E_{0i} - E_{0r}) \cos \theta_i = H_{0tx}, \tag{12-155}$$

$$\left(\frac{\epsilon_0}{\mu_0}\right)^{1/2} n_1(E_{0i} + E_{0r}) \sin \theta_i = H_{0tz}, \tag{12-156}$$

$$k_{2x}H_{0tx} + k_{2z}H_{0tz} = 0. \tag{12-157}$$

From the last equation,

$$H_{0tx} \gg H_{0tz}, \tag{12-158}$$

since k_{2x} is much smaller than k_{2z} (Eq. 12-151). The wave in the conductor therefore has its $\boldsymbol{E}$ vector parallel to the y-axis and its $\boldsymbol{H}$ vector parallel to the x-axis. It propagates in the negative direction of the z-axis.

Upon solving for E_{0r} and E_{0t} we find Eqs. 12-106 and 12-107. Solving for H_{0tx} shows that $\boldsymbol{E}$ and $\boldsymbol{H}$ in the conductor are related as in Eq. 11-106 and that $\boldsymbol{E}$ leads $\boldsymbol{H}$ by $\pi/4$ radian.

12.6 RADIATION PRESSURE AT NORMAL INCIDENCE ON A GOOD CONDUCTOR

We have now discussed at some length the reflection and the refraction of an electromagnetic wave at the interface between two dielectrics, and then at the interface between a dielectric and a good conductor. We also wish to investigate the same phenomena for the case of a dielectric and an ionized gas. Before going on with this, however, we shall study another phenomenon that is related to reflection from conductors, namely radiation pressure.

Let us consider an electromagnetic wave incident on a good conductor. We limit ourselves to normal incidence. We have seen above that the electric field intensity transmitted into the metal is small compared to that of the incident wave, but that it is not zero. We have also seen that the $\boldsymbol{E}$ and $\boldsymbol{H}$ vectors of the transmitted wave inside the conductor are orthogonal, just as in the incident wave.

According to Ohm's law, the electric field intensity $\boldsymbol{E}_t$ in the conductor gives a current density $\sigma \boldsymbol{E}_t$, where σ is the conductivity of the medium. This current, which is oriented like $\boldsymbol{E}_t$, is perpendicular to $\boldsymbol{H}_t$. It turns out, as we shall see presently, that under these conditions the conduction electrons are pushed by the magnetic force $Q\boldsymbol{u} \times \boldsymbol{B}$ in the direction of propagation of the wave. The electrons in turn push on the conductor in the process of colliding with the atoms in their path, giving rise to radiation pressure.

The situation would be very different if the conduction electrons were entirely free to move through the metal: the conductivity would then be

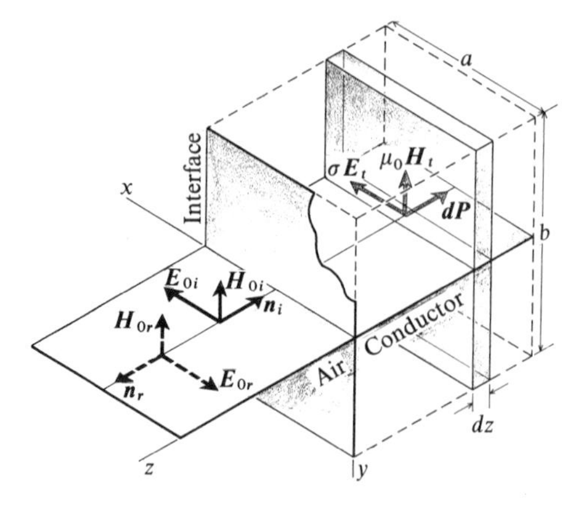

Figure 12-28. The conduction current σE_t and the magnetic induction $\mu_0 \boldsymbol{H}$ which give rise to radiation pressure in an element of volume of a reflector. The sides of the element of volume are parallel to the $\boldsymbol{E}$ and $\boldsymbol{H}$ vectors. The element has a thickness dz and is ab square meters in area.

imaginary, as in the low density gases that we studied in Section 11.6.1, and the radiation pressure would be zero.

Let us study the radiation pressure of a wave incident normally on a good conductor. We consider an element of volume that is parallel to the interface, of thickness dz as in Figure 12-28, and ab meter^2 in area. It carries a current $\sigma Eb\,dz$, and is submitted to a magnetic force $\sigma E\mu_0 Hab\,dz$ in the negative direction of the z-axis.

Note that the relative permeability μ_r of the conductor does not enter into the calculation. It has been shown that the magnetic force exerted on an electron within a magnetic material is proportional to $\mu_0 \boldsymbol{H}$ and not to $\mu \boldsymbol{H}$, for *slow* electrons like the ones we are considering.*

Then the pressure exerted by the wave on the element of thickness dz is

$$dP = \sigma E_t \mu_0 H_t dz. \tag{12-159}$$

With E_t and H_t chosen to be positive in the directions shown in Figure 12-28, a positive result will show that the wave pushes on the conductor.

From Eq. 12-107 for $\theta_i = 0$ and $\lambda_0 \gg \delta$,

$$E_t = \frac{2\mu_{r2} n_1 \delta}{\lambda_0 (1 - j)} E_{0i} \exp\left\{ j\left(\omega t + \frac{z}{\delta}\right) + \frac{z}{\delta} \right\}, \tag{12-160}$$

$$H_t = 2H_{0i} \exp\left\{ j\left(\omega t + \frac{z}{\delta}\right) + \frac{z}{\delta} \right\}. \tag{12-161}$$

We cannot continue using the exponential functions, however, since we must evaluate their product (see Appendix D). We must therefore rewrite them in the form

* G. H. Wannier, Phys. Rev. **72**, 304 (1947).

$$E_t = \frac{2\mu_{r2}n_1\delta}{2^{1/2}\lambda_0} E_{0i}e^{z/\delta} \cos\left(\omega t + \frac{z}{\delta} + \frac{\pi}{4}\right), \tag{12-162}$$

$$H_t = 2H_{0i}e^{z/\delta} \cos\left(\omega t + \frac{z}{\delta}\right). \tag{12-163}$$

Then

$$dP = \sigma\mu_0\delta E_t H_t \frac{dz}{\delta}, \tag{12-164}$$

$$= 4(2^{1/2})n_1 \frac{E_{0i}H_{0i}}{c} e^{2z/\delta} \cos\left(\omega t + \frac{z}{\delta}\right) \cos\left(\omega t + \frac{z}{\delta} + \frac{\pi}{4}\right) \frac{dz}{\delta}. \tag{12-165}$$

This is the *instantaneous* value of the pressure on the element of thickness dz at z. To find the *average* pressure we must replace the term between the the braces by its average value over one period $T = 2\pi/\omega$:

$$\frac{1}{T}\int_0^T \cos\left(\omega t + \frac{z}{\delta}\right) \cos\left(\omega t + \frac{z}{\delta} + \frac{\pi}{4}\right) dt = \frac{1}{2^{3/2}}. \tag{12-166}$$

Then

$$dP_{\text{av}} = 2n_1 \frac{E_{0i}H_{0i}}{c} e^{2z/\delta} \frac{dz}{\delta}. \tag{12-167}$$

The average pressure is in the same direction at all depths within the conductor.

Finally, integrating over all z within the conductor, from $-\infty$ to 0,

$$P_{\text{av}} = 2n_1 \frac{E_{0i}H_{0i}}{c} \int_{-\infty}^{0} e^{2z/\delta} \frac{dz}{\delta}, \tag{12-168}$$

$$= n_1 \frac{E_{0i}H_{0i}}{c}. \tag{12-169}$$

This is a positive quantity. As we saw at the beginning of this calculation, the fact that P_{av} is positive means that it is in the direction of propagation of the incident wave. The quantity P_{av} is the average pressure exerted on the conductor by the incident radiation, or the radiation pressure.

If the incident wave propagates in a vacuum,

$$P_{\text{av}} = 2\frac{\mathcal{S}_{i\,\text{av}}}{c}, \tag{12-170}$$

where $\mathcal{S}_{i\,\text{av}}$ is the average of the absolute value of the Poynting vector for the *incident* wave (Section 11.3).

We can ascribe this pressure to a change in momentum of $2\,\mathcal{S}_{\text{av}}/c$ per unit time and per unit area in the incident wave, the factor 2 being required because the wave is reflected with a momentum equal to its initial momentum, but of opposite sign.

An electromagnetic wave propagating in a vacuum can therefore be considered to involve a flux of momentum that is equal to $\mathcal{S}_{av}/c$, or to its energy density, according to Section 11.3; that is,

$$\text{(Flux of momentum)} = \text{(Momentum density)} \times \text{(Phase velocity)}, \tag{12-171}$$

$$= \text{(Energy density)}. \tag{12-172}$$

These results agree with those of atomic physics, where we consider an electromagnetic wave to involve photons of energy $\hbar\omega$($\hbar = 1.05 \times 10^{-34}$ joule-second is Planck's constant h divided by 2π) and of momentum $\hbar/\lambda\!\!\!^{-}$ traveling with a velocity u:

$$\frac{\hbar}{\lambda\!\!\!^{-}} u = \hbar\omega. \tag{12-173}$$

The radiation pressure of electromagnetic waves has been observed experimentally, and it has been found to agree with the above theory. This is a confirmation of our hypothesis that the effective magnetic induction acting on a *slow* electron inside a magnetic material is $\mu_0 \boldsymbol{H}$, and not $\mu \boldsymbol{H}$. It also is a demonstration of the fact that an electromagnetic field can have a momentum.

Examples

Radiation pressure is small for the usual radiation intensities. In sunlight the Poynting vector is approximately 1.4 kilowatts/meter^2 at the surface of the Earth, giving a radiation pressure of about 9.3×10^{-6} newton/meter^2 or 0.1 microgram/centimeter^2 on a metallic reflector. The radiation pressure P varies as the inverse square of the distance from the source, as does the Poynting vector $\mathcal{S}$.

To find P at the surface of the Sun we must multiply the above figures by

$$\left(\frac{\text{mean distance of Earth to Sun}}{\text{radius of Sun}}\right)^2 = \left(\frac{1.5 \times 10^{11}}{7 \times 10^8}\right)^2 = 4.6 \times 10^4, \tag{12-174}$$

which still gives only 4.4 milligrams/centimeter^2, or 4.3×10^{-6} atmosphere. Radiation pressure is unimportant even in the interior of the Sun, but it possibly plays an important role in the more luminous stars. As is well known, comet tails point predominantly away from the Sun. This phenomenon is explained in part by radiation pressure.

Much higher power densities and pressures are available in wave guides, where $\mathcal{S}_{av}$ can reach values of the order of 10^9 watts/meter^2. Then the radiation pressure is 0.1 gram/centimeter^2, assuming that the waves propagating within wave guides are plane. This is not correct, as we shall see, but the error is unimportant here, since we are concerned only with orders of magnitude.

In a laser beam with a power density of 10^{11} watts/meter^2 the radiation pressure is 10 grams/centimeter^2.

12.7 REFLECTION OF AN ELECTROMAGNETIC WAVE BY AN IONIZED GAS

Finally, we study the behavior of an electromagnetic wave that encounters an ionized gas. We assume, as in Section 11.6 that the electrons do not collide with the molecules of the gas; in other words, we assume that the pressure is low. Under those conditions the *phase* velocity is *larger* in the ionized gas than in free space, the index of refraction being

$$n = \frac{c}{u} = \left\{1 - \left(\frac{\omega_p}{\omega}\right)^2\right\}^{1/2}, \tag{12-175}$$

$$= \left\{1 - 80.5\frac{N_e}{f^2}\right\}^{1/2}, \tag{12-176}$$

from Eqs. 11-173 and 11-174, where c is the velocity of light, u is the phase velocity of the wave, ω_p is the plasma angular frequency of the wave, N_e is the number of free electrons per cubic meter, and f is the frequency of the wave.

If the ionized gas had a definite boundary and a uniform value of N_e throughout its volume, reflection and refraction at its surface would be simple to describe: the ionized gas would simply act as a dielectric with $n < 1$. As a rule, neither assumption is valid and the wave is reflected in much the same way that a light wave is reflected in a mirage.

It is possible to calculate the path of a ray by performing a numerical integration of the ray equation stated in Problem 12-2. However, one can deduce the main features of the reflection by simply using Snell's law.

We select coordinates as in Figure 12-29 and assume that the index of refraction n varies slowly with z, but not with the other two coordinates x and y. To be more specific, we assume that n varies by a negligible amount over one wavelength. If N_e gradually increases with z, a given ray gradually bends down as in Figure 12-29 to an angle θ, at a point where the index of refraction is n.

We can calculate θ in the following way. When refraction occurs at the interface between any two media n_1 and n_2, the quantity $n \sin\theta$ is conserved in going from one side of the interface to the other. This is Snell's law, Eq. 12-13. If the index of refraction varies continuously, the medium can be imagined to be stratified in infinitely thin layers, and $n \sin\theta$ is similarly conserved all along the ray. Thus

$$n \sin\theta = n_1 \sin\theta_i, \tag{12-177}$$

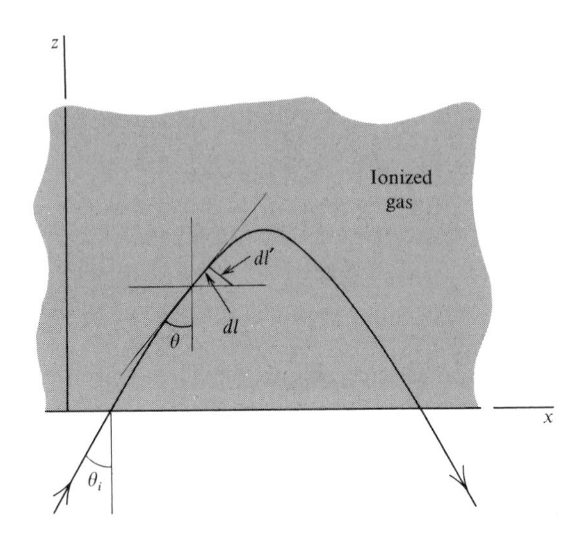

Figure 12-29. An electromagnetic wave incident on an ionized region at an angle θ_i is deflected at an angle θ after penetrating to a distance z.

where n_1 is the index of refraction at $z = 0$. If we set $n_1 = 1$, then

$$n \sin \theta = \sin \theta_i. \tag{12-178}$$

It is interesting to differentiate this equation with respect to the distance l measured along a ray. We find that

$$\frac{d\theta}{dl} = -\frac{1}{n}\frac{dn}{dl} \tan \theta. \tag{12-179}$$

If the ray penetrates into an ionized region where the ion density increases with z, the index of refraction n *de*creases with l and the derivative dn/dl is negative, so that the angle θ *in*creases with distance as in Figure 12-29. After some distance, if n becomes sufficiently large, θ becomes equal to 90°. At that point the tangent of θ becomes infinite, but dn/dl becomes zero. After this, $\tan \theta$ becomes negative, whereas the derivative dn/dl becomes positive, and θ keeps increasing until the ray escapes from the ionized region at an angle equal to the angle of incidence θ_i.

At the top of the trajectory,

$$\sin \theta = 1, \tag{12-180}$$

$$n_{90°} = \sin \theta_i. \tag{12-181}$$

This is the index of refraction required for reflection when the angle of incidence is θ_i.

From Eq. 12-175,

$$\sec \theta_i = \frac{\omega}{\omega_p}, \tag{12-182}$$

where ω is the angular frequency of the wave and ω_p is the plasma angular frequency at the value of z where $\theta = 90°$. For normal incidence, $\theta_i = 0$, $\sec \theta_i = 1$, and reflection occurs at a z sec $\omega_p = \omega$, that is, where n is zero and the phase velocity is infinite. At oblique incidence, however, $\sec \theta_i > 1$, and reflection occurs where $\omega_p < \omega$, or at a lower value of z, if we assume that the electron density N_e, and hence ω_p, increase with z.

We have found above the value of $d\theta/dl$. This is the reciprocal of the radius of curvature R, and thus

$$\frac{1}{R} = -\frac{1}{n}\frac{dn}{dl}\tan \theta. \tag{12-183}$$

A positive value of R correponds to a positive value of $d\theta/dl$ and to a trajectory which is concave downwards in this case. Also,

$$\frac{1}{R} = -\frac{1}{n}\frac{dn}{dl'}, \tag{12-184}$$

where l' is perpendicular to the direction of propagation, as in Figure 12-29. The ray therefore bends most sharply where the index of refraction n varies most rapidly in the direction perpendicular to the ray.

12.8 SUMMARY

The laws of reflection, Snell's law, and Fresnel's equations result from the continuity of the tangential components of $\boldsymbol{E}$ and of $\boldsymbol{H}$ at the interface between two media.

The *laws of reflection* are the following: (a) the angle of reflection is equal to the angle of incidence, (b) the normals to the wavefronts of the incident and reflected waves lie in a plane that also contains the normal to the interface and that is called the *plane of incidence.*

According to *Snell's law*,

$$\frac{\sin \theta_t}{\sin \theta_i} = \frac{n_1}{n_2}, \tag{12-13}$$

where θ_i and θ_t are the angles of incidence and of refraction respectively and where n_1 is the index of refraction of the first medium and n_2 is that of the second medium.

Fresnel's equations relate the amplitudes and phases of the reflected and transmitted waves to those of the incident wave. The subscripts N and P indicate that the $\boldsymbol{E}$ vector of the incident wave is either normal or parallel to the plane of incidence.

$$\left(\frac{E_{0r}}{E_{0i}}\right)_N = \frac{\dfrac{n_1}{\mu_{r1}}\cos\theta_i - \dfrac{n_2}{\mu_{r2}}\cos\theta_t}{\dfrac{n_1}{\mu_{r1}}\cos\theta_i + \dfrac{n_2}{\mu_{r2}}\cos\theta_t}, \tag{12-20}$$

$$\left(\frac{E_{0t}}{E_{0i}}\right)_N = \frac{2\dfrac{n_1}{\mu_{r1}}\cos\theta_i}{\dfrac{n_1}{\mu_{r1}}\cos\theta_i + \dfrac{n_2}{\mu_{r2}}\cos\theta_t}, \tag{12-21}$$

$$\left(\frac{E_{0r}}{E_{0i}}\right)_P = \frac{-\dfrac{n_2}{\mu_{r2}}\cos\theta_i + \dfrac{n_1}{\mu_{r1}}\cos\theta_t}{\dfrac{n_2}{\mu_{r2}}\cos\theta_i + \dfrac{n_1}{\mu_{r1}}\cos\theta_t}, \tag{12-25}$$

$$\left(\frac{E_{0t}}{E_{0i}}\right)_P = \frac{2\dfrac{n_1}{\mu_{r1}}\cos\theta_i}{\dfrac{n_2}{\mu_{r2}}\cos\theta_i + \dfrac{n_1}{\mu_{r1}}\cos\theta_t}. \tag{12-26}$$

At the *Brewster angle* for dielectrics there is no reflected wave if the incident wave is polarized with its $\boldsymbol{E}$ vector in the plane of incidence:

$$\frac{n_1}{n_2} = \cot\theta_{iB}. \tag{12-40}$$

The coefficient of reflection R gives the fraction of the incident power which is reflected. Similarly, the *coefficient of transmission* T is the fraction of the incident power which is transmitted. Thus $R + T = 1$.

Total reflection occurs when Snell's law gives $\sin\theta_t \geqq 1$. Then $R = 1$. The transmitted wave travels along the interface, and is attenuated exponentially in the direction perpendicular to the interface. The average energy flow across the interface is zero, but the instantaneous flow is alternately one way and then the other. Snell's law and Fresnel's equations give the correct results when applied to total reflection if we set

$$\cos\theta_t = -(1 - \sin^2\theta_t)^{1/2} = -\left\{1 - \left(\frac{n_1}{n_2}\right)^2 \sin^2\theta_i\right\}^{1/2}, \tag{12-54}$$

Good conductors have a coefficient of reflection R close to unity. The transmitted wave penetrates nearly perpendicularly into the conductor, whatever the angle of incidence, n_2 being much larger than n_1. It is attenuated by a factor e in one skin depth

$$\delta = \left(\frac{2}{\omega\sigma_2\mu_2}\right)^{1/2}. \quad (12\text{-}102)$$

The Fresnel equations again apply if we set

$$\cos\theta_t = +(1 - \sin^2\theta_t)^{1/2} = +\left\{1 - \left(\frac{n_1}{n_2}\right)^2 \sin^2\theta_i\right\}^{1/2}, \quad (12\text{-}104)$$

At normal incidence, *standing waves* are formed in front of a good conductor, the reflecting surface giving a node of $\boldsymbol{E}$ and a loop of $\boldsymbol{H}$. Nodes of $\boldsymbol{E}$ and of $\boldsymbol{H}$ are spaced $\lambda/4$ apart.

Radiation pressure is due to the magnetic force on the electrons oscillating under the influence of the electric field of the wave. It is shown that

$$P_{av} = 2\frac{\mathcal{S}_{i\ av}}{u}, \quad (12\text{-}170)$$

where P_{av} is the average value of the radiation pressure in newtons/meter^2, $\mathcal{S}_{i\ av}$ is the average value of the magnitude of the Poynting vector in watts/meter^2, and u is the phase velocity of the wave in meters/second. This leads to the concepts of flux of momentum and of momentum density for an electromagnetic wave. The forces due to radiation pressure are small and are usually negligible.

An electromagnetic wave is reflected by an ionized gas much as light waves are reflected in a mirage. If the ionization density does not vary in the direction parallel to the interface, we find that

$$n \sin\theta = n_1 \sin\theta_i, \quad (12\text{-}177)$$

where θ_i is the angle of incidence, n_1 is the index of refraction of medium 1, and θ gives the orientation of the ray at the point where the index of refraction is n (Figure 12-29). If $n_1 = 1$, then $\theta = 90°$ at a point where n is equal to $\sin\theta_i$. If N_e increases with z, the value of z where $\theta = 90°$ increases as the angle of incidence θ_i decreases and, at normal incidence, reflection occurs at a point where ω is equal to the plasma frequency ω_p and where the phase velocity is therefore infinite.

PROBLEMS

12-1. In a plane sinusoidal wave, the components of $\boldsymbol{E}$ and $\boldsymbol{H}$ are of the form

$$W = W_0 \exp j(\omega t - \boldsymbol{k}\cdot\boldsymbol{r}),$$
$$= W_0 \exp j(\omega t - k_x x - k_y y - k_z z),$$

where W_0 is a constant that can be complex. See Problem 11-11.

The vector $\boldsymbol{k}$, which is called the wave number (not to be confused with the unit vector along the z-axis), can also be complex. Then we can write that

$$\boldsymbol{k} = a\boldsymbol{n}_1 - jb\boldsymbol{n}_2,$$

where $\boldsymbol{n}_1$ and $\boldsymbol{n}_2$ are unit vectors, with $k_x = an_{1x} - jbn_{2x}$, and similarly for k_y and k_z.

(a) Show that surfaces of constant phase are normal to $\boldsymbol{n}_1$ and that surfaces of constant amplitude are normal to $\boldsymbol{n}_2$.
(b) Show that a is related to the wavelength, and b to the attenuation.
(c) Show that, in all operations involving $\boldsymbol{\nabla}$, this operator can be replaced by $-j\boldsymbol{k}$.
(d) Rewrite Maxwell's equations utilizing this fact.
(e) It does *not* follow that $\boldsymbol{E}, \boldsymbol{H}, \boldsymbol{k}$ are orthogonal when $\rho_t = 0$ and ϵ_r, μ_r are constants. To show this, use Maxwell's equations, with W expressed as a cosine function, and then find the conditions for orthogonality.

12-2. A wave travels in a stratified medium whose index of refraction is a function of the coordinate y.

(a) Show that the angle θ between a ray and the y-axis is given by:

$$\frac{d\theta}{ds} = -\frac{dn/dy}{n}\sin\theta,$$

where the distance s is measured along the ray.
(b) You can now verify the *ray equation*

$$\frac{d}{ds}(n\boldsymbol{t}) = \boldsymbol{\nabla} n,$$

where $\boldsymbol{t}$ is a unit vector tangent to the ray at a point where the index of refraction is n.‡

12-3. Show that

$$\left(\frac{H_{0t}}{H_{0i}}\right)_N = \frac{2\mu_{r1}n_2\cos\theta_i}{\mu_{r2}n_1\cos\theta_i + \mu_{r1}n_2\cos\theta_t},$$

$$\left(\frac{H_{0r}}{H_{0i}}\right)_N = -\frac{\mu_{r2}n_1\cos\theta_i - \mu_{r1}n_2\cos\theta_t}{\mu_{r2}n_1\cos\theta_i + \mu_{r1}n_2\cos\theta_t}.$$

12-4. Electromagnetic radiation is incident normally on a dielectric whose index of refraction is n.

Show that the reflected wave can be eliminated by covering the dielectric with a layer of a second dielectric whose index of refraction is $n^{1/2}$ and whose thickness is one-quarter wavelength.

12-5. Calculate the Brewster angles for the following cases:

(a) light incident on a glass whose index of refraction is 1.6,
(b) light emerging from the same type of glass,
(c) radio frequency wave incident on water ($n = 9$ at radio frequencies).

12-6. Show that there is no reflection from a *plate* of material illuminated at the Brewster angle when the $\boldsymbol{E}$ vector of the incident wave lies in the plane of incidence.

12-7. (a) Show that, for a wave incident in air on a nonconducting magnetic medium, $(E_{0r}/E_{0i})_P$ is zero for

$$\tan^2 \theta_i = \frac{\epsilon_r(\epsilon_r - \mu_r)}{\epsilon_r \mu_r - 1},$$

and hence that the Brewster angle exists only if $\epsilon_r > \mu_r$.
(b) Show that, similarly, $(E_{0r}/E_{0i})_N$ is zero for

$$\tan^2 \theta_i = \frac{\mu_r(\mu_r - \epsilon_r)}{\epsilon_r \mu_r - 1}.$$

In this case there is a Brewster angle only if $\mu_r > \epsilon_r$.

12-8. Show that the coefficient of reflection R and the coefficient of transmission T are both equal to 0.5 at normal incidence on the interface between two dielectrics if the ratio of the indexes of refraction is 5.83.

12-9. A 60-watt light bulb is situated in air one meter away from a water surface.

(a) Calculate the root mean square (rms) values of E and H for the incident, reflected, and refracted rays at the surface of the water directly under the bulb.
Assume that all the power is dissipated as electromagnetic radiation. The index of refraction of water is 1.33.
(b) Calculate the coefficients of reflection and of transmission.

12-10. Light is transmitted at normal incidence from a medium with index of refraction n to another of index $n + a$.

(a) Express the coefficient of transmission T as a power series in a/n, and show that the first two terms give an accuracy of 1% for $(a/n) \leqq \frac{1}{3}$, approximately.
(b) What is the value of T for light passing from air to glass ($n = 1.52$)?
(c) Discuss the reasons for using coated lenses in optical instruments.

12-11. A plane wave is reflected at the interface between two dielectrics. The wave is incident in medium 1, and $n_1/n_2 = 1 + a$.

(a) Show that the coefficient of reflection for waves polarized with their $\mathbf{E}$ vectors normal to the plane of incidence is

$$R_N = \left\{\frac{1 + a - A}{1 + a + A}\right\}^2$$

where

$$A^2 = 1 - a(a + 2) \tan^2 \theta_i,$$

θ_i being the angle of incidence.
(b) Show that $A = 0$ at the critical angle.

12-12. An electromagnetic wave polarized with its $\mathbf{E}$ vector normal to the plane of incidence is totally reflected at the interface between air and a dielectric whose index of refraction is 3.0. The angle of incidence is 75°.

(a) Calculate δ_z/λ_2 and δ_z/λ_1.
(b) Calculate the phases of the reflected and transmitted waves with respect to the incident wave at any point on the interface.

(c) Draw wave fronts for all three waves in the neighborhood of the interface, showing the phase shifts found above.

Draw parallel lines spaced 2π apart to represent the wave fronts.

(d) Check the continuity of E across the interface.

(e) Compare your results with Figure 12-16.

12-13. A plane-polarized electromagnetic wave is totally reflected at the interface between two media and the $\boldsymbol{E}$ vector of the incident wave has components parallel and normal to the plane of incidence.

Show that the reflected wave is elliptically polarized and that the component whose $\boldsymbol{E}$ vector is parallel to the plane of incidence leads the other component by

$$2 \arctan \frac{\cos \theta_i \left\{ \sin^2 \theta_i - \left(\frac{n_2}{n_1} \right)^2 \right\}^{1/2}}{\sin^2 \theta_i}.$$

12-14. In the case of the total reflection of a wave polarized with its $\boldsymbol{E}$ vector normal to the plane of incidence, we have found in Eqs. 12-96 and 12-97 that the x- and z-components of the $\boldsymbol{H}$ vector in the transmitted wave are out of phase by $\pi/2$ radians.

In what direction does the $\boldsymbol{H}$ vector rotate with respect to the direction of propagation of the transmitted wave: does it rotate like the wheels of a vehicle moving on a road or does it rotate in the opposite direction?

12-15. A *scintillator* is a substance that emits light when traversed by an ionizing particle, such as an electron. This light is detected by a photomultiplier, which produces electric pulses that actuate counters and other electronic equipment.

The scintillator has an index of refraction n_1 and is fixed to the face of the photomultiplier with a cement of index n_2. Light is emitted in all directions in the scintillator, a fraction F being transmitted out of the scintillator in the direction of the photomultiplier.

(a) Calculate the fraction F as a function of $\frac{n_1}{n_2}$, assuming that all the light is transmitted for angles of incidence smaller than the critical angle and that the scintillator is surrounded by a nonreflecting substance.

(b) Draw a graph of F for values of n_2/n_1 ranging from 0.1 to 1.0.

12-16. Construct $\sin(a + jb)$ and $\cos(a + jb)$, where a and b are both real, in the complex plane.

Over what ranges of values can these two functions vary?

12-17. Show that, at the surface of a good conductor in air,

$$\left|\frac{E_{0r}}{E_{0i}}\right|_N \approx 1 - \frac{\delta}{\lambda_0} \cos \theta_i,$$

if the conductor is nonmagnetic.

12-18. (a) Calculate $(E_{0r}/E_{0i})_P$ for reflection from the surface of a good conductor.

(b) Show that the reflected wave leads the incident wave by the angle

$$\varphi_P \approx \arc \tan \frac{2\alpha \cos \theta_i}{\alpha^2 - 2 \cos^2 \theta_i},$$

where

$$\alpha = n_1 \frac{\mu_{r2}}{\mu_{r1}} \frac{\delta}{\lambda_0} \ll 1.$$

12-19. A wave is incident at an angle θ_i on a plate of dielectric backed by a perfect conductor. Under what conditions can multiple reflections be avoided?

12-20. Draw two figures similar to the left-hand one in Figure 12-6c, showing E and H for an electromagnetic wave incident on a good conductor.

You will of course have to exaggerate the values of E_{0t} and of λ in the conductor.

Be sure to show the phases correctly. Show x, y, z axes on both figures so as to relate one to the other.

12-21. We have seen in the second example on page 535 that the standing wave pattern obtained by reflecting an electromagnetic wave at normal incidence on a good conductor gives nodes of E which are half way between the nodes of H.

Show that the electromagnetic energy density is constant throughout the standing wave pattern.

12-22. The *surface impedance* of a conductor is defined as the ratio E_t/H_t at the surface, where E_t and H_t are the tangential components of $\mathbf{E}$ and $\mathbf{H}$.

It was shown in the Problem 10-8 that H_t is numerically equal to the current per unit width in the conductor.

(a) Show that the surface impedance of a good conductor is

$$(1 + j)\left(\frac{\omega\mu}{2\sigma}\right)^{1/2}, \qquad \text{or} \qquad \frac{(1 + j)}{\sigma\delta}.$$

The quantity $1/\sigma\delta$ is called the *surface resistivity*.

Both the surface impedance and the surface resistivity are expressed in ohms/square. They give the impedance, or the resistance, between opposite sides of a square surface of any size. For example, the surface resistivity of copper is 8.25×10^{-3} ohm/square at 1 gigahertz.

(b) Show that the power dissipated per square meter in the conductor is

$$\frac{1}{2\sigma\delta} H_{t0}^2$$

where H_{t0} is the amplitude of H at the surface.

12-23. Calculate the radiation and gravitational forces exerted by the Sun on the Earth.

Assume that the radiation incident on the Earth is all absorbed. Then the radiation pressure is $\mathcal{S}_{av}/c$, and not $2\mathcal{S}_{av}/c$ as in Eq. 12-170.

The value of $\mathcal{S}_{av}$ is 1.4 kilowatts/meter^2.

12-24. (a) Compare the forces due to the gravitational attraction and to the radiation pressure of the Sun on a spherical particle of radius r and of specific gravity 5.

The Sun has a mass of 2.0×10^{30} kilograms, and it radiates 3.8×10^{26} watts in the form of electromagnetic radiation. The gravitational constant is 6.7×10^{-11} newton-meter²/kilogram².

In computing the radiation pressure, assume that the particle absorbs the radiation. See Problem 12-23.

(b) Calculate the value of r for which the two forces are equal.

12-25. A radio wave is incident at an angle θ_i on the ionosphere. One may assume that the electron density depends only on the altitude.

(a) Show that, at any given point in the ionosphere, the radius of curvature of a ray is given by

$$R = -\frac{n^2}{(dn/dz)\sin\theta_i}.$$

(b) Assuming that n decreases linearly with altitude, where does the ray bend most sharply, and what is the value of R at that point?

(c) Show that a ray would describe a circular arc of radius R if the index of refraction were given by

$$n = \frac{R\sin\theta_i}{z + R\sin\theta_i}.$$

(d) Draw a sketch of such a ray showing θ_i and the center of curvature. Is n equal to $\sin\theta_i$ at the top of the trajectory?

Hint

12-2. (b) Select your y axis along ∇n, and your x axis in the plane of incidence.

CHAPTER 13

PROPAGATION OF ELECTROMAGNETIC WAVES III

Guided Waves

In Chapter 11 we studied the propagation of plane electromagnetic waves in an unbounded region, first in a vacuum, and then in various media. Then, in Chapter 12, we investigated the reflection and the refraction of plane waves at the boundary between two different media. We shall now study the manner in which waves can be guided in prescribed directions by *metallic wave guides*. Various types of such wave guides are illustrated in Figures 13-1, 13-2, and 13-3.

We first investigate the propagation of electromagnetic waves in a straight line, along the z-axis, without making any assumption as to their dependence on the x- and y-coordinates. This discussion will therefore be more general than those in Chapters 11 and 12. After that, we shall study two relatively simple types of guided waves, which will serve as examples.

13.1 PROPAGATION IN A STRAIGHT LINE

We assume that the medium of propagation is isotropic, linear, and homogeneous; then $\boldsymbol{D} = \epsilon \boldsymbol{E}$ and $\boldsymbol{H} = \boldsymbol{B}/\mu$. We also set its conductivity equal to zero, since electromagnetic waves are strongly attenuated in conductors. This does not exclude metallic wave *guides* because the wave is propagated *outside* the conductors. Thirdly, we set the charge density ρ_t equal to zero in the medium of propagation. This makes $\boldsymbol{\nabla} \cdot \boldsymbol{E}$ equal to zero (Section 3.6). Fourthly, we assume zero attenuation. The method used for dealing with attenuation will be discussed in Section 13.3.4. Finally, we assume that propagation occurs in a straight line, namely in the positive direction along the z-axis.

Then, for a sinusoidal wave,

$$\boldsymbol{E} = (E_{0x}\boldsymbol{i} + E_{0y}\boldsymbol{j} + E_{0z}\boldsymbol{k}) \exp j(\omega t - k_g z) = \boldsymbol{E}_0 \exp j(\omega t - k_g z), \tag{13-1}$$

$$\boldsymbol{H} = (H_{0x}\boldsymbol{i} + H_{0y}\boldsymbol{j} + H_{0z}\boldsymbol{k}) \exp j(\omega t - k_g z) = \boldsymbol{H}_0 \exp j(\omega t - k_g z), \tag{13-2}$$

where the coefficients E_{0x}, E_{0y}, E_{0z}, H_{0x}, $\cdots$ are as yet unspecified functions of x and of y *only*. The dependence on t and on z appears in the exponential function, which is characteristic of a wave propagating in the positive direction of the z-axis.

The wave number for the guided wave,

$$k_g = \frac{1}{\lambda_g}, \tag{13-3}$$

is not necessarily equal to the wave number for a plane wave (Section 11.3) and is real if there is zero attenuation.

It is interesting to compare our present procedure with that of the preceding chapter. In discussing reflection and refraction, we used exponential functions to describe the dependence of the reflected and transmitted waves on all three coordinates and on the time, as for example in Eqs. 12-14 to 12-16. Here we use an exponential function only for z and t, whereas the dependence on x and on y is left unspecified. Let us write out Maxwell's equations for this field. Since $\rho_t = 0$, it follows that $\boldsymbol{\nabla} \cdot \boldsymbol{E} = 0$. Then

$$\frac{\partial E_{0x}}{\partial x} + \frac{\partial E_{0y}}{\partial y} = jk_g E_{0z}. \tag{13-4}$$

Similarly, since $\boldsymbol{\nabla} \cdot \boldsymbol{B} = 0$, and since μ is independent of the coordinates, $\boldsymbol{\nabla} \cdot \boldsymbol{H} = 0$ and

$$\frac{\partial H_{0x}}{\partial x} + \frac{\partial H_{0y}}{\partial y} = jk_g H_{0z}. \tag{13-5}$$

From $\boldsymbol{\nabla} \times \boldsymbol{E} = -\partial \boldsymbol{B}/\partial t$,

$$\frac{\partial E_{0z}}{\partial y} + jk_g E_{0y} = -j\omega\mu H_{0x}, \tag{13-6}$$

$$jk_g E_{0x} + \frac{\partial E_{0z}}{\partial x} = j\omega\mu H_{0y}, \tag{13-7}$$

$$\frac{\partial E_{0y}}{\partial x} - \frac{\partial E_{0x}}{\partial y} = -j\omega\mu H_{0z}, \tag{13-8}$$

and, from $\boldsymbol{\nabla} \times \boldsymbol{H} = \partial \boldsymbol{D}/\partial t$,

$$\frac{\partial H_{0z}}{\partial y} + jk_g H_{0y} = j\omega\epsilon E_{0x}, \tag{13-9}$$

$$jk_g H_{0x} + \frac{\partial H_{0z}}{\partial x} = -j\omega\epsilon E_{0y}, \tag{13-10}$$

$$\frac{\partial H_{0y}}{\partial x} - \frac{\partial H_{0x}}{\partial y} = j\omega\epsilon E_{0z}. \tag{13-11}$$

We can now show that the four transverse components E_{0x}, E_{0y}, H_{0x}, H_{0y} can be deduced from the two longitudinal components E_{0z} and H_{0z}. From Eqs. 13-7 and 13-9,

$$E_{0x} = \frac{-j\omega\mu}{\frac{1}{\lambda^2} - \frac{1}{\lambda_g^2}}\left(\frac{k_g}{\omega\mu}\frac{\partial E_{0z}}{\partial x} + \frac{\partial H_{0z}}{\partial y}\right) \qquad (\lambda_g \neq \lambda). \tag{13-12}$$

We have made the additional assumption that the radian length $\lambda = 1/\omega(\epsilon\mu)^{1/2}$ for a plane wave is different from the radian length of the guided wave, or that $\lambda_g \neq \lambda$. The case where $\lambda_g = \lambda$ will be discussed separately in Section 13.1.2.

Similarly,

$$E_{0y} = \frac{j\omega\mu}{\frac{1}{\lambda^2} - \frac{1}{\lambda_g^2}}\left(-\frac{k_g}{\omega\mu}\frac{\partial E_{0z}}{\partial y} + \frac{\partial H_{0z}}{\partial x}\right) \qquad (\lambda_g \neq \lambda), \tag{13-13}$$

$$H_{0x} = \frac{j\omega\epsilon}{\frac{1}{\lambda^2} - \frac{1}{\lambda_g^2}}\left(\frac{\partial E_{0z}}{\partial y} - \frac{k_g}{\omega\epsilon}\frac{\partial H_{0z}}{\partial x}\right) \qquad (\lambda_g \neq \lambda), \tag{13-14}$$

$$H_{0y} = \frac{-j\omega\epsilon}{\frac{1}{\lambda^2} - \frac{1}{\lambda_g^2}}\left(\frac{\partial E_{0z}}{\partial x} + \frac{k_g}{\omega\epsilon}\frac{\partial H_{0z}}{\partial y}\right). \qquad (\lambda_g \neq \lambda). \tag{13-15}$$

It is obvious, by inspection of the above four equations, that the wave is completely determined once E_{0z} and H_{0z} are known.

The wave equation 11-54,

$$\nabla^2 \mathbf{E} = \epsilon\mu \frac{\partial^2 \mathbf{E}}{\partial t^2} = -k^2 \mathbf{E}, \tag{13-16}$$

provides a differential equation for E_{0z}:

$$\frac{\partial^2 E_{0z}}{\partial x^2} + \frac{\partial^2 E_{0z}}{\partial y^2} - k_g^2 E_{0z} = -k^2 E_{0z}, \tag{13-17}$$

or

$$\frac{\partial^2 E_{0z}}{\partial x^2} + \frac{\partial^2 E_{0z}}{\partial y^2} = -\left(\frac{1}{\lambda^2} - \frac{1}{\lambda_g^2}\right) E_{0z}, \tag{13-18}$$

and an identical equation applies to H_{0z}:

$$\frac{\partial^2 H_{0z}}{\partial x^2} + \frac{\partial^2 H_{0z}}{\partial y^2} = -\left(\frac{1}{\lambda^2} - \frac{1}{\lambda_g^2}\right) H_{0z}. \tag{13-19}$$

The radian length λ_g of the guided wave is as yet unspecified; this is a constant which must be selected in such a way that E_{0z} and H_{0z} can satisfy the above differential equations, *and* the boundary conditions defined by the wave guide. It turns out, as we shall see in Section 13.3.1, that only certain discrete values of λ_g are possible. These values are called the *characteristic* or *eigen values* of the equation, and they depend (a) on the frequency, (b) on the geometry and on the electrical characteristics ϵ_r, μ_r of the medium of propagation, and (c) on the geometry and on the electrical properties σ, μ_r of the guiding structure.

The general procedure for calculating $\boldsymbol{E}$ and $\boldsymbol{H}$ is therefore the following. We first solve the above wave equations for E_{0z} and H_{0z}, using the boundary conditions for the wave guide under consideration. This gives not only E_{0z} and H_{0z}, but also λ_g. The other components of $\boldsymbol{E}$ and of $\boldsymbol{H}$ are then deduced from Eqs. 13-13 to 13-16. It is of course also possible to use Maxwell's equations directly, in conjunction with the appropriate boundary conditions.

13.1.1 TE and TM Waves

It is convenient to consider separately two types of wave: *Transverse Electric* (TE) waves, for which $E_{0z} = 0$, and *Transverse Magnetic* (TM) waves, for which $H_{0z} = 0$. We shall study one particular type of TE wave later on in Section 13.3.

Let us write out the transverse components of $\boldsymbol{E}_0$ and of $\boldsymbol{H}_0$ in vector form:

$$\boldsymbol{E}_{0t} = E_{0x}\boldsymbol{i} + E_{0y}\boldsymbol{j}, \tag{13-20}$$

$$\boldsymbol{H}_{0t} = H_{0x}\boldsymbol{i} + H_{0y}\boldsymbol{j}. \tag{13-21}$$

We can gain information on the relative orientations of these two vectors from their scalar product. We find that

$$\boldsymbol{E}_{0t}\cdot\boldsymbol{H}_{0tr} = E_{0x}H_{0x} + E_{0y}H_{0y} = 0 \tag{13-22}$$

for both TE and TM waves. *The transverse components of* **E** *and of* **H** *are everywhere mutually perpendicular*. This applies to *any* TE or TM wave propagating in a straight line. Our demonstration is valid only for $\lambda_g \neq \lambda$, but the result is correct for any value of λ_g, as we shall see in Section 13.1.2.

Let us calculate the ratio E_{0t}/H_{0t} for TE waves ($E_{0z} = 0$). We can select our axes in such a way that, at the point considered, $E_{0y} = H_{0x} = 0$. Then

$$\frac{E_{0t}}{H_{0t}} = \frac{E_{0x}}{H_{0y}} = \frac{\omega\mu}{k_g} = \left(\frac{\mu}{\epsilon}\right)^{1/2}\frac{\lambda_g}{\lambda}, \tag{13-23}$$

$$= 377\frac{\lambda_g}{\lambda_0} \text{ ohms} \qquad (\epsilon_r = 1,\ \mu_r = 1). \tag{13-24}$$

Similarly, for TM waves ($H_{0z} = 0$),

$$\frac{E_{0t}}{H_{0t}} = \frac{k_g}{\omega\epsilon} = \left(\frac{\mu}{\epsilon}\right)^{1/2}\frac{\lambda}{\lambda_g}, \tag{13-25}$$

$$= 377\,\frac{\lambda_0}{\lambda_g}\text{ ohms} \qquad (\epsilon_r = 1,\ \mu_r = 1). \tag{13-26}$$

The ratio E_{0t}/H_{0t} is called the *wave impedance* and is a real positive number.

Our demonstration is again valid only for $\lambda_g \neq \lambda$, but the result is really correct for any value of λ_g, as will also be shown in Section 13.1.2.

13.1.2 TEM Waves

Let us consider waves for which $\lambda_g = \lambda$. Since the factor

$$\frac{1}{\lambda^2} - \frac{1}{\lambda_g^2}$$

is zero, the parentheses on the right in Eqs. 13-12 to 13-15 must also be zero:

$$\frac{k}{\omega\mu}\frac{\partial E_{0z}}{\partial x} + \frac{\partial H_{0z}}{\partial y} = 0, \tag{13-27}$$

$$\frac{k}{\omega\mu}\frac{\partial E_{0z}}{\partial y} - \frac{\partial H_{0z}}{\partial x} = 0, \tag{13-28}$$

$$\frac{\partial E_{0z}}{\partial y} - \frac{k}{\omega\epsilon}\frac{\partial H_{0z}}{\partial x} = 0, \tag{13-29}$$

$$\frac{\partial E_{0z}}{\partial x} + \frac{k}{\omega\epsilon}\frac{\partial H_{0z}}{\partial y} = 0. \tag{13-30}$$

The last two equations are equivalent to the first two, since, by hypothesis, $\lambda_g = \lambda$ and $k/\omega\mu = \omega\epsilon/k = (\epsilon/\mu)^{1/2}$ in this case.

These equations can be satisfied by setting E_{0z} and H_{0z} both equal to zero, in which case we have a *Transverse Electric and Magnetic* (TEM) *wave*.

The TEM wave has some interesting characteristics. To begin with, $\lambda_g = \lambda$; hence the phase velocity $u = \omega\lambda$ is the same as that of a plane wave in the medium of propagation. This velocity is $1/(\epsilon\mu)^{1/2}$ and is independent of the frequency, insofar as ϵ and μ are themselves independent of the frequency.

If the wave propagates in a vacuum, its velocity is c, whatever the geometry of the wave guide, and whatever the frequency. Such a line is said to be *distortionless* because the various frequency components of a complex waveform are all transmitted at the same velocity.*

---

* This is only approximately correct in practice. Because of the finite conductivity of metal wave guides, there is both attenuation and dispersion.

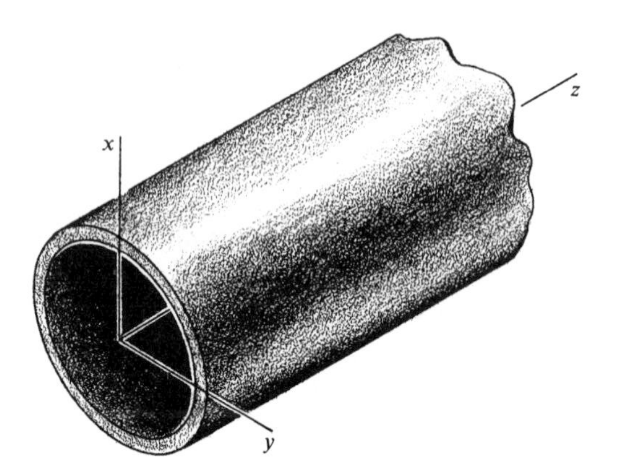

Figure 13-1. A hollow, perfectly conducting tube.

The TEM wave also has the following remarkable property. We have seen in Sections 6.5 and 8.2 that

$$\boldsymbol{E} = -\frac{\partial \boldsymbol{A}}{\partial t} - \boldsymbol{\nabla} V, \tag{13-31}$$

where the first term is associated with changes in the magnetic field, and the second with accumulations of charge. Now the currents are longitudinal, as we shall see in Section 13.1.3, and $\boldsymbol{A}$ must therefore also be longitudinal, as well as $\partial \boldsymbol{A}/\partial t$. Then $\boldsymbol{A} = A\boldsymbol{k}$, and

$$\boldsymbol{E} = -\frac{\partial V}{\partial x}\boldsymbol{i} - \frac{\partial V}{\partial y}\boldsymbol{j} - \left(\frac{\partial V}{\partial z} + \frac{\partial A}{\partial t}\right)\boldsymbol{k}. \tag{13-32}$$

But, since $\boldsymbol{E}$ is transverse, the longitudinal component of $\boldsymbol{\nabla} V$ must cancel that of $\partial A/\partial t$ exactly at all points,

$$\frac{\partial V}{\partial z} = -\frac{\partial A}{\partial t}, \tag{13-33}$$

and

$$\boldsymbol{E} = -\frac{\partial V}{\partial x}\boldsymbol{i} - \frac{\partial V}{\partial y}\boldsymbol{j}. \tag{13-34}$$

Since we are dealing with a wave, $\boldsymbol{E}$ and V must be of the form

$$\boldsymbol{E} = \boldsymbol{E}_0 \exp j\left(\omega t - \frac{z}{\lambda}\right), \tag{13-35}$$

$$V = V_0 \exp j\left(\omega t - \frac{z}{\lambda}\right), \tag{13-36}$$

where $\boldsymbol{E}_0$ and V_0 are functions of x and of y only, and thus

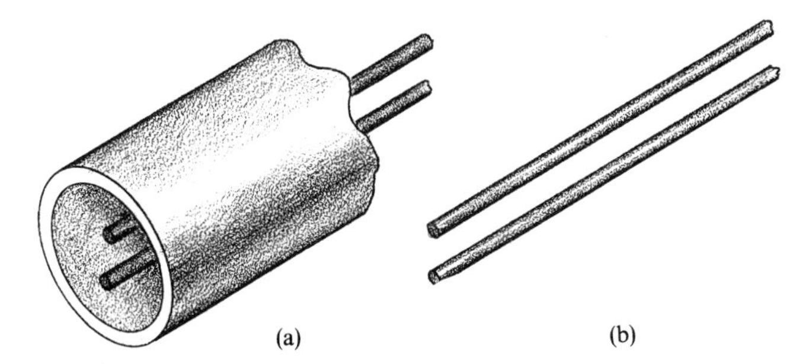

Figure 13-2. (a) Shielded-pair line. The signal is applied between the two wires; the outer cylindrical shield is grounded. (b) Parallel-wire line.

$$\boldsymbol{E} = -\left(\frac{\partial V_0}{\partial x}\boldsymbol{i} + \frac{\partial V_0}{\partial y}\boldsymbol{j}\right)\exp j\left(\omega t - \frac{z}{\lambda\!\!\!^-}\right), \tag{13-37}$$

$$= \boldsymbol{E}_0 \exp j\left(\omega t - \frac{z}{\lambda\!\!\!^-}\right), \tag{13-38}$$

with

$$\boldsymbol{E}_0 = -\frac{\partial V_0}{\partial x}\boldsymbol{i} - \frac{\partial V_0}{\partial y}\boldsymbol{j}. \tag{13-39}$$

This quantity $\boldsymbol{E}_0$ is the electric field intensity in a plane perpendicular to the direction of propagation (that is, in a plane parallel to the xy-plane) for corresponding values of z and t which make the phase angle $\omega t - (z/\lambda\!\!\!^-)$ equal to zero. This $\boldsymbol{E}_0$ is derivable from the potential V_0 in exactly the same manner as an electrostatic field.

If the wave guide is a hollow, perfectly conducting tube as in Figure 13-1, the tangential component of $\boldsymbol{E}$ at its surface is zero, V_0 is a constant all around the tube, and the only possible solution inside is V_0 = constant. Now, if V_0 is constant throughout the inside of the guide, $\boldsymbol{E}_0$ is zero, $\boldsymbol{E} = 0$, and, since $\boldsymbol{\nabla} \times \boldsymbol{E}$ is $-\partial \boldsymbol{B}/\partial t$, there is no $\boldsymbol{H}$ wave either. We therefore conclude that transverse electric and magnetic, or TEM, waves cannot be transmitted inside hollow conducting tubes. This is not rigorously true because TEM waves are transmitted if the wavelength is very much less than the cross-sectional dimensions. For example, light can of course be transmitted through a straight length of metal pipe. We shall see in Section 13.3.2 that the TEM wave is then a limiting case of a TE wave.

In a coaxial line, as in Figure 13-4, the inner conductor need not be at the same potential as the outer one, and $\boldsymbol{E}$ need not be zero. A field is therefore possible, and a TEM wave *can* be transmitted. Other lines with more than one conductor, such as the shielded pair or the parallel-wire lines illustrated in Figure 13-2, can also transmit TEM waves.

Let us rewrite Eqs. 13-4 to 13-11 for $E_{0z} = 0$, $H_{0z} = 0$. These eight equations are simply Maxwell's equations, as applied to a wave propagating in the positive direction along the z-axis. We now have only six distinct equations:

$$\frac{\partial E_{0x}}{\partial x} + \frac{\partial E_{0y}}{\partial y} = 0, \tag{13-40}$$

$$\frac{\partial H_{0x}}{\partial x} + \frac{\partial H_{0y}}{\partial y} = 0, \tag{13-41}$$

$$E_{0y} = -\left(\frac{\mu}{\epsilon}\right)^{1/2} H_{0x}, \tag{13-42}$$

$$E_{0x} = \left(\frac{\mu}{\epsilon}\right)^{1/2} H_{0y}, \tag{13-43}$$

$$\frac{\partial E_{0y}}{\partial x} - \frac{\partial E_{0x}}{\partial y} = 0, \tag{13-44}$$

$$\frac{\partial H_{0y}}{\partial x} - \frac{\partial H_{0x}}{\partial y} = 0. \tag{13-45}$$

We first note that, for a plane wave propagating along the z-axis, as in Chapter 11, the field vectors depend solely on z and on t, and the derivatives with respect to x and to y are zero. Then Eqs. 13-40, 13-41, 13-44, 13-45 become identities, and Eqs. 13-42 and 13-43 are equivalent to Eq. 11-61. Such a plane wave is the simplest form of TEM wave.

For any TEM wave, we can see, from the third and fourth of the above equations, that ***E** and **H** are mutually perpendicular*, just as for the transverse components of TE and TM waves. The wave impedance E_{0t}/H_{0t} can be calculated as in Section 13.1.1, choosing coordinate axes so that $E_{0y} = H_{0x} = 0$. Thus

$$\frac{E_{0t}}{H_{0t}} = \frac{E_{0x}}{H_{0y}} = \left(\frac{\mu}{\epsilon}\right)^{1/2}, \tag{13-46}$$

$$= 377 \text{ ohms} \qquad (\epsilon_r = 1,\ \mu_r = 1). \tag{13-47}$$

In addition, the electric and magnetic energy densities are equal,

$$\frac{1}{2}\epsilon E^2 = \frac{1}{2}\mu H^2, \tag{13-48}$$

and the total energy density is ϵE^2, or μH^2.

From Eq. 13-43, the average Poynting vector is

$$\mathbf{S}_{\text{av}} = \frac{1}{2}\,\text{Re}(\boldsymbol{E} \times \boldsymbol{H}^*) = \frac{1}{2}\left(\frac{\epsilon}{\mu}\right)^{1/2} E_{0x}^2\, \boldsymbol{k}, \tag{13-49}$$

$$= \left(\frac{\epsilon}{\mu}\right)^{1/2} E_{\text{rms}}^2\, \boldsymbol{k}, \tag{13-50}$$

$$= u\epsilon \boldsymbol{E}_{\text{rms}}^2 \boldsymbol{k}, \tag{13-51}$$

$$= u\mu H_{\text{rms}}^2\, \boldsymbol{k}. \tag{13-52}$$

This vector is directed in the positive direction of the z-axis, which is the direction of propagation of the wave, and its magnitude is equal to the phase velocity $u = 1/(\epsilon\mu)^{1/2}$ multiplied by the average energy density, just as for a plane wave (Section 11.2).

To find $\boldsymbol{E}$ and $\boldsymbol{H}$, one first chooses the static field $\boldsymbol{E}_0$ that corresponds to the mode of propagation desired. For example, with the coaxial line of Figure 13-4, $\boldsymbol{E}_0$ is radial. Then Eq. 13-37 gives $\boldsymbol{E}$ and Eqs. 13-42 and 13-43 give $\boldsymbol{H}$. We shall work out the field of the coaxial line in Section 13.2.

13.1.3 Boundary Conditions at the Surface of Metallic Wave Guides

For any type of wave the tangential component of $\boldsymbol{E}$ must vanish at the surface of an infinitely conducting guide. Then, close to the surface of the guide, $\boldsymbol{E}$ must be normal and, in particular, E_{0z} must be equal to zero. Real conductors, of course, do not have infinite conductivity unless they are superconducting, and we shall deal with this matter in Section 13.3.4.

There is also another boundary condition that applies to any type of wave guided by a perfect conductor and that is interesting from a qualitative point of view. Since the current density in the guide is tangent to its surface, we may use the results of Problem 10-8. Then, close to the guide, $\boldsymbol{H}$ must be (a) tangent to the surface, (b) perpendicular to the current density, and (c) equal in magnitude to the surface current density expressed in amperes/meter. For example, with TM and TEM waves, $\boldsymbol{H}$ is everywhere transverse and the currents in the guide are therefore longitudinal. We have already used this result in the preceding section.

Finally, in the case of a TE wave and a perfectly conducting guide, there is a third boundary condition that concerns H_{0z} and that we can find by writing out the value of $\boldsymbol{E}_0$:

$$\boldsymbol{E}_0 = E_{0x}\boldsymbol{i} + E_{0y}\boldsymbol{j}, \tag{13-53}$$

E_{0z} being equal to zero by hypothesis. Using Eqs. 13-12 and 13-13,

$$\boldsymbol{E}_0 = \frac{j\omega\mu}{\dfrac{1}{\lambda^2} - \dfrac{1}{\lambda_g^2}}\left(-\frac{\partial H_{0z}}{\partial y}\boldsymbol{i} + \frac{\partial H_{0z}}{\partial x}\boldsymbol{j}\right), \tag{13-54}$$

$$= \frac{j\omega\mu}{\dfrac{1}{\lambda^2} - \dfrac{1}{\lambda_g^2}}\boldsymbol{k} \times \nabla H_{0z}, \tag{13-55}$$

where $\boldsymbol{k}$ is the unit vector in the direction of the z-axis. The three vectors $\boldsymbol{E}_0$, $\boldsymbol{k}$, ∇H_{0z} are shown in Figure 13-3. It will be observed that ∇H_{0z} is necessarily

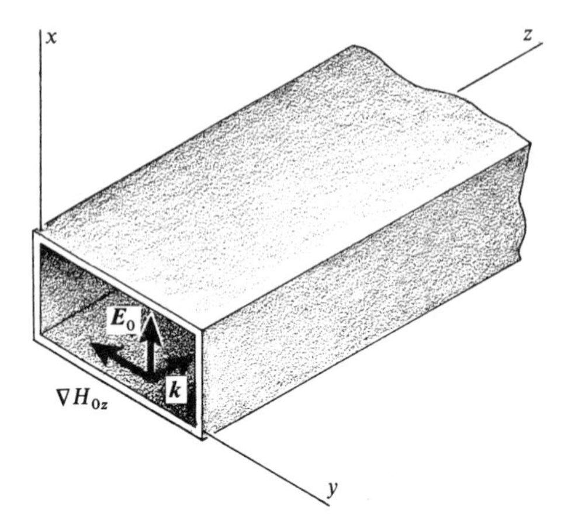

Figure 13-3. Portion of a rectangular wave guide. The electromagnetic wave propagates inside the tube. If the guide is perfectly conducting, the electric field intensity $\boldsymbol{E}$ is zero in the conductor, and $\boldsymbol{E}$ is either normal or zero at the surface. For a TE wave, it is shown that ∇H_{0z} is tangent to the wall.

tangent to the conducting wall and, therefore, that the rate of change of H_{0z} in the direction normal to the surface must be zero.

13.2 THE COAXIAL LINE

In the coaxial line illustrated in Figure 13-4 the electromagnetic wave propagates in the annular region between the two coaxial cylindrical conductors, and there is zero field outside. This type of wave guide is often used in electronic equipment. The medium of propagation is a low-loss dielectric.

We shall make several simplifying assumptions, namely, that (a) the guide is straight and its cross section is constant throughout its length, (b) the electrical conductivity of the walls is infinite, (c) the time dependence of the wave is described by a cosine function, and (d) the wave travels in the positive direction of the z-axis and there is no reflected wave traveling in the opposite direction. Assumptions (c) and (d) do not limit the generality of our calculation, since, according to the principle of superposition, the net field of a number of waves is simply given by the vector sum of their fields.

We have seen above that a TEM wave can be transmitted down such a guide; this is the mode that we shall study in this section. TE and TM modes are also possible, but they are much more complex and they are not used in practice.

Thus

$$E_{0z} = 0, \qquad H_{0z} = 0, \tag{13-56}$$

and, from Eq. 13-37,

$$\boldsymbol{E} = -\left(\frac{\partial V_0}{\partial x}\boldsymbol{i} + \frac{\partial V_0}{\partial y}\boldsymbol{j}\right)\exp j\left(\omega t - \frac{z}{\lambda}\right). \tag{13-57}$$

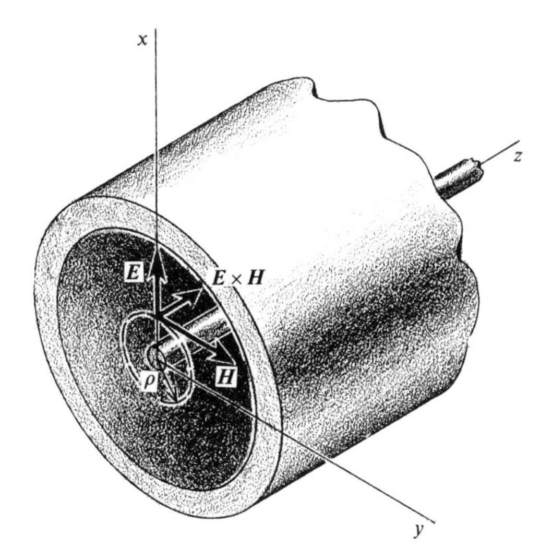

Figure 13-4. The $\boldsymbol{E}$ and $\boldsymbol{H}$ vectors inside a coaxial line. They are both transverse, $\boldsymbol{E}$ being radial and $\boldsymbol{H}$ azimuthal. The vector product $\boldsymbol{E} \times \boldsymbol{H}$ always points in the direction of propagation.

The dependence of $\boldsymbol{E}$ on x and y is the same as for a two-dimensional electrostatic field. For a given position z along the guide, and for a given time t, $\boldsymbol{E}$ is radial and varies as $1/\rho$, where ρ is the radial distance from the axis to the point considered. Then

$$\boldsymbol{E} = \frac{C}{\rho} \exp j\left(\omega t - \frac{z}{\lambda\!\!\!^{-}}\right) \boldsymbol{\rho}_1 \tag{13-58}$$

in cylindrical coordinates, $\boldsymbol{\rho}_1$ being the unit vector in the radial direction, and C being a constant.

Since we have a TEM wave, the radian length $\lambda\!\!\!^{-}_g$ of the guided wave is the same as that of an infinite plane wave, or $1/\omega(\epsilon\mu_0)^{1/2}$, and the velocity of propagation $\omega\lambda\!\!\!^{-} = 1/(\epsilon\mu_0)^{1/2}$ is the same at all frequencies.*

The line voltage $\mathcal{V}$, which is the potential of the inner conductor with respect to the outer conductor, is

$$\mathcal{V} = \int_{\rho_i}^{\rho_o} E\, d\rho = C \ln \frac{\rho_o}{\rho_i} \exp j\left(\omega t - \frac{z}{\lambda\!\!\!^{-}}\right). \tag{13-59}$$

From the previous section, $\boldsymbol{H}$ is orthogonal to $\boldsymbol{E}$, and

$$\boldsymbol{H} = \left(\frac{\epsilon}{\mu_0}\right)^{1/2} \frac{C}{\rho} \exp j\left(\omega t - \frac{z}{\lambda\!\!\!^{-}}\right) \boldsymbol{\varphi}_1. \tag{13-60}$$

The vectors $\boldsymbol{E}$ and $\boldsymbol{H}$ are oriented as in Figure 13-4 and the Poynting vector $\boldsymbol{E} \times \boldsymbol{H}$ points in the direction of propagation.

---

* Assuming that the line has infinite conductivity and that ϵ_r is independent of the frequency. In practice, the velocity of propagation does depend somewhat on the frequency. See Section 13.1.2.

The line current I flowing along the surface of the inner conductor can be calculated from the circuital law, Eq. 7-77:

$$I = \int H\rho_i \, d\varphi = 2\pi\rho_i \left(\frac{\epsilon}{\mu_0}\right)^{1/2} \frac{C}{\rho_i} \exp j\left(\omega t - \frac{z}{\bar{\lambda}}\right), \tag{13-61}$$

$$= \frac{C}{60} \exp j\left(\omega t - \frac{z}{\bar{\lambda}}\right) \qquad (\epsilon_r = 1,\ \mu_r = 1). \tag{13-62}$$

An equal current flows in the opposite direction along the inner surface of the outer conductor. Note that C is equal to 60 times the peak value of I if $\epsilon_r = 1$, $\mu_r = 1$.

The average transmitted power can be calculated by integrating the average Poynting vector over the annular area between the two conductors:

$$W_T = \int_{\rho_i}^{\rho_o} \mathcal{S}_{av}\, 2\pi\rho \, d\rho, \tag{13-63}$$

where

$$\mathcal{S}_{av} = \frac{1}{2} \operatorname{Re}(\boldsymbol{E} \times \boldsymbol{H}^*) = \left(\frac{\epsilon}{\mu_0}\right)^{1/2} \frac{C^2}{2\rho^2} \boldsymbol{k}. \tag{13-64}$$

Thus

$$W_T = \frac{C^2}{120} \ln \frac{\rho_o}{\rho_i} \text{ watts} \qquad (\epsilon_r = 1,\ \mu_r = 1). \tag{13-65}$$

The average transmitted power can also be calculated from $I_{rms}\mathcal{V}_{rms} = (1/2)I_{max}\mathcal{V}_{max}$.

The *characteristic impedance* of a coaxial line, which is the ratio $\mathcal{V}/I$ when there is no reflected wave traveling in the opposite direction, as we have assumed at the beginning, is

$$60 \ln \frac{\rho_o}{\rho_i} \text{ ohms} \qquad (\epsilon_r = 1,\ \mu_r = 1).$$

The *wave impedance* E/H is 377 ohms, as we saw earlier in Section 13.1.2.

13.3 THE HOLLOW RECTANGULAR WAVE GUIDE

Hollow wave guides are extensively used at microwave frequencies. These wave guides are simply metallic tubes inside which electromagnetic waves propagate by reflection on the inner surfaces, in much the same way that sound waves propagate through a tube. We shall consider hollow wave guides of rectangular cross-section because their guided waves are relatively simple and because they are the most widely used.

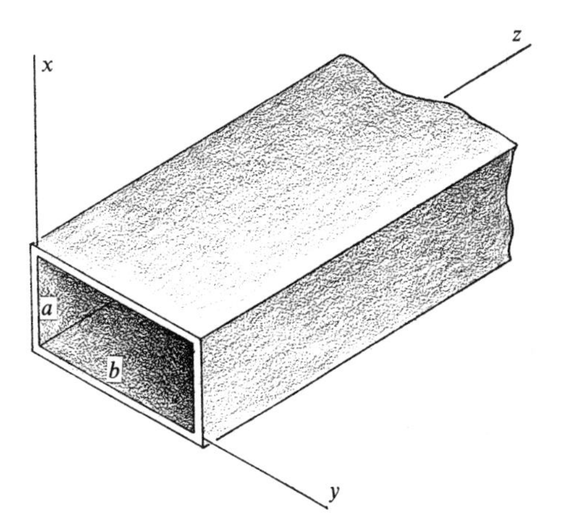

Figure 13-5. A hollow, rectangular wave guide. The wave propagates in the positive direction along the z-axis.

We make the same simplifying assumptions as for the coaxial line of the previous section. We also assume that the dielectric is air.

We have shown in Section 13.1.2 that TEM waves cannot propagate inside a hollow tube. Thus the radian length $\lambda\!\!\!^{-}_g$ of the guided TE or TM waves is different from $\lambda\!\!\!^{-}$, which is $\lambda\!\!\!^{-}_0$, or $1/\omega(\epsilon_0\mu_0)^{1/2}$ in this case.

We select axes as in Figure 13-5, with the wave propagating in the positive direction along the z-axis.

13.3.1 The TE Wave

We now consider the transverse electric (TE) wave ($E_z = 0$). This wave is simply a plane electromagnetic wave whose $\boldsymbol{E}$ and $\boldsymbol{H}$ vectors are oriented as in Figure 13-6 and which is reflected back and forth on the walls parallel to the xz-plane. We shall use this fact to simplify the analysis somewhat, but we could also find the TE wave by solving for H_{0z}, using the proper boundary conditions, without referring to the reflection.

We must find E_{0x}, E_{0y}, H_{0x}, H_{0y}, H_{0z}, $\lambda\!\!\!^{-}_g = 1/k_g$. From Figure 13-6 we can see immediately that

$$E_{0y} = 0, \qquad H_{0x} = 0. \tag{13-66}$$

To determine the remaining four quantities we shall proceed as indicated at the end of Section 13-1: we shall solve the wave equation for H_{0z} for the proper boundary conditions. This will give H_{0z} and $\lambda\!\!\!^{-}_g$. The values of E_{0x} and H_{0y} will then follow from Eqs. 13-12 and 13-15.

We have found in Eq. 13-19 that H_{0z} obeys the wave equation

$$\frac{\partial^2 H_{0z}}{\partial^2 x} + \frac{\partial^2 H_{0z}}{\partial y^2} = \left(\frac{1}{\lambda\!\!\!^{-2}_g} - \frac{1}{\lambda\!\!\!^{-2}_0}\right) H_{0z}. \tag{13-67}$$

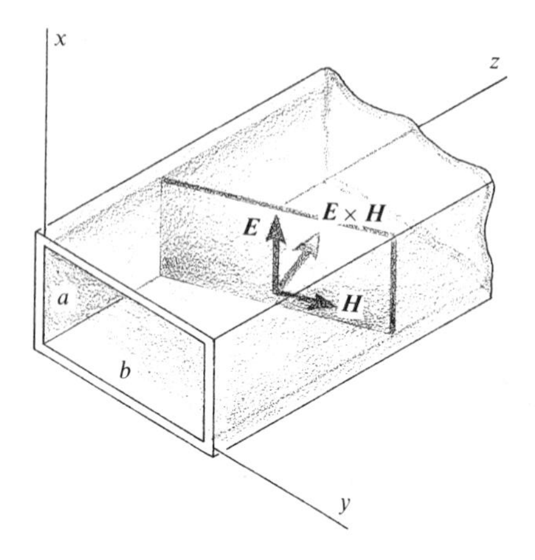

Figure 13-6. Typical plane wavefront with the $\boldsymbol{E}$, $\boldsymbol{H}$, and $\boldsymbol{E} \times \boldsymbol{H}$ vectors inside a rectangular wave guide. The reflection occurs on the faces parallel to the xz-plane. The $\boldsymbol{E}$ vector is transverse, that is, perpendicular to the direction of propagation Oz, and the wave is TE.

In the simple TE wave illustrated in Figure 13-6, the wave is independent of the x-coordinate, and

$$\frac{\partial^2 H_{0z}}{\partial y^2} = \left(\frac{1}{\lambda_g^2} - \frac{1}{\lambda_0^2}\right) H_{0z}. \tag{13-68}$$

Solving this equation, we find that

$$H_{0z} = F \sin Cy + G \cos Cy, \tag{13-69}$$

where

$$-C^2 = \frac{1}{\lambda_g^2} - \frac{1}{\lambda_0^2}. \tag{13-70}$$

We now apply the boundary condition discussed in Section 13.1.2:

$$\frac{\partial H_{0z}}{\partial x} = 0 \qquad \text{at } x = 0 \text{ and } x = a, \tag{13-71}$$

$$\frac{\partial H_{0z}}{\partial y} = 0 \qquad \text{at } y = 0 \text{ and } y = b. \tag{13-72}$$

The first condition is already satisfied, since H_{0z} is not a function of x. From the second condition,

$$\frac{\partial H_{0z}}{\partial y} = C(F \cos Cy - G \sin Cy) = 0 \qquad \text{at } y = 0 \text{ and } y = b. \tag{13-73}$$

To satisfy the boundary condition at $y = 0$, F must be zero. The condition $C = 0$ must be rejected because it implies that $\lambda_g = \lambda_0$, which is not compatible with a TE wave. At $y = b$,

$$\sin Cb = 0, \tag{13-74}$$

and thus

$$C = \frac{n\pi}{b}, \tag{13-75}$$

where n is an integer that cannot be zero, for then C would also be zero. Then

$$H_{0z} = G \cos \frac{n\pi}{b} y. \tag{13-76}$$

Let us find λ_g. This is easy now that we know H_{0z}. We can either substitute H_{0z} into the wave equation or use Eqs. 13-70 and 13-75 to obtain

$$-\frac{n^2\pi^2}{b^2} = \frac{1}{\lambda_g^2} - \frac{1}{\lambda_0^2} \qquad (n = 1, 2, 3, \cdots). \tag{13-77}$$

It will be noticed that λ_g can have only certain discrete characteristic values corresponding to $n = 1, 2, 3, \cdots$.

Since the left-hand side of the above equation is negative, $\lambda_g > \lambda_0$. This means that the wavelength in the guide is *longer* than that of a plane wave of the same frequency propagating in free space. Then the phase velocity u is *larger* than c. We shall return to this point later on.

Now that we have found H_{0z} and λ_g, we can calculate the two remaining unknowns E_{0x} and H_{0y} from Eqs. 13-12 and 13-15, setting $\lambda = \lambda_0$ since $\epsilon_r = 1$, $\mu_r = 1$, by hypothesis:

$$E_{0x} = \frac{j\omega\mu_0 bG}{n\pi} \sin \frac{n\pi}{b} y, \tag{13-78}$$

$$H_{0y} = \frac{jbG}{n\pi\lambda_g} \sin \frac{n\pi}{b} y, \tag{13-79}$$

or, setting

$$E_{00x} = \frac{j\omega\mu_0 bG}{n\pi}, \tag{13-80}$$

then

$$E_x = E_{00x} \sin\left(\frac{n\pi}{b} y\right) \exp j\left(\omega t - \frac{z}{\lambda_g}\right), \tag{13-81}$$

$$H_y = \frac{E_{00x}}{\omega\mu_0\lambda_g} \sin\left(\frac{n\pi}{b} y\right) \exp j\left(\omega t - \frac{z}{\lambda_g}\right), \tag{13-82}$$

$$H_z = \frac{E_{00x} n\pi}{j\omega\mu_0 b} \cos\left(\frac{n\pi}{b} y\right) \exp j\left(\omega t - \frac{z}{\lambda_g}\right), \tag{13-83}$$

where the values for λ_g are given by Eq. 13-77. The coefficient E_{00x} is the maximum value of E inside the guide. We have already found that $E_{0y} = 0$, $H_{0x} = 0$ (Eq. 13-66).

Let us consider the value of E_x. It can be understood qualitatively as follows. In any plane wave front such as that shown in Figure 13-6, $\mathbf{E}$ is independent of the x-coordinate; since the wave along the z-axis results from the superposition of such plane waves by multiple reflections, its $\mathbf{E}$ must also be independent of the x-coordinate. The same cannot be said about the y dependence, however. If we consider a single elementary plane wave progress-

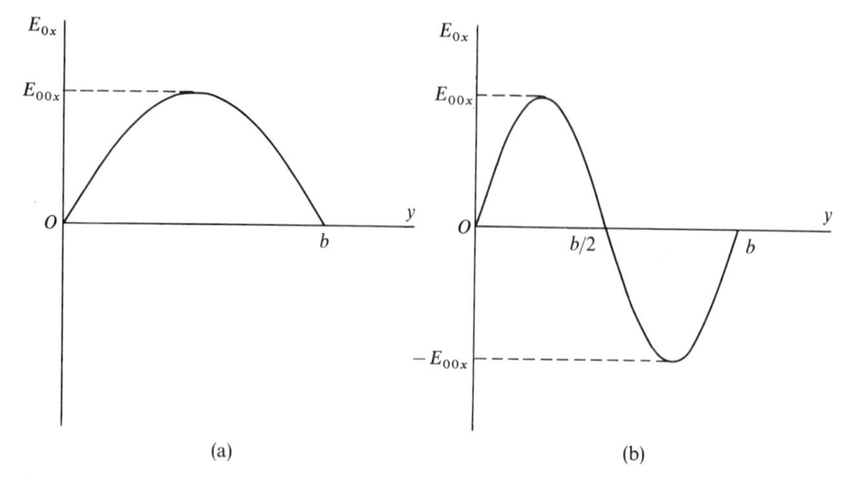

Figure 13-7. The amplitude of E for a TE wave in a hollow rectangular wave guide. The value of n of Eq. 13-78 is 1 in (a) and 2 in (b).

ing at an angle along the guide, the amplitude and the phase of $\boldsymbol{E}$ are constant over a wave front such as that shown in the figure. The superposition of such waves by multiple reflections gives an interference pattern, with the result that the amplitude varies with the y coordinate.

Let us now consider the meaning of the quantity n. For $n = 1$, E_{0x} varies from 0 at $y = 0$ to a maximum of E_{00x} at $y = b/2$ and to zero again at $y = b$ as in Figure 13-7a. For $n = 2$, E_{0x} is zero in the middle of the guide at $y = b/2$, and is of opposite sign on either side, as in Figure 13-7b. The different values of n thus correspond to different *modes of propagation* inside the guide.

Now let us return to Eq. 13-77 for λ_g. It can be rewritten as

$$\frac{\lambda_g}{\lambda_0} = \frac{1}{\left\{1 - \left(\dfrac{n\lambda_0}{2b}\right)^2\right\}^{1/2}}, \tag{13-84}$$

$$= \frac{1}{\left\{1 - \left(\dfrac{\lambda_0}{2b}\right)^2\right\}^{1/2}} \qquad \text{for } n = 1. \tag{13-85}$$

This equation shows that λ_g is real if $\lambda_0 < 2b$. If λ_g is real, the exponential functions of Eqs. 13-81 to 13-83 describe an *un*attenuated wave. Therefore, if the above inequality is satisfied, that is, if the frequency is high enough, and if the walls are perfectly conducting, a wave can propagate inside the hollow wave guide without attenuation.

If the above inequality is *not* satisfied, that is, if the frequency is too low, then λ_g is imaginary, and the exponential functions show that the field is attenuated exponentially with z. Then the phase does not vary with z, there is no wave, and there is zero energy flow into the guide once the field is established.

The attenuation is rapid. For example, if the frequency is too low by a factor of 2, $\lambda_0 = 4b$. Then, from Eq. 13-84,

$$\frac{1}{\lambda_g} = \pm j \frac{3^{1/2}}{\lambda_0}. \tag{13-86}$$

We must choose the negative sign, for otherwise the amplitude would increase exponentially with z. Then

$$\exp\left(-j\frac{z}{\lambda_g}\right) = \exp\left(-\frac{2\pi 3^{1/2} z}{\lambda_0}\right), \tag{13-87}$$

and the wave is attenuated in amplitude by a factor of 4×10^{-5} in one free-space wavelength λ_0!

The average Poynting vector $\mathbf{S}_{av}$ and the average transmitted power are attenuated by the square of this, or by a factor of 16×10^{-10} in one λ_0! The wave guide therefore acts like a high-pass filter, with the lower frequency limit determined solely by the width b, and not by a.

In the special case where the free-space wavelength λ_0 is $2b$, the guide wavelength λ_g becomes infinite, as does the phase velocity. This corresponds to $\omega = \omega_p$ in the case of propagation in an ionized gas (Section 11.6.3). The quantity $\lambda_c = 2b$ is called the *cut-off wavelength.*

Wave guides can therefore be used only if the free-space wavelength λ_0 is *shorter* than twice the distance between the reflecting walls for the $n = 1$ mode. They are inherently *high-frequency* devices. For example, if $b = 10$ centimeters, λ_0 must be shorter than 20 centimeters, and the frequency must be higher than $3 \times 10^8/0.2$, or 1.5 gigahertz.

Let us now rewrite the components of $\boldsymbol{E}$ and $\boldsymbol{H}$ for the mode $n = 1$, which is normally used:

$$E_x = E_{00x} \sin\left(\frac{\pi y}{b}\right) \exp j\left(\omega t - \frac{z}{\lambda_g}\right), \tag{13-88}$$

$$E_y = 0, \tag{13-89}$$

$$E_z = 0, \tag{13-90}$$

$$H_x = 0, \tag{13-91}$$

$$H_y = \frac{E_{00x}}{\omega\mu_0\lambda_g} \sin\left(\frac{\pi y}{b}\right) \exp j\left(\omega t - \frac{z}{\lambda_g}\right), \tag{13-92}$$

$$H_z = \frac{\pi E_{00x}}{\omega\mu_0 b} \cos\left(\frac{\pi y}{b}\right) \exp j\left(\omega t - \frac{z}{\lambda_g} - \frac{\pi}{2}\right), \tag{13-93}$$

$$= \frac{\pi E_{00x}}{\omega\mu_0 b} \cos\left(\frac{\pi y}{b}\right) \exp j\left\{\omega t - \frac{1}{\lambda_g}\left(z + \frac{\lambda_g}{4}\right)\right\}, \tag{13-94}$$

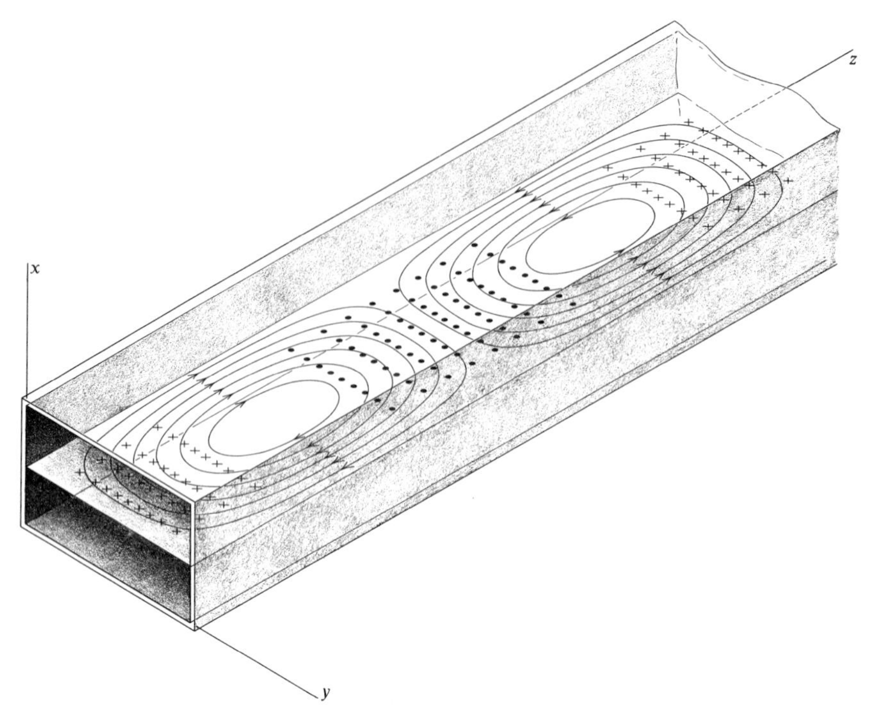

Figure 13-8. Lines of $\boldsymbol{E}$ and lines of $\boldsymbol{H}$ for a TE wave with $n = 1$ propagating in a hollow rectangular wave guide. The ovals are lines of $\boldsymbol{H}$. The lines of $\boldsymbol{E}$ are vertical straight lines and are represented by dots and crosses.

where the guide wavelength is

$$\lambda_g = \frac{\lambda_0}{\left\{1 - \left(\frac{\lambda_0}{2b}\right)^2\right\}^{1/2}} > \lambda_0, \tag{13-95}$$

and the phase velocity is

$$u_p = \frac{c}{\left\{1 - \left(\frac{\lambda_0}{2b}\right)^2\right\}^{1/2}} > c. \tag{13-96}$$

It will be observed that E_x and H_y are in phase, but that H_z has the same phase at $z + \dfrac{\lambda_g}{4}$ as have the two other vectors at z.

Figure 13-8 shows lines of $\boldsymbol{E}$ and of $\boldsymbol{H}$ for a TE wave with $n = 1$ propagating in a hollow rectangular wave guide.

13.3.2 Internal Reflections

Let us return to Figure 13-6, which shows a typical wave front for a wave zigzagging down the guide. It will be instructive to investigate the field by considering the interference resulting from the multiple reflections.

Figure 13-9 shows the multiple reflections in more detail. Let us assume that along the *fixed* line AB the electric field intensity of the wave which propagates upward and to the right is $E_0 \exp j\omega t$. The line AB is thus parallel to the wave fronts for this wave. The lines BC and DE are similarly fixed and parallel to wave fronts for the wave propagating to the right and downward.

These two waves must interfere at B to give zero E for all values of t at the perfectly conducting wall, since their electric field intensities are perpendicular to the paper and parallel to the wall. Then the electric field intensity along BC must be $E_0 \exp j(\omega t + \pi)$, or $-E_0 \exp j\omega t$. At C, interference must again give zero E for all t. Then, along CD, E is $E_0 \exp j(\omega t + 2\pi)$, or $E_0 \exp j\omega t$.

Thus, AB and CD are one free-space wavelength λ_0 apart, and, from the triangle CHI in the figure,

$$\cos \alpha = \frac{\lambda_0}{2b}. \tag{13-97}$$

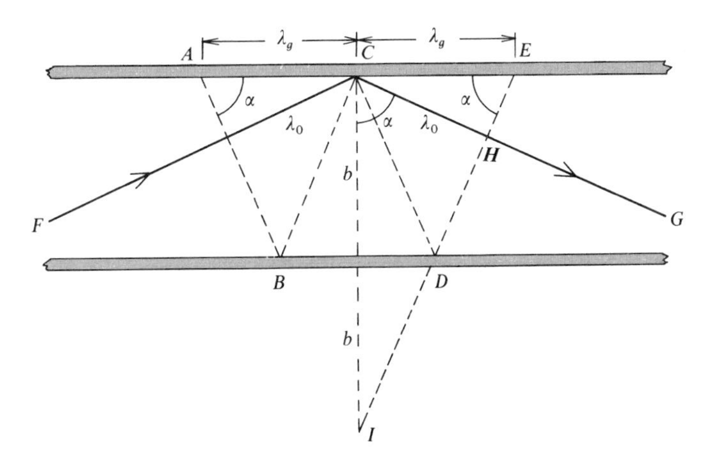

Figure 13-9. A plane electromagnetic wave propagating in a rectangular hollow wave guide along a zigzag path. The lines AB and CD are parallel to wave fronts for the wave propagating to the right and upward. Similarly, BC and DE are parallel to wave fronts traveling to the right and and downward. The angle α is the angle of incidence; the broken line FCG represents a ray reflected at C.

If the wavelength were n times smaller, the same lines AB and CD would be n wavelengths apart, and then we would have

$$\cos \alpha = \frac{n\lambda_0}{2b}. \tag{13-98}$$

There are obviously only certain discrete values of the angle α that permit destructive interference to occur at the walls of the guide.

We have therefore found a geometrical interpretation for the ratio $\lambda_0/(2b/n)$. From Eq. 13-84,

$$\frac{\lambda_g}{\lambda_0} = \frac{1}{\sin \alpha}, \tag{13-99}$$

where the guide wavelength λ_g is either AC, or CE, or BD.

At the critical wavelength, $\lambda_0 = 2b/n$, $\cos \alpha = 1$, $\alpha = 0$, and the wave fronts are parallel to the axis of the guide. The traveling wave then degenerates into a standing wave between the wave guide walls.

For $\lambda_0 \ll 2b/n$, $\cos \alpha \to 0$, and $\alpha \to \pi/2$. The TE wave then approaches a TEM mode as a limit.

The phase velocity is

$$u_p = \frac{\lambda_g}{\lambda_0} c = \frac{c}{\sin \alpha} > c. \tag{13-100}$$

This is the velocity at which the *phase* propagates along the guide. It is larger than c because the individual plane wave fronts are inclined at an angle with respect to the axis of the guide. This can be seen from Figure 13-9 as follows. Consider AB to be a wave front propagating parallel to itself and to the right at a velocity c. Then the point A moves along the z-axis at a velocity that is larger than c.

We can also consider the velocity at which a given *signal* progresses along the length of the guide. The z component of the velocity of an individual wave front is only $c \sin \alpha$ and is *smaller* than c. Thus, if we call this velocity u_s, then

$$u_s = c \sin \alpha, \tag{13-101}$$

$$= c\left\{1 - \left(\frac{n\lambda_0}{2b}\right)^2\right\}^{1/2} < c, \tag{13-102}$$

and

$$u_s u_p = c^2. \tag{13-103}$$

This is in agreement with our requirement that the signal velocity never exceed c (Section 5.10).

13.3.3 Energy Transmission

Let us consider the energy transmitted by a TE wave in a low-loss rectangular wave guide. We shall assume the usual $n = 1$ mode. The field is then completely described by Eqs. 13-78 to 13-83, with $n = 1$, or by Eqs. 13-88 to 13-94. The first set is slightly more convenient for our purpose.

From Eq. 11-90, the average value of the Poynting vector is

$$\mathbf{S}_{av} = \frac{1}{2}\,\mathrm{Re}(\boldsymbol{E} \times \boldsymbol{H}^*), \tag{13-104}$$

where $\boldsymbol{H}^*$ is the complex conjugate of $\boldsymbol{H}$. In the present case,

$$\mathbf{S}_{av} = \frac{1}{2}\,\mathrm{Re}\begin{vmatrix} \boldsymbol{i} & \boldsymbol{j} & \boldsymbol{k} \\ E_x & 0 & 0 \\ 0 & H_y^* & H_z^* \end{vmatrix} = \frac{1}{2}\,\mathrm{Re}(-E_x H_z^* \boldsymbol{j} + E_x H_y^* \boldsymbol{k}). \tag{13-105}$$

Substituting the values of E_x, H_y, H_z, we find that the first term in the parentheses is imaginary, whereas the second term is real.

The energy therefore flows only in the direction of the z-axis and

$$\mathbf{S}_{av} = \frac{E_{00x}^2}{2\omega\mu_0\lambda_g}\sin^2\left(\frac{\pi}{b}y\right)\boldsymbol{k}. \tag{13-106}$$

The value of $\mathbf{S}_{av}$ is independent of x, as expected, since the amplitude and phase of the wave are independent of the x coordinate. It is zero at the walls $y = 0$ and $y = b$, where $\boldsymbol{E}$ is zero, and it is maximum at $y = b/2$, again as expected.

The total average transmitted power is thus

$$W_T = \int_{y=0}^{y=b} \frac{E_{00x}^2}{2\omega\mu_0\lambda_g}\sin^2\left(\frac{\pi}{b}y\right) a\,dy, \tag{13-107}$$

$$= \frac{E_{00x}^2 ab}{4\omega\mu_0\lambda_g}, \tag{13-108}$$

$$= \frac{E_{00x}^2 ab}{4c\mu_0}\left\{1 - \left(\frac{\lambda_0}{2b}\right)^2\right\}^{1/2}. \tag{13-109}$$

Let us compare this transmitted power with the average electromagnetic energy per unit length within the guide. The instantaneous electric energy density is $(1/2)\epsilon_0 E^2$, and its average value is $(1/4)\epsilon_0 E_0^2$. The average electric energy per unit length is thus

$$\int_0^b \frac{1}{4}\epsilon_0 E_{00x}^2 \sin^2\left(\frac{\pi}{b}y\right) a\,dy = \frac{\epsilon_0}{8}\,ab\,E_{00x}^2. \tag{13-110}$$

To find the average magnetic energy content per unit length, we proceed similarly for both the y- and z-components of $\boldsymbol{H}$ and add the results, since

$$H^2 = H_y^2 + H_z^2. \tag{13-111}$$

Again the result is $(\epsilon_0/8)ab\, E_{00x}^2$, and the average electric and magnetic energies per unit length are equal. This is reasonable, since the plane electromagnetic waves that produce the field configuration by reflection at the side walls involve equal electric and magnetic energy densities. It is not obvious, however, because the interference effects tend to confuse the picture.

The total average electromagnetic energy content per unit length in the guide is therefore $\epsilon_0 ab\, E_{00x}^2/4$. Upon dividing the total average transmitted power by this quantity, we find that

$$E_{00x}^2 \frac{ab}{4\omega\mu_0\lambda_g} \frac{4}{E_{00x}^2 \epsilon_0 ab} = c\frac{\lambda_0}{\lambda_g} = c\left\{1 - \left(\frac{\lambda_0}{2b}\right)^2\right\}^{1/2}, \tag{13-112}$$

$$= u_s. \tag{13-113}$$

The average transmitted power is thus equal to the product of the average energy per unit length times the signal velocity u_s as we could have expected, intuitively.

13.3.4 Attenuation

We have assumed until now that the walls were perfectly conducting; let us now consider real wave guides of finite conductivity.

In the process of guiding electromagnetic waves, conductors dissipate part of the wave energy in the form of Joule losses. This is because the waves induce electric currents in the guide. A rigorous calculation of the field for a guide of finite conductivity is difficult, but fortunately unnecessary.

The procedure used for calculating the Joule losses is the following. We have performed a calculation on the assumption that the guide is perfectly conducting. This led to a field in which there is a tangential H at the surface of the guide. Since the tangential H must be continuous across any interface, we know the value of H inside the conductor. Then, using Maxwell's equations, we can find the corresponding tangential E inside, which is not zero unless the guide material is a perfect conductor. This small tangential E is then considered to be a perturbation of the ideal field obtained with perfect conductors. The method is entirely satisfactory because this E is so small that it hardly disturbs the wave. We thus have a tangential E, a tangential H, and a Poynting vector that is normal to the conducting surface and directed into the metal.

That both $\boldsymbol{E}$ and $\boldsymbol{H}$ vectors must exist inside the conducting walls can also be shown as follows. To begin with, we must have a tangential H just inside the wall. On the other hand, at some distance within the wall, there must be zero field, since the attenuation distance δ (Section 11.5) is quite short at frequencies which are high enough to propagate in wave guides. For example, if $b = 7.5$ centimeters, the frequency can be, say, 3000 megahertz and δ is then only 1.2 microns in copper (Table 11-1). The tangential component of $\boldsymbol{H}$ thus decreases rapidly with depth inside the conducting wall. Then $\nabla \times \boldsymbol{H}$ is not zero, and there is a current density $\boldsymbol{J}_f$ parallel to the surface and normal to $\boldsymbol{H}$ since $\nabla \times \boldsymbol{H} = \boldsymbol{J}_f$. We must therefore have both a tangential $\boldsymbol{E}$ to produce the tangential current density $\boldsymbol{J}_f$ and a tangential $\boldsymbol{H}$.

The average Poynting vector directed into the guide wall gives the average power W_L which is removed from the wave per meter of length. We then wish to calculate the attenuation constant k_{gi}. This constant must be such that, when both the E and the H of the transmitted wave are multiplied by $\exp(-k_{gi}z)$, the average Poynting vector for the transmitted wave and the average transmitted power W_T decrease by a factor of

$$\exp(-2k_{gi}\,\Delta z) \approx 1 - 2k_{gi}\,\Delta z \tag{13-114}$$

in a distance Δz. The approximation is excellent for ordinary types of wave guide. Then

$$W_L\,\Delta z = (2k_{gi}\,\Delta z)W_T, \tag{13-115}$$

or

$$k_{gi} = \frac{W_L}{2W_T}. \tag{13-116}$$

The real part k_{gr} of k_g can be taken to be the k_g obtained on the assumption of perfectly conducting walls.

It might be expected at first sight that the attenuation could be calculated from the reflection losses. It will be recalled from Section 12.5 that an electromagnetic wave reflected from a good conductor is slightly weaker than the incident wave. This method of calculation is incorrect because, as we shall see, there are also energy losses in the guide faces parallel to the yz- plane.

Let us calculate k_{gi}. The tangential H produces an electromagnetic wave that penetrates perpendicularly into the wall. Inside the conducting wall,

$$\frac{E}{H} = \left(\frac{\mu_0\omega}{\sigma}\right)^{1/2} e^{j\pi/4}, \tag{13-117}$$

as in Eq. 11-106.

We assume that the dielectric inside the guide is dry air, and is therefore lossless. We also assume that $n = 1$ in order that the field can be described, as a first approximation, by Eqs. 13-88 to 13-94.

Along the face that lies in the xz-plane,

$$H_z = \frac{\pi E_{00x}}{\omega\mu_0 b} \exp j\left(\omega t - \frac{z}{\lambda_g} - \frac{\pi}{2}\right) \qquad (y = 0). \tag{13-118}$$

Then E_x is *not* equal to zero for $y = 0$, as in Eq. 13-88, but is rather

$$E_x = \left(\frac{\mu_0\omega}{\sigma}\right)^{1/2} \frac{\pi E_{00x}}{\omega\mu_0 b} \exp j\left(\omega t - \frac{z}{\lambda_g} - \frac{\pi}{4}\right) \qquad (y = 0). \tag{13-119}$$

Note that, as the conductivity σ approaches infinity, E_x approaches zero. The average Poynting vector $(1/2)\,\mathrm{Re}\,(\boldsymbol{E} \times \boldsymbol{H}^*)$ is directed into the guide wall and is equal to

$$\left(\frac{\pi E_{00x}}{b}\right)^2 \frac{1}{\sigma^{1/2}(2\omega\mu_0)^{3/2}} \qquad (y = 0).$$

This is the average energy flowing into the wall at $y = 0$, per square meter and per second. It is interesting to note that this energy flux is the same at all points on the face $y = 0$. The power lost to the wall per meter of length is a times larger and, for the two faces parallel to the xz-plane,

$$W_{xz} = \left(\frac{\pi E_{00x}}{b}\right)^2 \frac{2a}{\sigma^{1/2}(2\omega\mu_0)^{3/2}}. \tag{13-120}$$

This is the average power lost by reflection.

At the face $x = 0$, $\boldsymbol{H}$ has y- and z-components as in Eqs. 13-92 and 13-94. Associated with H_y, we have an electric field intensity

$$E_z = \left(\frac{\mu_0\omega}{\sigma}\right)^{1/2} \frac{E_{00x}}{\omega\mu_0\lambda_g} \sin\left(\frac{\pi y}{b}\right) \exp j\left(\omega t - \frac{z}{\lambda_g} + \frac{\pi}{4}\right). \tag{13-121}$$

The magnitude of the corresponding average Poynting vector directed into the walls is now

$$\left(\frac{E_{00x}}{\lambda_g}\right)^2 \frac{1}{\sigma^{1/2}(2\omega\mu_0)^{3/2}} \sin^2\left(\frac{\pi y}{b}\right).$$

Similarly, the magnitude of the average Poynting vector corresponding to H_z is

$$\left(\frac{\pi E_{00x}}{b}\right)^2 \frac{1}{\sigma^{1/2}(2\omega\mu_0)^{3/2}} \cos^2\left(\frac{\pi y}{b}\right).$$

Integrating the sum of these two quantities from $y = 0$ to $y = b$ and multiplying by 2, we obtain the average power lost per meter in the two walls parallel to the yz-plane:

$$W_{yz} = \frac{\pi^2 E_{00x}^2}{b\sigma^{1/2}(2\omega\mu_0)^{3/2}} \left\{1 + \left(\frac{2b}{\lambda_g}\right)^2\right\} \tag{13-122}$$

or, using Eq. 13-85,

$$W_{yz} = \frac{\pi^2 E_{00x}^2}{b\sigma^{1/2}(2\omega\mu_0)^{3/2}} \left(\frac{2b}{\lambda_0}\right)^2. \tag{13-123}$$

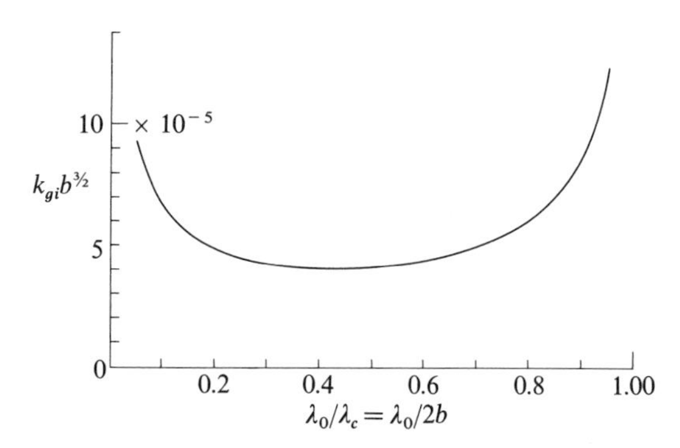

Figure 13-10. Dependence of $k_{gi}b^{3/2}$ on λ_0/λ_c for $a/b = 0.5$ and for copper.

The total average power loss per meter is then

$$W_L = W_{xz} + W_{yz}, \tag{13-124}$$

$$= \frac{\pi^2 E_{00x}^2}{b\sigma^{1/2}(2\omega\mu_0)^{3/2}} \left\{ \frac{2a}{b} + \left(\frac{2b}{\lambda_0}\right)^2 \right\}. \tag{13-125}$$

To find the attenuation constant k_{gi} for the wave, we now use Eqs. 13-116, 13-125, and 13-109:

$$k_{gi} = \frac{W_L}{2W_T}, \tag{13-126}$$

$$= \frac{1}{a(120\sigma\lambda_0)^{1/2}} \frac{1 + \dfrac{2a}{b}\left(\dfrac{\lambda_0}{2b}\right)^2}{\left\{1 - \left(\dfrac{\lambda_0}{2b}\right)^2\right\}^{1/2}}. \tag{13-127}$$

For an infinite guide height a, only the second term remains in the numerator and

$$k_{gi} = \frac{1}{b(30\sigma\lambda_0)^{1/2}} \frac{\left(\dfrac{\lambda_0}{2b}\right)^2}{\left\{1 - \left(\dfrac{\lambda_0}{2b}\right)^2\right\}^{1/2}} \qquad (a \to \infty). \tag{13-128}$$

This term comes from W_{xz}, and the losses then occur only on the guide faces parallel to the xz-plane. It is shown in Problem 13-22 that this value of k_{gi} can be accounted for entirely by the reflection losses.

In practice, the ratio $2a/b$ is close to unity, whereas $(\lambda_0/2b)^2$ is of the order of 1/2. The losses on the faces parallel to xz-plane are thus of the same order of magnitude as those on the other pair of faces, but smaller by a factor of about 2.

Figure 13-10 shows $k_{gi}b^{3/2}$ as a function of the ratio $\lambda_0/\lambda_c = \lambda_0/2b$ for

Table 13-1. *Characteristics of a Few Types of Rectangular Wave Guides* (TE *mode with* $n = 1$)

Inside Dimensions	Cut-off Frequency	Operating Range	Attenuation	Power Rating
inches	gigahertz		db/100 feet	megawatts
2.84 × 1.34	2.080	2.60– 3.95	1.10–0.75	2.2–3.2
1.872 × 0.872	3.155	3.95– 5.85	2.08–1.44	1.4–2.0
1.372 × 0.622	4.285	5.85– 8.20	2.87–2.30	0.6–0.7
0.900 × 0.400	6.56	8.20–12.4	6.45–4.48	0.2–0.3
0.622 × 0.311	9.49	12.4 –18.0	9.51–8.31	0.1–0.2

$2a/b$ equal to unity. According to this curve, the optimum value of $\lambda_0/2b$ is about 0.4, but the minimum is very broad; actual values of $\lambda_0/2b$ are larger so as to achieve strong attenuation for the $n = 2$ mode.

The attenuation is of the order of a tenth of a decibel per meter at frequencies of a few gigahertz, increasing as $f^{3/2}$ when the ratios a/b and $\lambda_0/2b$ are kept constant (Problem 13-21).

Table 13-1 shows the characteristics of a few types of hollow rectangular wave guides.

13.4 SUMMARY

In the general case of a sinusoidal electromagnetic wave propagating in the positive direction of the z-axis,

$$\boldsymbol{E} = (E_{0x}\boldsymbol{i} + E_{0y}\boldsymbol{j} + E_{0z}\boldsymbol{k}) \exp j(\omega t - k_g z) = \boldsymbol{E}_0 \exp j(\omega t - k_g z) \qquad (13\text{-}1)$$

$$\boldsymbol{H} = (H_{0x}\boldsymbol{i} + H_{0y}\boldsymbol{j} + H_{0z}\boldsymbol{k}) \exp j(\omega t - k_g z) = \boldsymbol{H}_0 \exp j(\omega t - k_g z) \qquad (13\text{-}2)$$

We have shown that, if the radian length $\bar{\lambda}_g$ of the guided wave is *not* equal to the radian length $\bar{\lambda}$ of a plane wave in the medium of propagation, then

$$E_{0x} = \frac{-j\omega\mu}{\dfrac{1}{\bar{\lambda}^2} - \dfrac{1}{\bar{\lambda}_g^2}} \left(\frac{k_g}{\omega\mu} \frac{\partial E_{0z}}{\partial x} + \frac{\partial H_{0z}}{\partial y} \right), \qquad (13\text{-}12)$$

$$E_{0y} = \frac{j\omega\mu}{\dfrac{1}{\bar{\lambda}^2} - \dfrac{1}{\bar{\lambda}_g^2}} \left(-\frac{k_g}{\omega\mu} \frac{\partial E_{0z}}{\partial y} + \frac{\partial H_{0z}}{\partial x} \right), \qquad (13\text{-}13)$$

$$H_{0x} = \frac{j\omega\epsilon}{\dfrac{1}{\bar{\lambda}^2} - \dfrac{1}{\bar{\lambda}_g^2}} \left(\frac{\partial E_{0z}}{\partial y} - \frac{k_g}{\omega\epsilon} \frac{\partial H_{0z}}{\partial x} \right), \qquad (13\text{-}14)$$

$$H_{0y} = \frac{-j\omega\epsilon}{\dfrac{1}{\lambda^2} - \dfrac{1}{\lambda_g^2}} \left(\frac{\partial E_{0z}}{\partial x} + \frac{k_g}{\omega\epsilon} \frac{\partial H_{0z}}{\partial y} \right). \tag{13-15}$$

The vectors $\boldsymbol{E}$ and $\boldsymbol{H}$ can therefore be calculated once E_{0z} and H_{0z} are known.

If $E_{0z} = 0$ we have a *TE wave*, and if $H_{0z} = 0$ we have a *TM wave*. If both E_{0z} and H_{0z} are zero, then we have a *TEM wave* and $\lambda_g = \lambda$.

For TE and TM waves we calculate E_{0z} and H_{0z} using the wave equations and the boundary conditions imposed by the guide. The wave equations for E_{0z} and H_{0z} are similar; that for E_{0z} is the following:

$$\frac{\partial^2 E_{0z}}{\partial x^2} + \frac{\partial^2 E_{0z}}{\partial y^2} = -\left(\frac{1}{\lambda^2} - \frac{1}{\lambda_g^2} \right) E_{0z}. \tag{13-18}$$

In *TEM waves,*

$$\frac{\partial V}{\partial z} = -\frac{\partial A}{\partial t}, \tag{13-31}$$

and

$$\boldsymbol{E} = -\left(\frac{\partial V_0}{\partial x} \boldsymbol{i} + \frac{\partial V_0}{\partial y} \boldsymbol{j} \right) \exp j \left(\omega t - \frac{z}{\lambda} \right), \tag{13-37}$$

$$= \boldsymbol{E}_0 \exp j \left(\omega t - \frac{z}{\lambda} \right). \tag{13-38}$$

The field $\boldsymbol{E}_0$ can be derived from the potential V_0 in exactly the same manner as for an electrostatic field. Then, inside a hollow perfectly conducting tube, $\boldsymbol{E}_0$ must be zero and TEM waves are impossible, except as a limiting case of TE and TM waves at very short wavelengths.

The *transverse components* of $\boldsymbol{E}$ and $\boldsymbol{H}$ are mutually perpendicular, whether we have TE, TM, or TEM waves. The *wave impedance* is

$$\frac{E_{0t}}{H_{0t}} = \left(\frac{\mu}{\epsilon} \right)^{1/2} \frac{\lambda_g}{\lambda} \text{ for TE waves,} \tag{13-23}$$

$$= \left(\frac{\mu}{\epsilon} \right)^{1/2} \frac{\lambda}{\lambda_g} \text{ for TM waves,} \tag{13-25}$$

$$= \left(\frac{\mu}{\epsilon} \right)^{1/2} \text{ for TEM waves.} \tag{13-46}$$

The *boundary conditions* at the surface of a perfectly conducting wave guide are as follows. (a) For any type of wave, the tangential component of $\boldsymbol{E}$ is zero. (b) For any type of wave, the $\boldsymbol{H}$ vector is tangent to the surface, perpendicular to the current density in the guide, and numerically equal to the surface current density expressed in amperes/meter. (c) For TE waves, $\boldsymbol{\nabla} H_{0z}$ is tangent to the wall, and the rate of change of H_{0z} in the direction normal to the surface is zero.

To illustrate how one can deal with guided waves, we studied two common types, namely, the TEM wave in a coaxial line and the TE wave with $n = 1$ in a rectangular wave guide. In a *coaxial line*,

$$\boldsymbol{E} = \frac{C}{\rho} \exp j\left(\omega t - \frac{z}{\bar{\lambda}}\right) \boldsymbol{\rho}_1, \qquad (13\text{-}58)$$

$$\boldsymbol{H} = \left(\frac{\epsilon}{\mu_0}\right)^{1/2} \frac{C}{\rho} \exp j\left(\omega t - \frac{z}{\bar{\lambda}}\right) \boldsymbol{\varphi}_1, \qquad (13\text{-}60)$$

and the average transmitted power is

$$W_T = \frac{C^2}{120} \ln \frac{\rho_o}{\rho_i} \text{ watts} \qquad (\epsilon_r = 1,\ \mu_r = 1). \qquad (13\text{-}65)$$

For the TE wave in a *rectangular wave guide*, we assumed a plane wave zigzagging down the guide, as in Figure 13-6. We then found H_{oz} from the wave equation and from the required boundary conditions. The other components followed immediately. The six components are shown in Eqs. 13-88 to 13-94 for $n = 1$. This parameter n is the number of "half cycles of $\boldsymbol{E}$" in the guide as in Figure 13-7.

Rectangular wave guides are high-frequency devices, the cut-off wave length λ_c being equal to twice the distance between the reflecting sides. The phase velocity u_p is larger than c, whereas the signal velocity u_s is smaller, and $u_p u_s = c^2$.

The average transmitted power is

$$W_T = \frac{E_{00x}^2 ab}{4c\mu_0} \left\{1 - \left(\frac{\lambda_0}{2b}\right)^2\right\}^{1/2}, \qquad (13\text{-}109)$$

where E_{00x} is the maximum value of E in the guide and the lengths a and b are as in Figure 13-5.

It is possible to take into account the finite conductivity of the guide by deducing the tangential E in the guide from the value of the tangential H calculated on the assumption of infinite conductivity as above. This leads to a Poynting vector that points into the guide material and that gives the Joule losses. The result is that the imagninary part of the wave number is

$$k_{gi} = \frac{1}{a(120\sigma\lambda_0)^{1/2}} \frac{1 + \dfrac{2a}{b}\left(\dfrac{\lambda_0}{2b}\right)^2}{\left\{1 - \left(\dfrac{\lambda_0}{2b}\right)^2\right\}^{1/2}}. \qquad (13\text{-}127)$$

It corresponds to losses in all four faces.

PROBLEMS

13-1. The transverse part of the vector $\boldsymbol{E}$ in a guided wave is written as a vector $\boldsymbol{E}_{tr}$.
What can you say about the orientation of this vector in space, as a function of the four variables x, y, z, t, in the case of TE waves? What if E_{0y}/E_{0x} is not real?

13-2. Sketch a rather large cross-sectional view of a coaxial line in a plane containing the axis.

(a) Show lines of $\boldsymbol{E}$ and of $\boldsymbol{H}$ at a given instant over at least one wavelength. The lines should be most closely spaced where the field is strongest. Indicate the directions of the fields by means of arrow heads. Show the direction of propagation.
(b) Add arrows at various points to represent Poynting vectors, using longer arrows where the power flow is larger. Assume that the length of the arrow represents the magnitude of the Poynting vector at its midpoint.
(c) How does the pattern change with time?
(d) Sketch a cross-sectional view of the coaxial line in a plane perpendicular to the axis and show lines of $\boldsymbol{E}$ and of $\boldsymbol{H}$ at a particular instant.
Relate this plane to the figure you drew under (a) above.
(e) Explain how this pattern changes with time.
(f) Add plus and minus signs to both figures to show the surface charges. The spacing between the signs should indicate qualitatively the relative magnitude of the surface charge density.
(g) How does the charge pattern change with time at a given z?
(h) Now add arrows of various lengths to your first figure to represent surface current densities.
(i) How does the current pattern change with time at a given z?

13-3. Show that, in the case of an idealized coaxial line of infinite conductivity, the current is given by the linear charge density multiplied by the velocity of propagation.

13-4. It is known, from transmission-line theory, that the characteristic impedance of a line is given by

$$Z_c = \left(\frac{L'}{C'}\right)^{1/2},$$

where L' and C' are, respectively, the inductance and capacitance per unit length.
Show that this is correct in the case of the coaxial line, by calculating L' and C', and then comparing with the value of the characteristic impedance given in Section 13.2.

13-5. (a) If the maximum allowed field strength in a coaxial line is E_m, show that the maximum allowed voltage is

$$V_m = \rho_o E_m \frac{\ln(\rho_o/\rho_i)}{\rho_o/\rho_i}.$$

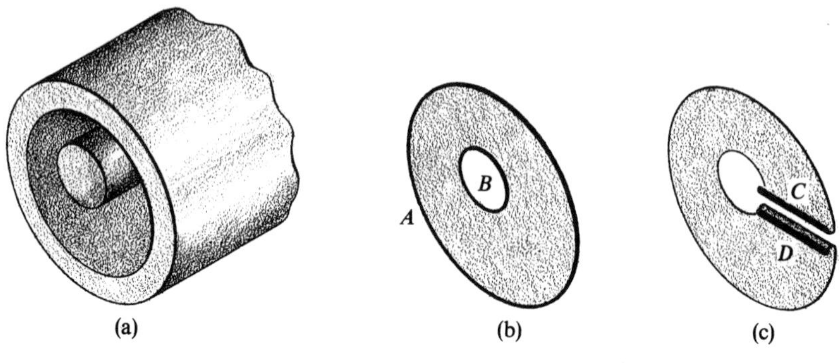

Figure 13-11.

(b) Show that, for a given value of ρ_o, V_m is greatest when ρ_o/ρ_i is equal to *e*.
(c) Show that the characteristic impedance is then 60 ohms if the line is air-insulated.

(d) Show that, under those conditions, the maximum allowable current is $\rho_o E_m/163$ ampere.

13-6. (a) Setting E_m to be the maximum allowed field strength in a coaxial line, use the result of Problem 13-5 to show that the maximum allowed power is

$$W_m = \rho_o^2 \frac{E_m^2}{120} \frac{\ln(\rho_o/\rho_i)}{(\rho_o/\rho_i)^2}$$

if the line is air-insulated.

(b) Show that, for a given value of ρ_o, this power is greatest when $\rho_o/\rho_i = 1.65$.

(c) Show that this ratio corresponds to a characteristic impedance of 30 ohms if the line is air-insulated.

13-7. One important parameter of a TEM transmission line is the *characteristic impedance.*

$$Z_c = \left(\frac{L'}{C'}\right)^{1/2},$$

where L' and C' are, respectively, the inductance and the capacitance per unit length of the line.

In designing such lines it is therefore essential to predict the values of L' and C'. If the geometry is such that L' and C' cannot be calculated analytically, as, for example, in Figure 13-11a, it is sometimes useful to perform the following measurements on a resistance-sheet analog.*

(a) We can find the value of C' by cutting out a sheet of resistive material in the shape of the cross section of the dielectric as in Figure 13-11b, and measuring the resistance between electrodes A and B.

* D. J. Epstein, Comment on a Theorem in the Field of Steady Current Flow, Proc. IEEE, **56**, 198 (1968).

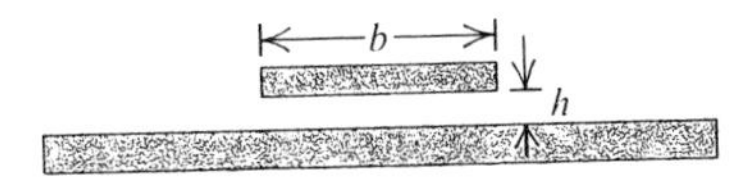

Figure 13-12.

Show that, if the material has a conductivity σ and a thickness s, and if the permittivity of the dielectric is ϵ, then

$$R_1 C' = \frac{\epsilon}{s\sigma},$$

where R_1 is the resistance between A and B.‡

(b) To measure L' we use a similar sheet with a radial cut and with electrodes C and D as in Figure 13-11c.

Show that, if μ is the permeability of the dielectric,

$$R_2 L' = \frac{\mu}{s\sigma},$$

where R_2 is the resistance between C and D.

Thus

$$Z_c = \left(\frac{\mu}{\epsilon}\right)^{1/2} \left(\frac{R_1}{R_2}\right)^{1/2}.$$

See also Problem 10-14.‡

13-8. Figure 13-12 shows a cross section of a *microstrip line*. The lower electrode is grounded and the wave is TEM.

The microstrip has the advantage of being much less costly than either the coaxial line or the rectangular wave guide. It is also particularly convenient for conveying high-frequency signals within printed and integrated circuits. Its main disadvantage lies in the fact that its field is not limited to the region immediately below the upper electrode. Microstrip lines can therefore interact with other elements in a circuit, unless they are either spaced or shielded properly. Shielding can be achieved by adding a second grounded plane above the strip, but this, of course, increases the complexity and cost of the line.

(a) Sketch lines of **E** and of **H**.

Use arrows to show the directions of **E** and **H** at a given time. Show the direction of propagation.

(b) In practice, the width b of the strip is much larger than its distance h to the grounded plane, and edge effects are small.

Show that the instantaneous value $I\mathcal{V}$ of the transmitted power is equal to the Poynting vector integrated over the cross section bh.

(c) Show that the characteristic impedance $\mathcal{V}/I$ is

$$\left(\frac{\mu}{\epsilon}\right)^{1/2} \frac{h}{b}.$$

(d) Show that one arrives at the same result if one defines the characteristic impedance as in Problems 13-4 or 13-7.

(e) Show that the addition of a second grounded plane placed symmetrically with the first reduces the characteristic impedance by a factor of two.

13-9. Figure 13-8 shows schematically the $\boldsymbol{E}$ and $\boldsymbol{H}$ fields inside a rectangular wave guide carrying a TE ($n = 1$) wave.

Draw sketches as in Problem 13-2 showing Poynting vectors $\boldsymbol{E} \times \boldsymbol{H}$, surface charge densities, and surface current densities.

To represent properly these last two quantities you will have to sketch two or three neighboring faces. Be sure to relate these faces to those shown in Figure 13-8.

13-10. If a wave guide is not properly connected to its load, part of the incident wave is reflected back towards the source and a standing wave is established along the line. Under such conditions the power fed to the load can become negligible.

It is therefore useful to be able to move a small probe along a longitudinal slot to sample the field inside the guide. The quantity that is usually measured is the *Voltage Standing Wave Ratio* (VSWR). This is the ratio of the maximum to the minimum time-averaged voltage picked up by the probe as it slides along the slot. Under ideal conditions of power transfer, there is no reflected wave and the VSWR is equal to unity.

The probe can be either a small loop that is coupled to the magnetic field, or a short length of wire that responds to the electric field. In both cases the projection into the guide is approximately one millimeter or less.

(a) In the case of a rectangular hollow wave guide and a TE mode, as in Figure 13-8, where should the slot be cut? The slot should, of course, disturb the wave as little as possible. Sketch a perspective view of the guide, showing both the $\boldsymbol{E}$ vector and the slot.

(b) If the probe is a small loop, how should it be oriented with respect to the guide?

(c) How would you use the movable probe to measure λ_g?

13-11. Ordinary wave guides are brass or copper tubes with wall thicknesses of the order of one or two millimeters. However, some microwave structures are quite complex and would be too expensive to fabricate either from tubing or from solid stock. Several techniques have therefore been developed for such cases.

One technique, called *electroforming*, utilizes a *former*, in the shape if the internal volume required, which is electroplated to a thickness of a few millimeters. The former is then removed, leaving the required structure.

In some cases the former is a plastic and is left in place. The resulting structure is thus dielectric-filled and is smaller (Problem 13-13), for a given operating frequency, than if it were air-filled.

In this latter case, what minimum thickness of copper would you recommend if the operating frequency is to be 3 gigahertz?

13-12. It is found that an electromagnetic wave propagating in the TE mode with $n = 1$ (Section 13.3.1) through the 2.84 inch $\times$ 1.34 inch wave guide of Table 13-1 has a λ_g of 13.8 centimeters.

Calculate its frequency.

13-13. (a) Show that, if a hollow rectangular wave guide is completely filled with a dielectric of relative permeability ϵ_r, its cut-off frequency is *lower* than if it were empty by a factor of $\epsilon_r^{1/2}$:

$$f_{cd} = \frac{1}{\epsilon_r^{1/2}} f_{cv},$$

where the subscripts d and v refer, respectively, to "dielectric" and to "vacuum."

Thus, for a given operating frequency, a dielectric-filled wave guide is *smaller* than an air-filled one. For example, if the dielectric is Teflon ($\epsilon_r = 2.1$), the cross-sectional dimensions a and b of Figure 13-5 are both reduced by a factor of 1.45.

(b) Consider two wave guides, one air-insulated and with a cross section a, b, and another that is dielectric-insulated and that has a cross-section $a/\epsilon_r^{1/2}$, $b/\epsilon_r^{1/2}$.

Show that, for a given frequency,

$$\lambda_{gd} = \frac{1}{\epsilon_r^{1/2}} \lambda_{gv}.$$

13-14. In Section 13.3.1 we showed that

$$\frac{1}{\lambda_0^2} - \frac{1}{\lambda_g^2} = \pi^2 \frac{n^2}{b^2}$$

for a TE wave propagating through a rectangular wave guide, as in Figure 13-6.

Show that, in the more general case of any TE or TM wave in a rectangular wave guide,

$$\frac{1}{\lambda_0^2} - \frac{1}{\lambda_g^2} = \pi^2 \left(\frac{m^2}{a^2} + \frac{n^2}{b^2}\right),$$

where m is the number of half cycles of $\boldsymbol{E}$ or of $\boldsymbol{H}$ in the direction of the x-axis, and similarly for n.

13-15. A plane electromagnetic wave is incident at an angle on a flat perfectly conducting surface, and $\boldsymbol{E}$ is normal to the plane of incidence.

(a) Draw carefully a set of equally spaced parallel lines representing the "crests" of E in the incident wave at a given instant, and draw broken lines for the "troughs." Draw similar lines for the reflected wave.

(b) Where is $\boldsymbol{E}$ always equal to zero?

(c) Can you relate this pattern to the $n = 1, 2$, etc., modes in a rectangular wave guide?

13-16. (a) Sketch a graph of λ_0/λ_g as a function of λ_0/λ_c for a rectangular wave guide.

In actual practice this curve is meaningful only for $\lambda_0/\lambda_c \approx 0.7$. At much smaller values of λ_0, higher modes of propagation ($n = 2, n = 3$, etc.), are possible, and the field configuration inside the guide becomes uncertain. At much larger values of λ_0 the attenuation becomes excessive.

(b) Sketch a curve of ω as a function of k_g.

For a given operating point, the phase velocity u_p is given by the ratio ω/k_g.

(c) Verify that the signal velocity is given by

$$u_s = \frac{1}{dk_g/d\omega} = \frac{d\omega}{dk_g},$$

and thus by the slope of this curve.

13-17. Check the power rating of the largest rectangular wave guide shown in Table 13-1. Assume that the operating frequency is 1.5 times the cut-off frequency.

The maximum permissible electric field intensity in air is 3×10^6 volts/meter. Allow a factor of safety of 2 to take into account the effects of irregularities in the inner surface of the guide.

13-18. A small single-turn loop of area A is situated at $x = 0$, $y = b/2$, just inside the hollow rectangular wave guide of Figure 13-5. The plane of the loop is parallel to the x-z plane, and the field inside the guide is as illustrated in Figure 13-8.

Show that the power flowing through the guide is given by

$$W_T = 3.36 \times 10^{-5} \frac{\lambda_0^2 ab}{\left\{1 - \left(\frac{\lambda_0}{2b}\right)^2\right\}^{1/2}} \frac{\mathcal{V}^2}{A^2},$$

where $\mathcal{V}$ is the rms electromotance induced in the loop.

13-19. It is suggested that one could measure the power transmitted by a rectangular wave guide by observing either the deflection or the energy gain of a beam of electrons crossing the guide.

Discuss the posibility of such measurements.

13-20. (a) Show that the average force per unit area exerted by the field on the face $x = 0$ of the rectangular wave guide shown in Figure 13-8 is given by

$$F_x = -\frac{\epsilon_0 E_{00x}^2}{4} \left(\frac{\lambda_0}{2b}\right)^2 \cos \frac{2\pi y}{b}$$

(b) Calculate F_x for the first guide listed in Table 13-1 at the maximum power rating and at a frequency of 3.00 gigahertz.

(c) Is it important to take this force into account in the design of rectangular wave guides?

(d) Could this force be used to measure the transmitted power?

13-21. Show that the attenuation constant k_{gi} for a TE($n = 1$) wave in a rectangular wave guide, as given in Eq. 13-127, varies as $f^{3/2}$ when the ratios a/b and $\lambda_0/2b$ are kept constant.

13-22. (a) Show that an electromagnetic wave is reduced in amplitude by a factor of approximately

$$1 - \left(\frac{2\omega\epsilon_0}{\sigma}\right)^{1/2} \cos \theta_i$$

upon reflection from a good conductor of conductivity σ. The angle of incidence is θ_i, the dielectric is air, and the $\boldsymbol{E}$ vector is normal to the plane of incidence.
(b) Show that this loss leads to the attenuation constant of Eq. 13-128 for rectangular wave guides.

13-23. (a) What is the surface current density in the face $y = 0$ of the wave guide illustrated in Figure 13-5 when it carries a TE($n = 1$) wave?
(b) Show that the power loss in the wall is the same as if the current were uniformly distributed in a thickness δ near the surface, δ being the attenuation distance or skin depth (Section 11.5).

This result is valid for curved surfaces, as long as the radius of curvature is much larger than the skin depth.

The quantity $1/\sigma\delta$ is called the *surface resistivity* and is expressed in ohms/square. A square sheet of given surface resistivity and of any size has a resistance equal to the surface resistivity between two parallel edges. See also Problems 11-20 and 12-22.

13-24. Use the result of Problem 11-20 to verify Eq. 13-125.

13-25. The degree of attenuation in a line can be expressed in *nepers/meter*, the number of nepers/meter being simply the numerical value of the attenuation constant k_{gi} in MKSA units.

The degree of attenuation is also expressed in *decibels/100 feet*. The number of decibels/100 feet is 20 times the logarithm to the base 10 of the ratio of the voltages (or the currents) at the two ends of a line 100 feet long.

Show that one neper/meter is equivalent to 264 decibels/100 feet.

13-26. (a) Check the attenuation of the largest type of rectangular wave guide listed in Table 13-1. Assume that the guide material is copper and that the operating frequency is 1.5 times the cut-off frequency.

See Problem 13-25 for transforming nepers/meter to decibels/100 feet.
(b) Calculate the power dissipated in the guide per meter of length at its maximum power rating of 2800 kilowatts.
(c) Calculate the rate at which the temperature of the guide would increase, at a maximum power rating, if it were thermally insulated. The walls have a thickness of 0.080 inch.
(d) How should the guide walls be cooled?

Would it be practical to transmit large amounts of electric energy over long distances in this way?

Wave guides are often used to transmit large amounts of power at low duty cycles. The average power dissipation is then much lower than in the above example.

13-27. Tabulate the values of the following quantities for the five types of rectangular wave guide of Table 13-1:

(a) b/a;
(b) $f_{\max}/f_{\min}$;
(c) $\lambda_0/2b$ at both limits of the operating range;

(d) the attenuation in nepers/meter (see Problem 13-25) for the $n = 2$ mode at both limits of the operating frequency.

13-28. (a) We are required to calculate the maximum power that can be transmitted at a frequency of 3.00 gigahertz through a short length of (a) coaxial line, and (b) rectangular wave guide.

The coaxial line has a diameter $2\rho_o$ of 2.50 centimeters and the wave guide has an inside cross section of 3.75×7.50 centimeters.

The coaxial line satisfies the condition for maximum power transfer, namely $\rho_o/\rho_i = 1.65$ (Problem 13-6). The radius ρ_0 is chosen to ensure attenuation of higher order modes.*

In both cases the dielectric is air and the current-carrying surfaces are silver plated. The maximum allowed electric field intensity is 1.5×10^6 volts/meter as in Problem 13-17.

It is assumed that there is no reflected wave.

(b) We are also required to calculate the power lost per meter of length in both cases. See Problem 13-29a.

(c) Finally, we wish to know the rms voltage and current at the input end of the coaxial line.

13-29. In calculating the resistance of a coaxial line it is correct to assume that the current is uniformly distributed throughout a thickness equal to the skin depth δ on both conductors (see Problem 13-23).

(a) Show that the resistance of a coaxial line per unit length is

$$R = \frac{1}{2\pi\sigma\delta}\left(\frac{1}{\rho_o} + \frac{1}{\rho_i}\right),$$

where σ is the conductivity of the material.

(b) Use the method of Section 13.3.4 to show that, for a coaxial line,

$$k_{gi} = 2.64 \times 10^{-6}\left(\frac{f}{\sigma}\right)^{1/2}\frac{1 + (\rho_0/\rho_i)}{\rho_o \ln(\rho_0/\rho_i)}.$$

(c) Show that, for a given value of ρ_o, k_{gi} is minimum when $\rho_o/\rho_i = 3.6$.

(d) Show that the characteristic impedance is then 77 ohms.

13-30. (a) Calculate the power rating of an air-insulated copper coaxial line whose outer conductor has an inside diameter $2\rho_o$ of 2.84 inches, when the ratio ρ_o/ρ_i is selected as in Problem 13-6 for maximum power transfer. The maximum electric field intensity in air is 3×10^6 volts/meter. Allow a factor of safety of 2 to take into account irregularities in the surfaces of the conductors.

(b) Use the results of Problem 13-29 to calculate the power dissipation in both the inner and outer conductors per meter of length at the power calculated under (a) above, and at a frequency of one gigahertz.

13-31. Let us calculate the attenuation constant k_{gi} for a coaxial line, taking into account energy dissipation in the dielectric.

---

* See, for example, A. F. Harvey, *Microwave Engineering*, Academic Press, New York, 1963, page 20.

Since the dielectric losses are normally small, we shall simply add their attenuation constant to that associated with resistive losses in the conductors (Problem 13-29) to obtain the value of k_{gi} for the line.

(a) A dielectric is said to be lossy when it dissipates energy in the course of the polarization process, or when it is slightly conducting.

Consider first a parallel-plate capacitor containing a lossy dielectric. The plates have an area A and are separated by a distance s. An alternating voltage voltage $\mathcal{V}$ is applied across the plates.

You should be able to show that, if there is energy dissipation in the dielectric, then one can either say that the dielectric has a conductivity σ, or that its relative permittivity is of the form

$$\epsilon_r = \epsilon' - j\epsilon_r'',$$

where ϵ_r' and ϵ_r'' are positive real quantities, with

$$\sigma = \omega\epsilon_0\epsilon_r''.$$

Energy dissipation in dielectrics is used extensively for welding plastics, gluing wood, heating food, etc.

(b) The calculation of k_g with a lossy dielectric is formally the same as with a loss-less medium, except that ϵ_r is complex.

Show that the part of k_{gi} that is due to the dielectric is

$$\frac{1}{2} k_{gr} \frac{\epsilon_r''}{\epsilon_r'}, \qquad \text{or} \qquad \frac{\omega}{2c} \frac{\epsilon_r''}{(\epsilon_r')^{1/2}}.$$

Thus, taking into account the losses in both the conductors and the dielectric, and using the result of Problem 13-29, the attenuation constant for a coaxial line is

$$k_{gi} = 2.64 \times 10^{-6} \left(\frac{f}{\sigma}\right)^{1/2} \frac{1 + (\rho_o/\rho_i)}{\rho_o \ln(\rho_o/\rho_i)} + 1.05 \times 10^{-8} f \frac{\epsilon_r''}{(\epsilon_r')^{1/2}} \text{ meters}^{-1},$$

where σ is the conductivity of the conductors, and where $\epsilon_r' - j\epsilon_r''$ is the relative permittivity of the dielectric.

13-32. A *gas lens* utilizes the fact that the index of refraction of a gas is dependent on its density, and therefore on its temperature. Thus, cool gas blown gently through a heated tube becomes hotter and lighter near the periphery than near the axis and acts as a weak converging lens.

After a certain distance the radial temperature gradient becomes negligible and, if further convergence is required, the hot gas must be evacuated and fresh cool gas injected in its place. Another possibility is to use the same gas flowing through a succession of alternately hot and cold tubes. Then the hot tubes act as focusing elements, while the cold tubes act as defocusing elements, and, under proper conditions, there is a net focusing effect. This technique is known as *alternate gradient focusing* and was developed for particle accelerators.

Such *beam wave guides* can be used for transmitting modulated light beams carrying, for example, telephone messages.

The index of refraction is related to the density D and to the temperature T as follows:

$$\frac{n-1}{n_0-1} = \frac{D}{D_0} = \frac{T_0}{T}.$$

Use the ray equation given in Problem 12-2 to show that

$$\frac{d^2\rho}{dz^2} \approx \frac{1}{n}\frac{\partial n}{\partial \rho},$$

where ρ is the radial position of a ray at a distance z along the guide.

Assume that

$$\frac{\partial}{\partial s} \approx \frac{\partial}{\partial z}$$

and neglect the term

$$\frac{\partial \rho}{\partial z}\frac{\partial n}{\partial z}.$$

Since n is very close to unity,

$$\frac{d^2\rho}{dz^2} \approx \frac{\partial n}{\partial \rho}.$$

Then

$$\frac{d^2\rho}{dz^2} \approx -(n_0-1)\frac{T_0}{T^2}\frac{\partial T}{\partial \rho},$$

and, finally, if we choose the reference temperature T_0 to be the average temperature, we can write that

$$\frac{d^2\rho}{dz^2} \approx -\frac{n_0-1}{T_0}\frac{\partial T}{\partial \rho}.$$

This equation can serve as a starting point for the study of such beam wave guides. Since the right-hand side is a complicated function of both ρ and z, the equation must be solved numerically.

Hints

13-7. (a) See Problem 4-7.

(b) See Problem 9-18.

CHAPTER 14

RADIATION OF ELECTROMAGNETIC WAVES

We have studied the propagation of electromagnetic waves in considerable detail. In Chapters 11 through 13 we have studied successively their propagation in free space and in various media, across an interface, and then along various guiding structures. The present chapter will now be devoted to the processes whereby these waves are produced. It may not seem logical to proceed in this order, but the reason is one of convenience and will become apparent after a while: the phenomenon of radiation is rather complex, and its discussion was best delayed until now. Indeed this chapter consists of a series of examples illustrating concepts we have developed in the preceding chapters.

We shall start with the electric dipole of Figure 14-1, which is the simplest type of source. Once we have mastered the electric dipole, we shall be able to study the radiation fields of the half-wave antenna and of antenna arrays. We shall then go on to magnetic dipole radiation, which is closely related to electric dipole radiation. After that it will be relatively easy to deduce the radiation fields of simple electric and magnetic quadrupoles. Finally, we shall prove a reciprocity theorem that is equally valid for antennas and for electric circuits.

14.1 ELECTRIC DIPOLE RADIATION

It will be recalled from Section 2.9 and from the example on page 428 that an electric dipole is formed of a pair of charges of equal magnitude and of opposite signs, as in Figure 14.1. Its dipole moment $\boldsymbol{p}$ is $Q\boldsymbol{s}$.

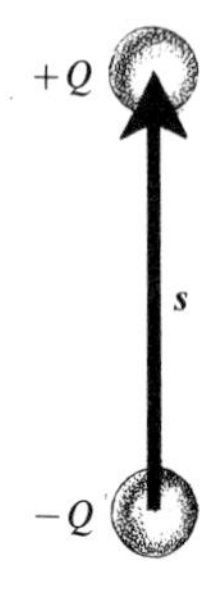

Figure 14-1. An electric dipole. The total charge is zero and the vector $\boldsymbol{s}$ is oriented from $-Q$ to $+Q$.

In the oscillating electric dipole,

$$Q = Q_0 e^{j\omega t}, \tag{14-1}$$

$$\boldsymbol{p} = Q\boldsymbol{s} = Q_0 \boldsymbol{s} e^{j\omega t} = \boldsymbol{p}_0 e^{j\omega t} \tag{14-2}$$

and an alternating current

$$I = I_0 e^{j\omega t} = j\omega\, Q_0 e^{j\omega t} \tag{14-3}$$

flows in the wire connecting the two charges as in Figure 14-2.

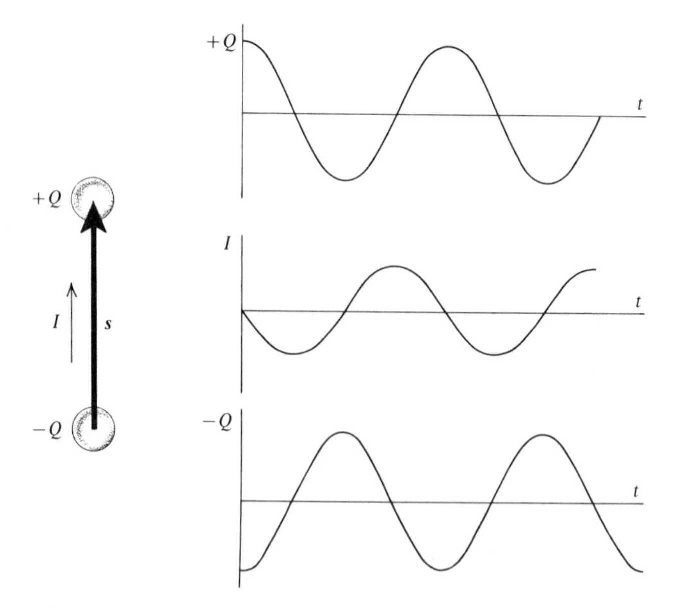

Figure 14-2. The charges $-Q$ and $+Q$ and the current I as functions of time in the oscillating dipole.

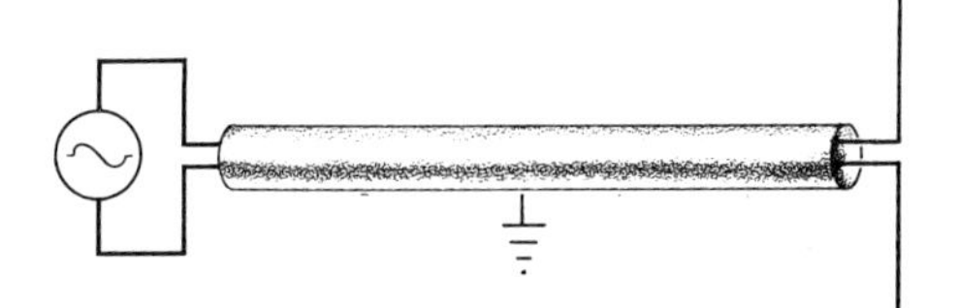

Figure 14-3. Electric dipole antenna fed by an oscillator.

We shall use this simple model, but electric dipole radiation is also produced by any charge distribution whose dipole moment

$$\boldsymbol{p} = \int \rho \boldsymbol{r} \, d\tau \tag{14-4}$$

is a sinusoidal function of the time. For example, charge oscillations can be imagined to occur on the surface of a conducting sphere in such a way that the electrons are driven alternately from one pole to the other. Figure 14-3 shows schematically an electric dipole antenna fed by an oscillator.

We shall calculate successively the potentials V and $\boldsymbol{A}$, then the field vectors $\boldsymbol{E}$ and $\boldsymbol{H}$, and finally the average Poynting vector $\mathbf{S}_{av}$ and the radiated power W.

14.1.1 The Scalar Potential V

We have already found in the example on page 428 the retarded V of the oscillating electric dipole in free space for $s \ll \lambda\!\!\!^{-}$ and $s \ll r$:

$$V = \frac{p_0 \exp j\omega[t]}{4\pi\epsilon_0 r \lambda\!\!\!^{-}} \left(\frac{\lambda\!\!\!^{-}}{r} + j\right) \cos\theta, \tag{14-5}$$

where r and θ are as in Figure 14-4 and $[t]$ is the retarded time $t - (r/c)$.

For a nonzero frequency, the exponential term appears to show that V propagates as a wave at a phase velocity c. This is not quite correct, because of the complex factor $(\lambda\!\!\!^{-}/r) + j$. Rewriting the complex factor in exponential form,

$$V = \frac{p_0}{4\pi\epsilon_0 r \lambda\!\!\!^{-}} \left(\frac{\lambda\!\!\!^{-2}}{r^2} + 1\right)^{1/2} \exp\left\{j\omega\left(t - \frac{r}{c} + \frac{1}{\omega}\arctan\frac{r}{\lambda\!\!\!^{-}}\right)\right\} \cos\theta. \tag{14-6}$$

For $r \gg \lambda\!\!\!^{-}$, $\arctan(r/\lambda\!\!\!^{-}) \approx \pi/2$ and is approximately independent of r. The phase velocity is then c. However, closer in to the dipole, where r is not much larger than $\lambda\!\!\!^{-}$, $\arctan(r/\lambda\!\!\!^{-})$ is not constant, and the effect of this term is to give a phase velocity that is *larger* than c.

The scalar potential V varies as $\cos\theta$ and is zero in the equatorial plane, where the fields of the two charges cancel exactly, just as in electrostatics. It varies as $1/r^2$ as in the static case, but only as $1/r$ when $r \gg \lambda\!\!\!^{-}$. Also, for

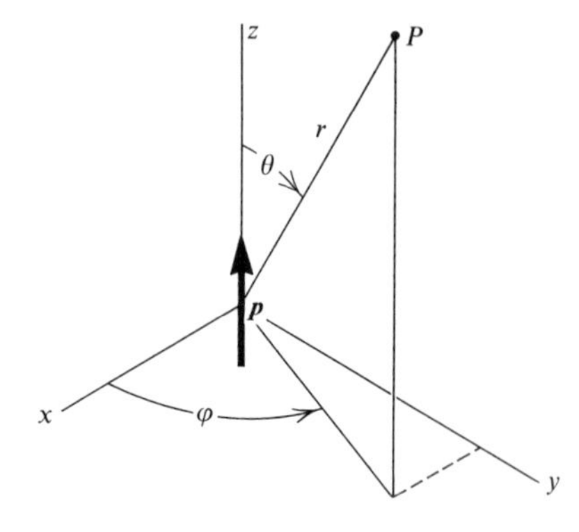

Figure 14-4. An oscillating electric dipole situated at the origin gives rise to a scalar potential V at the point $P(r, \theta, \varphi)$.

$r \gg \lambda$, V varies as $1/\lambda$ and is thus proportional to the frequency. Figure 14-5 shows a radial plot of V as a function of θ and φ.

14.1.2 The Vector Potential A and the Magnetic Field Intensity H

We shall calculate the vector potential $\boldsymbol{A}$ at the point $P(r, \theta, \varphi)$ and then the $\boldsymbol{H}$ vector from the curl of $\boldsymbol{A}$. We assume again that the dipole is in free space, that $s \ll r$, and that $s \ll \lambda$. The last condition eliminates standing-wave effects in the connecting wire, and the current is then the same throughout the length s of the dipole. We have already used these approximations in calculating V. We again make no assumption as to the relative magnitudes of r and λ.

The vector potential $\boldsymbol{A}$ is related to the current I in the dipole as in Eq. 10-36 and Figure 14-5 and 6:

$$\boldsymbol{A} = \frac{\mu_0}{4\pi r} I_0 \exp j\omega[t]\, \boldsymbol{s}. \tag{14-7}$$

It is independent of both θ and φ, and it depends only on the distance r to the dipole. It is also everywhere parallel to $\boldsymbol{s}$, and thus parallel to the polar axis. Expressing $\boldsymbol{s}$ in polar coordinates,

$$\boldsymbol{A} = \frac{\mu_0}{4\pi r} I_0 s\, (\exp j\omega[t])(\cos\theta\, \boldsymbol{r}_1 - \sin\theta\, \boldsymbol{\theta}_1) \tag{14-8}$$

$$= \frac{j\omega\mu_0 p_0}{4\pi r} (\exp j\omega[t])(\cos\theta\, \boldsymbol{r}_1 - \sin\theta\, \boldsymbol{\theta}_1) \tag{14-9}$$

where the polar unit vectors $\boldsymbol{r}_1$, $\boldsymbol{\theta}_1$, $\boldsymbol{\varphi}_1$ are as in Figure 1-19. The vector potential $\boldsymbol{A}$ propagates at a velocity c, even for $r \ll \lambda$.

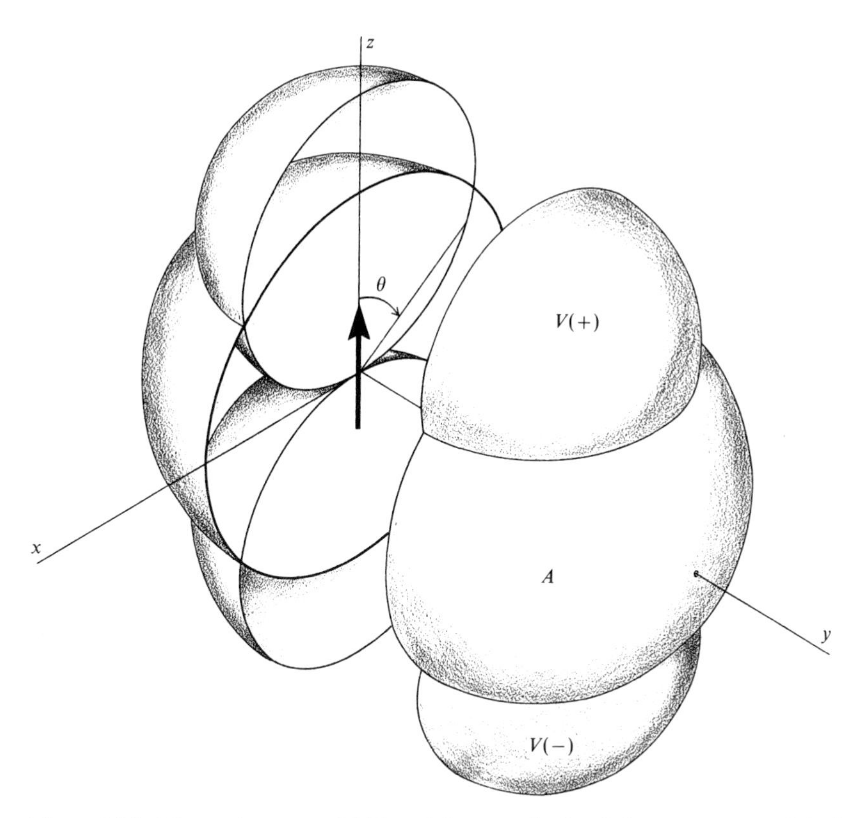

Figure 14-5. The scalar potential V and the magnitude of the vector potential $\boldsymbol{A}$ are shown here as functions of θ and φ about an oscillating electric dipole oriented as shown. The radial distance from the center of the dipole to the spheres marked V is proportional to the value of V in that particular direction. The scalar potential V is maximum at the poles; it vanishes at the equator, where the individual potentials of the charge $-Q$ and $+Q$ of the dipole cancel. It is positive in the northern hemisphere, where the field of $+Q$ is predominant, and negative in the southern hemisphere. The magnitude of $\boldsymbol{A}$ is similarly represented by the sphere marked $\boldsymbol{A}$. The vector potential is independent, both in magnitude and in direction, of the coordinates θ and φ.

We should really have integrated $\boldsymbol{da}$ over the length s of the dipole. However, it will be shown in Problem 14-1 that the integration leads to the above result when $s \ll \lambda$. The situation here is different from what it was for V, because V is the *difference* between the scalar potentials of the two charges $-Q$ and $+Q$. Since the two scalar potentials are very nearly equal in magnitude, their difference must be evaluated with care. For $\boldsymbol{A}$, however, the $\boldsymbol{da}$'s all add up, and the phase angle $\omega\{t-(r/c)\}$, or $\omega t - r/\lambda$, changes but slightly from one element to the next, since $s \ll \lambda$.

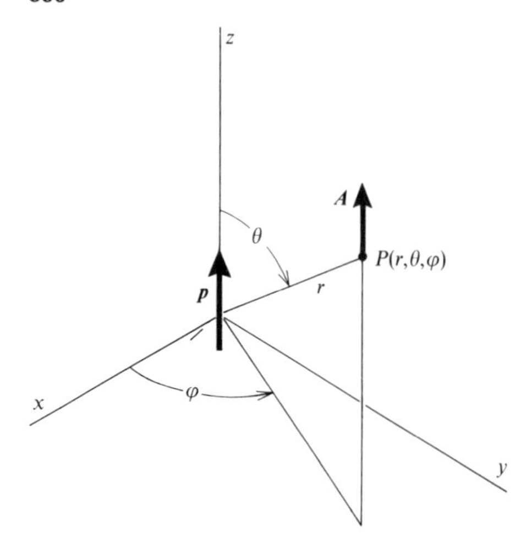

Figure 14-6. An oscillating electric dipole of moment $\boldsymbol{p}$ situated at the origin gives rise to a vector potential $\boldsymbol{A}$ parallel to $\boldsymbol{p}$ at the point $P(r, \theta, \varphi)$.

Thus

$$\boldsymbol{H} = \frac{1}{\mu_0} \boldsymbol{\nabla} \times \boldsymbol{A} = \frac{1}{\mu_0 r} \left\{ \frac{\partial}{\partial r} (rA_\theta) - \frac{\partial A_r}{\partial \theta} \right\} \boldsymbol{\varphi}_1. \tag{14-10}$$

The two other components of the curl are zero because $\boldsymbol{A}$ has no φ component, and because it is not a function of φ. Thus

$$\boldsymbol{H} = \frac{I_0 s}{4\pi r \lambda} (\exp j\omega[t]) \left(\frac{\lambda}{r} + j \right) \sin\theta \, \boldsymbol{\varphi}_1. \tag{14-11}$$

As usual, we compare this new result with a previously acquired one, namely the Biot-Savart law for steady currents, as stated in Eq. 7-8. According to this equation, the magnetic field intensity due to an element of current Is at the position of the dipole is

$$\boldsymbol{H} = \frac{I\boldsymbol{s} \times \boldsymbol{r}_1}{4\pi r^2} = \frac{Is \sin\theta}{4\pi r^2} \boldsymbol{\varphi}_1 \qquad (\omega = 0). \tag{14-12}$$

Thus, at zero frequency ($\omega = 0$, $\lambda \to \infty$), the two results agree.

For a nonzero frequency, $\boldsymbol{H}$ propagates at the velocity c when $r \gg \lambda$. Closer in, the phase velocity is larger than c, as it is for V.

It will be observed that, at zero frequency, the last parenthesis of Eq. 14-11 reduces to the first term, λ/r, and $\boldsymbol{H}$ varies as $1/r^2$.

The j term in the last parenthesis of Eq. 14-11 decreases only as $1/r$ and becomes predominant at large distances from the dipole where $r \gg \lambda$. This is the radiation term.

It is interesting to note how the amplitude of $\boldsymbol{H}$ changes from a $1/r^2$ dependence for a constant current to a $1/r$ dependence for a varying current. We have already observed a similar behavior for V.

Since both parts of $\boldsymbol{H}$ involve $\sin\theta$, they are both maximum in the equatorial plane and zero along the axis of the dipole. The magnitude of $\boldsymbol{H}$ as a function of θ and φ is plotted in Figure 14-8.

Since $I_0 s = j\omega p_0$,

$$\boldsymbol{H} = \frac{cp_0}{4\pi r \lambda\!\!\!^{-2}}(\exp j\omega[t])\left(j\frac{\lambda\!\!\!^{-}}{r} - 1\right)\sin\theta\, \boldsymbol{\varphi}_1. \tag{14-13}$$

For $r \gg \lambda\!\!\!^{-}$,

$$\boldsymbol{H} = -\frac{cp_0}{4\pi r \lambda\!\!\!^{-2}}\exp j\omega[t]\sin\theta\,\boldsymbol{\varphi}_1 \qquad (r \gg \lambda\!\!\!^{-}). \tag{14-14}$$

14.1.3 The Electric Field Intensity $\boldsymbol{E}$

To obtain $\boldsymbol{E}$, we require the time derivative of $\boldsymbol{A}$,

$$-\frac{\partial \boldsymbol{A}}{\partial t} = \frac{p_0}{4\pi\epsilon_0 r \lambda\!\!\!^{-2}}(\exp j\omega[t])(\cos\theta\,\boldsymbol{r}_1 - \sin\theta\,\boldsymbol{\theta}_1), \tag{14-15}$$

and the gradient of the scalar potential of Eq. 14-6,

$$-\boldsymbol{\nabla} V = -\left(\frac{\partial V}{\partial r}\boldsymbol{r}_1 + \frac{1}{r}\frac{\partial V}{\partial \theta}\boldsymbol{\theta}_1 + \frac{1}{r\sin\theta}\frac{\partial V}{\partial\varphi}\boldsymbol{\varphi}_1\right), \tag{14-16}$$

$$= \frac{-p_0}{4\pi\epsilon_0 r\lambda\!\!\!^{-2}}\exp j\omega[t]$$

$$\times\left\{\left(-2\frac{\lambda\!\!\!^{-2}}{r^2} - 2j\frac{\lambda\!\!\!^{-}}{r} + 1\right)\cos\theta\,\boldsymbol{r}_1 - \left(\frac{\lambda\!\!\!^{-2}}{r^2} + j\frac{\lambda\!\!\!^{-}}{r}\right)\sin\theta\,\boldsymbol{\theta}_1\right\}. \tag{14-17}$$

Then

$$\boldsymbol{E} = -\frac{\partial \boldsymbol{A}}{\partial t} - \boldsymbol{\nabla} V, \tag{14-18}$$

$$= -\frac{p_0}{4\pi\epsilon_0 r\lambda\!\!\!^{-2}}\exp j\omega[t]$$

$$\times\left\{\left\{(-1) + \left(-2\frac{\lambda\!\!\!^{-2}}{r^2} - 2j\frac{\lambda\!\!\!^{-}}{r} + 1\right)\right\}\cos\theta\,\boldsymbol{r}_1\right.$$

$$\left. + \left\{(+1) + \left(-\frac{\lambda\!\!\!^{-2}}{r^2} - j\frac{\lambda\!\!\!^{-}}{r}\right)\right\}\sin\theta\,\boldsymbol{\theta}_1\right\}. \tag{14-19}$$

At distances where $r \gg \lambda\!\!\!^{-}$, the terms in $\lambda\!\!\!^{-}/r$ and $\lambda\!\!\!^{-2}/r^2$ become negligible. In the coefficient of $\cos\theta$ the first and last terms are the largest, but they cancel.

Finally,

$$\boldsymbol{E} = \frac{p_0}{4\pi\epsilon_0 r\lambda\!\!\!^{-2}}(\exp j\omega[t])\left\{2\left(\frac{\lambda\!\!\!^{-2}}{r^2} + j\frac{\lambda\!\!\!^{-}}{r}\right)\cos\theta\,\boldsymbol{r}_1 + \left(\frac{\lambda\!\!\!^{-2}}{r^2} + j\frac{\lambda\!\!\!^{-}}{r} - 1\right)\sin\theta\,\boldsymbol{\theta}_1\right\}. \tag{14-20}$$

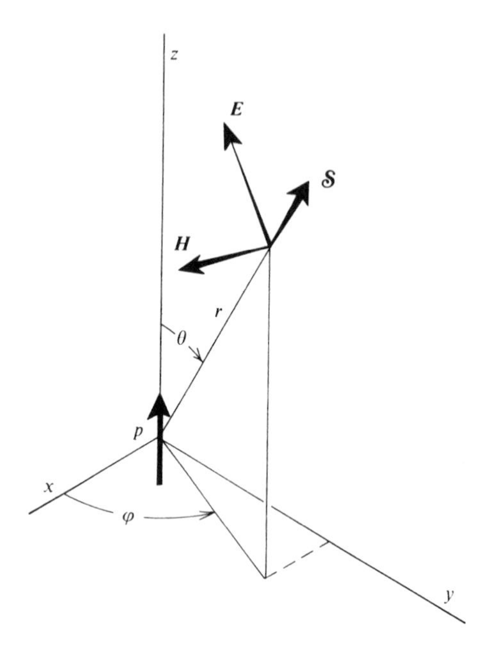

Figure 14-7. The $\boldsymbol{E}$, $\boldsymbol{H}$, $\mathbf{S}$ vectors for an oscillating dipole are oriented as shown at $r \gg \lambda\!\!\!^{-}$ when the phase angle $\omega[t]$ is zero.

Contrary to the $\boldsymbol{H}$ vector, which is entirely azimuthal, and thus transverse or perpendicular to the direction of propagation, $\boldsymbol{E}$ has a longitudinal component in the radial direction, at least close to the dipole where $r \not\gg \lambda\!\!\!^{-}$.

We can compare this $\boldsymbol{E}$ with that of a static electric dipole, as given in Section 2.9. For $\lambda\!\!\!^{-} \to \infty$, Eq. 14-20 becomes

$$\boldsymbol{E} = \frac{p_0}{4\pi\epsilon_0 r^3}(2\cos\theta\, \boldsymbol{r}_1 + \sin\theta\, \boldsymbol{\theta}_1) \qquad (\omega = 0), \qquad (14\text{-}21)$$

as required. These are the so-called *static terms.*

The electric field intensity $\boldsymbol{E}$ propagates through space with a velocity c for $r \gg \lambda\!\!\!^{-}$, as do V, $\boldsymbol{A}$, $\boldsymbol{H}$.

Close to the dipole, where r is not much larger than $\lambda\!\!\!^{-}$, $\boldsymbol{E}$ involves *five* terms. Two of these vary as $1/r^3$, two others lead them by $\pi/2$ radians and vary as $1/r^2$, and a fifth term varies as $1/r$ and leads the first pair by π radians.

Far from the source where $r \gg \lambda\!\!\!^{-}$, we are left with the radiation term

$$\boldsymbol{E} = -\frac{p_0}{4\pi\epsilon_0 r \lambda\!\!\!^{-2}} \exp j\omega[t] \sin\theta\, \boldsymbol{\theta}_1 \qquad (r \gg \lambda\!\!\!^{-}). \qquad (14\text{-}22)$$

The magnitude of $\boldsymbol{E}$ at $r \gg \lambda\!\!\!^{-}$ is plotted as a function of θ and φ in Figure 14-8.

In this case $\boldsymbol{E}$ changes from a $1/r^3$ dependence for a static field to a $1/r$ dependence for a varying field. We shall see later on that a $1/r$ dependence is required of both $\boldsymbol{E}$ and $\boldsymbol{H}$ for an oscillating dipole to ensure conservation of energy.

If we consider both $\boldsymbol{E}$ and $\boldsymbol{H}$ far from the source where $r \gg \lambda\!\!\!^{-}$, we notice that $\boldsymbol{E}$ lies in a plane passing through the polar axis, whereas $\boldsymbol{H}$ is azimuthal. The Poynting vector $\boldsymbol{E} \times \boldsymbol{H}$ is radial, as in Figure 14-7. We shall calculate the Poynting vector in the next section.

The ratio

$$\frac{E}{H} = \frac{1}{\epsilon_0 c} = \left(\frac{\mu_0}{\epsilon_0}\right)^{1/2} = 377 \text{ ohms} \qquad (r \gg \lambda\!\!\!^{-}), \tag{14-23}$$

just as for a plane wave in free space (Section 11.1). The electric and magnet energy densities $\epsilon_0 E^2/2$ and $\mu_0 H^2/2$ are equal.

We have therefore discovered that the $\boldsymbol{E}$ and $\boldsymbol{H}$ vectors at points remote from an oscillating electric dipole are related to each other exactly as in a plane electromagnetic wave.

14.1.4 The Average Poynting Vector and the Radiated Power

It is interesting to calculate the power radiated by an oscillating electric dipole. This can be found by integrating the average Poynting vector $\mathbf{S}_{av}$ over a spherical surface centered on the dipole. Let us first calculate $\mathbf{S}_{av}$:

$$\mathbf{S}_{av} = \frac{1}{2} \operatorname{Re} (\boldsymbol{E} \times \boldsymbol{H}^*), \tag{14-24}$$

$$= \frac{1}{2} \operatorname{Re} \{(E_r\, \boldsymbol{r}_1 + E_\theta\, \boldsymbol{\theta}_1) \times H_\varphi^*\, \boldsymbol{\varphi}_1\}. \tag{14-25}$$

Recalling that

$$\boldsymbol{r}_1 \times \boldsymbol{\varphi}_1 = -\,\boldsymbol{\theta}_1, \qquad \boldsymbol{\theta}_1 \times \boldsymbol{\varphi}_1 = \boldsymbol{r}_1, \tag{14-26}$$

$$\mathbf{S}_{av} = \frac{1}{2} \operatorname{Re} (-\, E_r H_\varphi^*\, \boldsymbol{\theta}_1 + E_\theta H_\varphi^*\, \boldsymbol{r}_1), \tag{14-27}$$

the components of $\boldsymbol{E}$ and of $\boldsymbol{H}$ being given by Eqs. 14-20 and 14-13. Thus

$$\mathbf{S}_{av} = \frac{c p_0^2 \sin^2 \theta}{32\pi^2 \epsilon_0 r^2 \lambda\!\!\!^{-4}}\, \boldsymbol{r}_1. \tag{14-28}$$

If we wish to express the average Poynting vector in terms of the current I_0, we must substitute $p_0 p_0^*$ for p_0^2, and then

$$\mathbf{S}_{av} = \frac{\mu_0 c I_0^2 s^2}{32\pi^2 r^2 \lambda\!\!\!^{-2}} \sin^2 \theta\, \boldsymbol{r}_1. \tag{14-29}$$

One striking feature of the Poynting vector is that it involves only the

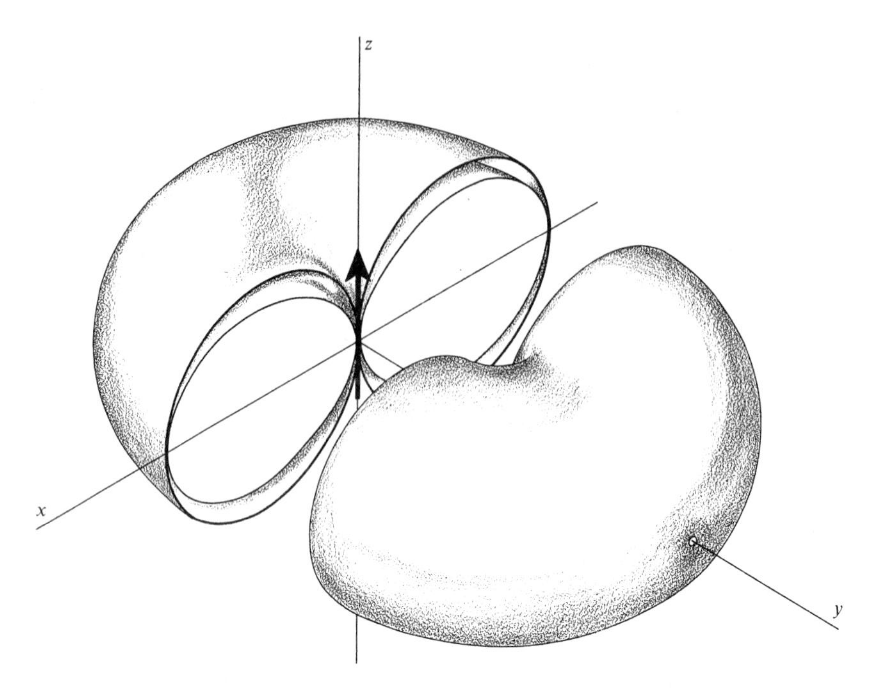

Figure 14-8. Polar diagrams of sin θ (outer surface) and of $\sin^2 \theta$ (inner surface) showing, respectively, the angular distributions of E or H, and of $\mathcal{S}_{av}$ at a distance $r \gg \lambda\!\!\!^-$ of an oscillating electric dipole situated at the origin. The radial distance from the dipole to one of the surfaces is proportional to the magnitude of these quantities in the corresponding direction. Most of the energy is radiated near the equatorial plane; *none* is radiated along the axis.

radiation terms, despite the fact that our calculation is valid even near the dipole, where r is not much larger than $\lambda\!\!\!^-$. In fact, it could have been calculated correctly by disregarding the terms that are unimportant far away from the dipole. The energy flow is everywhere purely radial, at least as long as the dipole length s is small compared to both r and $\lambda\!\!\!^-$.

The Poynting vector varies as $1/r^2$. This is required for conservation of energy since, under steady state conditions, the energy flow through any given solid angle must be the same for all r. This $1/r^2$ dependence results from the fact that the radiation terms for $\boldsymbol{E}$ and for $\boldsymbol{H}$ both vary as $1/r$.

Since the energy flow varies as $\sin^2 \theta$, it is zero along the axis of the dipole and maximum in the equatorial plane, as in Figure 14-8. *An electric dipole does not radiate energy along its axis.*

The total radiated power W is obtained by integrating $\mathcal{S}_{av}$. over the surface of a sphere of radius r:

$$W = \frac{cp_0^2}{32\pi^2\epsilon_0\lambda^4}\int_0^{2\pi}\int_0^{\pi}\frac{\sin^2\theta}{r^2}\,r^2\sin\theta\,d\theta\,d\varphi, \tag{14-30}$$

$$= \frac{c}{12\pi\epsilon_0}\frac{p_0^2}{\lambda^4}, \tag{14-31}$$

$$= 9.00 \times 10^{17}\frac{p_0^2}{\lambda^4}\text{ watts.} \tag{14-32}$$

The radiated energy varies as the *square* of the dipole moment $p_0 = Q_0 s$, and inversely as the *fourth* power of the wavelength, or directly as the *fourth* power of the frequency.

When the average Poynting vector is expressed in terms of I_0,

$$W = \frac{\mu_0 c}{12\pi}\left(\frac{s}{\lambda}\right)^2 I_0^2, \tag{14-33}$$

$$= 10.0\left(\frac{s}{\lambda}\right)^2 I_0^2\text{ watts,} \tag{14-34}$$

$$= 20.0\left(\frac{s}{\lambda}\right)^2 I_{\mathrm{rms}}^2\text{ watts.} \tag{14-35}$$

It will be observed that the energy radiated by the electric dipole is proportional to the square of the current flowing through it. The coefficient of I_{rms}^2 is called the *radiation resistance*:

$$R_{\mathrm{rad}} = 20.0\left(\frac{s}{\lambda}\right)^2\text{ ohms.} \tag{14-36}$$

This is the resistance that would dissipate in the form of heat the same power that the dipole radiates in the form of an electromagnetic wave, if it carried the same current. It will be recalled that we have assumed that $s \ll \lambda$.

14.1.5 The Electric and Magnetic Lines of Force

We have already found $\boldsymbol{E}$ in Eq. 14-20. To find the electric lines of force, we can proceed as in Section 2.9 and set

$$\frac{E_r}{dr} = \frac{E_\theta}{r\,d\theta}, \tag{14-37}$$

since an element of a line of force, having components dr and $r\,d\theta$, is parallel to the local $\boldsymbol{E}$, whose components are E_r and E_θ. The calculation is considerably simplified by using the vector

$$\boldsymbol{C} = \frac{p_0}{4\pi\epsilon_0 r\lambda}(\exp j\omega[t])\left(\frac{\lambda}{r} + j\right)\sin\theta\,\boldsymbol{\varphi}_1. \tag{14-38}$$

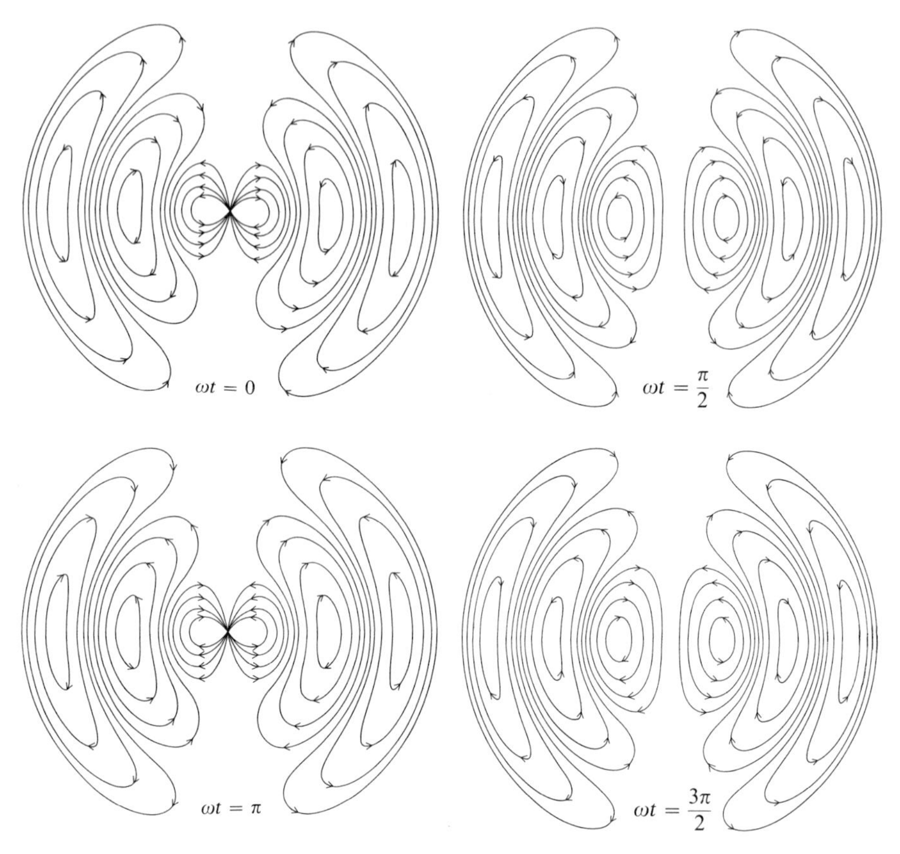

Figure 14-9. The electric lines of force of an oscillating dipole for $\omega t = 0, \pi/2, \pi$, and $3\pi/2$. The dipole is situated in the center and is oriented in the vertical direction. The decrease in wavelength with distance can be observed on these figures. The magnetic lines of force are circles perpendicular to the paper and centered on the axis of the dipole.

Then

$$\boldsymbol{E} = \boldsymbol{\nabla} \times \boldsymbol{C}, \tag{14-39}$$

$$E_r = \frac{1}{r \sin \theta} \frac{\partial}{\partial \theta} (C \sin \theta), \tag{14-40}$$

$$E_\theta = -\frac{1}{r} \frac{\partial (rC)}{\partial r}. \tag{14-41}$$

The differential equation for the lines of force is therefore

$$\frac{1}{\sin \theta} \frac{\partial}{\partial \theta} (C \sin \theta)\, d\theta = -\frac{1}{r} \frac{\partial}{\partial r} (rC)\, dr, \tag{14-42}$$

or

$$\frac{\partial}{\partial \theta}(Cr \sin \theta)\, d\theta + \frac{\partial}{\partial r}(Cr \sin \theta)\, dr = 0, \tag{14-43}$$

and the total differential of the quantity $Cr \sin \theta$ is zero. Then

$$Cr \sin \theta = \text{Constant}, \tag{14-44}$$

or, substituting the value of C and omitting the constant terms,

$$\frac{\sin^2 \theta}{\lambda}\left(\frac{\lambda}{r} + j\right) \exp j\omega\left(t - \frac{r}{c}\right) = \text{Constant}, \tag{14-45}$$

$$\sin^2 \theta \left(\frac{\lambda^2}{r^2} + 1\right)^{1/2} \cos\left(\omega t - \frac{r}{\lambda} + \arctan \frac{r}{\lambda}\right) = K\lambda, \tag{14-46}$$

where K is a real parameter that varies from one line of force to the next. The calculation is exact as long as $r \gg s$ and $\lambda \gg s$. All three factors determine the shape of the line of force as a function of r and θ, whereas the cosine term gives the radial motion. Figure 14-9 shows four families of lines of force.

For $r \gg \lambda$,

$$\sin^2 \theta \cos\left(\omega t - \frac{r}{\lambda} + \frac{\pi}{2}\right) = K\lambda, \tag{14-47}$$

and the lines of force travel outward at a velocity $\omega\lambda = c$. Closer in, however, the arc tan term varies with r, with the result that the velocity of the lines of force is larger than c.

The magnetic lines of force are much simpler: they are circles perpendicular to, and centered on, the axis of the dipole. This can be seen from Eq. 14-11, which shows that $\boldsymbol{H}$ is everywhere azimuthal.

14.1.6 The $K\lambda$ Surface*

It is instructive to represent the above equation in the form of a three-dimensional surface, as in Figure 14-10. This shows $K\lambda$ as a function of r and θ, as in Eq. 14-46, for $t = 0$. The loops drawn on the surface correspond to constant values of $K\lambda$ and are therefore lines of force. They are in fact the same lines of force as those shown in Figure 14-9, for $\omega t = 0$.

As time goes on, the angle $\omega t - (r/\lambda) + \arctan (r/\lambda)$ increases, and the result is that the ripples move out, somewhat like a damped wave, carrying the lines of force with them. Let us examine what happens to the lines of force.

Figure 14-11 shows the intersection of the $K\lambda$ surface, again for $t = 0$,

* You may wish to omit this rather detailed discussion of the electric lines of force of the oscillating electric dipole.

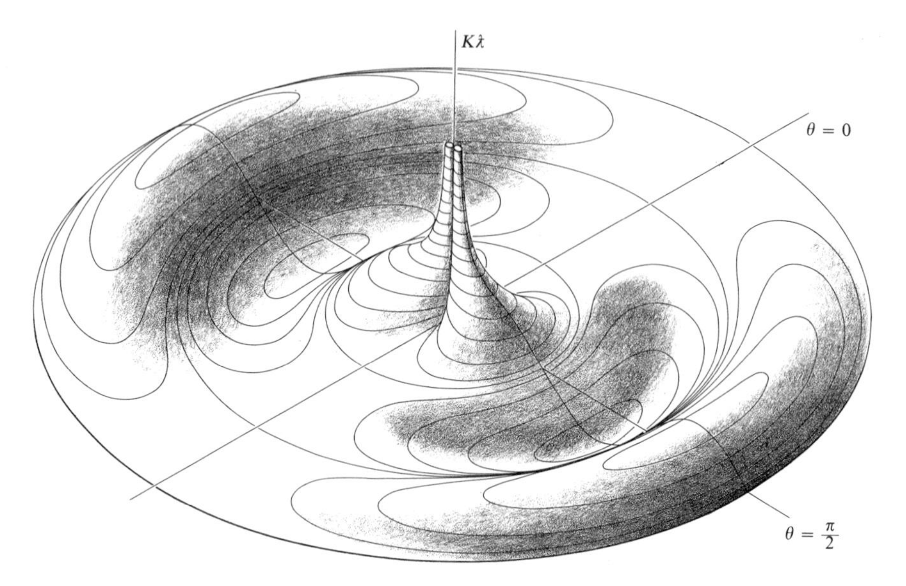

Figure 14-10. The parameter $K\lambda$ of Eq. 14-46 is plotted here as a function of the coordinates r and θ. The surface shown is that corresponding to $t = 0$; with increasing time t, the central peaks oscillate together from $-\infty$ to $+\infty$, and the ripples move out radially. The loops are electric lines of force corresponding to constant values of $K\lambda$.

with the plane $\theta = \pi/2$. This curve is situated inside its envelope, defined by the curves

$$K\lambda = \pm\left\{\left(\frac{\lambda}{r}\right)^2 + 1\right\}^{1/2} \qquad (\sin\theta = 1), \qquad (14\text{-}48)$$

which are shown by broken lines. These curves approach infinity as $r/\lambda \to 0$ and approach unity when $(r/\lambda)^2 \gg 1$. All values of $K\lambda$ are possible, from $-\infty$ to $+\infty$. The figure also shows a succession of curves for successive values of t.

As the ripples move out, their height decreases rapidly at first, and soon approaches unity. It is clear that lines of force with $K\lambda < 1$ can travel out to infinity. They give the *radiation field.* It is also clear that if $K\lambda > 1$, they cannot go far. Let us consider the case where $K\lambda$ is slightly larger than unity. The loop formed by the corresponding line of force shrinks until it reaches the top of the ripple and then disappears. This explains the relatively rapid decrease in the field intensity in the region where r is of the order of λ or less.

Closer to the dipole, which is situated at the origin, some lines of force, such as that for which $K\lambda = 5$, do not even get into a ripple. These lines of force simply pulsate in and out without ever escaping into space. This is the *static field.*

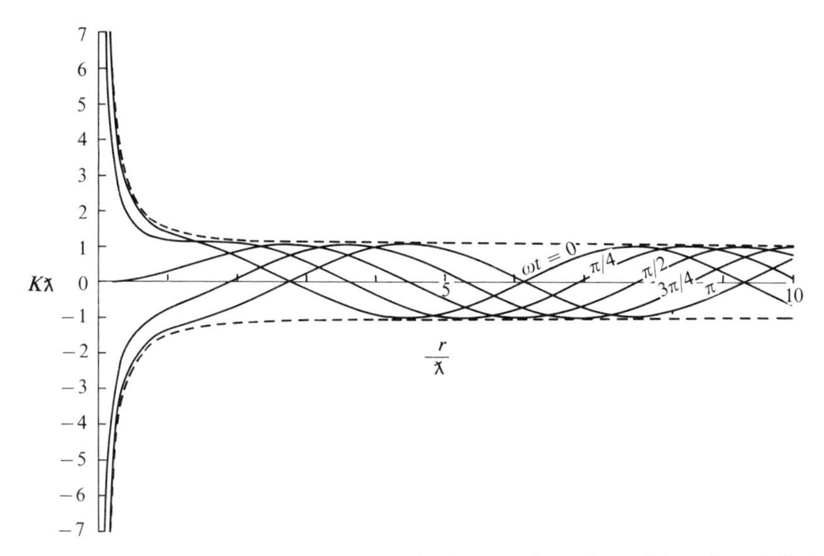

Figure 14-11. This figure illustrates how the intersection of the $K\lambda$ surface (Figure 14-10) with $\theta = \pi/2$ changes with time. The curves are all situated within the envelope shown; thus the amplitude is infinite at $r = 0$ and approaches unity for $r \gg \lambda$.

Let us analyze the motion of the lines of force quantitatively. We can set $\sin\theta = 1$ and consider their motion only in the $\theta = \pi/2$ plane. Then

$$\left(\frac{\lambda^2}{r^2} + 1\right)^{1/2} \cos\left(\omega t - \frac{r}{\lambda} + \arc\tan\frac{r}{\lambda}\right) = K\lambda \qquad (\sin\theta = 1). \quad (14\text{-}49)$$

This represents a family of curves such as those of Figure 14-11. To obtain the velocity of a line of force, we keep $K\lambda$ constant and calculate $u = \partial r/\partial t$. We find that

$$\frac{u}{c} = \frac{1}{c}\frac{\partial r}{\partial t} = \frac{\left(\frac{\lambda^2}{r^2} + 1\right)\sin X}{\sin X - \frac{\lambda^3}{r^3}\cos X}, \qquad (14\text{-}50)$$

where we have set

$$X = \omega t - \frac{r}{\lambda} + \arc\tan\frac{r}{\lambda}. \qquad (14\text{-}51)$$

If we eliminate the cosine term,

$$\frac{u}{c} = \frac{\left(\frac{\lambda^2}{r^2} + 1\right)\sin X}{\sin X - \frac{K\lambda}{\frac{r^2}{\lambda^2}\left(1 + \frac{r^2}{\lambda^2}\right)^{1/2}}}. \qquad (14\text{-}52)$$

The velocity u of the line of force becomes zero for $\sin X = 0$, or for $\cos X = 1$. This corresponds to a point where the ripple touches the envelope. The curve then "rolls" on the envelope and the line of force momentarily has zero velocity.

The velocity u also approaches plus or minus infinity near

$$\tan X = \frac{\lambda^3}{r^3} \tag{14-53}$$

or

$$\sin X = \frac{K\lambda}{\frac{r^2}{\lambda^2}\left(1 + \frac{r^2}{\lambda^2}\right)^{1/2}}. \tag{14-54}$$

Now this is precisely the condition that defines the top of a crest, or the bottom of a trough. The infinite velocity simply results from the fact that, as the ripple decreases in amplitude with increasing r, the line of force closes in with infinite velocity just before disappearing. This is interesting in that it tells us where the lines of force disappear.

Note that the top of a crest does *not* occur at $\cos X = 1$, or $\sin X = 0$, where the curve touches its envelope, but somewhat closer to the dipole at a value of X defined by the Eq. 14-112, since the cosine term in Eq. 14-49 is multiplied by another term that decreases with increasing r, namely

$$\{(\lambda/r)^2 + 1\}^{1/2}.$$

For large values of r/λ, $K\lambda$ is about unity, and the top of a crest, or the bottom of a trough, occurs at $X \approx 0$. For small values of r/λ, and for large values of $K\lambda$, these occur at $|\sin X| > 0$.

For $r \gg \lambda$, $K\lambda$ is about unity or less, and $u \rightarrow c$. The lines of force then travel outward at the velocity of light, as expected.

For $K\lambda \gg 1$ and $r \not\gg \lambda$,

$$\frac{u}{c} = -\frac{\left(\frac{r^2}{\lambda^2} + 1\right)^{3/2}}{K\lambda} \sin X, \tag{14-55}$$

and these lines pulsate in and out without ever escaping.

It is important to note that the $K\lambda$ surface and Figure 14-9 do *not* give the magnitude of $\mathbf{E}$, but only its direction. For example, in drawing a figure such as 14-9, one naturally selects equal intervals of $K\lambda$; this leads to a constant density of lines of force for $r \gg \lambda$, which *appears* to indicate that the amplitude of $\mathbf{E}$ does not decrease with r. In fact, $\mathbf{E}$ decreases as $1/r$ for $r \gg \lambda$, from Eq. 14-22.

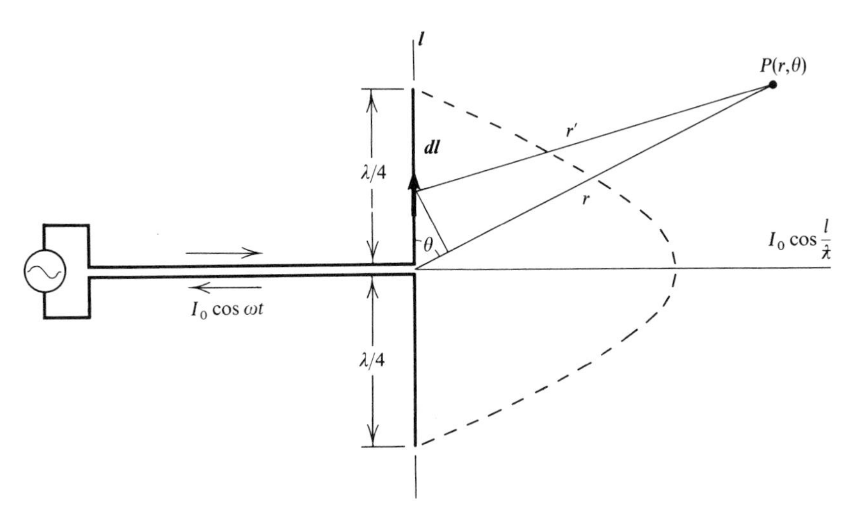

Figure 14-12. Half-wave antenna. The current distribution shown as a broken line is $I_o \cos (l/\lambda) \cos \omega t$. This is the standing wave pattern at some particular time when $\cos \omega t = 1$.

There is, of course, no such thing as a distinct line of force; the $K\lambda$ surface provides only the direction of $\boldsymbol{E}$ as a function of the position r, θ, and of the time t.

14.2 RADIATION FROM A HALF-WAVE ANTENNA

The half-wave antenna illustrated in Figure 14-12 is commonly used for radiating electromagnetic waves into space. It is simply a straight conductor whose length is half a free-space wavelength. When a current $I_0 \cos \omega t$ is established at the center by means of a suitable electronic circuit, a standing wave is formed along the conductor and the current I at l is

$$I = I_0 \cos \frac{l}{\lambda} \cos \omega t. \tag{14-56}$$

Each element $I\,dl$ of the antenna then radiates an electromagnetic wave similar to that of an electric dipole, and the field at any given point in space is the sum of all these fields.

In many cases the half-wave antenna is a one-quarter wave length mast set vertically on the ground, which then acts as a mirror (Problem 14-9). The mast and its image in the ground together form a half-wave antenna. Radio broadcast antennas are often of this type. To achieve good conductivity, the ground in the neighborhood of the antenna can be covered with a conducting screen.

This description of the half-wave antenna is really contradictory, because the standing wave along the wire can be truly sinusoidal only if there is zero energy loss, and hence no radiated wave. It turns out, however, that a rigorous calculation leads to nearly the same result as the approximate one. The current distribution is not quite sinusoidal, but the distortion has little effect on the field. It will therefore be sufficient for our purposes to assume a pure sinusoidal current distribution.

The standing wave can be expressed in exponential form as follows:

$$I = \mathrm{Re}\,\frac{I_0}{2}\left\{\exp j\left(\omega t - \frac{l}{\lambda}\right) + \exp j\left(\omega t + \frac{l}{\lambda}\right)\right\}, \tag{14-57}$$

where "Re" means, as usual, "Real part of." The right-hand side shows that the standing wave is the sum of two traveling waves, one in the positive direction, and one in the negative direction, with amplitudes $I_0/2$.

Then, from Eqs. 14-2 and 14-3, using the usual complex notation, we can express the electric dipole moment p of the element dl as

$$j\omega p = j\omega p_0 e^{j\omega t} = \frac{I_0}{2}\left\{\exp j\left(\omega t - \frac{l}{\lambda}\right) + \exp j\left(\omega t + \frac{l}{\lambda}\right)\right\} dl. \tag{14-58}$$

14.2.1 The Electric Field Intensity $\boldsymbol{E}$

In calculating the electric field intensity at the point (r, θ, φ), we shall assume that the distance r to the point of observation is much greater than λ. Then the electric field intensity $d\boldsymbol{E}$ from the element dl is given by Eq. 14-22, in which we must substitute the value of p_0 obtained from the above equation. Thus

$$d\boldsymbol{E} = -\frac{I_0}{j8\pi c\epsilon_0\lambda r'}\left\{\exp j\left(\omega t - \frac{l}{\lambda} - \frac{r'}{\lambda}\right) + \exp j\left(\omega t + \frac{l}{\lambda} - \frac{r'}{\lambda}\right)\right\} \sin\theta \, dl \, \boldsymbol{\theta}_1, \tag{14-59}$$

where r' is the distance between the element dl and the point (r, θ, φ), as in Figure 14-12,

$$r' \approx r - l\cos\theta, \tag{14-60}$$

and

$$\boldsymbol{E} = -\frac{I_0 \exp j\omega[t]}{j8\pi c\epsilon_0\lambda r} \times \sin\theta \int_{-\lambda/4}^{+\lambda/4} \left\{\exp\left(j\frac{l}{\lambda}(\cos\theta - 1)\right) + \exp\left(j\frac{l}{\lambda}(\cos\theta + 1)\right)\right\} dl \, \boldsymbol{\theta}_1. \tag{14-61}$$

We have removed the $1/r'$ from under the integral sign and set it equal to $1/r$, since it is assumed that $r \gg \lambda$. With this condition, the $d\mathbf{E}$'s can all be taken to be parallel for a given point of observation and can be assumed to have the same amplitude but different phases, and the integration can be limited to the phases. Integrating,

$$\mathbf{E} = -\frac{I_0 \exp j\omega[t]}{j8\pi c\epsilon_0 \lambda r}$$

$$\times \sin\theta \left(\frac{\lambda \exp\left\{ j\frac{l}{\lambda}(\cos\theta - 1)\right\}}{j(\cos\theta - 1)} + \frac{\lambda \exp\left\{ j\frac{l}{\lambda}(\cos\theta + 1)\right\}}{j(\cos\theta + 1)} \right)_{-\lambda/4}^{+\lambda/4} \boldsymbol{\theta}_1. \quad (14\text{-}62)$$

It is not permissible to expand the exponential functions between the main parentheses in series form, since l is not small with respect to λ. We have

$$\mathbf{E} = \frac{jI_0 \exp j\omega[t]}{4\pi c\epsilon_0 r} \sin\theta \left(\frac{\sin\left\{\frac{\pi}{2}(\cos\theta - 1)\right\}}{\cos\theta - 1} + \frac{\sin\left\{\frac{\pi}{2}(\cos\theta + 1)\right\}}{\cos\theta + 1} \right) \boldsymbol{\theta}_1. \quad (14\text{-}63)$$

The expression between the braces can be simplified by setting

$$\sin\left\{\frac{\pi}{2}(\cos\theta - 1)\right\} = -\cos\left(\frac{\pi}{2}\cos\theta\right) \quad (14\text{-}64)$$

$$\sin\left\{\frac{\pi}{2}(\cos\theta + 1)\right\} = +\cos\left(\frac{\pi}{2}\cos\theta\right) \quad (14\text{-}65)$$

and adding the two terms. Then

$$\mathbf{E} = \frac{j}{2\pi r c\epsilon_0} I_0 \exp j\omega[t] \frac{\cos\left(\frac{\pi}{2}\cos\theta\right)}{\sin\theta} \boldsymbol{\theta}_1, \quad (14\text{-}66)$$

$$= 60.0j \frac{I_0 \exp j\omega[t]}{r} \frac{\cos\left(\frac{\pi}{2}\cos\theta\right)}{\sin\theta} \boldsymbol{\theta}_1 \text{ volts/meter.} \qquad (r \gg \lambda). \quad (14\text{-}67)$$

At $\theta = 0$ and $\theta = \pi$, the above equation is indeterminate because the trigonometric term becomes 0/0. According to l'Hospital's rule, the limiting value of such a ratio is equal to the ratio of the derivatives of the two functions at the limit and

$$\lim_{\theta \to 0,\pi} \frac{\cos\left(\frac{\pi}{2}\cos\theta\right)}{\sin\theta} = \left\{ \frac{\frac{d}{d\theta}\cos\left(\frac{\pi}{2}\cos\theta\right)}{\frac{d}{d\theta}\sin\theta} \right\}_{\theta=0,\pi} \quad (14\text{-}68)$$

$$= \left\{ \frac{\sin\left(\frac{\pi}{2}\cos\theta\right)\frac{\pi}{2}\sin\theta}{\cos\theta} \right\}_{\theta=0,\pi} = 0. \quad (14\text{-}69)$$

Thus $\mathbf{E}$ is zero along the axis of the antenna.

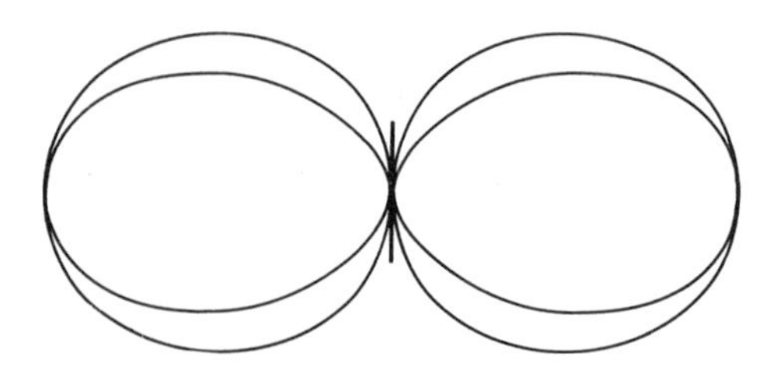

Figure 14-13. Polar diagrams the functions of $\cos\{(\pi/2)\cos\theta\}/\sin\theta$ and $\cos^2\{(\pi/2)\cos\theta\}/\sin^2\theta$ showing, respectively, the angular distributions of E or H and of $\mathcal{S}_{av}$ for the half-wave antenna at $r \gg \lambda$. The angular distributions are similar to those for the dipole, except that the half-wave antenna radiates a larger fraction of its power in the region of the equatorial plane.

It is interesting to note that the electric field intensity for a half-wave antenna is independent of the frequency for a given current I_0: the $\boldsymbol{E}$ for an elementary dipole is proportional to $1/\lambda$ for a given current amplitude, and the integration over the length of the antenna has introduced a factor of λ.

Figure 14-13 shows the radiation pattern for a half-wave antenna. It is quite similar to that for the dipole, except that a somewhat larger part of the field is radiated in the region of the equator. The reason for this similarity is that the phase difference between $d\boldsymbol{E}$'s from the elements of current along the antenna is small, except when the point of observation is near the polar axis, where the electric field intensity is zero in any case.

14.2.2 The Magnetic Field Intensity *H*

It is now a simple matter to find the $\boldsymbol{H}$ vector. For the electric dipole we found that $\boldsymbol{H}$ is azimuthal, as in Figure 14-7, and that

$$\frac{E}{H} = \left(\frac{\mu_0}{\epsilon_0}\right)^{1/2} = 377 \text{ ohms}, \tag{14-70}$$

as in Eq. 14-23. For the half-wave antenna, $\boldsymbol{H}$ is again azimuthal, and the above equation also applies; thus

$$\boldsymbol{H} = \frac{j}{2\pi r} I_0 \exp j\omega[t] \frac{\cos\left(\frac{\pi}{2}\cos\theta\right)}{\sin\theta} \boldsymbol{\varphi}_1. \tag{14-71}$$

14.2.3 The Average Poynting Vector and the Radiated Power

The average Poynting vector $\mathbf{S}_{av}$ gives the average flux of radiated energy:

$$\mathcal{S}_{av} = \frac{1}{2}\,\mathrm{Re}\,(\boldsymbol{E} \times \boldsymbol{H}^*), \tag{14-72}$$

$$= \frac{1}{8\pi^2 c\epsilon_0}\frac{I_0^2}{r^2}\frac{\cos^2\left(\dfrac{\pi}{2}\cos\theta\right)}{\sin^2\theta}\,\boldsymbol{r}_1, \tag{14-73}$$

$$= 9.55\,\frac{I_{rms}^2}{r^2}\frac{\cos^2\left(\dfrac{\pi}{2}\cos\theta\right)}{\sin^2\theta}\,\boldsymbol{r}_1 \text{ watts/meter}^2. \tag{14-74}$$

It points radially outward and varies as $1/r^2$, which ensures conservation of energy. The average Poynting vector is shown as a function of θ in Figure 14-13.

The radiated power is again obtained by integrating $\mathcal{S}_{av}$, over a sphere of radius r:

$$W = 9.55\,I_{rms}^2 \int_0^{2\pi}\int_0^{\pi} \frac{\cos^2\left(\dfrac{\pi}{2}\cos\theta\right)}{r^2\sin^2\theta}\,r^2\sin\theta\,d\theta\,d\varphi, \tag{14-75}$$

$$= 60.0\,I_{rms}^2 \int_0^{\pi} \frac{\cos^2\left(\dfrac{\pi}{2}\cos\theta\right)}{\sin\theta}\,d\theta. \tag{14-76}$$

To perform this integration, we set

$$\frac{\pi}{2}\cos\theta = \frac{\alpha}{2} - \frac{\pi}{2} \tag{14-77}$$

and

$$W = 60.0\pi I_{rms}^2 \int_0^{2\pi} \frac{1 - \cos\alpha}{\alpha(4\pi - 2\alpha)}\,d\alpha. \tag{14-78}$$

Then if we write

$$\frac{1}{\alpha(4\pi - 2\alpha)} = \frac{1}{4\pi}\left(\frac{1}{\alpha} + \frac{1}{2\pi - \alpha}\right), \tag{14-79}$$

the radiated power becomes

$$W = 15.0\,I_{rms}^2\left(\int_0^{2\pi}\frac{1 - \cos\alpha}{\alpha}\,d\alpha + \int_0^{2\pi}\frac{1 - \cos\alpha}{2\pi - \alpha}\,d\alpha\right). \tag{14-80}$$

Now the two integrals between the braces are equal, as can be seen from the curves for $1 - \cos\alpha$, for α, and for $2\pi - \alpha$, shown in Figure 14-14. Since the curve for $1 - \cos\alpha$ is symmetrical about $\alpha = \pi$, the area under the curve $(1 - \cos\alpha)/\alpha$ must be equal to that under $(1 - \cos\alpha)/(2\pi - \alpha)$ between the limits 0 and 2π. Thus

$$W = 30.0\,I_{rms}^2 \int_0^{2\pi}\frac{1 - \cos\alpha}{\alpha}\,d\alpha. \tag{14-81}$$

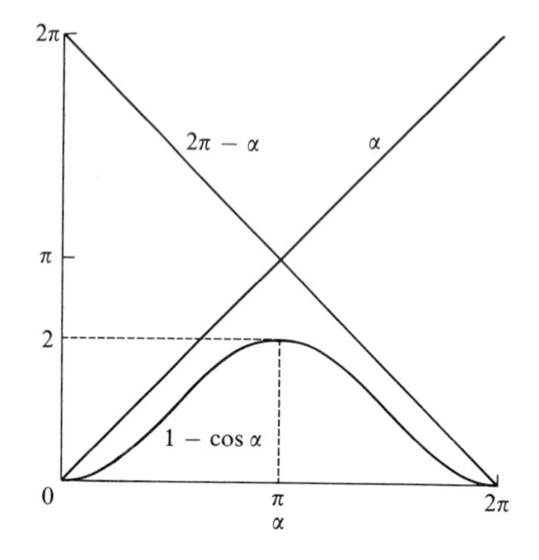

Figure 14-14. The three functions $1 - \cos\alpha$, α, and $2\pi - \alpha$ as functions of α for $\alpha = 0$ to 2π. The curves for $(1 - \cos\alpha)/\alpha$ and for $(1 - \cos\alpha)/(2\pi - \alpha)$ are symmetrical about $\alpha = \pi$; thus the areas under them must be equal for the interval shown.

This integration cannot be performed analytically, but tables are available* and the integral equals 2.4377. Thus

$$W = 73.1\, I_{\mathrm{rms}}^2 \text{ watts,} \tag{14-82}$$

and

$$R_{\mathrm{rad}} = 73.1 \text{ ohms.} \tag{14-83}$$

The radiation resistance of a half-wave antenna is therefore 73.1 ohms. This assumes that the current distribution on the antenna is sinusoidal, which is not quite correct, as we saw at the beginning of Section 14.2.

14.3 ANTENNA ARRAYS

It is often desirable to radiate energy predominantly in some given direction. This is achieved with arrays of antennas that utilize appropriate interference effects. Two simple examples are shown in Figures 14-15 and 14-16.

It is clear that by phasing and by spacing antennas properly a great variety of radiation patterns can be achieved.

Let us calculate the radiation pattern of Figure 14-15. We choose coordinates as in Figure 14-17 and assume that $r \gg \lambda$. Then the electric field intensity of each antenna is given by Eq. 14-67 and, if the antennas are excited in phase, the total field is

---

* See, for example E. Jahnke and F. Emde, *Tables of Functions* (Dover, New York, 1951), pp. 3 and 6.

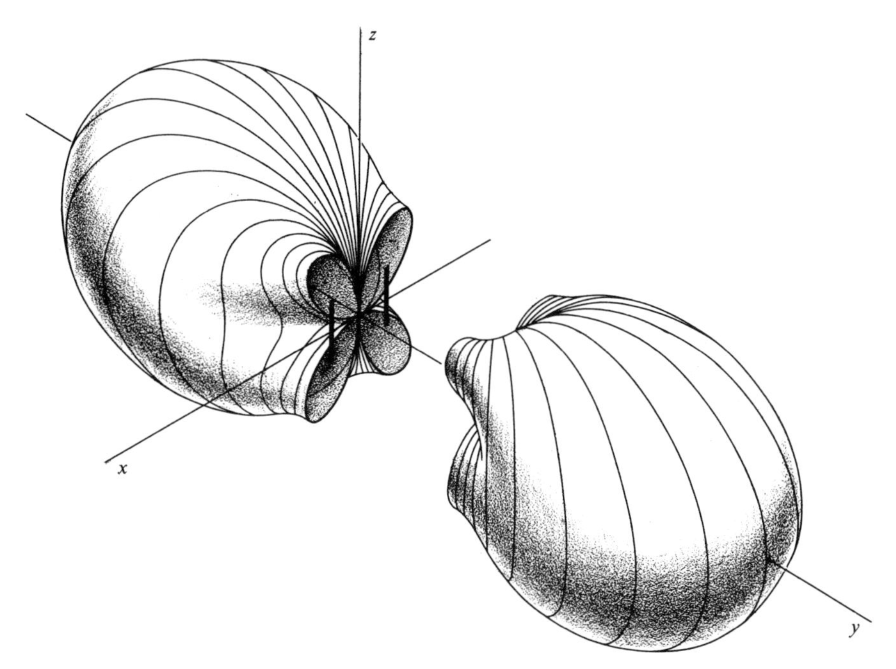

Figure 14-15. A simple antenna array, represented by the two vertical bars, and its radiation pattern for $r \gg \lambda$. The two half-wave antennas are spaced $\lambda/2$ apart and are excited in phase. The surface shows the magnitude of $\boldsymbol{E}$ plotted radially as a function of θ and φ. The curves shown are situated on the surface for constant values of φ chosen at 10° intervals. The surface is cut into two parts for clarity. Along the y-axis the two waves add, and the resulting electric field intensity is twice that produced by a single antenna. The same applies to all of the yz-plane, at least for $r \gg \lambda$. Along the x-axis, however, the two waves arrive in opposite phases and cancel. For other directions along the xz-plane the waves do not cancel completely, since the difference in path is smaller than $\lambda/2$. There is zero field along the z-axis for each of the antennas, and hence for the array.

$$\boldsymbol{E} = 60.0j\,\frac{I_0 \exp j\omega[t]}{r}\,\frac{\cos\left(\dfrac{\pi}{2}\cos\theta\right)}{\sin\theta} \times \left\{\exp\left(-j\frac{D\cos\psi}{2\lambda}\right) + \exp\left(j\frac{D\cos\psi}{2\lambda}\right)\right\}\boldsymbol{\theta}_1, \qquad (14\text{-}84)$$

where the quantity on the second line accounts for the fact that the two waves arrive out of phase, one having traveled a distance $r + (D/2)\cos\psi$, and the other a distance $r - (D/2)\cos\psi$. Thus

$$\boldsymbol{E} = 120.0j\,\frac{I_0 \exp j\omega[t]}{r}\,\frac{\cos\left(\dfrac{\pi}{2}\cos\theta\right)}{\sin\theta}\cos\left(\frac{D}{2\lambda}\cos\psi\right)\boldsymbol{\theta}_1. \qquad (14\text{-}85)$$

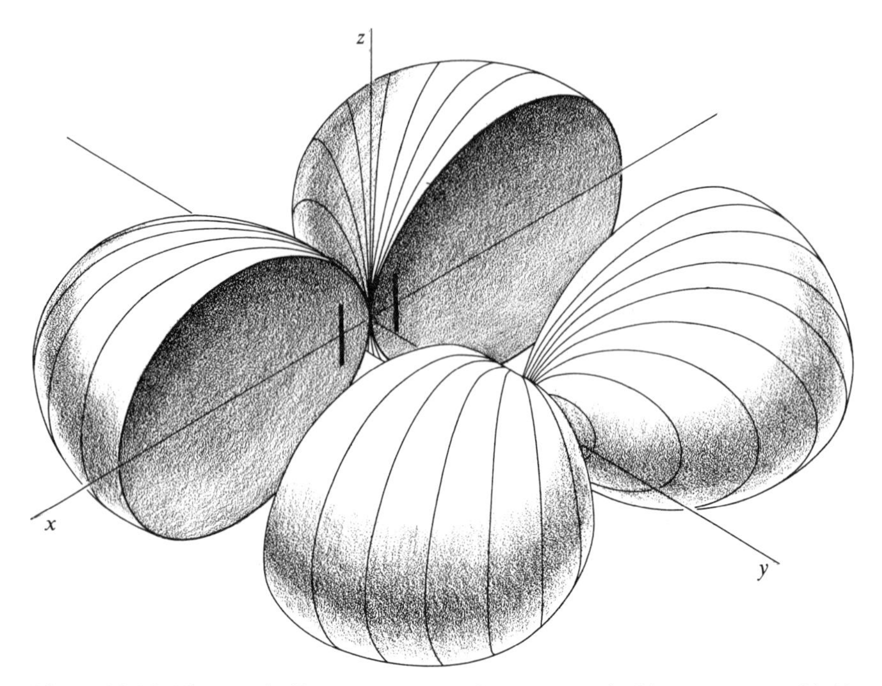

Figure 14-16. The two half-wave antennas shown as vertical bars are spaced $\lambda/2$ apart, but the one at $x = -D/2$ leads the other by π radians. See legend of Figure 14-15 for an explanation of the surface. The two waves now cancel everywhere on the yz-plane. All along the x-axis, the two waves arrive in phase to give twice the field of a single antenna. There is again no radiation in the direction of the z-axis.

We can replace the angle ψ by θ and φ, since

$$r \cos \psi = r \sin \theta \cos \varphi, \tag{14-86}$$

and then

$$\boldsymbol{E} = 120.0j \frac{I_0 \exp j\omega[t]}{r} \frac{\cos\left(\frac{\pi}{2} \cos \theta\right)}{\sin \theta} \cos\left(\frac{D}{2\lambda} \sin \theta \cos \varphi\right) \boldsymbol{\theta}_1 \text{ volts/meter.} \tag{14-87}$$

When the antennas are one-half wavelength apart, $D/(2\lambda) = \pi/2$. Then, in the xy-plane, where $\theta = \pi/2$, $\boldsymbol{E}$ varies as

$$\cos\left(\frac{\pi}{2} \cos \varphi\right).$$

At $\varphi = 0$ or π, this function is zero, while at $\varphi = \pi/2$ it is maximum. There is constructive interference along the y-axis and destructive interference along

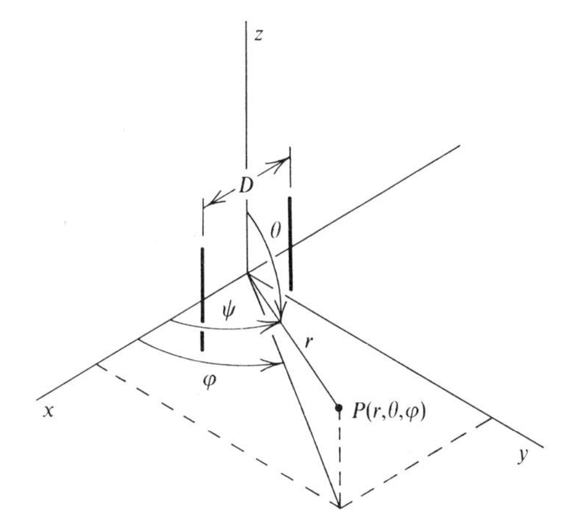

Figure 14-17. Pair of parallel half-wave antennas D meters apart. The point P is situated approximately at distances $r - (D/2)\cos\psi$ and $r + (D/2)\cos\psi$ from the centers of the antennas.

the x-axis, as must be expected. In the xz-plane, $\varphi = 0$ and $\boldsymbol{E}$ varies as

$$\frac{\cos\left(\frac{\pi}{2}\cos\theta\right)}{\sin\theta}\cos\left(\frac{\pi}{2}\sin\theta\right).$$

The first term is the angular distribution for a single half-wave antenna; it is zero at $\theta = 0$ and maximum at $\theta = \pi/2$. The second term comes from the interference between the two antennas; it is maximum at $\theta = 0$ and zero at $\theta = \pi/2$. The product of the two is zero both at $\theta = 0$ and $\theta = \pi/2$. Finally, in the yz-plane, $\varphi = \pi/2$, and $\boldsymbol{E}$ varies simply as

$$\frac{\cos\left(\frac{\pi}{2}\cos\theta\right)}{\sin\theta},$$

as does the $\boldsymbol{E}$ of a single half-wave antenna. This is to be expected, since the two waves arrive there in phase, and the total field is exactly twice that of a single antenna, at least for $r \gg \lambda$.

If the antenna centered at $x = -D/2$ has a phase lead of π, then the total electric field intensity is

$$\boldsymbol{E} = 60.0j\frac{I_0 \exp j\omega[t]}{r}\frac{\cos\left(\frac{\pi}{2}\cos\theta\right)}{\sin\theta} \times \left\{\exp\left(-j\frac{D\cos\psi}{2\lambda}\right) + \exp j\left(\frac{D\cos\psi}{2\lambda} + \pi\right)\right\}\boldsymbol{\theta}_1, \qquad (14\text{-}88)$$

which is the same as Eq. 14-84 except for the addition of $j\pi$ in the last exponent. Recalling that $e^{j\pi} = -1$,

$$E = 120.0 \frac{I_0 \exp j\omega[t]}{r} \frac{\cos\left(\frac{\pi}{2}\cos\theta\right)}{\sin\theta} \sin\left(\frac{D}{2\lambda}\sin\theta\cos\varphi\right) \boldsymbol{\theta}_1 \text{ volts/meter.} \tag{14-89}$$

The radiation pattern is shown in Figure 14-16.

14.4 ELECTRIC QUADRUPOLE RADIATION

We now go on to the more elaborate type of radiation produced by a linear electric quadrupole whose moment is a sinusoidal function of the time.

The linear electric quadrupole (Section 2.10) is composed of two dipoles of opposite polarity arranged in line to give three charges $+Q$, $-2Q$, $+Q$ as in Figure 14-18. The dipole moment of such a charge distribution is zero, but the quadrupole moment is

$$p_{zz} = \sum Qz^2 = 2Qs^2. \tag{14-90}$$

If

$$Q = Q_0 e^{j\omega t}, \tag{14-91}$$

then

$$p_{zz} = 2Q_0 s^2 e^{j\omega t}, \tag{14-92}$$

$$p_{zz0} = 2Q_0 s^2. \tag{14-93}$$

There is no dipole radiation, since the dipole moment $p = \sum Qz$ is always zero. Nevertheless, there must be some sort of radiation from the moving charges. This is what we shall investigate.

We shall use the above model to calculate the radiation field of a linear quadrupole, although *any charge distribution will produce exactly the same field if it oscillates in such a fashion that*

$$p_{zz} = \int \rho z^2 \, d\tau \tag{14-94}$$

is a sinusoidal function of time. This type of radiation would arise, for example, if currents were excited at the surface of a sphere in such a way as to drive the electrons alternately to both poles and then to the equator.

We could proceed exactly as for electric dipole radiation and calculate successively V, A, E, H. This will be done in Problem 14-11. It is easier, however, to add the fields of the two component dipoles as follows. For simplicity, we consider only the radiation field and set $r \gg \lambda$.

We now have two dipoles, one with moment $-p_0 e^{j\omega t}$ centered at $-s/2$, and another with moment $+p_0 e^{j\omega t}$ centered at $+s/2$, as in Figure 14-18. The electric field intensities of the two dipoles add vectorially, according to the

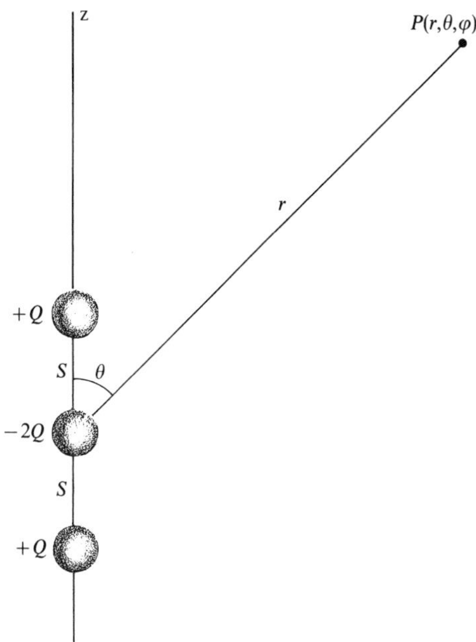

Figure 14-18. Linear electric quadrupole formed of two dipoles of opposite polarity, one above the other. The dipole centered at $-s/2$ has a moment of $-Qs$; the dipole centered at $+s/2$ has a moment of $+Qs$. The charges are assumed to pulsate with $Q = Q_0 e^{j\omega t}$.

principle of superposition (Section 2.2). From Eq. 14-22, the electric field intensity in the radiation field of an electric dipole situated at the origin is

$$\boldsymbol{E} = -\frac{p_0}{4\pi\epsilon_0 r \lambda^2} \exp j\omega[t] \sin\theta \, \boldsymbol{\theta}_1 \qquad (r \gg \lambda). \tag{14-95}$$

Since the two dipoles forming the quadrupole are centered some distance away from the origin, their electric field intensities at the point (r, θ, φ) will differ slightly in direction, in amplitude, and in phase. We may easily neglect the difference in direction and in amplitude for $r \gg s$, but not the difference in phase. Thus

$$\boldsymbol{E} = -\frac{p_0}{4\pi\epsilon_0 r \lambda^2} \exp j\omega[t] \times \sin\theta \left\{ \exp\left(j\frac{s}{2\lambda}\cos\theta \right) - \exp\left(-j\frac{s}{2\lambda}\cos\theta \right) \right\} \boldsymbol{\theta}_1. \tag{14-96}$$

The last two exponential functions can be expanded as power series in s/λ for $s \ll \lambda$, and their sum then reduces to $(js/\lambda)\cos\theta$. Thus

$$\boldsymbol{E} = -\frac{jp_{zz0}}{8\pi\epsilon_0 r \lambda^3} \exp j\omega[t] \sin\theta \cos\theta \, \boldsymbol{\theta}_1 \qquad (r \gg \lambda \gg s). \tag{14-97}$$

There can be no radiation along the axis $\theta = 0$ or π, where neither of the

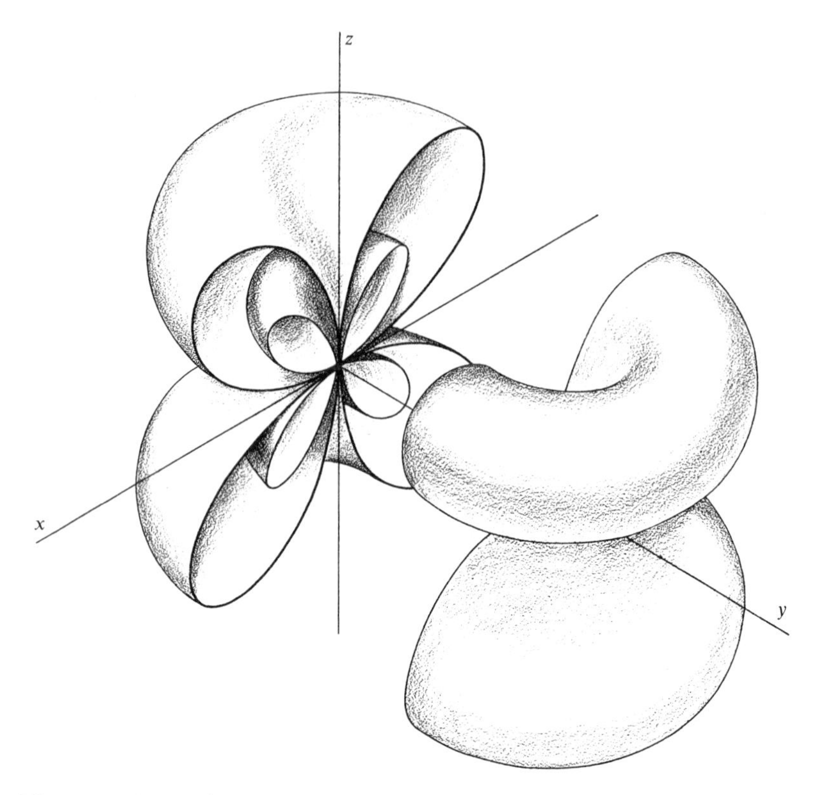

Figure 14-19. Radiation pattern for a vertical oscillating electric quadrupole at the origin. The amplitude of $\boldsymbol{E}$ or of $\boldsymbol{H}$ in any given direction is proportional to the distance between the origin and the outer surface in that direction. The inner surface is a similar plot of the magnitude of $\mathbf{S}_{av}$. There is no field along the axis or along the equator of the quadrupole. The maximum field intensity occurs along the surface of a cone at 45° to the axis.

dipoles forming the quadrupole radiate energy; there is also zero radiation along the equator $\theta = \pi/2$, where the two dipoles give equal and opposite fields.

The magnetic field intensity $\boldsymbol{H}$ is found in the same manner from Eq. 14-14 and

$$\boldsymbol{H} = \left(\frac{\epsilon_0}{\mu_0}\right)^{1/2} E\,\boldsymbol{\varphi}_1. \tag{14-98}$$

The $\boldsymbol{E}$ and $\boldsymbol{H}$ vectors for electric quadrupole radiation are therefore oriented in the same manner as those for electric dipole radiation. This was to be expected, since we have simply superposed the fields of two electric dipoles. The amplitudes of both $\boldsymbol{E}$ and $\boldsymbol{H}$ are inversely proportional to r and therefore decrease at the *same* rate as for dipole radiation. This makes the average Poynting vector (1/2) Re ($\boldsymbol{E} \times \boldsymbol{H}^*$) decrease again as $1/r^2$, which is necessary for conservation of energy.

There are two main differences between electric dipole and electric quadrupole radiation: the former field increases with the square of the frequency, whereas the latter increases as the cube of the frequency; the dipole field is zero along the polar axis, whereas the quadrupole field is zero both at the poles and at the equator.

Figure 14-19 shows the radiation pattern for an oscillating electric quadrupole at the origin.

14.5 MAGNETIC DIPOLE RADIATION

We studied the static magnetic dipole in Section 7.8, where a current loop of area $\boldsymbol{S}$ and current I was defined to have a magnetic moment

$$\boldsymbol{m} = I\boldsymbol{S}. \tag{14-99}$$

The vector $\boldsymbol{S}$ is perpendicular to the small area limited by the loop, and its direction is related to the current by the right-hand screw rule.

We shall consider a magnetic dipole as in Figure 14-20 carrying a current

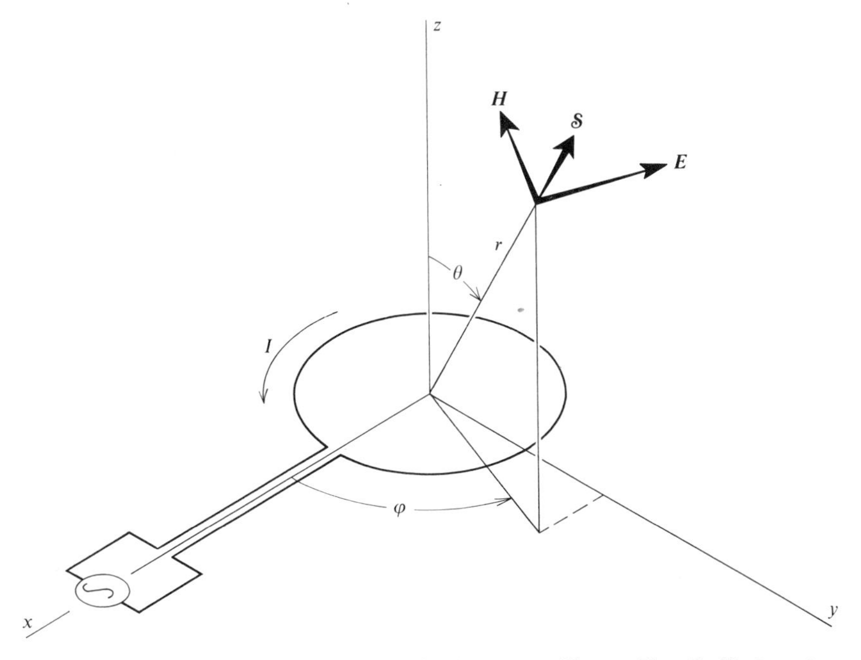

Figure 14-20. Magnetic dipole antenna fed by an oscillator. The $\boldsymbol{E}$, $\boldsymbol{H}$, $\boldsymbol{\mathcal{S}}$ vectors are oriented as shown when the phase angle $\omega\{t - (r/c)\}$ is zero.

$$I = I_0 e^{j\omega t}, \tag{14-100}$$

but *magnetic dipole radiation can also be produced by any current distribution whose magnetic dipole moment*

$$\boldsymbol{m} = \frac{1}{2}\int_\tau (\boldsymbol{r} \times \boldsymbol{J})\, d\tau \tag{14-101}$$

is a sinusoidal function of the time. This general definition of the magnetic dipole moment was stated in Eq. 7-116.

The conducting sphere can again be used to illustrate the oscillating magnetic dipole. In this case charge would flow parallel to the equator, alternately in one direction and then in the other.

We first calculate V and $\boldsymbol{A}$, and then $\boldsymbol{E}$, $\boldsymbol{H}$, $\boldsymbol{S}$, and W. We assume that the loop is small with respect to $\lambda\!\!\!^-$; if it were not, there would be wave effects along its circumference, and the current would not be the same all around it.

14.5.1 The Potentials V and A

If the impedance of the loop is small, all its points are at the same constant potential, which we can set equal to zero, since we are not interested in adding an electrostatic field to the radiation field. Then V is zero throughout space.

We have already found $\boldsymbol{A}$ in the example on page 430:

$$\boldsymbol{A} = \frac{\mu_0}{4\pi}\frac{\boldsymbol{m}_0 \times \boldsymbol{r}_1}{r\lambda\!\!\!^-}\left(\frac{\lambda\!\!\!^-}{r} + j\right)\exp j\omega[t]. \tag{14-102}$$

You will remember from Sections 6.7 and 10.3 that the Lorentz condition states that $\boldsymbol{\nabla}\cdot\boldsymbol{A}$ is $-\epsilon_0\mu_0$ times $\partial V/\partial t$. Since $V = 0$ for a magnetic dipole, the divergence of the above $\boldsymbol{A}$ must be zero. This will be shown in Problem 14-19.

14.5.2 The E and H Vectors

We can now find $\boldsymbol{E}$ and $\boldsymbol{H}$ for a magnetic dipole:

$$\boldsymbol{H} = \frac{1}{\mu_0}\boldsymbol{\nabla}\times\boldsymbol{A}, \tag{14-103}$$

$$= \frac{m_0}{4\pi r\lambda\!\!\!^{-2}}(\exp j\omega[t])\left\{2\left(\frac{\lambda\!\!\!^{-2}}{r^2} + j\frac{\lambda\!\!\!^-}{r}\right)\cos\theta\, \boldsymbol{r}_1 + \left(\frac{\lambda\!\!\!^{-2}}{r^2} + j\frac{\lambda\!\!\!^-}{r} - 1\right)\sin\theta\, \boldsymbol{\theta}_1\right\}, \tag{14-104}$$

$$= \frac{-m_0}{4\pi r\lambda\!\!\!^{-2}}\exp j\omega[t]\sin\theta\,\boldsymbol{\theta}_1 \qquad (r \gg \lambda\!\!\!^-); \tag{14-105}$$

$$\boldsymbol{E} = -\frac{\partial \boldsymbol{A}}{\partial t} - \boldsymbol{\nabla} V = -\frac{\partial \boldsymbol{A}}{\partial t}, \tag{14-106}$$

$$= -\left(\frac{\mu_0}{\epsilon_0}\right)^{1/2} \frac{m_0}{4\pi r \lambda\!\!\!^{-2}} (\exp j\omega[t]) \left(j\frac{\lambda\!\!\!^-}{r} - 1\right) \sin\theta\, \boldsymbol{\varphi}_1, \tag{14-107}$$

$$= \left(\frac{\mu_0}{\epsilon_0}\right)^{1/2} \frac{m_0}{4\pi r \lambda\!\!\!^{-2}} \exp j\omega[t] \sin\theta\, \boldsymbol{\varphi}_1 \qquad (r \gg \lambda\!\!\!^-). \tag{14-108}$$

The scalar potential V is zero, and the electric field intensity arises solely from the changing magnetic field. The vector $\boldsymbol{H}$ lies in a plane passing through the z-axis, whereas $\boldsymbol{E}$ is azimuthal, as in Figure 14-20.

At zero frequency ($\omega = 0$, $\lambda\!\!\!^- \rightarrow \infty$),

$$\boldsymbol{H} = \frac{m_0}{4\pi r^3}(2\cos\theta\, \boldsymbol{r}_1 + \sin\theta\, \boldsymbol{\theta}_1) \qquad (\omega = 0), \tag{14-109}$$

$$\boldsymbol{E} = 0 \qquad (\omega = 0). \tag{14-110}$$

This is the field of a static magnetic dipole at a distance r that is large compared to the radius of the loop. The value for $\boldsymbol{H}$ agrees with that found in Eqs. 7-112 to 7-114.

It is interesting to compare magnetic and electric dipole radiation. The $\boldsymbol{E}$ and $\boldsymbol{H}$ vectors for the latter are given by Eqs. 14-22 and 14-14 for $r \gg \lambda\!\!\!^-$. It will be observed that the two fields are quite similar, except that the expressions for $\boldsymbol{E}$ and $\boldsymbol{H}$ are interchanged. Also, the sign of $\boldsymbol{E}$ for the magnetic dipole is opposite that of $\boldsymbol{H}$ for the electric dipole. Such a change in sign is required to keep the Poynting vector directed outwards. The similarity between the fields of the electric and of the magnetic dipoles is a good illustration of the duality principle that was discussed in Section 10.8. It will be found that these two fields satisfy the symmetry conditions and that the constant of proportionality K is here equal to m_0/p_0 when the magnetic dipole field is chosen as the primed field.

The lines of force are similar to those of electric dipole radiation, except that, again, $\boldsymbol{E}$ and $\boldsymbol{H}$ are interchanged.

14.5.3 The Average Poynting Vector and the Radiated Power

The average Poynting vector and the radiated power for the magnetic dipole are calculated as in Section 14.1.4:

$$\boldsymbol{\mathcal{S}}_{av} = \frac{c\mu_0 m_0^2 \sin^2\theta}{32\pi^2 r^2 \lambda\!\!\!^{-4}}\, \boldsymbol{r}_1, \tag{14-111}$$

$$= 1.19\, \frac{m_0^2 \sin^2\theta}{r^2 \lambda\!\!\!^{-4}}\, \boldsymbol{r}_1 \text{ watts/meter}^2; \tag{14-112}$$

$$W = \frac{c\mu_0 m_0^2}{12\pi \lambda^4}, \tag{14-113}$$

$$= 10.0 \frac{m_0^2}{\lambda^4} \text{ watts.} \tag{14-114}$$

To find the radiation resistance of a circular loop of radius a, we set

$$m_0 = \pi a^2 I_0 \tag{14-115}$$

for the magnitude of the dipole moment, and then

$$W = 10.0\pi^2 \left(\frac{a}{\lambda}\right)^4 I_0^2 \text{ watts.} \tag{14-116}$$

The radiation resistance is the coefficient of $I_0^2/2$:

$$R_{\text{rad}} = 197 \left(\frac{a}{\lambda}\right)^4 \text{ ohms} \qquad (a \ll \lambda). \tag{14-117}$$

The radiation resistance of the magnetic dipole is proportional to the *fourth* power of the frequency, whereas that of the electric dipole was found to be proportional only to the second power of the frequency in Eq. 14-36.

14.6 MAGNETIC QUADRUPOLE RADIATION

In the previous section we studied an oscillating magnetic dipole consisting of a circular loop centered on the origin and carrying an alternating current.

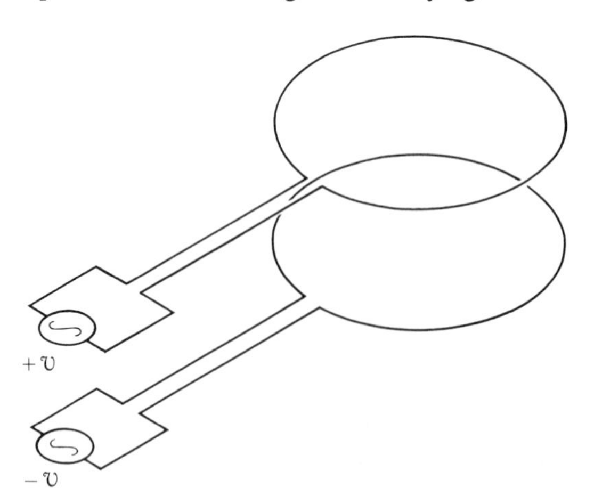

Figure 14-21. Simple magnetic quadrupole comprising two parallel loops of radius a separated by a distance a and excited in opposite phases, as indicated schematically by $+\mathcal{V}$ and $-\mathcal{V}$.

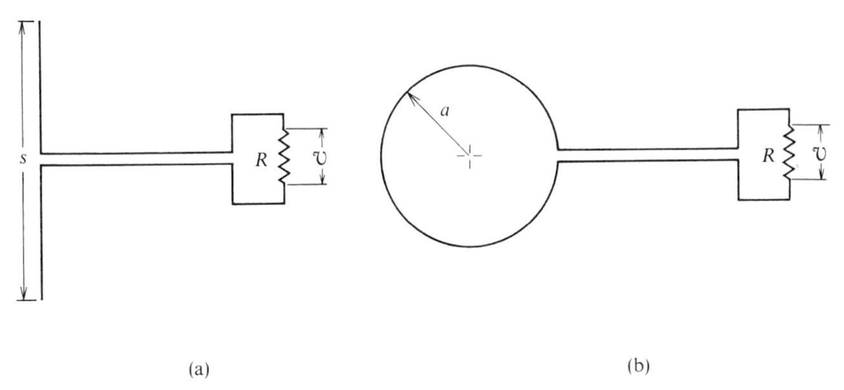

Figure 14-22. Electric and magnetic dipoles used as receiving antennas. The incident electromagnetic radiation induces an electric field in the wires, and the resulting voltage $\mathcal{V}$ across the resistance R is measured with a suitable electronic circuit.

We can now form an oscillating magnetic quadrupole with two such loops parallel to each other on either side of the origin and oscillating in opposite phases. We shall consider the simplest case, in which the distance between the loops is equal to the radius a of the loops, as in Figure 14-21, and we shall consider only the radiation field ($r \gg \lambdabar$). The lower dipole is centered at $z = -a/2$ and has a moment $-m_0 e^{j\omega t}$, whereas the upper dipole is centered at $z = +a/2$ and has a moment $+m_0 e^{j\omega t}$. This arrangement is purposely made similar to that of the linear electric quadrupole of Section 14.4.

As for the electric quadrupole, we may add the fields of the two magnetic dipoles, neglecting the differences in amplitude and in direction, but taking into account the difference in retardation. This is done by multiplying the dipole fields by plus and minus $\exp(ja/\lambdabar)\cos\theta$, as in Section 14.4. Thus,

$$\boldsymbol{E} = j\left(\frac{\mu_0}{\epsilon_0}\right)^{1/2} \frac{m_0 a}{4\pi r \lambdabar^3} \exp j\omega[t] \sin\theta\cos\theta\, \boldsymbol{\varphi}_1 \qquad (r \gg \lambdabar), \quad (14\text{-}118)$$

$$\boldsymbol{H} = -j\frac{m_0 a}{4\pi r \lambdabar^3} \exp j\omega[t] \sin\theta\cos\theta\, \boldsymbol{\theta}_1 \qquad (r \gg \lambdabar). \quad (14\text{-}119)$$

14.7 THE ELECTRIC AND MAGNETIC DIPOLES AS RECEIVING ANTENNAS

Figure 14-22 shows electric and magnetic dipoles used as receiving antennas. The tangential component of the incident electric field induces currents in the wires that (a) reradiate energy and (b) produce a voltage $\mathcal{V}$ across the load resistance R. It can usually be assumed that these currents have a negligible effect on those of the transmitting antenna. The voltage across R can be measured with some appropriate electronic circuit.

For the electric dipole, it can be shown that

$$\mathcal{V} = E_t s, \tag{14-120}$$

where E_t is the component of the incident electric field intensity that is tangent to the antenna and s is the length of the antenna. This assumes that $(s/2)^2 \ll \lambda\!\!\!^{-2}$ and that $R \to \infty$. A proper demonstration is quite elaborate and will not be gone into here.*

For the magnetic dipole, the induced electromotance is relatively easy to calculate:

$$\oint_C \boldsymbol{E} \cdot d\boldsymbol{l} = -\oint_C \left(\frac{\partial \boldsymbol{A}}{\partial t} + \boldsymbol{\nabla} V\right) \cdot d\boldsymbol{l}, \tag{14-121}$$

$$= -\int_S \left(\boldsymbol{\nabla} \times \frac{\partial \boldsymbol{A}}{\partial t} + \boldsymbol{\nabla} \times \boldsymbol{\nabla} V\right) \cdot d\boldsymbol{a}, \tag{14-122}$$

where S is any surface bounded by the circuit. The second term on the right vanishes, since the curl of a gradient is identically zero. Also, the order of the operations in the first term can be interchanged to give

$$\oint_C \boldsymbol{E} \cdot d\boldsymbol{l} = -\frac{\partial}{\partial t} \int_S \boldsymbol{\nabla} \times \boldsymbol{A} \cdot d\boldsymbol{a} = -\frac{\partial}{\partial t} \int_S \boldsymbol{B} \cdot d\boldsymbol{a}. \tag{14-123}$$

The induced electromotance in the loop is therefore equal to the rate of change of the flux linking the loop. It is maximum when the normal to the loop is parallel to the local $\boldsymbol{B}$.

The voltage $\mathcal{V}$ when $R \to \infty$ is not necessarily equal to the induced electromotance, because the circuit may also be excited in the electric dipole mode. For example, with a symmetrical loop such as that shown in Figure 14-22, if the $\boldsymbol{E}$ vector is parallel to the wires leading to R, charge oscillates from one end of the circuit to the other, $\mathcal{V}$ is not affected by the electric dipole oscillation, and the above relation is correct. On the other hand, if the $\boldsymbol{E}$ vector is in the plane of the loop but is perpendicular to the pair of wires, an extra voltage appears on R that comes from the dipole excitation and adds to the above induced electromotance.

---

* R. W. P. King, *Handbuch der Physik* (Springer-Verlag, Göttingen, 1958), Vol. XVI, p. 267.

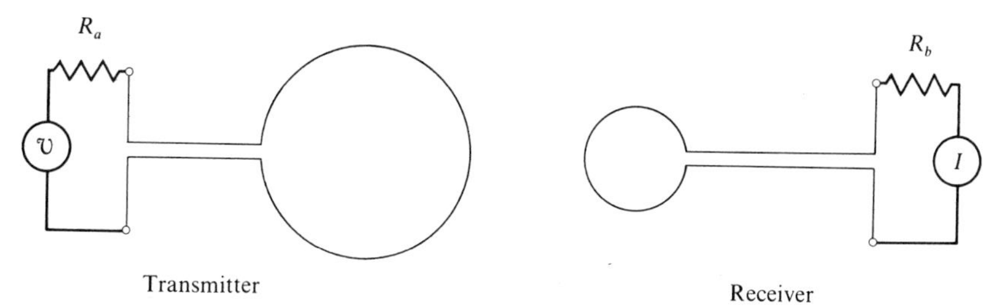

Figure 14-23. Pair of loop antennas. The one on the left is fed by an oscillator supplying a voltage $\mathcal{V}$; the other is connected to a load resistance R_b and to a zero impedance ammeter. The current I could also be measured with a high-impedance voltmeter across R_b. When a current I is drawn from a source such as an oscillator, the voltage drops at the output by an amount $\Delta\mathcal{V}$. The ratio $\Delta\mathcal{V}/I$ is called the output impedance of the source. It is shown here as the resistance R_a.

14.8 THE RECIPROCITY THEOREM

According to the reciprocity theorem, the *current* in a *detector* divided by the *voltage* at the *source* remains constant when source and detector are interchanged, as long as the frequency and all the impedances are left unchanged. This theorem is widely used for investigating both electric circuits and antennas. We shall prove it in the general case by using Maxwell's equations.

A pair of loop antennas, one of which is used as transmitter and the other as receiver, will serve to illustrate the discussion without restricting its generality. These are illustrated in Figure 14-23. The conductors and the medium of propagation are assumed to be isotropic. The source is connected to the left-hand antenna and supplies a voltage $\mathcal{V}$, while the detector is connected to the right-hand antenna and measures a current I. The reciprocity theorem states that the ratio $I/\mathcal{V}$ is not affected if source and detector are interchanged as in Figure 14-24.

We have seen in Section 10.9 that, for *any* two fields a and b of the same frequency,

$$\boldsymbol{\nabla}\cdot(\boldsymbol{E}_a \times \boldsymbol{H}_b) = \boldsymbol{\nabla}\cdot(\boldsymbol{E}_b \times \boldsymbol{H}_a) \tag{14-124}$$

at any point except within the sources. This is Lorentz's lemma. The more general case that does not exclude the sources is more complex, but it is also more interesting and more useful.

Let us consider the field a that is obtained when the antennas are used as in the top part of Figure 14-24, and the totally different field b that is obtained when the source is inserted in b and the current detector is in a as in the lower

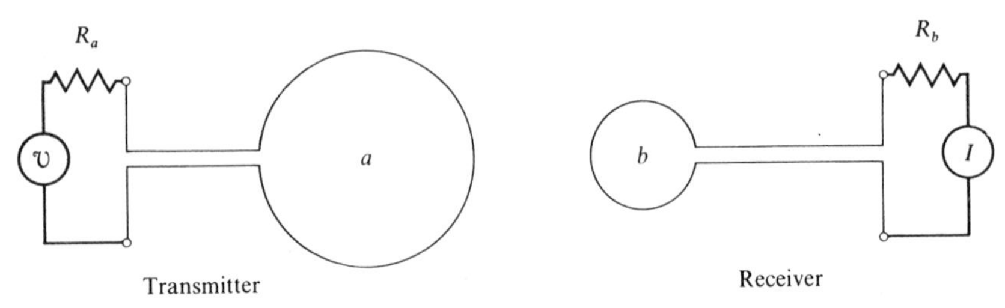

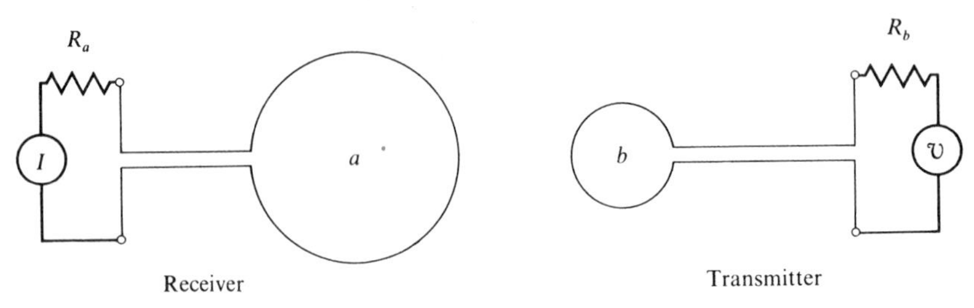

Figure 14-24. Pair of loop antennas with the source in a (top), and then in b (bottom). The frequency and the impedances are the same in both cases.

part of the same figure. The frequency and the impedances are assumed to be the same in the two cases.

At any point in space including the loops, and even including the source in loop a, we first have the mathematical identity

$$\begin{aligned}\nabla\cdot(\boldsymbol{E}_a \times \boldsymbol{H}_b - \boldsymbol{E}_b \times \boldsymbol{H}_a) \\ = \boldsymbol{H}_b\cdot\nabla \times \boldsymbol{E}_a - \boldsymbol{E}_a\cdot\nabla \times \boldsymbol{H}_b - \boldsymbol{H}_a\cdot\nabla \times \boldsymbol{E}_b + \boldsymbol{E}_b\cdot\nabla \times \boldsymbol{H}_a.\end{aligned} \tag{14-125}$$

Then, using Maxwell's equations,

$$\begin{aligned}\nabla\cdot(\boldsymbol{E}_a \times \boldsymbol{H}_b - \boldsymbol{E}_b \times \boldsymbol{H}_a) = -\boldsymbol{H}_b \cdot \frac{\partial \boldsymbol{B}_a}{\partial t} - \boldsymbol{E}_a \cdot \left(\boldsymbol{J}_{fb} + \frac{\partial \boldsymbol{D}_b}{\partial t}\right) \\ +\boldsymbol{H}_a \cdot \frac{\partial \boldsymbol{B}_b}{\partial t} + \boldsymbol{E}_b \cdot \left(\boldsymbol{J}_{fa} + \frac{\partial \boldsymbol{D}_a}{\partial t}\right),\end{aligned} \tag{14-126}$$

or, replacing the operator $\partial/\partial t$ by $j\omega$,

$$\nabla\cdot(\boldsymbol{E}_a \times \boldsymbol{H}_b - \boldsymbol{E}_b \times \boldsymbol{H}_a) = \boldsymbol{E}_b\cdot\boldsymbol{J}_{fa} - \boldsymbol{E}_a\cdot\boldsymbol{J}_{fb}. \tag{14-127}$$

For points outside the source, the current density $\boldsymbol{J}_f$ is simply $\sigma\boldsymbol{E}$, if we assume that Ohm's law applies. However, within the source, there is a further electric field intensity $\boldsymbol{E}_s$ (Section 8.5) and, in general,

$$\boldsymbol{J}_{fa} = \sigma(\boldsymbol{E}_a + \boldsymbol{E}_{sa}), \tag{14-128}$$

$$\boldsymbol{J}_{fb} = \sigma(\boldsymbol{E}_b + \boldsymbol{E}_{sb}). \tag{14-129}$$

The quantities $\boldsymbol{E}_{sa}$ and $\boldsymbol{E}_{sb}$ are the applied electric field intensities within the source when it is in loop a and when it is in loop b, respectively. The current density at any point in space is $\boldsymbol{J}_{fa}$ when the source is in a, and it is $\boldsymbol{J}_{fb}$ when the source is in b. Eliminating $\boldsymbol{E}_a$ and $\boldsymbol{E}_b$ on the right-hand side of Eq. 14-127,

$$\boldsymbol{\nabla}\cdot(\boldsymbol{E}_a \times \boldsymbol{H}_b - \boldsymbol{E}_b \times \boldsymbol{H}_a) = \boldsymbol{E}_{sa}\cdot\boldsymbol{J}_{fb} - \boldsymbol{E}_{sb}\cdot\boldsymbol{J}_{fa}. \tag{14-130}$$

In general, the right-hand side is *not* equal to zero. This relation applies to *any* pair of electromagnetic fields at *any* point in space, even inside the sources.

Let us integrate over all space:

$$\int_\infty \boldsymbol{\nabla}\cdot(\boldsymbol{E}_a \times \boldsymbol{H}_b - \boldsymbol{E}_b \times \boldsymbol{H}_a)\, d\tau = \int_\infty (\boldsymbol{E}_{sa}\cdot\boldsymbol{J}_{fb} - \boldsymbol{E}_{sb}\cdot\boldsymbol{J}_{fa})\, d\tau, \tag{14-131}$$

or, from the divergence theorem,

$$\int_\infty (\boldsymbol{E}_a \times \boldsymbol{H}_b - \boldsymbol{E}_b \times \boldsymbol{H}_a)\cdot d\boldsymbol{a} = \int_\infty (\boldsymbol{E}_{sa}\cdot\boldsymbol{J}_{fb} - \boldsymbol{E}_{sb}\cdot\boldsymbol{J}_{fa})\, d\tau. \tag{14-132}$$

If we now assume that the sources are limited to a finite volume, the surface of integration on the left-hand side is infinitely remote from them, and we have a plane wave with $\boldsymbol{E}$ and $\boldsymbol{H}$ orthogonal and transverse:

$$\boldsymbol{E} \times \boldsymbol{H} = EH\, \boldsymbol{r}_1, \tag{14-133}$$

where $\boldsymbol{r}_1$ is the unit radial vector. This is shown in Figure 14-25. Then

$$\boldsymbol{H}_a = \left(\frac{\epsilon_0}{\mu_0}\right)^{1/2} \boldsymbol{r}_1 \times \boldsymbol{E}_a, \tag{14-134}$$

and similarly for $\boldsymbol{H}_b$ and $\boldsymbol{E}_b$. Thus

$$\boldsymbol{E}_a \times \boldsymbol{H}_b - \boldsymbol{E}_b \times \boldsymbol{H}_a$$

$$= \left(\frac{\epsilon_0}{\mu_0}\right)^{1/2} \{\boldsymbol{E}_a \times (\boldsymbol{r}_1 \times \boldsymbol{E}_b) - \boldsymbol{E}_b \times (\boldsymbol{r}_1 \times \boldsymbol{E}_a)\} = 0 \tag{14-135}$$

at points infinitely remote from the sources. We can show that the quantity within braces is equal to zero merely by expanding it, recalling that $\boldsymbol{r}_1$ is perpendicular to both $\boldsymbol{E}_a$ and $\boldsymbol{E}_b$. Therefore the integral on the right-hand side of Eq. 14-132 must also be zero:

$$\int_\infty (\boldsymbol{E}_{sa}\cdot\boldsymbol{J}_{fb} - \boldsymbol{E}_{sb}\cdot\boldsymbol{J}_{fa})\, d\tau = 0. \tag{14-136}$$

The integration is extended to all space, but it can of course be limited to

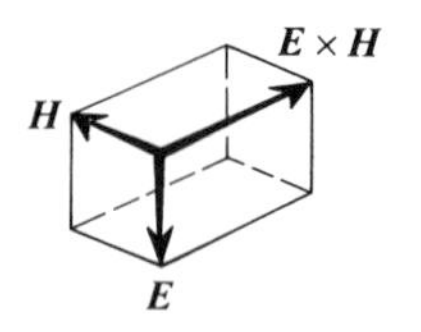

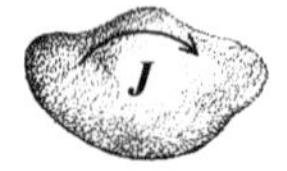

Figure 14-25. The field vectors $\boldsymbol{E}$ and $\boldsymbol{H}$ at a point infinitely remote from a source of radiation are orthogonal and transverse, as in a plane wave.

the sources, since $\boldsymbol{E}_{sa}$ and $\boldsymbol{E}_{sb}$ are zero everywhere else. Thus

$$\int_a \boldsymbol{E}_{sa} \cdot \boldsymbol{J}_{fb} \, d\tau = \int_b \boldsymbol{E}_{sb} \cdot \boldsymbol{J}_{fa} \, d\tau, \tag{14-137}$$

where the integrals are evaluated over the regions where $\boldsymbol{E}_{sa}$ and $\boldsymbol{E}_{sb}$ are nonzero.

The meaning of this equation can be illustrated by referring again to our pair of loop antennas. These are shown in Figure 14-24 for the field a, which obtains when the source is in a, and for the field b with the source in b.

For these antennas,

$$\int_a \boldsymbol{E}_{sa} \cdot \boldsymbol{J}_{fb} \, d\tau = \int_a \boldsymbol{E}_{sa} \cdot \boldsymbol{dl} \, J_{fb} \, da = \mathcal{V}_{sa} I_{b \text{ in } a}, \tag{14-138}$$

where $\mathcal{V}_{sa}$ is the voltage supplied by the source in a, and $I_{b \text{ in } a}$ is the current in the *same* loop *a when b is energized.* The other integral of Eq. 14-137 can be expressed similarly, and

$$\mathcal{V}_{sa} I_{b \text{ in } a} = \mathcal{V}_{sb} I_{a \text{ in } b}, \tag{14-139}$$

or

$$\frac{I_{a \text{ in } b}}{\mathcal{V}_{sa}} = \frac{I_{b \text{ in } a}}{\mathcal{V}_{sb}}. \tag{14-140}$$

Physically, this means that the current induced in b when a is energized, divided by the voltage applied on a, is the same as the current induced in a when b is energized, divided by the applied voltage on b, *as long as the frequency and the impedances remain unchanged. This is the reciprocity theorem.* We have illustrated it by referring to a pair of magnetic dipoles, but it is equally valid for any pair of antennas.

It must be kept in mind that this theorem is concerned solely with the ratio $I/\mathcal{V}$; it says nothing about the power expended by the source. This usually changes when the source is moved from one position to the other.

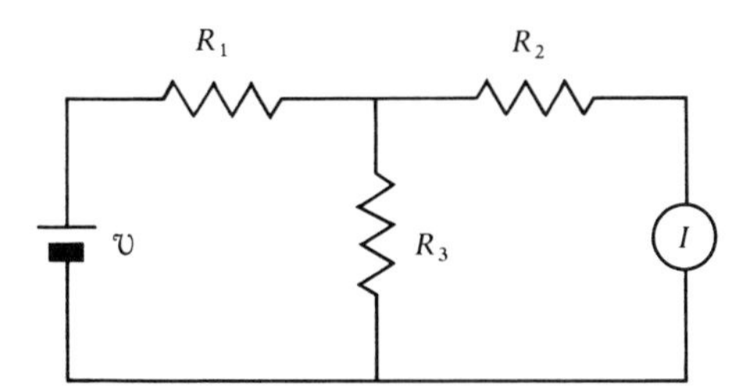

Figure 14-26. Simple electric circuit comprising a source of voltage $\mathcal{V}$ and an ammeter I. The reciprocity theorem applies: the ratio $I/\mathcal{V}$ remains constant when source and ammeter are interchanged, provided the frequency and the impedances remain unchanged.

One important consequence of the reciprocity theorem is that the radiation pattern of an antenna must have the same shape as the corresponding plot of the response of the antenna as a function of angle when it is used as a receiver. This fact is commonly used for the determination of radiation patterns.

Our demonstration of the theorem is general, except that the media have been assumed to be linear and isotropic. It therefore applies to ordinary electric circuits. It is a simple matter to check the reciprocity theorem for the circuit of Figure 14-26.

14.9 SUMMARY

Electric dipole radiation is produced by any charge distribution whose dipole moment $\boldsymbol{p}$ is a function of the time.

To obtain the field vectors $\boldsymbol{E}$ and $\boldsymbol{H}$, one first calculates the retarded potentials V and $\boldsymbol{A}$. One finds that, in free space, and for

$$\boldsymbol{p} = \boldsymbol{p}_0 e^{j\omega t}, \tag{14-2}$$

$$\boldsymbol{E} = -\frac{p_0}{4\pi\epsilon_0 r \lambda\!\!\!^{-2}} \exp j\omega[t] \sin\theta\, \boldsymbol{\theta}_1 \qquad (r \gg \lambda\!\!\!^{-}), \tag{14-22}$$

$$\boldsymbol{H} = -\frac{c p_0}{4\pi r \lambda\!\!\!^{-2}} \exp j\omega[t] \sin\theta\, \boldsymbol{\varphi}_1 \qquad (r \gg \lambda\!\!\!^{-}). \tag{14-14}$$

A transverse spherical wave therefore radiates away from the dipole, and the ratio E/H is the same as for a plane wave in free space. Closer in, the wave is more complex: there is a total of seven terms, and $\boldsymbol{E}$ has a radial component. The average Poynting vector is everywhere radial:

$$\mathbf{S}_{av} = \frac{cp_0^2 \sin^2 \theta}{32\pi^2 \epsilon_0 r^2 \lambda^4} \mathbf{r}_1. \qquad (14\text{-}28)$$

There is no energy radiated along the axis of the dipole.

The *half-wave antenna* is a conductor $\lambda/2$ long, which carries a standing wave of electric current. Its field is calculated by adding the radiation fields of the individual dipoles formed by the elements $\boldsymbol{dl}$ along its length. The field of the half-wave antenna is qualitatively similar to that of a dipole and, for $r \gg \lambda$,

$$\boldsymbol{E} = \frac{j}{2\pi r c \epsilon_0} I_0 \exp j\omega[t] \frac{\cos\left(\frac{\pi}{2}\cos\theta\right)}{\sin\theta} \boldsymbol{\theta}_1 \qquad (14\text{-}66)$$

$$\boldsymbol{H} = \frac{j}{2\pi r} I_0 \exp j\omega[t] \frac{\cos\left(\frac{\pi}{2}\cos\theta\right)}{\sin\theta} \boldsymbol{\varphi}_1. \qquad (14\text{-}71)$$

The radiated power is

$$W = 73.1\, I_{\text{rms}}^2 \text{ watts}, \qquad (14\text{-}82)$$

where I_{rms} is the rms current at the center of the antenna at the point where it is fed by a source of power. The coefficient 73.1 is the *radiation resistance* of the antenna.

Antenna arrays are sets of antennas properly spaced and phased so as to obtain, by interference, special radiation patterns.

Electric quadrupole radiation is produced by a charge distribution whose quadrupole moment is a function of the time. The simplest quadrupole source is the linear quadrupole. It is a simple matter to deduce its field from that of the electric dipole:

$$\boldsymbol{E} = -\frac{jp_{zz0}}{8\pi\epsilon_0 r \lambda^3} \exp j\omega[t] \sin\theta\cos\theta\, \boldsymbol{\theta}_1 \qquad (r \gg \lambda \gg s). \qquad (14\text{-}97)$$

The vectors $\boldsymbol{H}$ and $\boldsymbol{E}$ are related as with the electric dipole.

Magnetic dipole radiation is produced when the magnetic dipole moment of a current distribution is a function of time. If we assume a circular current loop of negligible impedance, then $V = 0$. From the retarded vector potential $\boldsymbol{A}$, we find values of $\boldsymbol{E}$ and $\boldsymbol{H}$ that are related to those of electric dipole radiation according to the duality principle of Section 10.8.

Magnetic quadrupole radiation is similarly related to electric quadrupole radiation.

Electric and magnetic dipoles can be used as *receiving antennas*. For a short electric dipole, the voltage on a high resistance load is equal to the tangential electric field multiplied by the length of the dipole. In the case of the magnetic dipole, this voltage is the electromotance induced by the changing magnetic flux linking the dipole, plus, in some cases, a voltage due to electric dipole excitation of the antenna.

The *reciprocity theorem* states that the current in a detector divided by the voltage applied at the source remains constant when source and detector are interchanged, as long as the frequency and the impedances are left unchanged. This applies to any electromagnetic field and, in particular, to electric circuits and to antennas.

PROBLEMS

14-1. (a) Calculate the vector potential A at a distance r on the axis of an oscillating electric dipole of length s.
(b) Show that the result is the same as it would be if the current were localized at the center of the dipole.

14-2. (a) Show that the phase velocity of the $\boldsymbol{H}$ field of an electric dipole is

$$u = \left(1 + \frac{1}{r'^2}\right) c,$$

where $r' = r/\lambda\!\!\!^-$.
(b) Draw a qualitative graph of u as a function of r'.

14-3. (a) Show that the phase velocities of the r and θ components of $\boldsymbol{E}$ for an oscillating electric dipole are, respectively,

$$u_r = \left(1 + \frac{1}{r'^2}\right) c,$$

$$u_\theta = \left\{\frac{r'^4 - r'^2 + 1}{r'^4 - 2r'^2}\right\} c,$$

where $r' = r/\lambda\!\!\!^-$.
(b) Draw qualitative graphs of u_r and of u_θ as functions of r'.

14-4. Calculate the ratio of the average electric energy density to the average magnetic energy density in the field of an electric dipole for (a) $r \ll \lambda\!\!\!^-$, (b) $r = \lambda\!\!\!^-$, (c) $r \gg \lambda\!\!\!^-$.

14-5. What fraction of the total power radiated by an electric dipole is radiated between $\pm 45°$ of the equatorial plane?

14-6. Show that the Poynting vector for the electric dipole at $r \gg \lambda\!\!\!^-$ is equal to the energy density multiplied by c.

14-7. (a) Show that, in the field of an electric dipole,

$$E_{rms} = 6.71\, W^{1/2} \frac{\sin\theta}{r},$$

where W is the radiated power and $r \gg \lambda\!\!\!^-$.

(b) Show also that, for the half-wave antenna,

$$E_{\rm rms} = 7.01\, W^{1/2} \frac{\cos\{(\pi/2)\cos\theta\}}{r\sin\theta}.$$

14-8. Calculate the electric field intensity in millivolts/meter at a distance of one kilometer in the equatorial plane of a half-wave antenna radiating one kilowatt of power. Set $\lambda \ll 1$ kilometer.

14-9. An antenna is normally situated near a conductor (the Earth, an airborne vehicle, a satellite, etc.) Energy radiated toward the conductor is reflected, and the total field is thus the vector sum of the direct wave plus the reflected wave.

It is convenient to consider that the latter is generated, not by reflection, but rather by an image of the antenna located below the surface of the conductor.

(a) Discuss the image of a *horizontal* half wave antenna over a perfectly conducting Earth, and show that *the current* in the image and the current in the antenna flow in *opposite* directions.‡

(b) Discuss the image of a *vertical* antenna over a perfectly conducting Earth, and show that the current in the image and that in the antenna flow in the *same* direction.

Both rules apply to oblique antennas.‡

(c) We have shown that the radiation resistance of a half-wave antenna is 73.1 ohms.

Show that the radiation resistance of a quarter-wave antenna perpendicular to a conducting plane is one half of this, or 36.6 ohms.

14-10. Identical parallel half-wave antennas are arranged in line with a uniform spacing D and are excited in phase.

(a) Show that the angular dependence of the electric field intensity E in the plane perpendicular to the antennas is given by

$$\frac{\sin\{(ND/2\lambda)\cos\varphi\}}{\sin\{(D/2\lambda)\cos\varphi\}},$$

where N is the number of antennas and φ is the angle between the direction of observation and the plane containing the antennas.‡

(b) Determine the angular positions of the maxima and minima of E for this radiation pattern. Differentiation of the above formula will yield only the maxima.

(c) Show that, for a given spacing D, the main lobe at $\varphi = \pi/2$ becomes narrower as the number of antennas N is increased.

(d) Show graphically, by considering the summation in the complex plane, that the main lobe is sharp and that the side lobes are small when N is large and $D < \lambda$.

(e) Plot the radiation pattern for an array composed of four parallel half-wave antennas spaced $\lambda/2$ apart.

14-11. Calculate the electromagnetic potentials V and A and the field vectors $\boldsymbol{E}$ and $\boldsymbol{H}$ for the oscillating linear electric quadrupole directly, without using the field of the oscillating electric dipole.

14-12. Show that the Lorentz condition applies to the radiation field of a linear electric quadrupole for $r \gg \lambda\!\!\!^{-}$.

14-13. Calculate the Poynting vector for the radiation field of an oscillating linear electric quadrupole, and show that the total radiated power is

$$1.8 \times 10^{17} \frac{Q_0^2 s^4}{\lambda\!\!\!^{-\,6}} \text{ watts.}$$

14-14. A sealed plastic box contains an antenna that radiates electromagnetic waves. How could you tell whether the antenna is an electric or a magnetic dipole?

14-15. Show that the electric dipole and magnetic dipole fields illustrate the duality property of electromagnetic fields (Section 10.8).

14-16. The gain G of an antenna is defined as the ratio of the Poynting vector at the maximum of the radiation pattern to the Poynting vector averaged over a spherical surface:

$$G = \frac{\mathcal{S}_{max}}{\frac{1}{4\pi} \int_0^{2\pi} \int_0^{\pi} \mathcal{S} \sin\theta \, d\theta \, d\varphi}.$$

The gain of an antenna is a measure of its directivity.

(a) Show that the gain of an electric or magnetic dipole is 1.5.
(b) Show that the gain of a half-wave antenna is 1.64.

14-17. (a) Show that the ratio $|E/H|$ for the field of an electric dipole is much larger close to the dipole than it is far away.
(b) Show that the inverse is true for the magnetic dipole.
(c) Which type of probe would you use to detect (*i*) electric dipole radiation, (*ii*) magnetic dipole radiation, near the source?

14-18. In the field of a magnetic dipole $V = 0$ and $A \neq 0$. Is it possible to have a radiation field such that $V \neq 0$ and $A = 0$?

14-19. (a) Show that $\nabla \cdot A = 0$ for the field of the magnetic dipole.
This is the Lorentz condition, since V is equal to zero.
(b) Show also that

$$\nabla^2 A - \epsilon_0 \mu_0 \frac{\partial^2 A}{\partial t^2} = 0$$

for the field of the magnetic dipole.
This is the wave equation for A.

14-20. Two identical magnetic dipoles are perpendicular to each other and have a common diameter.

(a) Show that the radiation pattern (amplitude as a function of θ) is a circle in the plane perpendicular to the common diameter if one dipole leads the other by $\pi/2$ radian.
(b) Explain the nature of the resulting field.
(c) How would you connect these antennas to a common source?
Such a pair of crossed coils can be used as an omnidirectional transmitting or receiving antenna.

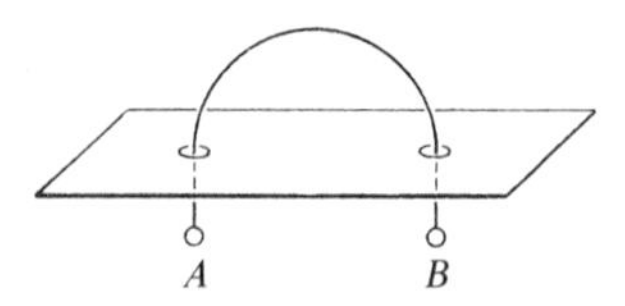

Figure 14-27.

14-21. Mobile radio receivers equipped with the usual vertical whip antennas are subject to signal fading when they move through the standing wave patterns that exist near electrically conducting buildings.

Let us see how this effect can be eliminated by using a half-loop antenna like the one shown in Figure 14-27. The $\boldsymbol{E}$ vector is vertical and the plane of the loop is normal to the local $\boldsymbol{H}$.

(a) Show that the sum of the signals at A and B is a measure of E, while their difference is a measure of H.

(b) Then, from Problem 12-21,

$$M(A+B)^2 + N(A-B)^2$$

will be a constant in the standing wave pattern if M/N is chosen properly.

Show that we must have that

$$\frac{M}{N} = \frac{\pi^4}{4c^2} R^2 f^2,$$

where R is the radius of the loop and f is the operating frequency.

This ratio is approximately equal to unity at 100 megahertz.

The single half-loop is not omnidirectional; to obtain an omnidirectional antenna one must use a pair of crossed half-loops connected as in Problem 14-20.

14-22. A loop antenna of inductance L feeds a load resistance R. The resistance of the loop is negligible compared to R.

Show that there will be maximum power transfer to the load when $R = \omega L$.

14-23. In the design of loop antennas for satellites, weight and size limitations are of foremost importance.

Show that, for a given loop diameter and for a given mass of copper, the ratio $\omega L/R$ is independent of the number of turns, L and R being, respectively, the inductance and resistance of the loop.

14-24. Compare the responses of the electric and of the magnetic dipole antennas when used as receivers (a) in air and (b) in sea water.

In (b), assume a frequency of 20 kilohertz. Assume that the loop antenna has a single turn, that its diameter is equal to the length l of the electric dipole, and that $l \ll \lambda$.

14-25. (a) Calculate the ratio $I/\mathcal{V}$ for the circuit of Figure 14-26, and show that the reciprocity theorem applies.

(b) Calculate the power expended by the source when the circuit is as shown and when $\mathcal{V}$ and I are interchanged.

14-26. Astronomers are interested in the following cosmological problem. Consider a particular class of objects, say quasars, and assume that they are all identical and distributed uniformly in a Euclidean universe. Then, if one plots $\log N$ as a function of $\log S$, where N is the number of objects giving a radio flux greater than S at the Earth, one should obtain a straight line whose slope is -1.5. The slope for quasars is in fact larger and possibly indicates the existence of cosmological evolution.*

Show that the expected slope is -1.5.

Hints

14-9. Think of the images of the charges flowing along the antennas.

14-10. (a) Perform the summation graphically in the complex plane.

* See, for example, A. Braccesi and L. Forniggini, *Astronomy and Astrophysics*, November 1969, page 364.

APPENDIXES

APPENDIX **A**

Conversion Table

Examples: One meter equals 100 centimeters. One volt equals 10^8 electromagnetic units of potential.

	SI	CGS SYSTEMS	
		esu	emu
Length	meter	10^2 centimeters	10^2 centimeters
Mass	kilogram	10^3 grams	10^3 grams
Time	second	1 second	1 second
Force	newton	10^5 dynes	10^5 dynes
Energy	joule	10^7 ergs	10^7 ergs
Power	watt	10^7 ergs/second	10^7 ergs/second
Charge	coulomb	3×10^9	10^{-1}
Electric potential	volt	$1/300$	10^8
Electric field intensity	volt/meter	$1/(3 \times 10^4)$	10^6
Electric displacement	coulomb/meter^2	$12\pi \times 10^5$	$4\pi \times 10^{-5}$
Displacement flux	coulomb	$12\pi \times 10^9$	$4\pi \times 10^{-1}$
Electric polarization	coulomb/meter^2	3×10^5	10^{-5}
Electric current	ampere	3×10^9	10^{-1}
Conductivity	mho/meter	9×10^9	10^{-11}
Resistance	ohm	$1/(9 \times 10^{11})$	10^9
Capacitance	farad	9×10^{11}	10^{-9}
Magnetic flux	weber	$1/300$	10^8 maxwells
Magnetic induction	tesla	$1/(3 \times 10^6)$	10^4 gausses
Magnetic field intensity	ampere/meter	$12\pi \times 10^7$	$4\pi \times 10^{-3}$ oersted
Magnetomotance	ampere	$12\pi \times 10^9$	$(4\pi/10)$ gilberts
Magnetic polarization	ampere/meter	$1/(3 \times 10^{13})$	10^{-3}
Inductance	henry	$1/(9 \times 10^{11})$	10^9
Reluctance	ampere/weber	$36\pi \times 10^{11}$	$4\pi \times 10^{-9}$

NOTE: We have set $c = 3 \times 10^8$ meters/second.

APPENDIX **B**

The Complex Potential

The complex potential provides a powerful method for calculating electrostatic fields. It is restricted, however, to two-dimensions, that is, to fields which are essentially constant in one direction. End-effects due to the finite length of the conductors in this particular direction are assumed to be negligible in the region considered. The method also assumes zero space charge density.

Similar methods of calculation are used in other fields of physics in the study of quantities which satisfy Laplace's equation. One important example is the field of hydrodynamics.

B.1 FUNCTIONS OF THE COMPLEX VARIABLE

Let us first consider the complex variable

$$z = x + jy, \tag{B-1}$$

where x and y are real numbers and where $j = (-1)^{1/2}$. This quantity can be represented by a point in the complex plane with x as abscissa and jy as ordinate, as in Figure B-1. This z must not be confused with the z-coordinate of Cartesian or cylindrical coordinates.

We can have functions $W(z)$ such as z^2, or $1/z$, or $\ln z$, and so on, and

$$W(z) = U(x, y) + jV(x, y) \tag{B-2}$$

with a real part U and an imaginary part jV, U and V being both *real* functions of x and of y. For example,

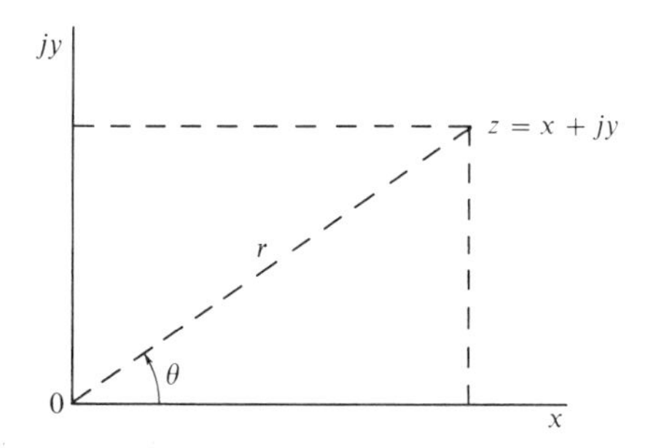

Figure B-1. Point $z = x + jy$ in the complex plane. The quantity r is called the modulus, and θ the argument, of the complex number z.

$$z^2 = (x + jy)^2 = (x^2 - y^2) + 2jxy, \tag{B-3}$$

in which

$$U = x^2 - y^2 \quad \text{and} \quad V = 2xy. \tag{B-4}$$

The function $W(z) = U + jV$ can be represented as a point on another complex plane with U as abscissa and jV as ordinate. We then speak of the z-plane and of the W-plane.

We shall consider functions $W(z)$ such that the derivative dW/dz exists in the region considered. This condition leads to an important pair of equations. First, let us examine the meaning of the derivative dW/dz by considering Figure B-2. The point W in the W-plane corresponds to the point z in the z-plane, according to some specified function $W(z)$. If z changes to $z + \Delta z$, W changes similarly to $W + \Delta W$, where the increments Δz and ΔW are complex numbers. The derivative dW/dz is the ratio of these increments at the limit $\Delta z \rightarrow 0$. The value of this derivative can take on different values for different values of z, but we wish to have a single value of dW/dz for a given value of z, that is, for a given point in the z-plane, *no matter how dz is chosen.*

We consider two particular values of dz: dx and $j\,dy$. In the first case, dz is parallel to the x-axis; in the second case, dz is parallel to the jy-axis. For both of these particular values of dz, we must have the same value of

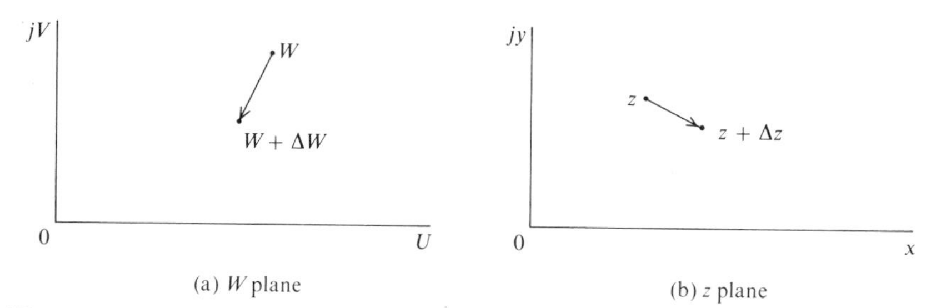

Figure B-2. The function $W(z) = U(x, y) + jV(x, y)$ shown in the W-plane. When z changes to $z + \Delta z$, W changes to $W + \Delta W$, and the ratio $\Delta W/\Delta z$ is equal to dW/dz for $\Delta z \rightarrow 0$.

dW/dz. For the first case, the value of dW/dz becomes $\partial W/\partial x$, and

$$\frac{\partial W}{\partial x} = \frac{\partial U}{\partial x} + j\frac{\partial V}{\partial x}. \tag{B-5}$$

For the second case, dW/dz becomes $\partial W/j\,\partial y$, and

$$\frac{\partial W}{j\,\partial y} = \frac{1}{j}\frac{\partial U}{\partial y} + \frac{\partial V}{\partial y}. \tag{B-6}$$

These two expressions must be equal:

$$\frac{\partial U}{\partial x} + j\frac{\partial V}{\partial x} = \frac{1}{j}\frac{\partial U}{\partial y} + \frac{\partial V}{\partial y}. \tag{B-7}$$

Then

$$\frac{\partial U}{\partial x} = \frac{\partial V}{\partial y} \quad \text{and} \quad \frac{\partial U}{\partial y} = -\frac{\partial V}{\partial x}. \tag{B-8}$$

These are the *Cauchy-Riemann equations.* The functions U and V are related to each other through these equations and are called *conjugate functions.*

A function $W(z)$ is said to be *analytic* if its four partial derivatives exist and are continuous throughout the region considered and, moreover, if they satisfy the Cauchy-Riemann equations.

B.2 CONFORMAL TRANSFORMATIONS

Consider now a point z and two neighboring points z' and z'' in the z-plane. These three points correspond to three other points in the W-plane: the point W and the neighboring points W' and W'', as in Figure B-3. Since dW/dz is unique at the point z, then

$$\frac{dW}{dz} = \lim_{z' \to z} \frac{W' - W}{z' - z} = \lim_{z'' \to z} \frac{W'' - W}{z'' - z}. \tag{B-9}$$

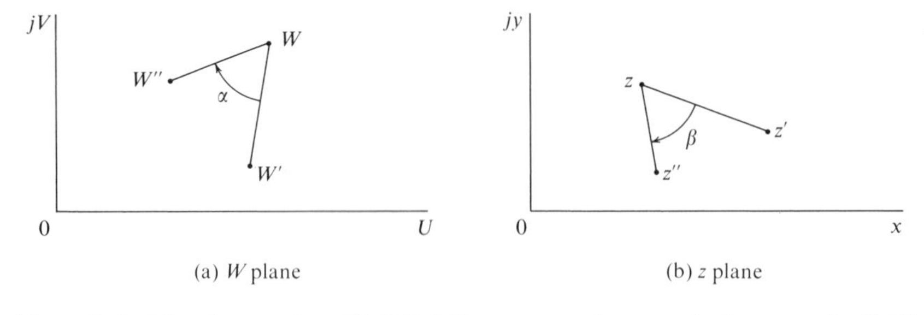

Figure B-3. The three points W, W', W'' correspond, respectively, to z, z', z''. The angles α and β are equal when $z' \to z$ and $z'' \to z$, as long as dW/dz exists and is not zero.

If $dW/dz \neq 0$, it can be written in the form $Ae^{j\varphi}$. (The argument φ has no meaning if $dW/dz = 0$.) Then, at the limits $z' \to z$ and $z'' \to z$, considering only the arguments of the various complex quantities,

$$\arg(W' - W) = \arg(z' - z) + \varphi, \tag{B-10}$$

$$\arg(W'' - W) = \arg(z'' - z) + \varphi, \tag{B-11}$$

$$\arg(W' - W) - \arg(W'' - W) = \arg(z' - z) - \arg(z'' - z), \tag{B-12}$$

or

$$\alpha = \beta, \tag{B-13}$$

where the angles α and β are as in Figure B-3. This result does not apply to points where $dW/dz = 0$.

The angle between two infinitesimal line segments is therefore conserved in passing from the z- to the W-plane, as long as the derivative dW/dz exists and is not zero. For example, the two families of straight lines represented by $U =$ Constant and $V =$ Constant in the W-plane are clearly orthogonal. Then, the corresponding curves $U =$ Constant and $V =$ Constant in the z-plane are also orthogonal.

Since there is a one-to-one correspondence between the points in the z-plane and those in the W-plane, we can imagine the W-plane to be distorted into the z-plane according to the function $W(z)$. Then a geometrical figure in the W-plane is "mapped" into a corresponding figure in the z-plane, and inversely. This process is called a *conformal transformation*.

B.3 THE FUNCTION $W(z)$ AS A COMPLEX POTENTIAL

Let us calculate the second derivatives of W with respect to x and to y:

$$\frac{\partial W}{\partial x} = \frac{dW}{dz}\frac{\partial z}{\partial x} = \frac{dW}{dz}, \tag{B-14}$$

$$\frac{\partial^2 W}{\partial x^2} = \frac{d^2 W}{dz^2}, \tag{B-15}$$

$$\frac{\partial W}{\partial y} = \frac{dW}{dz}\frac{\partial z}{\partial y} = j\frac{dW}{dz}, \tag{B-16}$$

$$\frac{\partial^2 W}{\partial y^2} = -\frac{d^2 W}{dz^2}, \tag{B-17}$$

and

$$\frac{\partial^2 W}{\partial x^2} + \frac{\partial^2 W}{\partial y^2} = 0. \tag{B-18}$$

The function $W(z)$ is thus a solution of Laplace's equation in two dimensions. Separating real and imaginary quantities, we find that

$$\frac{\partial^2 U}{\partial x^2} + \frac{\partial^2 U}{\partial y^2} = 0 \quad \text{and} \quad \frac{\partial^2 V}{\partial x^2} + \frac{\partial^2 V}{\partial y^2} = 0. \tag{B-19}$$

Thus both U and V independently satisfy Laplace's equation, and either one can be set to be the electric potential that also satisfies Laplace's equation, since we have assumed zero space charge density.

If the imaginary part of W is taken to be the electric potential, then the equipotentials are given by $V =$ Constant. Since the $U =$ Constant curves are orthogonal to the equipotentials, as we have seen above, they define the lines of force, from Section 2.3. The function V is then called the *potential function*, U is called the *stream function*, and W is called the *complex potential function*. It is shown in Problem B-2 that the electric field intensity is

$$E = \left|\frac{dW}{dz}\right| \tag{B-20}$$

at any point in the field.

A two-dimensional electrostatic field with zero space charge density is therefore completely determined once the complex potential function $W(z)$ is known.

It is essential to recall here the uniqueness theorem derived earlier in Section 4.2, according to which there is only one field configuration that satisfies given boundary conditions. Thus, if in one way or another we can find a satisfactory function $W(z)$, then that function is the proper one and the only one.

The determination of $W(z)$ is usually intuitive and empirical. However, much work has been done in this connection, and it is usually possible to determine the proper function for simple geometries.* One can usually arrive at $W(z)$ by using some function of the complex potential for some other known field.

Example

As an illustration, let us consider the field inside an infinite parallel-plate capacitor as in Figure B-4. The equipotentials and the lines of force are respectively $y =$ Constant and $x =$ Constant.

Setting V to be the electric potential,

$$V = \frac{V_1}{s} y. \tag{B-21}$$

* See, for example, E. Durand, *Electrostatique* (Masson et Cie, Paris, 1964), Vol. 1, and E. Kober, *Dictionary of Conformal Representations* (Dover, New York, 1952).

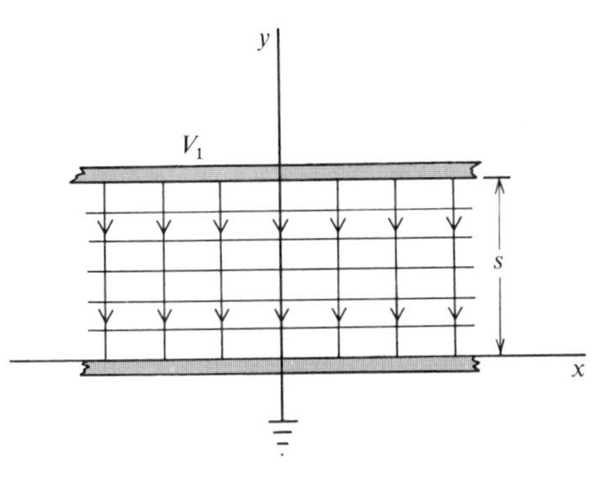

Figure B-4. A parallel-plate capacitor with its lower plate grounded and its upper plate at potential V_1. End effects are neglected. The horizontal lines are equipotentials $V = (V_1/s)y =$ constant; the vertical lines are lines of force $U = (V_1/s)x =$ constant.

This suggests that the function $W(z)$ must be

$$W(z) = U + jV = \frac{V_1}{s}(x + jy) = \frac{V_1}{s} z. \tag{B-22}$$

Thus,

$$U = \frac{V_1}{s} x. \tag{B-23}$$

We could also have determined the stream function U through the Cauchy-Riemann equations.

B.4 THE STREAM FUNCTION

We have seen above that the stream function is a constant along a line of force. We shall now see that it is quantitatively just as important as the potential function. We shall assume, as in Section B.3, that V is the potential function.

Figure B-5 shows three equipotentials and three lines of force in a portion of an electrostatic field. The vector $\boldsymbol{dn}$ is an element of length along the line of force normal to the equipotential at the point considered. The vector $\boldsymbol{ds}$ is an element of length along an equipotential and is oriented with respect to $\boldsymbol{dn}$ so that it points to the *left* when viewed along $\boldsymbol{dn}$. For convenience, we choose our x- and jy-axes as shown, so that the x-axis is parallel to $\boldsymbol{dn}$ and the jy axis is parallel to $\boldsymbol{ds}$ at the point considered.

According to the Cauchy-Riemann equations,

$$\frac{\partial U}{\partial s} = \frac{\partial U}{\partial y} = -\frac{\partial V}{\partial x} = -\frac{\partial V}{\partial n}. \tag{B-24}$$

Since $-\partial V/\partial n$ is the electric field intensity $\boldsymbol{E}$ at the point, the positive direction for $\boldsymbol{E}$ being along $\boldsymbol{dn}$, then, along an equipotential,

$$dU = E\,ds. \tag{B-25}$$

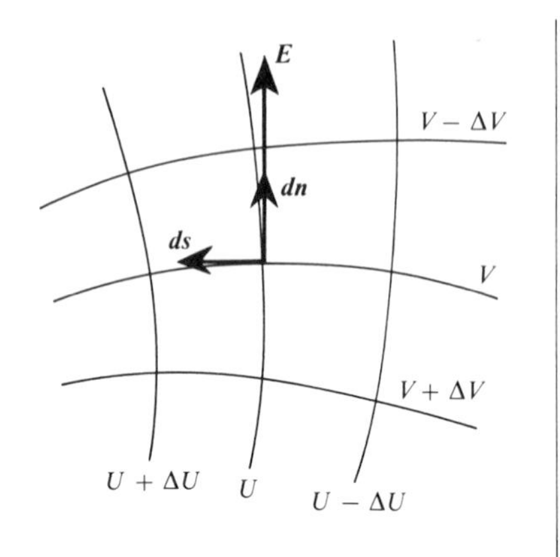

Figure B-5. A set of lines of force and a set of equipotentials in a portion of an electrostatic field. The elementary vectors ***dn*** and ***ds*** are directed, respectively, along the line of force and along the equipotential at the point considered. The vector ***dn*** points in the direction of ***E***, and the vector ***ds*** points to the *left* when one looks along ***dn***. For convenience, the axes are chosen to be parallel to these vectors as shown.

This relation is valid for any point in the field as long as V is chosen to be the potential function, and $\boldsymbol{E}$ and $\boldsymbol{ds}$ are oriented as in Figure B-5.

The stream function is thus related to $E\,ds$. This quantity $E\,ds$ is the flux of $\boldsymbol{E}$ crossing the equipotential in the direction of $\boldsymbol{dn}$ through an element of area on the equipotential surface that is ds wide and whose height, measured in the direction perpendicular to the paper, is the unit of length, namely one meter. Integrating Eq. B-25 along an equipotential between the lines of force U_1 and U_2,

$$U_2 - U_1 = \int_{U_1}^{U_2} \frac{dU}{ds}\,ds = \int_{U_1}^{U_2} E\,ds. \tag{B-26}$$

The line of force for which the stream function is zero is chosen arbitrarily, just as for the equipotential on which the potential is zero. Thus the charge density σ at the surface of a conductor is

$$\sigma = \epsilon_0 E = \epsilon_0 \frac{dU}{ds}, \tag{B-27}$$

where the vector $\boldsymbol{ds}$ points toward the *left* when one looks toward the *outside* of the conductor into the field.

The total charge Q per unit length on a cylindrical conductor whose axis is perpendicular to the paper is obtained by integrating $\sigma\,ds$ around the periphery of the conductor in the direction of increasing ds:

$$Q = \oint \sigma\,ds = \epsilon_0 \oint dU. \tag{B-28}$$

The stream function is thus useful for determining surface charge densities and total charges on conductors.

Example

THE PARALLEL-PLATE CAPACITOR

Let us return to the parallel-plate capacitor of Figure B-4, for which we found the complex potential function $W(z)$ in Eq. B-22. The charge density on the lower plate is obtained from Eq. B-22 with the vector $\mathbf{ds}$ pointing toward the *left* when one looks toward the outside of the conductor into the field:

$$\sigma_{\text{lower}} = \epsilon_0 \frac{dU}{ds} = -\epsilon_0 \frac{dU}{dx} = -\epsilon_0 \frac{V_1}{s}. \tag{B-29}$$

Similarly, the charge density on the upper plate is found to be $+\epsilon_0(V_1/s)$. Both of these charge densities can be verified to be correct by using Gauss's law.

The capacitance C' per unit area is

$$C' = \frac{\sigma}{V_1} = \frac{\epsilon_0}{s}, \tag{B-30}$$

and

$$E = \left|\frac{dW}{dz}\right| = \frac{V_1}{s}, \tag{B-31}$$

as expected.

We have thus verified our method of calculation by applying it to a well-known field.

Example

THE CYLINDRICAL CAPACITOR

In the case of the cylindrical capacitor, as in Figure B-6, the equipotentials are concentric circles, whereas the lines of force are radial straight lines.

We find empirically the following complex potential function, which is justified below:

$$W(z) = \frac{jV_1}{\ln \frac{r_2}{r_1}} \ln \frac{z}{r_1}, \tag{B-32}$$

or, writing

$$z = re^{j\theta}, \tag{B-33}$$

$$W(z) = \frac{jV_1}{\ln \frac{r_2}{r_1}}\left(\ln \frac{r}{r_1} + j\theta\right), \tag{B-34}$$

$$= -\frac{V_1}{\ln \frac{r_2}{r_1}}\theta + j\frac{V_1}{\ln \frac{r_2}{r_1}} \ln \frac{r}{r_1}. \tag{B-35}$$

Thus

$$U = -\frac{V_1}{\ln \frac{r_2}{r_1}}\theta, \tag{B-36}$$

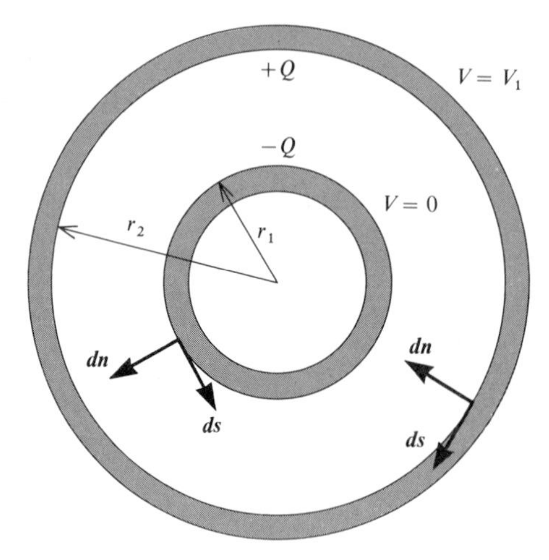

Figure B-6. A section through a cylindrical capacitor of internal radii r_1 and r_2. The inner cylinder is grounded, and the outer one is at a potential V_1, the charges being, respectively, $-Q$ and $+Q$.

$$V = \frac{V_1}{\ln \frac{r_2}{r_1}} \ln \frac{r}{r_1}. \tag{B-37}$$

The above expression for $W(z)$ is justified as follows: (a) the logarithm serves to give equipotentials and lines of force of the required form r = Constant and θ = Constant when the real and imaginary parts of $W(z)$ are set equal to constants; (b) the factor j serves to make V the potential; (c) the ratio z/r_1 makes V equal to zero at $r = r_1$; (d) finally, the factor $V_1/\ln(r_2/r_1)$ serves to make the potential equal to V_1 at $r = r_2$.

Let us find the charge on the inner cylinder. To do this we must integrate the stream function U in the direction of $\mathbf{ds}$, or of increasing θ, between $-\pi$ and $+\pi$, as in Figure B-6. Then

$$Q_{\text{inner}} = -\frac{2\pi\epsilon_0}{\ln \frac{r_2}{r_1}} V_1. \tag{B-38}$$

The charge on the outer cylinder is found by integrating U in the direction of decreasing θ, and $Q_{\text{outer}} = -Q_{\text{inner}}$, as expected.

The capacitance C' per unit length normal to the paper is

$$C' = \left|\frac{Q}{V_1}\right| = \frac{2\pi\epsilon_0}{\ln \frac{r_2}{r_1}}, \tag{B-39}$$

which is correct, and E is also given correctly by

$$E = \left|\frac{dw}{dz}\right| = \left|\frac{V_1}{\ln \frac{r_2}{r_1}} \frac{1}{z}\right|, \tag{B-40}$$

$$E = \frac{V_1}{r \ln \frac{r_2}{r_1}}. \tag{B-41}$$

Example

FIELD OF TWO PARALLEL LINE CHARGES OF OPPOSITE POLARITIES

We consider the field due to two parallel line charges of Q coulombs/meter and of opposite polarities, as in Figure B-7. The line charges shown at $-a$ and at $+a$ are presumed to be infinitely long in the direction perpendicular to the paper.

The potential V at any point due to this charge distribution can be found by integration, as in Problem 2-6:

$$V = \frac{Q}{2\pi\epsilon_0} \ln \frac{r_1}{r_2}, \tag{B-42}$$

where Q is the charge per unit length perpendicular to the paper, and r_1 and r_2 are as in Figure B-7. This equation shows that $V \to +\infty$ at $r_2 = 0$, $V \to -\infty$ at $r_1 = 0$, and $V = 0$ at $r_1 = r_2$, which is correct. Also, for a given value of r_1, V varies as the logarithm of r_2 and, similarly, for a given value of r_2, V varies as the logarithm of r_1, which is also correct.

We can also rewrite V as follows:

$$V = \frac{Q}{2\pi\epsilon_0} \ln \left|\frac{z+a}{z-a}\right|. \tag{B-43}$$

This suggests that

$$W(z) = j\frac{Q}{2\pi\epsilon_0} \ln \frac{z+a}{z-a}, \tag{B-44}$$

$$= j\frac{Q}{2\pi\epsilon_0} \ln \left(\left|\frac{z+a}{z-a}\right| \frac{e^{j\theta_1}}{e^{j\theta_2}}\right), \tag{B-45}$$

$$= j\frac{Q}{2\pi\epsilon_0} \left[\ln \left|\frac{z+a}{z-a}\right| + j(\theta_1 - \theta_2)\right], \tag{B-46}$$

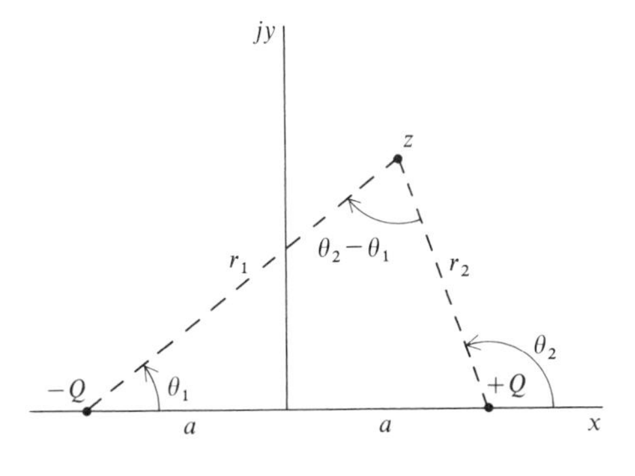

Figure B-7. Line charges of $-Q$ and $+Q$ coulombs/meter are situated, respectively, at $x = -a$ and $x = +a$.

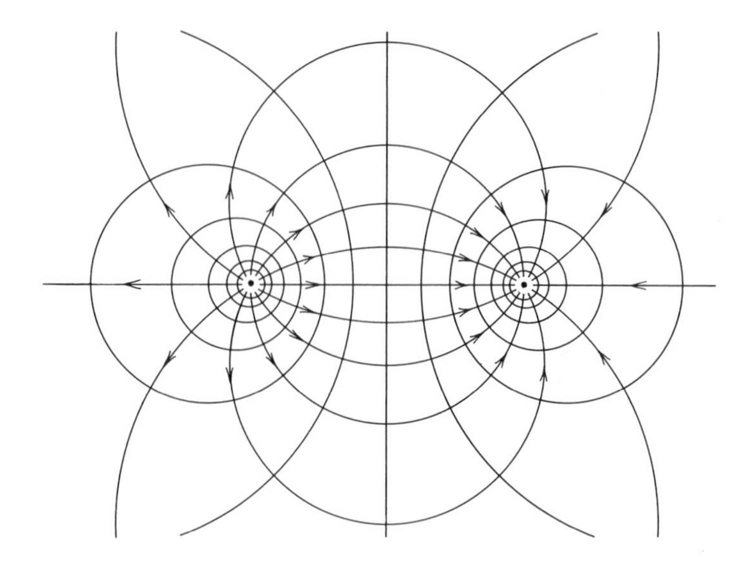

Figure B-8. Lines of force (indicated by arrows) and equipotentials for two infinite line charges perpendicular to the paper. All the curves are circles or arcs of circles. The equipotential surfaces are generated by sliding the figure along an axis perpendicular to the paper.

and

$$U = \frac{Q}{2\pi\epsilon_0}(\theta_2 - \theta_1), \tag{B-47}$$

$$V = \frac{Q}{2\pi\epsilon_0}\ln\left|\frac{z+a}{z-a}\right| = \frac{Q}{2\pi\epsilon_0}\ln\frac{r_1}{r_2}. \tag{B-48}$$

The lines of force are given by the equation $\theta_2 - \theta_1 = $ constant. They are thus arcs of circles passing through the line charges and centered on the jy-axis, both above and below the x-axis.

The equipotential V is determined by the equation

$$\frac{r_1}{r_2} = \exp(2\pi\epsilon_0 V/Q). \tag{B-49}$$

It is shown in Problem B-3 that this equipotential is a cylinder whose radius is $|a\,\mathrm{csch}\,(2\pi\epsilon_0 V/Q)|$ and whose axis is situated at $x = |a \coth 2\pi\epsilon_0 V/Q|$. For $V \to +\infty$, the equipotential surface reduces to a line perpendicular to the paper and situated at $x = a$, as expected. Similarly, for $V \to -\infty$, we have a line at $x = -a$.

The equipotentials and lines of force are shown in Figure B-8.

Example

FIELD OF TWO PARALLEL CONDUCTING CIRCULAR CYLINDERS OF OPPOSITE POLARITIES

The field investigated above was shown to have equipotentials in the form of circular cylinders whose axes are all parallel. We can place an *uncharged* conducting foil on any of these equipotentials without disturbing the field in any way. When we do this, charges migrate inside the foil so as to cancel the electrostatic field within it, and charges of opposite polarity appear on the two surfaces. In this way, the field remains everywhere exactly as it was, except for the conducting region inside the foil, where the field is zero. If the line charge surrounded by the foil is $+Q$ per unit length, a charge $-Q$ per unit length is induced on the inside surface of the foil, and a charge of $+Q$ per unit length is induced on the outside surface. These induced charges are of course due to the electrostatic field and are due to *both* of the line charges. We can cancel the $+Q$ and $-Q$ charges surrounded by the foil by shorting the line charge $+Q$ and the foil, leaving us with a net charge $+Q$ on the outside of the conducting foil and zero charge inside. The field in the internal region limited by the foil is then zero, but the field outside is unchanged. Instead of canceling the charges inside the foil as above, we could similarly have canceled the charges outside, making the field zero outside the region bounded by the conducting foil but leaving it intact inside.

We can now find the field due to a pair of parallel conducting circular cylinders by replacing two of the equipotentials with conducting cylinders carrying charges equal in magnitude but opposite in sign.

We consider two cylinders of radius R as in Figure B-9, with their axes separated by a distance D and carrying known charges $-Q$ and $+Q$, respectively. Their potentials $-V$ and $+V$ are unknown. These are the potentials with respect to that along the jy-axis, which is taken to be zero. The potential difference $2V$ is especially important, since it is required for the calculation of the capacitance per unit length of the system.

The function $W(z)$ for the pair of cylinders is the same as in Eq. B-44, except for the fact that now the quantity a is an unknown, with

$$R = a \operatorname{csch}\left(\frac{2\pi\epsilon_0 V}{Q}\right), \tag{B-50}$$

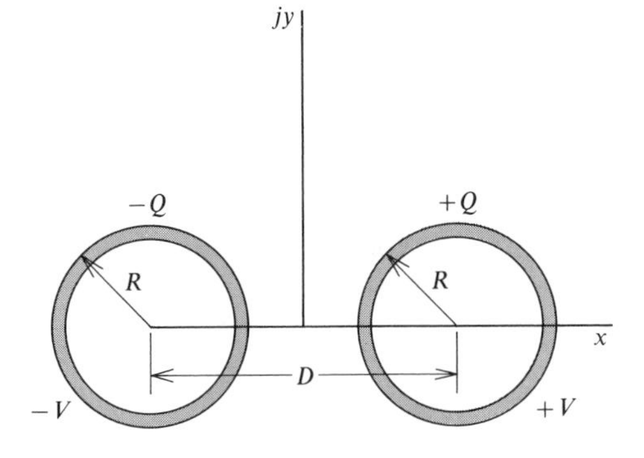

Figure B-9. A section through two parallel cylinders of radius R whose axes are separated by a distance D and which carry charges $-Q$ and $+Q$ coulombs/meter, respectively. The potentials are $-V$ and $+V$. The origin of coordinates is chosen midway between the axes.

$$\frac{D}{2} = a \coth\left(\frac{2\pi\epsilon_0 V}{Q}\right). \tag{B-51}$$

Since

$$\coth^2 x - \operatorname{csch}^2 x = 1, \tag{B-52}$$

then

$$a = \left[\left(\frac{D}{2}\right)^2 - R^2\right]^{1/2}. \tag{B-53}$$

Substituting this value of a in Eq. B-50 for R, recalling that the capacitance per meter $C' = Q/2V$, we find that

$$C' = \frac{\pi\epsilon_0}{\sinh^{-1}\left(\frac{D^2}{4R^2} - 1\right)^{1/2}}, \tag{B-54}$$

$$= \frac{\pi\epsilon_0}{\cosh^{-1}(D/2R)}. \tag{B-55}$$

It will be observed that C' depends only on the ratio D/R, and not on the actual dimensions, just as in the case of the cylindrical capacitor. When $D = 4R$, $C' = 21.2$ picofarads/meter.

PROBLEMS

B.1 Two cylinders of length l and radius a have their axes parallel and separated by a distance D in a liquid. A potential difference V is applied between the cylinders. Calculate the force of attraction for $l = 1.00$ meter, $a = 0.500$ centimeter, $D = 2.50$ centimeters, $\epsilon_r = 2.6$, and $V = 1.00 \times 10^4$ volts. Neglect end effects.

B.2 Show that the electric field intensity for a two-dimensional field in the xy-plane is

$$E = \left|\frac{dW}{dz}\right|,$$

where W is the complex potential and $z = x + jy$.

B.3 Show that the equipotentials defined by Eq. B-49 are the circular cylinders

$$\left[x - a \coth\left(\frac{2\pi\epsilon_0 V}{Q}\right)\right]^2 + y^2 = a^2 \operatorname{csch}^2\left(\frac{2\pi\epsilon_0 V}{Q}\right).$$

B.4 Consider an unknown two-dimensional field due to two charged conductors A' and B' in the complex z'-plane, and a known two-dimensional field due to the charged conductors A and B in the z-plane. Points in the z'-plane are related to those in the z-plane according to the transformation $z' = z'(z)$, and the corresponding potentials are equal: $V_A = V_{A'}$ and $V_B = V_{B'}$.

Show that there is conservation of charge under this transformation and that there is, in consequence, conservation of capacitance.

APPENDIX C

Induced Electromotance in Moving Systems*

To illustrate the meaning of Eq. 8-36,

$$\boldsymbol{\nabla} \times \boldsymbol{E} = -\frac{\partial \boldsymbol{B}}{\partial t} + \boldsymbol{\nabla} \times (\boldsymbol{u} \times \boldsymbol{B}), \tag{C-1}$$

in which the induced electric field intensity $\boldsymbol{E}$ is measured in one coordinate system and the magnetic induction $\boldsymbol{B}$ in another, let us consider the following experiments.

Example

EXPERIMENT 1

Figure C-1 shows a circular disk $D1$ rotating with an angular velocity ω about an axis perpendicular to its plane and parallel to a uniform magnetic field $\boldsymbol{B}$. The disk is assumed to be both nonconducting and nonmagnetic.

We now station two observers, one in the laboratory and one on the rotating disk, both equipped with a "curl-meter," a "$\boldsymbol{B}$-meter," and a stop watch. The "curl-meter" consists of a small loop of wire capable of orientation in any direction and connected in series with a sensitive, infinite-impedance voltmeter. By definition, the component of $\boldsymbol{\nabla} \times \boldsymbol{E}$ normal to the plane containing the path of integration is

$$(\boldsymbol{\nabla} \times \boldsymbol{E})_n = \lim_{S \to 0} \frac{\oint \boldsymbol{E} \cdot \boldsymbol{dl}}{S}. \tag{C-2}$$

* See Dale R. Corson, *Am. J. Phys.* **24,** 126 (1956).

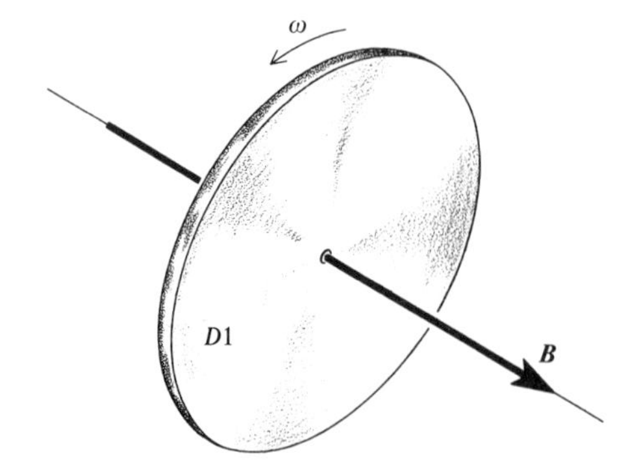

Figure C-1. A circular, nonconducting, nonmagnetic disk rotating with angular velocity ω about an axis perpendicular to its plane and parallel to a uniform magnetic field $\boldsymbol{B}$.

The voltmeter reading divided by the area of the loop is thus the component of the curl in the direction of the normal to the loop, if the loop is small enough. The "$\boldsymbol{B}$-meter" can be a cathode ray tube with the deflection of the electron beam on the tube face calibrated in teslas. The stop watch is used to measure the time rate of change of $\boldsymbol{B}$. The observers know nothing about the magnetic field except what they measure with their own instruments.

The laboratory observer measures a uniform magnetic field: $\boldsymbol{B}_L = \boldsymbol{B}$. He determines its direction and magnitude by observing the deflection of his electron beam for at least two mutually perpendicular orientations. He also observes that, for each orientation, the beam deflection is time-independent. Then $\partial \boldsymbol{B}_L/\partial t = 0$. Furthermore, his "curl-meter" reads zero for all orientations, since there is no changing flux through the loop, and thus

$$\boldsymbol{\nabla} \times \boldsymbol{E}_L = -\frac{\partial \boldsymbol{B}_L}{\partial t} = 0. \tag{C-3}$$

When the disk observer points his electron beam in the plane of the disk, he always records the same deflection, no matter where he is on the disk. When he points it parallel to the axis of the disk, he always records zero deflection. He therefore concludes that the magnetic induction is uniform and perpendicular to his disk. His value $\boldsymbol{B}_{D1}$ is the same as that of the laboratory observer: $\boldsymbol{B}_{D1} = \boldsymbol{B}_L = \boldsymbol{B}$. The "curl-meter" on the disk sees only a constant flux, and

$$\boldsymbol{\nabla} \times \boldsymbol{E}_{D1} = 0 \tag{C-4}$$

everywhere on the disk. Thus, the disk observer finds

$$\boldsymbol{\nabla} \times \boldsymbol{E}_{D1} = -\frac{\partial \boldsymbol{B}_{D1}}{\partial t} = 0. \tag{C-5}$$

Now let us consider a second similar nonconducting and nonmagnetic disk $D2$ rotating with an angular velocity ω about an axis in the plane of the disk and perpendicular to the direction of $\boldsymbol{B}$, as indicated in Figure C-2. An observer on this disk, if equipped with

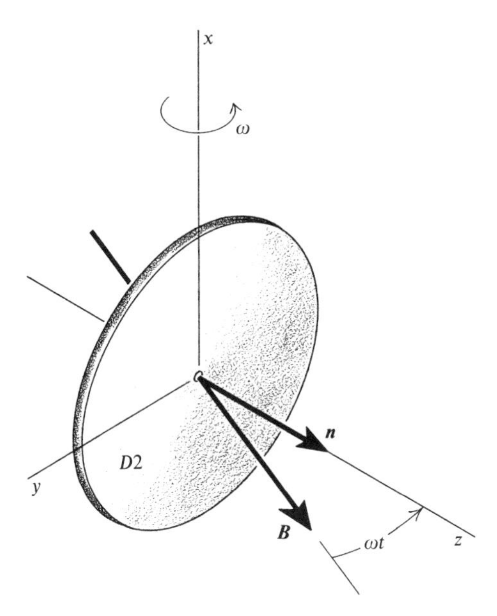

Figure C-2. A nonconducting, nonmagnetic disk rotating with an angular velocity ω about an axis in the plane of the disk and perpendicular to the direction of a uniform magnetic field $\boldsymbol{B}$.

the same instruments as the other observers, will ascribe entirely different properties to the field.

Let the observer on disk $D2$ establish a coordinate system with its z-axis perpendicular to the plane of the disk and its x-axis parallel to the axis of the rotation. When he points his electron beam in the z direction, the x-component of the beam deflection measures $(\boldsymbol{B}_{D2})_y$, and the y-component measures $(B_{D2})_x$. He finds that

$$(B_{D2})_x = 0, \tag{C-6}$$

$$(B_{D2})_y = B_0 \sin \omega t. \tag{C-7}$$

When he points his beam in the y direction, the x-deflection measures $(B_{D2})_z$, and the z-deflection measures $(B_{D2})_x$:

$$(B_{D2})_x = 0, \tag{C-8}$$

$$(B_{D2})_z = B_0 \cos \omega t. \tag{C-9}$$

He finds these same fields no matter where he measures on his disk. He can describe the field he measures as a uniform field $\boldsymbol{B}_0$ rotating with angular velocity ω about his x-axis. If he were to compare notes with the laboratory observer, he would find that $\boldsymbol{B}_0 = \boldsymbol{B}_L = \boldsymbol{B}$.

What about his "curl-meter"? When the disk observer points it so that the axis of the loop is in the x-direction, there is no flux through the loop, and the reading is zero. Then

$$(\boldsymbol{\nabla} \times \boldsymbol{E}_{D2})_x = -\frac{\partial (B_{D2})_x}{\partial t} = 0. \tag{C-10}$$

When he points the loop axis in the z direction, there is a changing flux through the loop, and

$$\left(\frac{d\Phi}{dt}\right)_{D2} = -BS\omega \sin \omega t, \tag{C-11}$$

where S is the area of the loop. Then, for this orientation,

$$\oint \boldsymbol{E}_{D2} \cdot d\boldsymbol{l} = BS\omega \sin \omega t, \tag{C-12}$$

and, from the definition of the curl,

$$(\boldsymbol{\nabla} \times \boldsymbol{E}_{D2})_z = B\omega \sin \omega t, \tag{C-13}$$

$$(\boldsymbol{\nabla} \times \boldsymbol{E}_{D2})_z = -\frac{\partial (\boldsymbol{B}_{D2})_z}{\partial t} = B\omega \sin \omega t. \tag{C-14}$$

If the "curl-meter" reading is calculated for the y-component,

$$(\boldsymbol{\nabla} \times \boldsymbol{E}_{D2})_y = -\frac{\partial (\boldsymbol{B}_{D2})_y}{\partial t} = -B\omega \cos \omega t. \tag{C-15}$$

From this experiment we can see that

$$\boldsymbol{\nabla} \times \boldsymbol{E} = -\frac{\partial \boldsymbol{B}}{\partial t} \tag{C-16}$$

whenever $\boldsymbol{E}$ *and* $\boldsymbol{B}$ are measured in the *same* coordinate system.

Example

EXPERIMENT 2

Let us now suppose that each observer is given some conducting wire and told to try to arrange it so as to induce an electromotance in a closed circuit at rest in his own coordinate system. For the laboratory observer,

$$\boldsymbol{\nabla} \times \boldsymbol{E}_L = -\frac{\partial \boldsymbol{B}_L}{\partial t} = 0 \tag{C-17}$$

everywhere and, from Stokes's theorem, the induced electromotance

$$\oint \boldsymbol{E}_L \cdot d\boldsymbol{l} = 0 \tag{C-18}$$

for any closed path.

The observer on disk $D1$ has the same experience, and if $D1$ is made of conducting material, there are no eddy currents induced in it.

Again the situation is different in $D2$. Let the conductor be a single loop around the rim of the disk, with a voltmeter connected in series. The voltmeter can either be on the disk or in a fixed position in the laboratory and connected to the loop by means of slip rings. The voltmeter is read both by the observer on the disk and by the laboratory observer. The former can calculate what the meter will read by invoking the Faraday law, since he knows $\partial \boldsymbol{B}_{D2}/\partial t$:

$$\oint \boldsymbol{E}_{D2} \cdot d\boldsymbol{l} = S \frac{\partial (\boldsymbol{B}_{D2})_z}{\partial t}, \tag{C-19}$$

$$= B_0 S\omega \sin \omega t. \tag{C-20}$$

He can also calculate the electromotance from Stokes's theorem, since he has measured $\nabla \times \boldsymbol{E}_{D2}$ for every point on the disk:

$$\oint \boldsymbol{E}_{D2} \cdot \boldsymbol{dl} = \int_S (\nabla \times \boldsymbol{E}_{D2}) \cdot \boldsymbol{da}, \tag{C-21}$$

$$= B_0 S \omega \sin \omega t. \tag{C-22}$$

The two results are the same, of course, since

$$\nabla \times \boldsymbol{E}_{D2} = -\frac{\partial \boldsymbol{B}_{D2}}{\partial t}. \tag{C-23}$$

Now let us calculate what the laboratory observer thinks the meter on the disk will read according to the Faraday law. He says that $d\Phi/dt$ differs from zero because the circuit is rotating in a time-independent field, whereas the observer on the disk says that $d\Phi/dt$ differs from zero because his circuit is in a time-varying field. The laboratory man calculates $d\Phi/dt$ through the rotating loop and gets

$$\oint \boldsymbol{E}_{D2} \cdot \boldsymbol{dl} = B_L S \omega \sin \omega t, \tag{C-24}$$

where his B_L is the same as the B_0 measured by the rotating observer.

The laboratory observer can also calculate the electromotance using Equation C-1:

$$\nabla \times \boldsymbol{E}_{D2} = -\frac{\partial \boldsymbol{B}_L}{\partial t} + \nabla \times (\boldsymbol{u} \times \boldsymbol{B}_L). \tag{C-25}$$

He must use this complete expression, since the path around which he calculates the electromotance is moving in his coordinate system. Since

$$\frac{\partial \boldsymbol{B}_L}{\partial t} = 0, \tag{C-26}$$

the equation

$$\nabla \times \boldsymbol{E}_{D2} = \nabla \times (\boldsymbol{u} \times \boldsymbol{B}_L) \tag{C-27}$$

holds everywhere on the disk or, excluding terms of zero curl,

$$\boldsymbol{E}_{D2} = \boldsymbol{u} \times \boldsymbol{B}_L. \tag{C-28}$$

The laboratory observer then says that

$$\oint \boldsymbol{E}_{D2} \cdot \boldsymbol{dl} = \oint (\boldsymbol{u} \times \boldsymbol{B}_L) \cdot \boldsymbol{dl}. \tag{C-29}$$

From Figure C-3, at an arbitrary point on the rim of the disk,

$$u = \omega r \sin \theta, \tag{C-30}$$

$$|\boldsymbol{u} \times \boldsymbol{B}_L| = B_L r \omega \sin \theta \sin \omega t. \tag{C-31}$$

Thus

$$\oint (\boldsymbol{u} \times \boldsymbol{B}_L) \cdot \boldsymbol{dl} = \int_0^{2\pi} B r \omega \sin \theta \sin \omega t \cos\left(\frac{\pi}{2} - \theta\right) r\, d\theta, \tag{C-32}$$

$$= S B \omega \sin \omega t, \tag{C-33}$$

as with the Faraday law.

The laboratory observer and the $D2$ observer therefore agree

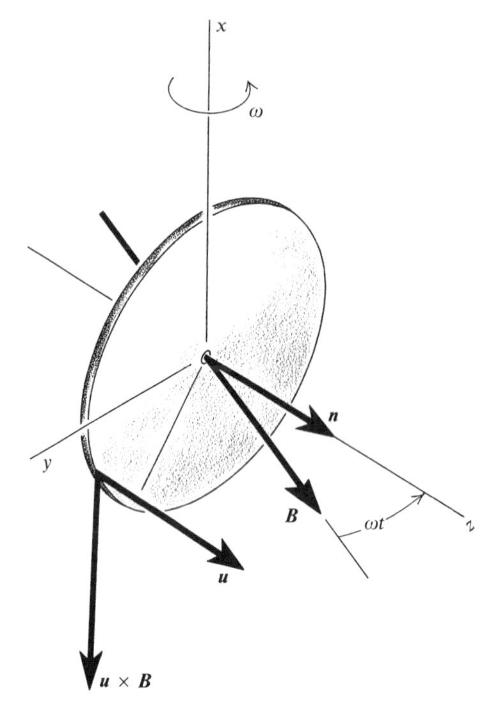

Figure C-3. The vector product $\boldsymbol{u} \times \boldsymbol{B}$ on the rim of a disk rotating about an axis lying on its plane and perpendicular to a uniform magnetic field $\boldsymbol{B}$. The vector $\boldsymbol{u}$ is perpendicular to the plane of the disk.

as to the voltmeter reading, but they disagree as to the reason for the induced electromotance. The laboratory observer says that the magnetic field is static but that the magnetic force ($\boldsymbol{u} \times \boldsymbol{B}$) on the free charges in the moving conductor produces the electromotance. The disk $D2$ observer, on the other hand, says that the conductor is at rest but that it is in a time-dependent magnetic field.

Example

EXPERIMENT 3

As a final example, let us consider the *Faraday disk* or, as it is also called, the *homopolar generator*. Let us return to the disk $D1$. We place a conducting ring around its circumference and use a conducting axle. The axle and the ring are connected by a radial conducting wire attached rigidly to the disk, and the circuit is completed by brushes and a stationary wire with a voltmeter in series, as indicated in Figure C-4.

According to the laboratory observer, $\partial \boldsymbol{B}_L/\partial t$ is everywhere zero, and

$$\oint \boldsymbol{E} \cdot d\boldsymbol{l} = \oint (\boldsymbol{u} \times \boldsymbol{B}_L) \cdot d\boldsymbol{l}. \tag{C-34}$$

The part of the circuit that is stationary in the laboratory has $\boldsymbol{u} = 0$. On the rim and on the axle, $(\boldsymbol{u} \times \boldsymbol{B}_L)$ is everywhere perpendicular

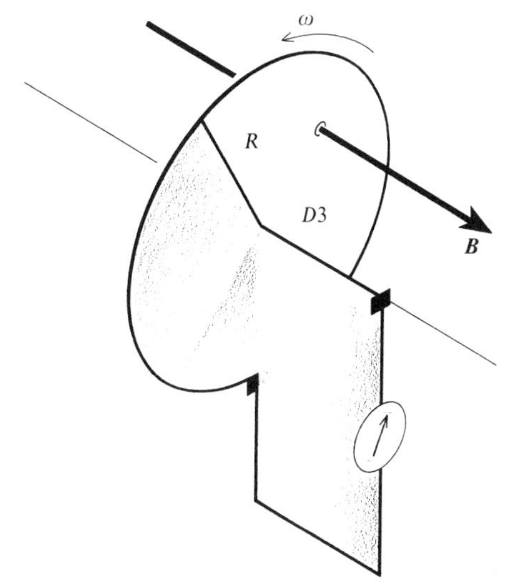

Figure C-4. Homopolar generator.

to the path of integration, thus there is no contribution to the integral. Along the radial conductor, the laboratory observer finds that

$$\oint \boldsymbol{E}\cdot d\boldsymbol{l} = \int_0^R r\omega B_L\, dr, \tag{C-35}$$

$$= \frac{R^2\omega B}{2}. \tag{C-36}$$

With the Faraday law, it is important to specify carefully the surface through which the flux is to be calculated. It can be any surface bounded by the path of integration in the electromotance calculation. For convenience we may choose a surface lying in two planes, as in Figure C-4. The only part of this surface in which the flux differs from zero is the part that lies on the disk. The rate of change of flux through this part of the surface is readily calculated.

If the observer on the disk calculates the electromotance, he finds that $\partial \boldsymbol{B}_{D3}/\partial t$ is also everywhere zero and that the only place where $(\boldsymbol{u} \times \boldsymbol{B}_{D3})$ differs from zero is in the portion of the circuit external to his disk. He sees this part of the circuit rotating with respect to the disk with angular velocity ω. Again, since the only contribution to the electromotance is in the radial parts of the circuit, his calculation also gives

$$\oint \boldsymbol{E}\cdot d\boldsymbol{l} = \frac{R^2\omega B}{2}. \tag{C-37}$$

The laboratory observer says there is an electromotance induced in the circuit because of the magnetic force on the moving charges of the disk, and the disk observer says there is an electromotance because of the magnetic force on the moving charges of the portion of the circuit external to the disk, but they always agree on the voltmeter reading.

If the *whole* disk is made of conducting material, the electromotance is calculated in exactly the same way. It makes no difference what integration path we choose from the axle to the rim, as long as the path is at rest relative to the disk. It is essential that this part of the path be at rest relative to the conductor, since the charges which experience the force resulting in the electromotance are, on the average, at rest with respect to the conductor. The electromotance is independent of the path in the conductor, since only the radial components of the path elements contribute to the integral of $(\boldsymbol{u} \times \boldsymbol{B}) \cdot \boldsymbol{dl}$.

The electromotance may also be calculated for the conducting disk from the Faraday law. Again we may choose any path in the moving conductor, as long as it is at rest relative to it. The result is the same as with the wire discussed above.

Our discussion is strictly valid only for large magnetic fields and small angular velocities, since centrifugal and Coriolis effects have been neglected. For practical laboratory purposes, however, this is not a limitation. The ratio of the magnetic force to the centripetal force on an electron in a disk rotating with angular velocity ω in a magnetic field B is

$$\frac{F_L}{F_C} = \frac{e}{m}\frac{B}{\omega}. \tag{C-38}$$

For an angular velocity of 1800 revolutions/minute and a magnetic field of one tesla, this ratio is about 10^9. This complete separation of electrical and mechanical effects is a consequence of the large value of the ratio e/m for an electron.

APPENDIX D

The Exponential Notation

The subject of this appendix is a mathematical technique for solving what is probably, from the point of view of the physicist, the most important class of differential equations.

The sine and cosine functions play a particularly important role in physics, mostly because of the relative ease with which they can be generated and measured by the ordinary types of instruments. They are also relatively easy to manipulate mathematically. All other periodic functions, such as square waves, for example, are vastly more complicated to use, both experimentally and mathematically.

The mathematical technique which we shall develop here is therefore widely used, despite the fact that it can apply only to functions of the form

$$x = x_0 \cos (\omega t + \theta), \tag{D-1}$$

where x_0 is the *amplitude* of x, ω is the *angular frequency*, and t is the time. The quantity $(\omega t + \theta)$ is the *phase*, or *phase angle*, θ being the phase at $t = 0$. We shall limit our discussion to the cosine function, since the sine can be transformed to a cosine by an appropriate choice of θ. We shall assume that the origin of time is chosen so that $\theta = 0$.

D.1 THE $j\omega$ OPERATOR

The procedure for differentiating $\cos \omega t$, although elementary, is rather inconvenient: the cosine function is changed to a sine, the result is multiplied by ω, and the sign is changed. To find the second derivative, the sine is changed back to a cosine, and the result is again multiplied by ω, but this time without changing sign, and so on.

Differentiation can be simplified if it is kept in mind that

$$e^{j\omega t} = \cos \omega t + j \sin \omega t, \tag{D-2}$$

where $j = \sqrt{-1}$. Then

$$x = x_0 \cos \omega t = \text{Re } x_0 e^{j\omega t}, \tag{D-3}$$

where Re is an operator which means "Real part of" whatever follows.

Let us calculate the first two derivatives:

$$\frac{dx}{dt} = \frac{d}{dt} \text{Re } (x_0 e^{j\omega t}), \tag{D-4}$$

$$= \text{Re } (j\omega x_0 e^{j\omega t}), \tag{D-5}$$

$$= -\omega x_0 \sin \omega t, \tag{D-6}$$

$$\frac{d^2x}{dt^2} = \text{Re } \{(j\omega)^2 x_0 e^{j\omega t}\}, \tag{D-7}$$

$$= -\omega^2 x_0 \cos \omega t. \tag{D-8}$$

The results are, of course, the same as by the usual method.

We now adopt the following convention: *we shall express the quantities x, dx/dt, d^2x/dt^2, and so forth, as exponential functions without writing the Re operator, but with the tacit understanding that only the real part must be used.* Then $x = x_0 \cos \omega t$ will be written as

$$x = x_0 e^{j\omega t}, \tag{D-9}$$

and

$$\frac{dx}{dt} = j\omega x_0 e^{j\omega t} = j\omega x, \tag{D-10}$$

$$\frac{d^2x}{dt^2} = (j\omega)^2 x_0 e^{j\omega t} = (j\omega)^2 x, \tag{D-11}$$

. . . .

With this convention, *the operator d/dt can be replaced by the factor $j\omega$.* This simplification is so useful that the exponential notation is almost invariably used to represent sinusoidally varying quantities, whether they are mechanical, acoustical, or electrical.

The coefficient before the exponential function can itself be complex. For example, $jx_0 e^{j\omega t}$ means

$$\text{Re } (jx_0 e^{j\omega t}) = -x_0 \sin \omega t \tag{D-12}$$

if x_0 is real. Since $j = e^{j\pi/2}$, one can also write that

$$\text{Re } (jx_0 e^{j\omega t}) = \text{Re } (x_0 e^{j(\omega t + \pi/2)}), \tag{D-13}$$

$$= x_0 \cos\left(\omega t + \frac{\pi}{2}\right), \quad \text{(D-14)}$$

$$= -x_0 \sin \omega t. \quad \text{(D-15)}$$

One common case is

$$(a + jb)e^{j\omega t} = \sqrt{a^2 + b^2} \exp j\left(\omega t + \arc\tan \frac{b}{a}\right), \quad \text{(D-16)}$$

where we have written

$$a + jb = \sqrt{a^2 + b^2} \exp j \arc\tan\left(\frac{b}{a}\right) \quad \text{(D-17)}$$

in the usual manner, on the assumption that a is positive, in order that the angle chosen lies either in the first or fourth quadrant. Thus

$$\text{Re}\,(a + jb)e^{j\omega t} = \sqrt{a^2 + b^2} \cos\left(\omega t + \arc\tan \frac{b}{a}\right). \quad \text{(D-18)}$$

Warning No. 1: If a is not positive, one must be careful to use the proper angle in the exponent. For example, the argument of $(-1 + 2j)$ is $(\pi - \arc\tan 2)$, not $\arc\tan(-2)$.

Space-dependent functions can also be represented with the exponential notation. For example, in an electric field, the vector $\boldsymbol{E}$ is oriented in some direction in space. It can also be a function of the time; for example

$$\boldsymbol{E} = \boldsymbol{E}_0 \cos(\omega t + \theta), \quad \text{(D-19)}$$

where $\boldsymbol{E}_0$ is a vector whose magnitude is the maximum value of $\boldsymbol{E}$. We then write

$$\boldsymbol{E} = \boldsymbol{E}_0 e^{j(\omega t + \theta)}, \quad \text{(D-20)}$$

$$= \boldsymbol{E}_0 e^{j\theta} e^{j\omega t}. \quad \text{(D-21)}$$

The coefficient of $\exp j\omega t$ can therefore be, at the same time, a vector and a complex quantity, the former property having to do with the orientation of $\boldsymbol{E}$ in space, whereas the second property is related to the phase $\boldsymbol{E}$ in the time dependence.

The exponential notation is used as follows. The sine or cosine functions are expressed in the form $x_0 e^{j(\omega t + \theta)}$, which is of course equal to $x_0\{\cos(\omega t + \theta) + j \sin(\omega t + \theta)\}$. We are concerned only with the real part, which is the first term; the second term can be considered as parasitic. Then, as long as the mathematical operations on the exponential functions are restricted to additions, subtractions, differentiations, and integrations, their real and imaginary parts do not mix. *The technique is thus useful for solving linear differential equations with constant coefficients.* Once the calculations are completed, the resulting expressions are often left in exponential form. However, if amplitudes and phases are required, the imaginary part is rejected, and the result is expressed again as a cosine function.

Warning No. 2: The exponential technique is valid only for mathematical operations that do not mix the real and the imaginary parts of the exponential functions. For example, for any two complex numbers A and B,

$$\text{Re}\{Ae^{j\omega t} + Be^{j(\omega t+\theta)}\} = \text{Re}\, Ae^{j\omega t} + \text{Re}\, Be^{j(\omega t+\theta)} \qquad \text{(D-22)}$$

but

$$\text{Re}\{Ae^{j\omega t}Be^{j(\omega t+\theta)}\} \neq \{\text{Re}\, Ae^{j\omega t}\}\{\text{Re}\, Be^{j(\omega t+\theta)}\} \qquad \text{(D-23)}$$

Whenever multiplications of $e^{j\omega t}$ terms are involved in the calculations, one must revert to the cosine functions and not use the exponential technique.

Example

SOLVING A LINEAR DIFFERENTIAL EQUATION WITH CONSTANT COEFFICIENTS, USING THE EXPONENTIAL NOTATION

As an illustration, let us consider the following differential equation, where all the terms are real:

$$\alpha \frac{\partial^2 x}{\partial t^2} + \beta \frac{\partial x}{\partial t} + \gamma x = X \cos \omega t. \qquad \text{(D-24)}$$

This could be the differential equation describing the motion of a mass α under the influence of a force $X \cos \omega t$, a restoring force $-\gamma x$ proportional to the displacement x from equilibrium, and a damping force $-\beta\, \partial x/\partial t$ proportional to the velocity $\partial x/\partial t$, as in Figure D-1a: the product of the mass α and the acceleration $\partial^2 x/\partial t^2$ is equal to the sum of the applied forces.

It could also be the differential equation for the charge x on a capacitor whose capacitance is $1/\gamma$ in series with an inductance α and a resistance β, for an applied voltage $X \cos \omega t$, as in Figure

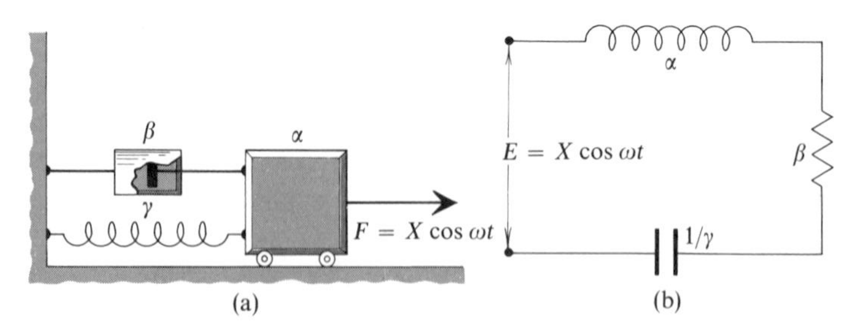

Figure D-1. Examples of oscillating systems.

D-1b: the applied voltage is equal to the sum of the voltages across α, β, and $1/\gamma$.

This equation can be solved, *without* using the exponential notation, as follows. We consider only the steady-state solution and neglect all transient effects obtained by setting the right-hand side equal to zero. There is, on the left, a sum of three terms involving the unknown function x, its first, and its second derivatives, with constant coefficients. Since the derivatives of the sine and cosine functions are themselves sine and cosine functions, it is plausible to try a function of the form

$$x = A\cos\omega t + B\sin\omega t. \tag{D-25}$$

Substituting this expression for x in Eq. D-24, and setting the coefficient of $\cos\omega t$ on the left-hand side equal to X and that of $\sin\omega t$ equal to zero, one can determine the coefficients A and B. The sum $(A\cos\omega t + B\sin\omega t)$ is then put into the form $C\cos(\omega t + \theta)$, and the result is as follows:

$$x = \frac{X}{\sqrt{(\gamma - \alpha\omega^2)^2 + \beta^2\omega^2}}\cos\left(\omega t - \arctan\frac{\beta\omega}{\gamma - \alpha\omega^2}\right). \tag{D-26}$$

We have assumed that $\gamma > \alpha\omega^2$ in writing out the arc tan term; otherwise this expression would not be correct. See Warning No. 1.

The procedure is much simpler with the exponential notation. Substituting $j\omega$ for the operator d/dt in Eq. D-24, we can write directly that

$$x = \frac{Xe^{j\omega t}}{-\alpha\omega^2 + j\omega\beta + \gamma}. \tag{D-27}$$

The amplitude and phase of x can be found easily:

$$x = \frac{X}{\sqrt{(\gamma - \alpha\omega^2)^2 + \beta^2\omega^2}}\,\frac{e^{j\omega t}}{\exp\left(j\arctan\dfrac{\beta\omega}{\gamma - \alpha\omega^2}\right)}, \tag{D-28}$$

$$= \frac{X}{\sqrt{(\gamma - \alpha\omega^2)^2 + \beta^2\omega^2}}\left\{\cos\left(\omega t - \arctan\frac{\beta\omega}{\gamma - \alpha\omega^2}\right) + j\sin\left(\omega t - \arctan\frac{\beta\omega}{\gamma - \alpha\omega^2}\right)\right\}. \tag{D-29}$$

If we reject the imaginary part, we obtain Eq. D-26.

The value of dx/dt corresponding to the velocity in Figure D-1a and to the current in Figure D-1b can be calculated easily:

$$\frac{dx}{dt} = j\omega x = \frac{j\omega Xe^{j\omega t}}{-\alpha\omega^2 + j\omega\beta + \gamma}, \tag{D-30}$$

or, rationalizing and again rejecting the imaginary part,

$$\frac{dx}{dt} = \frac{X}{\sqrt{\left(\dfrac{\gamma}{\omega} - \alpha\omega\right)^2 + \beta^2}}\cos\left\{\omega t + \arctan\frac{(\gamma - \alpha\omega^2)}{\beta\omega}\right\}. \tag{D-31}$$

APPENDIX E

Waves

A wave involves the propagation of a disturbance of some sort in space. One can think, for example, of the waves formed when a stretched string is fixed at one end and moved rapidly in a vertical plane in some arbitrary way at the free end.

If we call $y(t)$ the vertical position of the moving end, then it turns out that the vertical position y at a distance z along the string is $y\{t - (z/u)\}$, assuming no losses and a perfectly flexible string. The vertical position at z is thus given by the position at the moving end at a previous time $t - (z/u)$, the quantity z/u being the time required for the disturbance to travel through the distance z at the velocity u.

More generally, we can consider waves propagating in an extended region, such as acoustic waves in air or light waves in space. The quantity propagated can be either a scalar or a vector quantity. For example, in an acoustic wave, we can consider the propagation of pressure, which is a scalar. In an electromagnetic wave, we can consider the propagation of the electric field intensity vector $\boldsymbol{E}$ or of its components E_x, E_y, E_z.

E.1 PLANE SINUSOIDAL WAVES

If a certain quantity α propagating with a velocity u is given at $z = 0$ by

$$\alpha = \alpha_0 \cos \omega t, \tag{E-1}$$

then, for any position z in the direction of propagation of the plane wave,

$$\alpha = \alpha_0 \cos \omega \left(t - \frac{z}{u}\right). \tag{E-2}$$

This expression describes an *unattenuated plane sinusoidal wave*, since the amplitude α_0 is constant, and since α depends on z but not on x or on y. The *wave fronts*, which are surfaces of constant phase at a given time, are thus perpendicular to the z-axis. The quantity α_0 is called the *amplitude* of the wave. For a given position z, we have a sinusoidal variation of α with time. For a given time t, we also have a sinusoidal variation with z, as in Figure E-1.

The *phase angle* shown between brackets is a constant, that is

$$t - \frac{z}{u} = \text{Constant}, \tag{E-3}$$

for a point traveling with the velocity

$$\frac{dz}{dt} = u. \tag{E-4}$$

The quantity u is called the *phase velocity* of the wave, since it is the velocity with which the phase $\omega \left(t - \frac{z}{u}\right)$ is propagated in space.

We often write

$$\alpha = \alpha_0 \cos (\omega t - kz), \tag{E-5}$$

where

$$k = \frac{\omega}{u} \tag{E-6}$$

is called the *wave number*. It is important to note that *this wave number is 2π times that used in optics*. For this reason, the quantity k is also called the *circular wave number*.

The *wave length* λ is the distance over which kz changes by 2π radians,

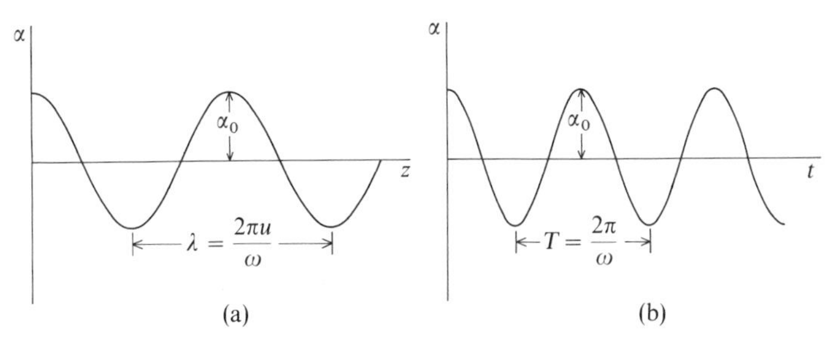

Figure E-1. The quantity $\alpha = \alpha_0 \cos (\omega t - kz)$ as a function of z and as a function of t.

as shown in Figure E-1a:

$$k\lambda = 2\pi. \tag{E-7}$$

If we write

$$\frac{\lambda}{2\pi} = \bar{\lambda}, \tag{E-8}$$

where $\bar{\lambda}$ is read as "lambda bar,"

$$k = \frac{1}{\bar{\lambda}}. \tag{E-9}$$

The quantity $\bar{\lambda}$ is called the *radian length*. It is the distance over which the phase of the wave changes by one radian; this is about $\lambda/6$, and is therefore considerably less than one-quarter wavelength. It turns out that the quantity that appears in nearly all the calculations is $\bar{\lambda}$, and not λ. We are already familiar with the fact that it is preferable to use the circular frequency ω instead of the frequency f. In both cases the intuitively simple quantity, namely, the wavelength or the frequency, is not the one that is "natural" from a mathematical standpoint.

With the exponential notation, the wave traveling in the positive direction along the z-axis is written*

$$\alpha = \alpha_0 \exp j(\omega t - kz). \tag{E-10}$$

Similarly, a wave traveling in the negative direction along the z-axis is described by

$$\alpha = \alpha_0 \exp j(\omega t + kz). \tag{E-11}$$

We shall have occasion to use plane sinusoidal waves traveling in some given direction specified by a unit vector $\boldsymbol{n}$. The wave fronts are then normal to $\boldsymbol{n}$, and such a wave is given by

$$\alpha = \alpha_0 \exp j\{\omega t - k(\boldsymbol{n}\cdot\boldsymbol{r})\}, \tag{E-12}$$

as can be seen in Figure E-2. When $\boldsymbol{n}$ coincides with the unit vector in the z direction, $\boldsymbol{n}\cdot\boldsymbol{r} = z$.

If a plane wave traveling along the z-axis is attenuated, its amplitude decreases exponentially, and

$$\alpha = \alpha_0 \exp \{j(\omega t - k_r z) - k_i z\}. \tag{E-13}$$

---

* We shall use the notation of Eq. E-10. Notice, however, that one could equally well write

$$\alpha = \alpha_0 \exp \{-j(\omega t - kz)\},$$

since the cosine is an even function. The latter notation is frequently used where it is convenient to omit the factor $e^{-j\omega t}$ for brevity. A wave traveling in the positive direction is then simply written

$$\alpha = \alpha_0' \exp jkz.$$

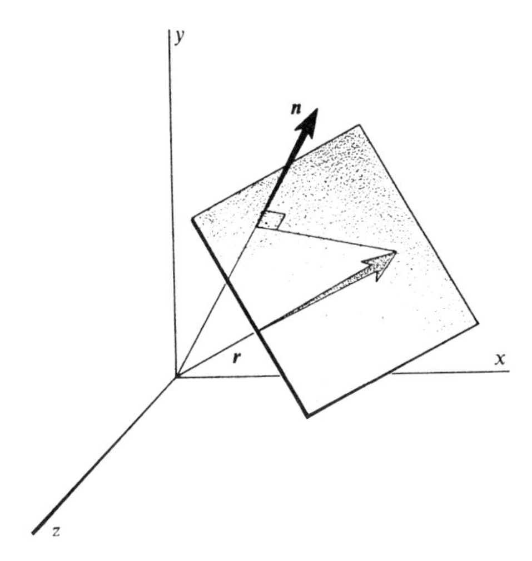

Figure E-2. The plane shown is defined by $\boldsymbol{n} \cdot \boldsymbol{r}$ = constant, the constant being the distance between the plane and the origin.

It is now k_r that is related to the wave length:

$$k_r = \frac{1}{\lambda}, \tag{E-14}$$

whereas the quantity k_i, which is called the *attenuation constant*, is such that the wave is reduced in amplitude by a factor of e in a distance

$$\delta = \frac{1}{k_i}. \tag{E-15}$$

called the *attenuation distance*. The phase velocity u of the wave is

$$u = \frac{\omega}{k_r}. \tag{E-16}$$

We can also rewrite Eq. E-13 in the form

$$\alpha = \alpha_0 \exp j(\omega t - kz), \tag{E-17}$$

where

$$k = k_r - jk_i. \tag{E-18}$$

It is important to note the *negative sign* in the above expression.

The quantity k is still called the wave number in this general case. However, it is complex, its *imaginary* part corresponding to absorption. It will be observed that an attenuated wave traveling in the positive direction along the z-axis requires that the real part of k be a positive quantity and that the imaginary part be a negative quantity. Otherwise, the wave would grow exponentially in amplitude with increasing z. The quantities k_r and k_i in Eq. E-18 are thus both positive quantities. In transmission line theory the quantity jk is written γ, and is called the *propagation constant*.

An attenuated wave traveling in the negative direction along the z-axis is given similarly by

$$\alpha = \alpha_0 \exp \{j(\omega t + k_r z) + k_i z\}, \quad \text{(E-19)}$$

$$= \alpha_0 \exp j(\omega t + kz), \quad \text{(E-20)}$$

where k_r and k_i are again positive quantities and where k is defined as above.

We shall meet with cases where $k_r = 0$. Then Eq. E-13 becomes

$$\alpha = \alpha_0 \exp(-k_i z) \exp j\omega t. \quad \text{(E-21)}$$

The phase angle ωt is then independent of z, all points are in phase, there is no traveling wave, and the amplitude decreases exponentially with z.

E.2 WAVES ON A STRETCHED STRING. THE DIFFERENTIAL EQUATION FOR AN UNATTENUATED WAVE

It is interesting to consider at this point the simple case of transverse waves propagating along a stretched string. We assume small transverse displacements on a flexible string of mass ρ per unit length stretched with a tension F, as in Figure E-3.

Figure E-3 shows an element of the string at some given time t. Both its displacement y and its angle θ are functions of both the position z along the string and the time t. We assume that there is no motion along the z-axis, in order that the stretching force F be constant all along the string.

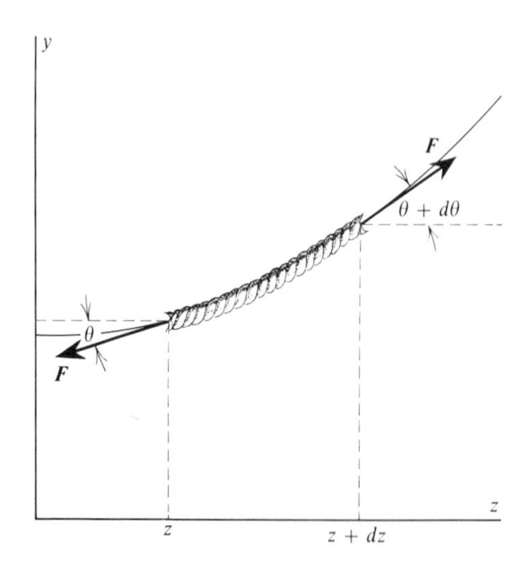

Figure E-3. Element of string stretched with a force $\boldsymbol{F}$. The angles and the displacement from the z-axis are grossly exaggerated for clarity.

The element of mass $\rho\, dz$ takes on an acceleration $\frac{\partial^2 y}{\partial t^2}$ under the action of the forces F at either end, and

$$\rho\, dz \frac{\partial^2 y}{\partial t^2} = F\{\sin(\theta + d\theta) - \sin\theta\}, \tag{E-22}$$

$$= F\, d(\sin\theta), \tag{E-23}$$

$$= F \cos\theta\, d\theta. \tag{E-24}$$

Then

$$\cos\theta \frac{\partial\theta}{\partial z} = \frac{\rho}{F}\frac{\partial^2 y}{\partial t^2}. \tag{E-25}$$

We have written a partial derivative for θ, since the $d\theta$ found above was for a given time t. For waves of small amplitude, we can set $\cos\theta = 1$ and $\theta = \partial y/\partial z$. Then

$$\frac{\partial^2 y}{\partial z^2} = \frac{\rho}{F}\frac{\partial^2 y}{\partial t^2}. \tag{E-26}$$

We can verify that the above differential equation does correspond to a wave motion by substituting for y *any* function of $\{t - (z/u)\}$ or of $\{t + (z/u)\}$. We assume, of course, that the second partial derivatives exist. We find that the phase velocity u is

$$u = \left(\frac{F}{\rho}\right)^{1/2}. \tag{E-27}$$

This result is general, and any *un*attenuated wave traveling with a velocity u along the z-axis is described by the following differential equation:

$$\frac{\partial^2 \alpha}{\partial z^2} = \frac{1}{u^2}\frac{\partial^2 \alpha}{\partial t^2}. \tag{E-28}$$

The differential equation for an attenuated wave will be discussed in Sections E.5 and E.6.

It will be observed that the above differential equation (E-28) is much more general than Eq. E-10. It does not involve the amplitude α_0, the angular frequency ω, or the wave number k, but only the velocity u. The differential equation therefore applies equally well to any wave form, periodic or not, and to any amplitude. Equation E-28 is also independent of the sign of u, since it involves only u^2. It can therefore represent waves traveling in either direction along the z-axis.

For a sinusoidal wave,

$$\frac{\partial^2 \alpha}{\partial z^2} = -\frac{\omega^2}{u^2}\alpha = -k^2\alpha, \tag{E-29}$$

$$\frac{\partial^2 \alpha}{\partial z^2} + k^2\alpha = 0. \tag{E-30}$$

E.3 SOLUTION OF THE DIFFERENTIAL EQUATION FOR AN UNATTENUATED WAVE BY THE SEPARATION OF VARIABLES

Equation E-28 can be solved formally by the method of separation of variables (Section 4.4). We set

$$\alpha = T(t)Z(z), \tag{E-31}$$

where T and Z are respectively functions of t and z only. Substituting this value of α in Eq. E-28 and dividing by TZ,

$$\frac{1}{Z}\frac{d^2Z}{dz^2} = \frac{1}{u^2T}\frac{d^2T}{dt^2}. \tag{E-32}$$

The term on the left is a function only of z, whereas that on the right is a function only of t. Then both sides of the equation can be equated to a separation constant $-k^2$:

$$\frac{1}{Z}\frac{d^2Z}{dz^2} = -k^2, \tag{E-33}$$

$$\frac{1}{T}\frac{d^2T}{dt^2} = -k^2u^2, \tag{E-34}$$

and

$$T = A\exp(jukt) + B\exp(-jukt), \tag{E-35}$$

$$Z = C\exp(jkz) + D\exp(-jkz). \tag{E-36}$$

Then

$$\alpha = AC\exp\{jk(ut + z)\} + BD\exp\{-jk(ut + z)\} + AD\exp\{jk(ut - z)\} + BC\exp\{-jk(ut - z)\}. \tag{E-37}$$

Now α is some physical quantity that must not increase or decrease exponentially with either z or t. This condition requires that k be a real number. The exponentials then reduce to sine and cosine functions.

Since the coefficients AC, BD, and so on, are as yet undetermined constants, we can set AC to be some complex number

$$AC = Ge^{j\varphi}, \tag{E-38}$$

where G and φ are real. Also, α must be a real quantity, and we must have

$$BD = Ge^{-j\varphi} \tag{E-39}$$

in order that the first two terms of Eq. E-37 can add up to give a cosine function.

Similarly, we can set

$$AD = Fe^{j\theta} \tag{E-40}$$

and

$$BC = Fe^{-j\theta}. \tag{E-41}$$

Finally, α can be written as follows:

$$\alpha = 2F\cos(kut - kz + \theta) + 2G\cos(kut + kz + \varphi), \tag{E-42}$$

where the angles θ and φ are constant. Comparing this with Eq. E-5, we find that the above equation determines a pair of waves traveling in opposite directions with a common velocity u. The separation constant k is the wave number, and

$$ku = \omega = 2\pi f. \tag{E-43}$$

Since α is a function only of z and of t, the waves are plane, with wave fronts parallel to the xy-plane.

It will be observed that the formal solution that we arrived at by separating the variables z and t has led us to a very special class of waves; namely, sine waves. We did not find a general function of $\{t - (z/u)\}$. This is quite disturbing at first sight, since our formal solution is presumably general. There is no contradiction, however, for the following reason. Since our differential equation is linear, that is, since its terms are all of the first degree in α or its derivatives, the sum of any number of solutions is also a solution. Any type of periodic wave form encountered in practice can be expressed as a Fourier series of sines and cosines of the fundamental frequency and of its harmonics (Section 4.4). Even individual nonperiodic pulses can be analyzed in a somewhat similar manner by means of Fourier integrals. Any wave form can thus be synthesized by combining terms of the form shown in Eq. E-42 with appropriate amplitudes and wave numbers.

E.4 REFLECTION OF A WAVE ON A STRETCHED STRING AT A POINT WHERE THE DENSITY CHANGES FROM ρ_1 TO ρ_2

If two strings of different densities ρ_1 and ρ_2 are tied together and stretched with a force F, as in Figure E-4, a wave traveling along the first section will

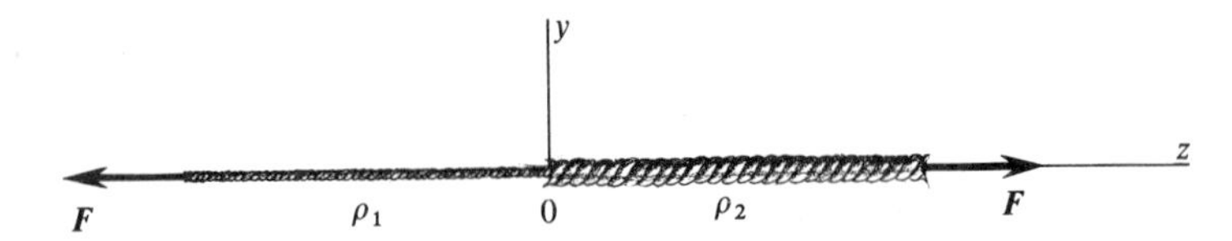

Figure E-4. Two strings of densities ρ_1 and ρ_2 fixed together at O and stretched with a force $\boldsymbol{F}$.

be partly reflected and partly transmitted at the knot. The phase velocities are respectively

$$u_1 = \left(\frac{F}{\rho_1}\right)^{1/2}, \qquad u_2 = \left(\frac{F}{\rho_2}\right)^{1/2}, \tag{E-44}$$

and the corresponding wave numbers are

$$k_1 = \frac{\omega}{u_1} = \omega\left(\frac{\rho_1}{F}\right)^{1/2}, \qquad k_2 = \frac{\omega}{u_2} = \omega\left(\frac{\rho_2}{F}\right)^{1/2}. \tag{E-45}$$

Let us assume that a wave travels to the right along string 1. We shall call this the incident wave and set

$$y_i = y_{0i} \exp j(\omega t - k_1 z), \tag{E-46}$$

y being the lateral displacement of the string and $z = 0$ being chosen at the knot. For the wave transmitted to string 2,

$$y_t = y_{0t} \exp j(\omega t - k_2 z). \tag{E-47}$$

Finally, for the wave reflected back at the knot,

$$y_r = y_{0r} \exp j(\omega t + k_1 z). \tag{E-48}$$

We assume, for simplicity, that there are no reflected waves originating at the supports.

In the above three expressions for y_i, y_t, y_r, the amplitude y_{0i} of the incident wave can be assumed to be known. Thus we have two quantities to determine: y_{0t} and y_{0r}.

It is possible to calculate the values of y_{0t} and y_{0r} in terms of y_{0i}, ρ_1, ρ_2 by considering the conditions of continuity which must be satisfied at the knot. First, there must, of course, be continuity of the displacement y: the value of y just to the left of the knot must be equal to its value just to the right. Then

$$y_i + y_r = y_t \qquad \text{at } z = 0,$$

or

$$y_{0i} + y_{0r} = y_{0t}. \tag{E-49}$$

Second, there must be continuity of the slope of the string dy/dz. The reason for this is as follows. We have assumed implicitly that the knot was weightless. Thus the sum of the forces acting on it must be zero, and the two opposing tension forces F at that point must be along the same line. Then

$$-k_1 y_{0i} + k_1 y_{0r} = -k_2 y_{0t}. \tag{E-50}$$

Solving these equations,

$$\frac{y_{0r}}{y_{0i}} = \frac{k_1 - k_2}{k_1 + k_2} = \frac{\rho_1^{1/2} - \rho_2^{1/2}}{\rho_1^{1/2} + \rho_2^{1/2}}, \tag{E-51}$$

$$\frac{y_{0t}}{y_{0i}} = \frac{2k_1}{k_1 + k_2} = \frac{2\rho_1^{1/2}}{\rho_1^{1/2} + \rho_2^{1/2}}. \tag{E-52}$$

Since the ratio y_{0t}/y_{0i} is always real and positive, the transmitted wave is always in phase with the incident wave. On the other hand, the ratio y_{0r}/y_{0i} can be either positive or negative. The reflected wave is in phase with the incident wave if $\rho_1 > \rho_2$, and π radians out of phase if $\rho_1 < \rho_2$. If $\rho_1 = \rho_2$ there is no discontinuity, no reflected wave, and $y_{0t} = y_{0i}$.

E.5 WAVES ON A STRETCHED STRING WITH DAMPING. THE DIFFERENTIAL EQUATION FOR AN ATTENUATED WAVE

Let us return to the case of the stretched string of uniform density ρ. We assume now that the string is in a viscous medium that provides a damping force that is proportional to the velocity. The damping force on the element of string dz is then

$$dF_D = -b\, dz \frac{\partial y}{\partial t}, \tag{E-53}$$

and, from Eq. E-24, we now have

$$\rho\, dz \frac{\partial^2 y}{\partial t^2} = F \cos\theta\, d\theta - b\, dz \frac{\partial y}{\partial t}, \tag{E-54}$$

or

$$\frac{\partial^2 y}{\partial z^2} = \frac{\rho}{F}\frac{\partial^2 y}{\partial t^2} + \frac{b}{F}\frac{\partial y}{\partial t}. \tag{E-55}$$

We can show that this is the differential equation for an attenuated wave by trying a solution of the form

$$y = y_0 \exp j(\omega t - k'z). \tag{E-56}$$

Substituting, we find that

$$k'^2 = \frac{\omega^2 \rho}{F}\left(1 - j\frac{b}{\omega\rho}\right), \tag{E-57}$$

and the wave number k' is now obviously complex. The above differential equation is therefore that of an attenuated wave. This is to be expected, since it is identical to Eq. E-26, except for the addition of the second term on the right-hand side. This term corresponds to a damping force which dissipates energy.

Equation E-55 can also be rewritten as

$$\frac{\partial^2 y}{\partial z^2} = -\frac{\omega^2 \rho}{F}\left(1 - j\frac{b}{\omega\rho}\right) y = -k'^2 y, \tag{E-58}$$

or

$$\left(\frac{\partial^2}{\partial z^2} + k'^2\right) y = 0. \tag{E-59}$$

This equation is similar to Eq. E-30, except that the wave number k' is now complex. It represents a pair of attenuated waves traveling in opposite directions along the z-axis.

It will be observed from Eq. E-57 that the differential equation is equally well satisfied by $+k'$ or by $-k'$, that is, by waves traveling in either direction along the z-axis.

We can find both k_r and k_i by recalling Eq. E-18 and substituting

$$k' = k'_r - jk'_i \tag{E-60}$$

in Eq. E-57, which gives

$$k'_r = \pm\omega\left(\frac{\rho}{2F}\right)^{1/2}\left\{1 \pm \left(1 + \frac{b^2}{\omega^2\rho^2}\right)^{1/2}\right\}^{1/2}, \tag{E-61}$$

$$k'_i = \pm\omega\left(\frac{\rho}{2F}\right)^{1/2}\left\{-1 \pm \left(1 + \frac{b^2}{\omega^2\rho^2}\right)^{1/2}\right\}^{1/2}. \tag{E-62}$$

Since, by definition, k_r and k_i are both positive and real (Section E.1), the $\pm$ signs must all be replaced by $+$ signs, and

$$k_r' = \omega \left(\frac{\rho}{2F}\right)^{1/2} \left\{ 1 + \left(1 + \frac{b^2}{\omega^2 \rho^2}\right)^{1/2} \right\}^{1/2}, \tag{E-63}$$

$$k_i' = \omega \left(\frac{\rho}{2F}\right)^{1/2} \left\{ -1 + \left(1 + \frac{b^2}{\omega^2 \rho^2}\right)^{1/2} \right\}^{1/2}. \tag{E-64}$$

It will also be recalled from Section E.1 that the velocity of the wave is ω divided by the real part of the propagation constant. This reduces to $(F/\rho)^{1/2}$ when $b = 0$, as expected.

E.6 SOLUTION OF THE DIFFERENTIAL EQUATION FOR AN ATTENUATED WAVE BY THE SEPARATION OF VARIABLES

The differential equation E-55 that we found above is general, and any attenuated wave traveling along the z-axis is described by

$$\frac{\partial^2 \alpha'}{\partial z^2} = g \frac{\partial^2 \alpha'}{\partial t^2} + h \frac{\partial \alpha'}{\partial t}. \tag{E-65}$$

We can solve this equation formally by separating the variables, as in Section E.3. We set

$$\alpha' = T'(t) Z'(z), \tag{E-66}$$

substitute in Eq. E-65, and divide by $T'Z'$. Then

$$\frac{1}{Z'} \frac{d^2 Z'}{dz^2} = \frac{g}{T'} \frac{d^2 T'}{dt^2} + \frac{h}{T'} \frac{dT'}{dt}, \tag{E-67}$$

where the left-hand side is a function only of z, whereas the right-hand side is a function only of t. Then

$$\frac{1}{Z'} \frac{d^2 Z'}{dz^2} = -k'^2, \tag{E-68}$$

$$\frac{g}{T'} \frac{d^2 T'}{dt^2} + \frac{h}{T'} \frac{dT'}{dt} = -k'^2. \tag{E-69}$$

From the first of these equations,

$$Z' = C' \exp(jk'z) + D' \exp(-jk'z), \tag{E-70}$$

and k' is the wave number.

The second equation takes on a more familiar form when rewritten as

$$g\frac{d^2T'}{dt^2} + h\frac{dT'}{dt} + k'^2T' = 0. \tag{E-71}$$

We can assume a sinusoidal wave without losing generality, as was noted in Section E.3, and set

$$T' = A'e^{j\omega t}. \tag{E-72}$$

The quantity ω must be real in order that the amplitude of the wave can decrease only with z. Substituting in Eq. E-71, we find that

$$k'^2 = \omega^2 g\left(1 - j\frac{h}{\omega g}\right). \tag{E-73}$$

This result is identical to that which we found in Eq. E-57, the coefficients g and h of the general differential equation for an attenuated wave (Eq. E-65) being equal, respectively, to ρ/F and to b/F in the differential equation for an attenuated wave on a stretched string (Eq. E-55).

If we again set

$$k' = k'_r - jk'_i, \tag{E-74}$$

we find that

$$k'_r = \omega\left(\frac{g}{2}\right)^{1/2}\left\{1 + \left(1 + \frac{h^2}{\omega^2 g^2}\right)^{1/2}\right\}^{1/2}, \tag{E-75}$$

$$k'_i = \omega\left(\frac{g}{2}\right)^{1/2}\left\{-1 + \left(1 + \frac{h^2}{\omega^2 g^2}\right)^{1/2}\right\}^{1/2}, \tag{E-76}$$

as in Eqs. E-61 and E-62, since $g = \rho/F$, and $h = b/F$.

Finally, from Eqs. E-66, E-70, and E-72,

$$\alpha' = A'C' \exp j(\omega t + k'z) + A'D' \exp j(\omega t - k'z), \tag{E-77}$$

where $k' = k'_r - jk'_i$. The first term represents a plane wave traveling in the negative direction along the z-axis, whereas the second term represents a similar wave traveling in the positive direction.

In this general case, the wave velocity u is ω/k_r, which reduces to $(1/g)^{1/2}$ when $h = 0$. The wave is attenuated by a factor of e in a distance $\delta = 1/k_i$, which approaches infinity as h approaches zero.

These results are the same as those of the preceding section.

E.7 WAVE PROPAGATION IN THREE DIMENSIONS

In the case of an attenuated wave propagating in space, the wave equation is similar to Eq. E-65, except that the second derivative with respect to z is replaced by the Laplacian

$$\nabla^2\alpha = g\frac{\partial^2\alpha}{\partial t^2} + h\frac{\partial\alpha}{\partial t}. \qquad \text{(E-78)}$$

The coefficients g and h again determine the wave number as in Eqs. E-75 and E-76. For sinusoidal waves,

$$\nabla^2\alpha = (-g\omega^2 + j\omega h)\alpha, \qquad \text{(E-79)}$$

or

$$\nabla^2\alpha + k^2\alpha = 0, \qquad \text{(E-80)}$$

where k is the wave number. We have omitted the primes that were used previously to identify the attenuated wave.

If there is no attenuation, $h = 0$, $g = 1/u^2$, and

$$\nabla^2\alpha = \frac{1}{u^2}\frac{\partial^2\alpha}{\partial t^2} \qquad \text{(E-81)}$$

for any waveform, u being the phase velocity.

E.8 WAVE PROPAGATION OF A VECTOR QUANTITY

As yet, we have only considered waves in which the quantity which is propagated is a scalar. Vector quantities, such as an electric field intensity $\boldsymbol{E}$, for example, can also propagate as a wave, and then, for a plane wave propagating in the positive direction of the z-axis,

$$\boldsymbol{E} = \boldsymbol{E}_0 \exp j(\omega t - kz). \qquad \text{(E-82)}$$

Since $\boldsymbol{E}_0$ is a vector, we may write that

$$\boldsymbol{E} = (E_{0x}\boldsymbol{i} + E_{0y}\boldsymbol{j} + E_{0z}\boldsymbol{k}) \exp j(\omega t - kz), \qquad \text{(E-83)}$$

where E_{0x}, E_{0y}, and E_{0z} are the components of $\boldsymbol{E}_0$. These components may conceivably depend on x and on y, but they do *not* depend on z, because a plane sinusoidal wave is characterized by the dependence on t and on z which is shown in the exponential function.

E.9 THE NONHOMOGENEOUS WAVE EQUATIONS

In the absence of attenuation, the scalar nonhomogeneous wave equation is of the form

$$\nabla^2\alpha - g\frac{\partial^2\alpha}{\partial t^2} = f(x', y', z', t), \qquad \text{(E-84)}$$

where the function f describes a disturbance that causes a wave to propagate in space. Thus, points where f is not zero are inside the source. The variable α is measured at a field point x, y, z, while a point inside the source is identified by the coordinates x', y', z'.

Outside the source we have the homogeneous wave equation

$$\nabla^2\alpha - g\frac{\partial^2\alpha}{\partial t^2} = 0. \tag{E-85}$$

Although a thorough mathematical discussion of the nonhomogeneous wave equation can be rather elaborate, its solution becomes intuitively quite obvious if we proceed as follows.

In the special case where f is not a function of the time,

$$\nabla^2\alpha = f(x', y', z'), \tag{E-86}$$

and α is some function of x, y, z only. Now this is Poisson's equation, Eq. 2-31, which we use in electrostatics, if α is the potential V and f is *minus* the charge density divided by ϵ_0. The solution of this equation is well known. It is

$$\alpha = -\frac{1}{4\pi}\int_\infty \frac{f(x', y', z')}{r}\,d\tau' \tag{E-87}$$

as in Eq. 2-19, where

$$r = \{(x - x')^2 + (y - y')^2 + (z - z')^2\}^{1/2} \tag{E-88}$$

is the distance between the source point x', y', z' and the field point x, y, z.

If now f is a function, not only of x', y', z' but also of t, we must take into account the time required for the disturbance to travel from x', y', z' to x, y, z at the velocity u, as explained in Section 10.2.1, and thus

$$\alpha = -\frac{1}{4\pi}\int_\infty \frac{f\left(x', y', z', t - \frac{r}{u}\right)}{r}\,d\tau'. \tag{E-89}$$

This is the solution of Eq. E-84.

The solution of the vector nonhomogeneous wave equation

$$\nabla^2\mathbf{E} - g\frac{\partial^2\mathbf{E}}{\partial t^2} = f(x', y', z', t)$$

is, similarly,

$$\mathbf{E} = -\frac{1}{4\pi}\int_\infty \frac{f\left(x', y', z', t - \frac{r}{u}\right)}{r}\,d\tau'. \tag{E-90}$$

ANSWERS

In many cases the answer is included in the statement of the problem; the list below contains one half of the remaining answers.

CHAPTER 1

1-10. $\boldsymbol{r} = 433t\boldsymbol{i} + (250t - 4.90t^2)\boldsymbol{j}$,
$\boldsymbol{v} = 433\boldsymbol{i} + (250 - 9.80t)\boldsymbol{j}$,
$\boldsymbol{a} = -9.80\boldsymbol{j}$.

1-16. Correct

1-26. $|\boldsymbol{\nabla} \times \boldsymbol{F}| = 1.4\, Ar^{-0.6}$.

CHAPTER 2

2-2. (a) 5.6×10^{-11} volt/meter.
(b) 5.1 meters.

2-4. There is zero field at $(-5.83a, 0, 0)$.

2-8. (a) 29.6 femtometers (1 femtometer $= 10^{-15}$ meter).
(b) 41.5 newtons.
(c) 6.24×10^{26} g.

2-11. $V = -\dfrac{\sigma}{2\epsilon_0} z + V_0$, where V_0 is an arbitrary constant.

2-13. (a) 2.32×10^{-7} coulomb/meter^3.
(b) Inside, $E = 1.31 \times 10^4\, r$ volts/meter.
Outside, $E = 1.31 \times 10^{-2}/r$ volt/meter.

(c,d)

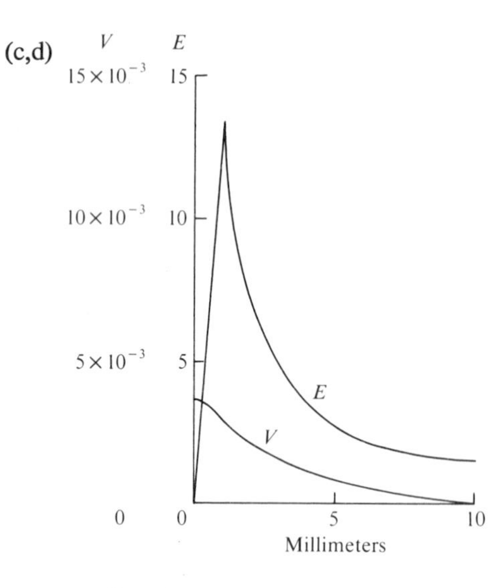

Figure A-1. Solution to Problem 2-13c,d.

Inside, $V = -6.6 \times 10^3(r^2 - 10^6) + 3.02 \times 10^{-2}$ volt.
Outside, $V = -(6.04 + 1.31 \ln r)10^{-2}$ volt.
See above figure.

2-19. (a) $\mathbf{F} = -\dfrac{e^2}{4\pi\epsilon_0 a^3}\mathbf{r}$.

(b) Simple harmonic motion.
(c) $f \approx 10^{15}$ hertz.
(d) Same order of magnitude. See Figure 11-1.

2-21. (a) $Q = (8/15)\pi\rho_0 a^3$.

(b) $E = \dfrac{2\rho_0 a^3}{15\epsilon_0 r^2}$.

$V = \dfrac{2\rho_0 a^3}{15\epsilon_0 r}$.

(c) $E = \dfrac{\rho_0 r}{\epsilon_0}\left(\dfrac{1}{3} - \dfrac{r^2}{5a^2}\right)$.

$V = \dfrac{\rho_0}{\epsilon_0}\left(\dfrac{a^2}{4} - \dfrac{r^2}{6} + \dfrac{r^4}{20a^2}\right)$.

(e)

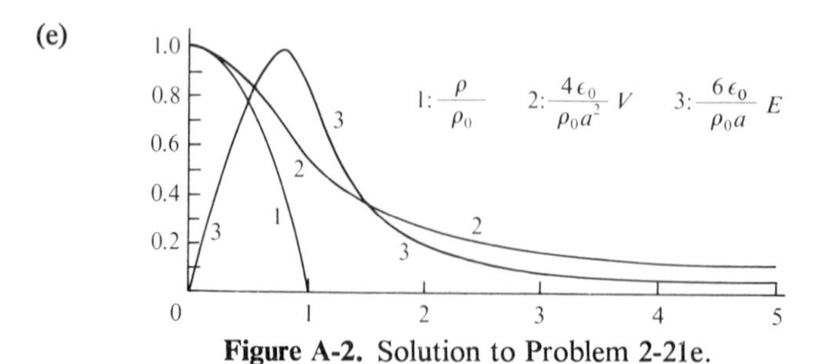

Figure A-2. Solution to Problem 2-21e.

2-24. $(4/3)\pi a^3\sigma_0$.

2-27. (a) $V = \dfrac{Q}{4\pi\epsilon_0 a} \ln \left\{ \dfrac{\left[r^2 - az + \dfrac{a^2}{4}\right]^{1/2} + \dfrac{a}{2} - z}{\left[r^2 + az + \dfrac{a^2}{4}\right]^{1/2} - \dfrac{a}{2} - z} \right\}.$

(b) $V_1 = \dfrac{Q}{4\pi\epsilon_0 r}.$

$V_2 = 0.$

$V_3 = \dfrac{Qa^2}{96\pi\epsilon_0 r^3}(3n^2 - 1).$

(c) Since $3n^2 - 1$ is of the order of unity,

$\dfrac{V_3}{V_1} \leqslant \dfrac{1}{100}$ for $r \geqslant 2a$.

2-29. $\dfrac{2\pi\epsilon_0}{\ln (b/a)}.$

2-31. (a) $\dfrac{(C_1Q_2 - C_2Q_1)^2}{2C_1C_2(C_1 + C_2)}.$

(b) The energy is dissipated as heat in the wires connecting the capacitors.

2-34. (a) $\dfrac{3e^2}{20\pi\epsilon_0 mc^2}.$ (b) $\dfrac{e^2}{8\pi\epsilon_0 mc^2}.$

2-36. (a) 150 kilovolts. (b) About 4×10^{-4} atmosphere.

2-39. (a) $\dfrac{\pi\epsilon_0}{\ln (b/a)} V^2.$ (b) The force is due to the fringing field.

2-42. (a) 20 centimeters. (b) 44 centimeters.

CHAPTER 3

3-1. (a) $p = 5.7 \times 10^{-37}$ coulomb-meter.
(b) $s = 5.9 \times 10^{-19}$ meter.

3-6.

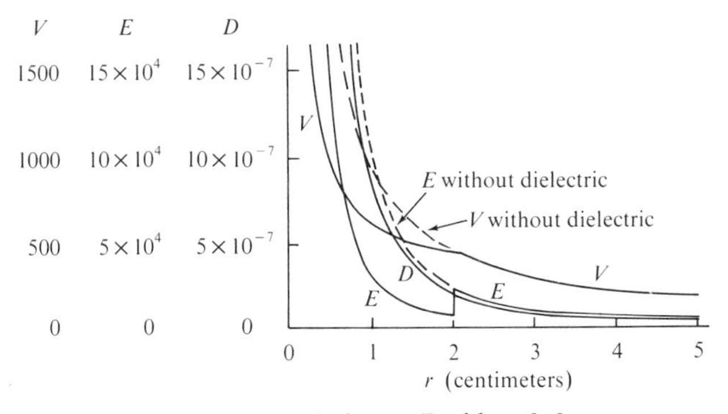

Figure A-3. Solution to Problem 3-6.

3-10. (a) $\epsilon_r = K/\rho$.
(b) $\rho_b = \lambda/2\pi K\rho$, λ being the linear charge density on the inner conductor.

3-13. (a) 1.33×10^{-5} coulomb/meter^2.

(b) $\epsilon AR \dfrac{d\sigma_e}{dt} = [t_d + \epsilon_r(t_{ao} + a \cos \omega t)]\sigma_e - t_d\sigma_d$, where A is the area of the film, t_d its thickness, and σ_d its charge density; σ_e is the charge density on the electrodes; and $t_{ao} + a \cos \omega t$ is the thickness of the air gap. Note that $AR(d\sigma_e/dt)$ is the output voltage.

(c) 1.53×10^{-3} coulomb/meter^2. (d) 6.1 volts.

3-15. 1.24×10^{-4} coulomb/meter^2.

3-17. (b) Assuming a 1 HP motor operating for one hour at 100% efficiency, the capacitor would have a volume of the order of 10 meters^3 and would weigh about 10 tons.

3-19. (a) $3.20 \times 10^{-3}/\rho^3$ newtons/meter^3.

3-22. (a) $\dfrac{\epsilon_r}{2s} CV^2$. (b) 11 tons. (c) 3.7 tons.

3-24. 44 micrometers.

CHAPTER 4

4-1. 4.6×10^5 coulombs.

4-4. (a) $r_2 \to \infty$. (c) For $r_1 \approx r_2$, $E \approx V/(r_2 - r_1)$ is large. For $r_1 \ll r_2$, E is large at the surface of the small inner sphere.
(e) $0.17 < r < 0.32$. (f) 4.15×10^6 volts/meter.

4-8. 2.4×10^4 volts/meter.

4-21. (a) Yes.
(b) $|\partial f/\partial r|$ becomes larger and the orbits can become unstable.
(c) Yes.

4-24. No.

4-28. (b) 3.3×10^{-3} newton, or one third of a gram-force.

(e)

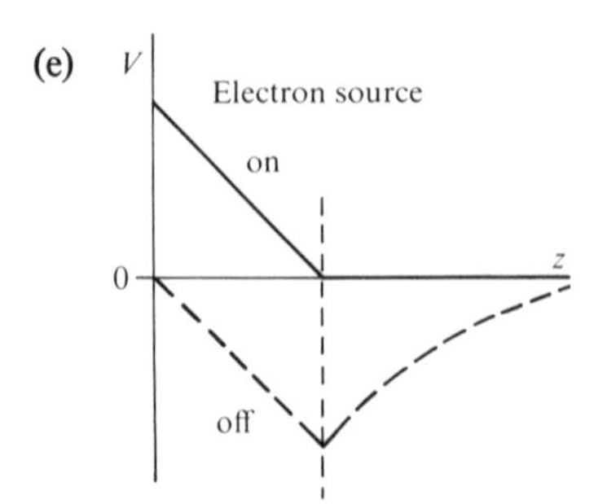

Figure A-4. Solution to Problem 4-28e.

(f) About 5.5 microseconds.

CHAPTER 5

5-1. $\mathcal{V} \approx c/7$.

5-4. O says that A and B emitted their light signals simultaneously.
O' says that B emitted his signal before A.

5-6. (a) $\tan \alpha_1 = \gamma \tan \alpha_2$.
(b) 90°.

5-8. (a) 0.363 c.
(b) 0.80 c.

5-10. Diameter $\leqslant 10^{13}$ meters.

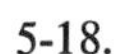
5-18.

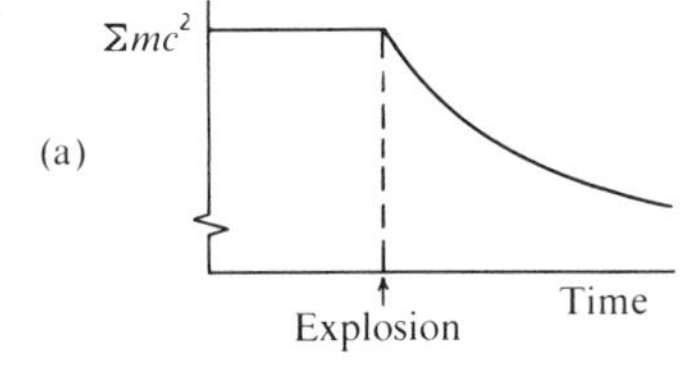

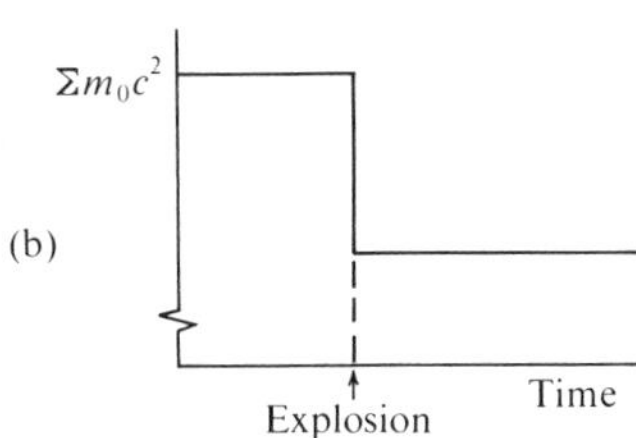

Figure A-5. Solution to Problem 5-18.

5-22. (a) 1.78×10^{-7} microgram.
(b) 30 seconds.

5-25. 9.34×10^8 electron-volts.

5-28. (b) For the Sun, $\Delta\nu/\nu = -2.1 \times 10^{-6}$.
For the Earth, $\Delta\nu/\nu = -7.0 \times 10^{-10}$.
(c) About 10^{11} kilograms/meter^3.
(d) $\pm 6.9 \times 10^{-6}$.
(e) The two shifts are of the same order of magnitude.

5-33. (a) $\Delta Q/Q = -4 \times 10^{-6}$.
(b) Yes. See the example in Section 5-20.

CHAPTER 6

6-3. (i) 2.31×10^{-22} newton.
(ii) 7.80×10^{-23} newton.

6-8. (d)

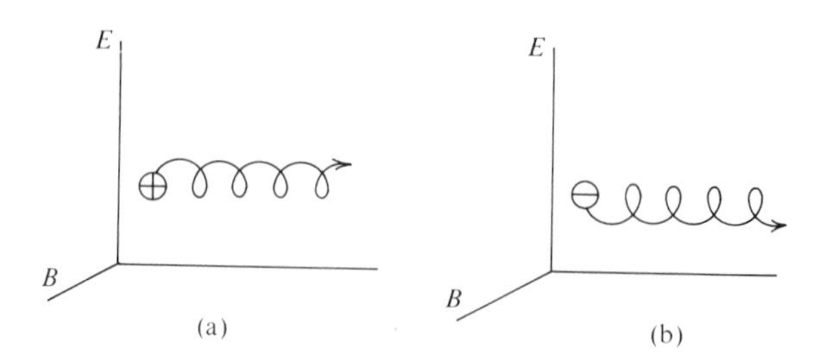

Figure A-6. Solution to Problem 6-8d.

(e) 2.5×10^{-3} meter/second eastward.
(f) Westward.

6-13. $\boldsymbol{E} = 1.45 \times 10^{-5} \dfrac{(-4.34 \times 10^{9} t\boldsymbol{i} + \boldsymbol{j})}{(19.2 \times 10^{18} t^{2} + 1)^{3/2}}$ volt/meter.

$\boldsymbol{B} = 7.00 \times 10^{-15} \dfrac{\boldsymbol{k}}{(19.2 \times 10^{18} t^{2} + 1)^{3/2}}$ tesla.

6-15. (a) Inside, $\boldsymbol{E}_1 = \gamma \mathcal{V} B_2 \boldsymbol{j}$, $\boldsymbol{B}_1 = \gamma B_2 \boldsymbol{k}$, if $\boldsymbol{B}_2 = B_2 \boldsymbol{k}$.
(b) Outside, $\boldsymbol{E}_1 = 0$, $\boldsymbol{B}_1 = 0$.
(e) Yes.
(f)

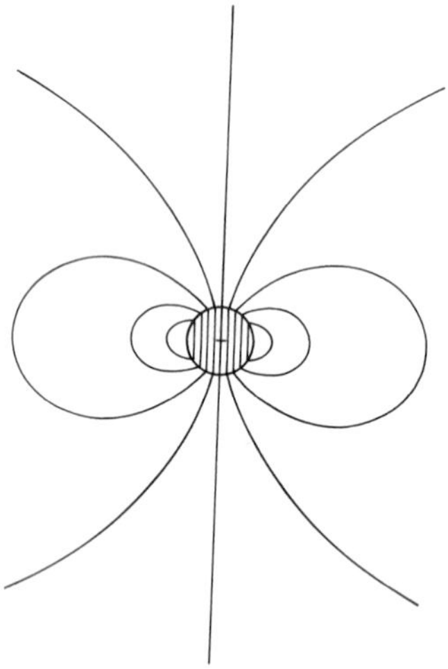

Figure A-7. Solution to Problem 6-15f.

6-19. $v = 0.27$ meter/hour.

6-23. (a) $h = 1.46$ meters.
$D = 5.0$ millimeters.
$\lambda = 3.28 \times 10^{-10}$ coulomb/meter.

(d) Yes;

$\dfrac{dr}{dt} = \dfrac{c}{3 \times 10^{4}}$ at $x = 400$ meters.

(e) No.
(f) We have neglected end effects on the cylinders of charge.

CHAPTER 7

7-4. 0.234 newton.

7-10. (a) Perpendicular to the plane passing through the two axes.
(b) Yes.

7-13. $\dfrac{2^{1/2}\mu_0}{\pi}\dfrac{I}{a}$.

7-16. (c) The electrons form part of a neutral plasma.

7-19. $V = (e/m)B^2(b^2 - a^2)^2/8b^2$, where e and m are the charge and mass of the electron, a is the radius of the cathode, and b that of the anode.

7-24. (a) $\epsilon_0\chi_e B\omega r$.
(b) $\rho_b = -2\epsilon_0\chi_e B\omega$
$\sigma_b = \epsilon_0\chi_e B\omega a$,
where a is the radius of the cylinder.

7-26. (a) $A = \dfrac{\mu_0 I}{4\pi}\ln\left\{\dfrac{H + (R^2 + H^2)^{1/2}}{-H + (R^2 + H^2)^{1/2}}\right\}$.

(b) It is only the axial component of the current which matters.
(c) No; to calculate the derivatives of $\mathbf{A}$ we require $\mathbf{A}$ as a function of ρ, φ, z.

7-29. (a) $\dfrac{\partial B_\rho}{\partial \rho} = \dfrac{3\mu_0 I a^2 z}{4(a^2 + z^2)^{5/2}}$.

(b) $B_\rho = \dfrac{3\mu_0 I a^2 z\rho}{4(a^2 + z^2)^{5/2}}$.

(c) $\dfrac{\partial B_z}{\partial \rho} = \dfrac{3\mu_0 I a^2 \rho}{4(a^2 + z^2)^{7/2}}(a^2 - 4z^2)$.

(d) $B_z = \dfrac{\mu_0 I a^2}{2(a^2 + z^2)^{3/2}}\left\{1 + \dfrac{3(a^2 - 4z^2)\rho^2}{4(a^2 + z^2)^2}\right\}$.

7-33. (c) 7.75×10^{-11} tesla.
(e) 4.83×10^{-8} ampere/meter^2.
(f) 6.14×10^{-6} ampere.

CHAPTER 8

8-1. 1.2 millivolts.

8-5. $\dfrac{\pi r^2\sigma L}{l}\dfrac{dE}{dt} + \dfrac{\pi r^2\sigma R}{l}E = \dfrac{V}{l}$,

where L is the inductance, σ is the conductivity of the wire, r is its radius, and l is its length.

8-8. (b)

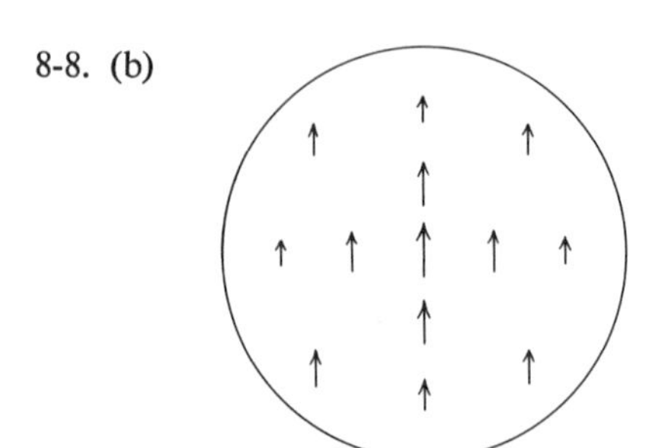

Figure A-8. Solution to Problem 8-8b.

(c)

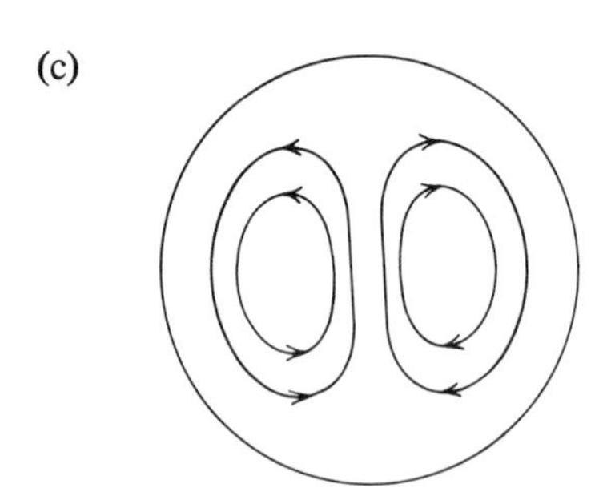

Figure A-9. Solution to Problem 8-8c.

(e)

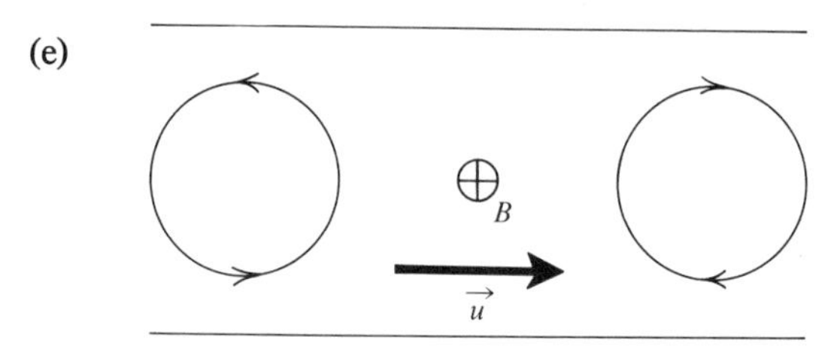

Figure A-10. Solution to Problem 8-8e.

8-11. (d) The transverse velocity decreases.

8-17. L decreases with increasing frequency until the skin depth becomes small compared to $b - a$.

8-21. $\dfrac{\mu_0 N^2 w}{4\pi} \ln\left(\dfrac{2R + w}{2R - w}\right) I^2.$

8-27. (b) 280 ampere-turns.

8-33. (c) Both are equal to the energy density and proportional to the square of the magnitude of the field.
The electric force is directed into the field; the magnetic force is directed away from the field.
The electric force per unit area is normal to the surface of a conductor and is equal to $\epsilon_0 E^2/2$ only if the field is static.
The magnetic pressure is normal to the surface of a conductor and equal to $B^2/2\mu_0$ only if there is zero field inside the conductor. This condition is achieved exactly with superconductors, and approximately if the frequency is high.

8-35. (a) 4.0×10^5 newtons/meter^2 = 4.0 atmospheres.
(b) The pressure would be unchanged since B is uniform inside a long solenoid.

8-37. (d) 3.2×10^4 teslas.
(e) 1.6×10^9 joules.

CHAPTER 9

9-5. 2.0.

9-8. (a) $\boldsymbol{E}$ is reduced.
$\boldsymbol{D}$ is unaffected.
(c) Minimum.
(d) $\boldsymbol{B}$ is increased.
$\boldsymbol{H}$ is unaffected.
(f) Minimum.

9-10. (a) $\dfrac{\mu l I}{2\pi} \ln \dfrac{c}{b}$.
(b) $\lambda_b = \chi_m I/2\pi b$, parallel to I.
$\lambda_c = \chi_m I/2\pi c$, opposite to I.
(c) Zero.
(d) $\mu_0 I/2\pi\rho$. Removing the iron cylinder would have no effect on this field.

9-15. Inside, $\boldsymbol{H} = \boldsymbol{M}/2$ in the direction opposite to $\boldsymbol{M}$.
$\boldsymbol{B} = \mu_0\boldsymbol{M}/2$ in the direction of $\boldsymbol{M}$.

Outside,

$$B = \frac{\mu_0 M t a^2}{2(a^2 + D^2)^{3/2}},$$

where D is the distance to the center, in the direction of $\boldsymbol{M}$.

9-20. Use a circuit similar to that of Figure 9-15 with A_m = 2.5 centimeters^2 and L_m = 8.84 centimeters (each magnet is 4.42 centimeters long).

CHAPTER 10

10-1. No. The relaxation time is of the order of 10^{-12} second.

10-4.
$$B = \mu_0 J_0/4a \qquad (z \leqslant 0),$$
$$= \mu_0 J_0 \left\{\frac{az^2}{2} - z + \frac{1}{4a}\right\} \qquad \left(0 \leqslant z \leqslant \frac{1}{a}\right),$$
$$= -\mu_0 J_0/4a \qquad \left(z \leqslant \frac{1}{a}\right).$$
$$B = 0 \quad \text{at} \quad z = 0.293/a.$$

10-7. $$\nabla^2 V - \epsilon\mu \frac{\partial^2 V}{\partial t^2} = -\frac{\rho_f}{\epsilon},$$

$$\nabla^2 \boldsymbol{A} - \epsilon\mu \frac{\partial^2 \boldsymbol{A}}{\partial t^2} = -\mu \boldsymbol{J}_f.$$

The other two equations ($\nabla \cdot \boldsymbol{B} = 0$ and $\nabla \times \boldsymbol{E} = -\partial \boldsymbol{B}/\partial t$) lead to identities.

10-9: (c) $B = \mu_0 \lambda$.
(d) The magnetic pressure is $\mu_0 \lambda^2/2$.
(f) Currents are induced which maintain Φ constant.

10-12. (c) $\boldsymbol{E}^{IV} = \boldsymbol{E}$; $\boldsymbol{H}^{IV} = \boldsymbol{H}$.

CHAPTER 11

11-3. 54.4 millivolts.

11-12. For $\mathcal{Q} = 1$, $k_r = 1.10(\epsilon_r \mu_r)^{1/2}/\lambda_0$, $k_i = 0.455(\epsilon_r \mu_r)^{1/2}/\lambda_0$, $\delta \approx 2\lambda$.
$|E/H|$ is smaller than for $\sigma = 0$ by a factor of 1.19.
$\boldsymbol{E}$ leads $\boldsymbol{H}$ by 0.393 radian or 22.5 degrees.
For $\mathcal{Q}^2 \gg 1$, $k_r \approx \omega(\epsilon\mu)^{1/2}\left(1 + \frac{1}{8\mathcal{Q}^2}\right)$, $k_i \approx \frac{\sigma}{2}(\mu/\epsilon)^{1/2}$, $\delta \approx 2\mathcal{Q}\lambda$.
$|E/H|$ is smaller than for $\sigma = 0$ by a factor of $\left(1 - \frac{1}{4\mathcal{Q}^2}\right)$.
$\boldsymbol{E}$ leads $\boldsymbol{H}$ by $1/2\mathcal{Q}$ radian.

11-19. Attenuation in db/meter:

	20 kilohertz	20 megahertz
Sea Water	5.4	170
Copper	1.9×10^4	5.9×10^5

11-25. (a) $\dfrac{c}{(f_2^2 - f_p^2)^{1/2} - (f_1^2 - f_p^2)^{1/2}}$.

(b) $u_g = c\left\{1 - \left(\frac{\omega_p}{\omega_1}\right)^2\right\}^{1/2}$.

(e) $u_1 = 3.24 \times 10^8$ meters/second.
$u_2 = 3.23 \times 10^8$ meters/second.
$u_g = 2.78 \times 10^8$ meters/second.
(f) 2.73×10^3 meters; 45 waves.

CHAPTER 12

12-1. (d) $j\boldsymbol{k} \cdot \boldsymbol{E} = \rho_t/\epsilon_0$, $\quad \boldsymbol{k} \cdot \boldsymbol{B} = 0$,
$\boldsymbol{k} \times \boldsymbol{E} = \omega \boldsymbol{B}$, $\quad -j\boldsymbol{k} \times \boldsymbol{B} = \mu_0 \boldsymbol{J}_m + j\omega \boldsymbol{E}/c^2$.

12-9. (a)

	E(volts/meter)	H(ampere-turns/meter)
Incident	42.4	0.115
Reflected	6.00	0.016
Refracted	36.4	0.131

(b) $R = 2.01\%$, $T = 98.0\%$.

12-12. (a) $\delta_z/\lambda_2 = 0.368$, $\delta_z/\lambda_1 = 1.11$.
(b) E_{0r} leads E_{0i} by 2.59 radians.
E_{0t} leads E_{0i} by 1.29 radians.

12-15. (a) $F = \frac{1}{2}\left\{1 - \left[1 - \left(\frac{n_2}{n_1}\right)^2\right]^{1/2}\right\}$.

(b)

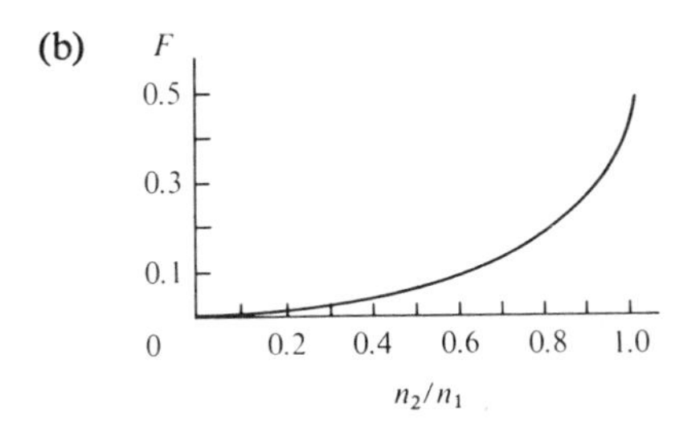

Figure A-11. Solution to Problem 12-15b.

12-18. $$\left(\frac{E_{0r}}{E_{0i}}\right)_P = \frac{-(1 - j)\cos\theta_i + n_1\mu_{r2}(\delta/\lambda_0)}{(1 - j)\cos\theta_i + n_1\mu_{r2}(\delta/\lambda_0)}.$$

12-20.

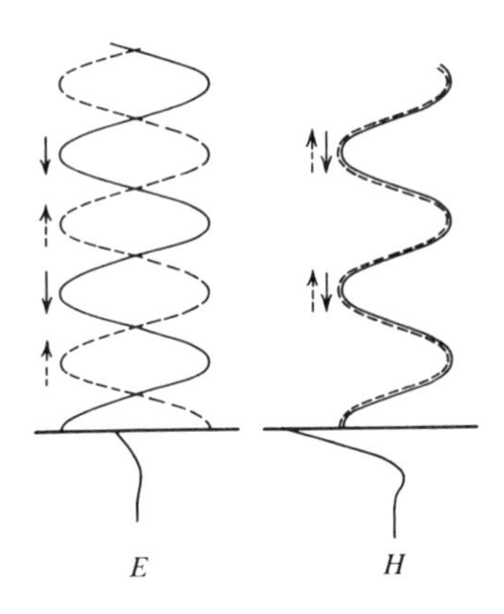

Figure A-12. Solution to Problem 12-20.

12-24. (a) $\frac{F_{\text{grav}}}{F_{\text{rad}}} = 8.86 \times 10^6\, r$.

(b) The two forces are equal for $r \approx 0.1$ micron.

CHAPTER 13

13-11. The skin depth is 1.5 microns; the coating should have a minimum thickness of about 5 microns.

13-15. (a)

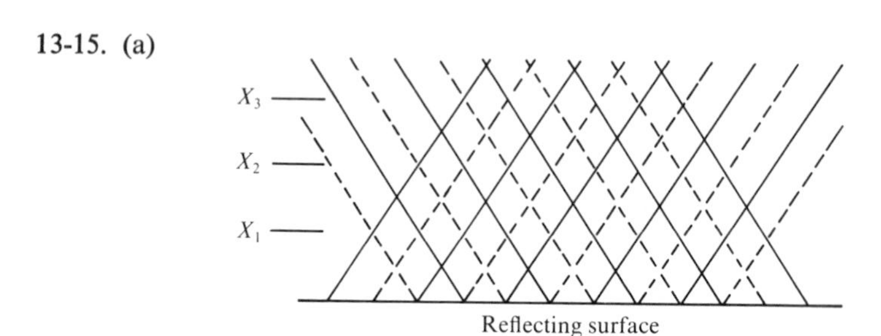

Figure A-13. Solution to Problem 13-15a.

(b) E is always zero in the planes marked x.
(c) If there were a reflecting surface at x_1, we would have the $n = 1$ mode. A reflecting surface at x_2 would give the $n = 2$ mode, etc.

13-19. Both methods can, in fact, be used. See, for example: Harvey, *Microwave Engineering* (New York: Academic Press), p. 144.

13-23. $\lambda = H_{y=0}$. This current flows inside the conductor, very close to the surface.

13-27.

Type	$\frac{b}{a}$	$\frac{f_{max}}{f_{min}}$	$\lambda_0/2b$		Attenuation ($n = 2$)	
			f_{min}	f_{max}	f_{min}	f_{max}
2.84 × 1.34	2.12	1.52	0.800	0.525	10.8	4.4
1.872 × 0.872	2.15	1.48	0.800	0.530	16.4	7.9
1.372 × 0.622	2.22	1.40	0.735	0.525	21.1	8.9
0.900 × 0.400	2.25	1.51	0.800	0.528	34.2	14.5
0.622 × 0.311	2.00	1.45	0.762	0.525	48.0	20.2

13-30. (a) 1.8 megawatts.
(b) 4.24 kilowatts/meter in the outer conductor.
7.00 kilowatts/meter in the inner conductor.

CHAPTER 14

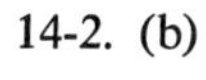

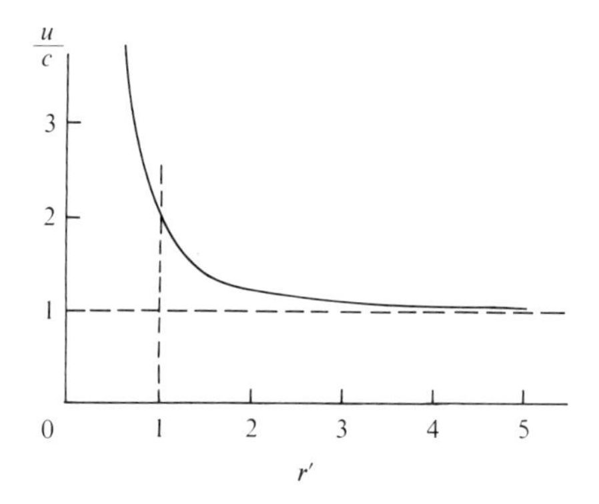

Figure A-14. Solution to Problem 14-2b.

14-4. (a) $\dfrac{1 + 3\cos^2\theta}{\sin^2\theta}\dfrac{\lambda\!\!\!^{-2}}{r^2}$.

(b) $\dfrac{1 + 7\cos^2\theta}{2\sin^2\theta}$.

(c) 1.

14-8. 222 millivolts/meter.

14-14. Explore the field with a small electric or magnetic dipole.

14-18. No.

14-24. In air, the ratio of the output voltages is

$$\frac{V_{el}}{V_{mag}} = \frac{4}{\pi}\frac{\lambda\!\!\!^{-}}{l} \gg 1.$$

In sea water,

$$\frac{V_{el}}{V_{mag}} = \frac{4}{\pi l\,(\sigma\mu_0\omega)^{1/2}}.$$

For a frequency of the order of 10^4 (see Section 12.5.1.

$$\frac{V_{el}}{V_{mag}} \simeq \frac{1}{l}.$$

APPENDIX B

B-1. 6.44×10^{-4} newton.

Index